WILDLIFE PROTECTION, DESTRUCTION AND EXTINCTION

BIOLOGY, EVOLUTION AND CONSERVATION OF RIVER DOLPHINS WITHIN SOUTH AMERICA AND ASIA

WILDLIFE PROTECTION, DESTRUCTION AND EXTINCTION

International Illegal Trade in Wildlife
Liana Sun Wyler and Pervaze A. Sheikh
2008. ISBN: 978-1-60456-757-1

Fishing, Hunting, and Wildlife Associated Recreation
Dustin N. Worley (Editor)
2009. ISBN: 978-1-60692-128-9

Wildlife: Destruction, Conservation and Biodiversity
John D. Harris and Paul L. Brown (Editors)
2009. ISBN: 978-1-60692-974-2

National Parks and Rivers: Background, Protection and Use Issues
Yolanda A. Reddy (Editor)
2009. ISBN: 978-1-60741-801-6

Protecting Marine Mammals
Marielle de Vries (Editor)
2010. ISBN: 978-1-60741-569-5

Wildlife Refuges: Factors and Concerns about Future Sustainability
Earl B. Taylor (Editor)
2010. ISBN: 978-1-60692-683-3

Conservation Research in Uganda's Forests: A Review of Site History, Research, and Use of Research in Uganda's Forest Parks and Budongo Forest Reserve
William Olupot and Andrew J. Plumptre
2010. ISBN: 978-1-60876-577-5

Biology, Evolution and Conservation of River Dolphins within South America and Asia
Manuel Ruiz-Garcia and Joseph Mark Shostell (Editors)
2010. ISBN: 978-1-60876-633-8

National Parks: Biodiversity, Conservation and Tourism
Angus O'Reilly and Doran Murphy (Editors)
2010. ISBN: 978-1-60741-465-0

National Parks: Vegetation, Wildlife and Threats
Grazia Polisciano and Olmo Farina (Editors)
2010. ISBN: 978-1-60876-742-7

Protection at the Wildlife - Urban Interface
Ernesto Di Bello (Editor)
2010. ISBN: 978-1-60876-093-0

Conservation Research in Uganda's Savannas: A Review of Park History, Applied Research, and Application of Research to Park Management
William Olupot, Luke Parry, Michelle Gunness and Andrew J. Plumptre
2010. ISBN: 978-1-60876-575-1

Are Animals in Zoos Rather Conspicuous Than Endangered?
D. Frynta, J. Marešová, E. Landová, S. Lišková, O. Šimková, I. Tichá, M. Zelenková and R. Fuchs
2010. ISBN: 978-1-61668-234-7
2010. ISBN: 978-1-61668-499-0 (E-book)

WILDLIFE PROTECTION, DESTRUCTION AND EXTINCTION

BIOLOGY, EVOLUTION AND CONSERVATION OF RIVER DOLPHINS WITHIN SOUTH AMERICA AND ASIA

MANUEL RUIZ-GARCIA
AND
JOSEPH MARK SHOSTELL
EDITORS

Nova Science Publishers, Inc.
New York

For permission to use material from this book please contact us:
Telephone 631-231-7269; Fax 631-231-8175
Web Site: http://www.novapublishers.com

LIBRARY OF CONGRESS CATALOGING-IN-PUBLICATION DATA

Biology, evolution, and conservation of river dolphins within South America and Asia / editors, Joseph Mark Shostell.
p. cm.
Includes index.
ISBN 978-1-60876-633-8 (hardcover)
1. River dolphins--Asia. 2. River dolphins--Evolution--Asia. 3. River dolphins--Conservation--Asia. 4. River dolphins--South America. 5. River dolphins--Evolution--South America. 6. River dolphins--Conservation--South America. I. Shostell, Joseph Mark.
QL737.C436B56 2009
599.53'8--dc22

2009048899

Published by Nova Science Publishers, Inc. † New York

CONTENTS

Preface		**xi**
Chapter 1	An Introduction to River Dolphin Species *Joseph Mark Shostell and Manuel Ruiz-García*	**1**
Chapter 2	Seasonal Ecology of Inia in Three River Basins of South America (Orinoco, Amazon, and Upper Madeira) *Tamara L. McGuire and Enzo Aliaga-Rossel*	**29**
Chapter 3	Conservation of the River Dolphin (*Inia boliviensis*) in Bolivia *Enzo Aliaga- Rossel*	**55**
Chapter 4	Mobility of the Axial Regions in a Captive Amazon River Dolphin (Inia geoffrensis) *Timothy D. Smith and Anne M. Burrows*	**71**
Chapter 5	The Application of Equation Models to Determine the Age of Pink River Dolphin Skulls *Luisa Fernanda Castellanos-Mora, Fernando Trujillo and Manuel Ruiz-García*	**83**
Chapter 6	Amazon River Dolphin: High Phylopatry due to Restricted Dispersion at Large and Short Distances *Juliana A. Vianna, Claudia Hollatz, Miriam Marmontel, Rodrigo A.F. Redondo and Fabrício R. Santos*	**101**
Chapter 7	Amazon River Dolphin Polymorphism and Population Differentiation of MHC Class II Peptides *María Martínez-Agüero, Sergio Flores-Ramírez and Manuel Ruiz-García*	**117**
Chapter 8	Micro-Geographical Genetic Structure of *Inia Geoffrensis* in the Napo-Curaray River Basin by Means of Chesser´S Models *Manuel Ruiz-García*	**131**

Chapter 9 Changes in the Demographic Trends of Pink River Dolphins (*Inia*) at the Microgeographical Level in Peruvian and Bolivian Rivers and Within the Upper Amazon: Microsatellites and Mtdna Analyses and Insights into *Inia's* Origin **161**
Manuel Ruiz-García

Chapter 10 Fossil Record and the Evolutionary History of Iniodea **193**
M. A. Cozzuol

Sotalia Fluviatilis-Sotalia Guianensis **219**

Chapter 11 Fishery Activity Impact on the *Sotalia* Populations from the Amazon Mouth **221**
Sandra Beltran-Pedreros and Miguel Petrere

Chapter 12 Dolphin-Fishery Interaction: Cost-Benefit, Social-Economic and Cultural Considerations **237**
Sandra Beltran-Pedreros and Ligia Amaral Filgueiras-Henriques

Chapter 13 Ethnoecology of *Sotalia Guianensis* (GERVAIS, 1853) in the Amazon Estuary **247**
Sandra Beltrán-Pedreros, Miguel Petrere and Ligia Amaral Filgueiras-Henriques

Chapter 14 Molecular Ecology and Systematics of *Sotalia* Dolphins **261**
H.A. Cunha, da Silva VMF and A.M. Solé-Cava

Chapter 15 Population Structure and Phylogeography of Tucuxi Dolphins (*Sotalia Fluviatilis*) **285**
Susana Caballero, Fernando Trujillo, Manuel Ruiz-García, Julianna A. Vianna, Miriam Marmontel, Fabricio R. Santos and C. Scott Baker

Pontoporia Blainvillei **299**

Chapter 16 Life History and Ecology of Franciscana, *Pontoporia Blainvillei* (Cetacea, Pontoporiidae) **301**
Eduardo R. Secchi

Chapter 17 Review on the Threats and Conservation Status of Franciscana, *Pontoporia Blainvillei* (Cetacea, Pontoporiidae) **323**
Eduardo R. Secchi

Asian River Dolphins **341**

Chapter 18 Detection of Yangtze Finless Porpoises in the Poyang Lake Mouth Area via Passive Acoustic Data-Loggers **343**
Songhai Li, Shouyue Dong, Satoko Kimura, Tomonari Akamatsu, Kexiong Wang and Ding Wang

Chapter 19 Population Status and Conservation of Baiji and the Yangtze Finless Porpoise **357**
Ding Wang and Xiujiang Zhao

Chapter 20 Failure of the Baiji Recovery Program: Conservation Lessons for other Freshwater Cetaceans **377**
Samuel T. Turvey

Chapter 21 High Level of MHC Polymorphism in the Baiji and Finless Porpoise, with Special Reference to Possible Convergent Adaptation to the Freshwater Yangtze River **395**
Shixia Xu, Wenhua Ren, Kaiya Zhou and Guang Yang

Chapter 22 Population Status and Conservation of the Ganges River Dolphin (*Platanista Gangetica Gangetica*) in the Indian Subcontinent **419**
R. K. Sinha, Sunil Kumar Verma and Lalji Singh

Chapter 23 The Evolutionary History and Phylogenetic Relationships of the Superfamily Platanistoidea **445**
Lawrence G. Barnes, Toshiyuki Kimura and Stephen J. Godfrey

Index **489**

PREFACE

True river dolphins as well as marine dolphins that frequent freshwater systems are large animals that have traditionally gone unnoticed by the general public and, in a certain sense, by marine mammal specialists as well. In fact, only a limited number of researchers have investigated the biology of these dolphin species. This is quite surprising given that these species are commonly the top predators in their habitats. Now for the first time, revolutionary molecular techniques are being applied to answer evolutionary reconstruction questions of many animals, including river dolphins. In addition, new paleontological records are dramatically changing our perspective about the relationships of these dolphins with each other and with other cetaceans. In this book, new census information and important ecological characteristics are provided of the river dolphins Inia, Sotalia, Pontoporia, Lipotes, Phocaena and Platinista. For the first time, molecular and genetic results of theses dolphin species are presented. A compilation of these data is essential if we are to present a strategic conservation plan for these animals. Upon being informed of critical evolutionary historical data, conservation biologists will now be able to tailor their conservation efforts for each threatened river dolphin species. Additionally, new morphological data and the new discoveries in the fossil record for river dolphins are examined. The major dolphin specialists in Colombia, Brazil, Bolivia, Argentina, the United States of America, China, England, India, Japan and New Zealand present their newest results within a single book that graduate students, professors, scientists, evolutionary ecologists, aquatic mammalogists, population ecologists, conservation ecologists, and marine biologists will all find valuable for the foreseeable future.

Chapter 1 - This chapter introduces nine dolphin species (*Neophocaena phocaenoides asiaeorientalis, Sotalia guianensis, Sotalia fluviatilis, Pontoporia blainvillei, Inia geoffrensis, Inia boliviensis, Platanista gangetica, Lipotes vexilliter* and *Orcaella brevirostris*) that are discussed in the succeeding chapters of this book and provides brief summaries on each species' population status, habitat condition and looming threats. There are commonalities among the threats for these dolphins and they are linked to human activities. Fishing, dams, and pollution generally affect all of the species with those species near the highest human densities being the most threatened and having the bleakest future. There are of course bright spots in the conservation efforts for these species and some dolphins, such as *Inia geoffrensis*, seem to be faring well and have a large population size and great distribution. Also, the authors discuss recent and new contributions of molecular, morphological, and paleontological data that tremendously help our understanding of phylogenetic relationships and evolutionary history of these graceful creatures.

Chapter 2 - Compared to most marine odontocetes, river dolphins live in an environment that is less-stable and more spatially complex than the ocean. Yearly seasonal fluctuations in river levels may be as great as 20 meters, and lead to seasonal extremes in quality and quantity of aquatic habitat available to river dolphins. Seasonal changes in water levels also affect the availability of dolphin prey due to seasonal patterns of fish reproduction and fish migrations. Human-induced threats to river dolphins, such as incidental net entanglement, vessel strikes, and deliberate killing appear to vary seasonally as well. In this chapter, the authors present their investigations of the seasonal ecology of *Inia* spp from three river basins of South America (*Inia geoffrensis humboldtiana* in Venezuela's Orinoco River Basin, *Inia geoffrensis geoffrensis* in Peru's Amazon Basin, and *Inia boliviensis* in Bolivia's Mamoré Basin). The authors provide results from their observational studies (which included boat-based surveys of groups and photo-identification of individuals) and the authors discuss these results in the context of other information about the seasonal ecology of *Inia*, including distribution, movement patterns, group size, age-class composition, and seasonality of reproduction. The authors conclude with a discussion of how seasonal ecology should be considered in the conservation of river dolphins and of the management of human activities that affect them.

Chapter 3 - The pink river dolphin genus *Inia*, is widely distributed in the Orinoco and Amazon basins. Locally called the bufeo (*Inia boliviensis*) in Bolivia, it is an endemic species to the region, geographically isolated from *Inia* populations within the Amazon's main stem by a series of rapids between Guayaramerin, Bolivia and Porto Velho, Brazil. In Bolivia, they are distributed in three main sub-basins: Abuna, Mamore and Itenes (Guapore). Despite bufeo being a native species and the only cetacean present in a land-locked country, its ecology and conservation status are poorly understood. Unfortunately, no conservation laws explicitly target this cetacean in Bolivia and consequently it only receives relatively minor legal protection when it resides in protected conservation areas. This chapter includes information on the studies that have been conducted in Bolivia; the conservation status; aspects related to the geographic distribution of the species, its behavior, ecology, population size, threats and possible means of protection. This information will lead to recommendations for the implementation of priorities in research programs and conservation for this species in Bolivia.

Chapter 4 - Here the authors analyze mobility of axial regions in a captive Amazon River dolphin (*Inia geoffrensis*), specifically regarding lateral movements of the neck and torso. Still images from video recordings of the swimming dolphin were extracted and analyzed using Scion Image software. Lateral movements of the neck can reach nearly a right angle (deviating from the thoracic region by up to at least 84 degrees). Much more lateral mobility is seen in the torso, with most occurring in the posterior torso (presumably at intervertebral joints in the caudal vertebrae). In sum, the lateral mobility allows this captive dolphin to touch rostrum to tail by lateral bending. Osteological correlates of lateral mobility in this species are also reviewed in this chapter. Based on behavioral descriptions in the literature, the extreme lateral mobility observed in this captive animal are likely representative of the species in general, and relates to locomotion in a complex environment. Further investigations must determine whether this mobility, and the morphological features that permit it, are unique adaptations or primitive features that characterized an ancestral condition.

Chapter 5 - In this chapter, the authors show that several biometric skull measurements in the pink river dolphin (*Inia geoffrensis*) are highly related to skull age, determined by teeth analysis. Out of 50 morphometric skull measurements, the maximum width between the zygomatic processes of the squamosal bones (V15) and the maximum width of the internal nares (V21) were highly correlated with age. Regression analyses such as linear and 24 other simple models, linear multiple regression, polynomial models, and distance multiple regression (Gower, Absolute value, Mahalanobis and Minkowski) had similar results. On average, these multiple regression equations demonstrated that the age of a dead pink river dolphin is determined by only four cranial measures which explained 64-65% of age variation.

Chapter 6 - The Amazon River Dolphin (*Inia geoffrensis*) is widely distributed along the Amazon and Orinoco basins, covering an area of about 7 million km^2. The authors have generated 519 base pair (bp) sequences of the control region (HVSI) and 1,140 bp of the Cytochrome B (Cyt-b) gene of mitochondrial DNA (mtDNA) for two populations from the Amazon basin in Brazil, separated by only 45 km. Six HVSI haplotypes were identified and the authors could detect a remarkable population structure despite of the short distance separating the localities. Compared to HVSI data from other South American countries, the Brazilian haplotypes occupy an intermediate position related to Colombian Amazon, Colombian Orinoco and Bolivian haplotypes. The Cyt-b data also detected a remarkable separation between both Brazilian locations, and the phylogenetic analysis indicated an association of Amazon and Orinoco haplotypes, separated from the Bolivian ones. This phylogeographic study emphasizes the outstanding population structure for the Amazon River Dolphin, considering both macro and microgeographic levels. These results suggest a strong phylopatry for this species due to gene flow restriction through long distances, as well as short distances by different water ecology characteristics. The studied Brazilian populations occur in close localities but are separated by the turbid fresh water environment of the Amazon River, a likely ecological barrier segregating *I. geoffrensis* populations.

Chapter 7 - *Inia*, the Amazon River dolphin, inhabits the three major basins of northern South America (Beni-Mamoré, Amazon and Orinoco). The authors analyzed class II DQB MHC gene peptide sequences in 60 dolphins from Bolivia (Beni-Mamoré), Peru (Amazon) and Colombia (Orinoco). Sixteen (16) peptide alleles were identified, generated by 17 polymorphic sites, most of them on the peptide binding region (PBR) residues. Four of the alleles were the most frequent of all the populations and several private alleles for each basin were found. A high level of polymorphism in the class II gene was determined, similar to those reported for the Chinese river dolphins, such as the Baiji and the finless porpoise. This polymorphism could be an adaptive response to the high level of pathogens in freshwater.

Chapter 8 - Thirty-three pink river dolphins (*Inia geoffrensis*) were caught at eight sampling places (one beach and seven lagoons; transect length of 280 km) in the Napo and Curaray rivers at the Peruvian Amazon. Nine microsatellites were applied to analyze the genetic structure of this species at the micro-geographic level and diverse population genetics procedures were used to determine if *Inia* is a solitary or a social reproductive species. Chesser's social model was used to determine asymptotic values of the F-statistics and showed that this species is a social reproductive one and the basic genetic lineages could be composed of: 1- seven reproductive females per lineage in each breeding period, 2- the number of reproductive males per linage is not important, 3- a reproductive male with four females, on average, within each lineage, and thus there is polygyny in this dolphin species, and 4- a probability of 0.30 that females of a same lineage have chosen the same male for breeding. This is the first work carried out at a micro-geographic level for a river dolphin species, where the basic social reproductive parameters are revealed.

Chapter 9 - More than 200 pink river dolphins (*Inia geoffrensis* and *Inia boliviensis*) were sampled in diverse rivers of Colombia, Peru, Brazil and Bolivia. Ten microsatellites and 400 bp of the mitochondrial control region (D-loop) gene were analyzed with special emphasis on three Peruvian rivers (Ucayali, Marañon and Napo-Curaray) and the Bolivian Mamoré River (and tributaries). Of the different evolutionary demographic tests applied to the microsatellite and mtDNA data, the tests of Kimmel et al., (2008) and Zhivotovsky et al., (2000) provided the most insights about the demographic history of the pink river dolphin. These tests showed that initial bottlenecks occurred prior to very recent population expansions in the diverse areas studied. Two tests (Zhivotovsky and Garza & Williamson) revealed a very strong bottleneck in the origin of the Bolivian population and not during its population expansion. Together, the microsatellite and mtDNA, analyses revealed a strong population expansion for the overall upper Amazon sample and supported that the population expansion and colonization of *Inia* throughout the Amazon, Orinoco and Beni-Mamoré basins occurred in the last 200,000 years ago (and in the majority of cases between 4,000-50,000 years ago) and not several millions of years ago as was claimed by other authors. Furthermore, the original population was the Amazon one, and not the Bolivian population as has been previously defended by several authors, such as Grabert (1984 a, b, c), Pilleri & Ghir (1977, 1980) and Pilleri et al. (1982).

Chapter 10 - This chapter discusses the possible phylogenetic relationships within the superfamily Inioidea (using fossil record data) and provides detailed descriptions of Brachydelphidae, Pontoporiidae and Iniidae (including *Goniodelphis, Ischyrhorhynchus, Saurocetes, Plicodontinia* and a possible new species of *Inia* that is estimated to have arisen approximately 45,000 years ago). Some previously related taxa to Iniidae are also discussed such as *Proinia patagonica.* Additionally, the chapter discusses the Lipotoidea and their relationship with Inioidea, the phylogenetic position of *Parapontoporia*, and the evolutionary process (and paths) that originated the inioid clades.

Chapter 11 - This chapter describes and analyzes the bycatch of *Sotalia guianensis*, in gillnets by an artisan fishing fleet within the Amazonian estuary during two time periods: 1996-1997 and 1999-2001. Number, size and gender data, as well as dolphin specimens were obtained from fishermen at Brazilian ports and analyzed. Fishing capacity and effort were determined via simple linear regression and bycatch, fishing trip and fishing effort data were analyzed between time-periods, among climatic (seasonal) periods and between strata (based on vessel length). Results indicated that the stratum two fishing fleet not only had larger vessels but longer fishing trips, used longer nets and had larger fishing crews compared to stratum one's fleet. Bycatch increased in both strata between periods but to a greater extent in stratum two. Although there was an increased percentage of fishing trips with bycatch across time, there was a reduced mean number of dolphins per bycatch. There were also differences in the bycatch by sexual maturity with an indiscriminately larger number of sexual-reproducing adults caught in stratum two. Collectively, these results in conjunction with other anthropogenic factors combined with dolphins being a k-selected species, suggest that dolphin mortality from bycatch may seriously affect *Sotalia guianensis* in the Amazonian estuary. Furthermore, the fishery-dolphin interaction was characterized and determined to be indirectly predatory.

Chapter 12 - The authors evaluated dolphin-fishery interaction dynamics as well as the social-economic and cultural aspects of this relationship on the artisan fleet of the city of Vigia (Para) that fished with gill nets in the Amazon estuary from 1996 to 2001. Simple regression analyses were used to define the fishing-effort components (fishing-power and time) and to correlate fishing-effort and dolphin bycatch. Economical system and fish marketing dynamics analyses were used to define the interaction cost-benefit. The fishing-effort unit was FN x Hr and, the correlation between it and the dolphin bycatch was low, decreasing in the second period of this study. The trade of whole dolphins or their parts does not represent an important revenue factor for the fishermen and therefore should not be considered dangerous to the dolphin population. A new model of relationships among the variables of the fishery-dolphin system is presented.

Chapter 13 - This chapter describes a study conducted on the ecology of *Sotalia guianensis* in the Amazon estuary from 1999 to 2001, using participatory research with methodology. Interviews of 150 fishermen across 11 towns as well as surveys of the estuary by boat were completed to obtain information regarding *S. guianensis* in relation to their group size, habitat fidelity and calf-dynamics. Interactions between the ecological variables were tested using a log linear analysis of frequency tables for three factors. The results indicate that the *S. guianensis* is a gregarious species, forming groups of two or three individuals. However, groups with more than 10 individuals and herds of up to 150 were not rare. Group size was related to the behavior and kind of habitat used. In this study dolphins were commonly observed in large groups, feeding and swimming in open water habitats, however they were rarely observed in ports and near human communities. Habitats such as "igarapés", lagoons and exposed coastal beaches were visited by the dolphins in the last hours of rising tide, high tide and the beginning of receding tide, when depth facilitated the exploration of the habitat.

Chapter 14 - Molecular markers have the potential to disclose genetic variation and provide clues on macro and microevolutionary issues. The taxonomic and phylogenetic status of species lie within the realm of macroevolution while intraspecific matters, such as geographic population structure, social organization and mating system, pertain to microevolution. This chapter describes the findings on the molecular systematics and ecology of Sotalia dolphins, and is divided in two sections, each focusing on one of those topics. The first section shows how molecular markers have helped to settle the issue of species composition within the genus Sotalia – a matter of debate for over 140 years. To explain the controversy, a brief history of taxonomic changes in the genus since the first species descriptions is included. In addition, the section also makes phylogenetic considerations and discusses the timing of the speciation between the two accepted Sotalia species. The second section deals with the molecular ecology of Sotalia, presenting results and prospects of studies on population structure, phylogeography and social structure. Although many studies are still underway, some important findings have already been produced. The section also includes comments on new analytical developments that promise to widen our knowledge on those issues. The two sections close with a discussion of the relevance of results for the conservation and management of Sotalia species. At least two important results stem from molecular systematics and ecology studies of Sotalia dolphins, both with immediate application to their conservation. At the end of the chapter there is a presentation of the prospects for new discoveries in these fields in the near future.

Chapter 15 - Here the authors consider the phylogeography and population structure of the tucuxi dolphin *Sotalia fluviatilis*, based on samples (n = 26) collected across the Peruvian, Colombian and Brazilian Amazon Regions. Fourteen control region (CR) and two cytochrome *b* (Cyt-*b*) haplotypes were identified among these samples. The Amazonian population units identified showed high mitochondrial haplotype diversity and relatively high female mediated gene flow when compared to *Sotalia guianensis* and another Amazonian dolphin species, *Inia geoffrensis* throughout the sampled regions of the main river and its tributaries. A Union of Maximum Parsimonious Trees analysis generated a CR haplotype genealogy reflecting connectivity among sampled regions and identified divergent haplotypes found in the extremes of the species distribution. These results indicate the need to maintain connectivity between populations along the Amazon River and its tributaries as a main objective of management and conservation programs for *Sotalia fluviatilis*.

Chapter 16 - In the current chapter, I discuss some aspects of the life history and ecology of the Franciscana (*Pontoporia blainvillei*). This is a small dolphin which inhabits the coasts of southern Brazil, Uruguay and Argentina. Disappointedly, this species suffers an extensive loss of individuals each year due to mortality in fishing nets. Therefore, all available knowledge about the ecology of this species is useful for its conservation.

Chapter 17 - Franciscana's restriction to shallow coastal waters makes it highly vulnerable to anthropogenic threats. Habitat degradation (noise pollution, chemical pollution and overfishing) and loss affect many coastal cetacean species around the world. Nonetheless, incidental catches in fishing gear are believed to be the main threat to franciscana conservation. This chapter aims at providing a review about the main conservation issues for franciscana with emphasis on bycatch in fishing gear. It also discusses the species conservation status, the potential alternatives for minimizing incidental mortality in fisheries and the constraints for the effective establishment and implementation of conservation measures.

Chapter 18 - This chapter presents preliminary results on the distribution pattern of Yangtze finless porpoises (*Neophocaena phocaenoides asiaeorientalis*) in the Poyang Lake mouth area by using passive acoustic data-loggers at four different stations. Porpoise sounds were detected at all stations but their abundance decreased as the distance from the Yantze River increased. Porpoises were detected swimming both upstream to the Poyang Lake and downstream to the Yangtze River as well as between railway and highway bridges at the end of the lake. They were detected 13.9% of the total time monitored, and detected less frequently between 05:00 and 10:00 and between 15:00 and 18:00 during heavier shipping traffic. Also, there were relatively vacant periods between July 12 and July 28, 2007, and between August 5 and August 22, 2007, when virtually no porpoises were detected while there was a reversal of water current or increased water turbulence in the mouth area. These results suggest that movement and genetic communication between porpoise groups in the Yangtze River section and Poyang Lake might still remain, and therefore, the groups should be considered collectively, as a uniform unit for conservation. Bridge construction, shipping traffic, and current (turbulence and direction), might have affected the presence or movement pattern of porpoises in the study area and should be included in future conservation plans.

Chapter 19 - The Yangtze River is home to two endemic cetaceans, the baiji or Yangtze River dolphin (*Lipotes vexillifer*) and Yangtze finless porpoise (*Neophocaena phocaenoides asiaeorientalis*). Both cetaceans suffered great abundance reduction and range contraction during the last three decades. Baiji had at one point been abundant in the river, but in 2006 was declared likely extinct because an extensive survey conducted by a team of international scientists throughout baiji's geographical range failed to observe a single baiji. The latest abundance estimate of the Yangtze finless porpoise, based on data collected in the same survey is approximately 1,800 which indicates that one half of the population has vanished since 1991. It is because the baiji and the Yangtze finless porpoise share the same river and almost the same habitat, they also have been facing the same kind of threats, i.e. over- and illegal fishing, heavy boat traffic, water constructions and water pollution. The authors provide an analysis of the effectiveness of our conservation methods over the last three decades regarding three measures (in situ, ex situ and captive breeding). The authors also provide suggestions for the future protection of the baiji and Yangtze finless porpoise including, forbidding fishing in the river or at least in the current reserves, expansion of the current Tian-e-Zhou Oxbow Reserve and establishing new similar ex situ reserves, and intensifying the captive breeding program.

Chapter 20 - The Yangtze River dolphin or baiji, a freshwater cetacean found in the mid-lower Yangtze River and neighboring lake and river systems, experienced a precipitous population decline throughout the late twentieth century driven by unsustainable by-catch in local fisheries and habitat degradation. An intensive survey in 2006 failed to find any evidence that the baiji still survives, and the species is now highly likely to be extinct. Although considerable protective legislation was put in place from the late 1970s onwards in China, notably laws banning harmful fishing practices and the establishment of a series of reserve sections in the main Yangtze channel, regulations were difficult or impossible to enforce and *in situ* reserves proved unable to provide adequate protection for baiji. More intensive species-specific recovery strategies also received considerable national and international attention, with extensive deliberation for over twenty years about an *ex situ* recovery program that aimed to establish a translocated breeding population of baiji under semi-natural conditions. However, minimal financial or logistical support for this active baiji conservation strategy was ever provided by the international conservation community. A more dynamic international response is required if other threatened river dolphin species are to be conserved in the future.

Chapter 21 - The authors surveyed the sequence variability at exon 2 of the MHC class I and class II (*DRA* and *DQB*) genes in the baiji (*Lipotes vexillifer*) and finless porpoise (*Neophocaena phocaenoides*). Little sequence variation was detected at the *DRA* locus whereas considerable variation was found at *DQB* and MHC-I. Three exon 2 MHC loci of the baiji revealed striking similarity with those of the finless porpoise. Some identical alleles shared by both species at the MHC-I and *DQB* loci suggest that convergent evolution as a consequence of common adaptive solutions to similar environmental pressures in the Yangtze River. As for the *DRA* locus, the identical alleles were shared not only by baiji and finless porpoise but also by some other cetacean species of the families Phocoenidae and Delphinidae, suggesting trans-species evolution of this gene.

Chapter 22 - Herein the authors discuss the Ganges River dolphin (*Platanista gangetica gangetica* or susu) which inhabits the Ganges-Brahmaputra-Meghna and Sangu-Karnaphuli river systems of India, Nepal and Bangladesh. The chapter begins with a discussion of the origin, evolution, and phylogeny of the Ganges River dolphin as well as river dolphins in general. Also included are descriptions of past and present distribution patterns of the Ganges River Dolphin along with its anatomical structure, including primitive characters and morphological characters of interest. In the second section of the chapter the authors elaborate on Ganges River dolphin population surveys the authors conducted within a 500 km section of the Ganges River in the state of Bihar during 2005 to 2007. Both upstream and downstream surveys were performed three times per year. A significantly greater number of Ganges dolphins were observed per kilometer upstream compared to downstream surveys (1.28 versus 1.0 respectively) and the mean number of dolphins observed per upstream survey ranged from 559 to 808. Their results also support spatial and temporal variation of the Ganges dolphin population with for example a greater number of animals in confluence areas. These survey results are similar to those obtained from other Ganges River surveys that used similar methods. The chapter concludes with a discussion on the Ganges River dolphin's conservation status and major threats to its existence. Direct catch, incidental catch, pollution, and habitat degradation are all serious threats.

Chapter 23 - The fossil record demonstrates that in the past the "river dolphin" superfamily, Platanistoidea, was much more widespread geographically, and more diverse ecologically and taxonomically than it is now, and that most of its early members lived in salt water, not fresh water. Families in the Platanistoidea comprise a significant initial radiation of dolphin-like toothed whales (suborder Odontoceti). Platanistoids were predominant odontocetes in some late Oligocene and early Miocene age fossil assemblages, from approximately 25 to 15 million years ago. However, the Platanistoidea gradually declined in abundance and diversity approximately 15 million years ago, and they were gradually replaced, largely by another rapidly diversifying odontocete superfamily, the Delphinoidea. During their evolutionary histories, these two superfamilies have had an inverse relationship of diversity and abundance. Among the archaic groups of Platanistoidea, the essentially cosmopolitan Oligocene and Miocene family Squalodontidae is the most primitive dentally, having heterodonty (teeth still recognizable as incisors, canines, premolars, and molars), large and projecting anterior teeth, and serrated and broad-crowned cheek teeth, but well-telescoped crania with their nares moved posteriorly, and relatively primitive body skeletons showing them to have been medium-size whales compared to living species. The Miocene family Allodelphinidae comprises strictly marine North Pacific odontocetes that had primitive braincases, with relatively small nares, and very long and dorsoventrally flattened rostra and symphyseal portions of their mandibles, which contained numerous small teeth. Late Oligocene and Early Miocene marine members of the family Waipatiidae from the South Pacific and northern hemisphere were smaller than squalodontids, and had smaller teeth with less recognizable heterodonty. Platanistoids in the more derived clade that ultimately culminated in the recent family Platanistidae have a modified zygomatic process of the squamosal that is compressed from side to side. Within this clade of Platanistoidea, the Atlantic and Southern Ocean family Squalodelphinidae includes the more primitive, small to medium-size species that have tuberosities superior to the orbits that are not invaded by the pterygoid sinuses, and teeth that still retained remnants of heterodonty. The more highly derived Miocene to Recent family Platanistidae includes two named subfamilies, the Miocene Pomatodelphininae, and the Miocene to Recent Platanistinae. Species of the North Atlantic subfamily Pomatodelphininae are relatively large, long-snouted dolphins that had many small teeth, rostra and symphyseal parts of the mandibles that are compressed dorsoventrally, and many species in this subfamily, but not all of them, have enlarged bony tuberosities over the orbits that are invaded by extensions of the pterygoid air sinuses. These dolphins have been found in near shore marine, estuarine, and fresh water deposits, and these are the first indications of any fresh water-dwelling Platanistoidea. The more derived species of

Platanistidae, those in the subfamily Platanistinae, have fenestrations within the supraorbital crest caused by invasion of the pterygoid sinus, and have a transversely flattened rostrum and symphyseal part of the mandible. Miocene members of the subfamily Platanistinae are known from North Pacific marine deposits, but the living members of the genus *Platanista* live only in rivers of south Asia. A cladistic analysis provides a framework for a classification of the Platanistoidea that is presented here.

In: Biology, Evolution and Conservation of River Dolphins… ISBN: 978-1-60876-633-8
Editors: M. Ruiz-García, J. Shostell. pp. 1-28

Chapter 1

An Introduction to River Dolphin Species

Joseph Mark Shostell[1]* and Manuel Ruiz-García[2]+
1-Biology Department, Penn State University-Fayette, Uniontown, USA
2-Laboratorio de Genética de Poblaciones Molecular-Biología Evolutiva. Unidad de Genética. Departamento de Biología. Facultad de Ciencias. Pontificia Universidad Javeriana, Bogotá DC, Colombia

Abstract

This chapter introduces nine dolphin species (*Neophocaena phocaenoides asiaeorientalis, Sotalia guianensis, Sotalia fluviatilis, Pontoporia blainvillei, Inia geoffrensis, Inia boliviensis, Platanista gangetica, Lipotes vexilliter* and *Orcaella brevirostris*) that are discussed in the succeeding chapters of this book and provides brief summaries on each species' population status, habitat condition and looming threats. There are commonalities among the threats for these dolphins and they are linked to human activities. Fishing, dams, and pollution generally affect all of the species with those species near the highest human densities being the most threatened and having the bleakest future. There are of course bright spots in the conservation efforts for these species and some dolphins, such as *Inia geoffrensis*, seem to be faring well and have a large population size and great distribution. Also, we discuss recent and new contributions of molecular, morphological, and paleontological data that tremendously help our understanding of phylogenetic relationships and evolutionary history of these graceful creatures.

Keywords: *Neophocaena phocaenoides asiaeorientalis, Sotalia guianensis, Sotalia fluviatilis, Pontoporia blainvillei, Inia geoffrensis, Inia boliviensis, Platanista gangética, Lipotes vexilliter, Orcaella brevirostris*

* Joseph.Shostell@gmail.com.
+ mruiz@javeriana.edu.co; mruizgar@yahoo.es.

INTRODUCTION

For having a pure heart, God told the little boy that he would grant him any request. The boy said "I want to be wild. I want to be free. I want to dance. I want to jump. I want to play." God replied, "then I will make you a dolphin."

Dolphins are one of the few groups of species that have demonstrated self awareness (Marten & Psarakos, 1995; Reiss & Marinon, 2001) and they have a high level of cephalization that is unparalleled relative to other animal species groups save for primates. They play a key role in ecosystems as top predators and are indicators of high water quality evidenced by their lower abundance in polluted waters. Because human health is also dependent on good water quality, their health is indirectly and positively linked to our own. The significance of dolphins to humans is not new, rather, the integration of their history with our own has been chronicled for at least the last two thousand years. Ancient Chinese records dated 200 B. C. (see Wang & Zhao's chapter 19), indicate that early Chinese settlers new the difference between the baiji river dolphin and the Yangtze finless porpoise and that they hunted cetaceans for lamp oil, caulking and medicine (Pilleri, 1979; Hoy, 1923). Similarly, other indigenous peoples, such as in the Amazon found value in dolphins using them for ornaments and charms and also incorporating them into cultural stories that have been communicated across generations to the present day (Yañez, 1999). Beginning more recently, cetaceans have added economic value as an integral part of the ecotourism industry (Woods-Ballard et al., 2003). Sadly, dolphins and their habitat are threatened by a number of factors, most of which are caused by man.

The historical increase of the Earth's human population and urbanization of the environment has been met with an increased demand for energy (fossil fuels, hydroelectric) and other resources, many of which are limited. Our growing technological power and rapid development of commercial fisheries combined with a lack of caution and unwillingness to perceive the failures of intelligent, well-minded scientists of the past will most likely have undesirable outcomes for dolphins in the not so distant future. Habitat degradation and pollution (chemical and noise) have been on the rise for many years and intensive deforestation, overfishing, and global warming are negatively affecting dolphins and the ecosystems they inhabit. Increases in pollution runoff from developed coastal areas create influxes of sewage, toxins, chemicals and plastics, all harmful to dolphins. Nutrient uptake by bacteria and phytoplankton results in eutrophication and low oxygen concentrations, at times causing anoxic dead zones. Toxins, such as heavy metals, bioaccumulate up the food chain and can have severe and deadly consequences for normal physiological functions of long-living top-predators like dolphins. Other links of human population growth to dolphins appear to be elevated pathogens along coastal waters. For example, morbillivirus, linked to pollution, has infected and killed thousands of dolphins in the early 1990s and again more recently (Raga et al., 2008; Osterhaus et al., 1995; Domingo et al., 1990). The largest and probably most obvious culprit of dolphin mortality is due to the fishing industry. Tens of thousands of dolphins are incidentally captured in fishing nets ending up as bycatch each year.

True river dolphins as well as marine dolphins that frequent freshwater systems are large animals that have traditionally gone unnoticed by the general public and, in a certain sense, by marine mammal specialists as well. In fact, only a limited number of researchers have investigated the biology of these dolphin species. This is quite surprising given that these species are commonly the top predators in their habitats. Unfortunately, river dolphins are extremely sensitive to environmental change, either natural or anthropogenic (Da Silva, 1995). The apparent extinction of the baiji, (see Wang & Zhao's and Turner chapters; chapters 19 and 20) the Chinese river dolphin, has been a wake up call for much of the general public as well as the scientific community and has intensified man's interest in these incredible and intelligent creatures. Much of our biological knowledge of river dolphins is based on a text published 20 years ago (Perrin et al., 1989). At that time, the major part of the biological knowledge concerning these animals was limited to basic ecological and census studies and some morphological and physiological data relating to taxonomy. But in the last 20 years, genetics and molecular biology have revolutionized our understanding of biology and evolution. Now for the first time, revolutionary molecular techniques are being applied to answer evolutionary reconstruction questions of many animals, including river dolphins (Cassens et al., 2000; Hamilton et al., 2001; Banguera-Hinestroza et al., 2002; Yang et al., 2002, 2005; Verma et al., 2004; Cunha et al., 2005; Caballero et al., 2007; Ruiz-García, 2007; Ruiz-García et al., 2007, 2008 and many chapters included in this book). In addition, new paleontological records are dramatically changing our perspective about the relationships of these dolphins with each other and with other cetaceans and yet, no book has incorporated these fascinating new discoveries (see the Cozzuol and Barnes et al., chapters; chapters 10 and 23). To meet this informational gap, we contacted the world's premier river dolphin specialists from Columbia, Brazil, Bolivia, Argentina, New Zealand, The United States of America, China, England, and India and asked them to contribute chapters to this updated river dolphin book. Moreover, this book provides new census information and important ecological characteristics of the river dolphins *Inia*, *Sotalia*, *Pontoporia*, and *Lipotes* and presents molecular and genetics results of these dolphin species. A compilation of these data is essential if we are to present a strategic conservation plan for these animals. Upon being informed of critical evolutionary history data, conservation biologists will be able to tailor their conservation efforts for each threatened river dolphin species. Additionally, new morphological data and the new discoveries in the fossil record for river dolphins are presented in this book that graduate students, professors, scientists, evolutionary ecologists, aquatic mammalogists, population ecologists, conservation ecologists, and marine biologists should find valuable for the foreseeable future. Introductory descriptions of each of the species covered in this book, along with field photos taken by the book's authors are provided in the following sections. We, the Authors and Editors, hope that you utilize the provided data, and, if not already, become active members in the conservation of river dolphin species.

Neophocaena phocaenoides asiaeorientalis (Finless Porpoise)

As its name states the finless porpoise does not possess a dorsal fin, but instead has a series of tubercles (Figure 1). Their flippers are relatively large and their overall body color is dark grey to black. It is also one of the smallest odontocetes with a maximum length for males of 1.9 m with males being slightly longer than females. Average adult weight ranges from around 30 to 45 kg (Wang et al., 2005). The subspecies of the finless porpoise in China (*Neophocaena phocaenoides asiaeorientalis*) is only found in the Main channel of the Yangtze River and Poyang Lake. Until recently, it seemed to prefer the confluence area of the Yangtze River and Poyang Lake where it historically had congregated in relatively large

A

B

Figure 1. Two images (a & b) of Neophocaena phocaenoides asiaeorientalis (copyright Dr. Wang).

numbers. The confluence can be a highly productive area with elevated nutrient concentrations and abundant dolphin prey. Observations by scientists support that the finless porpoise swims alone, in small groups, or even in larger groups of up to fifty individuals. Unfortunately, its numbers have dropped considerably to the current estimate of approximately 1,200 individuals and the species is listed as endangered by the International Union for the Conservation of Nature and Natural Resources (IUCN). The finless porpoise is affected by sand digging, sand-transport ships, fishing, pollution, bridges and dam construction. In the present book, Li et al.,(chapter 18) discuss how the population density decreases near construction areas and Wang and Zhao (chapter 19) provide an analysis of the possible effectiveness of several conservation methods over the last three decades regarding in situ, ex situ and captive breeding. Each of these negative factors contribute to group fragmentation and genetic isolation of this species.

Taxonomy: of *Neophocaena phocaenoides*

Kingdom: Animalia
Phylum: Chordata
Class: Mammalia
Order: Cetacea
Suborder: Odontoceti "toothed whales"
Family: Phocoenidae (porpoises)
Genus: Neophocaena
Species: *Neophocaena phocaenoides* (G. cuvier 1829)
Subspecies: *Neophocaena phocaenoides asiaeorientalis*

Sotalia guianensis (Marine Tucuxi Dolphin)

Marine tucuxi dolphins are small bodied delphinids with a maximum body length of about 2 meters (Figure 2) (Barros, 1991). Their dorsum is dark gray, while their ventral area is gray, white, or pinkish. They have a poorly developed lateral stripe, moderately long and slender beak, a triangular dorsal fin, and great muscle mass.

Figure 2. A *Sotalia guianensis* playing in Manzanillo, Costa Rica. (copyright Susana Caballero).

It's a K-selected species, with a slow growth rate, long life span, birth intervals of greater than one year (Fernando et al., 2002) and have a lengthy pre-sexual maturity period (6-7 years). *Sotalia guianensis* seems to prefer shallow waters (Edwards & Schnell, 2001) and has an Atlantic coastal distribution range that includes both Central and South America from Honduras to southern Brazil (da Silva & Best, 1996; Simões-Lopes, 1988), but may be also found hundreds of kilometers inland within the Orinoco River (Boher et al., 1995) and in the Amazon Estuary. They have relatively high densities for small cetaceans at 0.9 to 8.6 animals per squared kilometer (Vidal et al., 1997) and their group structure varies across its geographical range from small groups of one to three individuals (Monteiro-Filho, 2000; Filla & Monteiro-Filho, 2009) to extremely large groups in the hundreds (Lodi & Hetzel, 1998). Within the Amazon Estuary alone they are estimated to be over 92,000 tucuxi dolphins (see Beltran-Pedreros & Petrere's chapter). They seem to congregate in productive prey areas and tend to avoid areas of heavy human activity where their habitat has been degraded (Azevedo et al., 2007). Currently the IUCN classifies this species within the category of "insufficient data". The two largest threats for this species are the fishing industry and new dam construction. Bycatch accounts for losses of over 4,000 tucuxi dolphins every year just within the Amazon Estuary and seems to mostly affect solitary or small dolphin groups. All three of Beltran-Pedreros et al.'s chapters (11, 12, & 13) in the current book, discuss the bycatch incidence problems associated with marine tucuxi. It is easy to find tucuxi skulls, dried eyes and sexual organs (vaginas and penises) in some Amazon Estuary markets, such as the ver-O-peso market at Belém (Pará state, Brazil). For example, one of the authors (Ruiz-García) obtained more than 200 vaginas and penis samples in only a few hours (July 2005). One afro-Brazilian "witch" told him that she could obtain around 500 new vaginas and penises in one or two days if he was interested in additional samples. The author did not accept the proposal. These tissues are basically used for love charms and originate from tucuxi in bycatch.

Taxonomy: of *Sotalia guianensis*

Kingdom: Animalia
Phylum: Chordata
Class: Mammalia
Order: Cetacea
Suborder: Odontoceti "toothed whales"
Family: Delphinidae
Genus: *Sotalia*
Species: *Sotalia guianensis* (van Beneden, 1864)

Sotalia fluviatilis (Tucuxi Dolphin, Gray Dolphin, Bufeo-Negro, Bufeo-Gris)

Sotalia fluviatilis (Figure 3), the only exclusively freshwater dolphinid in the world (Cunha et al., 2005), has a structure similar to that of the coastal tucuxi dolphin (*Sotalia guianensis*), but is smaller in overall body size (Barros, 1991; da Silva & Best, 1996) with the maximum body length of *S. guianensis* being about 36% larger than that of *S. fluviatilis* (1.52 m). It wasn't until articles were published in 2002 (Monteiro-Filho et al.), 2005 (Cunha et.), and 2007 (Caballero et al.) were the marine (*S. guianensis*) and riverine *sotalia* (*S. fluviatilis*) forms classified as different species, a taxonomic decision that was based on morphological and molecular data. Cunha et al. (this book; chapter 14) suggests that this split occurred during the Pliocene, 2.5-5 million years ago (also see Caballero et al.'s phylogeographical analysis in chapter 15). The riverine form has a dark gray dorsum and a white or pinkish ventral area. It also possesses a moderately long slender beak, small rounded melon, large pectoral fins, stocky body and a triangular dorsal fin (Jefferson et al., 1993)

Figure 3. A *Sotalia fluviatilis* at the mouth of the Napo River in the Peruvian Amazon (copyright Pablo Escobar-Armel).

Sotalia fluviatilis is distributed throughout most of the Amazon River and its tributaries (Da Silva & Best, 1994) and have also been sighted in the Orinoco River (Borobia et al., 1991; Boher et al., 1995). There distribution extends from Brazil to Peru, Ecuador and Columbia (da Silva & Best, 1996). They demonstrate strong site fidelity but there appears to be high levels of gene flow among their Amazonian population units compared to those of *S. guianensis* (Caballero, 2006). *S. fluviatilis'* main threat is entanglement in gillnets (Trujillo et al., 2000) and is the most accidently captured dolphin in some Amazonian rivers (Barros & Teixeira, 1994; Siciliano, 1994). Other factors including habitat destruction, oil and pesticide pollution (Monteiro-Neto et al., 2003; Yogue et al., 2003), heavy metal contamination (Best & Silva, 1989), dam construction (da Silva & Best, 1996) and direct killing for specific organs (da Silva & Best, 1996; Siciliano, 1994) in combination with the problem of bycatch and growing coastal development have led some researchers (Barros & Teixeira, 1994) and the country of Ecuador (Tirira, 2001) to consider it endangered. Currently, IUCN lists *Sotalia fluviatilis* in the category of "data deficient". This species lives in the sympatric Amazonian area together with *Inia*, but does not have a mythological Indian tradition as developed as that of *Inia* and their tissues are not frequently sought after for love charms. Additionally, *Sotalia fluviatilis* is a much faster and more efficient swimmer than *Inia*, and also possesses amazing and superior agility. Therefore, fishermen find them comparatively, difficult to hunt and they are not used as bait to attract small catfishes, like the "mota" or "mapurito" (*Calophysus macropterus*) as in the Orinoco and in the Amazon rivers (Colombia and Brazil, mainly).

Ruiz-Garcia's anecdotal field observations support these claims. For example, during a population genetics and phylogeography study of river dolphins in the Amazonian river-system, Ruiz-Garcia intended to capture river dolphins. Even with expert Indian fishermen and with a 10-meter long wooden boat powered by a 40 horse power engine, only two *Sotalia* specimens were captured (in little lagoons of the Curaray and Samiria rivers at the Peruvian Amazon) compared to 200 *Inia*. In another example, Ruiz-Garcia's team used large nets to encircle mixed groups of botos and tucuxis (around seven to nine botos and four to six tucuxis) in the Napo and Curaray rivers. All the botos were caught, but incredibly, in the last instant, when the nets were perfectly closed, all the tucuxis escaped.

Taxonomy: of *Sotalia fluviatilis*

Kingdom: Animalia
Phylum: Chordata
Class: Mammalia
Order: Cetacea
Suborder: Odontoceti
Family: Delphinidae
Genus: *Sotalia*
Species: *Sotalia fluviatilis* (Gervais & Deville, 1853)

Pontoporia blainvillei (Franciscana, La Plata Dolphin)

Pontoporia blainvillei is a small dolphin species that has a gray, brown, to dark-yellowish dorsum that is comparatively darker than its flanks and ventral region (Figure 4). Key characteristics for this species include the fluke width to body length ratio (greater than 1:4) (Brownell, 1989) and an elongated, slender rostrum which is the longest of any dolphin relative to its body size. Both its broad flippers and dorsal fin have rounded tips and its dorsal fin is triangular and tall. They have small eyes, rounded head and a mouth that contains 250 small, sharp teeth (Bastida et al., 2007). And, similar to other river dolphin species it has unfused cervical vertebrae. It has a short life span (usually less than 12 years) (Pinedo, 1991) compared to other cetaceans and reaches sexual maturity quickly in two to three years (Kasuya & Brownell, 1979) and has birth intervals of approximately 1.5 years. Sexual dimorphism exists with female franciscanas slightly larger than males (1.53 versus 1.35 meters). Two geographical body forms exist for this species with smaller forms in the northern range and larger forms in the coastal waters (Pinedo 1995). There are no-overall population estimates at this time for this dolphin species, but limited surveys show that densities vary from 0.056 to 0.657 ind/km2. Their distribution range consists of the southern Brazil, Uruguay and Argentina coasts, estuaries (Santos et al. 2009) as well as the La Plata River and Babitonga Bay Estuaries in Uruguay (Cremer & Simoes-Lopes, 2005). Franciscanas are not continuously distributed across their range and studies indicate that their populations are genetically distinct from each other (Lazáro et al., 2004). They prefer turbid waters shallower than 30-35 meters deep (Pinedo et al., 1989; Danilewicz et al., 2009) and are considered an opportunistic predator that feed on small fish cephalopods and crustaceans (Danilewicz et al., 2002). The major threat for this dolphin species is mortality due to fishing, which exceeds population growth rates. There is mounting evidence that habitat degradation (Danilewicz et al., 2002) and chemical pollution (Kajiwara et al., 2004) have fueled the franciscana population decline, although further studies need to focus on this connection. *Pontoporia blainvillei* is listed as vulnerable by the IUCN because of its projected decline in numbers (Secchi 2006). Diminished fish stocks due to overfishing further exacerbate the situation (Haimovici, 1998; Bassoi & Secchi, 2000). The ecology, life history, threats and conservation status of franciscanas are reviewed by Secchi in chapters 16 and 17.

Figure 4. *Pontoporia blainvillei* (copyright Ricardo Bastida and Eduardo Secchi).

Taxonomy: of *Pontoporia blainvillei*

Kingdom: Animalia
Phylum: Chordata
Class: Mammalia
Order: Cetacea
Suborder: Odontoceti "toothed whales"
Family: Iniidae
Genus: *Pontoporia*
Species: *Pontoporia blainvillei* (Gervais and d'Orbigny, 1844)

Inia geoffrensis (Amazon Pink River Dolphin, Bufeo, Bugeo & Boto)

Inia geoffrensis is the largest river dolphin species reaching an average full body length of 2.6 m and weighing 160 kg. They have a long beak and may be a pink, bluish-gray or even white color (Figures 5a-f).

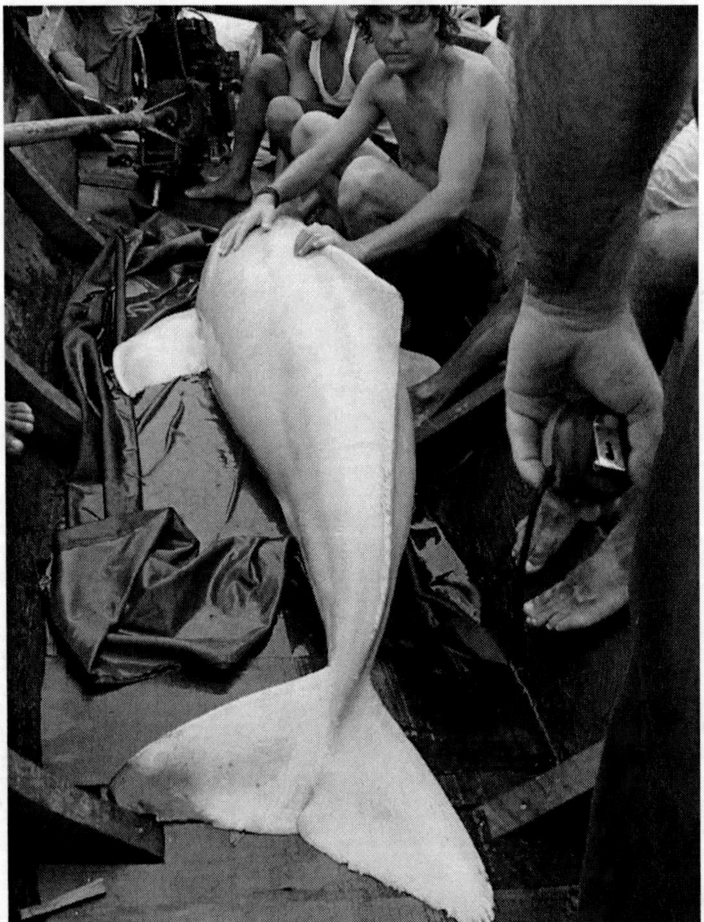

A

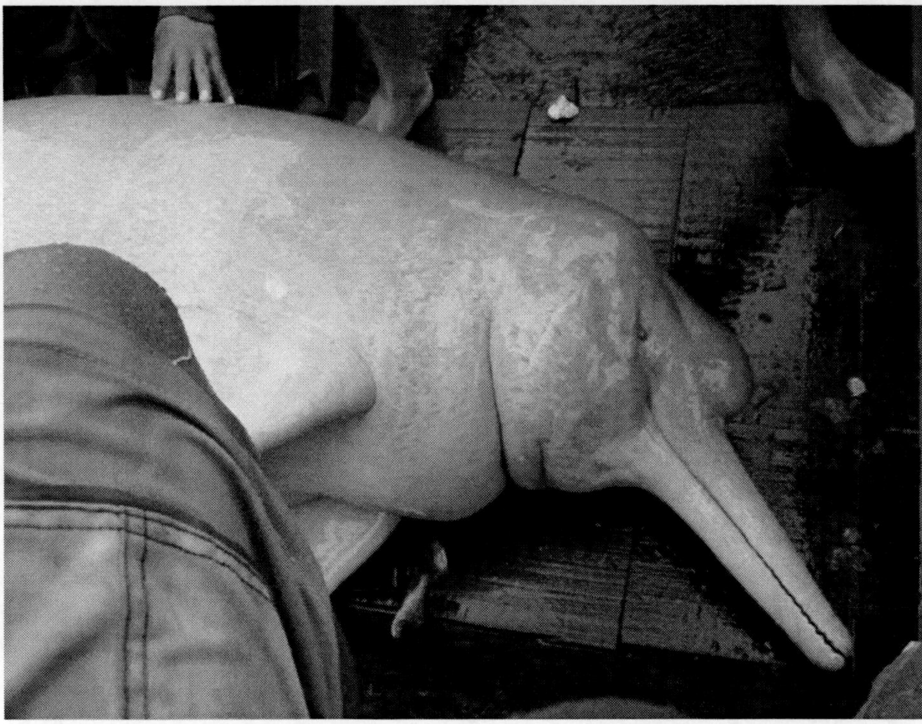

B

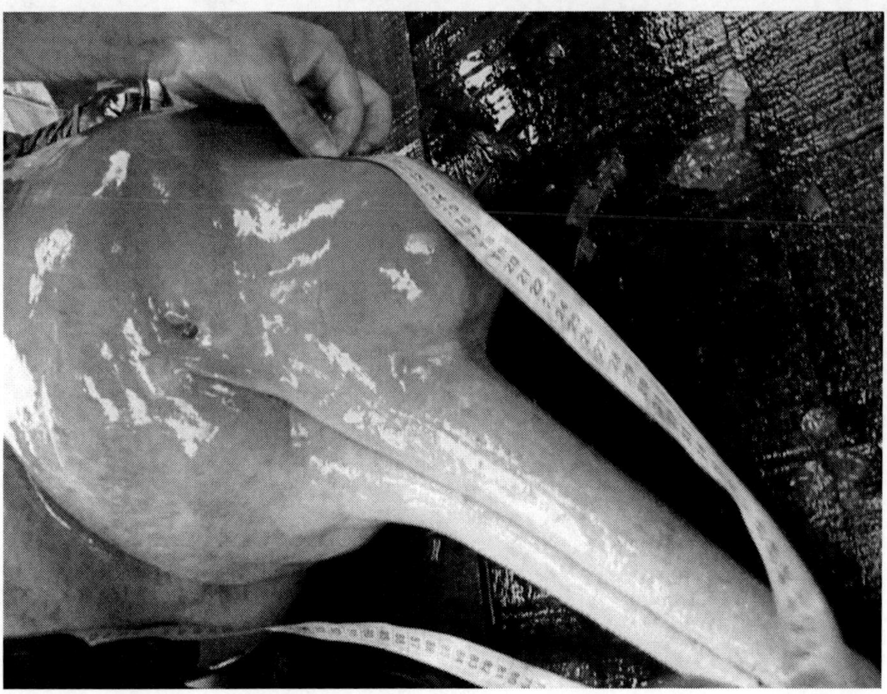

C

D

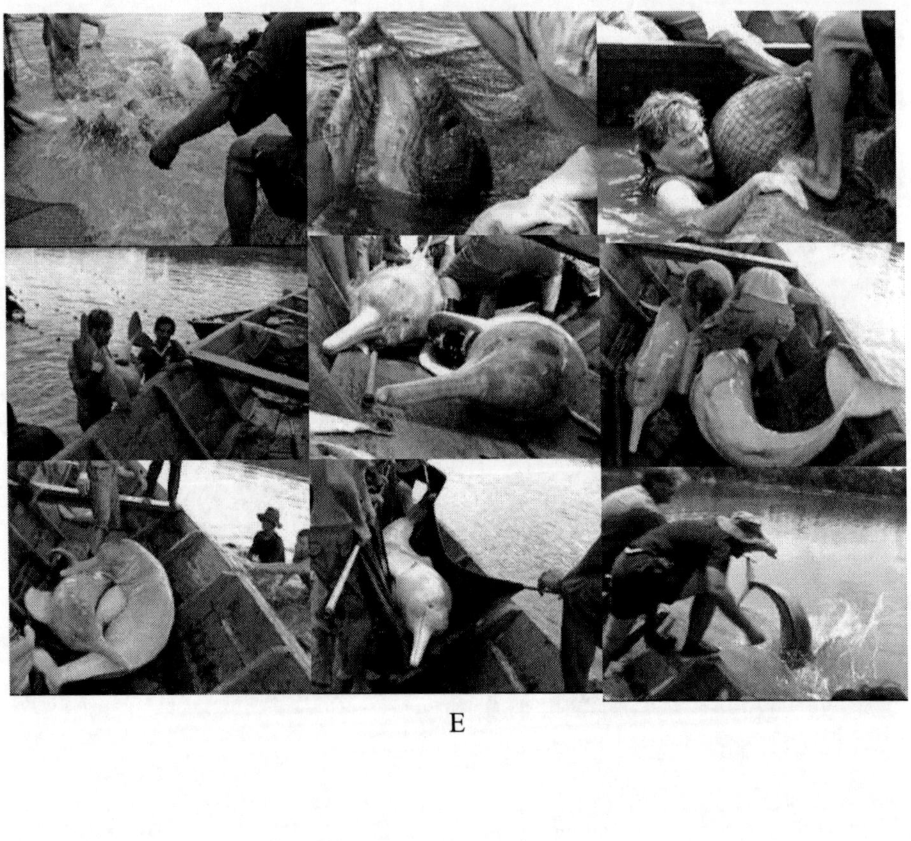

E

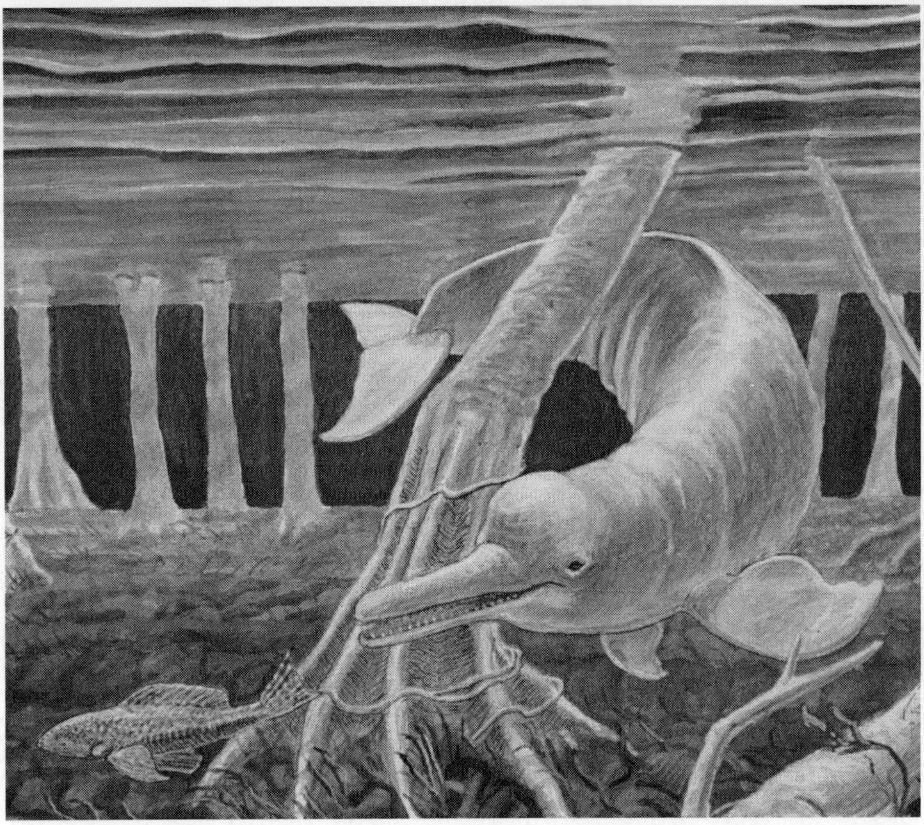

F

Figure 5. Four photos of *Inia geoffrensis* (A-D) captured at the Tapiche River and at Yanallpa cocha in the Ucayali River (Peru) by Ruiz-García's team in 2002-2003 and a mosaic of photos (E) of boto captures at the Napo, Curaray, Napo and Samiria Rivers in 2003 by Ruiz-García´s team (copyright Diana Alvarez & M. Ruiz-García). (F) An image of an *Inia* fishing a "carachama" (copyright Tim Smith).

Similar to the genera *Lipotes* and *Platanista*, *Inia geoffrensis* has unfused cervical vertebrae that can permit an increase of mobility in the vertebral area compared to other odontocetes (see Smith & Burrow's chapter 4). It has the largest population of all the river dolphins and is widely distributed in the Amazon and Orinoco basins (Best & da Silva, 1989) as well as in the upper Madeira River (da Silva, 1994), all in northern South America. Its range includes the countries of Bolivia, Brazil, Columbia, Ecuador, Peru, and Venezuela (and possibly some southern rivers of Guyana) (da Silva, 1994). Restricted to freshwater, the quality and quantity of the pink river dolphin's habitat is dependent on the season (rain versus dry) and will shrink considerably during the dry season. They have been observed commonly in main river channels, lagoons, and confluence areas (see McGuire & Aliaga-Rossel's chapter 2; Best & da Silva, 1993; Aliaga-Rossel, 2002) mostly alone or in pairs. At times they have been known to travel hundreds of kilometers (Martin & da Silva, 2004), but they seem to have a great fidelity and eventually return. Densities of pink river dolphins per linear km vary and range between 0.21 to 1.55 individuals (Magnusson et al., 1980; Schnapp & Howroyd, 1992; Aliaga-Rossel, 2002) with their greatest densities expected where their characin prey are abundant (da Silva, 1983). There appears to be little gene flow among different breeding populations and there are several different geographic genetic lineages of this species living together within larger populations (see Ruiz-García's chapter 8 & Vianna et al.'s chapter 6). In chapter 8, the bufeo is described for the first time in analytical detail, and the micro-genetic structure of populations within the Napo and Curaray rivers of the Peruvian Amazon are discussed with the help of DNA microsatellite analysis. The results support that this river dolphin has a social reproductive system. In fact, on January 1, 2003, Ruiz-García´s team captured a female in the Tiphisca Paucar lagoon (Canal del Puhinauva, Peruvian Amazon). Upon analysis of her constitution in a wooden research boat, the research team noticed that she exhibited some health problems and the team decided to liberate her in another section of the river that might be more conducive to her health. Immediately after she was brought on board, a group of five or six botos persecuted the boat with aggressive behavior and emitted intense vocalizations. They swam circles around the boat until the researchers released the female. The scene was incredible to experience and can be typical of a species that displays cooperative and social behavior. In chapter 9, Ruiz-García´s provides a new explanation of river colonization by *Inia* that has the central Amazon River as the probable origin of its expansion.

The pink river dolphin is probably the most secure of all the river dolphin species. Still, there are multiple threats to this species such as incidental catch, construction of dams (da Silva, 2002), direct hunting for genital organs (Best & da Silva, 1989), pesticides and mercury (Rosas & Lehti, 1996). Scientists are attempting to obtain new information to help conservation biologists by the analysis of carcasses that were the result of threats such as incidental capture. For example, Castellanos-Mora et al.'s chapter 5 addresses how the age of accidentally killed dolphins can be determined by craniometric and morphometric analyses, a relationship that has not been well-studied in cetacean species. Other threats to this species include freshwater pathogens. Martinez-Agüero et al.'s chapter 7 discusses the class II DQB major histocompatibility complex (MHC) gene of pink river dolphins and how it exhibits a high degree of polymorphism which may suggest an adaptive response to these pathogens. This species is classified as vulnerable by the IUCN and listed in appendix II of the Convention on International Trade in Endangered Species of Wild Flora and Fauna (CITES).

As mentioned previously, although this species of river dolphin has a wide distribution range, and is probably the least threatened of the strict river dolphins, it is never-the-less still threatened by several anthropogenic activities. Here are two brief field accounts by Ruiz-García's research team that attest to these threats. In October 2003, they obtained three decomposed carcasses of *Inia geoffrensis* at the Javarí River in the Peruvian and Brazilian Amazon frontier. The animals, three large males, were partially eaten by "mota" and had been deposited in the Peruvian bank of this river by Brazilian "colonos" fishermen. These dolphins were purposefully killed and left to rot to attract small catfishes (mota"-*Calophysus macropterus*), an event all too common in some Colombian and Brazilian Amazon rivers. In another incident, Ruiz-Garcia interviewed a "witch" who commented that *Inia* vaginas were employed for love charms. The witch stated "a little piece of the vagina is cut off and heated in water until ebullition (or in fire). During this process, the fat of the piece is obtained and mixed with aromatic alcoholic substances. A woman who desires to marry a specific man, a man not cooperating with the plan, touches the prepared concoction on three points of the man's face without his knowledge. In two or three months, the woman and the man will be married." Demand for dolphin parts, such as these by some indigenous people, bolster the dolphin market. Dolphin oil is even used as a form of pulmonary medicine (this practice extends into the Mamoré River in Bolivia).

It is noteworthy to remark that the myths around this genus are the same throughout the entirety of the Amazon basin with its thousands of kilometers. One such myth was recounted to Ruiz-Garcia in Alejandria and Exaltación (Mamoré River, Bolivia), Yarinacocha, near Pucallpa (Ucayali River, Peruvian Amazon), in Requena (Tapiche River; Peruvian Amazon), in Pompeya (Napo River, Ecuadorian Amazon), in San Francisco (Loreto-Yacu River, Colombian Amazon) and in villages near Manaus and Santarem (central Brazilian Amazon). People at each location stated their belief that pink river dolphins can transform themselves into elegantly dressed white men that don hats to cover their spiracles. They become excellent dancers and enamor young ladies at parties they crash. With cunning and a beautiful body, each "man" entices an infatuated young lady to the river and then transports the innocent female to the "dolphin city" under the water. The young lady is later returned to her village and nine months later, a baby is born. The father is unknown, but the new mother knows that the father was a bufeo. It is also interesting to remark on the intense sexual attraction that Indigenous people of the Amazon have for bufeos. The fishermen, that were part of Ruiz-García´s team in the areas where boto were captured, were very reluctant to touch the dolphins because they believed that if they touched the animals, they would incur illness. Nevertheless, they still looked for a dolphin's sexual area, especially in females, because their vulva is very similar to that of the human female. Also, in some areas of the Amazon, Ruiz-García listened to accounts of Indians sexually assaulting *Inia* females and raping them. Some of the *Inia* females were tied to trees and raped by a group of Indians for several hours.

In another belief, botos can sequester and kill their enemies because they are "yacurunas" or water spirits (in quechua) as is the Anaconda and they have the power to hold the spirit of their enemies in their sub-aquatic town. This myth, like some others, exists among the mixed Indian and Caucasian colonist communities ("colonos"). Ruiz-García interviewed two Caucasians, one in Bretaña (Canal del Puhinauva in the mouth of the Pacaya River, Peruvian Amazon) and another in San Francisco (Loreto-Yaku River at the Colombian Amazon). Both related the same story, that they personally observed a boto transform himself into a man during a moon-lit night. But, when he was discovered he returned to the river and again transformed himself into a boto. The first man even shot at the boto when it returned to the water, but missed. Therefore, certain species are not only important from a biodiversity or from a genetic diversity point of view, but also have cultural and mythical value. Bufeo is one of these species. Figure 6 contains a photo of bufeo's mythical form.

Taxonomy: of *Inia geoffrensis*

Kingdom: Animalia
Phylum: Chordata
Class: Mammalia
Order: Cetacea
Suborder: Odontoceti "toothed whales"
Family: Iniidae
Genus: *Inia*
Species: *Inia geoffrensis* (de Blainville 1817)

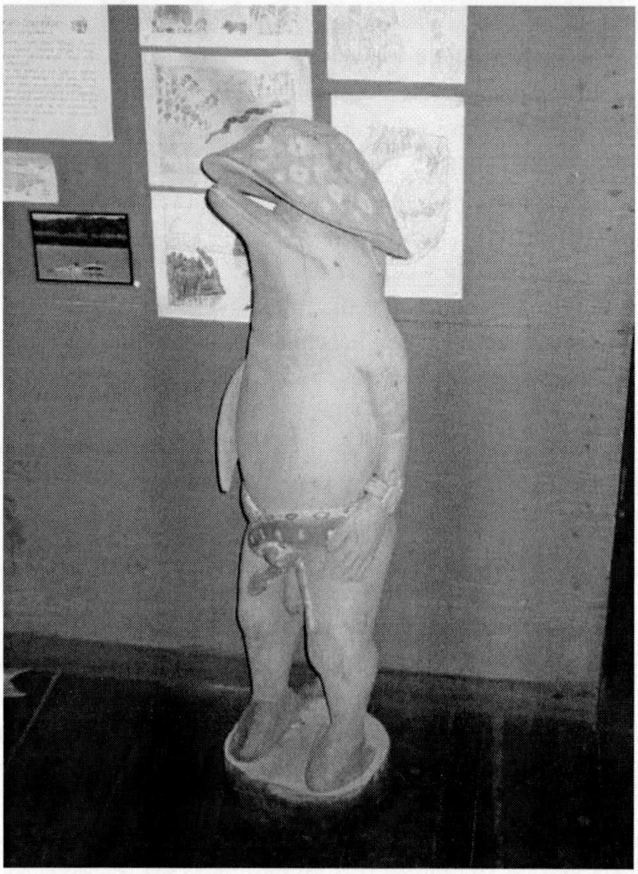

Figure 6. A mythic form of the *Inia geoffrensis* at the Colombian Amazon (copyright D.Alvarez & M. Ruiz-García).

Inia boliviensis (Bufeo, Pink River Dolphin)

Inia boliviensis and *Inia geoffrensis* were only recently classified as separate species based on molecular and morphological studies (Hamilton et al., 2001; Banguera-Hinestroza et al, 2002; Ruiz-García et al., 2006, 2008). Comparatively, *Inia boliviensis* (Figures 7a & b) is similar in color and in many characteristics to *Inia geoffrensis*, but it has more teeth, fewer phalanges per manus, a wider rostral incision on its sternum, as well as other different features (see Ruiz-Garcia's chapter 9). Ruiz-Garcia's chapter 9 also presents newly collected mitochondrial DNA data of *Inia geoffrensis* and *Ina boliviensis* that suggests that the two *Inia* forms separated from each other only approximately 150,000 years ago, much more recent than had been previously claimed. *Inia boliviensis* inhabit Bolivian rivers within the Cochabamba, Santa Cruz, Beni, and Pando areas of the Amazon (see Aliaga-Rossel's chapter 3) and is geographically isolated by 400 km of waterfalls and rapids from other *Inia* populations in the Amazon's main stem. Their highest densities occur during rising and high water (Aliaga-Rossel, 2002) and they have been observed in oxbow lakes, lagoons and rivers and move between these areas even in times of low water. During high water they swim into inundated areas and ephemeral rivers (Aliaga-Rossel, 2002; Aliaga-Rossel & Quevedo,

2008). Similar to *Inia geoffrensis*, *Inia boliviensis* is a pisivore and has a distribution that is based mainly on food availability (Aliaga-Rossel, 2003) and it can travel fairly long distances (60 km) in a single day, but has a tendency to remain in a single location. The waterfall barricade has supported a low genetic richness of the Bolivian population relative to other *Inia* populations (see Martínez-Agüero et al.'s chapter 7) and consequently it is more vulnerable.

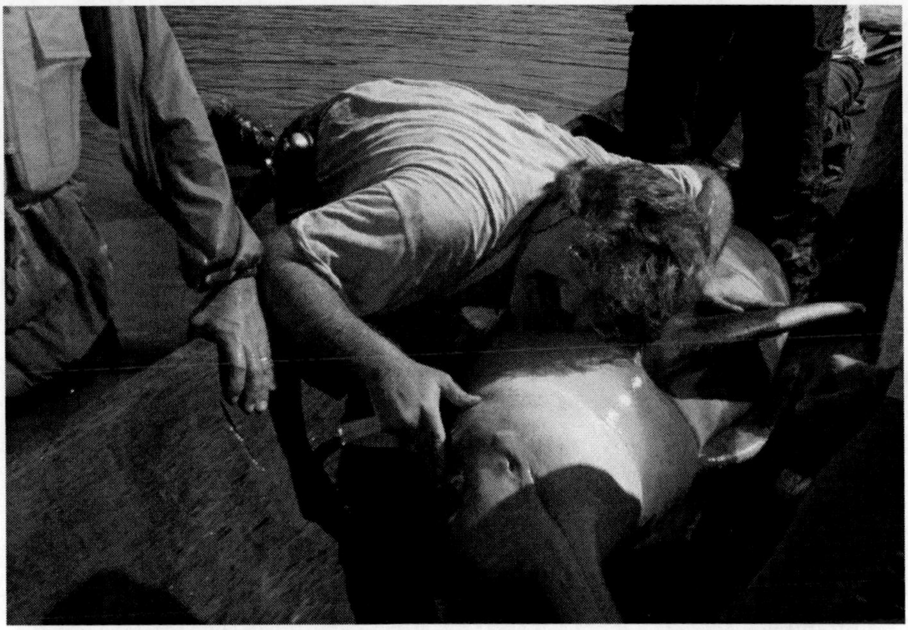

A

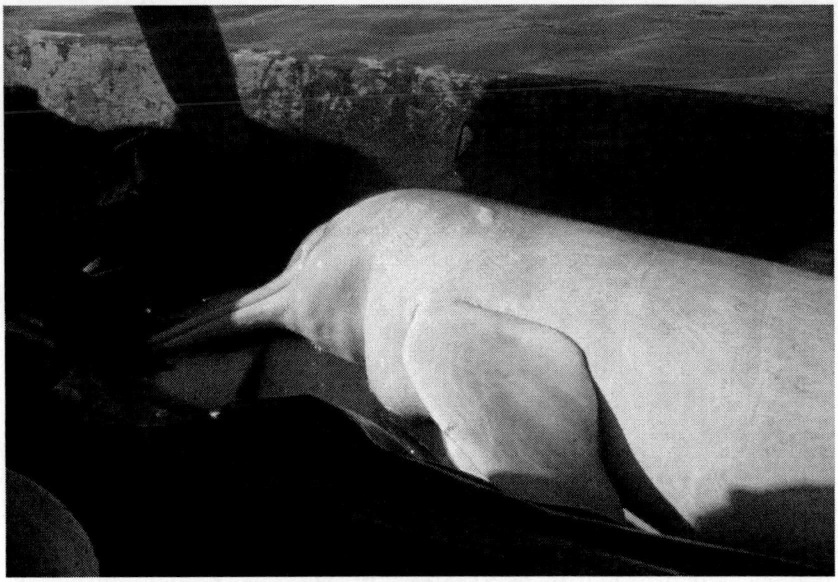

B

Figure 7. Two photos of *Inia boliviensis* captured at the Mamore River and affluents (Bolivia) by Ruiz-García´s team in 2003 (copyright D. Alvarez & M. Ruiz-García).

The major threat to *Inia boliviensis* is entanglement in fishing nets (Aliaga-Rossel, 2002), although commercial fishing is not as extensive as in adjacent countries. However, Ruiz-García observed that some Bolivian Indians occasionally consume *Inia* (California in one affluent of the Guapore River). Other threats include deforestation (Sayer & Whitmore, 1991), boat traffic (Pillieri & Gihr, 1977), phosphorus loading (Maurice-Bourgoin, 1999), mercury from mining activities (Dolbec et al., 2001), construction of hydroelectric dams (da Silva, 1995), gas exploration, and overfishing. The existence of a severe bottleneck determined by microsatellite analysis makes it difficult for *Inia boliviensis* to effectively deal with these anthropogenic factors. It's unfortunate that no conservation law in Bolivia specifically targets this species (although the Beni Department in the past year, 2008, declared the Bolivian pink river dolphin as emblematic and a protected species), but there are four protected areas such as the Beni Biological Biosphere Reserve that provide some shelter. As a group, *Inia* are listed as vulnerable by the International union for Conservation of Nature and Natural Resources. There is still of paucity a data for this species including a lack of population size estimates. Aliaga-Rossel's chapter 3 addresses research and conservation priorities for this species.

Taxonomy: of *Inia boliviensis*

Kingdom: Animalia
Phylum: Chordata
Class: Mammalia
Order: Cetacea
Suborder: Odontoceti "toothed whales"
Family: Iniidae
Genus: *Inia*
Species: *Inia boliviensis* (1817)

Platanista gangetica gangetica (Susa, Ganges River Dolphin, Water-Hog, Shushuk)

Platanista gangetica within the Platanistidae family is the only extant species of a once abundant and diversified Platanistoidea superfamily. Barnes et al.'s chapter 23 discusses the evolutionary history and phylogenetic relationships of this superfamily covering Platanistidae and the extinct families Allodelphinidae, Squalodontidae, Waipatiidae, and Squalodelphinidae. There are two subspecies of the Ganges River dolphin (*Platanista gangetica gangetica* and *Platanista gangetica minor*) but molecular phylogenetic studies support that they are quite similar (Guang & Kaiya, 1999) and therefore they are of the same species (see Sinha et al.'s chapter 22). General characteristics of the Ganges River dolphin are its brown color, long-pointed snout, and extremely small, pin-hole eyes (Figure 8). Adults have a body length of approximately 2.0-2.2 m and 2.4-2.6 m for males and females respectively, with an average weight of 70-90 kg. Similar to other river dolphins, their distribution pattern is partly due to the abundance of their prey: small fish, crustaceans, and snails (Sinha, 2006).

Their distribution is non-continuous within the Ganges-Brahmaputra-Meghna and Karnaphuli-Sangu River systems in India, Bangladesh and Nepal (Sinha et al., 2000) and have mostly been observed as solitary animals with surveyed densities of 0.76 to 1.36 dolphins per km (Smith et al., 2001). During the monsoon season they inhabit confluences and complex habitat areas while in the dry season they remain in river channels and deep tributary pools. Total estimated numbers for Ganges River dolphins are from 1,000 to 3,000, significantly less than the estimated 4,000-5,000 in 1986 (Mohan, 1989).

Figure 8. Some images of *Platanista gangetica gangetica* (copyright Dr. Singh).

Human developments, severe pollution, over fishing, habitat degradation, alteration of sedimentation, and hydrologic changes, all pose serious threats to the Ganges River dolphin (Dudgeon, 2000; Sinha, 2006). Elevated heavy metal and Butyltin compound concentrations, direct hunting (Sinha, 2002), and accidental bycatch are additional threats (Kannan et al., 1993; Kannan et al., 1997). Moreover, barrages and dams have confined and isolated Ganges River dolphins as well as reduced water flow (Mohan, 1989; Reeves & Leatherwood, 1994).

The Ganges River dolphin is listed as endangered by the IUCN and has been listed in the Indian Wildlife protection act of 1972.

Taxonomy: of *Platanista gangetica gangetica*

Kingdom: Animalia
Phylum: Chordata
Class: Mammalia
Order: Cetacea
Suborder: Odontoceti "toothed whales"
Family: Platanistidae (South Asian river dolphin)
Genus: *Platanista*
Species: *Platanista gangetica* (Roxburgh, 1801)
Subspecies: *Platanista gangetica gangetica* (Roxburgh, 1801)

Lipotes vexillifer (Baiji, Yangtze River Dolphin)

The baiji (*Lipotes vexillifer*), an endemic species in China, was once abundant in the Yangtze River, Qiantang River and Poyang and Dongting Lakes (Zhou et al., 1977). This blueish-gray (dorsal area) and white (ventral side) river dolphin has a long narrow beak like other river dolphins, small eyes, and average adult lengths of 2.3 m and 2.5 m for males and females respectively (Figure 9). Sadly, rising anthropogenic pressures from a human population now in excess of 400 million living within the Yangtze River watershed have taken its toll on the baiji (see Wang and Zhao's chapter 19). An influx of human sewage, construction explosives and mining have degraded the baiji's habitat. This along with overfishing and illegal fishing has led to a decline in fish production within the Yangtze River system (Wang et al., 2006; Wei et al., 2007) and concomitantly, a reduction in the abundance of baiji prey. Unselective fishing methods including the use of rolling hooks, electrofishing gear and gillnets have had direct and dire consequences for the baiji as well. Even as their numbers declined precipitously, Yangtze River dolphins continued to be entangled in fishing gear (Turvey et al., 2007). A rise in commerce transport to support a burgeoning human population presented other threats to *Lipotes vexillifer* through boat collisions and boat noise (Wang et al., 2006). The negative direct and indirect effects of these rising anthropogenic factors on the baiji are indicated by baiji population surveys conducted in the 1950s, 1970s, 1980s, and 1990s that document that their population dropped from 6,000 to 13 individuals (Chen et al., 1993; Yang et al., 2000). The last survey conducted in 2006 did not observe any animals (Barrett et al., 2006; Turvey et al., 2007). Anecdotal evidence suggests that at least one baiji was still alive in 2007, but more than likely, this species is either extinct or soon to be extinct. The IUCN has listed the baiji as critically endangered.

Wang and Zhao's chapter 19 as well as Turvey's chapter 20 discusses the development of conservation measures and failed attempts to save the baiji from extinction. River refuges and protection stations were established along with a captive breeding program to no avail. A critical analysis of the baiji recovery program may prove to be helpful for the conservation of cetaceans such as the Yangtze finless porpoise.

Figure 9. One of the last images of the technical extint *Lipotes vexillifer* (copyright Dr. Wang).

Taxonomy: of *Lipotes vexillifer*

Kingdom: Animalia
Phylum: Chordata
Class: Mammalia
Order: Cetacea
Suborder: Odontoceti "toothed whales"
Family: Lipotidae
Genus: *Lipotes*
Species: *Lipotes vexillifer* (Miller, 1918)

Orcaella brevirostris (Irrawady dolphin)

Orcaella brevirostris inhabits three river systems (Mekong, Mahakam, and Ayeyarwady) in southeastern Asia. Researchers of this dolphin species were still analyzing their most current data at the time of publication of this book and therefore we have not included its description, except to recognize that it does exist and that it is also threatened by human activities (a critically endangered species). We and other dolphin scientists look forward to observing contributions of new *Orcaella* papers to the scientific literature in the near future.

Taxonomy: of Orcaella brevirostris

Kingdom: Animalia
Phylum: Chordata
Class: Mammalia
Order: Cetacea

Suborder: Odontoceti “toothed whales”
Family: Delphinidae
Genus: Orcaella
Species: Orcaella brevirostris (Owen in Gray, 1866)

ACKNOWLEDGMENTS

Joseph Shostell gives special thanks to the constant support and love from Joelle and Sophia Shostell. Also, thanks to Mark Shostak who reviewed this manuscript. This work is dedicated to his daughter, Sophia, why dances and swims like a dolphin. Manuel Ruiz-García thanks Colciencias (Grant 1203-09-11239; Geographical population structure and genetic diversity of two river dolphin species, *Inia boliviensis* and *Inia geoffrensis*, using molecular markers) and the Fondo para la Accion Ambiental (US-Aid) (120108-E0102141; Structure and Genetic Conservation of river dolphins, *Inia* and *Sotalia*, in the Amazon and Orinoco basins) who financed these projects and allowed him to experience an exuberant, amazing, fascinating and incredible adventure through 10,000 km of Amazonian rivers for an European accustomed to another culture and different behaviors. Thanks to some special people that he knew in these rain forests (Isias in Requena, Angelito in Iquitos, Jose and Gabriel in Puerto Nariño, Alan in San Ramón, Ze Marubo in Atalaya do Norte, Javier Espiritu in Leticia). Special thanks to Pablo Escobar-Armel and Dr. Diana Alvarez who were both indispensable in all the facets of this large research project. Dr. Diana Alvarez, has also been an inseparable partner to Manuel for 10 years in all his travels and in his life… thank-you. Additional thanks to the many people of diverse Indian tribes in Peru (Bora, Ocaina, Shipigo-Comibo, Capanahua, Angoteros, Orejón, Cocama, Kishuarana and Alamas), Bolivia (Sirionó, Canichana, Cayubaba and Chacobo), Colombia (Jaguas, Ticunas, Huitoto, Cocama, Tucano, Nonuya, Yuri and Yucuna), Brazil (Marubos, Matis, Mayoruna, Kanaimari, Kulina, Maku and Waimiri-Atroari) and Ecuador (Kichwa, Huaorani, Shuar and Achuar) who provided mythical and practical information as well as thousands and thousands of samples of diverse mammal, bird and fish species. The company and help of Luisa Fernanda Castellanos-Mora during several Amazonian expeditions were important. Luisa helped to obtain pink river dolphins for population genetics works, and traveled to diverse locations in the Colombian, Brazilian and Peruvian Amazon to obtain *Inia*'s teeth and other samples as well as other species of genetics and taxonomic interest. Also, thanks goes to Alexandra Parra, who also dances as a beautiful tucuxi, for her continuous encouragement to produce this book. This work is dedicated to all the botos and tucuxis which were sampled and, especially, to I-16, III-1, *Inia* (a lovely *Lagothicha*, who was rescued in the third expedition), Copoazú (a two-toed sloth, also rescued in the third expedition) and all the feline friends which accompanied him throughout the decades (Spencer, Erik, Olaf, Twin, Chiqui, Talula, Odin, Thor, Yngwie, Tsunami, Shiva, Indra, Isis, Aymara and Yaku).

REFERENCES

[1] Aliaga-Rossel, E. (2002). Distribution and abundance of the river dolphin (*Inia geoffrensis*) in the Tijamuchu River, Beni, Bolivia. *Aquatic Mammals*, 28, 312-323.

[2] Aula, I., Braunschweiler, H., & Malin, I. (1995). The watershed flux of mercury examined with indicators in the Tucuri Reservoir in Pará, Brazil. *Science of the Total Environment*, 175, 97-107.

[3] Azevedo, A. F., Oliveira, A. M., Viana, S. C. & Van Sluys, M. (2007). Habitat use by marine tucuxis (*Sotalia guianensis*) (Cetacea: Delphinidae) in Guanabara Bay, south-eastern Brazil. *Journal of the Marine Biological Association of the United Kingdom*, 87, 201-205.

[4] Banguera-Hinestroza, E., Cardenas, H., Ruiz-García, M., García, Y. F., Marmontel, M., Gaitán, E., Vásquez, R. & García-Vallejo, F. (2002). Molecular identification of evolutionarily significant units in the Amazon River dolphin *Inia* sp. (Cetacea: Iniidae). *Journal of Heredity*, 93, 312-322.

[5] Barrett, L. A., Pfluger, A., & Wang, D. (2006). Successful pilot Yangtze freshwater dolphin expedition. *Oryx*, 40, 259-260.

[6] Barros, N. B. (1991). Recent cetacean records for southeastern Brazil. *Marine Mammal Science,* 7, 296-306.

[7] Barros, N., & Teixeira, R. (1994): Incidental catch of marine Tucuxi, *Sotalia fluviatilis*, in Alagoas, Northeastern Brazil. In Gillnets and Cetaceans. Reports to the International Whaling Commission (Special Issue 15, pp. 265-268). Cambridge, United Kingdom: International Whaling Commission.

[8] Bassoi, M. & Secchi, E. R. (2000). Temporal variation in the diet of franciscana Pontoporia blainvillei (Cetacea, Pontoporiidae) as a consequence of fish stocks depletion off southern Brazil. IV Encontro para a Coordenacao da Pesquisa e Conservacao da Fraciscana, Pontoporia blainvillei, no Atlantico Sul Ocidental, Working paper No. 9.

[9] Bastida, R., Rodriguez, D., Secchi, E. R., & da Silva, V. M. F. (2007). Mamíferos Acuáticos de Sudamérica y Antártida. Buenos Aires, Brazil: Vazquez Mazzini Editores.

[10] Best, R. C., & Silva, V. M. F. (1989). Biology, status and conservation of Inia geoffrensis in the Amazon and Orinoco River basin. In W. F. Perrin, R. L. Brownell, Jr., Z. Kaiya, & L. Jiankang (Eds.), Biology and conservation of the river dolphins (3, pp. 1-173). Gland, Switzerland: International Union for the Conservation of Nature and Natural Resources Species Survival Commission.

[11] Best, R. C. & da Silva, V. M. F. (1993). *Inia geoffrensis. Mammalian Species*, 426, 1-8.

[12] Boher, S. J. Bolaños, J. & Cova, L. J. (1995). Sobre un avistamiento del delfin estuarino o bufete (Sotalia fluviatilis) en el Orinoco Medio. *Acta Cientifica Venezolana,* 46, 217-218.

[13] Borobia, M., Siciliano, S., Lodi, L., & Hoek, W. (1991). Distribution of the South-American dolphin Sotalia fluviatilis. *Canadian Journal of Zoology*, 69, 1025-1039.

[14] Brownell, R. L. Jr. (1989). Franciscana, Pontoporia blainvillei (Gervais & d'Orbigny 1844). In S. H. Ridgway & R. J. Harrison, (Eds.), *Handbook of Marine Mammals* (volume 4, pp 45-67). London, United Kingdom: Academic Press.

[15] Cassens, I., Vicario, S., Waddell, V. G., et al., (2000). Independent adaption to riverine habitats allowed survival of ancient Cetacean lineages. *Proceedings of the National Academy of Sciences,* 97, 11343-11347.

[16] Caballero, S., Trujillo, F., Vianna, J. A., Barrios-Garrido, H., Montiel, M. G., Beltrán-Pedreros, S., Marmontel, M., Santos, M. C. O., Rossi-Santos, M., Santos, F. R., & Baker C. S. (2007). Taxonomic status of the genus *Sotalia*: Species level ranking for "tucuxi" (*Sotalia fluviatilis*) and "costero" (*Sotalia guianensis*) dolphins. *Marine Mammal Science*, 23(2), 358-386.

[17] Chen, P., Zhang, X., Wei, Z., Zhao, Q., Wang, X., Zhang, G., & Yang, J., (1993). Appraisal of the influence upon baiji, *Lipotes vexillifer* by the Three-Gorge Project and conservation strategy. *Acta Hydrobiologica Sinica, 11, 101-111.*

[18] Cremer, M. J. & Simoes-Lopes, P. C. (2005). The occurrence of Pontoporia blainvillei (Gervais & d'Orbigny) (Cetacea, Pontoporiidae) in an estuarine area in southern Brazil. *Revista Brasileira de Zoologia*, 22, 717-723.

[19] Cunha, H. A., da Silva, V. M. F., Lailson-Brito, J. Jr., Santos, M.C. O., Flores, P. A. C., Martin, A. R., Azevedo, A. F., Fragoso, A. B. L., Zanelatto, R. C., & Solé-Cava, A. M. (2005). Riverine and marine ecotypes of *Sotalia* dolphins are different species. *Marine Biology*, 148, 449-457.

[20] Danilewicz, D., Rosas, F., Bastida, R., Marigo, J., Muelbert, M., Rodriguez, D., Lailson-Brito Jr., J., Ruoppolo, V., Ramos, R., Bassoi, M., Ott, P. H., Caon, G., Rocha, A. M., Catão-Dias, J. L., & Secchi, E. R. (2002). Report of the Working Group on Biology and Ecology. In Secchi, E. R. (Ed.), *The Latin American Journal of Aquatic Mammals, Special Issue on the Biology and Conservation of Franciscana,* 1, 1-192.

[21] Danilewicz, D., Secchi, E. R., Ott, P. H., Moreno, I. B., Bassoi, M. & Borges-Martins, M. (2009). Habitat use patterns of franciscana dolphins (Pontoporia blainvillei) off southern Brazil in relation to water depth. *Journal of the Marine Biological Association of the United Kingdom*, doi: 10.1017/S002531540900054X.

[22] Da Silva, V. M. F. (1983). Ecología alimentar dos golfinhos da Amazonia. Master of Science Thesis. Manaus, Brazil: University of Amazonas.

[23] Da Silva, V. M. F. (2002). Amazon River Dolphin-*Inia geoffrensis*. In W. F. Perrin, B. Wursig, & J. G. M. Thewissen (Eds.), *Encyclopedia of Marine Mammals* (pp. 18-20). San Diego, CA: Academic Press.

[24] Da Silva, V. M. F. & Best, R. C. (1994). Tucuxi *Sotalia fluviatilis* (Gervais, 1853). In S. H. Ridgway, S. & R. Harrison, (Eds.), Handbook of Marine Mammals. (volume 5, pp. 43-69). San Diego, CA: Academic Press Inc.

[25] Da Silva, V. M. F. (1995). Conservation of the fresh wáter dolphin with special emphasis on Inia geoffrensis and Sotalia fluviatilis. In Delfines y otros mamíferos Acuáticos de Venezuela (pp. 71-92). Caracas, Venezuela: Fundación para el Desarrollo del Comercio Internacional (FUDECI).

[26] Da Silva, V. M. F. & Best, R. C. (1996). *Sotalia fluviatilis*. *Mammalian Species*, 527, 1-7.

[27] Dolbec, J., Mergler, D., Larribe, F., Roulet, M., Lebel, J., & Lucotte, M.(2001). Sequential analysis of hair mercury levels in relation to fish diet of an Amazonian population, Brazil. *Science Total Environment*, 23, 87-97.

[28] Domingo, M., Ferrer, L., Pumarola, M., Marco, A., Plana, J., & Kennedy, S. (1990). Morbillivirus in dolphins. *Nature*, 348, 21.

[29] Dudgeon, D. (2000). The ecology of tropical Asian rivers and streams in relation to biodiversity conservation. *Annual Review of Ecology and Systematics*, 31, 239-263.

[30] Edwards, H. H. & Schnell, G. D., (2001). Status and ecology of Sotalia fluviatilis in the Cayos Miskito Reserve, Nicaragua. *Marine Mammal Science*, 19, 445- 472.

[31] Filla, G. F. & Monteiro-Filho, E. L. A. (2009). Group structure of *Sotalia guianensis* in the bays on the coast of Parana State, south of Brazil. *Journal of the Marine Biological Association of the United Kingdom*, Published online by Cambridge University Press 05 Jun 2009 doi:10.1017/S0025315409002926

[32] Guang, Y. & Kaiya, Z. (1999). A study on the molecular phylogeny of river dolphins. *Acta Theriologica Sinica*, 19 (1), 1-19.

[33] Haimovici, M. (1998). Present state and perspectives for the southern Brazil shelf demersal fisheries. *Fishery Management and Ecology*, 5, 277-289.

[34] Hamilton, H., Caballero, S. & Collins, A. G. (2001). Evolution of river dolphins. *Proceedings of the Royal Society for Biological Sciences*, 268, 549-556.

[35] Hoy, C. M., (1923). The "white-flag" dolphin of the Tung Ting Lake. *The China Journal of Science & Arts*, 1, 154-157.

[36] Jefferson, T.A., Leatherwood, S., & Webber, M. A. (1993) *FAO species identification guide, Marine mammals of the world*, Rome, Italy: Food and Agriculture Organization of the United Nations (FAO).

[37] Kajiwara, N., Matsuoka, S., Iwata, H., Rosas, F. C. W., Fillmann, G., & Readman, J. W. (2004). Contamination by persistent organochlorines in cetaceans incidentally caught along Brazilian coastal waters. *Archives of Environmental Contamination and Toxicology*, 46, 124-134.

[38] Kannan, K., Sinha, R. K., Tanabe, S., Ichihashi, H. and Tatsukawa, R. (1993). Heavy metals and organochlorine residues in Ganges River dolphins from India. *Marine Pollution Bulletin* MPNBAZ, 26 (3), 159-162.

[39] Kannan, K., Senthilkumar, K., & Sinha, R. K. (1997). Sources and accumulation of butyltin compounds in Ganges River dolphin, *Platanista gangetica*. *Applied Organometallic Chemistry*, 11, 223-230.

[40] Kasuya, T. & Brownell, R. L. (1979). Age determination, reproduction and growth of franciscana dolphin, *Pontoporia blainvillei*. *Scientific Reports of the Whales Research Institute*, 31, 45-67.

[41] Lazáro, M., Lessa,E. & Hamilton, H. (2004). Geographic genetic structure in the franciscana dolphin (Pontoporia blainvillei). *Marine Mammal Science*, 20, 201-214.

[42] Lodi, L. & Hetzel, B. (1998) Grandes agregações do boto-cinza (Sotalia fluviatilis) na Baía da Ilha Grande, Rio de Janeiro. *Revista Bioikos*, 12, 26-30.

[43] Martin, A. R., & da Silva, V. M. F. (2004). River dolphins and flooded forest: seasonal habitat use and sexual segregation of botos (*Inia geoffrensis*) in an extreme cetacean environment. *Journal of the Zoological Society of London*, 263, 295-305.

[44] Maurice-Bourgoin (L.). Mercury pollution in Bolivian rivers. News-Scientific Bulletins 95. Institut de Recherche pour le Développement, 95 [online]. 1999 [cited 2009 June 29]. Available from: http://www.ird.fr/us/actualites/fiches/1999/95.htm

[45] Mohan, R. S. (1989). Conservation and management of the Ganges River dolphin, *Platanista gangetica*, in India. In W. F. Perrin, R. L.Brownell, Z. Kaiya & L. Jiankang (Eds.), *Biology and Conservation of the River Dolphins* (Species Survival Commission

Occasional Paper 3, pp. 64-69). Gland, Switzerland: International Union for Conservation of Nature.

[46] Monteiro-Filho, E. L. A. (2000). Group organization of the dolphin *sotalia fluviatilis* guianensis in an estuary of southeastern Brazil. Ciência e Cultura *Journal of the Brazilian Association for the Advancement of Science*, 52, 97-101.

[47] Monteiro-Filho, E. L. D. A., Rabello-Monteiro, L. & do Reis, S. F. (2002). Skull shape and size divergence in dolphins of the genus *Sotalia*: a morphometric tridimensional analysis. *Journal of Mammalogy*, 83, 125-134.

[48] Monteiro-Neto, C., Itavo, R. V. & do Souza-Moraes, L. E. (2003). Concentrations of heavy metals in Sotalia fluviatilis. *Environmental Pollution*, 123, 319-324.

[49] Osterhaus, A. D. M. E., de Swart, R. L., Vos, H. W., Ross, P. S., Kenter, J. H., & Barrett, T. (1995) morbillivirus infections of aquatic mammals: newly identified members of the genus. *Veterinary Microbiology*, 44, 219-227.

[50] Perrin, W. F., Brownell, R. L., Zhou, K., & Jiankang, L. (1989). *Biology and conservation of the river dolphins. Proceedings of the Workshop on Biology and Conservation of the Platanistoid Dolphin held at Wuhan, People's Republic of China, October 28-30, 1986. Occasional Papers of the IUCN species survival commission* (SSC), 3, 173.

[51] Pilleri, G., (1979). The Chinese river dolphin (Lipotes vexillifer) in poetry, literature and legend. *Investigations on Cetacea*, 10, 336-349.

[52] Pillieri, G. & Gihr, M. (1977). Observations on the Bolivian (*Inia boliviensis* d'Orbigny, 1834) and the Amazonian bufeo (*Inia geoffrensis* de Blainville, 1817) with a description of a new subspecies (*Inia geoffrensis humboldtiana*). *Investigations on Cetacea*, 8, 11-76.

[53] Pinedo, M. C. (1991). *Development and variation of the franciscana Pontoporia blainvillei. Ph.D. Thesis*, Santa Cruz, CA: University of California.

[54] Pinedo, M. C. (1995). Development and variation in external morphology of the franciscana, *Pontoporia blainvillei. Revista Brasileira de Biologia*, 55, 85-96.

[55] Pinedo, M. C., Praderi, R., & Brownell, R. Jr. (1989). Review of the biology and status of the franciscana Pontoporia blainvillei. In W. F. Perrin, R. L. Brownell, Z. Kaiya & L. Jiankang (Eds.), *Biology and Conservation of the River Dolphins, Occasional Papers* (International Union for Conservation of Nature SSC 3, pp. 46-51). Gland, Switzerland: International Union for Conservation of Nature.

[56] Raga, J. A., Banyard, Al, Domingo, M., Corteyn, M., Van Bressem, M. F., Fernandez, M., Aznar, F. J., & Barrett, T. (2008). Dolphin morbillivirus epizootic resurgence, Mediterranean Sea. *Emerging Infectious Diseases*, 14 (3), 471-473.

[57] Reeves, R. & Leatherwood, S. (1994). Dams and river dolphins: can they co-exist. Royal Swedish Academy of Sciences, 23 (3), 172-175.

[58] Rosas, F. C. W. & Lehti, K. K. (1996). Nutritional and mercury content of mild of the Amazon River dolphin, *Inia geoffrensis*. *Comparative Biochemistry and Physiology*, 115A, 117-119.

[59] Rosas, F. C. W. & Monteiro-Filho, E. L. A. (2002). Reproduction of the estuarine dolphin (Sotalia guianensis) on the coast of Parana, Southern Brazil. *Journal of Mammalogy*, 83, 507-515.

[60] Ruiz-García, M. (2007). Genética de Poblaciones: Teoría y aplicación a la conservación de mamíferos neotropicales (Oso andino y delfín rosado). *Boletín de la Real Sociedad Española de Historia Natural*, 102 (1-4), 99-126.

[61] Ruiz-García, M., Banguera, E. & Cárdenas, H. (2006). Morphological analysis of three *Inia* (Cetacea; Iniidae) populations from Colombian and Bolivia. *Acta Theriologica,* 51, 411-426.

[62] Ruiz-García, M., Murillo, A., Corrales, C., Romero-Aleán, N., & Alvarez-Prada, D. (2007). Genética de Poblaciones Amazónicas: La historia evolutiva del jaguar, ocelote, delfín rosado, mono lanudo y piurí reconstruida a partir de sus genes. *Animal Biodiversity and Conservation,* 30, 115-130.

[63] Ruiz-Garcia, M., Caballero, S., Martinez-Agüero, M., & Shostell, J. (2008). Molecular differentiation among *Inia geoffren*sis and *Inia boliviensis* (Iniidae Cetacea) by means of nuclear intron sequences. In V. P. Koven, (Ed.), Population Genetics Research Progress (pp. 177-223). New York, New York : Nova Science Publisher, Inc.

[64] Santos, M. C. O., Oshima, J. E. F. & Silva, E. (2009). Sightings of franciscana dolphins (*Pontoporia blainvillei*): the discovery of a population in the Paranaguá estuarine complex, southern Brazil. *Brazilian Journal of Oceanography,* 57(1), 57-63.

[65] Sayer, J. A. & Whitmore, T. C. (1991). Tropical moist forest: Destruction and species extinction. *Biological Conservation*, 55, 199-213.

[66] Schnapp, D. & Howroyd, J. (1992). Distribution and local range of the Orinoco dolphin (*Inia geoffrensis*) in the Rio-Apure, Venezuela. *Z. Saugetierkd*, 57(5), 313-315.

[67] Secchi E.R. (2006) *Modeling the population dynamics and viability analysis of franciscana (Pontoporia blainvillei) and Hector's dolphins (Cephalorynchus hectori) under the effects of bycatch in fisheries, parameter uncertainty and stochasticity. Doctoral thesis*. Dunedin, New Zealand: University of Otago.

[68] Siciliano, S. (1994). Review of small cetaceans and fishery interactions in the coastal waters of Brazil. In Gillnets and Cetaceans. Reports to the International Whaling Commission (Special Issue 15, pp. 241-250). Cambridge, United Kingdom: International Whaling Commission.

[69] Simões-Lopes, P. C. (1988). Sobre a amplição da distribuição do genero Sotalia Gray, 1866 (Cetacea, Delphinidae), para as águas do Estado de Santa Catarina, Brasil. *Biotemas*, 1, 58-62.

[70] Sinha, R. K. (2002). An alternative to dolphin oil as a fish attractant in the Ganges River system conservation of the Ganges River dolphin. *Biological Conservation*, 107, 253-257.

[71] Sinha, R. K. (2006). The Ganges River dolphin *Platanista gangetica gangetica*. *Journal of the Bombay Natural History Society*, 103, 254-263.

[72] Sinha, R. K., Smith, B. D., Sharma, G., Prasad, K., Choudhary, B. C., Sapkota, K., Sharma, R. K., & Behera, S. K. (2000). Status and distribution of the Ganges susu (*Platanista gangetica*) in Ganges River system of India and Nepal. In R. R. Reeves, B. D. Smith, T. Kasuya, (Eds.), *Biology and Conservation of Freshwater Cetaceans in Asia, Occasional Paper of the IUCN Species Survival Commission* (No. 23, pp. 42-48). Gland, Switzerland and Cambridge, United Kingdom: International Union for Conservation of Nature.

[73] Smith, B. D., Ahmed, B., Ali, M. E. & Braulik, G. (2001). Status of the Ganges River dolphin or shushuk Platanista gangetica in Kaptai Lake and the southern rivers of Bangladesh. *Oryx,* 35 (1), 61-72.

[74] Tirira, D. (2001). Libro Rojo de los Mamíferos de Ecuador. Serie Libros Rojos del Ecuador, Tomo 1. Publicación Especial Sobre los Mamíferos del Ecuador. Quito, Ecuador: Sociedad para la Investigación y Monitoreo de la Biodiversidad Ecuatoriana (SIMBIOE)/Ecociencias/Ministerio del Ambiente.

[75] Trujillo, F., Garcia, C., & Avila, J. M. (2000). *Status and conservation of the tucuxi Sotalia fluviatilis (Gervais, 1853): marine and fluvial ecotypes in Columbia.* (SC/52SM11/2000). Adelaide, Australia: International Whaling Commission.

[76] Turvey, S. T., Pitman, R. L., Taylor, B. L., Barlow, J., Akamatsu, T., Barrett, L. A., Zhao, X., Reeves, R. R., Stewart, B. S., Pusser, L. T., Wang, K., Wei, Z., Zhang, X., Richlen, M., Brandon, J. R. & Wang, D., (2007). First human-caused extinction of a cetacean species? *Biology Letters*, 3, 537-540.

[77] Verma SK, Sinha RK, & Singh L. Phylogenetic position of *Platanista gangetica*: insights from the mitochondrial cytochrome *b* and nuclear interphotoreceptor retinoid-binding protein gene sequences. *Molecular Phylogenetics and Evolution*, 2004, 33, 280-288.

[78] Wang D., Hao Y., Wang K., Zhao Q., Chen D., Wei Z. & Zhang X., (2005). The first Yangtzefinless porpoise successfully born in captivity. *Environmental Science and Pollution Research,* 12, 247-250.

[79] Wang, D., Zhang, X., Wang, K., Wei, Z., Würsig, B., Braulik, G. T., & Ellis, S. (2006). Conservation of the baiji: no simple solution. *Conservation Biology*, 20, 623-625.

[80] Woods-Ballard, A., Parsons, E. C. M., Hughes, A., Velander, K. A, Lakle, R. J., Warburton, C. A, (2003). The sustainability of whale-watching in Scotland. *Journal of Sustainable Tourism*, 11(1), 40-54.

[81] Yañez, M. (1999). *Etología, ecología y conservación del delfin Inia geoffrensis en los ríos Itenez y Paragua del Parque Nacional Noel Kempf Mercado (Masters Thesis)*. La Paz, Bolivia: Universidad Mayor de San Andres.

[82] Yang, J., Xiao, W., Kuang, X., Wei, Z., & Liu, R., (2000). Studies on the distribution, population size and the activity of *Lipotes vexillifer* and *Neophocaena phocaenoides* in Dongting Lake and Boyang Lake. *Resources and Environment in the Yangtze Basin*, 9, 444-450.

[83] Yang, G., Yan, J., Zhou, K.Y. & Wei, F.U. (2005). Sequence variation and gene duplication at MHC DQB loci of Baiji (*Lipotes vexillifer*), a Chinese river dolphin. *Journal of Heredity,* 96, 310–317.

[84] Yang, G., Zhou, K.Y., Ren, W.H., Ji, G.Q., Liu, S., Bastida, R., & Rivero, L. (2002). Molecular systematics of river dolphins inferred from complete mitochondrial cytochrome b gene sequences. *Marine Mammal Science,* 18, 20–29.

[85] Yogue, G. T., Santos, M. C. D. O., & Montone, R. C. (2003). Chlorinated pesticides and polychlorinated biphenyls in marine tucuxi dolphins (Sotalia fluviatilis) from the Cananéia Estuary, southeastern Brazil. *The Science of the Total Environment*, 312, 67-78.

[86] Zhou, K., Qian, W. & Li, Y. (1977). Studies on the distribution of baiji, *Lipotes vexillifer Miller. Acta Zoologica*, 23, 72-79.

In: Biology, Evolution and Conservation of River Dolphins… ISBN: 978-1-60876-633-8
Editors: M. Ruiz-García, J. Shostell. pp. 29-53

Chapter 2

SEASONAL ECOLOGY OF INIA IN THREE RIVER BASINS OF SOUTH AMERICA (ORINOCO, AMAZON, AND UPPER MADEIRA)

***Tamara L. McGuire*[1] *and Enzo Aliaga-Rossel*[2]**
[1]LGL Alaska Research Associates, Anchorage, AK, USA
[2]Instituto de Ecología, Universidad Mayor de San Andres, La Paz, Bolivia and University of Hawaii, Honolulu, Hawaii, USA

ABSTRACT

Compared to most marine odontocetes, river dolphins live in an environment that is less-stable and more spatially complex than the ocean. Yearly seasonal fluctuations in river levels may be as great as 20 meters, and lead to seasonal extremes in quality and quantity of aquatic habitat available to river dolphins. Seasonal changes in water levels also affect the availability of dolphin prey due to seasonal patterns of fish reproduction and fish migrations. Human-induced threats to river dolphins, such as incidental net entanglement, vessel strikes, and deliberate killing appear to vary seasonally as well. In this chapter, we present our investigations of the seasonal ecology of *Inia* spp from three river basins of South America (*Inia geoffrensis humboldtiana* in Venezuela's Orinoco River Basin, *Inia geoffrensis geoffrensis* in Peru's Amazon Basin, and *Inia boliviensis* in Bolivia's Mamoré Basin). We provide results from our observational studies (which included boat- based surveys of groups and photo-identification of individuals) and we discuss these results in the context of other information about the seasonal ecology of *Inia*, including distribution, movement patterns, group size, age-class composition, and seasonality of reproduction. We conclude with a discussion of how seasonal ecology should be considered in the conservation of river dolphins and of the management of human activities that affect them.

Keywords: seasonal ecology, seasonality, distribution, movement patterns, group size, age-class composition, reproduction, river dolphins, South American, *Inia boliviensis*, *Inia geoffrensis*.

INTRODUCTION

River dolphins live in an environment that is less-stable and more spatially complex than the ocean, where most odontocetes (toothed whales) occur. The Amazon River dolphin *Inia,* occurs in the Neotropics, where water temperature and photoperiod remain almost constant throughout the year; seasonal differences are primarily due to changes in precipitation (wet and dry seasons) which in turn corresponds to river water levels (low, rising, high, and falling).Yearly seasonal fluctuations in river levels may be as great as 20 meters and these differences result in seasonal extremes in quality and quantity of aquatic habitat available to river dolphins. Dolphins restricted to deep waters of river channels, confluences, and seasonally isolated oxbow lakes during the dry season are free to swim through submerged primary rainforest and across the llanos (grassland plains) during the height of the rainy season.

Changes in water levels affect not only the amount and type of aquatic habitat available to *Inia*, but also to their prey. *Inia* are piscivores (fish-eaters), and therefore prey biomass and availability are largely determined by seasonal water levels, via seasonal patterns of fish reproduction and fish migrations. Fish reproduction and migrations are highly seasonal, although the timing of these events varies according to species (Goulding, 1980). Ease of prey capture may also be affected by water depth and aquatic habitat. Fish may be easier to catch in shallow confined waters, than in areas of deep open water, or when they are dispersed and hidden in the structure provided by flooded vegetation. Such seasonal extremes in habitat and prey availability would be expected to be reflected in seasonal patterns of other aspects of river dolphin ecology, such as distribution, habitat association, movement and residency patterns, group size, reproduction, and mortality.

The Amazon River Dolphin, *Inia geoffrensis,* occurs in freshwaters of the Amazon, Orinoco, and upper Madeira River basins of South America, in the countries of Bolivia, Brazil, Colombia, Ecuador, Peru, and Venezuela (da Silva, 1994). *Inia* is listed by the International Whaling Commission and the International Union for the Conservation of Nature as a single species with three subspecies: *Inia geoffrensis humboldtiana* in the Orinoco River Basin, *Inia g. geoffrensis* in the main Amazon River Basin, and *Inia g. boliviensis* in the Bolivian sub-basin of the Amazon. The taxonomic status of *Inia* is in question however, and many researchers have proposed classifying *Inia boliviensis* as a separate species, based on genetic and morphologic differences (D'Orbigny, 1834; Pillieri & Gihr 1977; da Silva 1994; Hamilton et al,. 2001; Banguera-Hinestroza et al., 2002; Ruiz-García et al., 2008).

In this chapter, we present results from our investigations of the seasonal ecology of *Inia* from three river basins of South America: *Inia geoffrensis humboldtiana* in Venezuela's Orinoco River Basin; *Inia geoffrensis geoffrensis* in Peru's Amazon Basin; and *Inia boliviensis* in Bolivia's Mamore Basin (upper Madeira Basin). We also discuss seasonal patterns in natural and anthropogenic mortality, including strandings, intra-specific aggression, entanglement in fishing nets, vessel strikes, and deliberate killing by fishermen. Although this chapter focuses on *Inia*, the discussion of seasonal ecology can be applied to river dolphin species presented in other chapters of this book.

METHODS

Study Areas

River dolphin surveys were conducted in freshwaters of Venezuela, Peru, and Bolivia between 1993 - 2001 (Figure 1; Table 1). All three study areas contained main stem rivers, tributaries, confluences, and oxbow lakes (i.e., former river channels than may become seasonally isolated from the main channel), and contained whitewater and blackwater habitats (differentiated by turbidity, nutrient load, pH, and origin; Sioli, 1984). Four seasons were assigned according to relative water levels: high water, falling water, low water, and rising water. Seasons were not correlated to month of the year as this differed among study areas. Water levels of Neotropical river systems are influenced by local rains, snowmelt from the Andes, and water levels both up- and downstream. The Peruvian study area is classified as lowland tropical rainforest, whereas the Bolivian and Venezuelan study areas are characterized as tropical savannas with gallery forests (Cox & Moore, 1993).

The Venezuelan study area along the Cinaruco River is 3 km^2 in area and contains 20 km of water courses. This area lies within the Santos Luzardo National Park. The blackwater Cinaruco River flows east to the Orinoco River across the llanos (lowland plains) of Apure state. The lower portion of the Cinaruco River is particularly sinuous and forms a complex flood plain with numerous channels and oxbow lakes. The river floods the surrounding plains from May-November, then returns to the main channel during the dry season (usually February-April).

The Peruvian study area is located within Peru's Pacaya-Samiria Reserve, 93 km upriver from the city of Iquitos, Department of Loreto, in the far western Amazon Basin. The Reserve is bounded by the Marañón and Ucayali rivers, which are the parent rivers of the Amazon River. Major tributaries of these rivers are, respectively, the Samiria River (400 km long), and the Pacaya River (380 km long). The Reserve has over 10,000 km of linear waterways (blackwater, whitewater, and mixed), comprised of main stem rivers, tributaries, confluences, channels, and lakes. A protected area since 1940, the Pacaya-Samiria Reserve is the largest reserve in Peru at 2,150,700 ha (INRENA-CTARL, 2000). Water levels are usually lowest during September, although this varies yearly. Water levels can fluctuate greatly during the overall rise between October and May, and peak waters generally occur in May or June.

The Bolivian study area is located along the Tijamuchi River, Moxos province, Department of Beni, and contains 185 linear km of waterways, upriver from and including the confluence of the Tijamuchi and Mamoré rivers. Three oxbox lakes adjoining the Tijamuchi were also surveyed. Highest water levels occur between December and April, and lowest water levels are generally from June to October. The lower Tijamuchi is a mixed white- and blackwater river.

Field Sampling

Data collection consisted of boat-based surveys of dolphins and habitat, photo-identification of individual dolphins, necropsies of dolphins, and interviews with local people. Environmental surveys consisted of measurements of water depth, channel width, water temperature, water turbidity, and characterization of water type (i.e., white water, black water, or mixed). Aquatic habitat was defined as main river channel, tributary, oxbow lake, or confluence. Surveys were assigned to seasons based on the relative depth of water and the month of the year (Table 1). Surveys in Bolivia and Peru were conducted in all seasons. Surveys in Venezuela were conducted during falling, low, and rising water seasons, but not during the high water season.

Table 1. Dates, locations, and seasons of field observations

Country	Location	River Basin	Long/ Lat	Year	Seasons sampled	Peak water level	Reference
Venezuela	Cinaruco River, Santos Luzardo National Park	Orinoco	67°W, 6°N	1993-1994	low rising falling	July	McGuire 1995 McGuire and Winemiller 1998
Peru	Pacaya-Samiria National Reserve	Amazon	74°W, 5°S	1996-2000	all	May	McGuire 2002
Bolivia	Tijamuchi River	Amazon, upper Madeira, Mamoré Sub-basin	65°W, 14°S	1998 1999 2001	all	Feb	Aliaga-Rossel 2000, 2002

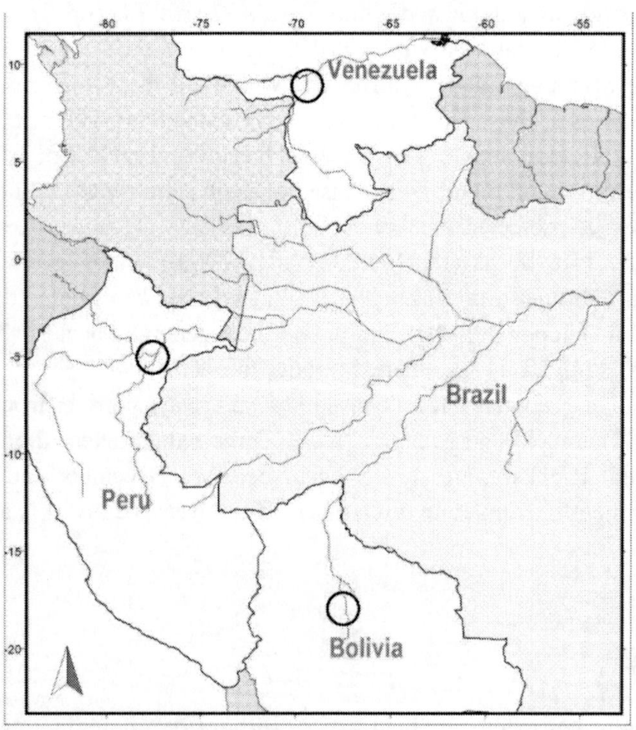

Figure 1. Location of study sites, Bolivia, Peru and Venezuela.

During boat-based surveys in rivers, tributaries, lakes, and confluences (for detailed methods see McGuire, 1995; Aliaga-Rossel, 2000; McGuire, 2002) group size, age composition and GPS position were recorded for each dolphin sighting and results were used to determine encounter rates. Age classifications of *Inia* were based on visual estimation of total body length and divided into two categories: *neonates* (<1 m), and *other* (>1 m). Neonates were further identified by their uncoordinated swimming and surfacing behavior, and fetal folds (when possible). The *other* category included older calves, juveniles, subadults, and adults, and was not further subdivided as it was often difficult to visually differentiate size-class of intermediate sized animals. Age, sexual maturity, and sex are not clearly differentiated based on length alone in these dolphins (da Silva, 1994). Mating behavior and intra-specific aggression were opportunistically recorded. An interaction was categorized as mating if ventral to ventral contact between two or more dolphins was observed.

Fishing activity by humans was noted opportunistically in Venezuela and Bolivia, and was systematically recorded during dolphin surveys in Peru's lake San Pablo de Tipishca in 2000. Observers counted and recorded the position of all fishing nets (seines and gill nets), temporary fishing camps, and fish cages. Fishing nets were counted regardless of if they were deployed or in canoes. Fish cages were used to hold live fish that had been caught in nets.

In Venezuela, potential prey fish were collected with seines and an experimental gill net (see McGuire & Winemiller, 1998 for details), and catch per unit effort (CPUE) was calculated by dividing the total number of fishes caught by either the total number of meters the seine was pulled or by the total number of minutes the gill net was in the water.

Rates of Travel

Rates of travel were calculated for individual dolphins seen in multiple times during the course of a survey trip (which was typically 7-12 days in Bolivia and Peru, and 6 months in Venezuela). Photo-identification techniques were used to identify individual *Inia* by cuts and nicks to the dorsal fin and back, pigmentation patterns on the back and head, scars, tooth-rake marks, and abnormally shaped beaks (McGuire, 1995; Aliaga-Rossel, 2000; McGuire, 2002; McGuire & Henningsen 2007). Photo-identification results from Peru were supplemented by photo-catalogs from the same area assembled by Leatherwood (1996), Henningsen (1998), and Zúñiga (1999), and range maps and sighting histories were created from the compiled data, which spanned the period 1991-2000.

Mortality

All dead dolphins encountered were examined for length, girth, body condition, sex, stomach contents, pregnancy and/or lactation in females, tooth eruption of neonates and calves, and possible signs of death.

Literature Review

We reviewed the published and unpublished literature about the seasonal ecology of *Inia*; the unpublished literature consisted of project reports, bachelor and master theses, doctoral dissertations, and conference abstracts.

RESULTS

Distribution

In Venezuela, *Inia* sighting rates throughout the study area were significantly associated with season ($\chi^2 = 48.65$, df = 2, $P = 0.001$, N = 489) and dolphins were observed most often during the falling water season and least often during rising water (Figure 2). Sighting rates varied seasonally within oxbow lake and river habitats ($\chi^2 = 52.09$, df = 2, $P = 0.001$, N = 258 and $\chi^2 = 10.21$, df = 2, $P = 0.006$, N = 63, respectively). Sighting rates within lakes were greatest during falling water, then declined during low and rising water. Fewer dolphins were seen in rivers during falling water than during any other season. Sighting rates in confluences were not associated with season ($\chi^2 = 0.470$, df = 2, P = 0.79, N=156).

Inia sighting rates in Peru did not differ significantly according to season in two of the three surveyed lakes (Table 2), or four of the six surveyed rivers (Table 3), as the variation in sighting rates was greater within-seasons than among-seasons. When all sightings were pooled across all lakes, rivers, and confluences surveyed and standardized for survey effort, sighting rates were greatest during low water and lowest during rising water (Figure 2).

Table 2. Seasonal *Inia* abundance in lakes of the Pacaya-Samiria Reserve, Peru. SSDWL (significant seasonal differences within a lake*)

Lake	Season	# transects	Mean *Inia*/km	CV	SSDWL
San Pablo	All	25	10.9 (0.7 *Inia*/km^2)	0.73	Yes P = 0.02*
	Falling	5	7.2	1.28	
	High	3	17.3	0.35	
	Low	4	1.0	0.82	
	Rising	13	13.5	0.41	
Atun Cocha	All	23	9.2 (1.5 *Inia*/km^2)	0.79	No P=0.43
	Falling	3	7.7	1.04	
	High	3	9.7	0.43	
	Low	6	5.5	1.55	
	Rising	11	11.6	0.59	
Tipishca del Samiria	All	29	50.7 (3.5 *Inia*/km^2)	0.45	No P= 0.84
	Falling	5	57.6	0.44	
	High	7	45.9	0.32	
	Low	5	53.6	0.56	
	Rising	12	49.4	0.49	

* Single factor ANOVAS were used to compare means among seasons. When data were not normally distributed, the Kruskal-Wallis test was used to compare sample medians.

Table 3. Encounter rates of *Inia* in rivers of the Pacaya-Samiria Reserve, Peru. S.S.D.W.R=significant seasonal differences within a river*

River (water type)	Season	# surveys	Mean *Inia*/km	CV	S.S.D.W.R
Marañón (white)	All	39	0.3	1.06	yes P = 0.01*
	Falling	9	0.5	0.56	
	High	7	0.2	1.90	
	Low	4	0.4	1.40	
	Rising	19	0.1	1.90	
Samiria section 1 (mixed)	All	11	1.5	1.92	no P = 0.75
	Falling	4	0.7	0.75	
	High	3	0.6	0.12	
	Low	3	3.5	1.6	
	Rising	1	0.2	0	
Samiria section 2 (black)	All	32	0.4	0.67	no P = 0.16
	Falling	9	0.6	0.38	
	High	7	0.4	0.45	
	Low	5	0.5	0.74	
	Rising	11	0.3	0.97	
Yanayaquillo (black)	All	25	0.4	1.65	no P = 0.92
	Falling	5	0.5	1.18	
	High	7	0.4	1.23	
	Low	3	0.6	1.32	
	Rising	10	0.4	1.95	
Atun Caño (black)	All	26	0.3	1.30	no P = 0.32
	Falling	4	0.3	1.10	
	High	7	0.3	1.50	
	Low	5	0.1	2.10	
	Rising	10	0.4	1.05	
Yanayacu (mixed)	All	15	1.1	1.00	yes P = 0.01*
	Falling	0	x	x	
	High	5	0.3	1.23	
	Low	2	2.0	0.32	
	Rising	8	1.3	1.00	
Pucate (black)	All	9	0.4	1.65	no P = 0.98
	Falling	1	0.4	1.42	
	High	4	0.2	1.45	
	Low	1	0.3	1.40	
	Rising	3	0.6	1.83	

* Single factor ANOVAS were used to compare means among seasons. When data were not normally distributed, the Kruskal-Wallis test was used to compare sample medians.

Inia occurred in lakes as shallow as 1.5-m mean depth and in rivers as shallow as 2.4-m mean depth (Table 4). There were significant seasonal differences in the mean and maximum number of *Inia* in confluences (Table 5). *Inia* in confluences were seen most often during low water, least often during high water, and in the largest aggregations during low water.

Inia sighting rates in Bolivia varied significantly among seasons ($F = 80.55$, $df = 3,4$, $P = 0.0005$; Figure 2). Sighting rates were greatest during the low water season (30.4% of the total), and lowest during high water (21.5% of the total). Sampling effort did not vary by season. In the rivers, dolphins were most often seen during falling and low waters (Figure 3). In the oxbow lakes, dolphins were most often seen during rising and high waters. The number of dolphins in the oxbow lakes declined during low water, sometimes to the point where they were absent.

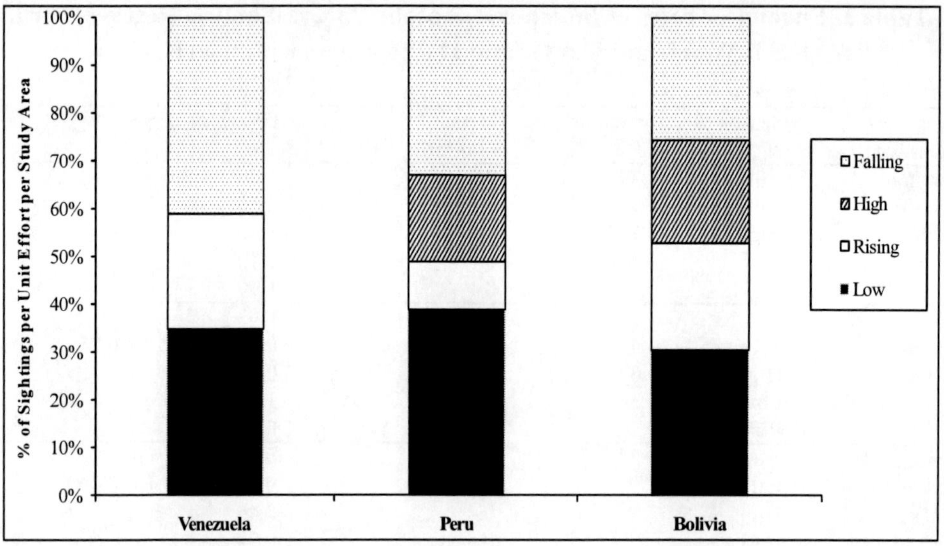

Figure 2. Percent of *Inia* sightings per unit effort of survey per study area. Surveys were not conducted in Venezuela during high water.

Table 4. Seasonal depths of rivers, lakes, and Inia depth thresholds in the Pacaya-Samiria Reserve, Peru. * indicates dolphins were always seen, regardless of depth

Location	Mean maximum depth (m)	Mean minimum depth (m)	Flux (m) (maximum-minimum)	Shallowest water (m) with *Inia*	Deepest water (m) without *Inia*
Rivers					
Marañón	11.7	5.2	6.5	5.2	*
Samiria (Section 1)	7.4	3.0	4.4	3.0	*
Samiria (Section 2)	8.2	3.0	5.2	3.0	*
Yanayaquillo	10.8	4.7	6.1	4.7	*
Atun Caño	9.2	2.2	7.0	2.4	2.2
Yanayacu	15.4	4.5	10.9	4.5	*
Pucate	13.7	4.6	9.1	4.7	4.6
Lakes					
San Pablo	7.9	3.2	4.7	3.2	3.0
Atun Cocha	7.2	0.9	6.3	1.5	1.4
Tipishca del Samiria	7.4	3.8	3.6	3.8	*

Table 5. Seasonal differences in *Inia* occurrence and numbers in confluences of the Pacaya-Samiria Reserve, Peru

Comparison	Falling	High	Low	Rising	Kruskal-Wallace Test statistic and P value	Significant differences as indicated by the Bonferroni test
Mean # *Inia* by season	3.4 (± 2.20 SD)	1.7 (± 0.72 SD)	5.6 (± 2.47 SD)	3.5 (± 1.91 SD)	H = 11.87 P = 0.0007	high and low high and rising
Maximum # *Inia* by season	8.5 (± 4.08 SD)	5.8 (± 1.91 SD)	13.0 (± 4.24 SD)	9.2 (± 2.98 SD)	H = 13.01 P = 0.004	high and low high and rising

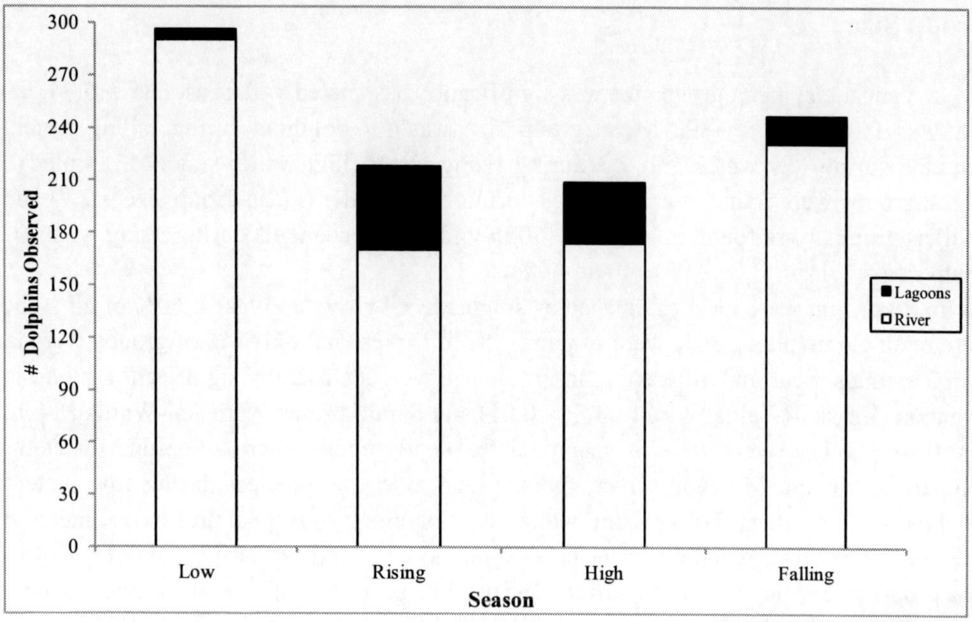

Figure 3. Number of river dolphins observed by season in Tijamuchi River, Bolivia (survey effort was equally distributed among seasons).

Movement Patterns of Individuals

In Venezuela, six *Inia* were photo-identified and resighted (McGuire, 1995; McGuire &Winemiller, 1998). Four dolphins had fewer than ten days between their initial identification and final sightings and were not included in the seasonal movement analysis. One individual was sighted eight times over a period of 186 days and another was seen seven times over a 156 day period. Both of these dolphins were initially identified during falling water, and were not resighted until the end of the late low water/early rising water period (with 4-5 month gaps between the first and second sightings).

Within the Peruvian study area, 25 *Inia* were photo-identified and resighted (McGuire, 2002; McGuire & Henningsen, 2007). Some identified *Inia* frequently moved 40-60 km within a 24-h period, while other individuals remained in the same location for several days. These relatively high rates of travel occurred during high and falling water levels, although sample size was insufficient to detect any statistically significant relationships between seasons and distance traveled.

In Bolivia, two individuals were identified and resighted (Aliaga-Rossel, 2000, 2002). One dolphin was only resighted once, the day after it was first identified. The other animal was resighted four times, the first time during high water, then three months later during falling water, 60 km away, then months later in high water. There was a 239-day range between the initial and final sighting.

Group Size

In Venezuela, mean group size was significantly associated with season ($F = 6.74$, $df = 2, 478$, $P < 0.0129$, $N = 489$). Mean group size was 1.7 dolphins during falling water, 1.9 dolphins during low water, and 2.3 during rising water (high water was not sampled). The largest groups were found in confluences during rising water (mean group size = 2.7) and the smallest groups were found in narrow (<50 m wide) river channels during rising water (mean group size = 1.1).

In Peru, *Inia* were most often seen as singles in all rivers and lakes: 60% of all sightings were of single dolphins, 30% were of pairs, 9% of triples, and only 1% of groups were larger than 3 animals. Seasonal differences in group size were statistically significant for *Inia* in the Marañón (Kruskal-Wallis $H = 11.03$, $P = 0.01$) and Samiria rivers (Kruskal-Wallis $H = 12.48$, $P = 0.006$); other rivers were not sampled frequently enough across seasons for statistical comparison. In the Marañón River, mean group size was greatest during low water (1.7 dolphins), and smallest during rising water (1.5 dolphins). In the Samiria River, mean group size was largest during falling water (1.9 dolphins) and smallest during rising (1.5). Median *Inia* group size in lakes did not differ significantly according to season in two of the three lakes surveyed. In lake Atun Cocha, the seasonal differences were significant (Kruskal-Wallis $H = 20.45$, $P = 0.0002$) and the largest groups occurred during falling water (mean group size of 2.1 dolphins). Regardless of season or habitat, mean *Inia* group size was never greater than 2.1 dolphins.

In Bolivia, solitary individuals comprised 41% of all observations. The frequency of group size varied significantly by season ($\chi^2 = 73.67$; $df = 15$; $P = 0.00$). Solitary individuals were seen most frequently in all seasons, with the exception of low water, during which time pairs were seen more often than other group sizes. Mean group size was 2.7 during falling water, 2.9 during low water, 2.3 during rising water, and 2.1 during high water.

Seasonality of Reproduction

Inia neonates were observed year-round in Peru and Bolivia, although diffuse seasonal peaks occurred during falling water in both of these study areas. In Venezuela, neonates were never observed during falling water, but began appearing at the end of the low water season and peaked during rising water (high water was not sampled due to logistical constraints). Significant seasonal differences in neonate sightings existed in all three study areas, even after accounting for seasonal differences in sampling efforts and overall dolphin abundance (Figure 4 & Figure 5; Peru $\chi^2 = 135$, $df = 3$, $P < 0.001$; Bolivia $\chi^2 = 17$, $df = 3$, $P < 0.001$; Venezuela $\chi^2 = 34$, $df = 2$, $P < 0.001$) .

One *Inia* neonate was necropsied 13 August 1997 during low water in Peru. The animal was a female, 82.5 cm total length (straight length, tip of snout to tail notch), with fetal folds. The teeth had not yet erupted but were visible just below the surface of the gums. The stomach was empty, although the intestinal contents indicated a milk diet. Based on its total length and da Silva's (1994) estimates of neonate growth rates, it was estimated to be < 1month old, and born between falling and low water. Dead females that were pregnant or lactating were not encountered.

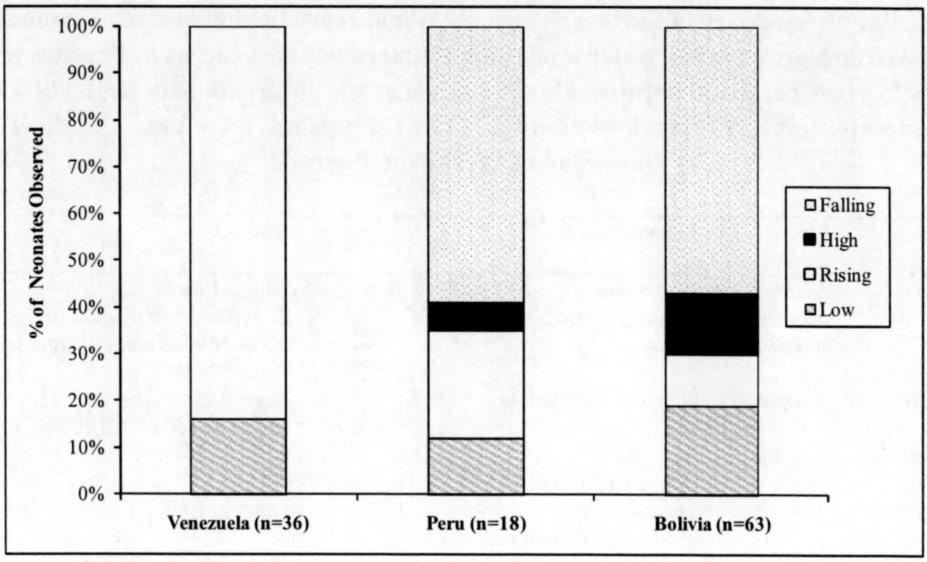

Figure 4. Neonates per season, as percent of all neonates observed within a study area (n= number of neonates observed). Venezuela was not sampled during high water (McGuire and Aliaga-Rossel 2007).

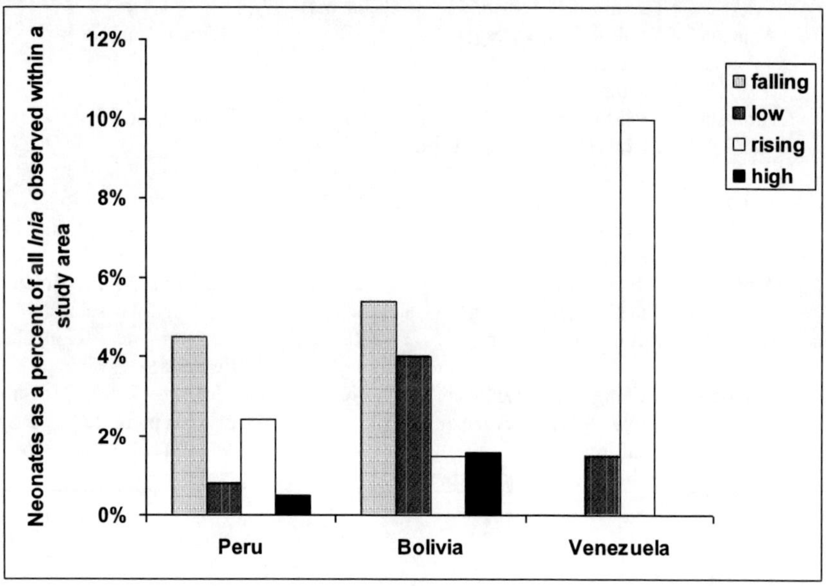

Figure 5. Neonates as a percent of all *Inia* observed within a season and study area. Venezuela was not sampled during high water (McGuire and Aliaga-Rossel 2007).

Inia mating behavior in Peru was observed in all seasons and was not significantly associated with season ($\chi^2 = 3$, df = 3, $P = 0.40$, $n = 11$). Mating behavior was observed only during falling and low water in Bolivia ($n=14$). Mating was not observed in Venezuela.

A review of published and unpublished studies from Bolivia, Brazil, Colombia, Ecuador, Peru and Venezuela indicate geographic variation in seasonality of *Inia* reproduction (Table 6).

Table 6. Summary of the literature on seasonal reproduction in *Inia*. Seasonal comparisons are based on water levels and not month of the year, as peak water levels vary by river basin and latitude. Flux is defined as the difference between highest and lowest water levels. Mating: F =Falling, L = Low, R = rising, YR = Year round, NI = no information, NO = Not observed

Country	River Basin	Neonates/calves	Mating	Authors	Flux (m)
Bolivia	Amazon (Mamoré sub-basin)	year-round (peaks during falling water)	F & L	Aliaga-Rossel 2000, 2002; Aliaga-Rossel *et al.* 2006, McGuire & Aliaga-Rossel 2007	7
Brazil	Amazon	high and early falling	NI	Best 1984, Best & da Silva 1984, da Silva 1994	13
Colombia	Amazon	year-round (few seen during rising)	NI	Hurtado 1996	10-15
Colombia	Amazon	year-round, (peaks during high water)	NI	Galindo 1998	10
Colombia	Amazon & Orinoco	falling (Amazon Basin) no info (Orinoco Basin)	L (Amazon and Orinoco basins)	Beltrán & Trujillo 1993. Trujillo *et al.* 1998	no info.
Ecuador	Amazon	falling/low water (other seasons not sampled)	NI	Herman *et al.* 1996	no info.
Peru	Amazon	year-round (fewest during high water)	NI	Henningsen 1998	7
Peru	Amazon	rising and falling water (other seasons not sampled)	NI	Leatherwood 1996	7
Peru	Amazon	year-round peak during falling	YR	McGuire 2002, McGuire & Aliaga-Rossel 2007	7
Venezuela	Orinoco	end of low water	R	Caranto & Gonzalez-Fernandez 1998	no info.
Venezuela	Orinoco	rising water and end of low water never during falling water (high water not sampled)	NO	McGuire 1995, McGuire & Winemiller 1998, McGuire & Aliaga-Rossel 2007	5

Seasonal Mortality

Necropsy results from Venezuela, Peru, and Bolivia are summarized in Table 7. It was rare for us to encounter dead dolphins, and in remote tropical areas, carcasses were usually in advanced stages of decomposition with significant loss to scavengers. Furthermore, local people in Peru reported that dolphins that had been intentionally killed by fishermen were sometimes hidden from survey teams (Reeves et al., 1999).

Table 7. Necropsies of *Inia* from Peru, Venezuela, and Bolivia, 1993-2000 field seasons (McGuire 1995, Aliaga-Rossel 2000, McGuire 2002). Sex (F=female, M=male, U=undetermined), and total length (T.L; snout to notch of tail) data are included. An X indicates that a body was too decomposed to measure total length. Potential cause of death comments are also included (P = poisoning, HT = Head trauma, Intra-specific aggression = ISA, VS= Vessel-strike, & U = undetermined)

Date	Location	Sex *	T.L. (cm)	Season (relative water level)	Comment
28-October-96	Peru	U	230.00	Low	P
28-October-96	Peru	U	217.05	Low	P
28-October-96	Peru	U	X	Low	P
29-October-96	Peru	U	232.50	Low	P
29-October-96	Peru	U	X	Low	P
29-October-96	Peru	U	X	Low	P
21-January-97	Peru	M	193.04	Rising	P
13-August-97	Peru	F	82.50	Low	HT
5-December-98	Peru	M	185.00	Rising	U
12-September-00	Peru	U	X	Falling	U
6-March-94	Venezuela	M	168.0	Low	U
4-September-98	Bolivia	M	200.0	Low	ISA
9-July-99	Bolivia	M	215.0	Falling	VS

Strandings

Stranded dolphins were not observed in any of the study areas. In Bolivia, live *Inia* were observed in an oxbow lake that had been isolated from the main river channel during extreme low water. During the extreme low water season in Bolivia and Venezuela, *Inia* were observed to transit between water bodies via very shallow channels (0.5 m depth), but they always appeared to move quickly and gave the impression of increased wariness as they did so.

Intra-Specific Aggression

Although tooth-rake marks believed to be made from con-specifics were often seen on live *Inia* in all three study areas, in Peru and Venezuela none of the necropsied *Inia* (Table 7) had tooth-rake marks on the body that might have indicated aggression from other dolphins.

During extreme low water in Venezuela, an adult *Inia* (sex undetermined) was observed for approximately ten hours as it appeared to try to maintain its position in very shallow water (0.25-0.5 m) along a sandy beach, while as many as five adult dolphins patrolled the deeper water nearby, occasionally approaching the beached dolphin as singles, pairs, and trios, and biting, prodding, and grabbing it by the tail in apparent attempts to drive or drag it to deeper water. The beached dolphin vigorously resisted these efforts, and active bouts of snorting, chuffing, tail slapping, pectoral fin slapping, and thrashing occurred throughout the day. This dolphin was not seen in the study area again (McGuire, 1995).

A dead *Inia* encountered during the low water season in Bolivia may have died as a result of intra-specific aggression. One day before the carcass was discovered, reproductive activity among approximately 12 *Inia* was observed along a sandy beach at a confluence. There was a great deal of splashing, chasing, snorting/chuffing, and contact between the dolphins. This activity continued throughout the night, accompanied by abnormally loud exhalations and labored breathing by at least one dolphin. The following morning, the body of a freshly-dead adult male *Inia* was encountered on the beach. Its pectoral and caudal fins displayed numerous deep tooth scars from other *Inia* (determined by scar depth, shape, and spacing), bleeding from the anal slit (from internal hemorrhaging), and heavy bruises and swelling around the blowhole (Aliaga-Rossel, 2000).

Vessel Strikes

In Venezuela, vessel strikes of *Inia* were not observed, nor were any dead *Inia* encountered that appeared to have been victims of vessels strikes. Two photo-identified *Inia* in Peru had deep dorsal scars that may have been caused by boat propellers, and a third dolphin had a broken upper beak that might have resulted from a boat strike. A dead dolphin encountered during rising water in Bolivia appeared to have been struck by an outboard motor, as the left side of the beak was heavily cut and the lower left jaw was broken (Aliaga-Rossel, 2000).

Interactions with Fisheries

Aliaga-Rossel (2000, 2002) reported that the major threat to *Inia* in Bolivia's Mamoré River Basin is entanglement in nets during fishing operations, especially during low water when fishing effort by humans is greatest. Interviews with fishermen indicated that the season of greatest fishing success is between May and September (period of falling and low water). During this time, the fishermen place their nets at the mouths of the lagoons and tributaries in order to catch fish heading for the main river, and they frequently encounter dolphins trapped in their nets. They are more likely to trap the younger dolphins, as perhaps these animals are the least experienced with nets and the most curious. Dolphins may be released alive by fishermen, but often they are left to die, or are killed with machetes. The fishermen may use dolphin meat as bait to attract fish, or use the oil to cure lung ailments. Local people report that during one year in the Bolivian study area, six dolphins died in fishing nets; of these, two calves, one juvenile, and one adult were captured between May to September, which correspond to the seasons of falling and low water. The nets were located across the mouth of a lagoon.

The major threats to river dolphins in Peru have been reported as entanglement in fishing gear (especially gill nets), and also capture in drop traps designed to catch paiche fish (*Arapaima gigas*) and manatees (*Trichechus inunguis*; Leatherwood, 1996). A local fisherman estimated that as many as 50 dolphins a year were being killed in nets or drop traps in the oxbow lake Tipishca del Samiria (Leatherwood, 1996), although Zúñiga (1999) conducted almost daily surveys of this lake for nearly a year and did not encounter fishery-related dolphin mortalities. Taboos that prevented the intentional harm of river dolphins once existed among the native people of the region (Best & da Silva, 1989; Slater 1994), but such beliefs are disappearing in some places, and more recent settlers often regard the dolphins as competitors for fish (Leatherwood, 1996; Reeves et al., 1999). On occasion, dolphins are deliberately killed by fishermen attempting to protect their nets or reduce the dolphins' take of fish (Leatherwood, 1996; Reeves et al., 1999; McGuire 2002).

Of the ten dead *Inia* encountered in Peru between 1996 and 2000, seven were encountered during low water, two during rising water, and one during falling water (Table 7). Of the seven dead *Inia* encountered during low water, six were believed to have died from fisheries-related poisoning (for more details see McGuire, 2000 and Reeves et al., 1999). Dead *Inia* were never encountered during high water, perhaps because sources of mortality are reduced during this season, and/or carcasses would have been more difficult to detect in the submerged vegetation. In Peru's lake San Pablo de Tipishca, nets, traps, and fishing camps were most common during the period between late falling to rising water (Figure 6). Fishing effort was minimal during high water. The inverse pattern was observed for dolphin abundance.

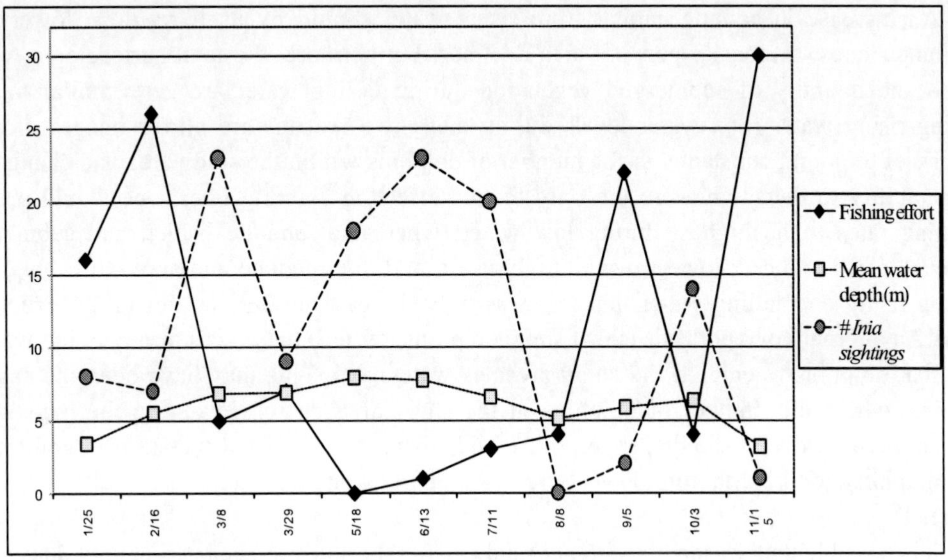

Figure 6. Seasonal pattern of number of dolphins, human fishing effort, and water depth in lake San Pablo, Peru in the year 2000 (McGuire 2002).

Venezuela's Santos Luzardo Park is closed to commercial fishing, and subsistence fishing is minimal in this sparely populated area. A minor threat from sports fishing does exist, as some sports fishermen shoot at *Inia* to drive them away from game fish (McGuire, 1995; McGuire & Winemiller, 1998). Sports fishing is most common during the low water season (especially during the Holy Week vacation), and does not generally occur during the high water season. The mean CPUE of seine and gill net samples in Venezuela varied by habitat and season. Maximum gill net CPUE occurred during low water in lakes, and minimum gill net CPUE occurred during low water in confluence areas. Maximum seine CPUE was obtained during low water in confluences, and minimum CPUE occurred during rising water in lakes.

Discussion

Distribution

Consistent seasonal differences in encounter rates from the three study areas had been predicted, but were not observed. Although seasonal differences were observed in Bolivia and Venezuela, the seasonal patterns between the two sites were different. In Peru, encounter rates in rivers and lakes generally did not differ among seasons once differences in seasonal sampling effort and the high variation within seasons were taken into account. Other studies have reported seasonal changes in river dolphin distribution and abundance, although the patterns of seasonal changes have varied. For example, *Inia* were observed most often in the Amazon River in Brazil during low water (Martin et al., 2004), but during high water in the Marañón River in Peru (Leatherwood, 1996). Maximum *Inia* encounter rates were observed during high and rising water in the Ecuadorian Amazon (Utreras, 1996), in the Marañón River of Peru (Leatherwood, 1996), and Rio Caquetá in Colombia (Galindo, 1998).

An increase in aquatic habitat from rising water should by itself result in lower *Inia* encounter rates, as density per unit area declines. Furthermore, the total surface area, water depth and quantity of submerged vegetation during falling water are very similar to that during rising water (i.e., water levels during these two seasons are mirror images of each other). If temporal constancy in the number of dolphins within the study area and changes in vulnerability to detection were the only factors affecting sighting rates, one would predict sighting rates to be highest during low water, when total aquatic habitat and submerged vegetation are reduced, lowest during high water, and intermediate and approximately equal during rising and falling water, but this was only the case in Bolivia, not in Venezuela or Peru. An alternative hypothesis is that some dolphins left the study areas during rising waters, and that dolphins re-entered the study areas as waters fell. This may have been the case in Bolivia, where the number of dolphins in the study area decreased during the high water season, perhaps due to the dispersion of individuals in the inundated forest, or migrations of the dolphins to the Mamoré River and its numerous lagoons and seasonally inundated channels.

It is possible that seasonal differences did exist in the study areas, but were not detected if the four broad categories of seasons that were used did not accurately represent the complex seasonality of the study area. In general, within-season variability was higher than between-season variability; for example during one 5-day period in Peru, the water level of a lake dropped by 2 m, and the number of *Inia* decreased by 32. Dolphins may well respond to changes in water depths and fish abundance, but perhaps not on the scale likely to be detected with only four categories of season. Movements and spawning of many fishes are triggered by small changes in water level, rather than by season (Lowe-McConnell, 1975). Short-term rises in water level are associated with changes in fish migrations (Goulding, 1980). Local fishermen respond to these fluctuations with changes in fishing location and effort, and it is likely that dolphins would do the same.

In general, fish abundance declines during the dry season as young of the year and other fishes are either eaten or strand in shallow, oxygen-poor stagnant waters, while fish biomass achieves its maximum during the late rainy and falling water periods (Lowe-McConnell, 1975, 1979). The dispersal of fishes across the flooded forest and floodplain decreases fish densities per unit area during the rainy season. Fish biomass approaches its maximum at the end of the rainy season and beginning of falling water. As waters recede, young-of-the-year fish move out of lakes and into the rivers. Thus, we had predicted maximum dolphin encounter rates in rivers during falling and low water, and lowest encounter rates in rivers during rising and high water as fish travel up tributaries and disperse onto the floodplain. Why this pattern was observed in some rivers and not all is unclear. Dolphins likely respond to not only the relative abundance of fishes in an area but also the relative ease of capturing them. During the rising and high water period, flooded vegetation permits fish dispersal and provides greater refuge from predators, likely making prey more difficult to locate and capture. Although fish abundance declines during the low water period, fishes are presumably easier for predators to catch because the reduced water volume spatially concentrates them. In Venezuela, maximum fish CPUE for gill nets and seines occurred during low water. This index reflects not only the relative abundance of fishes caught, but also the ease of capture.

The association between dolphin abundance and absolute water depth was examined separately from season because there was substantial variation in depth within seasons, and seasons were classified according to a combination of relative depth and month of the year. The flexible body of *Inia* allows them to maneuver in shallow water, and their long beaks allow them to extract fish from submerged vegetation (da Silva & Best, 1996), thus equipping them to exploit extremes in water depth. In general, dolphin abundance in rivers was not related to absolute water depth. The relationship between *Inia* abundance and depth was not significant in any of the rivers surveyed in Peru. Other factors such as depths of nearby rivers and lakes, river currents, human activity, and distribution and abundance of prey probably affect dolphin abundance as well.

The relationship between seasonal lake depth and dolphin sighting rates appeared to vary according to study area and lake. Maximum dolphin densities in lakes would be expected during transition seasons, when dolphins enter lakes from rivers with the rising water and migratory fish, or leave the floodplain and forest with the receding waters and enter lakes; however, these seasonal patterns of dolphin distribution were not observed in the three study areas. In Venezuela, *Inia* densities in lakes were greatest during falling water, and lowest in rising water (high water season not sampled; McGuire & Winemiller, 1998). In the Bolivian Amazon, Aliaga-Rossel (2002) recorded the highest density of *Inia* in lakes during rising and high water, and lowest densities during the dry season. In one lake in Peru, *Inia* encounter rates were constant year-round, regardless of season or water depth. The findings from this lake suggest that large lakes may be year-round dolphin habitat due to their buffering capacity, as they do not experience the same reduction in surface areas with depth changes as do shallower lakes, and therefore may provide refuge to fish and dolphins during low water periods when shallower lakes and rivers become dry. If dolphins track not only water levels, but also fish movements, as some have suggested (da Silva, 1994; McGuire & Winemiller, 1998; da Silva and Martin, 2000a), dolphin densities in lakes should be minimal during low water as fish migrate to river habitats (Lowe-McConnell, 1975; Goulding 1980), and also during high water, as fish disperse onto the floodplain and into the flooded forest. In the central Brazilian Amazon, da Silva & Martin (2000) observed that *Inia* densities in lakes were greatest during rising water, and that *Inia* moved from lakes to rivers during the dry season, and from rivers to the flooded forest during high water. Trujillo (1990) reported that dolphins in Colombia were found in lakes during the rainy season, and then moved to the Amazon River during falling water, although some *Inia* remained in the lakes during this season. In summary, seasonal dolphin abundance in lakes can be explained only partially by water depth, and other factors such as distance to nearby rivers, lake volume, human activity, seasonal variability, and distribution and abundance of prey probably affect dolphin abundance as well.

As with other river dolphins, confluences appear to be important habitat for *Inia*. There were significant seasonal differences in occurrence of dolphins in confluences in Peru. During low water, *Inia* persisted longer in the confluences and occurred at higher densities than in any other season. Dolphins were least likely to be in confluences during high water relative to other seasons, and when present, were in lower densities than in other seasons. Confluences, with their deep water and high density of prey, appear to provide refuge to dolphins during low water periods, when many lakes and small tributaries are dry or shallow, and fish have left the shallow areas for deep water. Conversely, during high water, dolphins probably leave the confluence areas and follow the rising waters into lakes, tributaries and inundated forests, where fish density is high. Confluence areas in Venezuela however appeared to be favored *Inia* habitat year-round; dolphin densities in confluences were higher than in lakes or rivers, and did not change seasonally. Confluences throughout the Neotropics are known to provide year-round availability of deep water and high densities of fish, as fishes pass through confluences while leaving oxbow lakes and tributaries during falling water, and pass again during rising water when they return (Lowe-McConnell, 1975).

Rates of Travel

Maximum rates of re-sightings of identified dolphins might have been expected during the low water season, when *Inia* were predicted to be concentrated in deep water areas and relatively easy to locate and photograph. In all three study areas however, identified dolphins were more likely to be resighted during transition seasons (i.e., falling and rising water). Although sample sizes are too small to allow for anything more than speculation at this point, the high rates of travel observed for a few identified dolphins during the high and falling water seasons in Peru are consistent with the idea that some dolphins may extend their ranges during this period as they search for dispersed prey.

Group Size

Mean group size remained less than two animals for most habitats and seasons in Peru and Venezuela. Although seasonal differences were statistically significant, in Venezuela, they were numerically small and may have been biologically irrelevant. Mean group size in Bolivia did increase slightly during the low water season, but was never more than three dolphins. A significant seasonal change in group size might indicate seasonal changes in prey distribution and abundance. A distinct seasonal change in group size might also be indicative of the onset of a distinct mating season and/or calving season.

Few studies have examined seasonal changes in *Inia* group size, and fewer still accounted for seasonal differences in sampling effort. It is interesting to note that among the few studies that examined seasonality of *Inia* group size, seasonality existed in the Orinoco River Basin and in the Mamoré River Basin, but rarely in the main Amazon River Basin. The Amazon River is just a few degrees south of the equator, and its seasonal flux is generally not as extreme as those rivers located at higher latitudes (Lowe-McConnell, 1979; Lewis et al., 1995). Compared to the rest of South America, seasonality is minimal in the northwestern Amazon Basin (Lewis et al., 1995), in which the Pacaya-Samiria Reserve is located. The Orinoco River has a more pronounced dry season than the Amazon (Meade & Koehnken, 1991), and it is likely that a similar situation exists for the Mamoré River system, located at about15° S latitude. Perhaps seasonal differences in dolphin group size are more pronounced where seasonal differences in aquatic habitat are most predictable and extreme.

Seasonality of Reproduction

Our field observations and review of the published and unpublished literature from throughout the approximately 8 million km^2 range of *Inia* indicates geographic variation in reproductive seasonality, with some areas exhibiting year-round reproduction with diffuse seasonal peaks (McGuire, 1995; McGuire & Winemiller, 1998; Aliaga-Rossel, 2002; McGuire & Aliaga-Rossel, 2007). Seasonality of peaks in births varied according to study area, and may be more closely associated with local environmental and prey conditions than with relative seasonal differences between high and low water levels of the study areas, or distribution according to river basin or latitude.

Inia births in the central Brazilian Amazon have been reported as highly seasonal, occurring during the local period of high water and the beginning of falling water (Best, 1984; Best & da Silva, 1984; Brownell, 1984; Best & da Silva, 1989; da Silva, 1994; da Silva & Best, 1996). Best & da Silva (1989) hypothesized these births coincided with increased access to prey fish, which become vulnerable as habitat and cover from previously inundated vegetation decrease with falling water levels.

Because the taxonomy of *Inia* is determined by distribution according to river basin, the possible effects of differences in phylogeny are essentially indistinguishable from differences in broad geographic distribution. When we associated study area with species (*i.e., I. boliviensis* in Bolivia, *I.g. humboltiania* in Venezuela, and *I.g.geoffrensis* in Peru and Brazil) no clear patterns emerged with respect to synchrony of reproduction. In fact, there is more similarity in the patterns of seasonality of reproduction in dolphins from the Peruvian and Bolivian study areas (*I.g. geoffrensis* and *I.bolivensis*, respectively) than between *I.g.geoffrensis* from the Peruvian and Brazilian study areas from the same river basin. Although sample sizes are small, two separate studies from Venezuela (McGuire, 1995; Caranto & Gonzalez-Fernandez, 1998) suggest *I.g.humboltania* may differ from the other subspecies in that the calving peaks during low water, and does not occur year-round.

Based on patterns of reproductive seasonality found in other odontocetes, we were interested in exploring if degree of seasonality of reproduction in *Inia* increases with distance from the equator. If so, we would predict little to no seasonality in the da Silva Brazilian study site (3°S) and strong seasonality in the Bolivian study site (14°S), with intermediate levels in Peru and Venezuela (5°S and 6°N, respectively); however this was not the case, and no clear patterns emerged.

Is degree of reproductive seasonality in *Inia* related to relative differences between high and low water levels? Are differences in reproductive seasonality more pronounced where seasonal differences in aquatic habitat are most extreme? In general, differences between high and low water levels not only increase with latitude, but also along the headwater-to-confluence course of a river. Compared to the rest of South America, seasonality is minimal in the northwestern Amazon Basin (Lewis et al., 1995). The Amazon River is just a few degrees south of the equator, where its seasonal flux is generally not as extreme as those rivers located at higher latitudes. In addition, there is an east/west gradient of seasonal changes in water level. For example, the difference between maximum and minimum waters in the Peruvian study area is 7 m, but downriver in central Brazil it is almost twice that amount (Martin & da Silva, 2004a,b); *Inia* reproduction is year-round in the first study site and seasonal in the second. This model does not readily apply to our other study sites however, as Venezuela had the smallest difference between high and low waters, yet *Inia* reproduction was seasonal. Further testing of this hypothesis requires more data from throughout the vast geographic range of *Inia*, and a refinement of our definition of seasonal changes in habitat (*e.g.,* measure not only differences in water depths, but the amount and type of habitat lost and gained throughout a complete flood cycle).

Observations of mating offer little insight into seasonal reproduction of *Inia*, because mating was rarely observed and difficult to classify. Mating by *Inia* was observed in all seasons in Peru, during falling and low water in Bolivia, and not at all in Venezuela. In general, categorizing an activity as mating was subjective. Little of the animals was visible above the surface of the water and this made classification and description of behavior difficult. It was impossible to determine how much of the behavior described as mating was truly reproductive in nature, and how much was other social behavior such as play, aggression, sexual behavior without copulation, or dominance displays. Caldwell et al., (1989) reported that heterosexual and homosexual behaviors were very common in captive *Inia*, regardless of season. Based on examination of dead animals, male *Inia* do not appear to exhibit seasonality in reproductive condition (Best & da Silva, 1984).

Reproduction in odontocetes ranges from highly seasonal to year-round. Extreme latitudes may influence reproduction via extreme seasonal differences in photo-period, water temperature, and prey abundance. In many species, degree of reproductive seasonality is related to geographic distribution; for example tropical *Stenella* reproduce year-round (Barlow, 1984), while at high latitudes, reproduction in *Phocoena, Monodon,* and *Delphinapterus* is highly seasonal (Leatherwood & Reeves, 1983; Read 1990). Reproductive seasonality in some odontocetes however, such as *Tursiops,* has been shown to be flexible and inherent to populations, rather than correlated with latitude (Urian et al., 1996). There is little information on the seasonality of reproduction in other river dolphin species. *Lipotes* is reported to mate in the spring and calve in the winter and spring (Zhou & Zhang, 1991). *Platanista* is thought to have a bimodal calving and reproductive season (Kasuya, 1972; Shresta, 1989). *Pontoporia* is reported to give birth during the austral spring and early summer of higher latitudes (Harrison et al., 1981; Danilewicz, 2003), with year-round births at slightly lower latitudes (Ramos, 1997).

Seasonal Mortality

Inia are listed as vulnerable by the International Union for the Conservation of Nature (IUCN, 2007). Mortality rates in the wild are unknown, and causes of natural mortality in the wild include parasites and respiratory infections (Best & da Silva, 1993; da Silva & Best, 1996). There are no records of non-human predation on river dolphins, although caiman, piranha, jaguars, and bull sharks occur throughout their range. In recent years *Inia* in Brazil and Colombia reportedly have been killed for use as fish bait (Martin & da Silva, 2004b), but we did not witness or hear of this occurring in any of the study sites during the time we were there.

The low water season appears to be the period in which *Inia* are most vulnerable to mortality from stranding, vessel strikes, intra-specific aggression, and fishery interactions. Although strandings due to shallow water were not observed in the study areas, this undoubtedly occurs on occasion, and has been reported in Brazil (Best, 1984) and Venezuela (Kirk Winemiller, Texas A&M University, personal communication). Vessel strikes, although also not observed directly, would seem more likely to occur during extreme low water seasons, when dolphins and vessels are navigating the same restricted waterways. Fishing and tourism activities are primarily vessel-based, and peak during the low water season. In addition to the potential threats from strikes, boat motors produce noise, which may cause disruption of dolphin behavior and habitat use. The two instances of presumed intra-specific aggression were observed during the low water season, and may have been sexual in nature (McGuire, 1995; Aliaga-Rossel, 2000).

Although much of the area of distribution of *Inia* is unprotected, river dolphins do occur in protected areas of countries throughout their range (e.g., Santos Luzardo National Park in Venezuela, Pacaya-Samiria National Reserve in Peru, Mamirauá Reserve in Brazil, Noel Kempff Mercado National Park in Bolivia). Better understanding of the seasonal ecology of *Inia* could contribute to the conservation of this vulnerable species, primarily by enhancing reserve and fisheries management. Consideration of restrictions on vessel traffic and certain fishing practices in critical seasons and habitats may be warranted in some areas.

ACKNOWLEDGMENTS

We thank our families and the following: the Willie May Harris Fellowship, the Mendon B. Krischer Scholarship, Fundación Fluvial de los Llanos, Laguna Larga Lodge, the Cinaruco Fishing Club, Kirk Winemiller, Bernd Würsig, David Jepsen, Don Taphorn, the Garcia family, the volunteers and staff of Earthwatch, Elderhostel and the Oceanic Society, the crew of the Miron Lento and Delfín, INRENA, ProNaturaleza, the Tenazoa family, Elizabeth Zúñiga, Gerónimo Vega Quevare, the Virtual Explorers, Dulcie Powell, Fremen Tours, Ramiro Cuellar and family, Healy Hamilton, the Kramarae/Kramer family, Michael Link and LGL Alaska, the American Cetacean Society, and Cetacean Society International. Research was conducted under the following permits in Peru: INRENA-DGANPFS-DNAP #53-97, #27-99, and #02-S/C2000. For Venezuela, one permit was needed (SARPA #0493) while no permits were necessary for Bolivia (unprotected area).

REFERENCES

[1] Aliaga-Rossel, E. (2000). Distribución y abundancia del delfín de río, bufeo (Inia geoffrensis) en el río Tijamuchi, Beni, Bolivia. Lic. Thesis, La Paz, Bolivia: Universidad Mayor de San Andres.

[2] Aliaga-Rossel, E. (2002). Distribution and abundance of the river dolphin (Inia geoffrensis) in the Tijamuchu River, Beni, Bolivia. *Aquatic Mammals*, 28, 312-23.

[3] Aliaga- Rossel, E., Mc Guire T. L. & Hamilton, H. (2006). Distribution and encounter rates of the river dolphin (Inia geoffrensis boliviensis) in the central Bolivian Amazon. *Journal of Cetacean Research and Management,* 8, 87–92.

[4] Banguera-Hinestroza, E., Cardenas, H., Ruiz-García, M., Marmontel, M., Gaitán, E., Vásquez, R. & García-Vallejo, F. (2002). Molecular identification of evolutionarily significant units in the Amazon River Dolphin Inia sp. (Cetacea: Iniidae). *Journal of Heredity,* 93, 312-22.

[5] Barlow, J. (1984). Reproductive seasonality in pelagic dolphins (Stenella spp.): implications for measuring rates. Reports of the International Whaling Commission, Special Issue, 6, 191-198.

[6] Beltrán, S., & F. Trujillo, F. (1993). Mating behavior of the fresh water dolphin Inia geoffrensis (de Blainville, 1817), in an Orinoco tributary of Colombia. In: Abstracts of the 10th Biennial Conference on the Biology of Marine Mammals, November 11-15, Galveston, Texas: Society for Marine Mammalogy.

[7] Best, R.C. (1984). The aquatic mammals and reptiles of the Amazon. In H. Sioli (Ed.), The Amazon. Limnology and landscape ecology of a mighty tropical river and its basins (pp. 371-412). Dordrecht, the Netherlands: Dr. W. Junk Publisher.

[8] Best, R.C. & da Silva, V.M.F. (1993). Inia geoffrensis. Mammalian Species, 426, 1-8.

[9] Best, R.C. & da Silva, V.M.F. (1984). Preliminary analysis of reproductive parameters of the boutu, *Inia geoffrensis*, and the tucuxi, *Sotalia fluviatilis*, in the Amazon River system. Report International Whaling Commission, Special issue 6, 361-9.

[10] Best, R.C. and da Silva V.M.F. 1989b. Biology, status and conservation of *Inia geoffrensis* in the Amazon and Orinoco river basins. In Perrin,W.F., Brownell, Jr. R.L., Zhou Kaia & Jiankang, L. (Eds.), Biology and Conservation of the River Dolphins (paper 3, pp. 24-34). Gland, Switzerland; International Union for Conservation of Nature.

[11] Brownell,R.L. Jr. (1984). Review of reproduction in platanistid dolphins. Reports of the International Whaling Commission, Special Issue 6, 149-158.

[12] Caldwell, M.C., D.K. Caldwell & R. L. Brill. (1989). Inia geoffrensis in captivity in the United States. In W. F. Perrin, R. L. Brownell, Jr., Z. Kaiya & L. Jiankang (Eds.), Biology and conservation of the river dolphins. Species Survival Commission, International Union for Conservation of Nature and Natural Resources (Occasional Paper 3, pp. 35-40). Gland, Switzerland: International Union for Conservation of Nature and Natural Resources

[13] Caranto, T.M., & A.J. Gonzalez- Fernandez. (1998). Reproduction of the river dolphin (*Inia geoffrensis*) in the refuge of wildlife Caño Guaritico, Apure State, Venezuela. In: Abstracts of the World Marine Mammal Science Conference 1998 (January 20-24, pp. 23). Monaco, Monaco: The Society of Marine Mammalogy.

[14] Cox, B.C. & P.D. Moore. (1993). Biogeography. An ecological and evolutionary approach. Oxford, UK: Blackwell Scientific Publications.

[15] D'Orbigny, M. A. (1834). Notice sur un nouveau genre de cetacé des rivieres du centre de l'Amerique meridionale. Noveau annals du Musee d'Histoire Naturel de Paris, 3, 28-36.

[16] da Silva, V.M.F. (1994). Aspects of the biology of the Amazonian dolphins, *Inia geoffrensis* de Blainville, 1817 (Cetacean, Iniidae) and *Sotalia fluviatilis* Gervais, 1853 (Cetacea, Delphinidae). Doctoral Thesis. Cambridge, United Kingdom: University of Cambridge.

[17] da Silva, V.M.F. & Martin, A.R. (2000). A study of the boto, or Amazon River dolphin *Inia geoffrensis* in the Mamirauá Reserve, Brazil: operation and techniques. In: R.R. Reeves, B.D. Smith & T. Kasuya (Eds.), Biology and Conservation of Freshwater Dolphins in Asia (pp. 121-31). Gland, Switzerland and Cambridge, United Kingdom: International Union for Conservation of Nature.

[18] da Silva, V.M.F., & R.C. Best. (1996). *Sotalia fluviatilis*. Mammalian Species, 527, 1-7.

[19] Danilewicz, D. (2003). Reproduction of female franciscana (*Pontoporia blainvillei*) in Rio Grande do Sol, Southern Brazil. *Latin American Journal of Aquatic Mammals*, 2(2), 67-78.

[20] Galindo, M.A. (1998). Estimación de abundancia y distribución de los delfines de agua dulce *Inia geoffrensis* y *Sotalia fluviatilis* en el Río Caquetá (la Pedrera-Depto. Amazonas). Bachelor Thesis. Cali, Colombia: Universidad del Valle.

[21] Goulding, M. (1980). The Fishes and the forest: explorations in Amazonian natural history. Berkeley, CA, University of California Press.

[22] Hamilton, H., Caballero, S. & Collins, A.G. (2001). Evolution of river dolphins. Proceedings of the Royal Society of Biological Sciences, 268, 549-56.

[23] Harrison, R.J., M.M. Bryden, D.A. McBreardy, & R. L. Brownell, Jr. (1981). The ovaries and reproduction in *Pontoporia blainvillei* (Cetacea: Platanistidae). *Journal of Zoology*, 193, 563-80.

[24] Henningsen, T. (1998). Zur Verbreitung, Habitatwahl und Verhaltensökologie der Delphine Inia geoffrensis und Sotalia fluviatilis im Oberlauf des Amazonas. Doctoral Thesis, Zentrum für Marine Tropenökologie, Bremen, Germany: University of Bremen.

[25] Herman, L.M., Vonfersen, L. & Solangi, M. (1996). The bufeo *Inia geoffrensis* in the Rio Lagarto Cocha of the Ecuadorian Amazon. Marine Mammal Science, 12 (1), 118-25.

[26] Hurtado Clavijo, L.A. (1996). Distributión, uso del habitat, movimientos y organización social del bufeo colorado *Inia geoffrensis* (Cetacea: Inidae) en el alto Río Amazonas. Master Thesis, Guaymas, México: Instituto Tecnológico y de Estudios Superiores de Monterrey.

[27] Instituto Nacional de Recursos Naturales (INRENA-CTARL). (2000). Plan maestro para la conservación y el desarrollo sostenible de la Reserva Nacional Pacaya Samiria y su zona de amortiguamiento (unpublished, available from INRENA, Iquitos, Peru).

[28] IUCN (2007). IUCN Red List of Threatened Species. Available from: www.iucnredlist.org.

[29] Kasuya, T. (1972). Some information on the growth of the Ganges dolphin with a comment on the Indus dolphin. The Scientific Reports of the Whales Research Institute, Tokyo, 24, 87-108.

[30] Leatherwood, J.S. (1996). Distributional ecology and conservation status of river dolphins (*Inia geoffrensis* and *Sotalia fluviatilis*) in portions of the Peruvian Amazon. Doctoral Thesis, College Station, Texas: Texas A & M University.

[31] Leatherwood, J.S., & R.R. Reeves. (1983). The Sierra Club handbook of whales and dolphins. San Francisco, CA: Sierra Club Books.

[32] Lewis, W.M., Jr., S.K. Hamilton, & J.F. Saunders III. (1995). Rivers of northern South America. In C. E. Cushing, K. W. Cummins, & G. W. Minshall (Eds.), Ecosystems of the world. Vol. 22, River and stream ecosystems (pp.219-256). Amsterdam, The Netherlands: Elsevier.

[33] Lowe-McConnell, R. H. (1975). Fish communities in tropical freshwaters. London, United Kingdom: Longman.

[34] Lowe-McConnell, R. H. (1979). Ecological aspects of seasonality in fishes of tropical waters. *Symposium Zoological Society London*, 44, 219-241.

[35] Martin, A. R., & da Silva, V. M. F. (2004a). River dolphins and flooded forest: seasonal habitat use and sexual segregation of botos (*Inia geoffrensis*) in an extreme cetacean environment. *Journal of the Zoological Society of London*, 263, 295-305.

[36] Martin, A.R., & da Silva, V.M. 2004b. Number, seasonal movements, and residency characteristics of river dolphins in an Amazonian floodplain lake system. *Canadian Journal of Zoology*, 82, 1307–1315.

[37] McGuire, T.L. (1995). The ecology of the river dolphin, *Inia geoffrensis*, in the Cinaruco River, Venezuela. Masters Thesis, College Station, Texas, Texas A & M University.

[38] McGuire, T.L. (2002). Distribution and abundance of river dolphins in the Peruvian Amazon. Doctoral Thesis, College Station, Texas: Texas A & M University.

[39] McGuire, T.L. & Winemiller, K.O. (1998). Occurrence patterns, habitat associations and potential prey of the river dolphin, Inia geoffrensis, in the Cinaruco River. Venezuala. Biotropica, 30, 625-38.

[40] McGuire, T. L., & Aliaga-Rossel, R. (2007). Seasonality of reproduction in Amazon River dolphins (*Inia geoffrensis*) in three major river basins of South America. Biotropica, 39(1), 139-135.

[41] McGuire,T.L. & T. Henningsen. (2007). Movement patterns and site fidelity of river dolphins (*Inia geoffrensis* and *Sotalia fluviatilis*) in the Peruvian Amazon as determined by photo-identification. Aquatic Mammals, 33, 359-367.

[42] Pilleri, G. & Gihr, M. (1977). Observations on the Bolivian (*Inia boliviensis* d'Orbigny, 1834) and the Amazonian Bufeo (*Inia geoffrensis* Blainville, 1817) with description of a new subspecies (*Inia geoffrensis humboldtiana*). In: G. Pilleri (Ed.), Investigations on Cetacea. (Volume 8, pp. 11-76). Berne, Switzerland: Brain Anatomy Institute.

[43] Meade, R.H. & Koehnken, L. (1991). Distribution of the river dolphin, tonina *Inia geoffrensis*, in the Orinoco River Basin of Venezuela and Colombia. Interciencia, 16, 300-12.

[44] Ramos, R. M. A. (1997). Determinacao de idade e biología reprodutiva de *Pontoporia blainvillei* e da forma marinha de *Sotalia fluviatilis* no litoral norte do Rio de Janiero. Dissertacao de mestrado. Rio de Janiero, Brazil: Univesidade estadual do Norte Fluminense.

[45] Read, A. J. (1990). Reproductive seasonality in harbour porpoises, *Phocoena phocoena*, from the Bay of Fundy. *Canadian Journal of Zoology*, 68, 284-288.

[46] Reeves, R.R., McGuire, T.L. & Zúñiga, E.L. (1999). Ecology and conservation of river dolphins in the Peruvian Amazon. International Marine Biological Research Institute IBI Reports, 9, 21-32.

[47] Ruiz-García, M., Caballero, S., Martinez-Agüero, M & Shostell, J. M. (2008). Molecular differentiation among *Inia geoffrensis* and *Inia boliviensis* (Iniidae, Cetacea) by means of nuclear intron sequences. In: Koven, V (Ed.), Population Genetics Research Progress.(pp 177-211). New York, NY: Nova Science Publisher, Inc.

[48] Shrestha, T.K. (1989). Biology, status and conservation of the Ganges river dolphin, *Platanista gangetica*, in Nepal. In W. F. Perrin, R. L. Brownell, Jr., Z. Kaiya & L. Jiankang (Eds.), Biology and conservation of the river dolphins (Species Survival Commission, Occasional Paper 3, pp. 70-76). Gland, Switzerland: International Union for Conservation of Nature and Natural Resources.

[49] Sioli, H. (1984). The Amazon. In: H. Sioli (Ed.), The Amazon: limnology and landscape ecology of a mighty tropical river and its basin (pp. 1-25). Hague, The Netherlands: Dr. W. Junk Publishers.

[50] Slatter (1994). The dance of the dolphin: transformation and disenchantment in the Amazonian imagination. University of Chicago Press, Chicago, Illinois, USA.

[51] Trujillo (1990). Aspectos ecologicos y etologicos de los delfines *Inia geoffrensis* y *Sotalia fluviatilis* en la Amazonia Colombiana. Trabajo de grado Biologo Marino. Bogotá, Colombia: Universidad de Bogotá Jorge Tadeo Lozano.

[52] Trujillo, F., S. Kendall, M.C. Diazgranados, M.A. Galindo & L. Fuentes. (1998). Mating behavior of the freshwater dolphin *Inia geoffrensis* (de Blainville, 1817) in the Amazon and Orinoco River systems in Colombia. In: Abstracts of the World Marine Mammal Science Conference (January 20-24, pp. 136). Monaco, Monaco: Marine Mammal Science.

[53] Urian, K.W., D.A. Duffield, A.J. Read, R. S. Wells & E. D. Shell. (1996). Seasonality of reproduction in bottlenose dolphins, *Tursiops trucatus*. *Journal of Mammalogy*, 77(2), 394-403.

[54] Utreras, V. M. (1996). Estimación de la abundancia, aspectos ecológicas y etológicos del delfín Amazónico *Inia geoffrensis geoffrensis* (Cetacea: Iniidae) en el Río Lagartococha, Amazonia Ecuatoriana. Bachelor Thesis, Quito, Ecuador: Pontifica Universidad Católica del Ecuador.

[55] Zhou, K. & X. Zhang. (1991). Baiji: the Yangtze River dolphin and other endangered animals of China. Washington, D.C.: Stone Wall Press.

[56] Zúñiga, E. L. (1999). Seasonal distribution of freshwater dolphins in Tipishca del Samiria, Peru, Master of Science Thesis. College Station, Texas: Texas A&M University.

In: Biology, Evolution and Conservation of River Dolphins… ISBN: 978-1-60876-633-8
Editors: M. Ruiz-García, J. Shostell. pp. 55-70

Chapter 3

CONSERVATION OF THE RIVER DOLPHIN (*INIA BOLIVIENSIS*) IN BOLIVIA

Enzo Aliaga-Rossel
Instituto de Ecología, Universidad Mayor de San Andres – Bolivia.
University of Hawaii at Manoa, EECB program.

ABSTRACT

The pink river dolphin genus *Inia*, is widely distributed in the Orinoco and Amazon basins. Locally called the bufeo (*Inia boliviensis*) in Bolivia, it is an endemic species to the region, geographically isolated from *Inia* populations within the Amazon's main stem by a series of rapids between Guayaramerin, Bolivia and Porto Velho, Brazil. In Bolivia, they are distributed in three main sub-basins: Abuna, Mamore and Itenes (Guapore). Despite bufeo being a native species and the only cetacean present in a land-locked country, its ecology and conservation status are poorly understood. Unfortunately, no conservation laws explicitly target this cetacean in Bolivia and consequently it only receives relatively minor legal protection when it resides in protected conservation areas. This chapter includes information on the studies that have been conducted in Bolivia; the conservation status; aspects related to the geographic distribution of the species, its behavior, ecology, population size, threats and possible means of protection. This information will lead to recommendations for the implementation of priorities in research programs and conservation for this species in Bolivia.

Key words: *Inia boliviensis*, bufeo, pink River dolphin, conservation, Bolivia

INTRODUCTION

The only cetacean species of the South American continent that lives exclusively in freshwaters is the pink River Dolphin (genus *Inia*), locally called bufeo in Bolivia. The group is listed as Data Deficient by The International Union for Conservation of Nature and Natural Resources (IUCN) (IUCN, 2009). Although the population is in better condition than other freshwater dolphin taxa, such as the endangered South Asian River dolphins (*Platanista gangetica*) and the critically endangered baiji (*Lipotes vexillifer*), which it is doubtful will survive as a species (IUCN, 2009; Reeves et al., 2003; Zhanga et al., 2003). Multiple and potentially adverse anthropogenic pressures occur in the major river basins of South America including mining, logging, dam construction, oil and gas exploration, and use of toxic contaminants.

The River Dolphin is distributed widely throughout the Orinoco River basins of Colombia and Venezuela as well as the Amazon basin in Brazil, Colombia, Ecuador, Peru, Guyana and Bolivia (Best & Da Silva, 1993). Little is known about the status of the *Inia* population, and published studies that refer to their ecology, behavior, social structure and biology are scarce. Our knowledge of the basic ecology of the *Inia* comes from research conducted in Brazil (Magnusson et al., 1980; Best & da Silva, 1984; 1989; 1993; da Silva, 1994; da Silva & Martin, 2000; Martin & da Silva 2004), Colombia (Layne, 1958; Trujillo, 1992; Hurtado-Clavijo, 1996; Vidal et al., 1997), Ecuador (Utreras, 1995; Herman et al., 1996; Deniker, 1999), Peru (Leatherwood, 1996; Reeves et al., 1999; Zúñiga, 1999; Leatherwood, S. et al., 2000; McGuire, 2002; McGuire & Henningsen, 2007), Venezuela (Trebbau & Van Bree, 1974; Trebbau, 1978; Meade & Koehnken, 1991; Schnapp & Howroyd, 1992; McGuire, 1995; McGuire & Wienemiller, 1998;McGuire & Aliaga-Rossel, 2007), and Bolivia (Aliaga-Rossel, 2000, 2002; Aliaga-Rossel et al., 2006). However, information from Ecuador and Venezuela generally comes from short term studies, using different methods, which makes comparisons among studies difficult. Long term studies in South America are currently being carried out in Brazil by Vera Da Silva and by the Omacha Foundation in Colombia (McGuire, pers. com., 2008; Trujillo, pers. com., 2008) but no long term studies have been initiated for the bufeo in Bolivia. The Faunagua Foundation does have a program that studies aquatic mammals and water quality in Northern Bolivian Amazonian Rivers. Data garnered from this program might prove to be useful and be incorporated into future dolphin studies. This chapter reviews the current information on the taxonomic situation, ecology, distribution, threats and conservation status of the River Dolphin in Bolivia, and identifies research and conservation priorities.

Taxonomy and Morphology

The Pink River Dolphin belongs to the Order Cetacea, Suborder Odontoceti; superfamily Platanistoidea; Family Iniidae, with a single Genus *Inia*, located only in South America (Reeves et al., 2003).

The Bolivian River Dolphin is geographically isolated from *Inia* populations in the Amazon's main stem, by a series of waterfalls and rapids between Guayaramerin, Bolivia and Porto Velho, Brazil (ca. 400 km). This isolation formed during the late Pliocene (5- 6 millions ago), which might be the cause of the allopatric separation from the other *Inia* populations in the Amazon basin. Comparative mitochondrial DNA sequence analysis has been used to investigate and clarify the taxonomic relationships within *Inia* (Hamilton et al., 2001; Banguera-Hinestroza et al., 2002). These studies found substantial sequence divergence between Bolivian *Inia* and *Inia geoffrensis* in the Amazon and Orinoco Rivers. Banguera et al., (2002) provided additional and stronger evidence from an *Inia* population in Bolivia that warranted its status as a separate species (*Inia boliviensis)*. Their results indicated that two mitochondrial genes (control region and Cyt-b) in the Bolivian form did not share any haplotype in comparison to the other two populations. Furthermore, Ruiz-García et al., (2008b), analyzing autosomal and Y-chromosome intron sequences, presented additional molecular evidence that favored an independent evolutionary history of the Bolivian population. This reinforces the results of Banguera et al., (2002) which subdivided *Inia* into two evolutionary units (Moritz, 1994), although the divergence may not be as temporally distant as was previously claimed.

Similar to molecular genetics studies, morphological studies, have indicated that the Bolivian population is a separate species because of its greater number of teeth, smaller body size but more robust, smaller skull size, larger flipper size and greater tail length (relative to body length) (Pilleri & Gihr, 1977; da Silva, 1994; Ruiz-García et al., 2006). These morphologic and molecular data clearly indicate the uniqueness of the Bolivian *Inia*, highlighting the importance of obtaining further knowledge of its biology, distribution, abundance and ecology.

Distribution and Abundance

In Bolivia, the bufeo is located in rivers of the Amazon basin, within the Cochabamba, Santa Cruz, Beni, and Pando areas (Figure 1).

Their population and distribution within these Rivers is determined by the availability of food rather than by the type of water (white waters, black, clear or mixed), pH or physicochemical characteristics (Aliaga-Rossel, 2003; da Silva, 1994; McGuire, 2002). The bufeo has a preference for river confluences, lagoons, river bends/curves. During high water season (November to April), they expand their area of search for food by swimming inside flooded areas (forests) and small or ephemeral tributary rivers (Best & Da Silva, 1993; Aliaga-Rossel, 2002; Aliaga-Rossel and Quevedo, in prep). They can stay in the same area up to a year upon which they move to other habitats. A paucity of seasonal movement information makes it impossible to completely describe their movement patterns. McGuire & Henningsen (2007) indicated that *Inia* frequently move 40 to 60 km within a 24-h period. Although some individuals can remain in the same location for several days, there are anecdotal reports of *Inia* traveling in excess of 1,000 km (McGuire & Henningsen, 2007). In contrast, Aliaga-Rossel (2000) suggested that *Inia boliviensis* had a more conservative maximum range of 60 km which they can potentially travel in less than a week.

The bufeo is located in the sub-basin of the Madeira River (Fig 1), in the department of Pando, in the Negro and Abuná rivers, which flow into the Madeira River. However, there are no published registers of these occurrences (Anderson, 1997).

They also inhabit the Itenez's sub-basin (Guapore in Brazil) in the Iténez, and its respective tributaries including the Baures, Paraguá, Pauserna, Verde, Blanco, San Luis (Yañez, 1999), San Martín (Salinas 2007, Tapia 1995), Santa Rosa, Machupo (Anderson, 1997) and Irupururu.

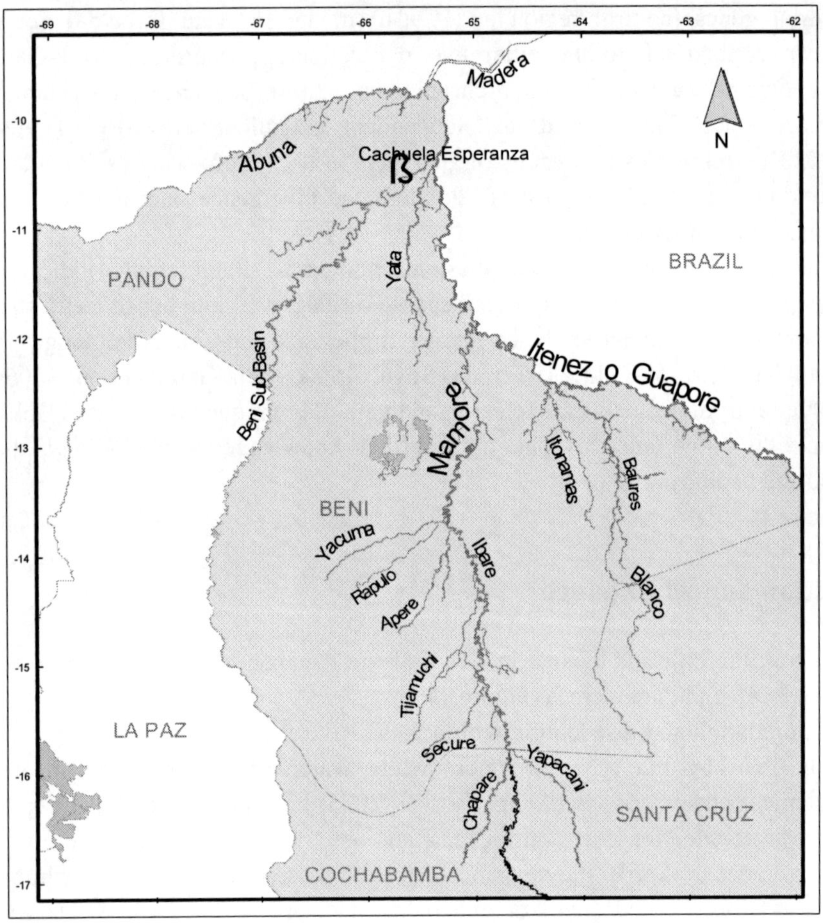

Figure 1. Distribution of the river dolphin, bufeo (*Inia boliviensis*) in the three sub-basins in Bolivia. In the Abuna sub-basin, there is a possible presence of *I. geoffrensis.* "Cachuela Esperanza" is the beginning of the rapids that extends to Brazil and isolates *Inia boliviensis*.

In the Mamore sub-basin, the bufeo is located all along the Mamore River, from the Ichilo River (in Cochabamba) to the Guayanamerin in Beni, including most of their tributaries, such as Sécure, Ibare, Tijamuchi, Apere, Abuná, Rapulo, Itonamas, Yacuma, Yata, Mariquipiri, Baures, and Chapare rivers close to the confluence of the Rio Grande . The indigenous group, the Yuracaré, along the Chapare River have reported the presence of the bufeo close to the community of San Antonio (16º 56'45.4''S- 65º 22'42,9''W, approximately 600 m above sea level) only during the rainy season when the river current is rapid and there is strong turbulence (J. Flores, I. Soria, N. Chavez 2007 com. pers.). This might be the highest elevation for the species, the previous highest elevation being 380m, close to the Puerto Villarroel on the Ichilo River, in the same sub-basin (Pilleri & Ghir, 1977).

The bufeo is not registered in the Madre de Dios and Beni sub-basins, probably because these rivers are isolated by rapids, principally in the area called "Cachuela Esperanza" that form a natural barrier to the Mamore River (Figure 1). Pilleri & Ghir, (1977) and Tello, (1986) indicated the occurrence of the bufeo in the Beni River, but these reports may be inaccurate reports because small tributaries of the Mamore sub-Basin are only a few kilometers from the reported area (Anderson, 1997). For example, in a town close to the Beni

River (Beni Sub-Basin), tourist companies offer swimming experiences with River dolphins, taking tourists to the Yacuma River (Mamore Sub-Basin). Informal interviews with fisherman and local people by the author indicate that there are no bufeos in the Rivers of these Sub-Basins.

There are studies determining the abundance of the bufeo in Bolivia (Table 1). In 2007 the Omacha Foundation (Colombia) and Faunagua (Bolivia) started a program to determine the abundance of bufeo in South America, which is the largest survey carried out in Bolivia. However to date, these data are not yet published.

Table 1. Population Density (PD) of *Inia* as individuals per square kilometer within rivers, sub-basins and departments of Bolivia

Rivers	Sub-Basin	Department	PD (ind/km²)	Study
Ibaré	Mamoré	Beni	1.0 0.2	Pilleri 1969; Pilleri & Gihr 1977 Aliaga-Rossel & Quevedo (in prep)
Ichilo	Mamoré	Cochabamba	0.25	Pilleri & Gihr 1977
Ichilo-Mamore	Mamoré	Cochabamba-Beni	0.88	Omacha-Faunagua unpublished data 2008
Mamoré	Mamoré	Beni	1.6 (dry season)	Aliaga-Rossel et al. 2006
Tijamuchi	Mamoré	Beni	1.12 (all seasons); 5.8 (dry Season)	Aliaga-Rossel 2000, 2002; Aliaga-Rossel et al. 2006
Apere	Mamoré	Beni	2.9 (dry season)	Aliaga-Rossel et al. 2006
Rapulo	Mamoré	Beni	2.6 (dry season)	Aliaga-Rossel et al. 2006
Yacuma	Mamoré	Beni	2.4 (dry season)	Aliaga-Rossel et al. 2006
Iténez	Iténez	Santa Cruz	1.57	Omacha–Faunagua unpublished data 2008
Irupurupuru	Iténez	Beni	1.17	Pilleri & Gihr 1977
Blanco	Iténez	Beni	1.62	Salinas 2007
San Martin	Iténez	Beni	0.74	Salinas 2007

Research on *Inia* in Bolivia

Only a handful of Bolivian bufeo studies have been conducted and published. The first study was conducted by Pilleri, (1969) in the Mamore River, and described the morphology of collected individuals, estimated abundance and compared the behavior of individuals collected in Bolivia with those in Venezuela. Van Bree & Robineau, (1973) compared the morphology of the River dolphins *I.g. geoffrensis* and *I.g. boliviensis*. They suggested the presence of *I.g. geoffrensis* in the Abuná sub-basin, and the other subspecies (now *Inia boliviensis*) in the Mamore and Itenez sub-basins. However, this proposal was not accepted due its lack of comprehensive morphometric and genetic data (Anderson, 1997). Pilleri & Gihr, (1977) surveyed the Ichilo River from Puerto Villarroel to Mamore and morphologically compared Venezuelan and Bolivian individuals and suggested that the Bolivian bufeo was a new species (*Inia boliviensis*). Also, Anderson's book of Bolivian mammals (1997) provides a limited description of the river dolphin distributions based on collections.

Yáñez, (1999) worked in Noel Kempf Mercado National Park, on the Itenez and Paragua Rivers, in the Santa Cruz area. His descriptive study included distribution patterns, dolphin sounds, and possible interactions with the giant otter (*Pteronura brasiliensis*) as well as incorporated legends and myths related to bufeos in the area.

The studies conducted by Aliaga-Rossel, (2000, 2002) evaluated the distribution and abundance of river dolphins during four heavy precipitation seasons across three different water types (white, black and mixed waters) in the Tijamuchi River. Aliaga-Rossel documented one of the highest population densities relative to other areas surveyed. This work constituted the first study in Bolivia using standardized methodology.

Aliaga-Rossel et al., (2006) reported distribution and abundance survey results collected during the dry season in the central Mamore River and four of its tributaries: the Tijamuchi, Apere, Yacuma and its tributary, the Rapulo (Table 1). More recently, Aliaga-Rossel & Quevedo, (in prep) surveyed the Tijamuchi and Ibare Rivers and compared their findings with earlier surveys of the same areas (Aliaga-Rossel 2002; Pilleri 1977). They concluded that the abundance of river dolphins varied temporally with the lowest values for both rivers in the most recent surveys. Although both rivers possessed similar water characteristics, the authors noted a decreased abundance of river dolphins in the Ibare River compared to Pilleri & Gihr, (1977) and the Tijamuchi River. The authors also suggested that this spatial variation may reflect increased human activities in the Ibare River. This work was conducted during the rainy season of a Niña year, when garbage accumulates along the shores and sewage overflows into the river. Salinas (2007) worked in the Blanco and San Martin Rivers of the Itenez sub-basin, covering an area of 56 km with a single observer in a boat (non standardized method). This constituted the first study of relative abundance in this area.

In 2007, Omacha-Faunagua (unpublished data), initiated a long survey project titled "First evaluation of the abundance of the three River dolphin species (*Inia geoffrensis, I. boliviensis*, and *Sotalia fluviatilis*) in the Orinoco and Amazon River Basins, South America". This extensive survey was carried out during the dry season in Bolivia, travelling more than 500 km along the Mamore River, and 598 km along the Itenez River.

Several expeditions were carried out to capture samples for different genetic studies (Banguera *et al.,* 2002; Hamilton et al., 2001; Ruiz-García et al., 2006, 2007, 2008a, b; Ruiz-García, 2010), all of them supporting the status of the Bolivian bufeo as a separate species. In the genetic expedition carried out by Ruiz-García and collaborators in 2003, more than 1.300 km of the Mamoré, Yacumo, Iruyañez, Yata, Itenez and Beni rivers were traveled.

Many other completed studies simply provide lists or distribution data, but these reports are not directly related to bufeos. Other less reliable or non standardized reports such as grey literature, casual observation and posters in scientific conferences also exist (Aramayo, 2008; Tapia, 1995; Tello, 1986). They are considered questionable sources because the methods and/or results were poorly presented and unclear. A common problem that easily leads to the collection of inaccurate data is the use of non-standard methods.

Conservation Status and Threats

The conservation status of the bufeo in Bolivia is incompletely known and laws that explicitly protect the bufeo do not currently exist in Bolivia. Some protection is afforded by the Veda General Indefinida (D.S. 25458), enacted in July of 1999, which is a general decree that prohibits the harassment, capture, harvest, and training of wild animals and their derivatives. Debate exits over the effectiveness of this law, but it continues to be in effect in the absence of more stringent or specific laws (Aliaga-Rossel, 2002).

The IUCN has categorized the bufeo (*Inia geoffrensis*) as a Data Deficient (IUCN, 2009) and it is listed under CITES Appendix II. In the 2008 workshop for the "Red List of Vertebrates of Bolivia", *Inia boliviensis* was categorized as *Vulnerable*, but there are no direct actions or conservation plans from the government. It is important to highlight that the Beni Department declared the bufeo as a "Natural Heritage" of the Beni area (Law evicted on March 2008) (Aliaga-Rossel, 2009).

Although much of the area of distribution of *Inia* is unprotected, River Dolphins do occur in five protected areas: Noel Kempff Mercado National Park, Indigenous territory and National Park Isiboro-Securé, Beni Biological Biosphere Reserve, Itenez Protected Area and the Elsner Espiritu Private Wildlife Refuge. However even in these areas water quality can be reduced because of contamination or other factors, from outside the area.

Genetic studies by Ruiz-García et al. (2008a), indicated that the bufeo population (*Inia boliviensis*) has the lowest genetic richness compared to the other populations, indicating that in some lagoons there are no genetic flows and that there is a very limited inter-connection among populations. Consequently the species seems to be vulnerable to any threat that may affect it.

The direct threats to the bufeo population are identified as habitat degradation, contaminate loads, hunting, over-fishing, navigation by aquatic vessels, and dams. Each of these factors is discussed next.

Habitat Degradation, Decreased Water Quality and Contaminants

The main threat is the deterioration and degradation of the aquatic habitat due to high phosphorus loads and influxes of toxins such as heavy metals, DDT, and chlorine compounds (Maurice-Bourgoin, 1999; Reeves & Kasuya, 2000; Reeves *et al.,* 2003). That comes from pesticides, and agricultural activities. Heavy metals (mercury, arsenic, and lead) originate from gold mine areas, and are used to separate gold from other elements. Later these contaminate bi-products are dumped into rivers where they can magnify in the food chain (Best & Da Silva, 1993; Maurice-Bourgoin et al., 1999).

There is uncontrolled gold exploitation along the Beni River, where there are no controls or specific laws regarding the contaminants that are dumped into the water. Studies carried on the Beni River and among human populations along the upper Madeira River in Brazil have shown that the mercury level is above regulatory limits (Barbosa et al., 1998; Dorea et al., 1998; Maurice-Bourgoin et al., 1999, 2000; Hacon et al., 2000; Dolbec et al., 2001). These authors indicate that these elevated contamination concentrations are affecting not only the people directly engaged in or near mining activities, but also affecting populations located 150 km downriver, where mercury has been found in the hair of people that have a high consumption of fish. Mercury is associated with lesions in the neurological system, motor dysfunction and ocular problems. Studies show the direct effects of the bio-accumulation of contaminants in fish and humans (Hacon et al., 2000; Senthilkumar et al., 1999). This clearly indicates how mining contaminant bi-products might be affecting the bufeo population, constituting a serious threat for the species located at the top of the food chain.

Hunting and Fishing

There is no strong evidence in Bolivia that people living in the area kill bufeos as a source of protein in their diet. However, in Brazil, Colombia, and Peru a main human cause of mortality comes from the death of individuals trapped in fishing nets. Although some of these trapped bufeos are released, in many cases the fishermen let them die or kill them with machetes to avoid damages to their catch and nets. In such cases it is common for the fishermen to use the remains as bait (Best & Da Silva, 1989; Reeves et al., 1999; Aliaga-Rossel, 2002). However, interviews with fishermen in the central and downriver portions of the Mamore River indicate that the fishermen do not consider the bufeo to be a competitor for their fish (Aliaga- Rossel, 2002), as is the case in the Pacaya-Samiria National Reserve in Peru. Here, commercial fishing is a problem, and there have been reports that the dolphins have been intentionally poisoned using methyl-parathion (Reeves *et al.,* 1999). Also a report from near of Tefé, in the Brazilian Amazon, indicates the intentional kills to use as bait, where 1650 bufeos were estimated to be killed per year (V. da Silva 2008 com.pers).

There are only a few studies that have focused on the effects of commercial fishing and the possible impact on aquatic populations. Commercial fisheries in Bolivia are not as extensive compared to neighboring counties. There are reports that some members of the indigenous group, the Yucararés, occasionally hunt bufeos to sell the fat in towns around the area of Trinity City. This seems to be an income-generating practice. Bufeos are also shot for amusement or by sport hunters practicing their aim (R. Cuellar 2008 com. Pers; Aliaga-Rossel 2003).

Navigation

The increase in boat traffic on the rivers has had a negative effect on the bufeos and other aquatic fauna (Constantine et al., 2004; King & Heinen, 2004; Lusseau 2003). A similar observation was made in the study of Pilleri & Gihr (1977), indicating that the increase in boat traffic and the construction of new roads are harmful for the species, were the construction of new roads lead to river bank erosion problems. The increase in traffic is a real threat for the bufeos causing acoustic contamination which can affect communication, orientation and stress for the bufeos, but also the effects of the leakage of oil from boats.

Bufeos can also be injured or killed by boat propellers. A necropsy was performed on a bufeo from the Tijamuchi River (Aliaga-Rossel, 2000) and it was found that the animal had probably been struck by an outboard motor propeller. The left side of its beak was heavily cut and the lower left jaw was broken. Although this was an isolated registered event, deaths by boats have been documented in the Colombian Amazon (Trujillo, 1992). This danger increases during the low water season when more neonates are observed (McGuire and Aliaga-Rossel, 2007) and dolphins use shallow river channels that are also used by fishing boats and other vessels. In some areas (i.e. the Yacuma River), an increase in unorganized tourism and the offer of swimming with dolphins increases boat traffic and results in pollution and increased stress and other possible negative effects on dolphin and other wildlife populations.

Dams

The construction of dams is a direct threat to the bufeo populations due to habitat fragmentation (IUCN, 2009) which influences movement and migration, resulting in isolated populations, as has happened in Asian Rivers and the Brazilian Amazon (Reeves et al. 2000; Reeves et al. 2003), and having a major impact on fish movement and migration, affecting the food chain (Best & da Silva 1989). Currently, there is no official information in Bolivia to build dams on the main rivers in the basins where bufeos are present. However, there are plans to build two dams (Jirau and Santo Antonio) on the Madeira River in Brazil near the border with Bolivia. While there is a lot of concern about the project by the Bolivian scientific community, it seems it will proceed. The argument against the project is that this dam might affect water levels (eventually flooding Bolivian territory and therefore affecting the entire ecosystem), fish populations, migration, and several ecologic factors. Brazil anticipates that the construction of channels will facilitate navigation between Porto Velho and Guayaramerín which will allow boat traffic all year around. Furthermore, the construction of a fish bypass will also allow dolphins to pass the rapids with unknown consequences for the isolated *I. boliviensis* and therefore, the construction of this dam might constitute a direct threat for this population. To date, there is no further information about this construction, and the studies and controversy continue.

Traditions, Myths and Local Use

Traditional myths and legends in Bolivia related to bufeo are rare and relatively unknown, as is the case in Colombia and Peru. For example, some indigenous communities in Bolivia, such as the Itonamas or Baurenses believe that the bufeos were people that were punished by gods by converting them into animals (Ribera 2000). Other groups believe that bufeos transform themselves into humans and seduce young woman in the surrounding villages (Yañez 1999). However, most of the beliefs are simply erroneous perceptions, such as dolphins come out of the water to reproduce; they can have up to 5 neonates at the same time or that they have similar mammary glands (breasts) and reproductive organs as humans. These particular beliefs do not result in respect for or fear of dolphins, as do beliefs common in neighboring countries.

The indigenous group "the Yuracaré", located on the central Mamore, occasionally hunt for subsistence commerce and consume dolphin meat. There are no registers of the number of dolphins killed by the Yuracaré. However, in general, the dolphin meat is considered too greasy and smelly for regular consumption (I. Soria; N. Chavez, com. pers.2006). In the north of the country (Costa Marquez) people believe that the meat is unhealthy and even poisonous (Anderson 1997). People in several villages along the rivers and also in the main towns agree that the fat can be used as a very effective traditional medicine to cure respiratory problems (tuberculosis) and some lung infections. In the local market of Riberalta, they sell bufeo teeth valued as good luck charms or as sexual attractors. There are some indigenous witches who use dolphin genitals, eyes or other body parts for various purposes. This however is not very common.

Natural Mortality

There are no registers on the natural predation of the bufeo. Best & Da Silva (1993) indicated that the black caiman (*Melanusuchus niger*) and the jaguar (*Panthera onca*) are potential predators. There is an anecdotic report in Bolivia that a crocodile (*Caiman yacare*) was on the shore of a river with a bufeo calf in its jaws, but there is no certainty that the crocodile killed the bufeo or if it was already dead (Leonardo Cuellar pers. com. 2009). The author of the chapter observed a dead bufeo possibly caused by intra-species sexual aggression in the Mamore River; the day before encountering the corpse, there was a lot of behavior activity, plays, splashes and contact between them. These behaviors were observed and heard the whole night. The next morning the dead individual was found, with bites marks on the fins, cuts and lesions in the blow hole and internal bleeding. Similar sexual aggression has also been observed in the Colombian Orinoquia (Fuentes *et al.* 2000).

Research Priorities

It is a priority to conduct long term studies on the abundance, distribution, social organization, migration patterns, behavior and mortality of bufeos in different areas where they are present. Standardized techniques should be used in these studies so that the results can be replicated and easily compared, to identify threats to bufeo populations in the different areas of their distribution.

It is important to conduct in-depth studies on the distribution and morphometrics and genetics of bufeos in the Abuna Sub-basin, to evaluate Van Bree & Robineau's (1973) proposition of the possible presence of *Inia g. geoffrensis.* If this is confirmed, Bolivia would have two different river dolphin species. It would also be interesting to monitor the northern area around the rapids or cachuelas, to verify the number of bufeos present and the possible barriers that the rapids represent, because it is probable that during high water season they can overcome the rapids.

Studies need to be carried out on the impacts of deforestation, as well as human settlements along the rivers and the pollution caused by gold mines and the effects of the industries and pesticides on water quality and fish populations. More comprehensive studies on the effect of fishing and the impact on fish communities need to be undertaken.

A monitoring system that records natural and human caused mortalities among the bufeos should also be initiated.

Conclusion and Conservation Priorities

It is necessary to improve the administration and the effectiveness of protected areas where bufeos are located. It is also important to create laws that regulate water use in the head basins, since pollution and contaminates travel well-beyond their sources. It is a priority to not only stop habitat degradation but to also remediate those habitat areas already negatively impacted.

Increased educational awareness of the importance of the Bufeo is critical to a successful long-term conservation plan. Bufeos are clean-water species and consequently their presence and health indicate the general health of the river ecosystem. Therefore the conservation of the bufeo is correlated to the conservation of all aquatic species which in turn is related to water quality. Human pressures on this resource are increasing, therefore activities such as environmental education campaigns can be started, to teach people and make them aware of the importance of conserving and protecting biological diversity and natural resources, as well as recovering local traditions.

It would be important to monitor bufeo populations, to detect natural fluctuations, which could be a basis to understand genetic exchanges.

An environmental education campaign should be started that highlights the indiscriminate use of mercury and its effects on water quality. Mercury regulated levels need to be created and enforced.

Its recommended to stipulate clear regulatory guidelines regarding the use of nylon fishing nets (monofilaments) and to reduce the use of nets that do not discriminate fish size.

The Mamore River should be a priority area to implement the proposed environmental education plans because it is heavily used by boats and fishermen, and the high biodiversity along this river.

There is also a need to regulate tourism activities and tour guides. The activity of swimming with dolphins should be discouraged because this activity causes stress in the animals and changes their behaviors, as has also been determined for other aquatic mammals (Constantine et al. 2004; King & Heinen 2004). As indicated by Wilson et al. (1999) in a species like the dolphin, that is long-lived and slow-breeding, the long-term effects of reduced rest on fitness, individual reproductive success and population size could take decades to detect, which reinforces the need for long term studies.

It is essential to increase public awareness of the River Dolphin across Bolivia. The charismatic or emblematic species can stimulate people to be engaged in conservation issues and to understand conservation problems. In this way, by using this species, it will be possible to start long term educational campaigns similar to the Omacha Foundation's in Colombia or in Brazil. An efficient conservation plan that incorporates a strong educational component can even transform factors that negatively affect river dolphins into positive ones. For example, a well-organized and regulated tourism based on cetacean observation can be a good incentive for protection of these areas, as occurs in Brazil. This activity can contribute to sustainable use and provide another employment option that helps the economies of the communities along these rivers.

Acknowledgments

Thanks to my boat drivers Ramiro and Leonardo Cuellar and their families; and to Tamara McGuire for suggestions to the manuscript; to Moses Montgomery for his review of grammar. Thanks to the Hawaii regional chapter of the Society of Marine Mammals. Thanks to my family.

References

[1] Anderson, S. (1997). *Mammals of Bolivia - Taxonomy and distribution* (231). New York, New York: Bulletin of the American Museum of Natural History.

[2] Aliaga-Rossel, E. (2000). Distribution and Abundance of the River Dolphin, Boto (*Inia geoffrensis*) in the Tijamuchi River in Beni, Bolivia (Licencitura Thesis). La Paz, Bolivia: San Andres University.

[3] Aliaga-Rossel, E. (2002). Distribution and abundance of the river dolphin (*Inia geoffrensis*) in the Tijamuchu River, Beni, Bolivia. *Aquatic Mammals* 28, 312-323.

[4] Aliaga-Rossel, E., Mc Guire T. L. & Hamilton, H. (2006). Distribution and encounter rates of the river dolphin (*Inia geoffrensis boliviensis*) in the central Bolivian Amazon. *Journal of Cetacean Research and Management*, 8, 87–92.

[5] Aliaga-Rossel, E. (2009). *Inia boliviensis*. Pp: 534-535. In: Aguirre, L. F., Aguayo, R., Valderrama, J., Cortez, C., & Tarifa, T (Eds). *Red book of the wildlife Vertebrates of Bolivia*. Ministerio de Medio Ambiente y Agua, La Paz, Bolivia.

[6] Aramayo, P. (2008). Identification of tourist activity impacts on the bufeo (*Inia boliviensis*) in the Yacuma River, Bolivia. *Abstracts of the III meeting of mammalogy in Bolivia.* Santa Cruz. October 20-24, 54.

[7] Barbosa A.C., Silva S.R., & Dorea, J.G. (1998). Concentration of mercury in hair of indigenous mothers and infants from the Amazon basin. *Archives of Environment Contamination and Toxicology*, 34, 100-105.

[8] Banguera-Hinestroza, E., Cardenas, H., Ruiz-García, M., García, Y.F., Marmontel, M., Gaitán, E., Vásquez, R. & García-Vallejo, F. (2002). Molecular identification of evolutionarily significant units in the Amazon River Dolphin *Inia* sp. (*Cetacea: Iniidae*). *The Journal of Heredity*, 93, 312-322.

[9] Best, R.C. & da Silva, V.M.F. (1984). Preliminary analysis of reproductive parameters of the boutu, *Inia geoffrensis*, and the tucuxi, *Sotalia fluviatilis*, in the Amazon River system. *Report international Whale Communication* (special issue), 6, 361-369.

[10] Best, R.C. & da Silva V.M.F. (1989). Biology, status and conservation of *Inia geoffrensis* in the Amazon and Orinoco river basins. In W. F. Perrin, R. L. Brownell, Jr., K. Zhou & L. Jiankang, (Eds.), *Biology and Conservation of the River Dolphins*. (3, pp 23- 34). Gland, Switzerland and Cambridge, United Kingdom: International Union for Conservation and Natural Resources (IUCN).

[11] Best, R.C. & da Silva, V.M.F. (1993). *Inia geoffrensis*. *Mammal Species,* 426, 1-8.

[12] Constantine, R., Brunton, D.H. & Dennos, T. (2004). Dolphin watching tour boats change dolphin (*Tursiops truncatus*) behavior. *Biological conservation*, 117, 299-307.

[13] da Silva, V.M.F. (1994). Aspects of the biology of the Amazonian dolphins, *Inia geoffrensis* de Blainville, 1817 (Cetacean, Iniidae) and *Sotalia fluviatilis* Gervais, 1853 (Cetacea, Delphinidae). (Ph.D. Thesis), University of Cambridge.

[14] da Silva, V.M.F. & Martin, A.R. (2000). A study of the boto, or Amazon River dolphin *Inia geoffrensis* in the Mamiraua Reserve, Brazil: operation and techniques. In: R.R. B.

D. Reeves, B.D. Smith and T. Kasuya (Eds.) *Biology and Conservation of Freshwater Dolphins in Asia*. (pp. 121-131). Gland, Switzerland and Cambridge, United Kingdom: IUCN.

[15] Deniker, J. (1998). A comparison of the abundance and group composition of the Amazon river dolphin (*Inia geoffrensis*) in two blackwater rivers of the Reserva de Produccion Faunistica Cuyabeno, Ecuador. *Abstracts of the World Marine Mammal Science Conference,* January 20-24, 35.

[16] Dolbec, J., Mergler, D., Larribe, F., Roulet, M. Lebel J., and Lucotte, M. (2001). Sequential analysis of hair mercury levels in relation to fish diet of an Amazonian population, Brazil. *Science Total Environ*ment, 23, 87-97.

[17] Dorea, J.G., Moreira, M.B. East, G., & Barbosa A.C. (1998). Selenium and mercury concentrations in some fish species of the Madeira River, Amazon basin, Brazil. *Biologic Trace Element Research*, 65, 211-220.

[18] Fuentes, L., Diezgranados, M. & Trujillo, F. (2000). Reproductive Strategies of *Inia geoffrensis* in the Orinoquia colombiana. Resúmenes, 9° Reunión de Especialistas en Mamíferos Acuáticos de América del Sur. Oct/Nov. 2000. Buenos Aires.

[19] Hacon, S., Yokoo, E., Valente, J., Campos, R.C., Da Silva V.A, de Menezes, A. C., de Moraes L.P., & Ignotti, E. (2000). Exposure to mercury in pregnant women from Alta Floresta-Amazon basin, Brazil. *Environmental research*, 84(3), 204-210.

[20] Hamilton, H., Caballero, S. & Collins, A.G. (2001). Evolution of river dolphins. *Proceedings of the Royal Society for Biological Sciences,* 268, 549-556.

[21] Herman, L.M., Vonfersen, L. & Solangi, M. (1996). The bufeo *Inia geoffrensis* in the Rio Lagarto Cocha of the Ecuadorian Amazon. *Marine Mammal Science*, 12, 118-125.

[22] Hurtado Clavijo, L.A. (1996). Distributión, uso del habitat, movimientos y organización social del boto colorado *Inia geoffrensis* (*Cetacea: Inidae*) en el alto Río Amazonas (Master Thesis), Guaymas, México: Instituto Tecnológico y de Estudios Superiores de Monterrey.

[23] International Union for Conservation of Nature and Natural Resources. IUCN Red List of Threatened Species. [online] 2009 [cited 2010 January 09]. Available from: www.iucnredlist.org

[24] King J.M. & Heinen, J. T. (2004). An assessment of the behaviors of overwintering manatees as influenced by interactions with tourists at two sites in central Florida. *Biological Conservation*, 117, 227-234.

[25] Layne, J.N. (1958). Observations on freshwater dolphins in the upper Amazon. *Journal of Mammalogy*, 39, 1-22.

[26] Leatherwood, J.S. (1996). Distributional ecology and conservation status of river dolphins (*Inia geoffrensis* and *Sotalia fluviatilis*) in portions of the Peruvian Amazon (Ph.D. Thesis). College Station, Texas: Texas A & M University.

[27] Leatherwood, J.S., Reeves, R.R., Wursig, B. & Shearn, D. (2000). Habitat preferences of river dolphins in the Peruvian Amazon. In: R.R. Reeves, B. D. Smith & T. Kasuya (Eds.), Occasional Paper of the IUCN Species Survival Commission, Biology and Conservation of Freshwater Cetaceans in Asia. (23, pp.131-144). Gland, Switzerland: IUCN.

[28] Lusseau, D. (2003). Effects of tour boats on the behavior of bottlenose dolphins: using Markov chains to model anthropogenic impacts. *Conservation Biology*, 17, 1785-1793.

[29] Magnusson, W., Best, R.C. & da Silva V. M. (1980). Numbers and behavior of Amazonian dolphins, *Inia geoffrensis* and *Sotalia fluviatilis*, in the Solimoes River, Brazil. *Aquatic mammals*, 8, 27-32.

[30] Martin, A.R., da Silva, V.M. (2004). Number, seasonal movements, and residency characteristics of river dolphins in an Amazonian floodplain lake system. *Canadian Journal of Zoology,* 82, 1307–1315.

[31] Maurice-Bourgoin (L.) Mercury pollution in Bolivian rivers. News- Scientific Bulletins 95. IRD, 95. [online]. 1999 [cited 2009 March 13]. Available from: http://www.ird.fr/us/actualites/fiches/1999/95.htm

[32] Maurice-Bourgoin, L., Quiroga, I., Chincheros, J., & Courau, P. (2000). Mercury distribution in waters and fishes of the upper Madeira Rivers and mercury exposure in riparian Amazonian populations. *The Science of the Total Environment*, 260. 73-86.

[33] McGuire, T.L. (1995). The ecology of the river dolphin, *Inia geoffrensis*, in the Cinaruco River, Venezuela (Masters Thesis). College Station, Texas: Texas A & M University.

[34] McGuire, T.L. (2002). Distribution and abundance of river dolphins in the Peruvian Amazon (Ph.D. Thesis). College Station, Texas: Texas A & M University, USA.

[35] McGuire, T.L. & Wienemiller, K.O. (1998). Occurrence patterns, habitat associations and potential prey of the river dolphin, *Inia geoffrensis*, in the Cinaruco River, Venezuela. *Biotropica*, 30, 625-638.

[36] McGuire, T. L. & Aliaga-Rossel, E. (2007). Seasonality of Reproduction in Amazon River Dolphins (*Inia geoffrensis*) in Three Major River Basins of South America. *Biotropica.* 39, 129-135.

[37] McGuire,T. L. & Henningsen, T. (2007). Movement patterns and site fidelity of river dolphins (*Inia geoffrensis* and *Sotalia fluviatilis*) in the Peruvian Amazon as determined by photo-identification. *Aquatic Mammals,* 33, 359-367.

[38] Meade, R. H. & Koehnken, L. (1991). Distribution of the river dolphin, Tonina *Inia geoffrensis*, in the Orinoco River Basin of Venezuela and Colombia. *Interciencia,* 16, 300-312.

[39] Moritz, C. (1994). Defining 'Evolutionarily Significant Units' for conservation. *TREE,* 9, 373-375.

[40] Pilleri, G. (1969). On the behavior of the Amazon dolphin, *Inia geoffrensis*, in Beni Bolivia. *Revue Suisse de Zoologie,* 76, 57-74.

[41] Pilleri, G. & Gihr, M. (1977). Observations on the Bolivian (*Inia boliviensis* d'Orbigny, 1834) and the Amazonian Bufeo (*Inia geoffrensis* Blainville, 1817) with description of a new subspecies (*Inia geoffrensis humboldtiana*). In G. Pilleri (Ed.) Vol. 8. *Investigations on Cetacea*, (pp. 11-76). Berne, Switzerland: Brain Anatomy Institute.

[42] Reeves, R. R., McGuire, T. L. & Zúñiga, E. L. (1999). Ecology and conservation of river dolphins in the Peruvian Amazon. *IBI Reports,* 9, 21-32.

[43] Reeves, R. R., Smith, B. D., & Kasuya, T. (2000). Biology and conservation of freshwater Cetacean in Asia. Gland, Switzerland and Cambridge, United Kingdom: IUCN.

[44] Reeves, R. R., Smith, B. D., Crespo, E. A. & Notarbartolo di Sciara, G. (2003). Dolphins, Whales and Porpoises: 2002–2010 Conservation Action Plan for the World's Cetaceans. IUCN/SSC Cetacean Specialist Group. Gland, Switzerland and Cambridge. United Kingdom: IUCN.

[45] Ribera, J. (2000). *Patrimonio espiritual de los pueblos amazónicos bolivianos*. Trinidad: Vicariato del Beni- Comisión de pastoral indígena.

[46] Ruiz-García, M. (2010). Changes in the demographic trends of Pink River Dolphins (*Inia*) at the microgeographical level in Peruvian and Bolivian rivers and within the Upper Amazon: Microsatellites and mtDNA analyses and insights into *Inia's* origin. *In* Ruiz-García, M., & J. Shostell (Eds.), *Biology, Evolution, and Conservation of River Dolphins Within South America and Asia: Unknown Dolphins In Danger*. Hauppauge, NY: Nova Science Publishers., Inc.

[47] Ruiz-García, M., E. Banguera-Hinestroza, H. Cardenas. (2006). Morphological analysis of three *Inia* (Cetacea: Innidae) populations from Colombia and Bolivia. *Acta Theriologica*, 51, 411-426.

[48] Ruiz-García, M., Murillo, A., Corrales, C., Romero-Aleán, N., & Alvarez-Prada, D. (2007). Genética de Poblaciones Amazónicas: La historia evolutiva del jaguar, ocelote, delfín rosado, mono lanudo y piurí reconstruida a partir de sus genes. *Animal Biodiversity and Conservation,* 30, 115-130.

[49] Ruiz-García, M., Escobar-Armel, P., Caballero S. & Secchi, E. (2008a). Determination of microsatellite mutation rates and effective numbers in four Cetacean species (*Inia boliviensis, Inia geoffrensis, Pontoporia blainvillei* and *Sotalia fluviatilis*): comparisons with terrestrial mammals. *Molecular Biology and Evolution* (in press).

[50] Ruiz-García, M., Caballero, S., Martínez-Agüero, M., & Shostell, J. (2008b). Molecular differentiation among *Inia geoffrensis* and *Inia boliviensis* (Iniidae, Cetacea) by means of nuclear intron sequences. In Koven, V. P (Ed.). *Population Genetics Research Progress* (pp. 177-223). Hauppauge, New York: Nova Science Publisher, Inc.

[51] Salinas, A. (2007). Distribución y estado poblacional del bufeo (*Inia boliviensis*) en los ríos Blanco y San Martín (Cuenca del Río Iténez) (Licenciatura Thesis). Cochabamba-Bolivia: Universidad Mayor de San Simón.

[52] Schnapp, D. & Howroyd, J. (1992). Distribution and local range of the Orinoco dolphin (*Inia geoffrensis*) in the Rio-Apure, Venezuela. *Z. Saugetierkd*, 57(5), 313-315.

[53] Senthilkumar, K., Kannan, K., Sinha, R.K, Tanabe, S. & Giesy, J.P. (1999). Bioaccumulation profiles of polychlorinated biphenyl congeners and organochlorine pesticides in Ganges river dolphins. Environmental Toxicology and Chemistry, 18: 1511–1520.

[54] Tapia, C. (1995). Mamíferos acuáticos de la Reserva de Vida Silvestre Ríos Blanco y Negro. Santa Cruz. 41 p. Informe para el Museo de Historia Natural Noel Kempf Mercado.

[55] Tello, L. (1986). La situación de los gatos salvajes (Felidae) en Bolivia. Reporte preparado para CITES. pp. 1-63.

[56] Trebbau, P. (1978). La tonina on *Inia geoffrensis* de agua dulce. *Natura*, 65, 27-28.

[57] Trebbau, P. & Van Bree, P.J.H. (1974). Notes concerning the freshwater dolphin, *Inia geoffrensis* (de Blainville, 1817), in Venezuela. *Z.Saugetierkd*, 39, 50-57.

[58] Trujillo, F. (1992). Estimación poblacional de las especies dulceacuícolas de delfines *Inia geoffrensis y Sotalia fluviatilis* en el sistema lacustre de Tarapoto y El Correo, Amazonia *Colombiana Special Report*. Bogotá, Colombia: Centro de Investigaciones Científicas, Universidad Jorge Tadeo.

[59] Utreras, V. (1995). Abundance estimation, ecological and ethological aspects of the Amazon River dolphin *Inia geoffrensis* in eastern Ecuador. *Abstracts of the Biennial Conference on the Biology of Marine Mammals*, 11, 117.

[60] Van Bree, P. & Robineau, D. (1973). Notes sur les holotypes *Inia geoffrensis geoffrensis* (de Blainville, 1817) et de *Inia geoffrensis boliviensis* D'Orbigny,1834 (Cetacea, Platanistidae). *Mammalia*, 37, 658-668.

[61] Vidal, O., Barlow, J., Hurtado, L., Torre, J, Cendon, P. & Ojeda, Z. (1997). Distribution and abundance of the Amazon River Dolphin (*Inia geoffrensis*) and the tucuxi (*Sotalia fluviatilis*) in the Upper Amazon River. *Marine Mammal Science*, 13, 427-445

[62] Wilson, B., Hammond, P. S. & Thompson, P. M. (1999). Estimating size and assessing trends on a coastal bottlenose dolphin population. *Ecological Applications*, 9, 288-300.

[63] Yañez, M. (1999). Etología, ecología y conservación del delfín *Inia geoffrensis* en los ríos Itenez y Paragua del Parque Nacional Noel Kempf Mercado (Master Thesis). La Paz, Bolivia: Universidad Mayor de San Andres.

[64] Zhanga, X., Wanga, D., Liua, R., Weia, Z., Huab, Y., Wangc, Y., Chend Z., & Wang, L. (2003). The Yangtze River dolphin /or baiji (*Lipotes vexillifer*): population status and conservation issues in theYangtze River, China. *Aquatic Conservation of Marine and Freshwater Ecosystems* 13, 51–64

[65] Zúñiga, E.L. (1999). Seasonal distribution of freshwater dolphins in Tipishca del Samiria, Peru (Masters Thesis). College Station, Texas: Texas A & M University.

In: Biology, Evolution and Conservation of River Dolphins… ISBN: 978-1-60876-633-8
Editors: M. Ruiz-García, J. Shostell. pp. 71-81

Chapter 4

MOBILITY OF THE AXIAL REGIONS IN A CAPTIVE AMAZON RIVER DOLPHIN (*INIA GEOFFRENSIS*)

***Timothy D. Smith*[1, 3*] *and Anne M. Burrows*[2, 4]**

[1]School of Physical Therapy, Slippery Rock University, Slippery Rock PA, 16057 USA
[2]Department of Anthropology, University of Pittsburgh, Pittsburgh, PA, USA
[3]Section of Mammals, Carnegie Museum of Natural History; Pittsburgh, PA, USA
[4]Department of Physical Therapy, Duquesne University, Pittsburgh, PA, USA

ABSTRACT

Here we analyze mobility of axial regions in a captive Amazon River dolphin (Inia geoffrensis), specifically regarding lateral movements of the neck and torso. Still images from video recordings of the swimming dolphin were extracted and analyzed using Scion Image software. Lateral movements of the neck can reach nearly a right angle (deviating from the thoracic region by up to at least 84 degrees). Much more lateral mobility is seen in the torso, with most occurring in the posterior torso (presumably at intervertebral joints in the caudal vertebrae). In sum, the lateral mobility allows this captive dolphin to touch rostrum to tail by lateral bending. Osteological correlates of lateral mobility in this species are also reviewed in this chapter. Based on behavioral descriptions in the literature, the extreme lateral mobility observed in this captive animal are likely representative of the species in general, and relates to locomotion in a complex environment. Further investigations must determine whether this mobility, and the morphological features that permit it, are unique adaptations or primitive features that characterized an ancestral condition.

Keywords: caudal, cervical, *Inia geoffrensis*, lumbar, thoracic, vertebra.

---

* tdsmith@gmail.com

INTRODUCTION

Cetacean evolution has emphasized morphological adaptations to a fully aquatic lifestyle (e.g., Thewissen, 1998; Reidenberg, 2007; Uhen, 2007). In particular, studies of odontocete axial anatomy emphasize features that create a stable, relatively rigid vertebral column and associated connective tissues for rapid swimming (Pabst, 1996; Fish, 1997; Long et al., 1997). The vertebral morphology of many toothed whales (odontocetes) have some fusion of vertebrae in the cervical region and rostrocaudally short vertebral bodies (or centra), characteristics which limit mobility (Buchholtz, 2001). Some axial flexibility is clearly required for propulsion alone, as in dorsoventral fluke movements (Figure, 1a; mobility in the sagittal plane). Axial mobility also affects swimming posture (e.g., in banking), which is important for maneuverability (Fish, 1997). Some lateral flexion (that is, bending in the horizontal plane; Figure 1b) is used by all cetaceans. However, since morphology constrains lateral mobility in odontocetes (see below), lateral flexibility has not been subject to much scrutiny.

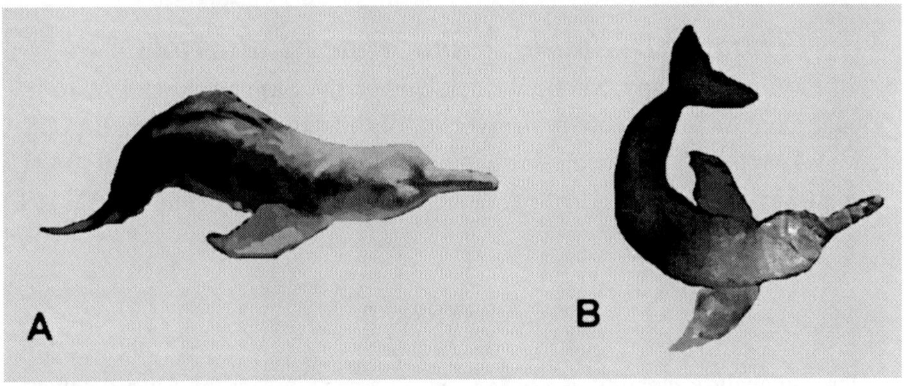

Figure 1. Mobility of a cetacean in the sagittal (a) and horizontal planes (b). Lateral view of a river dolphin showing dorsoventral bending of the body during propulsive movements of the flukes (a) and dorsal view of a river dolphin showing lateral bending of the torso (b).

Some dolphins possess relatively greater vertebral mobility than others. Certain extant river dolphins, such as those of the genera *Inia*, *Lipotes*, and *Platanista*, have unfused cervical vertebrae and other morphologic characteristics that increase mobility (Klima et al., 1980; Fish, 1997; Buchholtz, 2001). Captive or wild *Inia geoffrensis* (Amazon River dolphin) are well known for extreme mobility during locomotion (Schreib et al., 1994; Fish, 1997). For example, the ability to touch the rostrum to the tail by lateral bending of the neck and peduncle has been observed in a captive *Inia* (Schreib et al., 1994). The same animal was observed by Fish (1997), who noted turns could be accomplished by lateral flexion rather than by rotating (banking) the body away from the horizontal plane. A great mobility of the pectoral limbs has been observed in captive *Inia* (Layne and Caldwell, 1964; Klima et al., 1980), in which the forelimbs are frequently used in "oar-like" movements to assist in turns (Klima et al., 1980; Smith et al., 1994) and a captive *Inia* used such movements to maintain a stationary location in an enclosure (T.D. Smith, personal observations).

Numerous studies have utilized video imaging to determine critical aspects of locomotory behavior in aquatic or semiaquatic mammals (e.g., Smith et al., 1976; Fish, 1994; Buchholtz,

2001). This chapter briefly reports unpublished observations on the axial mobility of an adult captive, male *Inia* that was housed at the Pittsburgh Zoo (Figure 2), and examines the mobility of vertebral regions in this mammal based on measurements from video recorded images. This dolphin was previously studied by Fish (1997) and Buchholtz (2001). Herein, movements from the dorsal perspective are examined in the first attempt to quantify the degree of lateral axial flexibility of this species. Anatomical attributes of river dolphins generally, and *Inia geoffrensis* in particular, are reviewed in the context of our findings.

Figure 2. Illustration of a captive (Pittsburgh Zoo) Amazon River dolphin (*Inia geoffrensis*) pursuing live fish (© 2008, Timothy D. Smith). This image emphasizes the extreme mobility of the species, afforded by unfused cervical vertebrae, relatively long vertebral centra throughout the torso, and large paddle-shaped pectoral flippers.

Observations of a Captive River Dolphin: Measurement Techniques

Despite some chronic health problems (Bonar & Wagner, 2003), the male Amazon River dolphin at the Pittsburgh Zoo lived for more than three decades. This dolphin was fed a diet of live brown trout and had daily interactions with trainers or zoo staff. On four separate days, mobility at the head and neck (atlantooccipital joint; cervical vertebrae) and torso (primarily intervertebral joints of lumbar and caudal vertebrae) was examined by videotaping the dolphin during training or feeding sessions. These sessions were taped simultaneously from lateral and dorsal perspectives. In order to synchronize the two videotapes for analysis, a trainer initiated the session by targeting the dolphin or dropping a fish into the tank. The lateral videotapes were used to determine when the body axis of the animal was roughly parallel to the floor of the tank, thus ensuring little distortion on measurements taken from the dorsal view. The dorsal videotapes were then used to obtain measurements of body mobility, but only when a parallel position had been confirmed based on synchronized portions of the lateral videotape. These particular moments were captured and saved as bitmap files for analysis using Scion Image software (NIH).

The captured images were used to assess the axial region in which lateral flexion was occurring. Figure 3a shows a superimposition of a full skeleton of *Inia geoffrensis* (drawn after Flower, 1869, and de Miranda-Ribeiro, 1943) over an illustration of the captive dolphin from the Pittsburgh Zoo. This superimposition shows that dorsal ridge and peduncle are post-thoracic regions, corresponding to the "torso" region as defined by Buchholtz (2001). Herein, we refer to the region of the dorsal ridge, which include the lumbar and parts of the adjacent vertebrae, as the anterior torso. The peduncle, which includes only caudal vertebrae, corresponds to the posterior torso. The anterior torso is the point of greatest dorsoventral bending during swimming in *Inia*, as is also likely true for many extinct cetaceans (Buchholtz, 2001). The lateral mobility in the two parts of the torso is a focus in this chapter.

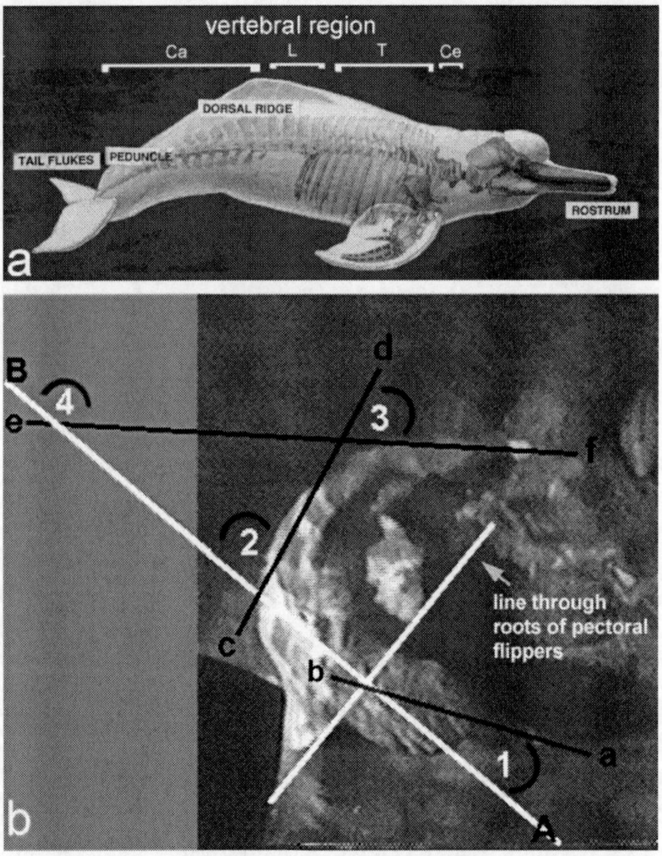

Figure 3. a) Superimposition of the skeleton of *Inia geoffrensis* (drawn after Flower, 1869; and de Miranda-Ribeiro, 1943) over an illustration of the captive dolphin from the Pittsburgh zoo (drawn after a photograph). b) Dorsal still image from video tape of the captive river dolphin, with lines approximating the thorax (A-B, used as midline for calculating lateral flexibility) orientation of neck (based on a line drawn through the rostrum - a-b) and torso (c-d and e-f). Angles of these regions are indicated. The thorax line was identified by drawing a line transecting the roots of the two pectoral flippers and then drawing A-B perpendicular to this line. Angles: 1 = cervical; 2 = anterior torso; 3 = posterior torso; 4 = total torso angle (anterior + posterior). (© 2008, Timothy D. Smith).

The presumptive cervical and torso regions were assessed relative to a thoracic region (see below) in which lateral flexibility is presumed to be more constrained due to costal articulations. The results for each of these regions may include a small amount of mobility afforded by thoracic articulations. Moreover, flexibility in the cervical region could presumably include some movement at the atlanto-occipital joint, although this too is presumed to be rather minimal due the ovoid and transversely broad shape of this joint.

Measurements were obtained from captured still dorsal images as follows. First, a straight line was drawn across the longitudinal (i.e., midline) body axis and measuring the maximum divergence of the neck and torso from this line (recorded in degrees). A line through the thoracic region (Figure 3a) was selected as the midline axis, since this is the least mobile region due to costal articulations. The thoracic line was established as follows: a line was drawn across the flippers, specifically where their anterior edge meets the body wall. The line perpendicular to this line (A-B, Figure 3b) was used to estimate the central line of the thoracic region. Flexion occurring in the cervical and atlanto-occipital regions was estimated by measuring the angle between A-B and a line drawn through the rostrum (a-b, Figure 3b). To estimate the degree of bending in the anterior torso region, a line was drawn through the dorsal ridge (c-d, Figure 3b). The angle between A-B and c-d was interpreted as the angle of lateral flexion formed in the anterior torso region. Finally, a line was drawn longitudinally through the center of the flukes, in line with the peduncle ridge (e-f, Figure 3b). The angle formed by c-d and e-f was interpreted as the angle of lateral flexion in the posterior torso region.

RANGE OF MOBILITY

Table 1 shows the ranges of lateral flexion observed in each presumptive region of the vertebral column as measured from captured images. Also included are ranges for training behaviors versus feeding sessions. The most acute angles, indicating the highest degrees of flexion, are seen in the torso region. The greatest magnitude of lateral flexion is observed in the posterior torso region (about 125 degrees), although the anterior torso region itself has a high degree of lateral flexibility (as much as 95 degrees). Cervical flexibility was more limited, reaching a maximum of 83.8 degrees.

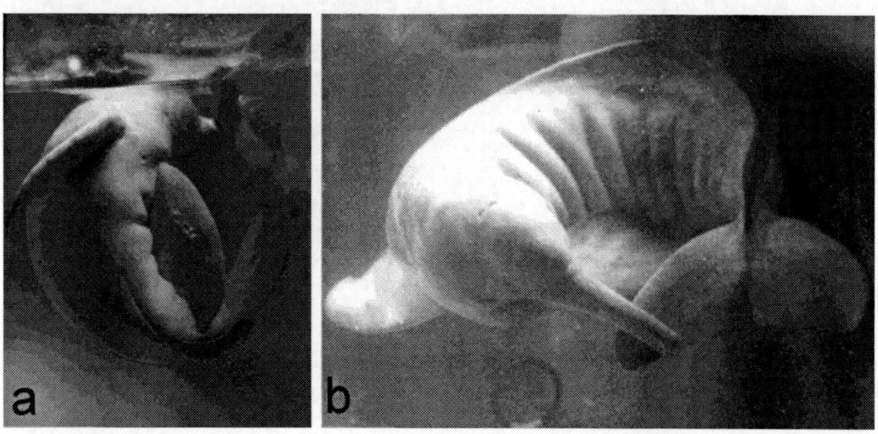

Figure 4. A trained river dolphin touches its rostrum to its tail (a & b). (photographs by TD Smith).

Table 1. Observed Ranges of Lateral Flexibility of Axial Regions (in degrees)

Activity	Cervical[1]	Anterior torso	Posterior torso	Total lateral flexibility[2]
Training	5.7-80.1	32.2-96.9	42-124.8	137.3-314.7
Feeding	14.5-83.8	40-52.4	60.8-95.6	148.7-158.0
All	5.7-83.8	32.2-96.9	42-124.8	137.3-314.7

[1]for definitions of angles see text and figure 2b. [2]includes only observations where all three regions were quantified in the same still image.

The most acute torso angles were measured from training sessions (Table 1). In one trained behavior, this dolphin touched his rostrum to his tail (Figures. 4a, b). The most acute cervical angle was measured from the feeding-only session, almost reaching a right angle relative to the thorax. It is noteworthy that the feeding-only sessions resulted in few measurement opportunities because the dolphin was chasing live fish. Therefore, in many instances he was positioned outside of the horizontal plane. Archived photographs of this dolphin chasing trout suggest he would capture fish using lateral flexion in all axial regions (Figure 5).

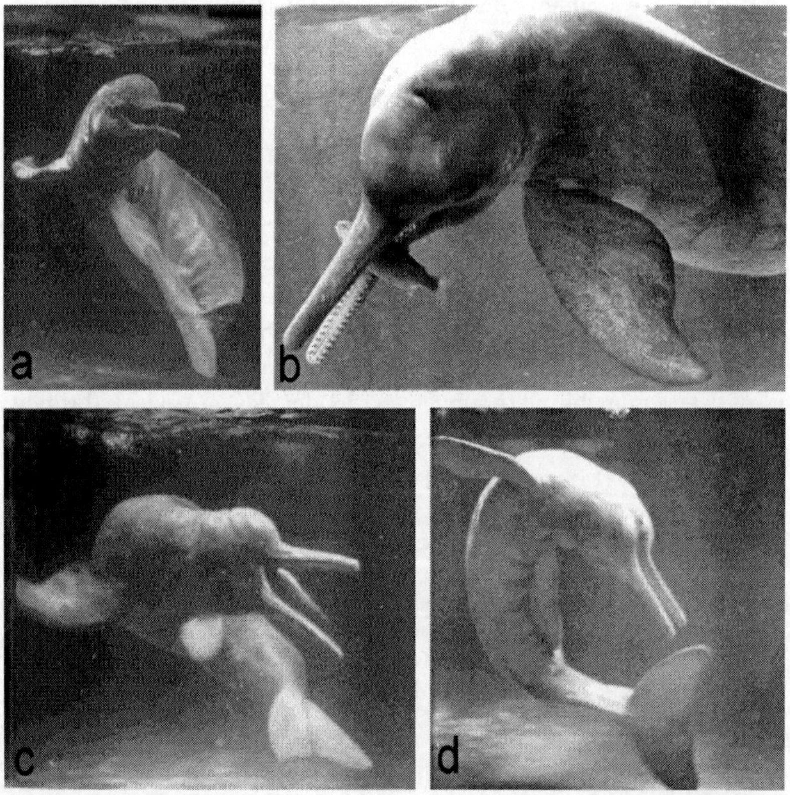

Figure 5. Photographs (a-d) of the captive river dolphin showing lateral flexion in the neck and torso during the pursuit of live trout (photographs by TD Smith).

These measurements are the first attempt to quantify lateral axial mobility in dolphins. The interpretation of regional mobility is made with caution, as it is inferred from skeletal proportions in reference to the external anatomy. Skeletal proportions do not account for certain variables, such as intervertebral disc thickness, which could differentially affect proportions of the axial regions. That said, two deviations from the thoracic line (the cervical and torso angles) provide valid assessments of atlantooccipital/cervical and overall torso mobility. A second caveat is that there is no means to verify that the upper ends of these ranges represent true limits in axial mobility. Nonetheless, the acute lateral flexion measured in these regions in the captive Amazon River dolphin support qualitative descriptions of mobility in captive *Inia* (Schreib et al., 1994; Fish, 1997).

Dolphin Axial Mobility and Navigation

Since no comparable data exist for other cetaceans, it cannot unequivocally be said that *Inia* are unique in lateral mobility. Certainly, lateral mobility in other riverine dolphins would provide advantages in relatively complex, contained environments, in comparison to oceanic environments. Published accounts of swimming behavior in other riverine species are scarce, however. Dolphins of the genus *Platanista* habitually "side-swim," thus favoring dorsoventral bending for navigating turns. In captive *Lipotes vexillifer,* banking during high speed turns has been described ("body bent inward and turned to one side") by Zhou & Zhang (1991). These authors provide little description of lateral mobility, except to indicate the head may be moved "side to side".

Pelagic species have been subject to far more scrutiny (e.g., Smith et al., 1976; Fish, 1997; Buchholtz, 2001). Fish (1997) notes that lateral flexion of the neck is used to initiate turns in pelagic odontocetes. In unpowered turns (i.e., gliding turns without propulsive fluke movements), the caudal vertebrae in the peduncle may be laterally flexed. Thus, lateral mobility is of some importance to other odontocetes, at least at slow speeds. In high speed swimming, however, lateral movements are minimized in oceanic dolphins. Instead, body posture (i.e., banking) and flipper position are used to navigate turns (Fish, 1997).

Fish (1997) observed that *Inia* have no reliance on banking for turning. Aside from some side-swimming trained behaviors, this was borne out by our observations (albeit on the same individual dolphin as observed by Fish). This relates in part to swimming speed as *Inia* have slow speed and a small turning radius. In most marine dolphins, turns occur at a faster speed at a high radius (Fish, 2002). The lateral mobility in *Inia* also may be of very important to navigation in extremely complex environments, specifically one that includes seasonally flooded forests (Best & da Silva, 1989). Data on lateral mobility in other dolphins inhabiting environments that differ in complexity would be of great interest.

Axial Anatomy and Dolphin Mobility

Generally, cetaceans differ anatomically from other mammals in a manner that restricts vertebral mobility. Cetaceans have an extreme length reduction of the vertebral centrum (or body; Figure 6) in the cervical region (see vertebral regions in Figure 3) compared to semiaquatic mammals (e.g., pinnepeds and otters). This decreases mobility while enhancing axial stability (Buchholtz, 2001). Fully aquatic mammals living in the "most extreme aquatic habits" (Buchholtz, 2001 - p 179) have the shortest relative centrum lengths of unfused cervical vertebrae. Buchholtz graphically illustrates this as a trend, revealing that river dolphins have the longest centrum lengths compared to pelagic odontocetes, though still widely separated from the range for pinnepeds or otters. *Inia* also have a relatively low vertebral count in the "torso," the axial region from the end of the thorax to the fluke (Buchholtz, 2001). Although the torso vertebral count is higher in pelagic species, relative centrum length is low. Short centrum length is related to column stability, whereas long centrum length is related to column flexibility (Long et al., 1997). Moreover, shorter neural and transverse processes are related to column flexibility (Long et al., 1997).

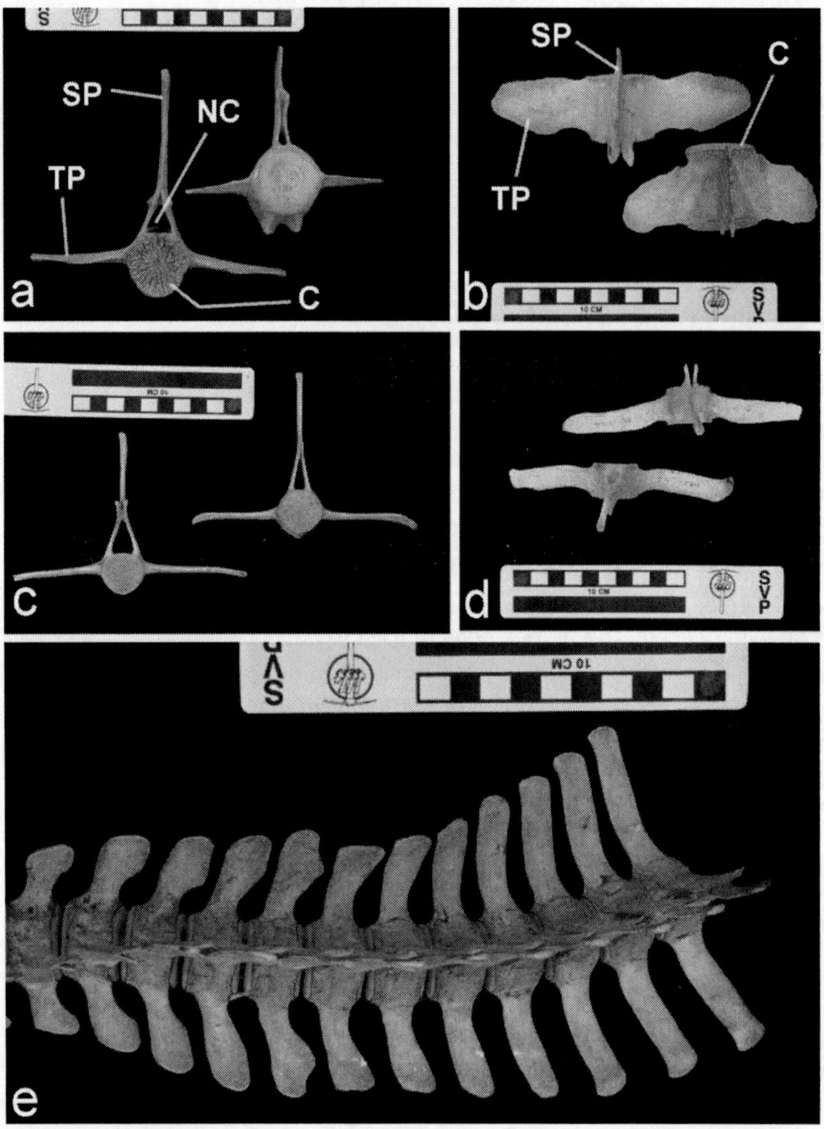

Figure 6. The lumbar vertebrae of *Inia geoffrensis* compared to pelagic odontocetes: photographs of rostral (a, c) and dorsal views (b, d, e) of vertebra from the anterior torso in *Inia geoffrensis* (top) and *Delphinus delphis* (middle), and dorsal view of vertebrae from the anterior torso in the common porpoise (*Phocoena phocoena*)(bottom). Abbreviations: C = centrum (body), nc = neural canal, SP = spinous process, TP = transverse process. Scale in cm.

In more rapidly swimming odontocetes, relatively elongated transverse processes of the vertebrae constrain lateral bending of the vertebral column (Long et al., 1997). Shortened rostrocaudal dimensions of cervical vertebrae and varying degrees of fusion are additional limitations on mobility. These features simultaneously reduce vertebral mobility while increasing stability. In oceanic odontocetes, these limitations significantly affect the anterior torso, where transverse processes and centra are long (Rommel, 1990; Buchholtz, 2001). In these respects, the anatomy of *Inia* departs. Figure 6 shows the lumbar vertebrae of *Inia geoffrensis* compared to pelagic odontocetes. *Inia* has proportionally wider (transversely) and longer (rostrocaudally) centra compared to the pelagic species. Also note the proportionally shorter lateral projection of the transverse processes. These features presumably allow *Inia* to retain lateral mobility in the torso region, while unfused cervical vertebrae (Buchholtz, 2001) allow neck mobility.

Whether such features are retained from ancestors or secondarily evolved as an adaptation to riverine environments is a crucial question. Although the ancestry of modern river dolphins is much debated (de Muizon, 1994; Messenger, 1994), it seems highly plausible that the morphological features enabling mobility of the anterior torso are primitive features. This is the most undulatory axial region during dorsoventral swimming movements in *Inia* (Buchholtz, 2001), as was probably the case for many fossil cetaceans (Buchholtz, 1998). Miocene delphinoids likely had the greatest amount of undulation in the anterior torso as well, based on relative centrum lengths (Buchholtz, 2001).

However, there are ways in which *Inia* stand out among extant odontocetes. *Inia* have uniformly high relative centrum length throughout the torso (Buchholtz, 2001). Since greater centrum length augments mobility (Buchholtz, 1998), this probably accounts for the great lateral flexibility in both anterior and posterior portions of the torso observed in the captive *Inia*. A question that seems unresolved at present is whether the morphological features that facilitate this nearly uniformly mobile torso are primitive or derived. If *Inia* and some of the other river dolphins also derive from a delphinoid lineage, disparate trends for oceanic and riverine dolphins would not be surprising. Vertebral morphology in some known fossil delphinoids departs in significant ways from some extant delphinoids. For example, they do not exhibit a relative increase in centrum length in the posterior torso (as seen especially in the Delphinidae – Buchholtz, 2001). With more detailed knowledge of their ancestry, derived features of the vertebral morphology of river dolphins may become apparent.

SUMMARY AND CONCLUSSIONS

In this chapter, extreme lateral bending in cervical and torso axial regions are documented. Further investigations must determine whether this mobility, and the morphological features that permit it, are unique adaptations as opposed to primitive features that characterized an ancestral condition. In either case, the "flexible body design" (Fish, 2002) of *Inia* is apparent both behaviorally and in vertebral morphology. The evolutionary trend for these dolphins includes the "sacrifice (of) speed for maneuverability..." (Fish, 2002).

ACKNOWLEDGMENTS

We thank the trainers at the Pittsburgh zoo, especially S. Schreib and K. Kello, who helped us immensely with the simultaneous filming sessions. We are also grateful to J.R. Wible and S.B. McLaren of the Carnegie Museum of Natural History (Pittsburgh, PA, USA) for access to odontocete skeletal specimens and photographic equipment.

REFERENCES

[1] Best, R.C., & da Silva, V.M.F. (1989). Biology, status and conservation of *Inia geoffrensis* in the Amazon and Orinoco river basins. In W. F. Perrin, R. L. Brownell, K. Zhou, & J. Liu (Eds). *Biology and Conservation of the River Dolphins Occasional Papers of the IUCN Species Survival Commission* (No. 3, pp 23-34). Gland, Switzerland: International Union for the Conservation of Nature (IUCN).

[2] Bonar, C.J., & Wagner, R.A., (2003). A third report of "golf ball disease" in an Amazon River dolphin (*Inia geoffrensis*) associated with *Streptococcus iniae*. *Journal of Zoo Wildlife Medicine,* 34, 296-301.

[3] Buchholtz, E.A., (1998). Implications of vertebral morphology for locomotor evolution in early Cetacea. In J. G. M. Thewissen (Ed.), *The Emergence of Whales* (pp. 325-351). Plenum, New York: Springer Life Sciences.

[4] Buchholtz, E.A., (2001). Vertebral osteology and swimming style in living and fossil whales (Order: Cetacea). *Journal of Zoology, London,* 253, 175-190.

[5] Fish, F.E., (1994). Association of propulsive swimming mode with behavior in river otters (*Lutracanadensis*). *Journal of Mammalogy,* 75, 989-997.

[6] Fish, F.E., (1997). Biological designs for enhanced maneuverability: Analysis of marine mammal performance. In: *Tenth International Symposium on Unmanned Untethered SubmersibleTechnology: Special Section on Bio-Engineering research Related to Autonomous Underwater Vehicles* (pp 109-117). Autonomous Undersea Systems Institute. Lee, New Hampshire.

[7] Fish, F.E., (1998). Biomechanical perspectives on the origin of cetacean flukes. In J. G. M. Thewissen (Ed.), *The Emergence of Whales* (pp. 303–324). Plenum, New York: Springer Life Sciences.

[8] Fish, F.E., (2002). Balancing Requirements for Stability and Maneuverability in Cetaceans. *Integrative and Comparative Biology,* 42, 85-93.

[9] Flower, W.H., (1869). Description of the skeleton of *Inia geoffrensis* and the skull of *Pontoporia blainvillei*, with remarks on the systematic position of these animals in the order Cetacea. *Transactions Zoological Society London,* 6, 87-116.

[10] Klima, M., Oelschläger, H.A., & Wünsch, D, (1980). Morphology of the pectoral girdle in the Amazon dolphin *Inia geoffrensis* with special reference to the shoulder joint and the movements of the flippers. *Zeitschrifte fur Säugetierkunde,* 45, 288-309.

[11] ***Layne***, J. N., and ***Caldwell, D.K.,*** ***(1964)***. Behavior of the Amazon dolphin, **Inia** *geoffrensis* (Blainville), in captivity. *Zoologica*, 49, 81-108.

[12] Long, J.H., Pabst, D.A., Shepherd, W.R., & McLellan, W.A., (1997). Locomotor design of dolphin vertbebral columns: Bending mechanics and morphology of Delphinus delphis. *Journal of Experimental Biology,* 200, 65-81.

[13] de Miranda Ribeiro, A. (1943). *Inia geoffrensis* (Blainville). *Archivos do Museu Nacional do Rio de Janeiro*, 37, 23-58.

[14] Messenger, S.L., (1994). Phylogenetic relationships of Platanistoid river dolphins: Assessing the significance of fossil taxa. In Berta, A. and Demere, T. (Eds.), Contributions in Marine Mammal Paleontology Honoring Frank C. Whitmore, *Junior. Proceedings of the San Diego Society of Natural History* (Volume 29, pp. 125-134). San Diego, CA: San Diego Society of Natural History.

[15] De Muizon, C., (1994). Are the squalodonts related to platanistoids? In Berta, A. & Demere, T. (Eds.), *Contributions in Marine Mammal Paleontology Honoring Frank C. Whitmore, Jr. Proceedings of the San Diego Society of Natural History* (Volume 29, pp 135-146). San Diegao, CA: San Diego Society of Natural History

[16] Pabst, D.A., (1996). Springs in swimming animals. *American Zoologist,* 36, 723-735.

[17] Reidenberg, J.S., (2007), Anatomical adaptations of aquatic mammals. *Anatomical Record,* 290, 507-513.

[18] Rommel, S., (1990). Osteology of the bottlenose dolphin. In S. Leatherwood, S., & R. R. Reeves (Eds.), *The Bottlenose Dolphin* (pp 29-49). San Diego, CA: Academic Press.

[19] Schreib, S., Burrows, A., & Smith, T., (1994). The Amazon River dolphin (*Inia geoffrensis*). *IMATA Soundings,* 19, 5.

[20] Smith, G.J., Brown, K.W., & Gaskin, D.E., (1976). Functional mycology of the harbor porpoise, *Phocoena phocoena* (L.). *Canadian Journal of Zoology,* 54, 716-729.

[21] Smith, T.D., Mooney, M.P., Siegel, M.I., Taylor, A.B., & Burrows, A.M., (1994). Shape of scapular fossae in freshwater and marine dolphins. Journal of Mammalogy, 75, 515-519.

[22] Thewissen, J.G.M., (1998). *The Emergence of Whales*. Plenum, New York: Springer.

[23] Uhen, M.D., (2007). Evolution of marine mammals: Back to the sea after 300 million years. *Anatomical Record,* 290, 514-522.

[24] Zhou, K. & Zhang, X., (1991). *The Yangtze River dolphin and other endangered animals of China*. Washington DC: The Stone Wall Press.

In: Biology, Evolution and Conservation of River Dolphins… ISBN: 978-1-60876-633-8
Editors: M. Ruiz-García, J. Shostell. pp. 83-99

Chapter 5

The Application of Equation Models to Determine the Age of Pink River Dolphin Skulls

Luisa Fernanda Castellanos-Mora*[1,2]*;
Fernando Trujillo*[2] *and Manuel Ruiz-García*[1]****

[1]Laboratorio de Genética de Poblaciones Molecular-Biología Evolutiva. Unidad de Genética. Departamento de Biología. Facultad de Ciencias. Pontificia Universidad Javeriana. Bogotá DC. Colombia.
[2]Fundación Omacha, Bogotá DC. Colombia.

Abstract

In this chapter, we show that several biometric skull measurements in the pink river dolphin (*Inia geoffrensis*) are highly related to skull age, determined by teeth analysis. Out of 50 morphometric skull measurements, the maximum width between the zygomatic processes of the squamosal bones (V15) and the maximum width of the internal nares (V21) were highly correlated with age. Regression analyses such as linear and 24 other simple models, linear multiple regression, polynomial models, and distance multiple regression (Gower, Absolute value, Mahalanobis and Minkowski) had similar results. On average, these multiple regression equations demonstrated that the age of a dead pink river dolphin is determined by only four cranial measures which explained 64-65% of age variation.

Keywords: *Inia*, ages by teeth analysis, craniometric analyses, regression models.

* mruiz@javeriana.edu.co , mruizgar@yahoo.es

INTRODUCTION

For several decades now, society has known about the negative impact of fishery activities on the world's Cetacean populations. The main cause of their death is fish nets, a problem, which also affects other animals such as turtles and is happening in both marine and in river ecosystems (Northridge & Hofman, 1999).

The case of the pink river dolphin (*Inia geoffrensis*) is no exception as there are regular cases of pink river dolphins accidentally caught and found dead in monofilament nets. Several studies have shown the negative impact of these nets on the populations of this river dolphin (Trujillo, 2000; Trujillo et al., 2006). Additionally, the fishermen of large catfish species believe that this dolphin is a competitor of their resources and consequently, sometimes they resort to guns and harpoons to kill the dolphins. More recently, another threat developed that is affecting the pink river dolphins. The dolphins are purposefully killed and left to rot to attract some small catfishes, like the "mota" or "mapurito" (*Calophysus macropterus*) in the Orinoco and in the Amazon (Colombia and Brazil, mainly). Some indigenous people have a demand for dolphin parts which bolsters a dolphin market. For example, dolphin oil is used as a form of medicine against pulmonary illness (for instance, this practice is extended into the Mamoré River within Bolivia, Ruiz-García, unpublished observation), whereas teeth and genitalia are utilized as love charms by Indian "witches" (for instance, the "pasaje Paquita" in the Mercado de Belén at Iquitos, Peru; Ruiz-García, unpublished observation).

However, when a dolphin is accidentally killed by a net or fisherman, the real age of the animal is unknown. Only the size of the animal has sometimes been recorded and thus we only know if they are an adult, young or calf. For this reason, the main objective of this chapter is to explain a method of how to determine the age of a dead pink river dolphin through the analyses of craniometric measurements and different mathematical methodologies. We will show the existence of a relationship between the age of dead pink river dolphins estimated by means of teeth analysis and several craniometrical measures. This relationship has not yet been demonstrated with this species, although several craniometrical and morphometric studies have been carried out with *Inia* (Da Silva, 1994; Ruiz-García et al., 2006). Therefore, if one specimen is found dead, some skull variables could be measured and the age of this animal could be estimated. It could be extremely useful to determine which age fraction of the pink river dolphin populations are caught in nets or killed by firearms or harpoons or, if the dead animals represent a random sample of a population. Ages of dead animals were determined through the analysis of teeth growth layer groups (GLGs).

The first works determining the age of mammals by means of GLGs were those carried out on pinnipeda by Scheffer (1950) and Laws (1953) in *Callorhinus ursinus* and *Mirounga leonina,* respectively. Subsequently, other research on this topic was accomplished by Laws (1957, 1958), Mc Laren (1958), Nishiwaki & Yagi (1953) and Sergeant (1959) in pinnipidea and odontocetes. During 1968, in Oslo, The International Whaling Commission (IWC) accepted that the formation of one GLG was equal to one year of life in odontocetes. Later works were in agreement with this finding (Nielsen 1972) with *Phocoena phocoena*, Hohn (1980) and Hohn et al., (1989) with *Tursiops truncates*, Collet (1981) with *Delphinus delphis*, Manzanilla (1989) with *Lagenorhynchus obscurus* and Lockyer (1993) with *Globicephala macrorynchus*. Also the Platanistoid dolphins show an annual GLGs. This was demonstrated by Kasuya (1972) for *Platanista gangetica* and by Anli & Zhou (1992) for *Lipotes vexillifer*.

Pinedo (1991, 1994) and Pinedo & Hohn (2000) showed that most individuals of *Pontoporia blainvillei* caught in nets in Southern Brazil and Uruguay were between 0 and 4 years old by using GLGs, with females of one year old being the most common. Chivers (2002) found that the age range most affected by nets for *Stenella longirostris orientalis* in the eastern tropical Pacific Ocean was between 12 and 17 years old without significant differences between males and females. Archer & Chivers (2002) analyzed a sample of 1,160 dead spinner dolphins obtained from 1973 to 2000, with the highest frequency between 0 and 5 years old. On the contrary, Rosas (2002) did not find any age trend in the *Sotalia guianensis* studied in the Paraná coast of southern Brazil. A unique study of *Inia geoffrensis* conducted by Da Silva (1994) determined the ages of several animals and reported that the oldest individual found was a male of 36 years.

Studies that determine the relationships between age and skull´s measurements have not been numerous in Cetacean species. One example is Mead and Potter (1990) who studied the relationship of age with the pre-maxilla -maxilla coalition (distal fusion) in *Tursiops truncatus*. They concluded that this coalition begins to occur at 7 years of age, before they reach sexual maturity. Perrin (1993) did not find any relationship between skull maturity and the age in the common dolphin, *Delphinus delphis*. Stevick (1999) employed the von Bertalanffy equation to determine a relationship between the age of the animal *Megaptera novaeangliae* and its length. Kemper (1999) determined, that for a whale species (*Caperea marginata*), several post-cranial measures were highly correlated with the age of the animals. Finally, Nummela et al. (2001) determined the existence of a significant relationship between two measures of the lateral-mandibular grosor and the age in *Tursiops truncatus*.

In this chapter, we present the first evidence of a significant relationship among age and several craniometric traits in a river dolphin species.

Material and Methods

The ages of 71 dead river dolphins were determined by means of GLGs using the laboratory procedures described by Da Silva (1995) and Castellanos-Mora (2007). Some *Inia* teeth and GLGs are shown in Figure 1.

Descriptions of the geographical origin of these exemplars are as follow: Twelve (12) samples were from the Ucayali, Marañón, Napo and Amazon rivers at the Peruvian Amazon; nine (9) samples were from the Yavarí river, which follows the frontier among Peru and Brazil; twenty-nine (29) samples proceeding from the Colombian Amazon (mainly from Leticia to the Loreto-Yacu River); three (3) samples from the Negro river at the central Amazon in Brazil; two (2) samples from the Mamoré River in Bolivia and, finally, sixteen (16) samples from diverse rivers of the Colombian Orinoquia (10 from the Orinoco river, 2 from Meta river, 2 from Tomo River, one from Bita River and other from the Inirida River).

Of these 71 exemplars, 45 had complete skulls (28 from the Colombian Amazon, one from the Bolivian Amazon and 16 from the diverse rivers of the Colombian Orinoquia). Fifty (50) variables were recorded from each skull; 42 were skull measurements and eight were teeth counts (Figure 2 and Table 1).

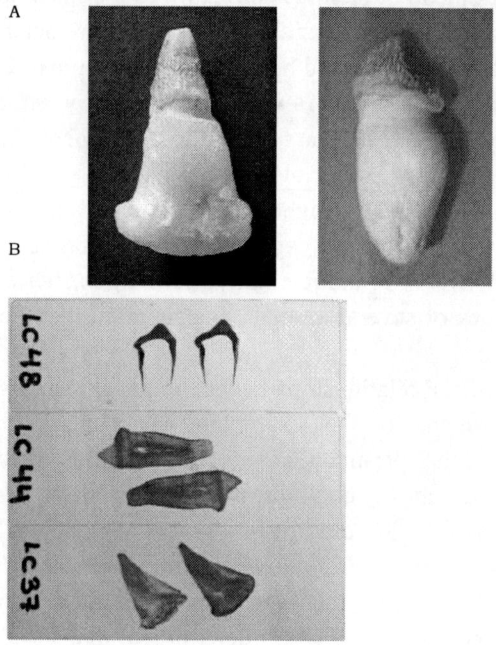

Figure 1. A/ *Inia geoffrensis* teeth and B/ Analysis of teeth growth layer groups (GLGs).

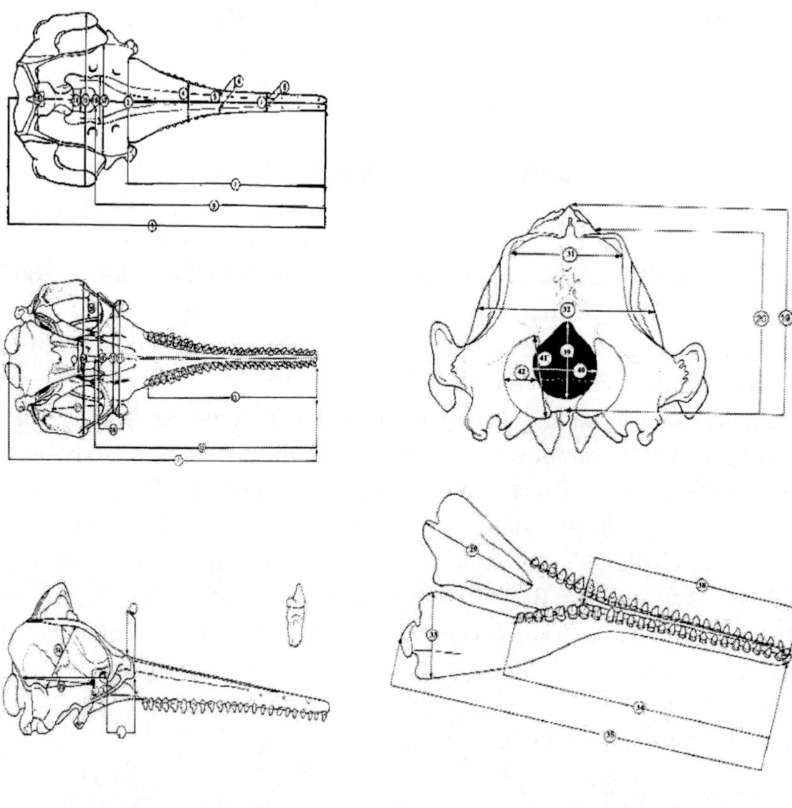

Figure 2. Morphometric skull measures of *Inia geoffrensis* (Modified from da Silva, 1994).

Table 1. Morphometric skull variables analyzed in *Inia*. VN = variable number

VN	Morphometric Measurements
1.	Condylobasal length
2.	Length of rostrum
3.	Width of rostrum at base
4.	Width of rostrum 60 mm anterior to line across hindmost limits of antorbital notches
5.	Width of rostrum at mid length
6.	Width of premaxillaries at mid length of rostrum
7.	Width of rostrum at ¾ of this length
8.	Premaxilla width at ¾ of this length
9.	Distance from tip of rostrum to external nares
10.	Distance from tip of rostrum to internal nares
11.	Greater preorbital width
12.	Greatest postorbital width
13.	Least suproaorbital width
14.	Greatest width of external nares
15.	Greatest width between zygomatic processes of squamous bones
16.	Greatest width of premaxillaries
17.	Greatest width of maxillaries
18.	Greatest parietal width within post temporal fossae measured in the suture of the bone
19.	Vertical external height of braincase with supraoccipital crest
20.	Vertical external height of braincase without supraoccipital crest
21.	Internal length of braincase
22.	Greatest length of left post temporal fossa from post orbital bone
23.	Greatest length of left post temporal fossa from zygomatic extremity
24.	Greatest width of left temporal fossa at right angles to greatest length
25.	Major diameter of left temporal fossa proper
26.	Minor diameter of left temporal fossa proper
27.	Length of left orbit from apex of preorbital process of frontal to apex of post orbital process
28.	Length from edge of lacrimal to end of post orbital
29.	Length of antorbital process of left lacrimal
30.	Greatest width of internal nares
31.	Width of supraoccipital area
32.	Width of braincase across parietals
33.	Length of upper left tooth row
34.	Length of lower left tooth row
35.	Greatest length of left ramus
36.	Greatest height of left ramus
37.	Length of left mandibular fossa
38.	Length of mandibular symphysis
39.	Height of foramen magnum
40.	Width of foramen magnum
41.	Left condyle length
42.	Left condyle width
43.	Number of teeth-upper left
44.	Number of teeth-upper right
45.	Number of teeth- lower left
46.	Number of teeth-lower right
47.	Number of non-conical teeth-upper left
48.	Number of non-conical teeth-upper right
49.	Number of non-conical teeth-lower left
50.	Number of non-conical teeth-lower right

Statistical Analyses

The first statistical analysis performed was a linear regression between the age and each one of the 50 morphometric variables studied. The relationship was expressed as $y = a + bx$, where age was represented by the dependent variable (y). For those variables which showed a significant relationship with age, a multiple regression analysis was undertaken with the expression $y= \mu+b_1 x_1 + ...+b_m x_m$ (Arenas *et al,* 1991), were y was the dependent variable (age) and $x_1 .... x_m$ were the skull variables. Furthermore, the square of multiple correlation was calculated, which represented the goodness of fit between x_i, the observed value of each genetic variable. The better linear predictor, X_i, is a function of m x_i and the expression $R^2 = \Sigma_I (x_i - X_i)^2 / \Sigma_I (x_i - X)^2$, where X is the average of the recorded dependent variable values. The residual variances were calculated as well; these were the mean square distances of the points to the regression hyperplanes, R_o^2 / n, with $R_o^2 = \Sigma_I (x_i - X_i)^2$, where X_i was the prediction of x_i as a function of the regression hyperplane. The number of points compared is abbreviated as (n) and variance was calculated with the equation var $(x_i) = \sigma_i^2 (1 - R^2)$, where σ_i^2 is the variance of the observed distribution of each one of the skull measurements and R^2 is the square multiple correlation coefficient. The square root of this value is the typical error of the estimated x_i value. These analyses were performed with the statistical programs NCSS and Multicua.

Also, a regression analysis employing distances was carried out. Five different distances were used (Gower, Absolute value, Mahalanobis, and Minkowski with q exponent equals to 2 and 4; Cuadras & Arenas, 1990). Their mathematical expressions are shown in Table 2.

Table 2. Distances employed in the regression analysis

Distance	Equation
Gower	$d_i^2 = 1-s_{ij}$
Absolute value	$d_{ij}^2 = \sum_{h=1}^{p} \lvert x_{ih} - x_{jh} \rvert$
Mahalanobis	$d_{ij}^2 = (x_i - x_j)' C^{-1} (x_i - x_j)$
Minkowski with q exponent (2 and 4)	$d_{ij} = \left[\sum_{h=1}^{p} \lvert x_{ih} - x_{jh} \rvert^q \right]^{1/q}$

The coefficient of determination (based on the function of the number of principal coordinates selected), the residual sum of squares (SSQ = $\Sigma_I (x_i - X_i)^2$, where X_i is the prediction of x_i) and the cross-validation coefficient (C) (calculated with the equation (C = $1/n \, \Sigma_I (x_i - X_i)^2$, where X_i is the prediction of x_i obtained to eliminate this individual from the original matrix data) were also estimated.

Additionally for the linear and distance models, we analyzed 24 other regression models (Table 3) as well as polynomial regressions of third, fourth and fifth order with the variables which showed significant correlations with the linear regression. Finally, we obtained a multiple regression with the four variables which presented the highest correlations with the linear regressions, along with the four variables with the highest correlations with the polynomial regressions combined with the four variables with the highest correlations independent of the 25 regression models employed (including the linear one).

Table 3. Twenty four different regression models applied to detect which could best estimate the ages of pink river dolphins

24 different regression models
1- Age = bX
2 - Age = 1/(a + b X)
3 - Age = a + b X + c/X
4 - Age = a + b/X
5 - Age = X/(a + b)
6 - Age = a + b/X + c/X^2
7 - Age = a + b X + c X^2
8 - Age = a X + b X^2
9 - Age = a X^b
10 – Age = a b^X
11 – Age = a b $^{(1/X)}$
12 – Age = a X$^{(bX)}$
13 – Age = a X$^{(b/X)}$
14 – Age = a e$^{(bX)}$
15 – Age = a e$^{(b/x)}$
16 – Age = a + b ln X
17 – Age = 1/(a + b ln X)
18 – Age = a b^X X^c
19 – Age = a b$^{(1/X)}$ X^c
20 – Age = a e$^{(X-b)\,(2/C)}$
21 – Age = a e$^{(\ln X - b)(2/C)}$
22 – Age = a X$^{b(1-X)c}$
23 – Age = a (X /b)$^{ce(X/b)}$
24 – Age = 1/(a (X + b)^2 + c)

RESULTS

Simple Lineal Regression

Out of 50 simple lineal regressions carried out, 24 skull morphometric variables showed significant relationships with age ($p \leq 0.05$) (Table 4). Five of these variables yielded probabilities of 0.00001. Thus, this indicated a very high correlation with age. These characters were as follows: Width of the face at the skull base (V3), maximum width through the zygomatic process (V15), internal longitude of the skull (V21), maximum width of the internal nares (V30), and width of the left condole (V42). The variable, that explained the highest proportion of the age, was the maximum width between the zygomatic processes of the squamosal bone, with R^2= 0.4641. The width of premaxillaries at midlength of rostrum (V6) was the morphometric trait which showed a significant relationship with age, but with lower correlation.

Table 4. Linear regression equations *Y*= *a* + *bX*. V = Significant variable. P = significant probability. R^2 = coefficient of determination

Y =	*a* +	*b*	X	P R^2
Age	-38.8895	2.6687	V15	0.0000 0.4641
Age	-27.1907	5.1523	V21	0.0000 0.4275
Age	-30.1253	16.5529	V42	0.0000 0.3854
Age	-39.108	13.1322	V30	0.0000 0.3739
Age	-36.7558	5.9261	V3	0.0000 0.3637
Age	-30.5325	2.7541	V12	0.0001 0.3555
Age	-23.2609	2.9087	V11	0.0001 0.3314
Age	-29.8114	3.6238	V37	0.0002 0.3160
Age	-14.0205	6.5668	V29	0.0003 0.2986
Age	-5.9999	5.1261	V26	0.0004 0.2816
Age	-12.3174	3.9934	V27	0.0005 0.2730
Age	-18.6417	5.1957	V23	0.0005 0.2726
Age	-16.961	1.5761	V38	0.0019 0.2498
Age	-33.0639	3.9577	V22	0.0014 0.2371
Age	-48.0576	14.9523	V41	0.0015 0.2291
Age	-21.0493	1.2048	V2	0.0016 0.2233
Age	-20.2478	1.0683	V9	0.0039 0.1950
Age	-21.7943	0.8342	V1	0.0053 0.1783
Age	-17.6538	1.2612	V33	0.0059 0.1745
Age	-7.3988	1.942	V13	0.0092 0.1697
Age	-19.1031	8.4843	V28	0.0091 0.1657
Age	-19.8644	2.665	V19	0.0133 0.1438
Age	-3.7023	0.5223	V35	0.0328 0.1307
Age	-3.1229	9.093	V6	0.0441 0.0975

Multiple Lineal Regressions

Out of the 24 variables employed (those which were significant in the linear regression analysis), 8 skull measurements turned out to be highly significant in the multiple regression (Table 5). That is to say, these were the variables that correlated the best (compared to other variables) with age. The maximum width between zygomatic processes of squamosal bones again turned out to be the variable that better correlated with the age, with p = 0.00001 and R^2 = 0.0861. In this case, the smaller diameter of the left temporal fossa proper was the variable which showed the smallest correlation of the significant variables. R^2 was 0.8171 for the 24 variables employed, which means that these variables explained 81 % of the age variation.

In Table 6, the observed age estimates for the 71 exemplars analyzed were compared with those estimated throughout the multiple regression equation calculated.

Table 5. Significant variables (V) in the Multiple regression analysis. Probability (P) and coefficient of determination (R^2) values are listed

V	Description	P	R^2
15	Greatest width between zygomatic processes	0.0000	0.0861
22	Greatest length of left post temporal fossa from post orbital bone	0.0009	0.331
28	Length from edge of lacrimal bone to end of postorbital	0.0039	0.0241
30	Greatest width of internal nares	0.0051	0.0044
13	Least supraorbital width	0.0067	0.021
11	Least supraorbital width	0.019	0.0154
37	Length of left mandibular fossa	0.0236	0.0143
26	Minor diameter of left temporal fossa proper	0.0251	0.014

The multiple regression equation obtained was:

Age = 0.0887 (V1)+0.3624(V2)+1.976(V3)+7.091(V6)+0.1471(V9)-5.040(V11)+2,616(V12)+3.601(V13)-7.608(V15)-3.021(V19)-1.608(V21) +6.898(V22)-0.4529(V23)+5.767(V26)-0.1168(V27)+14.12(V28) 0.09582(V29)-11.33(V30)+1.407(V33)+0.01818(V35)-5.521(V37)+0.4650(V38)-3.964(V41)+1.577(V42)+ 46.24

Table 6. Observed and estimated age values by means of the multiple regression equation

Individual	Observed age	Estimated age	Residual	Absolute error
1	31	28.845	2.155	6.952
2	44.67	41.559	3.111	6.964
3	15.33	15.762	-0.432	2.82
4	16.67	20.342	-3.672	22.029
5	19.33	28.158	-8.828	45.67
6	14	10.54	3.46	24.717
7	36	29.579	6.421	17.837
8	30	18.059	11.941	39.802
9	15	11.669	3.331	22.204
10	1	-2.272	3.272	327.238
11	23,33	22.429	0.901	3.86
12	6	15.073	-9.073	151.209
13	17.33	16.934	0.396	2.287
14	17.33	21.461	-4.131	23.834
15	27	28.769	-1.769	6.553
16	5	6.532	-1.532	30.64
17	37	34.651	2.349	6.348
18	1	3.16	-2.16	215.964
19	4	3.978	0.022	0.561
20	21	32.793	-11.793	56.157
21	20.33	20.642	-0.312	1.537
22	3	4.163	-1.163	38.782

Table 6. (Continued)

23	6.33	10.529	-4.199	66.341
24	8.33	8.901	-0.571	6.857
25	14.33	10.927	3.403	23.748
26	6.33	8.484	-2.154	34.021
27	1	-9.448	10.448	1044.777
28	10	11.895	-1.895	18.948
29	9.67	8.101	1.569	16.228
30	9.67	8.101	1,569	16,228
31	32.67	31.025	1.645	5.036
32	6	8.244	-2.244	37.394
33	3	1.347	1.653	55.095
34	2	4,765	-2,765	138,236
35	42	30.041	11.959	28.474
36	1	5.122	-4.122	412.186
37	20	20.567	-0.567	2.835
38	20	14.553	5.447	27.237
39	32.33	37.033	-4.703	14.548
40	22	27.197	-5.197	23.625
41	34.67	27.685	6.985	20.146
42	27.67	25.258	2.412	8.716
43	2	7.335	-5.335	266.734
44	18.33	24.353	-6.023	32.857
45	17.67	15.313	2.357	13.338
46	19	17.835	1.165	6.129
47	20.33	20.798	-0.468	2.301
48	0.5	2.582	-2.862	416.348
49	5	7.862	-2.862	57.233
50	40	38.21	1.79	4.475
51	42.67	38.932	3.738	8.761
52	20.67	20.065	0.605	2.925
53	14	13.759	0.241	1.723
54	1	4.75	-3.75	374.997
55	2.67	4.132	-1.462	54.774
56	6	5.994	0.006	0.095
57	0.5	-2.177	2.677	535.352
58	14	12.487	1.513	10.806
59	0.5	-0.201	0.701	140.194
60	6	5.959	0.041	0.686
61	2	11.229	-9.229	461.451
62	17.33	17.429	-0.099	0.571
63	37.67	36.594	1.076	2.857
64	12.33	14.406	-2.076	16.839
65	24	22.772	1.228	5.118
66	18	13.747	4.253	23.629
67	27.33	29.354	-2.024	7.404
68	0.5	-0.201	0.701	140.194
69	0.5	-0.201	0.701	140.194
70	16	15.409	0.591	3.696
71	19.33	18.472	0.858	4.439

If we observed the absolute error values in table 6, we can see that the ages with the highest errors were those near zero. For instance, the 27 and 57 individuals ages one and 0.5 years, respectively, present the highest absolute errors, with values of 1044.7 and 535.3.

Regression Using Distances

Such as in the previous analysis, 24 variables were employed for this procedure (those significant for the linear regression analysis). For each one of the 5 distances used (Gower, absolute Value, Mahalanobis, Minkowski with exponent 2 and 4), the coefficient of determination, the residual sum of squares and the cross-validation statistic were considered. These values can be observed in Table 7.

Table 7. Values of the coefficient of determination, sum of residual squares, and cross-validation statistic for each one of the distances employed in the distance regression analysis

Distance	Coefficient of determination	Sum of residual squares	Cross-validation statistic
Gower	0.81911	1938.2	58.815
Absolute value	0.93293	719.01	70.57
Mahalanobis	0.87981	1288.3	192.61
Minkowski exp 2	0.87984	1321.9	193.5
Minkowski exp 4	0.82288	2012.2	5580.6

The absolute distance was the best estimator of the real age. The determination coefficient showed that 93 % of age variation was determined by the 24 variables employed. This distance also yielded the lower sum of residual squares. Nevertheless, Gower distance had the best cross-validation coefficient. The observed and estimated ages with the absolute value distance are shown in table 8.

There was an excellent agreement between the observed and estimated ages for many individuals. For instance, the individuals 10, 12, 31, 38, 42, 50, 55 and 64 the observed and estimated ages were exactly the same. The individuals 5, 11, 16, 21, 24, 26, 29, 30, 33, 44, 49, 51, 56, 60, 63, 71 only differed in one year between the observed and the estimated ages. The worst agreement was found in the set of individuals between the age range of less than one year to two years. The estimated values for these cases were negative. However, these animals were easier to recognize because of their calf or juvenile characteristics.

In addition to the linear and distances models, we used 24 other regression models and polynomial regressions of third, fourth and fifth order to analyze the 24 variables which showed the significant correlations with the linear regression. The different variables presented diverse behavior with regard to these 24 regression models. Those which were up to $r = 0.50$ are described here. Variable one (V1) showed an important correlation with age using model 24 ($r = 0.8324$; $r^2 = 0.6928$) while V15 yielded a significant relationship with model 21 ($r = 0.6501$; $r^2 = 0.4226$). Significant relationships were also found for other variables with model 21 including: as V11 ($r = 0.5887$; $r^2 = 0.3466$), V12 ($r = 0.5826$; $r^2 = 0.3394$), V23 ($r = 0.5088$; $r^2 = 0.2589$), V26 ($r = 0.5838$; $r^2 = 0.3408$), V30 ($r = 0.6313$; $r^2 =$

0.3985), V37 ($r = 0.5325$; $r^2 = 0.2835$), V38 ($r = 0.5197$; $r^2 = 0.2700$), and V42 ($r = 0.5847$; $r^2 = 0.3419$). Variable 21 (V21) showed a significant relationship with age using model 7 ($r = 0.5758$; $r^2 = 0.3315$), as did V29 ($r = 0.5246$; $r^2 = 0.2752$).

Table 8. Observed and expected age estimates by means of multiple regression using absolute value distances

	Observed Age	Estimated Age		Observed Age	Estimated Age
1	31.000	28.428	37	20.000	22.134
2	44.670	38.046	38	20.000	20.323
3	15.330	18.473	39	32.330	27.850
4	16.670	18.223	40	22.000	17.985
5	19.330	20.868	41	34.670	31.994
6	14.000	11.934	42	27.670	27.613
7	36.000	30.818	43	2.0000	-.48393
8	30.000	25.542	44	18.330	17.983
9	15.000	10.431	45	17.670	14.392
10	1.0000	1.5782	46	19.000	21.181
11	23.330	22.058	47	20.330	22.732
12	6.0000	6.8494	48	.50000	-1.2580
13	17.330	24.649	49	5.0000	4.2978
14	17.330	19.082	50	40.000	40.648
15	27.000	33.746	51	42.670	41.517
16	5.0000	6.4392	52	20.670	22.744
17	37.000	35.251	53	14.000	11.629
18	1.0000	9.6655	54	1.0000	5.9686
19	4.0000	6.4875	55	2.6700	2.9932
20	21.000	26.903	56	6.0000	5.2628
21	20.330	21.555	57	.50000	1.1774
22	3.0000	1.9518	58	14.000	11.504
23	6.3300	13.319	59	.50000	-.95871
24	8.3300	7.3142	60	6.0000	5.6430
25	14.330	11.715	61	2.0000	4.6141
26	6.3300	7.1177	62	17.330	21.102
27	1.0000	-2.2043	63	37.670	36.941
28	10.000	12.592	64	12.330	12.194
29	9.6700	10.826	65	24.000	26.619
30	9.6700	10.826	66	18.000	20.924
31	32.670	32.426	67	27.330	20.690
32	6.0000	4.4641	68	.50000	-.95871
33	3.0000	4.3992	69	.50000	-.95871
34	2.0000	-3.0196	70	16.000	19.630
35	42.000	37.568	71	19.330	18.645
36	1.0000	-.72749			

The polynomial regressions with the higher correlations coefficients were those using V11 (fifth order: $r = 0.5435$; $r^2 = 0.2954$), V12 (fifth order: $r = 0.5507$; $r^2 = 0.3033$), V15 (fifth order: $r = 0.6441$; $r^2 = 0.4148$), V21 (fifth order: $r = 0.6219$; $r^2 = 0.3868$), V23 (fifth order: $r = 0.6094$; $r^2 = 0.3715$), V26 (fifth order: $r = 0.5606$; $r^2 = 0.3142$), V27 (fifth order: $r =$

0.5449; r^2 = 0.2969), V29 (fifth order: r = 0.5757; r^2 = 0.3314), V30 (fifth order: r = 0.6013; r^2 = 0.3616), V35 (fifth order: r = 0.5542; r^2 = 0.3071) and V42 (fifth order: r = 0.5545; r^2 = 0.3074).

Furthermore, we obtained a multiple regression that included the four variables which presented the highest correlations with the linear regressions (V15, V21, V30 and V42; Y = -61.2597 + 2.251799 (V15) + 2.336477 (V21) + 4.37399 (V30) – 2.52504 (V42), r = 0.8059 , r^2 = 0.6493), the four variables with the highest correlations with the polynomial regressions (V15, V21, V23 and V30; Y = -61.9367 + 1.783891 (V15) + 2.176527 (V21) + 3.460791 (V23) + 1.24999 (V30), r = 0.8080 , r^2 = 0.6529) and the four variables with the highest correlations independent of the 24 regression models employed (V15, V1, V30 and V11; Y = -64.3776 + 4.3223 (V15) – 0.2514 (V1) + 6.350545 (V30) – 1.75691 (V11), r = 0.8088 , r^2 = 0.6541). Therefore, for these multiple regression equations the age of a dead pink river dolphin is determined by only four cranial measurements which explained 64-65% of age variation.

DISCUSSION

Two morphometric traits seem to be extremely important in determining the skull age of a dead animal. These variables are the maximum width between zygomatic processes of the squamosal bones (V15) and the maximum width of the internal nares (V21). The maximum width between the zygomatic squamosal processes offered the highest correlation values practically in all the analyses undertaken. In relation to other dolphins, *Inia geoffrensis* is characterized to have a zygomatic process quite conspicuous, and that it increases its size notably with increase of years. Keeping in mind that the masseter muscle (in charge of the mastication) has part of its insertion on the zygomatic process. Its conspicuous size can be related to a particular diet and heterodonty (cusped posterior teeth) of this species, which allows for crushing of even armored catfishes (for instance, Loricariidae such as *Hypostomus plecostomus, Pterygoplichthys multiradiatus* or Doradidae as *Oxydoras niger*), and to have the capacity to manage a diverse diet. Da Silva (1983) identified 43 fish species in 22 *Inia* stomachs including those from benthic, littoral and pelagic environments. Several analyses indicated that two other morphometric traits; the internal longitude of the skull and the left condyle width; were also important in the determination of age. Therefore, of all the variables tested, these four were the most useful in determining the ages of pink river dolphin skulls. It is interesting to note that no dental variables showed any relationship with age.

The obtained regression equations estimated the age with accurate precision. The biggest errors between the observed and estimated ages were in the individuals of less than one year old as well as one year olds. This is probably due to the hindrance of counting dental layers in calves and young individuals which have a thin cement width, and which have incompletely differentiated layers. At about two years of age the layers in the cement begin to differ from each other more and are more clearly defined and then stay this way more or less through age 27. In this age range (ages 2-27) we found the best fit between the observed and estimated ages. Post 27 years, we observed some differences among the observed and expected ages with the linear multiple regression equation. According to our experience in the laboratory, this can be due to that teeth of mature individuals present numerous lines in a very limited space, which can complicate the reading of the layers. Additionally, the growth of the skull of these dolphins could be more or less linear from two until 27 years but not from zero to two years and after 27 years. However, this is not a grave problem, because the animals younger than two years of age or very old are those easiest to determine in relationship of their ages. The multiple regression equation with the absolute value distance also offered comparable results with the linear multiple regression procedure.

Therefore, the four aforementioned skull measurements and the linear and distance regression methods seem to be adequate and precise tools to determine the age of pink river dolphins using their skulls. Nevertheless, some more sophisticated procedures could be applied to obtain more accurate estimations especially in the youngest and oldest animals. For example, Stevick (1999) applied the von Bertalanffy equation showing that is a good descriptor of the growth in marine species and it has been applied to describe the relationships between longitude and age in several mysticete species, such as hunchback whales (Chittleborouh 1965,) and some odontoces (Kasuya, 1972; Martin, 1980; Bloch et al., 1993). The equation, $L_t=L^{\infty}(1-e^{-k(t+to)})$, (where L_t is the longitude of the body at certain t moment, L^{∞} is the corporal longitude at the sexual maturity or the asymptotic longitude of the individuals in one given population, k is a growth constant, t is the age at that time and t_o is a time constant), could be adapted to the growth of several skull morphometric traits and through it, the age could be obtained. However, in this case, it should be necessary a constantly monitor the growth of the implied individuals.

In the current chapter, we showed the usefulness of several skull measures to determine accurate age estimations in pink river dolphins.

ACKNOWLEDGMENTS

The authors want to thank different peoples and institutions for their collaboration in carrying out the present research. Thanks to Dr. Vera da Silva and Dr. Fernando Rosas for sharing their research experiences on this topic to the first author during the tooth analysis in the INPA (Manaus, Brazil). Thanks to Sandra Beltrán-Pedreros for her hospitality in Manaus. Also, a thank-you goes to the von Humboldt Institute (at Villa de Leyva, Colombia) for providing access to its *Inia* skull collection. Finally, thanks go to diverse Indian communities which helped to obtain pink river dolphin material in Colombia (Ticunas and Cocamas in diverse communities such as Puerto Nariño, Patrulleros or in the Tarapoto lake), in Perú (Ticuna, Ocaina, Capanahua, Angoteros, Orejones, Cocamas and Alamas) and in Brazil (Marubos, Kanaimari and Kulina in diverse points of the Yavarí river).

REFERENCES

[1] Anli, G. & Zhou, K. (1992). Sexual dimorphism in the baiji, *Lipotes vexillifer*. *Canadian Journal of Zoology,* 70, 484-493.

[2] Archer, F. & Chivers, S. (2002). *Age Structure of the Northeastern Spotted Dolphin Incidental Kill By Year, for 1971 to 1990 and 1996 to 2000. Administrative Report LJ-*

02-12. La Jolla Shores, CA: Southwest Fisheries Science Center. National Marine Fisheries.

[3] Arenas, C., Cuadras, C.M. & Fortiana J. (1991). *Multicua. Publicaciones del Departamento de Estadística N4.* Barcelona, Spain: Universidad de Barcelona.

[4] Bloch, D., Lockyer, C. & Zachariassen, M. (1993). Age and growth parameters of the long-finned pilot whale off the Faroe Island. *Report of the International Whaling. Commission (Special Issue)*, 14, 163-207.

[5] Castellanos-Mora, L. (2007). *Evaluación del impacto por mallas de pesca en función de la edad y relaciones craneométricas en poblaciones de delfín de río Inia geoffrensis en el Amazonas y Orinoco. Bachelor of Science Thesis.* Bogotá DC., Colombia: Pontificia Universidad Javeriana.

[6] Chittleborough, R.G. (1965). Dynamics of two populations of the humpback whale. *Australian Journal of Marine and Freshwater Research*, 16, 13-128.

[7] Chivers, S. J. (2002). *Age structure of female eastern spinner dolphins (Stenella longirostris orientalis) incidentally killed in the eastern tropical Pacific tuna purse-seine fishery. Administrative Report LJ-02-11.* La Jolla Shores, CA: Southwest Fisheries Science Center. National Marine Fisheries.

[8] Collet, A. (1981). *Biologie du dauphin commun Delphinus delphis L. en Atlantique Nord-Est. Doctoral Thesis.* Poitiers, France: L' Université de.

[9] Cuadras, C. M. & Arenas, C. (1990). A distance based regression model for prediction with mixed data. *Communications* in *Statistics - Theory* and Methods*.,* 19, 2261-2279.

[10] Da Silva, V. (1983). *Ecologia alimentar dos golfinhos da Amazónia. Master of Science thesis.* Manaus. Brazil: University of Amazonas,.

[11] Da Silva, V. (1994). Aspects of the Biology of the Amazonian Dolphins Genus *Inia* and *Sotalia fluviatilis*. Doctoral Thesis. Cambridge, UK: The University of Cambridge.

[12] Da Silva, V. (1995). Age estimation o the Amazon dolphin, *Inia geoffrensis*, using laminae in the teeth. *Report of the International Whaling. Commission. (special issue*), 16, 532-543.

[13] Hohn, A. A. (1980). Analysis of growth layers in the teeth of *Tursiops truncates* using light microscopy, microradiography, and SEM. *Report of the International Whaling. Commission. (Special issue)*, 3, 155-160.

[14] Hohn, A. A., Scott, M. D., Wells, R. S., Sweeney, J. C. & Irvine A. B. (1989). Growth layers in teeth from known-age, free-ranging bottlenose dolphins. *Marine Mammal Science,* 5, 315-342.

[15] Kasuya, T. (1972). Growth and reproduction of *Stenella coeruleoalba* based on age determination by means of dentinal growth layers. *Scientific Reports of the Whales Research Institute*, Tokyo., 24, 57-79.

[16] Kemper, C. (1999). Estimating body length of Pygmy right whales (*Caperea marginata*) from measurements of the skeleton and baleen. *Marine Mammal Science*, 15, 683-700.

[17] Laws, R.M. (1953). A new method of age determination for mammals with special reference to the elephant seal (*Mirounga leonine* Linn). Falkland Is. *Depend. Surv. Sci. Rep.*, 2, 1-11

[18] Laws, R.M. (1957). On the Growth rates of the leopard seal, *Hydrurga leptonyx*. (De Blainbille, 1820). *Saeugetierkunde Mitteilung*, 5, 49-55.

[19] Laws, R.M. (1958). Growth rates and ages of crabeater seal, *Lobodon carcinophagus* Jacquinot and Pucheran. *Proceedings of the Zoological. Society of London,* 130, 275-88.

[20] Lockyer, C. (1993). A report on patterns of deposition of dentine and cement in teeth of pilot whales, genus *Globicephala. Report of the International Whaling. Commission (special issue),* 14, 137-163.

[21] Manzanilla, S. R. (1989). The 1982-1983 El niño event recorded in dentinal growth layers in teeth of Peruvian dusky dolphins (*Lagenorhynchus obscurus*). *Canadian Journal of Zoology*, 67, 2120-2125.

[22] Martin, A.R. (1980). An examination of sperm age and length data from the 1949-78. Icelandic catch. *Report of the International Whaling. Commission*, 30, 227-230.

[23] McLaren, I.A. (1958). The biology of the ringed seal (*Phoca hispida* Schereber) in the eastern Canadian Arctic. *Bulletin of the Fisheries Research* Board of *Canada*, 118, 1-97.

[24] Mead. J. G & Potter. C. W. (1990). Natural history of Bottlenose dolphins along the central Atlantic coast of the United States. In S. Leatherwood & R. Reevs. *The Bottlenose dolphin* (pp 165-195). San Diego, CA: Academic Press.

[25] Nielsen, H. G. (1972). Age determination of the harbor porpoise *Phocoena phocoena* (L.) (Cetacea). *Vidensk. Meddr dansk naturh. Foren*, 135, 61-84.

[26] Nishiwaki, M & Yagi, T. (1953). On the Age and the growth of teeth in a dolphin (*Prodelphinus caeruleo-albus*). *Sci. Rep. Whales. Rest. Inst.* Tokyo, 8, 133-46.

[27] Northridge, S & Hofman, R. (1999). Marine mammal interactions with fisheries. In J. R. Twiss & R. Reevs (Eds.), Conservation and management of marine mammals (pp 99-119). London, UK: Smithsonian institution press.

[28] Nummela. S., Kosove J., Lancaster, T & Thewissen, T. (2004). Lateral Mandibular Wall Thickness in *Tursiops truncatus*: Variation due to Sex and Age. *Marine Mammal Science*, 20, 491-497.

[29] Perrin, W. F. (1993). Rostral Fusion as a Criterion of Cranial Maturity in the Common Dolphin, *Delphinus delphis*. *Marine Mammal Science*, 9, 195-1997.

[30] Pinedo, M. C. (1991). *Development and variation of the franciscana, Pontoporia blainvillei. Doctoral Thesis*. Santa Cruz, CA: University of California.

[31] Pinedo. M. C. (1994). Impact of Incidental Fishery Mortality on the Age Structure of *Pontoporia blainvillei* in Southern Brazil and Uruguay. *Report of the International Whaling Commission Special Issue*, 15, 261-264.

[32] Pinedo. M. C & Hohn. A. A. (2000). Growth layer patterns in the teeth from the Franciscana, *Pontoporia blainville*: Developing a model for precision in age estimation. *Marine Mammal Science*, 16, 1-27.

[33] Rosas, F. (2002). Age and Growth of the estuarine dolphin (*Sotalia guanensis*) (Cetacea, Delphinidae) on the Paraná Coast, southern Brazil. *Fishery Bulletin*, 101, 377-383.

[34] Ruíz-García, M., Banguera, E. & Cárdenas, E. (2006). Morphological analysis of three *Inia* (Cetacea; Iniidae) populations from Colombian and Bolivia. *Acta Theriologica*, 51, 411-426.

[35] Sergeant, D. E. (1959). Age determination in odontocete whales from dentinal growth layers. *Norsk Hvalfangsttid*, 48, 273-88.

[36] Scheffer, V. B. (1950). Winter injury to young fur seals on the northwest coast. *California Fish and Game*, 34, 378-379.

[37] Stevick, P. (1999). Age- Length relationships in Humpback whales: A comparison of strandings in the western north Atlantic with commercial catches. *Marine Mammal Science*, 15, 725-737.

[38] Trujillo, F. (2000). *Habitat Use and Social Behavior of the Freshwater dolphin Inia geoffrensis (De Blainville, 1817) in the Amazon and Orinoco Basins. Doctoral Thesis*. Aberden, Scotland: University of Aberdeen.

[39] Trujillo, F., Diazgranados, M., Galindo, A. & Fuentes, L. (2006). Delfín Rosado *Inia geoffrensis*. pp: 285-290. In J. V. Rodríguez, M. Alberico, F. Trujillo, & J. Joergenson (Eds.), *Libro Rojo de los Mamíferos de Colombia. Serie de Libros Rojos de Especies Amenazadas de Colombia* (pp 285-290). Bogotá, Colombia: Conservación Internacional, Universidad Nacional, Ministerio del Medio Ambiente.

In: Biology, Evolution and Conservation of River Dolphins… ISBN: 978-1-60876-633-8
Editors: M. Ruiz-García, J. Shostell. pp. 101-116

Chapter 6

AMAZON RIVER DOLPHIN: HIGH PHYLOPATRY DUE TO RESTRICTED DISPERSION AT LARGE AND SHORT DISTANCES

Juliana A. Vianna[1,3], Claudia Hollatz[1], Miriam Marmontel[2], Rodrigo A.F. Redondo[1], and Fabrício R. Santos[1]*

[1]Departamento de Biologia Geral, ICB, Universidade Federal de Minas Gerais, ICB – UFMG, Belo Horizonte, MG, Brazil.
[2]Instituto de Desenvolvimento Sustentável Mamirauá, R. Augusto Correa No.1 Campus do Guamá, Belém, PA, Brazil.
[3]Universidad Andrés Bello, Escuela Medicina Veterinária, Santiago, Chile.

The Amazon River Dolphin (*Inia geoffrensis*) is widely distributed along the Amazon and Orinoco basins, covering an area of about 7 million km^2. We have generated 519 base pair (bp) sequences of the control region (HVSI) and 1,140 bp of the Cytochrome B (Cyt-b) gene of mitochondrial DNA (mtDNA) for two populations from the Amazon basin in Brazil, separated by only 45 km. Six HVSI haplotypes were identified and we could detect a remarkable population structure despite of the short distance separating the localities. Compared to HVSI data from other South American countries, the Brazilian haplotypes occupy an intermediate position related to Colombian Amazon, Colombian Orinoco and Bolivian haplotypes. The Cyt-b data also detected a remarkable separation between both Brazilian locations, and the phylogenetic analysis indicated an association of Amazon and Orinoco haplotypes, separated from the Bolivian ones. This phylogeographic study emphasizes the outstanding population structure for the Amazon River Dolphin, considering both macro and microgeographic levels. These results suggest a strong phylopatry for this species due to gene flow restriction through long distances, as well as short distances by different water ecology characteristics. The studied Brazilian populations occur in close localities but are separated by the turbid fresh water environment of the Amazon River, a likely ecological barrier segregating *I. geoffrensis* populations.

* Corresponding author. fsantos@icb.ufmg.br

Keywords: Phylopatry, *Inia geoffrensis*, mtDNA, phylogeography.

INTRODUCTION

The Amazon River Dolphin or Pink Dolphin (*Inia geoffrensis*) is endemic to the Amazon and Orinoco rivers and their tributaries, distributed in a total range area of seven million km^2 (Best & da Silva, 1989). The species is classified as vulnerable by The World Conservation Union (IUCN) and listed in Appendix II of the Convention on International Trade in Endangered Species of Wild Flora and Fauna (CITES). The major threats include the incidental catch of *I. geoffrensis* by fishing nets and the recent construction of dams (da Silva 2002). Presently, the Amazon River Dolphin is also hunted for the extraction of its genital organs which are sold as magical amulets (Best & da Silva, 1989). Furthermore, the species seems to also be affected by chemical contamination, such as pesticides and mercury (Rosas & Lehti, 1996).

Some previous studies (Cassens et al., 2000; Hamilton et al., 2001; Yan et al., 2005) have focused on the phylogeny of the river dolphins (*Inia*, *Pontoporia*, *Lipotes*, and *Platanista*), suggesting a paraphyletic relationship among the existent river species, indicating a high adaptive plasticity for these cetaceans. However, few studies have investigated the population structure and phylogeography of these species (Banguera-Hinestroza et al., 2002; Yang et al., 2003).

A recent study (Banguera-Hinestroza et al., 2002) analyzed the mtDNA control region and Cytochrome b gene haplotypes of *Inia geoffrensis*, supporting the proposal of a subdivision of the *Inia* genus into at least two evolutionarily-significant units. It has also showed a strong phylogeographic pattern at a macro-geographic scale. However, it did not take into account the micro-geographic level to evaluate the species' capacity to migrate through short distances.

In this chapter we have analyzed mtDNA control region sequences of *Inia geoffrensis* from two close but ecologically different populations in Brazil to assess the spatial distribution of genetic diversity and likely restrictions to gene flow. Besides, we compared the population genetic structure in a macrogeographic scale with populations from other South-American regions.

MATERIALS AND METHODS

Sample Collection and DNA Extraction

Tissue samples were collected from 21 pink dolphins (*Inia geoffrensis*) from three different locations within the central Amazon basin, Brazil. Eleven samples were collected near the lake on the Mamirauá Sustainable Development Reserve (MSDR) referred here simply as "Mamirauá", nine in the Tefé River and Lake, referred as "Tefé", and one at an intermediate area between the previous locations (Figure 1). These samples were collected from captured and dead animals over a nine year period. Muscle and skin tissues were preserved in 70% ethanol. DNA was extracted with a standard phenol-chloroform protocol (Sambrook et al., 2001) or DNeasy® DNA extraction kit (Qiagen®).

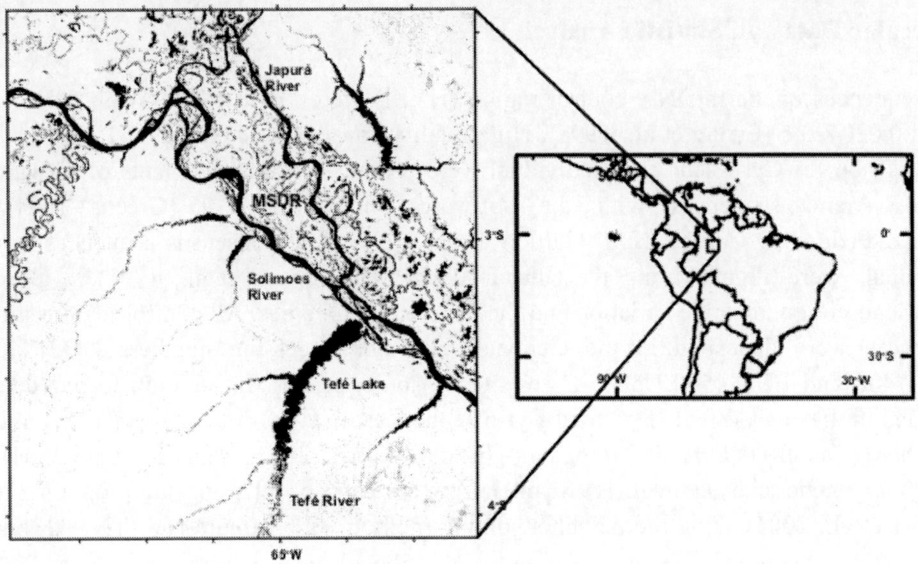

Figure 1. Map showing the two sampled areas in Brazil: the Tefé River and Lake and the Mamirauá Reserve located between the Japurá and Solimões (Amazon) rivers in the central part of the Amazon basin.

DNA Amplification and Sequencing

The mtDNA control region (HVSI) was amplified via the polymerase chain reaction (PCR) using primers modified from Kocher et al. (1989): L15926 5' TCAAAGCTTACACCAGTCTTGTAAACC 3' and H16498 5' CCTGAAGTAGGAACCAGATG 3'. Each PCR mix contained 20 ng of genomic DNA, 1X Taq reaction buffer 1B (Phoneutria® - 1.5 mM $MgCl_2$, 10 mM Tris-HCl (pH 8.4), 50 mM KCl, 0.1% Triton X-100), 200 µM dNTPs, 0.5 µM of each primer, and 1 unit of Taq DNA polymerase (Phoneutria®). The PCR cycling was performed in an Eppendorff Gradient thermocycler at a temperature of 94°C for 5 min, followed by 35 cycles at 47°C for 45 s, 72°C for 1 min and 94°C for 30 s, and a final extension period at 72°C for 10 min.

The complete mtDNA Cytochrome *b* (Cyt-b) gene (1140 bp) was amplified using primers L14121, H15318 and XL14733 (Redondo et al., 2008) and MVZ4 (Kocher et al., 1998). The cycling profile used for this gene was the same as for the control region, but using 52° C as the annealing temperature.

PCR products were purified using a 1:1 volume ratio of 10 units/µl of Exonuclease I and 1 unit/µl Shrimp Alkaline Phosphatase (GE Healthcare®) at the temperature of 37°C for 45 min and 80°C for 15 min. Both strands were sequenced using one of the primers described above and the DYEnamic ET® dye terminator kit (GE Healthcare®) following the temperature cycling profile: 94°C for 1 min, 35 cycles of 94°C at 30 s, annealing at 55°C for

30 s, and extension at 60°C for 1 min. Resulting fragments were sequenced in a MegaBACE 1000 DNA Analysis System with protocols indicated by the fabricant (GE Healthcare®).

Molecular Data and Statistic Analysis

Sequences of the mtDNA control region (HVSI) were "base called" with the software Phred 0.020425.c (Ewing et al., 1998). High quality consensus sequences (519 bp for HVSI and 1140 bp for Cyt-b) for each individual were produced from alignments of forward and reverse strand sequences with the software Phrap 0.990319 (Green, 1994) and visualized/edited in Consed 12.0 (Gordon et al., 1998). The consensus sequences from all individuals were aligned using the Clustal-X 1.83 (Thompson et al., 1997) to allow the identification of nucleotide variation and the different haplotypes. All control regions mtDNA sequences were deposited in the Genbank with the accession numbers DQ217403 to DQ217408 and EU55456/EU554562 for the Cyt-b sequences. In addition to our data, we included in the analyses all HVSI and Cyt-b sequences of *I. geoffrensis* specimens published elsewhere (Cassens et al., 2000; Banguera-Hinestroza et al., 2002; Arnasson et al., 2004).

Phylogenetic analyses with HVSI mtDNA were carried out using the program Mega 4 (Kumar et al., 2004) using the neighbor-joining method with Tamura-Nei (TN – Tamura & Nei, 1993) substitution model, and gamma distributed rates among sites (α = 0.88) as suggested by the Modeltest approach (Posada & Crandall, 1998). For Cyt-b data, we used the neighbor-joining algorithm and Kimura 2-parameters genetic distance with (α = 0.5. Branch support tests were calculated by bootstrap using 1000 replicates with Mega. The sequence of *Pontoporia blainvillei* (Genbank accession number AY644451) was used as the outgroup for the phylogenetic analyses.

Haplotype networks were constructed by the median-joining algorithm (Bandelt et al., 1999) using the program Network v 4.5 (www.fluxusengineering.com) to infer the relationship among the haplotypes for the populations in a geographic context.

Population genetics parameters such as haplotype diversity (*h*) and nucleotide diversity (π) were estimated using either the average number of pairwise differences (K) or alternative models of evolution (see above) in the Arlequin 3.1 software (Schneider et al., 2000). We used Mega to calculate average sequence pairwise divergences within and between populations using TN + Γ (α = 0.88) model.

The analysis of molecular variance (AMOVA), a test to calculate the distribution of genetic variation in a particular hierarchical grouping of populations, was performed with 1,000 permutations, using either F_{ST} or Φ_{ST} calculated with TN + Γ (α = 0.88) in Arlequin. Tajima's D-test (Tajima, 1989) and Fu's Fs test (Fu, 1997) were calculated for each population to evaluate the possibility of recent population expansion using 1,000 bootstrap replicates. Mismatch distributions (Schneider et al., 2000) were also used to evaluate the past occurrence of population bottlenecks and expansion. An exact test of population differentiation was performed with 10,000 Markov Chain steps to test significance as implemented in Arlequin. Population pairwise distances were calculated as F_{ST} and Φ_{ST} (TN, α=0.88). The number of female migrants between populations was estimated as $N_{mf} = \frac{1-\Phi_{ST}}{2\Phi_{ST}}$ following Slatkin's (1991) formula. Mantel tests were performed to test the correlation between genetic (F_{ST} or Φ_{ST}) and geographic distances. Geographic distances among samples were calculated following the river's course in the Amazon basin using the MapSource (Garmin®) software.

We also used the SAMOVA (Dupanloup et al., 2002) approach to identify groups of geographically homogeneous populations and to maximally differentiate them from each other. Groupings suggested by SAMOVA were further tested by AMOVA in Arlequin as described above. Besides, we also used the software Barrier (Manni et al., 2004) to identify likely barriers to gene flow using the haplotype distribution in populations spatially defined by geographic coordinates. These tests were used to identify significant discontinuities in the population structure of *I. geoffrensis*.

RESULTS

Populations Phylopatry

A fragment of 519 base pairs (bp) of the mtDNA control region (HVSI) was sequenced in 21 Brazilian individuals. Seven polymorphic transitions were identified in the *Inia geoffrensis* sequences from all the locations and we found a total of six control region mtDNA haplotypes restricted to Brazil. Three haplotypes were found exclusively in individuals from the Mamirauá Lake, and two in the Tefé River and Lake (Tefé). Another haplotype (BrC) was identified in both locations and in one additional specimen in an intermediate area between the two studied locations (Table 1).

The Mamirauá population shows a higher haplotype diversity than the Tefé population (Table 1). Although the two localities are only about 45 km apart (straight line between their edges), a high differentiation was found between the two studied populations (Φ_{ST} (TN+Γ) = 0.66, F_{ST} = 0.38), sharing only one haplotype (BrC). AMOVA analysis (using Φ_{ST} (TN)) shows that most of the variation is observed between the localities (65.7%), but there is still a considerable amount of variation within each locality (34.3%). Despite the observation of negative values, significant expansion could not be detected for Mamirauá or Tefé populations with Tajima's and Fu's tests. Mismatch distributions analyses also showed no sign of population expansion.

The Median Joining Network (MJN) shows two population clusters in Brazil (Figure 2). The star-like tree found for the Mamirauá population showing an excess of rare haplotypes originated from the most frequent haplotype (BrC) could suggest population growth, however it has been considered non significant by previous expansion tests. The haplotypes from the Tefé population are more differentiated. This is also supported by the mismatch distribution analysis that presented unimodal distribution for Mamirauá and bimodal for Tefé (Figure 3), and by the slightly higher nucleotide diversity observed in Tefé (Table 1).

The complete Cyt-b gene (1140 bp) was also sequenced in 20 individuals from the Brazilian Amazon. Two different haplotypes were distinguished, differentiated by one transversion (non-synonymous). One haplotype (BrM) appears in 13 individuals from Mamirauá and another one appears in 7 individuals from Tefé (BrT). Thus, there is a specific Cyt-b haplotype found in Tefé and another found in Mamirauá, even though they are only 45 Km apart. This emphasizes the remarkable population structure that was already suggested by the control region, but presenting no haplotype sharing between both regions. The haplotypic diversity is 0.479 ± 0.072 and the nucleotide diversity is 0.00042, a low value suggesting a short period of separation.

Table 1. Absolute haplotype frequencies, and gene (Nei 1978) and nucleotide (Tamura & Nei 1993) diversities for the Brazilian subpopulations using the control region data

Population		Haplotypes						Gene	Nucleotide
	n	BrA	BrB	BrC	BrD	BrE	BrF	diversity (*h*)	diversity (π)
Mamirauá	11	3	0	5	1	0	2	0.7455 +/- 0.0978	0.0024 +/- 0.0019
Tefé	9	0	7	1	0	1	0	0.4167 +/- 0.1907	0.0031 +/- 0.0024

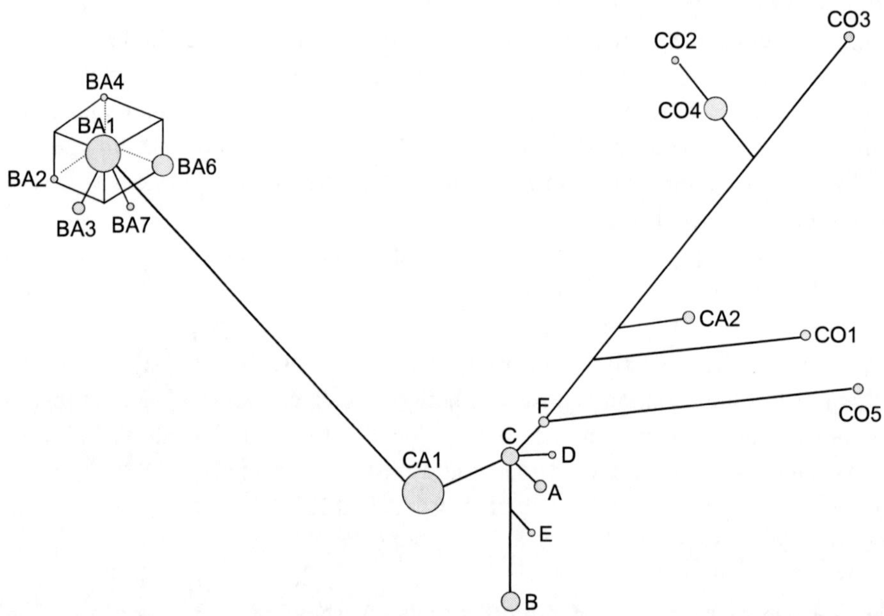

Figure 2. Median joining network (Bandelt et al., 1999) constructed with control region mtDNA haplotypes for *I. geoffrensis* individuals from four populations, including two subpopulations from Brazil. Circle areas are proportional to haplotype frequencies and branch lengths are proportional to mutation events. Haplotypes are from the Brazilian Amazon (BR, A-F), Bolivian Amazon (BA), Colombian Orinoco (CO) and Colombian Amazon (CA) populations. BA1 = BA5 (Banguera-Hinestroza et al., 2002).

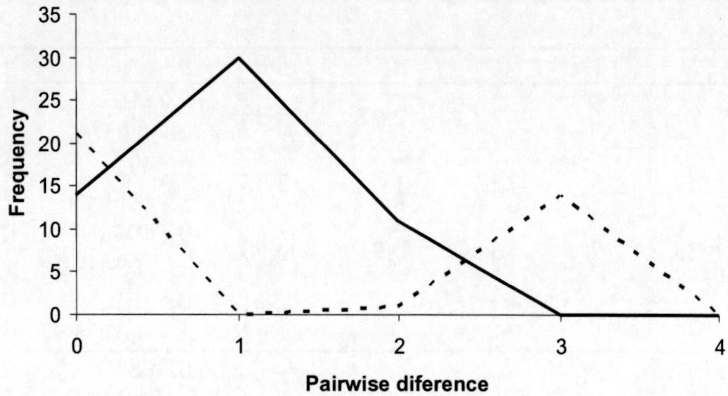

Figure 3. Graphic mismatch distributions of pairwise differences for Mamirauá (continuous line) and Tefé (broken line) populations, using the control region data.

Comparative Phylogeography

All polymorphic sites found on the 519 bp HVSI fragment from the Brazilian Amazon sequences were also part of the 400 bp studied by Banguera-Hinestroza et al. (2002). Comparing our data with 96 control region sequences (Table 2) from three different locations (Banguera-Hinestroza et al., 2002), a total of 20 haplotypes were identified, six from Brazil (n = 21), six from Bolivia (n = 41), two from Colombian Amazon (n = 38) and five from Orinoco (n = 17). The haplotypes show 50 polymorphic sites (36 informative, 14 singletons), 47 transitions and three transversions, and one position containing both a transition and a transversion.

Among the four localities, Brazil displays the highest haplotype diversity and Colombian Amazon the lowest one, but Bolivia presents the lowest nucleotide diversity while Colombian Orinoco has the highest one (Table 2). Tajima's and Fu's tests revealed no significant results for expansion in all populations (data not shown), although MJN analysis (Figure 2) shows a star-like pattern for Brazil and Bolivia while a mismatch distribution displays a significant unimodal shape only for the Bolivia population.

Inter-population comparisons (Table 3) reveal significant distances between all pairs of populations. Indeed, using a more appropriate exact test of population differentiation (Schneider et al., 2000), very significant statistical values ($P < 0.001$) were observed for all comparisons, including between the two Brazilian subpopulations (data not shown).

AMOVA analysis (Table 4); considering the four main localities, reveals that 90.4% of the variation occurs among populations, showing an extremely reduced gene flow among localities. The low N_{mf} values (Table 3) also reflect the relative isolation observed among *Inia* populations.

Table 2. Population genetic parameters for the four studied populations

Populations	*N*	HT	*S*	*d*	*S/d*	*h*	π
Brazilian Amazon	21	6	7	2.07	3.38	0.7810 +/- 0.0562	0.0053 +/- 0.0034
Colombia Orinoco	17	5	27	11.48	2.35	0.6471 +/- 0.1185	0.0211 +/- 0.0115
Colombian Amazon	38	2	6	0.91	6.59	0.1494 +/- 0.0739	0.0023 +/- 0.0018
Bolivian Amazon	41	6*	5	0.64	7.81	0.5561 +/- 0.0755	0.0016 +/- 0.0014
Total	117	20	50			0.8083 +/- 0.0259	0.0357 +/- 0.0178

Table 3. Population pairwise distances (Φ_{ST}(TN) -below diagonal) and estimated female migrants (N_{mf}-above diagonal) per generation among the four Inia populations. All Φ_{ST} values are significant at P < 0.0001)

	Brazilian Amazon	Colombian Orinoco	Bolivian Amazon	Colombian Amazon
Brazilian Amazon	-	0.11026	0.02003	0.11730
Colombian Orinoco	0.67167	-	0.03904	0.09024
Bolivian Amazon	0.95820	0.91464	-	0.01523
Colombian Amazon	0.62408	0.76368	0.96856	-

Pairwise Φ_{ST} comparisons (Table 3) suggest that a significant part of the interpopulation diversity is due to the inclusion of the Bolivian population. Excluding the Bolivian population from AMOVA (Table 4) causes more than a 20% reduction of the variance and Φ_{ST} values, although heterogeneity still remains high among the remnant localities (Φ_{ST} ~0.7). However, when the same test was performed for each population subtraction, we did not observe any significant reduction in Φ_{ST} (0.90-0.94).

To test the partition of the genetic variability in relation to the possible geographic barriers proposed by Banguera-Hinestroza et al., (2002), a three level AMOVA was made using three groups: Brazil+Colombian Amazon, Bolivia, and Colombian Orinoco (Table 4). The AMOVA shows 82.3% of variation occurring among groups, a very high value compared to other tested groupings.

The SAMOVA tests with five populations (including both Brazilian subpopulations) indicate three and four population groupings (Table 4) as the ones maximizing the variation among groups (F_{CT}). Surprisingly, the four groups' structure is the one presenting the highest F_{CT}, and separates both Brazilian subpopulations leaving a cluster formed by the Colombian Amazon and Brazilian Tefé. This result emphasizes the distinctiveness of all populations as already shown by the pairwise F_{ST}'s comparison (Table 3) and the exact test of differentiation. SAMOVA also indicates that Bolivia presents the most divergent population, separating from the others in structures formed by two groups (Table 4). Using the Barrier software (Manni et al., 2004) based in the pairwise F_{ST}'s distances between populations represented by geographic coordinates, the most likely barrier to gene flow suggested by the algorithm is the one separating Bolivia from the other populations (data not shown).

Table 4. AMOVA results for several groupings of populations using Φ_{ST} calculated with Tamura-Nei (TN) model and α = 0.88, or pairwise number of differences (K) – the later model is the one used in the SAMOVA test. Distinct groupings, including the ones suggested by SAMOVA analysis*, were tested

Structure of populations	Among groups	Among populations	Within populations		F_{CT}	F_{ST}
5 populations/1 group:		90.7%	9.3%	Φ_{ST} (TN)		0.907
BA+CA+CO+BT+BM		90.2%	9.8%	Φ_{ST} (K)		0.902
5 populations/2 groups*:	68.8%	24.3%	6.9%	Φ_{ST} (TN)	0.688	0.931
BA x CA+CO+BT+BM	66.9%	25.7%	7.4%	Φ_{ST} (K)	0.669	0.926
5 populations/3 groups*:	81.2%	10.9%	7.9%	Φ_{ST} (TN)	0.811	0.920
BA x CO x CA+BT+BM	79.4%	12.1%	8.5%	Φ_{ST} (K)	0.794	0.915
5 populations/4 groups*:	82.9%	8.5%	8.6%	Φ_{ST} (TN)	0.829	0.914
BA x CO x CA+BT x BM	81.4%	9.5%	9.1%	Φ_{ST} (K)	0.814	0.908
4 populations/1 group:		90.4%	9.96%	Φ_{ST} (TN)		0.900
BA+CA+CO+BR		89.4%	10.6%	Φ_{ST} (K)		0.894
4 populations/2 groups:	68.7%	23.8%	7.5%	Φ_{ST} (TN)	0.687	0.926
BA x CA+CO+BR	66.9%	25.1%	8%	Φ_{ST} (K)	0.669	0.919
4 populations/3 groups:	82.3%	9.1%	8.6%	Φ_{ST} (TN)	0.823	0.913
BA x CO x CA+BR	80.7%	10.1%	9.2%	Φ_{ST} (K)	0.807	0.908
3 populations/1 group:		69.8%	30.2%	Φ_{ST} (TN)		0.698
CA+CO+BR		69.5%	30.5%	Φ_{ST} (K)		0.695
2 Brazilian populations:		65.7%	34.3%	Φ_{ST} (TN)		0.657
BT + BM		65.4%	34.6%	Φ_{ST} (K)		0.654

Populations: BA (Bolivian Amazon), CA (Colombian Amazon), CO (Colombian Orinoco), BT (Brazilian Tefé), BM (Brazilian Mamirauá), BR (Brazilian Tefé + Mamirauá).

The MJN (Figure 2) shows that among the Brazilian Amazon haplotypes, BrC and BrF are basal. The haplotype BrC is closely related to the Colombian Amazon haplotype CA1 and the other Brazilian haplotypes. The haplotype BrF connects with the remnant haplotypes from the Colombian Amazon (CA2) and Orinoco (CO1 to CO5). The Brazilian haplotypes fill a phylogenetic gap between Colombian Orinoco and Bolivian haplotypes, as would be expected due to the intermediate geographic position between both localities, considering the connectivity of rivers. The MJN also indicates some heterogeneity in the Colombian Orinoco population.

Phylogenetic analysis of the mtDNA control region (Figure 4) shows that the Brazilian Amazon haplotypes are more related to the Colombian Amazon and Orinoco when compared to the Bolivian Amazon ones. The Brazilian-Colombian group and the Bolivian group were both supported by 91% bootstrap values (Figure 4). Colombian Amazon haplotypes appear as basal in this clade, and although Colombian Orinoco ones appear as a cluster, there is no significant bootstrap support, and this cluster is also inconsistent using maximum parsimony (data not shown).

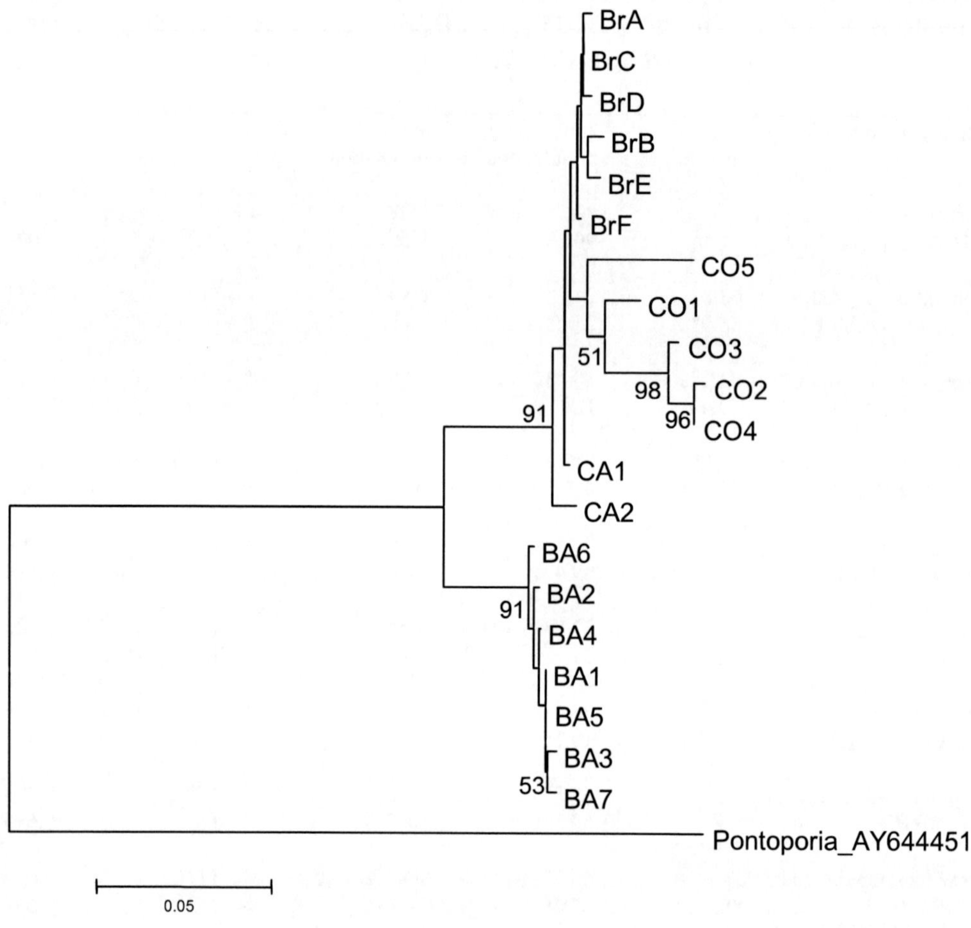

Figure 4. Neighbor-joining tree for HVSI mtDNA haplotypes. Bootstrap supporting values (percents of 1,000 replicates) are depicted on branches; only values above 50% are shown. CO (Colombian Orinoco), CA (Colombian Amazon), BA (Bolivian Amazon) and Br (Brazilian Amazon) haplotypes. We used a *Pontoporia blainvillei* control region sequence, retrieved from Genbank as an outgroup.

To address the systematic position of Brazilian *Inia* sp population, we included the 600 bp Cyt-b sequences reported by Banguera-Hinestroza et al., (2002). Phylogenetic reconstruction of all *Inia* sp haplotypes shows that the Brazilian Amazon population represents the most frequent haplotype discovered in the Colombian Amazon (CA2), in agreement with the results obtained by these authors. However, it was not possible to discriminate the two Brazilian haplotypes (Mamiraua-Tefe) in the tree, once the position of the polymorphic site that separates them was not included in the 600 bp by Banguera-Hinestroza et al., (2002) (data not shown).

The numbers of haplotypes (HT), polymorphic sites (S), mean pairwise difference (d), gene diversity (h), nucleotide diversity (π) are displayed. Standard deviations are included for *h* and π ($p < 0.05$). Asterisks (*) indicate Bolivian haplotypes reported by Banguera-Hinestroza et al., (2002) that were originally 7 haplotypes, but they have published (see Figures 2 and 3 of their paper) and deposited in Genbank the haplotypes BA1 and BA5 as identical sequences.

CONCLUSION

Some general aspects can be retrieved from the diversity data of the Brazilian populations. The higher haplotype diversity found on the Mamirauá location may be a result of greater habitat diversity found on the reserve, as compared to the non protected Tefé area. The Mamirauá reserve (MSDR, 64°45'W; 03°35'S) is a flooded land with many lakes and intermittent channels, comprising 11.240 km^2. It includes the Mamirauá Lake system (225 km^2) where many samples were collected. The reserve is situated between the juncture of the Japurá and Solimões (Amazon) rivers, two large whitewater tributaries of the Amazon basin. The MSDR is a floodplain area with a seasonal flood of 10-15 m in amplitude. Furthermore, the reserve has been established since 1990, preserving a high diversity (>300 *spp*) and biomass of fishes (Crampton, 1999), which attracts a diverse community of piscivorous predators, including *Inia,* a generalist fish predator (da Silva, 1983). The Tefé population occurs in a dark transparent water environment, which is less common in that Amazon area where the turbid water environment is widespread. The Tefé is a large, black water lake (125 km^2) but shallow, with a maximum depth of 20 m. The lake is connected to the Amazon River by a narrow channel. Although the two Brazilian locations are close to each other, the maternal lineages reveal a relatively low gene flow detected by the mtDNA HVSI (Φ_{ST}=0.66, Table 4), and no shared Cyt-b haplotypes. The distance between the mouth of the Tefé River and the beginning of the Mamirauá Reserve (MSDR) is only 45 km and the results reveal a low (or presently absent) female migration between these areas. A different situation is found in other sympatric aquatic Amazonian mammals. The Amazonian manatee displays haplotypes widely spread and a reduced population structure (Φ_{ST} = 0.10 to 0.22) (Vianna et al., 2006). Furthermore, the Amazon River dolphin *Sotalia fluviatilis* also displays a significantly lower population structure (Φ_{ST} = 0.29) (Caballero et al., 2007). The higher level of population structure detected in *Inia geoffrensis* should be seriously considered in future strategies for population management and species conservation, particularly those involving translocation of animals.

The phylogeographic pattern observed in the analysis of control region mtDNA haplotypes from Brazilian, Colombian and Bolivian populations described by Banguera-Hinestroza et al., (2002) is still very remarking when adding our data and analyses. Our results also support the occurrence of two ESUs, through the use of network (MJN) and tree (NJ) reconstructions, as well as using SAMOVA (Dupanloup et al., 2002) and the Barrier software (Manni et al., 2004). The separation of the Bolivian ESU can be likely due to a barrier consisted by 200 km of rapids between Guajara-mirim and Porto Velho in Brazil (da Silva, 1994), isolated since the end of the Pliocene and the beginning of the Pleistocene (Pilleri et al., 1984).

The Brazilian haplotypes occupy an intermediate position between haplotypes of Colombian Orinoco and Bolivian populations (Figure 2). Interestingly, our MJN analyses

(Figure 2) also indicates that the Orinoco basin has been settled by a diverse set of mtDNA haplotypes. Although the NJ tree (Figure 4) could suggest the monophyly of Orinoco haplotypes, this clade is not well supported by bootstrap and it does not remain clustered in some maximum parsimony analyses (data not shown). The Orinoco basin has a unique communication with the Amazon basin in the Casiquiari Channel (Venezuela) and, if we assume that *Inia* has an origin in the Amazon as suggested by our phylogenetic analysis (Figure 4), it could indicate a previous colonization event of the Orinoco bringing together distinct haplotype groups in one or more dispersion waves.

Although the phylogeographic pattern reveals a strong geographic correlation (Figures 2 & 4, Table 4), all Mantel tests using pairwise F_{ST}'s and distances along the rivers revealed no significant correlation (data not shown). Thus, geographic distances may not be the only factor that is significantly affecting the present distribution of mtDNA haplotypes.

The strong phylogeographic structure observed in *Inia* indicates a high phylopatry, at least for the females. In mammals the females are usually more phylopatric than males. *Inia* probably does not often migrate long distances, or the migrations may be only made by males. However, the significant differentiation observed between the two closely located (45 km) Brazilian populations (potentially in contact), requires another explanation. Sampling bias is not a likely explanation as capture of individuals has not involved members of a single family, and has occurred in different times. Although small, sample sizes look representative of the sampled areas considering that a population around 250-300 resident individuals was estimated for the Mamirauá Lake system (Martin & da Silva, 2004a). According to the authors, there is an average of 1.16 *Inia* per km^2 of *várzea* (only the flooded area) or 0.84 per km^2 in the total area. Extrapolating this value to the 125 km^2 of Tefé Lake system, we could predict a total population of 105 to 145 *Inia* individuals, although lower values would be expected in the Tefé Lake due to higher human impact and because it is less productive when compared to Mamirauá. Thus, our current sample sizes would correspond to about 4.5% (n = 11) and 8.5% (n = 9) of the total resident population, in Mamirauá and Tefé respectively. Besides, our conclusions are based on the presence of significant differentiation observed between our samples, not on the lack of data that is expected to produce non-significant comparisons.

Considering the significant differentiation between the two adjacent Brazilian subpopulations, our explanatory alternative is based on the observation that Mamirauá and Tefé are ecologically very distinct water environments, typically found in the Amazon Basin (Sioli, 1984). *Inia* populations may have local adaptations to dark, transparent, low pH water that are different from a turbid water environment such as the Amazon (Solimões) River, which separates Mamirauá and Tefé. Locally adapted populations tend to isolate from their neighbors adapted to other environments, thus these *Inia* populations may have been accumulating genetic differentiation separated by ecological barriers.

Indeed, unique adaptations related to the water environment are present in *Inia*. The flexible vertebral column and neck (nonfused cervical vertebrae) and large, broad, and paddle-like flippers which are capable of circular movements, allowing to move between trees and submerged vegetation to search for food in the flooded forest (da Silva, 2002). Furthermore, like other river dolphins, *Inia* are endowed with a sophisticated sonar system, but also have good vision, different from Indus and Ganges' dolphins, which are functionally blind (Reeves, 2002). All adaptive characteristics would be very dependent on the water characteristics in the Amazon basin. The Amazon rivers are characterized by very distinct physical and chemical properties of the water divided in three main types: turbid white water (depth of the Secchi disc: 0.10-0.50 m; pH 6.2-7.2), clear water (1.10-4.30 m; pH 4.5-7.8) and black water (1.30-2.90 m; pH 3.8-4.9) (Sioli, 1984). Our results present a significant differentiation in mtDNA variation between populations from different environments separated by the Amazon River with turbid water. This is an indication of post-isolation differentiation, which could be initially promoted by adaptive segregation of *Inia* populations from diverse water environments. Besides, mtDNA divergence tends to accumulate by drift much faster than autosomal loci because it presents a quarter of the effective population size of diploid markers, such as microsatellites. Thus, it is not an expectation to also find significant differentiation in autosomal markers, unless they are linked or directly involved with putative genes associated to the adaptive response to different water environments.

In a previous long-term study of dolphins in the Mamirauá system, Martin & da Silva (2004a) identified a high degree of fidelity, where 90% of the pink dolphin population sighted in the system was considered to be resident of the area. These authors never recorded marked dolphins at distances over 10 km from the area. However, they believe these animals use the continuous "várzea" system extensively, and can move tens to hundreds of kilometers, which would grant them a high mixing potential. On the other hand, Martin and da Silva (2004b) observed marked sexual segregation in which males would preferentially remain in the main rivers' area, whereas females would tend to inhabit the more remote and protected portions of the habitat. Even though Mamirauá is annually flooded with the turbid water of the Amazon and Japurá rivers, Mamirauá and Tefé can be regarded as relatively isolated habitats, separated by the Amazon River.

Taking into account the significant differentiation observed in *Inia*, with a unique spatial distribution shaped by geography and ecological barriers, we suggest a specific conservation management. To minimize anthropogenic impact, the reintroduction and translocation of the animals in nature should consider the maintenance of the *Inia* phylopatry, managing local populations as semi-independent evolutionary units. It will be a particularly important consideration if translocations should be necessary in the future, involving individuals from populations coming from farther regions and from different water environments. Finally, we suggest further studies using several neutral markers and adaptation related genes in many different Amazon habitats to investigate in detail the differentiation process among *Inia geoffrensis* populations.

ACKNOWLEDGMENTS

This work received a research grant from CNPq (National Research Council of Brazil). JAV and CH were supported by CAPES (Coordenação de Aperfeiçoamento do Pessoal de Ensino Superior), FRS and RAFR by CNPq, and MM by MCT (Ministry of Science and Technology) from Brazil. Thanks to the Cetacean Society International for their financial support. All Brazilian samples were collected with the government permit IBAMA 130/2004. This work was also performed according to the special authorization for access to genetic resources in Brazil # 03/2004 issued by IBAMA/CGEN.

REFERENCES

[1] Arnason U., Gullberg A., & Janke A. (2004). Mitogenomic analyses provide new insights into cetacean origin and evolution. *Gene,* 333, 27-34

[2] Bandelt, H-J., Forster, P., & Röhl, A. (1999). Median-joining networks for inferring intra-specific phylogenies. *Molecular Biology and Evolution,* 16, 37-48

[3] Banguera-Hinestroza E, Cárdenas H, Ruiz-García M, Marmontel M, Gaitán E, Vázquez R., & García-Vallejo, F. (2002). Molecular Identification of Evolutionarily Significant Units in the Amazon River Dolphin Inia sp. (Cetacea: Iniidae). *The Journal of Heredity*, 93, 312-322

[4] Best R., & da Silva, V.M.F. (1989). Amazon River dolphin, Boto. Inia geoffrensis, de Blainville, 1817. In S. H. Ridgway & S. R. Harrison (Eds.), *Handbook of marine mammals*. London, New England: Academic Press.

[5] Caballero, S., Trujillo, F., Vianna, J.A., Barrios, H., Montiel, M.G., Beltrán, S., Marmontel, M., Santos, M.C., Rossi, M., Santos, F.R., & Baker, S.C. (2007). Taxonomic status of South American coastal and riverine dolphins Sotalia spp., with insights into cetacean molecular systematics. *Marine Mammal Science,* 23 (2), 358–386.

[6] Cassens, I., Saverio, V., Waddell, V.G., Balchowsky, H., Van Belle, D., Ding, W., Fan,

[7] C., Mohan Lal, R.S., Simões-Lopes, P.C., Bastida, R., Meyer, A., Stanhope, M.J., & Milinkovitch, M.C. (2000). Independent adaptation to riverine habitats allowed survival of ancient cetacean lineages *Proceedings of the National Academy of Sciences USA,* 97, 11343-11347.

[8] Crampton, W.G.R. (1999). Os peixes da Reserva Mamiruá: diversidade e história natural na planície alagável da Amazônia. In Quiroz HL & Crampton WGR (Eds), *Estratégias para Manejo de Recursos Pesqueiros em Mamirauá, Brasília* (pp. 10-36), Include city here. Brazil: Sociedade Civil Mamirauá/CNPq.

[9] da Silva, V. M. F. (1983). *Ecologia Alimentar dos Golfinhos da Amazônia. Master Thesis*, Manaus, Brazil: University of Amazonas.

[10] da Silva, V.M.F. (1994). *Aspects of the biology of the Amazonian dolphin genus Inia and Sotalia fluviatilis (Doctorate Dissertation)*. Cambridge, United Kingdom: Cambridge University.

[11] da Silva, V. M. F. (2002). Amazon River Dolphin - Inia geoffrensis. In W. F. Perrin, B. Würsig, & J. G. M. Thewissen (Eds.), *Encyclopedia of marine mammals* (pp. 18-20). San Diego, CA: Academic Press.

[12] Dupanloup, I., Schneider, S., & Excoffier, L. (2002). A simulated annealing approach to define the genetic structure of populations. *Molecular Ecology,* 11, 2571-2581.

[13] Ewing, B., Hillier, L., Wendl, M., & Green, P. (1998). Basecalling of automated sequencer traces using phred. I. Accuracy assessment. *Genome Research,* 8, 175-185.

[14] Fu, Y-X. (1997). Statistical tests of neutrality mutations against population growth, hitchhiking and background selection. *Genetics,* 147, 915-925.

[15] Gordon, D., Abajian, C., & Green, P. (1998). Consed: A graphical tool for sequence finishing. *Genome Research,* 8, 195-202.

[16] Green, P. Phrap [online]. 1994 [cited 2009, July 17]. Available from: URL: www.genome.washington.edu/UWGC/analysistools/phrap.htm.

[17] Hamilton, H., Caballero, S., Collins, A.G., & Brownell-Jr, R.L. (2001). Evolution of river dolphins. *Proceedings of the Royal Society of London Biological*, 268, 549-556.

[18] Kocher, T.D., Thomas, W.K., Meyer, A., Edwards, S.V., Paabo, S., Villablanca, F.X. & Wilson, E.A.C., (1989). Dynamics of mitochondrial DNA evolution in animals: amplification and sequencing with conserved primers. *Proceedings of the National Academy of Sciences USA,* 86, 6196-6200.

[19] Kumar, S., Tamura, K., & Nei, M. (2004). MEGA3: Integrated software for Molecular Evolutionary Genetics Analysis and sequence alignment. *Briefings in Bioinformatics,* 5, 150-163.

[20] Manni, F., Guérard, E., & Heyer, E. (2004). Geographic pattern of (genetic, morphologic, linguistic) variation: how barriers can be detected by "Monmonier's algorithm". *Human Biology,* 76, 173-190.

[21] Martin, A.R., & da Silva, V.M.F. (2004a). Numbers, seasonal movements and residency characteristics of river dolphins in an Amazonian floodplain lake system. *Canadian Journal of Zoology,* 82, 1307-1315.

[22] Martin, A.R., & da Silva, V.M.F. (2004b). River dolphins and flooded forest: seasonal habitat use and sexual segregation of botos Inia geoffrensis in an extreme cetacean environment. *Journal of Zoology,* 263, 295-305.

[23] Pilleri, G., Marcuzzi, G., & Pilleri, O. (1984). Speciation in the Platanistoidea: systematic, zoogeographical and ecological observations on recent species. *Investigations on Cetacea*. 14:15-46.

[24] Posada, D., & Crandall, K.A. (1998). Modeltest: testing the model of DNA substitution. *Bioinformatics*, 14, 817-818.

[25] Redondo, R. A. F., Brina, L. P. S., Silva R. F., & Ditchfield A. D. (2008). Molecular systematics of genus *Artibeus* (Chiroptera: Phyllostomidae) (*in press*).

[26] Reeves, R. R. (2002). River Dolphins. In W. F. Perrin, B. Würsig, & J. G. M. Thewissen (Eds.), *Encyclopedia of Marine Mammals* (pp. 18-20). San Diego, CA: Academic Press.

[27] Rosas, F. C. W. & Lehti, K. K. (1996). Nutritional and mercury content of milk of the Amazon River Dolphin, Inia geoffrensis. *Comparative Biochemistry and Physiology,* 115A, 117-119.

[28] Sambrook, J., Russell, D.W., & Sambrook, J. (2001). *Molecular Cloning: A Laboratory Manual. 3rd ed.* Woodbury, NY: Cold Spring Harbor Laboratory Press.

[29] Schneider, S., Roessli, D., & Excoffier, L. (2000). *Arlequin ver. 2.000. A Software for Population Genetics Data Analysis*. Geneva, Switzerland: Genetics and Biometry Laboratory, University of Geneva.

[30] Sioli, H. (1984). The Amazon and its main affluent: hydrology, morphology of the river courses, and river types. In H. Sioli & J. W. Dordrecht (Eds.), *The Amazon: Limnology and landscape ecology of a mighty tropical river and its basins* (pp. 127-165).

[31] Slatkin, M. (1991). Inbreeding coefficients and coalescence times. *Genetical Research,* 58, 167-175.

[32] Tajima, F. (1989). Statistical method for testing the neutral mutation hypothesis by DNA polymorphism. *Genetics,* 123, 585-595.

[33] Tamura, K., & Nei, M. (1993). Estimation of the number of nucleotide substitutions in the control region of mitochondrial DNA in humans and chimpanzees. *Molecular Biology and Evolution,* 10, 512-526

[34] Thompson, J.D., Gibson, T.J., Plewniak, F., Jeanmougin, F., & Higgins, D.G. (1997). The Clustal-X windows interface: flexible strategies for multiple sequence alignment aided by quality analysis tools. *Nucleic Acids Research,* 24, 4876-4882.

[35] Vianna, J. A., Bonde, R. K., Caballero, S., Giraldo, J. P., Lima, R. P., Clark, A. M., Marmontel, M., Morales-Vela, B., Souza, M. J., Parr, L., Rodriguez-Lopez, M. A., Mignucci-Giannoni, A. A., Powell, J., & Santos, F. R. (2006). Phylogeography, phylogeny and hybridization in trichechid sirenians: Implications on manatee conservation. *Molecular Ecology,* 15, 433-447.

[36] Yan, J., Zhou, K., & Yang, G. (2005). Molecular phylogenetics of 'river dolphins' and the baiji mitochondrial genome. *Molecular Phylogenetics and Evolution,* 37, 743-757.

[37] Yang, G., Liu, S., Ren, W.H., Zhou, K., & Wei, F. (2003). Mitochondrial control region variability of baiji and the Yangtze finless porpoises, two sympatric small cetaceans in the Yangtze River. *Acta Theriologica*, 48, 469–483.

In: Biology, Evolution and Conservation of River Dolphins... ISBN: 978-1-60876-633-8
Editors: M. Ruiz-García, J. Shostell. pp. 117-130

Chapter 7

AMAZON RIVER DOLPHIN POLYMORPHISM AND POPULATION DIFFERENTIATION OF MHC CLASS II PEPTIDES

María Martínez-Agüero[1], Sergio Flores-Ramírez[2] and Manuel Ruiz-García[1*]

[1]Unidad de Genética (Grupo de Genética de Poblaciones Molecular y Biología Evolutiva). Departamento de Biología. Facultad de Ciencias. Pontificia Universidad Javeriana, Bogotá DC, Colombia

[2]Laboratorio de Ecología Molecular y Genética de la Conservación, Universidad Autónoma de Baja California Sur, México

ABSTRACT

Inia, the Amazon River dolphin, inhabits the three major basins of northern South America (Beni-Mamoré, Amazon and Orinoco). We analyzed class II DQB MHC gene peptide sequences in 60 dolphins from Bolivia (Beni-Mamoré), Peru (Amazon) and Colombia (Orinoco). Sixteen (16) peptide alleles were identified, generated by 17 polymorphic sites, most of them on the peptide binding region (PBR) residues. Four of the alleles were the most frequent of all the populations and several private alleles for each basin were found. A high level of polymorphism in the class II gene was determined, similar to those reported for the Chinese river dolphins, such as the Baiji and the finless porpoise. This polymorphism could be an adaptive response to the high level of pathogens in freshwater.

Keywords: *MHC, DQB, Inia geoffrensis,* gene variability, adaptative response, Bolivia, Peru, Colombia.

* mruiz@javeriana.edu.co , mruizgar@yahoo.es.

INTRODUCTION

Inia, the Amazon pink river dolphin, is the largest river dolphin and has the most stable and largest population of all the world's freshwater dolphins. Its distribution includes three major basins in northern South America: the Beni-Mamoré, in the Bolivian Amazon, and the Amazon and Orinoco. Traditionally, *Inia* was considered a monospecific genus with only one species, *Inia geoffrensis* (de Blainville, 1817), and three subspecies, *Inia geoffrensis boliviensis*, *I. g. geoffrensis*, and *I. g. humboldtiana*, distributed within each of the major basins. Nevertheless, some molecular and morphological studies suggest that, at least, the Bolivian morphotype could be a different and an allopatric species, separated by 400 km of falls and rapids in the Madeira River in Bolivia and Brazil (Pilleri & Gihr 1977; Banguera-Hinestroza et al., 2002; Ruiz-García, 2007, 2010; Ruiz-García et al., 2006, 2007, 2008).

The waters that flow throughout these rivers in the overall Amazon and Orinoco basins have two different origins. The white rivers have their origin in the Andean mountains and have high levels of suspended sediments and organic resources, while the black rivers originate in the lowlands and have low productivity because of their high concentration of tannins that produce low pH values and few organic resources, although some black waters, such as those from the Negro River watershed in Brazil can have astonishing biological richness (Goulding et al., 1988).

Traditionally, it was believed that *Inia* could not live in black rivers, but some studies in Venezuela suggest that the dolphins even inhabit the Casiquiare Channel (Pilleri & Pilleri, 1982; Borobia 1990, Meade & Koehnken, 1991; Romero et al., 2001) that has been considered as a barrier between *Inia geoffrensis humboldtiana* (Orinoco subspecies) and *I. g. geoffrensis* (Amazon subspecies) due to the extreme low pH and low productivity. So, these dolphins are exposed to different environments and, probably, the pathogen pressure and parasite virulence in these habitats is different too, because the habitats are quite distinct displaying their own seasonal flooding and specific ecological settings, specific to black or white tributaries. Furthermore, the amount of white and black water for any of the basins is different; therefore we expect that the different *Inia* forms have different and specific adaptations to live in those basins.

To solve this question we may use different molecular markers, but neutral polymorphisms (as mitochondrial DNA, nuclear microsatellites, and autosome or Y chromosome introns) do not provide precise information on how selection operates on individual interaction with the environment nor their ability to adapt to environmental change, two relevant issues for conservation (Meyers & Bull 2002; van Tienderen et al., 2002; Crandall et al., 2000). In such cases, it may turn profitable to analyze genes potentially subjected to selection (Cohen 2002; Koskinen et al., 2002), such as those of the major histocompatibility complex (*MHC*). *MHC* genes are the most variable genes among vertebrates (Robinson et al., 2003) due to high non-synonymous substitution rates that change the amino acid sequence and physiochemical traits of *MHC* proteins (Hughes & Nei, 1988).

Studies on *MHC* polymorphism and evolution in humans and certain primate populations are numerous (*e.g.* Klein et al., 1993; Bergström et al., 1999; Bontrop et al., 1999) and point out the benefits of *MHC* variation in humans and macaques (Black & Hedrick 1997; Dorak et al., 2002; Sauermann et al., 2001) and the importance of certain *MHC* alleles for eliciting efficient immune responses (*e.g.* Hill et al., 1991; Jepson et al., 1997; Hill 1999; Thursz et al.,

1999; Carrington et al., 1999). This last concept has motivated studies on endangered mammal species living in small populations, that due to their expected low *MHC* variation, were also expected to be more susceptible to infectious disease, and thus to extinction (Evermann et al., 1988; Lyles & Dobson 1993; Mikko et al., 1999).

These, and other recent studies, have helped to built the widely accepted, but somewhat contentious perception, that *MHC* variation is related to disease resistance in wild vertebrates (Hedrick & Miller 199; Gutierrez-Espeleta et al., 2001; Garrigan & Hedrick 2003; Aguilar et al,. 2004; Hedrick 2004). A major consensus is that, with rare exceptions, *MHC* variation is strongly influenced by selection, albeit it is more common to detect that selection if it has operated at some point throughout evolutionary time rather than in the extant generations. Most importantly, contentious perceptions point out that we know little on the evolution of host parasite molecular interactions in wild vertebrates as a whole and less on the vertebrate-sub-group of marine and freshwater cetaceans.

Early studies found low *MHC* polymorphism in fin and blue whales, which was attributed to the weak pressure of marine pathogens (Trowsdale et al. 1989). However, more recent studies found a considerable amount of *MHC*-II polymorphisms in Beluga whales (*Delphinapterus leucas*), Narwhals (*Monodon monoceros*) and other cetaceans like the finless porpoise (*Neophocaena phocaenoides*) and southern minke whales (*Balaenoptera bonaerensis*) (Murray et al., 1995; Murray & White 1998; Hayashi et al., 2003).

On the other hand, the unique freshwater dolphin studied from this point of view showed a high variability in the class II MHC DQB gene. This was the case of the recently extinct Yangtze River dolphin *Lipotes vexillifer* (Yang et al., 2005). Eighteen animals were analyzed and these authors found the highest level of polymorphism at this gene comparing with other cetacean species. They proposed that the high polymorphism could be a consequence of the adaptation to freshwater, although environmental studies are insufficient to conclude that the freshwater systems have higher levels of pathogens than the marine ones (Yang et al., 2005).

Additionally, the *MHC* genes are related to sexual selection because female mammals could detect males with adequate *MHC* genotypes, which could increase the fitness and survival of their litters by augmenting their heterozygosity levels (and therefore an augment of protection against external pathogens) at these *MHC* genes. This has been demonstrated for rodents (Potts et al., 1991) as well as for humans (Wedeking et al., 1995; Wedeking & Seebeck, 1996). In an identical sense, *MHC* genes are related to kin recognition (Beauchamp et al., 1985; Manning et al., 1992; Potts & Wakeland, 1993; Brown & Eklund, 1994). Furthermore, homozygosity in some *MHC* genes is correlated with lower fertility and higher rates of spontaneous abortion (Weckstein et al., 1991; Ober et al., 1992; Ober, 1995; Balasch et al., 1993; Ho et al., 1994).

The main goal of this study was to infer, using peptide sequence analyses, the evolutionary and adaptative potential of river dolphins populations in each of the three aforementioned *Inia* populations, because their different river environments suggest that they have been submitted to different pathogen pressures and parasite virulences.

MATERIALS AND METHODS

We analyzed pink river dolphin skin biopsies from the main rivers or tributaries of the three major basins in Northern South America. Twenty-three (23) animals were from the Beni-Mamoré basin in the Bolivian Amazon (Mamoré, Iruyañez, Iténez-Guaporé and Ipurupuru rivers). Another 27 were from the Peruvian Amazon (Napo-Curaray, Marañón-Samiria, Ucayali-Tapiche rivers) as well as from the Colombian Amazon (Putumayo River), and 10 were from the Orinoco River and some of its tributaries (Guaviare and Inirida rivers) (Figure 1).

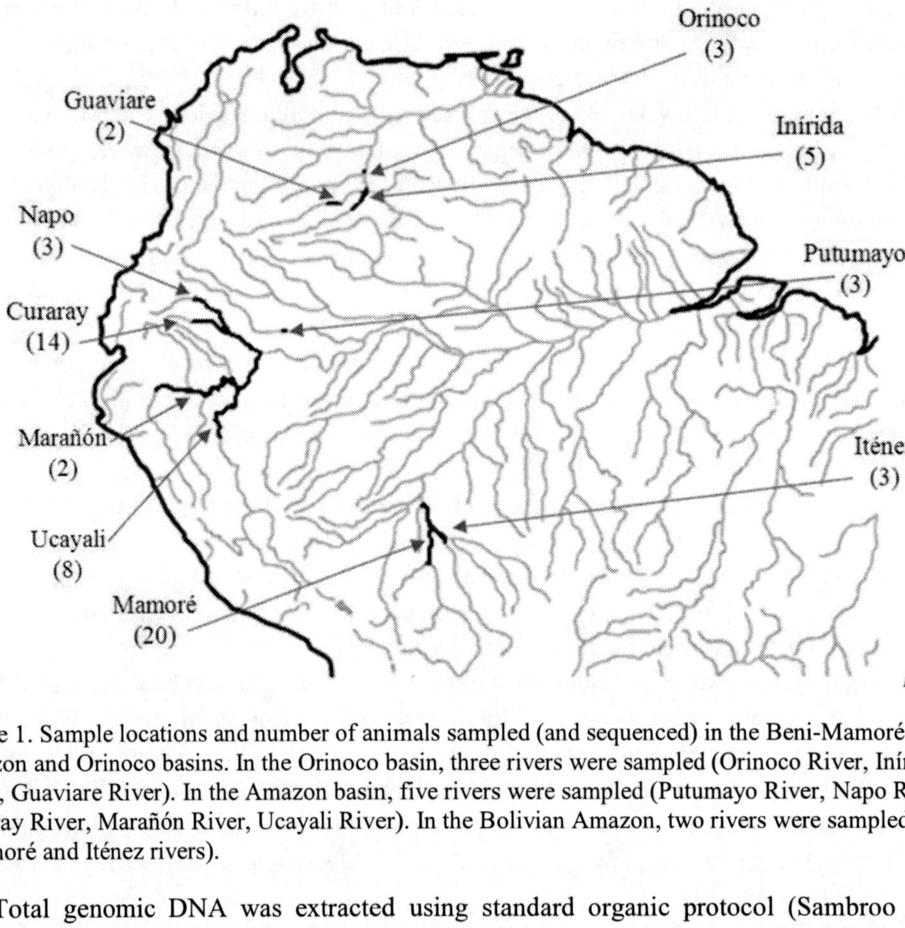

Figure 1. Sample locations and number of animals sampled (and sequenced) in the Beni-Mamoré, Amazon and Orinoco basins. In the Orinoco basin, three rivers were sampled (Orinoco River, Inírida River, Guaviare River). In the Amazon basin, five rivers were sampled (Putumayo River, Napo River, Curaray River, Marañón River, Ucayali River). In the Bolivian Amazon, two rivers were sampled (Mamoré and Iténez rivers).

Total genomic DNA was extracted using standard organic protocol (Sambroo et al., 1989) and used to amplify via PCR the second exon of MHC class II DQB using primers previously proved in other Cetacea (Murray et al. 1995): 5'-CAT GTG CTA CTT CAC CAA CGG-3' (forward) and 5'-CTG GTA GTT GTG TCT GCA CAC-3' (reverse). The amplification reactions were conducted on a Techne Genius Thermocycler or in a BioRad Cycler in a 25 μl total volume using 0.6 μM of each primer, 1X buffer reaction, 3.5 mM $MgCl_2$, 0.4 μM dNTP, 1 U Taq polymerase (Invitrogen) and 25 to 50 ng of DNA. Thermal cycling conditions comprised an initial denaturation at 95°C for 5 min, followed by 30 cycles of 94°C for 1 min, 55°C for 1 min, 72°C for 2 min, and a final extension at 72°C for 10 min. PCR products of expected size from the first assayed individual were cloned and sequenced on both strands (as described in Martínez-Agüero et al., 2006) in order to confirm their identity as a MHC locus throughout a BLAST homology search (Altschul et al,. 1990). The remaining individuals were assayed for their DQB polymorphism by applying a previously described PCR protocol and single strand conformational polymorphism (SSCP) methods (Orita et al., 1989) following previously reported protocols (Nigenda-Morales et al., 2008). After electrophoresis, the SSCP gel was silver stained using the protocol of Basam et al., (1991) or incubated for 30 min in darkness with SYBR Gold© 0.5X (Molecular Probes, Invitrogen) and visualized by UV illumination for SSCP scoring. All PCR products displaying unique SSCP patterns were cloned using a TOPO TA cloning kit (Invitrogen), and 5 positive clones per sample were sequenced on both strands using the BigDye Terminator Cycle Sequencing Ready Reaction Kit and an ABI 373 automated sequencer (Macrogen Inc., Seoul, Korea).

Unique sequences were aligned using Clustal X (Thompson et al., 1997) and corrected by hand, and later translated using BioEdit 7.0.5.2 (Hall, 1999). Amino acid sequence divergence was estimated by means of the MEGA 4.0 program by using different algorithms (Kumar et al., 2004).

RESULTS

We successfully obtained amplification of DQB exon 2 from all individuals analyzed. Eighteen different alleles were identified. None of the sequences had shown in/del mutations and no more than two different sequences for each animal were obtained, thus suggesting that only one locus was amplified. The nucleotide sequences were translated on the three possible reading frames and we obtained just one open reading frame (Table 1) and 16 different alleles, all of them with 57 amino acids. We highlighted the binding peptide sites in red, and the lateral chains of the PBR in green (Table 1).

Three major groups of alleles were found by comparing the similarity of the sequences between them and with a consensus sequence using a median-joining algorithm (Bandelt et al., 1999): *Inge* DQB1-01, with eight different peptides, *Inge* DQB1-02, with five peptides, and *Inge* DQB1-03, with three different peptides. The third group was the most divergent one.

Fifteen animals were homozygous and 45 were heterozygous; Table 2 shows the results for each river and basin indicating the alleles and number of copies of each allele found in the different populations. The Bolivian population showed nine different alleles. Seven animals were homozygous and 16 were heterozygous; the alleles *Inge* DQB1-0104, -0107, and -0205 were only found in the Mamoré River population. At the Orinoco basin, nine different alleles were found; these specimens yielded the major levels of homozygosity; five animals were homozygous and five were heterozygous; the alleles *Inge* DQB1-0302 and -0303 were only found in the Inirida River. Finally, in the upper Amazon we found 11 different alleles, two of these, *Inge* DQB1-0105 and -0202, were only in the Peruvian population. In this basin, three homozygous and 24 heterozygous animals were determined.

The amino acid sequences have 17 polymorphic sites (Table 1), most of them on the PBR (peptide binding region) and the variability at the amino acid level is 29.8% (17/57). These peptide mutations were analyzed using PAM 1 and Blossum 80 similarity matrices trying to identify the most important changes for the structure of the peptide comparing the alleles with the consensus sequence. The major divergence was found comparing the consensus sequence with *Inge* DQB1-0302, a private allele from the Inírida River in the Orinoco Basin. Furthermore, the allele *Inge* DQB1-0102 was the most similar sequence to the consensus one.

The major physical-chemical divergence was produced changing tyrosine-asparagine (position 37), an aromatic amino acid for a long linear acid (hydrophobic one), and also changing arginine-glycine (positions 5 and 51).

The highest polymorphism was found in the Orinoco Basin although that basin showed the major proportion of homozygous animals, meanwhile the Bolivian animals showed the lowest polymorphism, which is consistent with that observed with other molecular markers. The average evolutionary divergences were calculated between and within the populations and the results confirmed the major differences between the Orinoco samples (0.135 – 0.140).

Table 1. Segregate sites on theoretical amino acid sequences for DQB-1 exon 2 locus identified in *Inia*

		*	20	*	40	*		
Inge 0101	DQB1	TERVRFVSSY	IYNREEYVRF	DSDVGEYRAV	TELGRPYAEY	WNRQKDILEQ	TRAELDT	57
Inge 0102	DQB1	........R.	....	....	....	....	R......	57
Inge 0103	DQB1	........R.	....	....	....	.........R	....	57
Inge 0104	DQB1	.....Y..R.	....	....	....	..S.......	....	57
Inge 0105	DQB1	........R.	....	....	....	....	R...V..	57
Inge 0106	DQB1	......M.R.	....	....	S.........	....	R...V..	57
Inge 0107	DQB1	....GY..R.	....	....	....	....	....V..	57
Inge 0108	DQB1	....	....	....	......N...	..S...L..R	G......	57
Inge 0201	DQB1	.......TR.	....	....	....	........R.	R......	57
Inge 0202	DQB1	........R.	....	....	....	........R.	R......	57
Inge 0203	DQB1	......M.R.	......FA..	....	....	........R.	K......	57
Inge 0204	DQB1	.....YM.R.	....	....	....	........RR	R...V..	57
Inge 0205	DQB1	.....Y..R.	....	....	....	........RK	....	57
Inge 0301	DQB1	.....Y..R.	....	....	.....RA...	..S.......	....V..	57
Inge 0302	DQB1	.....Y.TR.	......FA..	....	.....RT.K.	..S...L...	R......	57
Inge 0303	DQB1	.....Y.TR.	....	....	......T...	..S.......	....V..	57

Table 2. Number of copies with each *Inge DQB1* allele per river analyzed in the three basins considered. We show the subspecies and the tributary rivers on each basin

Subspecies	River	allele	No. copies	Homozygous
I. g. boliviensis	Mamoré and Iruyáñez	0101	11	2
		0102	4	0
		0103	1	0
		0104	1	0
		0107	2	0
		0201	2	2
		0203	3	0
		0204	6	1
		0205	2	0
	Iténez-Guaporé and Ipurupuru	0101	3	1
		0103	2	1
		0201	2	0
		0204	1	0
I. g. humboldtiana	Orinoco	0101	2	1
		0108	2	0
		0204	1	0
		0301	1	0
	Inírida	0108	2	1
		0203	2	1
		0204	2	1
		0301	1	0
		0302	2	1
		0303	1	0
	Guaviare	0103	1	0
		0106	1	0
		0108	1	0
		0203	1	0
I. g. geoffrensis	Napo-Curaray	0101	2	0
		0103	1	0
		0105	2	0
		0106	1	0
		0108	1	0
		0201	1	1
		0202	3	0
		0203	7	1
		0204	10	1
	Marañón-Samiria	0101	1	0
		0201	1	0
		0203	1	0
		0204	1	0
	Putumayo	0101	2	0
		0103	1	0
		0106	2	0
		0203	1	0
	Ucayali-Tapiche	0102	2	0
		0103	1	0
		0105	2	0
		0106	1	0
		0108	1	0
		0201	1	0
		0202	1	0
		0203	3	0
		0204	3	0
		0301	1	0

DISCUSSION

In the present study, 16 different alleles were identified from the theoretical translation of the exon 2 *MHC DQB* locus in 60 *Inia* dolphins. The sequences do not show in/del mutation or stop codons. Thus, they might be functional "in vivo". Additional evidence of functionality is the existence of high mutation spots on PBR creating variation in the binding site (Edwards & Hedrick 1998) that may be the result of environmental selective pathogenic pressure (Klein et al., 1993; Hedrick 1994; Bergstrom & Gyllensten, 1995).

Ten out of 17 polymorphic sites, 10 sites (5, 9, 18, 36, 37, 39, 43, 49, 50 and 51) were non-conservatives due the physicochemical properties of the residues, which would indicate changes in the recognition of foreign peptides by the PBR. These hotspots show the effect of balancing natural selection through over-dominance or frequency-dependent selection (Hughes & Nei, 1989; Garrigan & Hedrick, 2003)

Site 49 changed from a long acid hydrophilic residue in group 01 to a basic residue on group 02. In position 43, a small hydrophobic residue, characteristic of group 03, was replaced by a basic one in groups 01 and 02. And site 37 changed from a small hydrophilic residue on group 03 to an aromatic one on groups 01 and 02. These changes could be evidence of distinct antigen recognition between the allele groups.

It is important to note that, although there were many private (unique alleles found in only one population of that studied) alleles, most of the amplified products (60.8%) were the four most frequent alleles: *Inge* DQB -0101, -0201, -0203 y -0204. These alleles could give some kind of adaptive advantage to animals that carry them or reflect trans-specific polymorphism of MHC genes (Munguia-Vega et al., 2007; Xu et al., 2008).

The results of the exon 2 *MHC DQB* locus revealed basic aspects of the evolutionary and demographic history which agree quite well with that revealed by other molecular markers. The Bolivian populations showed the lowest polymorphic level of the three basin populations studied. By using two mtDNA genes (Banguera-Hinestroza et al., 2002; Ruiz-García et al., 2009a), nuclear DNA microsatellites (Ruiz-García, 2007, 2010; Ruiz-García et al., 2010b), autosome and Y chromosome intron sequences (Ruiz-García et al., 2008) and RAPDs (Ruiz-García et al., 2007), we demonstrated that the Bolivian populations was also the population with the lowest gene diversity. In parallel, the Bolivian basin individuals yielded the highest percentage of private alleles (33%) when compared with the number of private alleles found in the upper Amazon (18%) and Orinoco basins (22%). These results show that the Bolivian population came from a strong founder event that implied a bottleneck and that these original animals proceeded from the Amazon basin. The existence of the highest level of private alleles in the Bolivian rivers could show that the physical-chemical water characteristics and the number and types of pathogens of the Beni-Mamoré waters are note worthily different from those of the Amazon. Other interesting points revealed by molecular markers, were the highest levels of polymorphism and internal genetic differences found within the Orinoco basin. Ruiz-García (2010) and Ruiz-García et al., (2010a) showed that there are two different mitochondrial lineages in the same rivers of the Orinoco basin and that proceeded from the Amazon in two different epochs.

This could explain why in the Orinoco basin, we found the highest polymorphic level and the highest internal divergence among the individuals. The existence of a lower percentage of private alleles in the Orinoco than in the Bolivian rivers could be because the separation of the Orinoco's lineages are temporally more recent than the Bolivian divergence from the Amazon population (which was demonstrated by Ruiz-García, 2010, using mtDNA) and/or the physical-chemical water characteristics and the number and types of pathogens of the Orinoco basin waters are less divergent from the Amazon ones than those present in the Bolivian waters.

Bear in mind that most of the alleles are represented by animals of the three basins, and that the differences are in allele frequencies among these populations. The results suggest that the most frequent genotypes could provide the dolphins some selective advantage to specific physicochemical and microbiological conditions, as a result of the diverse selective pathogen pressures in the different water characteristics (white, black and mixed waters) and physicochemical conditions of the rivers. It could be interesting in the future to compare the microbiological and physicochemical characteristics of these basins and their main tributaries with the current results. However, the data are in the meantime insufficient for this kind of comparison.

Comparing the levels of *MHC* genetic diversity of this work with other cetacean species, we see that the genetic variability and allelic diversity of the Amazon River dolphin are very high, similar to that found in the Yangtze river dolphin (Yang et al., 2005) and the finless porpoise population in the Yangtze River (Xu et al., 2007) and clearly higher than what is found in many marine species (Slade, 1992; Murray et al., 1995; Nigenda-Morales et al., 2008). This could be due to the low environmental pathogenic pressure in marine systems, as has been proposed before (Trowsdale et al., 1989; Slade, 1992) and the higher pathogenic pressure in freshwater systems which are similar to terrestrial ones. However, several studies have shown many diseases and epizootic episodes in marine mammal populations (Acevedo-Whitehouse et al., 2006, 2003; Raga et al., 1997; van Bressem et al., 1998, 1999), and therefore an explanation of why freshwater dolphins present such high levels of variability at this gene is not yet clear.

ACKNOWLEDGMENTS

Economic resources to carry out this study were obtained from the Administrative Department of Science, Technology and Innovation (Colciencias) (1203-09-11239; Geographical population structure and genetic diversity of two river dolphin species, *Inia boliviensis* and *Inia geoffrensis*, using molecular markers) and Fondo para la Acción Ambiental (120108-E0102141; Structure and Genetic Conservation of river dolphins, *Inia* and *Sotalia*, in the Amazon and Orinoco basins). Additional thanks go to Dr. F. Trujillo (Omacha Foundation), H. Gálvez (Peru), M. Escovar and Dr. J. Vargas (Bolivia), Dr. D. Alvarez, A. Rodríguez, and P. Escobar-Armel (Colombia), A. Castellanos (Ecuador) and Isaias and his sons (Requena, Peru). Many thanks go to the Bolivian, Peruvian, Ecuadorian Ministry of Environment for their role in facilitating the obtainment of the CITES and collection permits. Thanks also go to the Pontificia Universidad Javeriana (Colombia) and to the Universidad de Baja California Sur (México) for respective economic assistance.

REFERENCES

[1] Acevedo-Whitehouse K., Gulland F., Greig D. & Amos, W. (2003). Disease Susceptibility in California sea lions. *Nature*, 422,35.

[2] Acevedo-Whitehouse K., Spraker T.R., Lyons E., Melin S. R., Gulland F., Delong R.L. & Amos, W. (2006). Contrasting effects of heterozygosity on survival and hookworm resistance in California sea lion pups. *Molecular Ecology*, 15, 1973-1982.

[3] Altschul, S.F., Madden, T. L., Schäffer, A. A., Zhang, J., Zhang, Z., Miller, W., & Lipman, D. J. (1997). Gapped BLAST and PSI-BLAST: a new generation of protein database search programs. *Nucleic Acids Research*, 25, 3389-3402.

[4] Balash, J., Jov'e, I., Martorell, J., Gay's, A., & Vanrell, J. A. (1993). Histocompatibility in in vitro fertilization couples. *Fertility and Sterility*, 59, 456-458.

[5] Bandelt, H.J., Forster, P. & Röhl, A. (1999). Median-Joining Networks for Inferring Intraspecific Phylogenies. *Molecular Biology and Evolution*, 16, 37-48.

[6] Banguera-Hinestroza, E., Cárdenas, H., Ruiz-García, M., Marmontel, M., Gaitán, E., Vázquez, R., & García-Vallejo, F. (2002). Molecular identification of evolutionarily significant units in the Amazon River dolphin *Inia sp.* (Cetacea: Iniidae). *The Journal of Heredity*, 93, 312-322.

[7] Bassam, B. J., Caetano-Anollés, G. & Gresshoff, P. M. (1991). Fast and Sensitive Silver Staining of DNA in Polyacrylamide Gels. *Analytical Biochemistry*, 196, 80-83.

[8] Bergström, T., & Gyllenstein, U. (1995). Evolution of Mhc class II polymorphism: The rise and fall of class II gene function in primates. *Immunological reviews,* 143, 13–32.

[9] Bergström, T. F., Erlandsson, R., Engkvist, H., Josefsson, A., Erlich, H. A. & Gyllensten, U. (1999). Phylogenetic history of hominoid DRB loci and alleles inferred from intron sequences. *Immunological reviews,* 167, 351-365.

[10] Beauchamp, G. K., Yamazaki, K., Wysocki, C. J., Slotnick, B. M., Thomas, L., & Boyse, E. A. (1985). Chemosensory recognition of mouse major histocompatibility types by another species. *Proceedings of the National Academy of Sciences, USA*, 82, 4186-4188.

[11] Black, F. L. & Hedrick, P. W. (1997). Strong balancing selection at HLA loci: evidence from segregation in South Amerindian families. *Proceedings of the National Academy of Science of the USA*, 94, 12452–12456.

[12] Bontrop R. E., Otting, N., de Groot, N. G., & Doxiadis, G. G. (1999). Major histocompatibility complex class II polymorphisms in primates. *Immunological reviews,* 167, 339-350.

[13] Borobia, M. (1990). Survey of *Sotalia fluviatilis* in the Casiquiare Canal, Venezuela (Project). In S. Leatherwood & C.L. Hill (Eds.), *Newsletter of the Cetacean Specialist Group*, 6, 9.

[14] Brown, J. L., & Eklund, A. (1994). Kin recognition and the major histocompatibility complex: An integrative review. *The American Naturalist*, 143, 435-461.

[15] Carrington, M., Nelson, G. W. and Martin M. P. (1999). HLA and HIV-1: heterozygote advantage and B*35-Cw*04 disadvantage. *Science*, 28, 1748–1752.

[16] Cohen, S. (2002). Strong positive selection and habitat-specific amino acid substitution patterns in MHC from an estuarine fish under intense pollution stress. *Molecular Biology and Evolution*, 19, 1870–1880.

[17] Crandall, K. A., Bininda-Emonds, O. R. P., Mace, G. M. & Wayne, R. K. (2000). Considering evolutionary processes in conservation biology. *Trends in Ecology and Evolution*, 15, 290–295.

[18] Dorak, M.T., Lawson, T., Machulla, H. K., Mills, K. I., & Burnett, A K. (2002). Increased heterozygosity for MHC class II lineages in newborn males. *Genes and Immunity*, 3, 263–269.

[19] Edwards, S. V. & Hedrick, P. W. (1998). Evolution and Ecology of MHC Molecules: From Genomics to Sexual Selection. *Trends in Ecology and Evolution*, 13, 305-311.

[20] Evermann, J.F., Heeney, J. L., Roelke, M. E., McKeirnan, A. J., & O'Brien, S. J. (1988) Biological and pathological consequences of feline infectious peritonitis virus infection in the cheetah. *Archives of Virology*, 102, 155–171.

[21] Garrigan, D., Hedrick, P.W. (2003). Perspective: detecting adaptive molecular polymorphism: lessons from the MHC. *Evolution*, 58, 1707-1722.

[22] Goulding, M., Carvalho, M. L., & Ferreira, E. G. (1988). *Río Negro, Rich Life in Poor Water*. The Hague, Netherlands: SPB Academic Publisher.

[23] Gutierrez-Espeleta, G.A., Hedrick, P.W., Kalinowski, S. T., Garrigan, D., & Boyce, W. M. (2001). Is the decline of desert bighorn sheep from infectious disease the result of low MHC variation? *Heredity*, 86, 439–450.

[24] Hall, T. BioEdit: A User-Friendly Biological Sequence Alignment Editor and Analysis [online]. 1999. [cited 2009, July 19] Available from URL:http://www.mbio.ncsu.edu/BioEdit/bioedit.html

[25] Hayashi, K., Nishida, S., Yoshida, H., Goto, M., Pastene, L. A. & Koike, H. (2003). Sequence variation of the DQB allele in the cetacean MHC. *Mammal Study*, 28, 89-96.

[26] Hedrick, P. W. (1994). Evolutionary Genetics of the Major Histocompatibility Complex. *The American Naturalist*, 6, 945-964.

[27] Hedrick, P. (2004). Evolutionary genomics: foxy MHC selection story. *Heredity*. 89, 237-8.

[28] Hedrick, P. W. & Miller, P. S. (1991). MHC polymorphism and the design of captive breeding programs — simple solutions are not the simple answer. *Conservation Biology*, 5, 556–558.

[29] Hill, A.V. (1999). The immunogenetics of resistance to malaria. *Proceedings of the Association of American Physicians*, 111, 272–277.

[30] Hill, A.V., Allsopp, C. E. & Kwiatkowski, D. (1991). Common west African HLA antigens are associated with protection from severe malaria. *Nature*, 352, 595–600.

[31] Ho, H. N., Yang, Y. S., Hsieh, R. P., Lin, H. R., Chen, S. U. (1994). Sharing of human leukocyte antigens in couples with unexplained infertility affects the success of *in vitro* fertilization and tubal embryo transfer. *American Journal of Obstetrics and Gynecology*, 170, 63-71.

[32] Hughes, A.L. & Nei, M. (1988). Pattern of nucleotide substitution at major histocompatibility complex class I loci reveals overdominant selection. *Proceedings of the National Academy of Sciences of USA*, 86, 958-962.

[33] Hughes, A.L. & Nei, M. (1989). Nucleotide Substitution at Major Histocompatibility Complex Class II Loci: Evidence for overdominant Selection. *Proceedings of the National Academy of Sciences*, 86, 958-962.

[34] Jepson, A., Banya, W., Sisay-Joof, F. (1997) Quantification of the relative contribution ofmajor histocompatibility complex (MHC) and non-MHC genes to human immune responses to foreign antigens. *Infection and Immunity*, 65, 872–876.

[35] Klein J., O'Huigin, C., Figueroa, F., Mayer, W. E., & Klein. D. (1993). Different modes of MHC evolution in primates. *Molecular Biology and Evolution*, 10, 48–59.

[36] Koskinen, M.T., Haugen, T. O. and Primmer, C. R. (2002). Contemporary fisherian life-history evolution in small salmonid populations. *Nature*, 419, 826–830.

[37] Kumar, S., Tamura, K. & Nei, M. (2004). MEGA3: Integrated Software for Molecular Evolutionary Genetics Analysis and Sequence Alignment. *Briefing in Bioinformatics*, 5, 150-163.

[38] Lyles, A.M., & Dobson, A. P. (1993). Infectious disease and intensive management: population dynamics, threatened hosts, and their parasites. *Journal of Zoo and Wildlife Medicine*, 24, 315–326.

[39] Manning, C. J., Wakeland, E. K., & Potts, W. K. (1992). Communal nesting patterns in mice implicate MHC genes in kin recognition. *Nature*, 360, 581-583.

[40] Martínez-Agüero, M., Flores-Ramírez, S. & Ruiz-García, M. (2006). First Report of Major Histocompatibility Complex Class II Loci from the Amazon Pink River Dolphins (Genus *Inia*). *Genetics and Molecular Research*, 5, 421-431.

[41] Meade R. H., & Koehnken, L. (1991). Distribution of the river dolphin, tonina *Inia geoffrensis*, in the Orinoco river basin of Venezuela and Colombia. *Interciencia,* 16, 300-312.

[42] Meyers, L.A. & Bull, J. J. (2002). Fighting change with change: adaptive variation in an uncertain world. *Trends in Ecology and Evolution*, 17, 551–557.

[43] Mikko, S., Roed, K., Schmutz, S. & L. Andersson. (1999). Monomorphism and polymorphism at Mhc DRB loci in domestic and wild ruminants. *Immunological Reviews*, 167, 169–178.

[44] Munguía-Vega, A., Esquer-Garrigos, Y., Rojas-Brancho, L., Vázquez-Juárez, R., Castro-Prieto, A. & Flores-Ramírez, S. (2007). Genetic Drift vs. Natural Selection in a Long-Term Small Isolated Population: Major Histocompatibility Complex Class II Variation in the Gulf of California Endemic Porpoise (*Phocoena sinus*). *Molecular Ecology*, 16, 4051-4065.

[45] Murray, B.W., Malik, S., & White, B. N. (1995). Sequence variation at the major histocompatibility complex locus DQB in beluga whales (*Delphinapterus leucas*). *Molecular Biology and Evolution*, 12, 582-593.

[46] Murray, B.W. & White, B.N. (1998). Sequence variation at the major histocompatibility complex DRB loci in beluga (*Delphinapterus leucas*) and narwhal (*Monodon monoceros*). *Immunogenetics*, 4, 242-52.

[47] Nigenda-Morales, S., Flores-Ramírez, S., Urbán-Ramírez, J. & Vázquez-Juárez, R. (2008). MHC DQB-I Polymorphism in the Gulf Of California Fin Whale (*Balaenoptera physalus*) Population. *Journal of Heredity*, 99, 14-21.

[48] Ober, C. (1995). Current topic: HLA and reproduction: Lessons from studies in the Hutterites. *Placenta*, 16, 569-577.

[49] Ober, C., Elias, S., Kostyu, D. D., & Hauck, W. W. (1992). Decreased fecundability in Hutterite couples sharing HLA-DR. *American Journal of Human Genetics*, 50, 6-14.

[50] Orita, M., Iwahana, H., Kanazawa, H., Hayashi, K., & Sekiya, T. (1989). Detection of Polymorphisms of Human DNA by Gel Electrophoresis as Single-Strand Conformation Polymorphisms. *Proceedings of the National Academy of Sciences*, 86, 2766-2770.

[51] Pilleri, G., & Ghir, M. (1977). Observations on the Bolivian, *Inia boliviensis* (D'Orbigny,1834) and the Amazonian bufeo, *Inia geoffrensis* (Blainville 1817), with a description of a new subspecies (*Inia geoffrensis humboldtiana*). *Investigations on Cetacea,* 8, 11-76.

[52] Pilleri, G., & Pilleri, O. (1982). *Zoologische Expedition zum Orinoco un Brazo Casiquiari 1981*. Bern, Switzerland: Hirnanatomisch Institute, University of Bern.

[53] Potts, W. K., Manning, C. J., & Wakeland, E. K. (1991). Mating patterns in seminatural populations of mice influenced by MHC genotype. *Nature*, 352, 619-621.

[54] Potts, W. K., & Wakeland, E. K. (1993). Evolution of MHC genetic diversity: A tale of incest, pestilence and sexual preference. *Trends in Genetics*, 9, 408-412.

[55] Raga J. A., Balluena J. A., Aznar J., & Fernandez M. (1997). The impacts of parasites on marine mammals: a review. *Parassitologia*, 39, 293–296.

[56] Robinson J., Waller, M. J., & Parham, P., (2003). IMGT/HLA and IMGT/MHC: equence databases for the study of the major histocompatibility complex. *Nucleic Acids Research*, 31, 311–314.

[57] Romero, A., Agudo, A. I., Green, S. M., & di Sciara, G. N., (2001). Cetaceans of enezuela: Their distribution and Conservation Status. *NOAA Technical Report NMFS*, 151.

[58] Ruiz-García, M. (2007). Genética de Poblaciones: Teoría y aplicación a la conservación de mamíferos neotropicales (Oso andino y delfín rosado). *Bol. Real Soc. Esp. Hist. Nat.*, 102 (1-4), 99-126.

[59] Ruiz-García, M. (2010). Changes in the demographic trends of Pink River Dolphins (*Inia*) at the microgeographical level in Peruvian and Bolivian rivers and within the Upper Amazon: Microsatellites and mtDNA analyses and insights into *Inia*'s origin. *In* Ruiz-García, M., & J. Shostell (Eds.), *Biology, Evolution, and Conservation of River Dolphins Within South America and Asia: Unknown Dolphins In Danger*. Hauppauge, NY: Nova Science Publishers., Inc.

[60] Ruiz-García M, Banguera-Hinestroza E, & Cardenas H. (2006). Morphological analysis of three *Inia* (Cetacea: Iniidae) populations from Colombia and Bolivia. *Acta Theriologica*, 51, 411-426.

[61] Ruiz-García, M., Murillo, A., Corrales, C., Romero-Aleán, N., & Alvarez-Prada, D. (2007). Genética de Poblaciones Amazónicas: La historia evolutiva del jaguar, ocelote, delfín rosado, mono lanudo y piurí reconstruida a partir de sus genes. *Animal Biodiversity and Conservation,* 30, 115-130.

[62] Ruiz-García, M., Caballero, S., Martínez-Agüero, M., & Shostell, J. (2008). Molecular differentiation among *Inia geoffrensis* and *Inia boliviensis* (Iniidae, Cetacea) by means of nuclear intron sequences. In Koven, V. P (Ed.), *Population Genetics Research Progress* (pp. 177-223). Hauppauge, NY: Nova Science Publisher, Inc.

[63] Ruiz-García, M., Martinez-Agüero, M., Escobar-Armel, P., Gaviria, M. & Alvarez, D. (2010a). Historical demographic evolution in the pink river dolphin (*Inia geoffrensis* and *Inia boliviensis*) and comparative demographic trends with other Latin American dolphins related with freshwater systems (*Sotalia fluviatilis* and *Pontoporia blainvillei*). *Evolution,* (in press).

[64] Ruiz-García, M., Gaviria, M., Martinez-Agüero, M., Escobar-Armel, P., & Alvarez, D. (2010b). Genetic structure of *Inia geoffrensis* and *Inia boliviensis* by means of heterologous DNA microsatellites. *The Journal of Heredity* (submitted).

[65] Sambrook, J., Fritsch, E. F. & Maniatis, T. (1989). *Molecular Cloning: A LaboratoryManual. 2nd Ed.* Cold Spring Harbor, NY: Cold Spring Harbor Laboratory Press.

[66] Sauermann, U., Nürnberg, P. and Bercovitch, F. B. (2001). Increased reproductive uccess of MHC class II heterozygous males among free-ranging rhesus macaques. *uman Genetics*, 108, 249–254.

[67] Slade RW. (1992). Limited MHC polymorphism in the southern elephant seal: mplications for MHC evolution and marine mammal population biology. *Proceedings of the Royal Society of London B.*, 249, 163–171.

[68] Thompson, J.D., Gibson, T. J., Plewniak, F., Jeanmougin F., & Higgins, D. G.,. (1997). The ClustalX windows interface: flexible strategies for multiple alignment aided by quality analysis tools. *Nucleic Acids Research*, 25, 4876-4882.

[69] Thursz, M., Yallop, R., Goldin, R., Trepo, C., & Thomas, H. C., (1999). Influence of HC class II genotype on outcome of infection with hepatitis C virus. The HENCORE group. Hepatitis C European Network for Cooperative Research. *Lancet*, 354, 2119–2124.

[70] Trowsdale, J., Groves, V., & Arnason, A., (1989). Limited MHC polymorphism in whales. *Immunogenetics*, 29, 19-24.

[71] Van Bressem, M. F., Jepson, P. & Barrett, T. (1998). Further insight on the epidemiology of cetacean morbillivirus in the northeastern Atlantic. *Marine Mammal Science*, 14, 605–613.

[72] Van Bressem M. F., Van Waerebeek K, & Raga J. A. (1999). A review of virus infections of cetaceans and the potential impact of morbilliviruses, poxviruses and papillomaviruses on host population dynamics. *Diseases of Aquatic Organisms*, 38, 53–65.

[73] Van Tienderen, P. H., A. A. de Haan, G. van der Linden & B. Vosman. (2002). Biodiversity assessment using markers for ecologically important traits. *Trends in Ecology and Evolution*, 17, 577–582.

[74] Weckstein, L. N., Patrizio, P., Balmaceda, J. P., Asch, R. H., & Branco, D. W. (1991). Human leukocyte antigen compatibility and failure to achieve a viable pregnancy with assisted reproductive technology. *Acta Europaea Fertilitatis*, 22, 103-107.

[75] Wedekind, C., & Seebeck, T. (1996). MHC and mate selection in humans?., reply. *Trends in Ecology and Evolution*, 11, 24-25.

[76] Wedekind, C., Seebeck, T., Bettens, F., & Paepke, A. (1995). MHC dependent mate preferences in humans. *Proceedings of the Royal Society of London. B.*, 260: 245-249.

[77] Xu, S., P. Sun, K. Zhou & G. Yang. (2007). Sequence Variability at Three MHC Loci of Finless Porpoises (*Neophocaena phocaenoides*). *Immunogenetics*, 59, 581-592.

[78] Xu, S., B. Chen, K. Zhou & G. Yang. (2008). High Similarity at Three MHC Loci between the Baiji and Finless Porpoise: Trans-Species or Convergent Evolution? *Molecular Phylogenetics and Evolution*, 47, 36-44.

[79] Yang, G., J. Yan, K. Zhou & F. Wei. (2005). Sequence Variation and Gene Duplication at MHC DQB Loci of Baiji (*Lipotes vexillifer*), a Chinese River Dolphin. *The Journal of Heredity*, 96, 310-317.

In: Biology, Evolution and Conservation of River Dolphins… ISBN: 978-1-60876-633-8
Editors: M. Ruiz-García, J. Shostell. pp. 131-160

Chapter 8

MICRO-GEOGRAPHICAL GENETIC STRUCTURE OF *INIA GEOFFRENSIS* IN THE NAPO-CURARAY RIVER BASIN BY MEANS OF CHESSER´S MODELS

Manuel Ruiz-García*

Laboratorio de Genética de Poblaciones Molecular-Biología Evolutiva. Unidad de Genética. Departamento de Biología. Facultad de Ciencias. Pontificia Universidad Javeriana, Bogotá DC. Colombia.

ABSTRACT

Thirty-three pink river dolphins (*Inia geoffrensis*) were caught at eight sampling places (one beach and seven lagoons; transect length of 280 km) in the Napo and Curaray rivers at the Peruvian Amazon. Nine microsatellites were applied to analyze the genetic structure of this species at the micro-geographic level and diverse population genetics procedures were used to determine if *Inia* is a solitary or a social reproductive species. Chesser's social model was used to determine asymptotic values of the F-statistics and showed that this species is a social reproductive one and the basic genetic lineages could be composed of: 1- seven reproductive females per lineage in each breeding period, 2- the number of reproductive males per linage is not important, 3- a reproductive male with four females, on average, within each lineage, and thus there is polygyny in this dolphin species, and 4- a probability of 0.30 that females of a same lineage have chosen the same male for breeding. This is the first work carried out at a micro-geographic level for a river dolphin species, where the basic social reproductive parameters are revealed.

Keywords: *Inia geoffrensis*; micro-genetic structure; social structure; Chesser's models; DNA microsatellites; Napo and Curaray rivers; Peru.

---

* mruiz@javeriana.edu.co , mruizgar@yahoo.es.

INTRODUCTION

The micro-genetic structure in mammals, and other vertebrates, is very difficult to analyze because the sampling level for it is not very easy to obtain by the researcher. In the case of mammals, only a few examples of micro-geographic genetic structure have been published considering the larger number of species the category of mammals contains and the importance of micro-genetic structure (*Alouatta seniculus*, Pope, 1992; *Balaenoptera borealis,* Kanda et al., 2006; *Cercopithecus aethiops*, Cheney & Seyfarth, 1983; *Cynomys ludovicianus*, Chesser, 1983; *Dypodomys spectabilis,* Waser & Elliot, 1991; *Felis catus*, Ruiz-García, 1998; 1999, Say et al., 2003; *Macaca mulatta*, Melnick et al., 1984; *Marmota flaviventris,* Schwartz & Armitage, 1980; Armitage, 1988; *Microtus pennsylvanicus*, Sheridan & Tamarin, 1986; *Mus musculus musculus*, Dallas et al., 1995; *Oryctolagus cuniculus*, Daly, 1981; *Petrogale xanthopus*, Pope et al., 1996; *Phocoena phocoena*, Andersen et al., 1997; *Rangifer tarandus*, Cote et al., 2002; *Spermophilus richardsonii*, van Staaden et al., 1994). This was not a comprehensive list, but it does name the more outstanding studies. No micro-geographical genetics studies have been carried out with a river dolphin because it is not easy to obtain samples of these organisms. However, we caught, 33 pink river dolphins (*Inia geoffrensis*) with large nets in a 280 km transect across seven lagoons, and in one beach, of the Napo-Curaray river basin within the Peruvian Amazon. This represented a unique opportunity to study the possible genetic structure of this species at a very fine geographic level incorporating social reproductive parameters. To do this, nine DNA microsatellites (short tandem repeat polymorphisms; STRPs) were applied to all the pink river dolphins captured in the Napo and Curaray rivers. These markers are based on the polymerase chain reaction (PCR) methodology. Microsatellites are composed of tandem, repetitive units of two to six base pairs in length (Weber and May, 1989). STRPs are randomly distributed, highly polymorphic and frequently found inside eukaryotic genomes. Therefore, they are an important tool in the study of population biology dynamics (Bruford & Wayne, 1993). All of these facts have contributed to the use of STRPs to construct genetic recombination maps in different species and to apply them to diverse population and systematic studies. Several examples are as follows: 1- To determine social structure in several wild species (for instance, in some cetaceans as *Megaptera novaeangliae,* Amos et al., 1993); 2- To determine paternity diagnoses in a vast range of animal species (from chimpanzee populations, Morin et al., 1993, to *Drosophila* populations, Noor, 1995); 3- To determine possible bottleneck events in wild species (for example in *Canis simensis,* Gottelli et al., 1994, in the Australian wombat, *Lasiorhinus krefftii,* Taylor et al., 1994 or in the upper Amazon populations of Humboldt woolly monkey, *Lagothrix lagotricha*, Ruiz-García, 2005); 4- To determine pure lineages or even possible new species (for instance, the existence of pure lineages of the Mexican gray wolf, *Canis lupus bayileyi,* Garcia-Moreno et al., 1996, or possible different chimpanzee species, Morin et al., 1994). Herein, I applied STRPs to accrue new evidence about the possibility of a consistent social reproductive system in *Inia geoffrensis* and to determine its possible genetic structure at a micro-geographical level within a determined river basin. This is important because cetacean ecologists are trying to determine if pink river dolphins have a social system or if they really are solitary animals without social structure. The ethological or ecological studies are ambiguous in this sense. The pioneer works of Trebbau & Van Bree (1974) and Pilleri & Gihr (1977), suggested that groups of these dolphins showed territoriality and were solitary. In a 550 km transect of the Amazon River, Magnusson et al. (1980), observed that the majority of pink river dolphins were solitary (81 %), followed by couples (15 %), whereas other Amazon dolphin, *Sotalia fluviatilis*, presented greater numbers of groups than did *Inia* (55 % in pairs or more). Similarly, this aforementioned work showed that the groups of *Sotalia* were more clumped than a Poisson (random) distribution, whereas the groups of *Inia* did not differ significantly from random. This was interpreted that *Sotalia* need more specialized food and are less adaptable to a variety of river conditions than *Inia* as well as showing that *Inia* are more territorial than *Sotalia*. But it was not clear if this territoriality was in favor of a social structure or in favor of an individual and solitary species. Schnapp & Howroyd (1992) tried to study whether *Inia* is solitary and if grouping behavior is only the result of aggregation in favorable habitats. They found that individual dolphins were in patch distributions and the groups did not differ significantly from a Poisson distribution. Fifty-eight of their sightings were clusters of two or more individuals (2 to 7 dolphins forming groups). They found a significant correlation between group size and stream velocity. Most large groups were found in slow running sections of the rivers and lagoons. These authors reached the conclusion that *Inia* occupied an undefended home-range. Vidal et al., (1997) reported a mean group size of 2.9 dolphins in approximately 120 km of the Amazon River bordering Colombia, Brazil and Peru that they analyzed. The highest mean group size was in the tributaries (4.1), whereas the smaller mean group was determined in lakes (2.0). McGuire & Winemiller (1998) determined an average group size of two individuals and that the number of sightings were higher in the habitat with the higher degree of heterogeneity and that the frequency of sightings most frequently occurred during the dry season than during the rainy season (41 % vs. 24 %). The mean group and the relative frequency of group size changed by season. The largest group they observed was composed of eight individuals. Denkinger et al., (2000) showed population density estimations in the Cuyabeno River at the Ecuadorian Amazon ranging from 0.01 dolphins/linear km to 0.47 dolphins/linear km depending on the rainy or dry seasons, respectively. Aliaga-Rossel (2002) found on average, 1.12 dolphins per linear km in 185 km transect conducted along the Tijamuchí River in Bolivia. He found 41 % of the animals to be solitary animals, 32 % were in pairs, 15 % were in groups of three animals and that the maximum group size was 19 animals, which suggests that there is a certain sociability within this species. Martin et al. (2004) determined that 99 % of pink river dolphins groups observed consisted of one to four specimens with a mean of 1.42 individuals in the confluence area of the Amazon and Japurá rivers in the Brazilian Amazon. However, the work of Martin & da Siva (2004b, 2006) showed strong evidence in favor of social components in the behavior and ecology of the pink river dolphins. One difficulty here is that a continuous distribution of this species in many Amazon rivers can mask any population structure that may exist (Martin & da Silva, 2004a). Additionally, since this is a top predator in the Amazon and Orinoco river basins and because many fish-nets converge on them (Ruiz-García et al., 2007), population extinctions of this species could result and consequently reveal extreme alteration in the ecology of the aforementioned rivers. Denkinger et al. (2000), for example, determined an *Inia* population density decline in the Cuyabeno and Lagartococha rivers within the Cuyabeno Reserve in the north east of the Ecuadorian Amazon from 1996 to 1998. This declination seemed to be due to the contamination caused by six oil spills and waste water effluent from oil fields. In fact, since 1990, the other Amazon dolphin (*Sotalia fluviatilis*) has disappeared from the Lagunas Grandes in the upper Cuyabeno River, an affluent of the Aguarico River.

Therefore, the main aims of this chapter to determine a possible social reproductive genetic system and a possible existence of genetic structure at a microgeographical level for *Inia geoffrensis* are as follows:

1- To determine the levels of genetic diversity in the population of *Inia geoffrensis* in a 280 km section of the Napo-Curaray rivers and within each one of the eight lagoons sampled in that transect. 2- To estimate possible existence, or not, of Hardy-Weinberg equilibrium inside each one of the lagoons studied. 3- To establish how the microsatellite genetic variance is divided into several sampling units (rivers, lagoons, individuals, genes) by means of an analysis molecular of variance (AMOVA). 4- To utilize F-statistics (Chesser 1991a,b, 1993) and analyze if a social reproductive system affects pink river dolphins in the Peruvian Amazon.

Material and Methods

During October and December 2003, we traveled 550 km of the Napo River, in Peru, from its mouth at Francisco de Orellana until Arica in the Curaray River (close to the Ecuadorian frontier), which is one of the main tributaries of the Napo River. Within these 550 km, 33 pink river dolphins were caught in eight locations distributed within 280 km of this transect. The points, where the dolphins were caught, were as follows: 1- A little beach on the left band of the Napo river near Santa Marta (S 03° 05' 00.1"; W 73° 10' 32") where two exemplars were captured. 2- Cocha Zapote (S 02° 57' 04.9"; W 73° 18' 44.6"), a lagoon of the Napo river, where two animals were caught. 3- Cocha Echevarría (S 2° 20' 58.4"; W 74° 07' 16.6"), a lagoon of the Curaray river, where one specimen was sampled. 4- Aucacocha (S 2° 18' 28.8"; W 74° 12' 12.6"), a lagoon in the Curaray River, where six individuals where captured. 5- Tiphisca Santa María (S 2° 17' 25.3"; W 74° 15' 38.3"), a lagoon in the Curaray river, where two (2) dolphins were caught. 6- Tiphisca Loro (S 2° 11' 55.6"; W 74° 32' 11.8"), a lagoon of the Curaray River, where eight exemplars were sampled. 7- Tiphisca Avispa (S 1° 58' 07.1"; W 74° 57' 31.8"), a lagoon of the Curaray River, where five individuals were captured. 8- Tiphisca Chambirá (S 1° 54' 26.5''; W 75° 00' 16.5''), a lagoon of the Curaray river, where seven dolphins were caught. The geographic sample points are shown in Figure 1.

The dolphins were captured using special fishing nets with lengths of 400 meters and widths of 10 meters, taking special care to ensure the physical safety of each dolphin. Five Indian fishermen and three biologists worked together to capture each animal (including the author). One 10-meter long wooden boat powered by a 40 horse power engine was employed to capture the animals. The individuals were brought on board and I personally took a small biopsy from the caudal fin of each dolphin captured. After the biopsy, the wound was covered with an antibiotic. Later, the animals were measured for different biometric characteristics and safely released after 5-8 minutes of manipulation. The animals were marked to avoid any recapture. The biopsies were stored in absolute alcohol until DNA extraction.

Each capture reflected a large fraction of all the animals that were present in the lagoons at the moment of capture. A total of 48 animals were present when we arrived and 33 animals were captured in the eight sampling points. Therefore, 69 % of the individuals present "in situ" were captured with the nets.

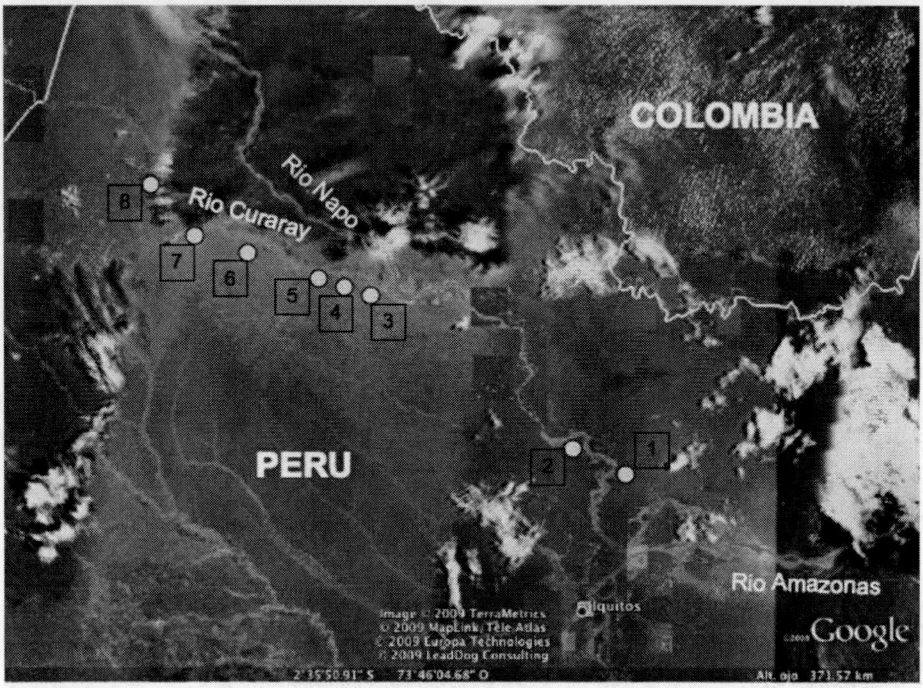

Figure 1. The eight pink river dolphin sampling points in the Napo (1-2) (affluent of the Amazon River) and Curaray rivers (3-8) (affluent of the Napo River) within the Peruvian Amazon.

DNA extraction from the caudal fin biopsies was performed by the phenol-chloroform method (Sambrock et al., 1989). The nine microsatellite markers studied in the *Inia geoffrensis* samples were Ev14, Ev37, Ev76, Ev94 (Valsecchi & Amos, 1996), MK5 (Krutzen et al., 2001), PPHO137 (Rosel et al., 1999), KWM2a, KWM2b and KWM12a (Hoelzel et al., 1998, 2002). These citations and Ruiz-García et al., (2007, 2010a) explain the amplification conditions of these microsatellites.

The amplification products were kept at 4 °C until they were used. The PCR amplification products were run in denatured 6 % polyacrilamide gels in a Hoefer SQ3 sequencer vertical chamber with 35 W as a constant and the gels were stained with $AgNO_3$ (silver nitrate) after 2-3 hours of migration. We used the molecular weight markers ϕX174 DNA digested with *Hind* III and *Hinf* I. A molecular weight marker was loaded every eight lanes.

Population Genetics Analyses

The mean number of alleles and the observed and expected heterozygosities (H_o and H, respectively), and their respective standard errors, were estimated for each one of the point samples and for the eight locality samples taken as a whole representing the 280 km of the Napo-Curaray rivers sampled.

To determine possible deviations from the Hardy-Weinberg equilibrium (HWE), in each lagoon as well as globally, we used an exact test using Markov chains, with forecasted chain length of 1,000,000 and 100,000 demorization steps. The average value of the F_{is} statistic (which measures the deviations from HWE inside each point sample) was obtained for each one of the microsatellites analyzed as well as the number of positive and negative F_{IS} values in each one of the localities sampled.

A likelihood assignation analysis of individual genotypes in all the sampling localities was carried out by using the procedures of Paetkau et al., (1995, 1997) and Waser & Strobeck (1998).

Two AMOVAs were undertaken. In the first AMOVA, three hierarchical levels were taken as the source of variation: among sampling localities, among individuals within the sampling localities and within individuals (F_{ST}, F_{IS}, F_{IT}). In the second AMOVA, four hierarchical levels were taken as the source of variation: among rivers (Napo and Curaray), among sampling locations within the rivers, among individuals within the sampling localities and within individuals (F_{CT}, F_{SC}, F_{IS}, F_{IT}). Two out of nine microsatellites were excluded from this analysis (KWM12a and PPHO 137) because of their low gene variability. The probability significances were estimated by means of 1023 bootstrap permutations. For these analyses, the Arlequin 3.1 software was employed (Excoffier et al., 2006)

Finally, the models of Chesser (1991a,b; 1993) were applied to the molecular data obtained. For this, some definitions are necessary and based on terminology developed by Cockerham (1967, 1969, 1973):

γ = coancestry of parents in the same lineage
θ = coancestry of random offspring in the same lineage
φ = coancestry of parents of different lineages
α = coancestry of random offspring of different lineages
F = coancestry of genes within random individuals
S = lineage number in the overall population
n = number of females for lineage
m = number of reproductive males for lineage
b = number of females within a determined lineage which bred with an specific male
ϕ = probability that females of the same lineage have chosen the same male for breeding, when more than one male breeds within a lineage.

In this case, I followed the Chesser's nomenclature for F-statistics, with individual lineages in one overall subpopulation:

$F_{LS} = (\theta - \alpha)/(1 - \alpha)$; $F_{IL} = (F - \theta)/(1 - \theta)$; $F_{IS} = (F - \alpha)/(1 - \alpha)$, where F_{LS} is the proportion of genetic variance found among the lineages which integrated the population studied, F_{IL} is the gene correlation within individuals in regard to those within the same lineage. F_{IS} is the correlation of genes within the individuals in regard to those within the overall population. Henceforth, F_{LS} is equivalent to F_{ST}, F_{IL} is equivalent to F_{IS}, and F_{IS} is equivalent to F_{IT} by using the original definition of Wright (1951). The parameters, constants and probabilities defined above were used to estimate transitions of coancestries over successive generations. The four transitions of coancestries in generation *t + 1* we are interest in are listed below.

1. $\alpha_{\tau+1} = (\xi\theta_{mmt})/4 + (\psi\theta_{mft})/2 + ((4 - \xi - 2\psi)\, \alpha_t/4)$, where $\xi = (n - 1)/(ns - 1)$, $\psi = 1/s$,

θ_{mmt} is the gene correlation of any given male with the n – 1 males of the same lineage at the *t* generation, θ_{mft} is the gene correlation of a given female with the n males of her own lineage at the *t* generation, whereas α_t is the average of the gene correlation of any given male with the n (s – 1) males of the all other lineages and the gene correlation of a given female to all the other n (s – 1) males of all the other lineages.

2. $F_{t+1} = \psi\theta_{mft} = (\psi\theta_{mft}) + (1 - \psi)\alpha_t$.

All of the terms within this equation are defined above.

3. $\theta_{mmt+1} = (\phi (1 + F_t)/8) + ((1 + \xi(1 - \phi)\theta_{mmt})/4) + ((\psi\theta_{mft})/2) + (((2 (1-\psi) + (1 - \xi) (1 - \phi))\alpha_t /4)$.

Again, all of the terms within this equation have been previously defined.

4. $\theta_{mft+1} = ((2n (1 - \psi) + (n - 1) (1 - \xi) (1 - \phi))\alpha_t /4) + (((n - 1) (1 + \xi (1 - \phi)) \theta_{mmt})/4) + ((\psi\theta_{mft})/2) + ((1 + F_t)(\phi(n - 1) + 2)/8n)$.

These four expressions were developed for the case where only the males migrate and the females were strictly phylopatric. Therefore, a column vector termed **S** was constructed in the t generation.

$\mathbf{S}_t = (\alpha_t , F_t , \theta_{mmt} , \theta_{mft})'$. This column vector was employed to obtain the same vector in the next generation (*t + 1*), by means of the following expression: $\mathbf{S}_{t+1} = T\mathbf{S}_t + C$, where

$$T = \begin{pmatrix} (4 - \xi - 2\psi)/4 & 0 & \xi/4 & \psi/2 \\ 1 - \psi & 0 & 0 & \psi \\ (2(1 - \psi) + (1 - \xi)(1 - \phi))/4 & \phi/8 & (1 + \xi(1 - \phi))/4 & \psi/2 \\ (2n(1 - \psi) + (n - 1)(1 - \xi)(1 - \phi))/4n & (\phi(n - 1) + 2)/8n & ((n - 1)(1 + \xi(1 - \phi))/4n & \psi/2 \end{pmatrix}$$

and $C = (0, 0, \phi/8, (\phi(n - 1) + 2)/8n)'$

For this specific case, the asymptotic values for θ, α and F used to estimate the asymptotic F-statistics were $\theta = (\theta_{mm} + \theta_{mf} + F)/2$, $\alpha = ((\xi\theta_{mm} + 2\psi\theta_{mf})/4)$, $F = \psi\theta_{mf}$, where $\theta_{mm} = (\phi/(6 - 2\xi(1 - \phi))$ and $\theta_{mf} = ((n - 1)\ \theta_{mm}/n) + (1/4n))$

If the females and the males are all dispersers, then, some equations change:

$\alpha_{\tau+1} = (\xi\theta_{mmt})/2 + (\psi\theta_{mft})/2 + ((2 - \xi - \psi)\ \alpha_t/2)$ and

$\theta_{mmt+1} = (\phi\,(1 + F_t)/8) + ((\xi(2 - \phi)\theta_{mmt})/4) + ((\psi\theta_{mft})/2) + (((2\,(1-\psi) + (1 - \xi)\,(2 - \phi))\alpha_t/4)$ and the T matrix to estimate $\mathbf{S}_{t+1}$ is now

$$T = \begin{pmatrix} (2-\xi-\psi)/2 & 0 & \xi/2 & \psi/2 \\ 1-\psi & 0 & 0 & \psi \\ (2(1-\psi)+(1-\xi)(2-\phi))/4 & \phi/8 & \xi(2-\phi)/4 & \psi/2 \\ (2n(1-\psi)+(n-1)(1-\xi)(1-\phi))/4n & (\phi(n-1)+2)/8n & ((n-1)(1+\xi(1-\phi))/4n & \psi/2 \end{pmatrix}$$

To estimate the asymptotic F-statistics, only $\theta_{mm} = (\phi/(8 - 2\xi(2 - \phi))$ needs to be modified.

A third model that I used was for the case where both sexes could simultaneously migrate with different migration rates (Chesser, 1991b). Then, the relevant equations were as follows:

1- $\alpha_{\tau+1} = (\xi(\delta_\mu + \delta_\phi)\,\theta_{mmt})/4 + (\psi(\delta_\mu + \delta_\phi - \delta_\mu\delta_\phi)\,\theta_{mft})/2 + ((4 - \xi(\delta_\mu + \delta_\phi) - 2\psi(\delta_\mu + \delta_\phi - \delta_\mu\delta_\phi))\,\alpha_t/4)$, where δ_μ is the male migration rate and δ_ϕ is the female migration rate, respectively;

2- $F_{t+1} = (1 - (1 - \psi)(\delta_\mu + \delta_\phi - \delta_\mu\delta_\phi))\theta_{mft} + (1 - \psi)(\delta_\mu + \delta_\phi - \delta_\mu\delta_\phi)\,\alpha_t$;

3- $\theta_{mmt+1} = (\phi\,(1 + F_t)/8) + ((2 - \phi - (1 - \xi)(\delta\mu\,(1 - \phi) + \delta_\phi)\,\theta_{mmt})/4) + (1 - (1 - \psi)(\delta_\mu + \delta_\phi - \delta_\mu\delta_\phi)\,\theta_{mft})/2) + (((4 - 2\psi(\delta_\mu + \delta_\phi - \delta_\mu\delta_\phi) - \xi(\delta_\mu + \delta_\phi))\alpha_t/4)$;

4- $\theta_{mft+1} = ((1 + F_t)(\phi(n - 1) + 2)/8n) + ((n - 1)(2 - \phi - (1 - \xi)(\delta_\mu(1 - \phi) + \delta_\phi)\theta_{mmt})/4n + ((1 - (1 - \psi)(\delta_\mu + \delta_\phi - \delta_\mu\delta_\phi)\theta_{mft})/2 + ((2n\,(1 - \psi)(\delta_\mu + \delta_\phi - \delta_\mu\delta_\phi) + (n - 1)(1 - \xi)(\delta_\mu(1 - \phi) + \delta_\phi)\alpha_t/4n$, where all the terms were previously explained. With these equations, a new transition T matrix was constructed to obtain $\mathbf{S}_{t+1} = TS_t + C$,

$$T = \begin{pmatrix} (4-\xi(\delta_\mu+\delta_\phi)-2\psi A)/4 & 0 & \xi(\delta_\mu+\delta_\phi)/4 & \psi A/2 \\ (1-\psi)A & 0 & 0 & 1-(1-\psi)A \\ ((2(1-\psi)A)+B)/4 & \phi/8 & (2-\phi-B)/4 & (1-(1-\psi)A)/2 \\ (2n(1-\psi)A+(n-1)B)/4n & (\phi(n-1)+2)/8n & ((n-1)(2-\phi-B))/4n & (1-(1-\psi)A)/2 \end{pmatrix}$$

The asymptotic values for θ , α and F to estimate the asymptotic F-statistics were, for this case: $\theta = ((\theta_{mm} + \theta_{mf})(1 - F)/2) + F$, $\alpha = ((\xi(\delta_\mu + \delta_\phi)\theta_{mm})/4 + (\psi(\delta_\mu + \delta_\phi - \delta_\mu\delta_\phi)\theta_{mf})/2$, $F = (1 - (1 - \psi)(\delta_\mu + \delta_\phi - \delta_\mu\delta_\phi)\,\theta_{mf}$, where

$$\theta_{mm} = (\phi/(4 + 2\phi + 2(1 - \xi)(\delta_\mu(1 - \phi) + \delta_\phi)) \text{ and } \theta_{mf} = ((n - 1)\,\theta_{mm}/n) + (1/4n)).$$

In fact, the reality could be more complex, because here I consider that generations did not overlap and adult females were replaced by their female progeny each generation within each lineage. But, these could be the first simulations to obtain vital data to understand the genetic and reproductive structure of pink river dolphins. To make these simulations a program was written by the author with the assistance of O. Suescun.

RESULTS

The three sampling points showing the highest average allele number were the lagoons 6, 7 and 8 (3.111 $\pm$ 0.875, 3$\pm$ 1.491, 3.111 $\pm$ 1.523, respectively), while lagoon 3 presented the lowest average allele number (1.222 $\pm$ 0.629). However, the average allele number is dependent on the sample size, while the expected heterozygosity is not. By using this last statistic, lagoons 2, 6 and 8 (H = 0.6852 $\pm$ 0.3963, 0.5599 $\pm$ 0.2178, 0.5526 $\pm$ 0.2606, respectively) presented the highest levels of gene diversity, while sampling localities 1 and 3 registered the lowest levels (H_e = 0.3333 $\pm$ 0.3143, 0.333 $\pm$ 0.4714, respectively). Therefore, no spatial pattern was found for the geographical disposition of the gene diversity in that Napo-Curaray river transect. Table 1 displays the different gene diversity statistics by lagoon. Globally, the average allele number (5 $\pm$ 2.539) and H_e (0.5271 $\pm$ 0.2537), values could be considered as medium for molecular markers such as microsatellites.

Table 1. Average number of alleles (ana), observed heterozygosity (H_{obs}) and expected heterozygosity (H_{exp}) for the nine microsatellites employed in each one of the sampling points studied and overall

Population	ana	H_{obs}	H_{exp}
beach1	1.67 ± 0.67	0.278 ± 0.342	0.333 ± 0.314
lagoon2	1.89 ± 0.87	0.667 ± 0.408	0.658 ± 0.396
lagoon3	1.22 ± 0.63	0.333 ± 0.471	0.333 ± 0.471
lagoon4	2.67 ± 1.05	0.459 ± 0.303	0.538 ± 0.239
lagoon5	1.89 ± 0.99	0.278 ± 0.342	0.389 ± 0.377
Lagoon6	3.11 ± 0.87	0.459 ± 0.172	0.559 ± 0.218
lagoon7	3.00 ± 1.49	0.294 ± 0.267	0.489 ± 0.316
lagoon8	3.11 ± 1.52	0.456 ± 0.281	0.552 ± 0.260
Total	5.00 ± 2.54	0.405 ± 0.166	0.527 ± 0.254

By means of an exact test using Markov chains, I analyzed possible HWE deviations within each geographic point sampled (Table 2). Only a very limited number of cases showed a significant homozygote excess.

Table 2. Hardy-Weinberg equilibrium exact tests for each point sampled for the nine microsatellites studied. * = significant homozygote excess at 0.05 probability level

Population	EV76	EV14	KWM2b	KWM2a	KWM12a	MK5	PPHO137	EV37	EV94
beach1	1 ± 0	-	-	1 ± 0	-	1 ± 0	-	0.332 ± 0.004	1±0
lagoon2	1 ± 0	1 ± 0	-	1 ± 0	-	1 ± 0	1 ± 0	1 ± 0	0.333±0.004
lagoon3	1 ± 0	-	-	-	-	1 ± 0	-	-	1±0
lagoon4	0.203 ±0.004	0.111 ±0.003	1 ± 0	1 ± 0	1 ± 0	0.127 ± 0.000	-	0.048 ± 0.000*	0.201±0.000
lagoon5	1 ± 0	1 ± 0	-	-	-	1 ± 0	-	0.334 ± 0.000	0.332±0.000
lagoon6	0.072 ±0.002	0.086 ±0.002	1 ± 0	1 ± 0	1 ± 0	0.315 ± 0.004	1 ± 0	0.367 ± 0.004	0.048±0.002*
lagoon7	0.111 ±0.003	1 ± 0	1 ± 0	0.111 ±0.003	-	0.029 ± 0.001*	-	0.034 ± 0.000*	0.465±0.005
lagoon8	0.441 ±0.000	0.091 ±0.000	1 ± 0	0.134 ±0.000	-	1 ± 0	1 ± 0	0.008 ± 0.001*	0.563±0.004

These cases were as follows: In lagoon 4, Ev37 showed a homozygote excess ($p = 0.0478 \pm 0.0002$); in the lagoon 7, MK5 and Ev37 yielded homozygote excess ($p = 0.0290 \pm 0.0015$ and $p = 0.0337 \pm 0.0018$, respectively) as well as did Ev37 in lagoon 8 ($p = 0.0089 \pm 0.0010$). Thus, out 52 HWE tests only four were significant at the level of $p < 0.05$, which represents 7.69 % and did not significantly deviate from the 5 % error type. In fact, if I apply the Bonferroni criteria, the significant probability level was 0.00096. In this case, no significant homozygote excess cases were found in the lagoons sampled. Therefore, no evidence of endogamy for *Inia geoffrensis* was found within any lagoon analyzed. When the F_{IS} statistic was considered in each sample point, some fraction was negative. In sample point 1, 20 % (1/5) of polymorphic microsatellites presented a negative F_{IS} (heterozygote excess). In lagoon 2, 16.67 % (1/6) of microsatellites analyzed yielded a negative F_{IS} meanwhile in lagoon 4, 42.86 % (3/7), and lagoons 6, 7 and 8, 28.57 % (2/7), also showed negative F_{IS} values. These results were incompatible with endogamy in the little pink river dolphin populations studied within the lagoons. If the nine microsatellites are studied for all the lagoons collectively, the picture obtained is different (Table 3).

Table 3. Overall Hardy-Weinberg equilibrium exact probability tests (P) and standard deviations (SD) for the nine DNA microsatellites studied. Six loci (*) showed homozygote excess

Marker	P	SD	
EV76	0.00604*	0.00008	Homozygote excess
EV14	0.00342*	0.00005	Homozygote excess
KWM2b	1	0	
KWM2a	0.00246*	0.0005	Homozygote excess
KWM12a	1	0	
MK5	0.03316*	0.0016	Homozygote excess
PPHO 137	1	0	
EV37	0.00000*	0	Homozygote excess
EV94	0.00447*	0.0006	Homozygote excess

Six microsatellites showed significant homozygote excess with positive F_{IS}, Ev76 ($p = 0.0060 \pm 0.00008$), Ev14 ($p = 0.0034 \pm 0.00005$), KWM2a ($p = 0.0025 \pm 0.0005$), MK5 ($p = 0.00332 \pm 0.0016$), Ev37 ($p = 0.0000 \pm 0.0000$) and Ev94 ($p = 0.0045 \pm 0.00006$). This means, and it is the first indication, that each lagoon had a pink river dolphin genotype composition different from the other lagoons. Therefore, the homozygote excess detected at six out of nine loci was caused by Wahlund effect (= population subdivision) and the gene flow among the lagoons is not extensive enough to totally homogenize the gene composition of all the lagoons analyzed.

Similarly, the assignation analysis showed the following situation. The individuals sampled in the first sample point were grouped in its own cluster. The same was true for the lagoons 2, 3, 5 and 8. All the individuals were correctly classified in their own lagoons. Nevertheless, in lagoon 4, two individuals were more intensely associated with lagoons 6 and 7, respectively. In lagoon 6, one individual was significantly more clustered with lagoon 7 as well as in lagoon 7 one individual was clearly associated with the individuals of lagoon 8.

Therefore of 33 individuals, four exemplars (12.12 %) could be migrants, or descendents of migrants, of other lagoons.

The two completed AMOVA showed interesting and complementary results to those generated with other analyses.

Table 4. Two different AMOVAs carried out with the pink river dolphin populations analyzed in the Napo-Curaray Rivers at the Peruvian Amazon. A. With three hierarchical geographic levels. B. With four geographical levels. * = significant probabilities at the 0.05 significance levels

A.
AMOVA

Sourceof variation	Percentage of variation	Significance	
Among populations (lagoons)	-0.4549	1.000	F_{ST} = - 0.0045
Among individuals, within populations	26.0672	0.0000*	F_{IS} = 0.2595
Within Individuals	74.3877	0.0000*	F_{IT} = 0.2561
B			
Among rivers (Napo & Curaray)	2.5115	0.1730	F_{CT} = 0.0251
Among populations within rivers	-1.1034	0.5894	F_{SC} = -0.0113
Among individuals within populations	25.5838	0.0000*	F_{IS} = 0.2595
Within Individuals	73.0081	0.0000*	F_{IT} = 0.2699

In the first analysis, three hierarchical geographical levels were employed (Table 4). The differences among the lagoons were basically inexistent (Percentage of variation = -0.00455 %, p = 1.0000). However, a significant variance fraction was estimated among the individuals within the lagoons (26.0672 %, p = 0.0000), although the highest variance fraction was within the individuals (74.3877 %, p = 0.0000). Thus, although in the previous analyses certain genetic differences were found among the pink river dolphins of diverse neighbor lagoons, this analysis showed that all the little populations sampled, in this 280 km transect belonged to a unique global gene pool and that the main genetic differences were among the individuals independently of their geographical origins. The previous analyses and the current one could be conciliated if we consider the following. All the animals sampled in the eight points belonged to the gene pool for the nine microsatellites analyzed. Therefore, no strong genetic differences could be found among these samples (AMOVA). However, the population size of each lagoon was quite small. No more than 10-12 animals and, in many cases, we sampled all the animals present in those lagoons (for instance, lagoons 6, 7 and 8). With these population sizes, genetic drift could be intense. Significant genetic differences could exist among dolphin groups of different lagoons, when they were taken together, although they belonged to the same gene pool (exact tests and F_{IS} statistics for the different microsatellites employed). If the action of gene drift is stochastic with regard to spatial patterns, then the major fraction of the genetic differences was among the individuals and not among other geographic levels considered. The second AMOVA considered a fourth hierarchical level, the existence of two different rivers, Napo and Curaray (two sampling points in the Napo River and six sampling points in the Curaray River). This analysis revealed that the differences of the two rivers and the differences of the lagoons within these two rivers were practically insignificant (Percentage of variation = 2.5115 %, p = 0.1730 and -1.1035 %, p = 0.5894). Again, the variance fraction among individuals within lagoons and, especially, the differences within individuals irrespective of their geographical origin were highly significant (25.5838 %, p = 0.00000 and 73.0081 %, p = 0.00000, respectively). This result ratified that rivers and lagoons were composed of a unique pink river dolphin gene pool for the markers studied and that there were genetic differences, especially among the individuals.

But the main population genetics results presented in this study were those obtained with the procedures of Chesser (1991a,b) to estimate F statistics for social reproductive systems and not demic models, as are usually applied. In the first step, some gene correlation or coancestry parameters were estimated: $\theta = 0.595203$, $\alpha = 0.455277$ and $F = 0.569179$ as well as the current F-statistics: $F_{LS} = 0.25687$, $F_{IL} = -0.06429$, $F_{IS} = 0.20910$. These values satisfied the relationship $(1 - F_{IS}) = (1 - F_{IL})(1 - F_{LS})$. As I explained in the Materials and Methods section, these F-statistics were assumed to reach an asymptotic value, because the pink river dolphins at the Napo-Curaray rivers are old enough to reach this condition, and then several simulations, with diverse constant population reproductive parameters, were constructed to obtain possible asymptotic F-statistic values which agreed with the F-statistics obtained from the data. Henceforth, those population parameters which generated simulated F-statistics very similar to those directly observed throughout the real data, could be used to decide if the pink river dolphin populations have a social reproductive system or if they behave as demic units. Some of the different Chesser's simulations were as follows (Figure 2): The first (A) simulation had values of s = 8, n = 5, m = 1, b = 5, g = 1250 (and migration rates, for males $d_m = 1$, and for females $d_f = 0$ with ϕ calculated throughout the values of n, m and b employed in the simulation) and yielded the asymptotic F-statistic values $F_{LS} = 0.17426$, $F_{IL} = -0.20221$, and $F_{IS} = 0.00729$, which did not agree with the calculated F-statistic values ($F_{LS} = 0.25687$, $F_{IL} = -0.06429$, $F_{IS} = 0.20910$). Thus, the population parameters used in this simulation did not seem be those present in the real pink river dolphin populations that produced the observed F-statistic values. (B) The simulation values of s = 8, n = 12, m = 4, b = 3, g = 1250 (and migration rates, for males of $d_m = 1$, and for females, $d_f = 0$) yielded the asymptotic F-statistic values $F_{LS} = 0.01772$, $F_{IL} = -0.01611$, and $F_{IS} = 0.00152$, which agreed even less with F-statistic values obtained than the previous simulation. (C) With the different values of s = 8, n = 5, m = 4, b = 3, and g = 1250 (and the migration rates, for males $d_m = 1$, and for females $d_f = 0$), the simulation provided the asymptotic F-statistic values $F_{LS} = 0.07308$, $F_{IL} = -0.07089$, and $F_{IS} = 0.00439$. These did not agree with the F-statistic values obtained, although the F_{IL} value is similar to that obtained directly from the molecular data. (D) With the values of s = 8, n = 20, m = 1, b = 20, and g = 1250 (and migration rates, for males $d_m = 1$, and for females $d_f = 0$) the simulation yielded the asymptotic F-statistic values of $F_{LS} = 0.16666$, $F_{IL} = -0.19305$, and $F_{IS} = 0.00572$, which did not agree with the F-statistic values obtained. (E) When the values were s = 8, n = 25, m = 3, b = 25, and g = 1250 (and the migration rates, for males $d_m = 0.7$, and for females $d_f = 0.3$) the simulation provided the following asymptotic F-statistic values: $F_{LS} = 0.15196$, $F_{IL} = -0.16999$, and $F_{IS} = 0.04290$, which also did not agree with the F-statistic values obtained. Nevertheless, the next simulations showed a considerably better fit between the simulated and the real F-statistic values. (F) The simulation values of: s = 8, n = 8, m = 2, b = 2, and g = 1250 (and the migration rates, for males $d_m = 0.1$, and for females $d_f = 0$) yielded the asymptotic F-statistic values $F_{LS} = 0.21821$, $F_{IL} = -0.01182$, and $F_{IS} = 0.20897$. The F_{IS} value was almost identical to the real value obtained. The F_{LS} value was relatively similar to the real one, but the excess

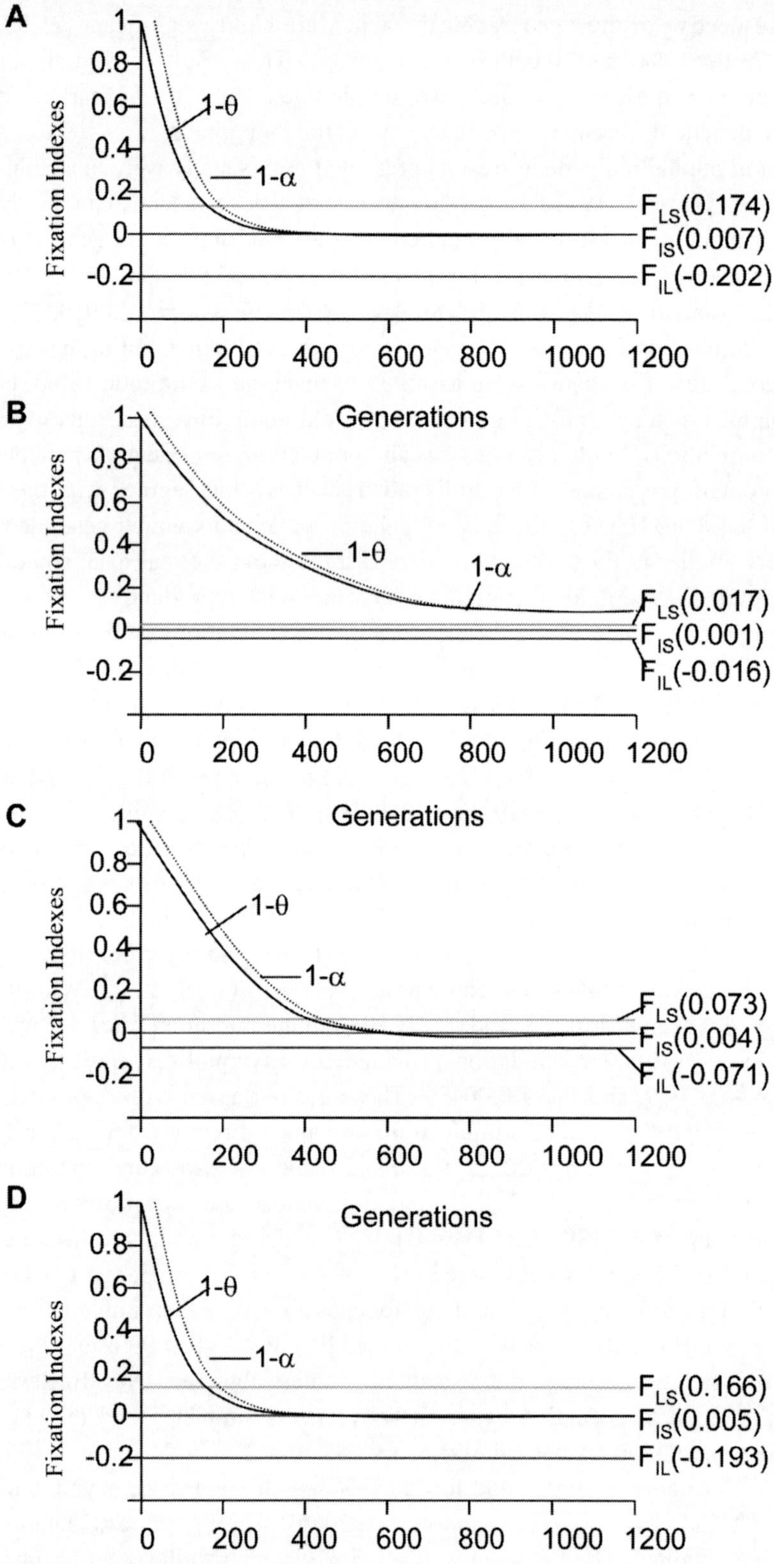

Figure 2. Continued on next page.

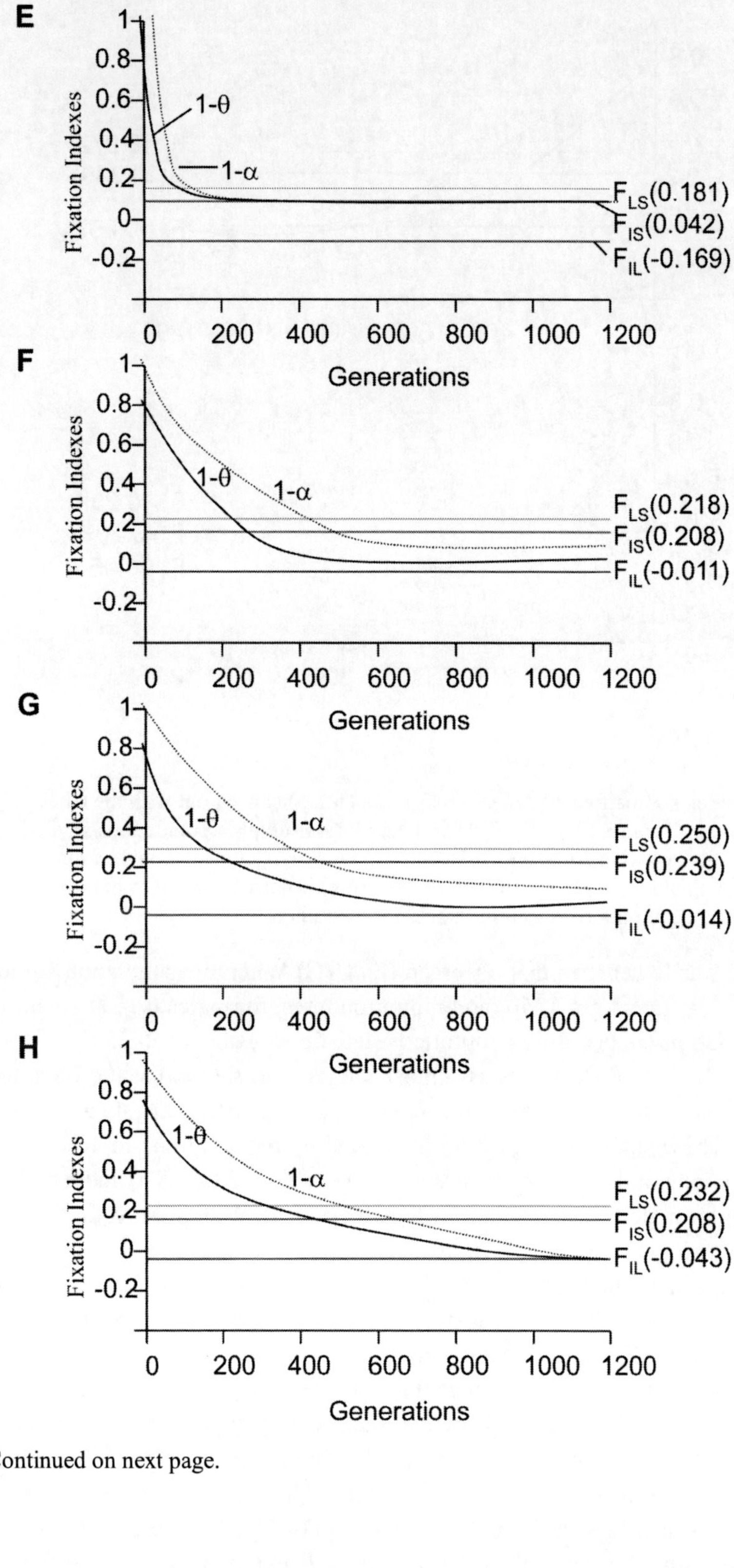

Figure 2. Continued on next page.

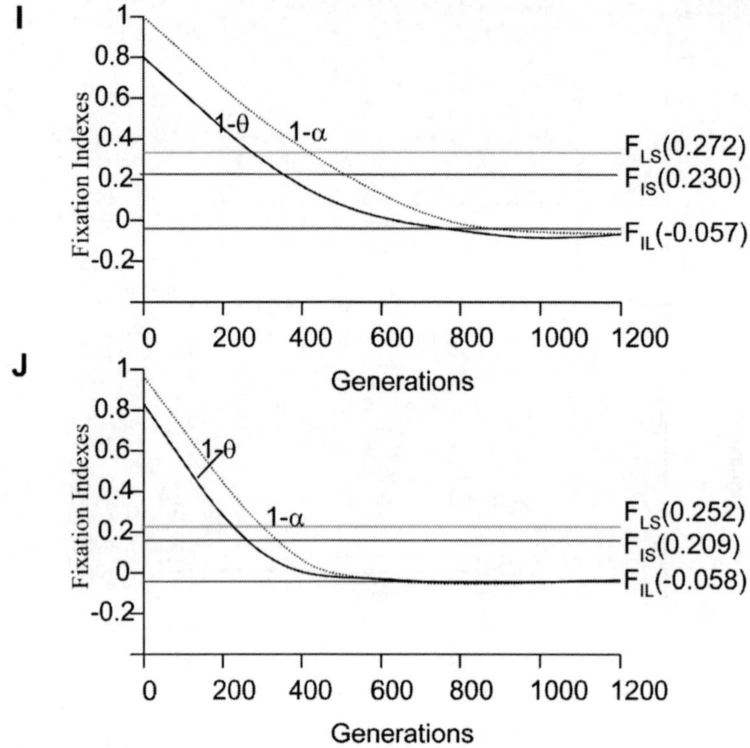

Figure 2. The first ten simulations (A, B, C, D, E, F, G, H, I, J) carried out with the Chesser's model to determine asymptotic F values (F_{LS}, F_{IL}, F_{IS}) for the pink river dolphin populations studied in the Napo-Curaray Rivers at the Peruvian Amazon. Different *s*, *n*, *m*, *b*, *g*, d_m and d_f variable values were employed in the simulations to obtain asymptotic F values virtually identical to the real F values observed among the eight pink river dolphin populations studied.

of heterozygote was lower than that observed (F_{IL}). (G) When the simulation values were s = 8, n = 7, m = 2, b = 2, and g = 1250 (and migration rates, for males d_m = 0.1, and for females d_f = 0) the simulation yielded the asymptotic F-statistic values F_{LS} = 0.25011, F_{IL} = -0.01478, and F_{IS} = 0.23902. The F_{LS} value was almost identical to the real value obtained. The F_{IS} value was relatively similar to the real one, but the excess of heterozygote was lower than that observed (F_{IL}). These small changes were motivated by reducing one reproductive female in each lineage. Simulation H used the values s = 8, n = 7, m = 2, b = 3.5, and g = 1250 (and the migration rates, for males d_m = 0.2, and for females d_f = 0). With these values, this simulation yielded the following asymptotic F-statistic values: F_{LS} = 0.23248, F_{IL} = -0.04375, F_{IS} = 0.19891. The agreement between the simulated and the real data begin to be very high. For simulation I, the values were: s = 8, n = 7, m = 1, b = 4, and g = 1250 (and the migration rates, for males d_m = 0.2, and for females d_f = 0). With these values, this simulation yielded the following asymptotic F-statistic values: F_{LS} = 0.27232, F_{IL} = -0.05771, F_{IS} = 0.23032. The agreement between the simulated and the real data was again very high. In simulation J, the values were: s = 8, n = 7, m = 1, b = 4, and g = 1250 (and the migration rates, for males d_m = 0.22, and for females d_f = 0). With these values, this simulation yielded the following asymptotic F-statistic values: F_{LS} = 0.25293, F_{IL} = -0.05780, F_{IS} = 0.20975. These simulated F-statistics were virtually identical to the real ones. It is noteworthy to mention that in this simulation one could change one, two, three until 100 males by lineage and the simulated F-statistics did not vary. In an identical sense, the simulated results did not change if there was one sex or other for migration ($d_m = 0.22$ or $d_f = 0.22$, but not both sexes simultaneously). Simulation K had values s = 8, n = 7, m = 1, b = 4.1, and g = 1250 (and the migration rates, for males $d_m = 0.22$, or for females $d_f = 0.22$). With these values, this simulation yielded the following asymptotic F-statistic values: $F_{LS} = 0.26083$, $F_{IL} = -0.06088$, $F_{IS} = 0.21584$. These results were also practically identical to the real ones.

Henceforth, the following primary foci were determined from these simulations:

1. The negative value of F_{IL} suggests that *Inia* conforms to a social reproductive system, with philopatric sex and with some migrant sex, although the migrant potential of this sex is very low.
2. The number of lineages, s, affects the values of $1 - \alpha$ and $1 - \theta$ but not the F-statistic values.
3. The optimal number of reproductive females, n, in each lagoon was seven. If the reproductive female number will be eight, nine or ten (higher than seven), the F_{IS} and F_{LS} will be lower than the real values, while the F_{IL} value will be higher than the real one. On the contrary, if the reproductive female number will be six or five (lower than seven), the F_{IS} and F_{LS} will be higher than the real values, while the F_{IL} value will be lower than the real one.
4. The number of reproductive males in each lineage, m, did not influence the calculated F-statistic values. For example, values from one to 100 will offer the same results.
5. The optimal number of females in a lineage which bred with the *ith* male, b, was around 4. If this parameter is higher (five, six, etc), the F_{IS} and F_{LS} will be higher than those estimated directly from the molecular data, while F_{IL} will be lower. If b is lower (three, two, etc), the contrary will be true.
6. The specific number of generations chosen (g = 1,250) is really irrelevant because in a very few number of generations, the asymptotic values of the simulated F-statistics was reached.
7. The optimal migration rates for males, d_m, or for females, d_f, should be around 0.22 for one sex or for the other, but not for both simultaneously. If d_m or d_f will be higher, F_{IS} and F_{LS} will be lower than the observed values. If the migration rates were multiples of d_m and d_f, then F_{IS} and F_{LS} would also lower than the observed F values.
8. The optimal probability that females selected the same male with which to breed, ϕ, was around 0.30.

DISCUSSION

General Aspects

The pink river dolphins sampled in the Napo and the Curaray rivers were captured in a river margin (sandy beach) and in seven different lagoons. As many studies have previously demonstrated, pink river dolphins are seen more often in lagoons, river margins and river confluences than in the main river channels (Alliaga-Rossel 2002; Martin et al., 2004). Therefore, our captures were carried out in the areas where the dolphins were more frequently concentrated, with the exception of river confluences where the currents were so intense that no nets could be used to capture animals. As I will briefly comment, these are the sampling points where the social reproductive structure, if it exists in this species, can be recorded.

Wright (1940) was the first to consider the dispersal of animals from their natal areas. But this author considered this process to be random with regard to both sex and distance. Nonetheless, many studies have demonstrated that sexual differences in dispersal trends exist (Chesser 1991a,b; Chesser et al., 1993; Prout, 1981) and that this affects the genetic structure of the species. For example, Chesser (1991a,b) and Chesser et al. (1993) demonstrated that migrant sex is the vector of gene flow and a factor of global genetic homogeneity, while philopatric sex introduces internal homogeneity within the lineages (at least, promotes gene correlation between individuals in the same lineage or close geographic lineages) and genetic heterogeneity among the most distant lineages. The genetic heterogeneity among the lineages in the population considered (F_{LS} = 0.26) was higher than those usually published for other mammal species (*Mus musculus* from different farms, F_{ST} = 0.047, Selander & Kaufman, 1975; *Macaca mulatta* among different social groups, F_{ST} = 0.035, Melnick et al., 1984; *Cynomys ludovicianus,* between wards within populations, F_{ST} = 0.045-0.065, Chesser, 1983, to cite a few cases). However, this genetic heterogeneity was of a similar magnitude to the differences found by Chesser (1983), for the quoted rodent, among coteries within wards (F_{ST} = 0.227). However, we must take into account, following the theoretical models and simulations made by Rothman et al. (1974) and Fix (1978) that the generation and splitting of new lineages have an inflationary effect on the F_{ST} estimations, even when the population was a single reproductive one and the gene flow is extensive. However, the F_{LS} obtained was similar to that obtained for other molecular marker set (RAPD), we applied to the different lagoon dolphin samples that we obtained in the Napo-Curaray, Ucayali and Marañón rivers (F_{ST} = 0.2302; Ruiz-García et al., 2007). Lewontin (1972) was the first researcher that determined that a large proportion of population genetic variance exists within small population units in humans. The same has been demonstrated for other wild primates (Melnick, 1987), rodents (Chesser, 1983; van Staadent et al., 1994) or in small or large cats (*Felis catus*, Ruiz-García, 1998, 1999; *Panthera onca* and *Puma concolor*, Ruiz-García et al., 2006a, 2009).

Philopatry and Gene Flow

The results obtained showed that all the animals studied belonged to a unique gene pool (AMOVA). Although the gene flow among different lagoons existed, this gene flow was highly restricted (assignation analysis and Chesser's social structure analysis) and only one sex was responsible for this gene flow. Martin & da Silva (2004a) annually determined the existence of around 260 pink river dolphins in 225 km^2 of the main Mamirauá lake system, in the Japurá (Caquetá) river in Brazil, where half were permanent residents. However, 90 % of the dolphin sightings that these authors marked within the lake system between November 2001 and November 2002 were of permanent residents, or their dependent offspring. They also found a possible cline in site fidelity between those that always lived in the lake system and those where dolphins occasionally visited the lake system. These observational results agree with our genetics results in the sense that the fidelity for lagoons seems to be important for a significant fraction of pink river dolphins. In fact, the negative correlation between the monthly proportion of marked pink river dolphins in the Mamirauá lake system and on the river within two kilometers could indicate that many resident animals traveled small distances when forced to go to the river during the low water season. On the other hand, Martin & Da Siva (2004a) detected individuals which moved from tens to hundreds (even up to 1000) of kilometers along the rivers, but without broad-scale seasonal migration. It is possible that some animals travel many kilometers along the rivers, but this individual movement does not mean “gene movement”, because these exemplars can return to their original lagoons to breed. In fact, although a substantial amount of movement in the Mamirauá area was demonstrated on any particular day when a large fraction of the marked resident animals were outside of the lake system, almost all of them returned after periods of days, weeks or even years. Moreover, these authors only found a low percentage of marked animals on rivers outside the influence area of Mamirauá (1.3 %), which agrees quite well with a strong phylopatry for this species for both sexes. For instance, I detected that 12 % of the individuals (4/33) had genetic profiles which more probably belonged to other lagoons from where they were captured. If I analyze in detail the sex of these animals the picture was as follows: The two animals in lagoon 4 with more genetic resemblance to individuals from lagoons 6 and 7 were adult females as well as the animal of lagoon 7, which grouped with the animals of lagoon 8. Contrarily, the individual from the lagoon 6, which was classified together with the individuals from lagoon 7, was an adult male. Furthermore, in all the cases, these possible migrants, or descendents of migrants, were coming from other lagoons in the upper basin of the river. It seems that these animals descended the river to integrate themselves into other lagoons. Nevertheless, we cannot know if these animals were migrants (or migrant descendents), which were reproductively integrated in the new lagoons, or were simply occasional visitors. Thus, I cannot emphatically decide if the very low gene flow found among the lagoons were caused by males or females (d_m or d_f = 0.22), but the Chesser´s simulations demonstrated that it cannot be a simultaneous contribution of both sexes. If the exemplars, with different genetic profiles in regards to the lagoons where they were sampled, were reproductively integrated into these lagoons, then, it seems more probably that females are better contributors for gene flow. Data from Martin & da Silva (2004a) agree quite well with this possibility. They found that the proportion of residents that were males (52 %) was higher than that of non residents (43 %), although this difference was not statistically significant. Additionally, the data of these authors showed that it is probable that males presented a similar degree of site fidelity to the lagoons as females, although they simply spent less time in the lagoons and more on the rivers near the lagoons. However, I detected that females and their calves used lagoons preferentially, such as Martin & da Silva (2004b) determined in Mamirauá, and where males were numerically dominant in the main rivers. Therefore, the probability to catch females was higher than for males. This could also explain why three of the four animals, with genetic profiles that did not belong to the lagoons where they were sampled, were females. If the males spend more time in the main river channel, they have a greater possibility of traveling to distant lagoons compared to females. Moreover, adult males are more robust and strong than females and this could enhance the possibility for swimming greater distances along rivers. If this is true, then, males could be the decisive sex that carries out gene flow among the different lagoons. But, philopatry is extremely important for both sexes, although males probably contribute more than females to gene flows because they are polygynous (Chesser 1991a, Chesser et al., 1993).

Main Reproductive Repercussions of the Chesser's Simulation Data Obtained in the Biology of *Inia*

The simulation social reproductive population structure results showed relevant data (listed as four points here) to help to determine if *Inia geoffiensis* is a social or a solitary species: 1- seven reproductive females per lineage are active in each breeding period, 2- the number of reproductive males per linage are not important, 3- a reproductive male breeds with four females on average within each lineage, and thus there is polygyny in this dolphin species, and 4- the probability that females of the same lineage choose the same male for breeding is 0.30. Additionally, the negative values of F_{IL} are clearly symptomatic of a philopatric sex and other migrant sex. As the negative magnitude of F_{IL} was small, this means something that I earlier discussed: the migrant sex is also highly philopatric (site tenacity) and gene flow is low and occasional. The negative F_{IL} values indicating heterozygote excess within lineages agree quite well with expectations for sex-dependent migration as earlier discussed for theoretical models (Prout, 1981) and for social structured populations (Cockerham, 1969, 1973; Chesser, 1991a,b; Chesser et al., 1993) in which social genetic reproductive units have been accurately identified. Thus, a demic structure seems to be inexistent in the pink river dolphins (panmictically breeding units relatively isolated from other demes) and the data support social lineages for areas with a single population (or mixed populations in an area that conserve their breeding integrity) wherein related individuals may remain philopatric and/or within which mate choice is complex and not at random. Philopatry is usually employed by only one sex, and typically, in mammals, it is the female, which could be the case for pink river dolphins. Although no isolation by distance was detected in the current study with STRPs, the behavioral factors could result in a population that was genetically structured in a non-random manner. Ruiz-García et al. (2007) showed a significant correlation coefficient (r = 0.6743, P = 0.014) among the genetic distances of lagoon pairs and among geographical distances of lagoon pairs for diverse lagoons sampled in the Mamoré River at Bolivia with nine RAPD markers. This means that 45.47 % of the genetic differences among the dolphins of different lagoons were caused by the geographic distance among the lagoons. Therefore, the social reproductive structure in *Inia* could create a non-random distribution of the genetic variants. For my simulations, two data were irrelevant. First, as the samples proceeded from eight sampling points, I considered eight lineages. It's entirely possible that in the 280 km transect sampled more than eight lineages could exist and that we did not observe, detect or capture them. However, the number of lineages did not affect the F-statistic estimations, although it affects the extant gene diversity within lineages and the overall population. Secondly, I used 1,250 generations to reach asymptotic F-statistics. I chose this value because it represented 10,000 years (considering eight years per generation in this species) and that is the time after the last glacial period. Nevertheless, the asymptotic F-statistic values were reached in less than 8-10 generations. This means that no importance was given to the number of generations that I originally took. These last two results were also obtained by Chesser (1991a,b) in his theoretical simulations. I consider that these important reproductive parameters obtained through simulations were probably not affected as much by several seasonal changes because the animals sampled were in the reproductive epoch. Martin & da Silva (2004a) demonstrated that there is no clear influx or efflux of pink river dolphins in Mamirauá at any time of the year. In fact, although no animal spent the entire year within this lake system (Martin & da Silva, 2004a,b), an important fraction of them probably traveled no farther than the area near to the entrance of the lake system. Nonetheless, Alliaga-Rossel (2002) showed that group size frequency significantly varied by season (31 %; p = 0.001) in the Tijamuchí river in Bolivia. Solitary individuals were more frequent in all the seasons with exception of low water, where pairs were seen more frequently. But in falling (19 individuals) and low waters (14 individuals) (from April to September), this author observed the highest groups of pink river dolphins, coinciding with the greatest fish abundance. He even commented that he observed more than 30 individuals in the same lagoon. I saw similar sized groups in Peruvian and Bolivian rivers. In fact, the same has been observed by other authors such as Vidal et al. (1993), Best & da Silva (1993) or Schmidt-Lynch (1994), who observed aggregation of 20-35 individuals. In the falling and low waters is the reproductive time for *Inia* and it is when the group size increases and probably reflects more precisely, which is the social and reproductive size of the *Inia*'s lineages. This author observed a reproductive activity season day with 12 individuals involved in the event. One day after, one animal appeared dead probably due to sexual aggression (a male of 2.25 m). In my experience, I have witnessed many reproductive activities (and many seasons) for this species and participation from 5 to 15 animals was quite common. The most frequent number of animals ranged from 7 to 10. The social Chesser's model simulations yielded seven females, as the most probably number within each lineage, and one male breeding with four females. Therefore, 2 or 3 males could be the fathers of all the new offspring. If I sum these quantities, 9-10 animals are the effective lineage sizes for the minimal social reproductive unit for *Inia* which coincides with those numbers seen in other reproductive seasons by Aliaga-Rossel (2002) and me. The polygyny for this species, obtained by the Chesser's simulations, agrees quite well with certain sexual dimorphism determined in *Inia*. Although some females could reach a total length similar to the largest males (Ruiz-García et al., 2006b; Ruiz-García et al., 2010b), the oldest and largest males are more robust than the oldest and longest females (for instance, girth in front of pectoral flippers, width of the pectoral flippers or weight) (da Silva, 1994; Martin & da Silva, 2006; Ruiz-García et al., 2010b). Therefore, Chesser's social simulations offer the possibility of polygyny in *Inia,* which supports sexual dimorphism in favor of males because they actively fight for the sexual access to females (Clutton-Brock, 1989). If so, this is one of the clearer cases of polygyny in cetaceans, together with the possible case of *Physeter catodon.* A possible close kinship within maternal lineages, could generate kin clusters identified by a greater geographic proximity between the members of these clusters than between members of clusters far away. Although greater geographic physical overlap of home ranges (lagoons within Várzea, for instance) could occur among kin females than with non-kin females. Kin female lineages could not develop cooperative social behavior with other close kin female lineages, with each female lineage retaining the best possible area in floodplain lagoons (although the best core habitat for this species is dependent on fish movements and the rainy season). Such behaviors could enhance a population consisting of small philopatric groups, creating genetic patches within the overall population. Henceforth, the comparisons of my molecular results and the observational data can indicate that most solitary animals could be part of social reproductive groups and that they regroup for fishing or breeding and later they again disperse (like a continuous fission-fusion process within a social lineage). Additionally, the social system in *Inia* is reinforced with other data. Kendall (1994) reported the existence of "nurseries", where calves were cared by one or several adults in the Colombian Amazon, although this event has not been observed in other areas of *Inia's* geographic distribution. Another finding which could help to demonstrate the existence of a social reproductive system in *Inia* was that determined by Martin & da Silva (2004b). They found that, at the Mamirauá Reserve, the proportion of males was highest in the main rivers and diminished towards the innermost parts of lagoons within the flooded forest, reaching the highest male proportion at mid-rising water of the main rivers. Males always presented higher numbers than females in main rivers and their margins, meanwhile females greatly outnumbered males in low lying, remote open areas within Várzea comprising temporary and permanent shallow lagoons (they are called chavascal). The exception was during September-November in the lower water season where the sex-ratio was identical in all habitats. They explained this possible sexual segregation by three potential explanations. First, calves begin to take fish within their first year of life. The abundance and diversity of small fishes is greater in shallow lagoons within várzea. Therefore, calves and young could easily obtain more food. Secondly, lactating females can receive shelter and seek refuge from strong riverine currents, and thus maintain low energetic cost. A refuge would provide an easier situation for resting, and nursing, as well as provide an area for calves and young to learn how to catch fish. This explanation has been proposed for *Platanista gangetica* (Smith, 1993) and it is possible for *Inia*. Thirdly, sexual segregation may protect calves from the attacks of adult males, which can be extremely aggressive with other males. These activities frequently wound young and potentially lactating females and their calves. Nonetheless, we observed a different situation in some Peruvian lagoons. For instance, we appreciated in the tiphisca Urarinas at the Puhinauva channel (Ucayali River), how an adult male protected a female and her young (another male) when we captured these animals during high water. In identical sense, in Cocha Paucar, another lagoon in the Puhinauva channel (Ucayali River), we assisted in the spectacle of several adult and young individuals emitting agonistic display against us and following our wood boat when we captured one exemplar and transported it for 30-40 meters to release in another river area. In whatever case, these behaviors agree quite well with Chesser's simulations showing a social reproductive system within and between small reproductive size units in this species but does not support the claims of those authors which consider this as a solitary species. All of these attributes found in the pink river dolphins could enhance the possibility of both the interdemic selection model (Wright, 1982) and trait-group model (Wilson, 1977) acting on this and other cetacean species. Genetic differences over small geographic distances could be the rule rather the exception (Smith et al., 1978).

Could the Social Reproductive Parameters found in this Study be Typical for all the *Inia* Populations?

It is recognized that some lake systems contain numerically important populations of pink river dolphins (for example, Tipischa San Pablo in Marañón River in Perú or Mamirauá in the Japurá (Caquetá) River in Brazil). Martin & da Silva (2004a) estimated, at least, 100 individuals used the main Mamirauá lake as the center of their residence. This depends on the density, the temporal and spatial distribution, and fish species diversity in these lakes. But it's likely that several different geographic genetic lineages of pink river dolphins are living together within these large populations. However, no such large populations were observed in the Napo-Curaray River transect analyzed. Furthermore, it's unclear if the genetic structure found in this study could be applied to these large pink river dolphin populations. It is logical to think, as was determined by Martin et al. (2004) in rivers of central Brazilian Amazon, that the highest densities of the dolphins occur close to the margin, the location where characins, a significant component of *Inia*'s diet, migrate (da Silva, 1983, migrate). Here, sediment-rich white waters meet acidic black waters, creating a high productive area and an abundant number and variety of fish. This water-meeting area has a reduced current, the lowest in the river's center and is a much preferred area by the dolphins. A relatively high density of pink river dolphins was only found in the river center due to annual minimum water levels, probably, because fish were forced to migrate by them (Martin et al., 2004). It's likely that different genetic lineages of pink river dolphins converge on these high productive areas where these two river channels meet (Magnusson et al., 1980; Leatherwood, 1996; McGuire & Winemiller, 1998). But their genetic integrities are maintained throughout this space no matter if the animals are dispersed as solitaries or as different lineages converging on points of high fish density when, for example, fish movements are produced by the Amazon's seasonal changes in water level and dissolved oxygen. (Martin & da Silva, 2004b). Martin & da Silva (2004b) showed that an exodus of pink river dolphins is produced from the floodplain to the main rivers at Mamirauá during the low water season which results in the highest reported density for any cetacean worldwide (18 dolphins/km^2 or 4.2 dolphins/linear km in the várzea). However, the climatic, physical and ecological conditions of the Napo-Curaray rivers, where the samples of dolphins for this study were obtained were different to those found by these authors in Mamirauá. The animals were caught during the low-falling water period of these rivers in October-November and almost 69 % of the animals present in the diverse lagoons and in the beach were caught with our nets (33 out 48 animals). Therefore, in the eight sampling points, a total of 15 animals escaped from our nets. In the main rivers (Napo, Curaray and their affluents), 169 animals were counted (with the help of M. F. Gómez and P. Escobar-Armel). Thus, a total number of 217 pink river dolphins were documented in seven lagoons, one beach 280 km away and in the main rivers between these lagoons. This yielded a minimal density of 0.775 dolphins/linear km. This density is within the range of other densities found for *Inia* (for instance, 0.22 dolphins/linear km in 490 km from Manaus to Jutica in the Amazon river, Magnusson et al., 1980; 0.56 dolphins/linear km in the Apuré River and 1.15 dolphins/linear km in the Apurito River within Venezuela, Schnapp & Howroyd, 1992; 1.12 dolphins/linear km in the Tijamuchí river, an affluent of the Mamoré river in Bolivia, Aliaga-Rossel, 2002; a mean of 0.66 dolphins/linear km along margins of Japurá (Caquetá) and Amazon River near to Tefé, Martin et al., 2004; 0.47-0.68 dolphins/linear km and 0.73-1.47 dolphins/linear km were determined for the Samiria and the Tapiche rivers, affluents of the Ucayali River in the Peruvian Amazon, Henningsen, 1998; 0.38 dolphins/linear km in low water and 0.44 dolphins/linear km in high water in the lower part of the Lagartococha river in the Ecuadorian Amazon, Utreras, 1996; or even lower densities between 0.06 to 0.21 dolphins/linear km during three years in the Cuyabeno River and 0.21-0.27 dolphins/linear km during three years in the Lagartococha River, in Ecuador, Denkinger et al., 2000) and was substantially lower than what was determined for low water within the Mamirauá Reserve by Martin & da Silva (2004b). Although the captures were made in the falling-low water season, a significant fraction of the animals were living in the lagoons (22 %) and not in the main river, probably because these lagoons do not completely dry-out during the low water season. On the other hand, the sex-ratio (45.8 % males and 54.2 % females) of these captured animals, did not significantly differ from a 1:1 sex-ratio. Many females were with calves, and there was no segregation of sexes as was observed by Martin & da Silva (2004b). Although our collections occurred during low-falling water season, the same time of year when Martin & da Silva observed the low level of segregation of sexes in Mamirauá. Nonetheless, we did not observe sexual segregation in diverse lagoons of the Ucayali and Marañón rivers in the Peruvian Amazon during high rising-high water for this species, which could be in disagreement with that postulated by Martin & da Silva (2004b). Although, I have no evidence that the sexual segregation was present in the Peruvian rivers analyzed, this does not necessarily affect the social reproductive system obtained by the Chesser's simulations because the reproductive social system is really important in low water areas where males and females search for each other to breed if there is fidelity for the reproduction place. However, such as I commented before, in areas of extreme density or with enormous population size, such as Mamirauá (Martin & da Silva, 2004a, estimated about 13,000 pink river dolphins in the 11,240 km^2 of the Mamirauá reserve), specific genetic analyses must be developed to determine if the genetic social reproductive parameters are similar to those estimated in the major part of *Inia's* geographic distribution, where population sizes are not as large (such as I presented here), or are clearly different. Additionally, all the areas where observational and genetics results were obtained correspond to white-water rivers (that is, rivers that drain the Andes mountains; Meade & Koehnken, 1991). It would be of interest to carry out micro-genetic structure analyses of pink river dolphins from clear-water rivers (draining, for instance, the Colombian and Venezuelan Llanos, like the Cinaruco River, or rivers from the upland Guyana Shield, such as the Ventuari River) and acid black-waters (as the Negro River in the Brazilian Amazon or the Atabapo River in Venezuela; Meade & Koehnken, 1991) to know if these diverse ecological systems have similar social reproductive parameters that I determined for the pink river dolphin population from a main white-water affluent (Napo/Curaray River) of the upper Amazon River. It seems that pink river dolphin populations in black-water rivers are scarce compared to those in white-water rivers. For instance, Pilleri & Pilleri (1982) affirmed that the density of *Inia* was 50 times less in black-water systems such as in the upper Orinoco and the Casiquiari channel compared to a white-water river like the Apure River. In fact, some black water tributaries, for example, of the Napo River in Ecuador do not contain *Inia* populations (Tagliavini & Pilleri, 1984). It would be of additional interest, and decisive for conservation programs, to determine how anthropogenic activities including fishing with diverse types of nets (da Silva & Best, 1996), use of motorized boats and colonization of river banks alter the micro-genetic structure of these amazing dolphins.

ACKNOWLEDGMENTS

Economic resources to carry out this study were obtained from Colciencias (Grant 1203-09-11239; Geographical population structure and genetic diversity of two river dolphin species, *Inia boliviensis* and *Inia geoffrensis*, using molecular markers) and the Fondo para la Accion Ambiental (US-Aid) (120108-E0102141; Structure and Genetic Conservation of river dolphins, *Inia* and *Sotalia*, in the Amazon and Orinoco basins). Main thanks go to P. Escobar-Armel (Colombia), C. Vergara (Colombia), M. F. Gómez (Colombia), N. Romero (Colombia), Juanito and Angelito (Iquitos, Perú), and, especially to Isaias and his sons (Requena, Perú) to be part in the capture of the pink river dolphins herein studied. Also diverse Peruvian Indian communities collaborated with our pink river dolphins captures throughout the Peruvian rivers (Bora, Ocaina, Shipigo-Comibo, Capanahua, Angoteros, Orejón, Cocama, Kishuarana and Alamas). Additional thanks go to Hugo Gálvez (Iquitos, Perú) and Armando Castellanos (Quito, Ecuador) to collaborate in collection permits in both countries. Also, many thanks go to PRODUCE, Dirección Nacional de Extracción and Procesamiento Pesquero from Perú for their role in facilitating the obtainment of the collection permits. Thanks also to the Fundación Sociedad Portuaria de Santa Marta (Colombia) for its logistical support.

REFERENCES

[1] Alliaga-Rossel, E. (2002). Distribution and abundance of the river dolphin (*Inia geoffrensis*) in the Tijamuchí river, Beni, Bolivia. *Aquatic Mammals,* 28, 312-323.

[2] Amos, B., Schlotterer, C and Tautz, D. (1993). Social structure of pilot whales revealed by analytical DNA profiling. *Science,* 260, 670-672.

[3] Andersen, L.W., Holm, L. E., Siegismund, H. R., Clausen, B., Kinze, C. C., and Loeschcke, V. (1997). A combined DNA-microsatellite and isozyme analysis of the population structure of the harbor porpoise in Danish waters and west Greenland. *Heredity,* 78, 270-276.

[4] Armitage, K. B. (1988). Resources and social organization of ground-dwelling squirrels. In C. N. Slobodchikoff (Ed.). *The ecology of social behavior.* (pp 131-155). San Diego, California: Academic Press.

[5] Best, R. C., and da Silva V. M. F. (1993). *Inia geoffrensis*. *Mammalian species,* 426, 1-8.

[6] Bruford, M. W., and Wayne R, K. (1993). Microsatellites and their application to population genetics studies. *Current Opinion in Genetics and Development,* 3, 939-943.

[7] Cheney, D. L. and Seyfarth, R. M. (1983). Nonrandom dispersal in free-ranging vervet monkeys: social and genetic consequences. *The American Naturalist,* 122, 392-412.

[8] Chesser, R. K. (1983). Genetic variability within and among populations of the black-tailed prairie dog. *Evolution,* 37, 320-331.

[9] Chesser, R. K. (1991a). Gene diversity and female philopatry. *Genetics,* 127, 437-447.

[10] Chesser, R. K. (1991b). Influence of gene flow and breeding tactics on gene diversity within populations. *Genetics,* 129, 573-583.

[11] Chesser, R. K., Sugg, D. W., Rhodes Jr, O. E., Novak, J. N., & Smith, M. H. (1993). Evolution of mammalian social structure. *Acta Theriologica,* 38, 163-174.

[12] Clutton-Brock, T. H. (1989). Mammalian mating systems. *Proceedings of the Royal Society of London B,* 236, 339-372.

[13] Cockerham, C. C. (1967). Group inbreeding and coancestry. *Genetics,* 56, 89-104.

[14] Cockerham, C. C. (1969). Variance of gene frequencies. *Evolution,* 23, 72-84.

[15] Cockerham, C. C. (1973). Analysis of gene frequencies. *Genetics,* 74, 679-700.

[16] Cote, S.D., Dallas, J. F., Marshall, F., Irvine, R. J., Langvatn, R., and Albon, S. D. (2002). Microsatellite DNA evidence for genetic drift and philopatry in Svalbard reindeer. *Molecular Ecology,* 11, 1923-1930.

[17] Dallas, J.F., Dod, B., Boursot, P., Praguer, E. M., and Bonhomme, F. (1995) Population subdivision and gene flow in Danish house mice. *Molecular Ecology,* 4, 311- 320.

[18] Daly, J. C. (1981). Effects of social organization and environmental diversity on determining the genetic structure of a population of the wild rabbit *Oryctolagus cuniculus*. *Evolution,* 35, 689-706.

[19] Da Silva, V. M. F. (1983). *Ecologia alimentar dos golfinhos da Amazónia. Master of Science Thesis.* Manaus, Brazil: University of Amazonas.

[20] Da Silva, V. M. F. (1994). Aspects of the Biology of the Amazonian Dolphins Genus *Inia* and *Sotalia fluviatilis Doctoral Thesis.* Cambridge, United Kingdom: The University of Cambridge.

[21] Da Silva, V. M. F., & R. C. Best. (1996). Freshwater dolphin/fisheries in the Central Amazon (Brazil). *Amazoniana,* 14, 165-175.

[22] Denkinger, J., Campos, F., & von Fersen L. (2000). Status of the bufeo (*Inia geoffrensis*) and the tucuxi (*Sotalia fluviatilis*) in the Cuyabeno Reserve, Ecuador. (unpublished manuscript).

[23] Excoffier, L., Laval, G., & Schneider, S. (2006). Arlequin ver 3.1. An integrated software package for population genetics data analysis. http://cmpg.unibe.ch/software/arlequin3.

[24] Fix, A. G. (1978). The role of kin-structured migration in genetic microdifferentiation. *Annals of Human Genetics,* 41, 329-339.

[25] García-Moreno, J., Roy, M.S, Geffen, E & Wayne, R.K. (1996). Relationships of genetic purity of the endangered Mexican wolf based on analysis of microsatellite loci. *Conservation Biology,* 10, 376-384.

[26] Gottelli, D., Sillero-Zubiri, C, Applebaum, G. D, Roy, M. S, Girmann, D. J, García-Moreno, J., Ostrander, E. A & Wayne, R. K. (1994). Molecular genetics of the most endangered canid: the Ethiopian wolf, *Canis simensis*. *Molecular Ecology,* 3, 301-312.

[27] Henningsen, T. (1998). *Zur Verbreitung, Habitatwahl und Verhaltensoekologie der delphine Inia geoffrensis und Sotalia fluviatilis im oberlauf des Amazonas. PhD Dissertation.* Bremen, Germany: Faculty of Sciences, University of Bremen.

[28] Hoelzel, A. R., Dahlheim, M & Stern, S. J. (1998). Low genetic variation among killer whales (*Orcinus orca*) in the Eastern North Pacific and genetic differentiation between foraging specialists. *The Journal of Heredity,* 89, 121-128.

[29] Hoelzel, A. R., Natoli, A., Dahlheim, M. E., Olavarria, C., Baird, R. W & Black, N. (2002). Low worldwide genetic diversity in the killer whale (*Orcinus orca*): implications for demographic history. *Proceedings of the Royal Society of London,* 269, 1467-1473.

[30] Kanda, N., Goto, M., & Pastene, L. A. (2006). Genetic characteristics of western North Pacific sei whales, *Balaenoptera borealis*, as revealed by microsatellites. *Marine Biotechnology* (NY), 8, 86-93.

[31] Kendall, S. (1994). *Los delfines de la Amazonía y la Orinoquía.* Bogotá, Colombia: Fundación Omacha.

[32] Krutzen, M., Valsecchi, E., Connor, R. C., & Sherwin, W. B. (2001). Characterization of microsatellite loci in *Tursiops aduncus*. *Molecular Ecology,* 1, 170-172.

[33] Leatherwood, S. (1996). *Distributional ecology and conservation status of river dolphins (Inia geoffrensis and Sotalia fluviatilis) in portions of the Peruvian Amazon.* PhD dissertation. College Station, Texas: Texas A&M University.

[34] Lewontin, R. C. (1972). The apportionment of human diversity. *Evolutionary Biology,* 6, 381-398.

[35] Magnusson, W. E., Best, R. C & da Silva, V. M. F. (1980). Numbers and behavior of Amazonian dolphins, *Inia geoffrensis* and *Sotalia fluviatilis* in the Rio Solimoes, Brazil. *Aquatic Mammals,* 8, 27-41.

[36] Martin, A.R., & da Silva, V. M. F. (2004a). Number, seasonal movements, and residency characteristics of river dolphins in an Amazonian floodplain lake system. *Canadian Journal of Zoology,* 82, 1307-1315.

[37] Martin, A.R., & da Silva, V. M. F. (2004b). River dolphin and flooded forest: seasonal habitat use and sexual segregation of botos (*Inia geoffrensis*) in an extreme cetacean environment. *Journal of Zoology,* 263, 295-305.

[38] Martin, A.R., & da Silva, V. M. F. (2006). Sexual dimorphism and body scarring in the boto (Amazon river dolphin), *Inia geoffrensis*. *Marine Mammal Science*, 22, 25-33.

[39] Martin, A.R., da Silva, V. M. F., & Salmon, D. L. (2004). Riverine habitat preferences of botos (*Inia geoffrensis*) and tucuxis (*Sotalia fluviatilis*) in the central Amazon. *Marine Mammal Science,* 20, 189-200.

[40] McGuire, T. L., and Winemiller, K. O. 1998. Occurrence patterns, habitat association, and potential prey of the river dolphin, *Inia geoffrensis*, in the Cinaruco River, Venezuela. *Biotropica* 30: 625-638.

[41] Meade R. H., & Koehnken, L. (1991). Distribution of the river dolphin, tonina *Inia geoffrensis*, in the Orinoco river basin of Venezuela and Colombia. *Interciencia,* 16, 300-312.

[42] Melnick, D. J. (1987). The genetic consequences of primate social organization: a review of macaques, baboons and vervet monkeys. Genetica, 73, 117-135.

[43] Melnick, D. J., Jolly, C. J., & Kidd, K. K. (1984). The genetics of a wild population of rhesus monkeys (*Macaca mulatta*). I. Genetic variability within and between social groups. *American Journal of Physical Anthropology,* 63, 341-360.

[44] Morin, P. A., Wallis J., Moore, J., Chakraborty, R & Woodruff, D. (1993). Non-invasive sampling and DNA amplification for paternity exclusion, community structure, and phylogeography in wild chimpanzees. *Primates,* 34, 347-356.

[45] Morin, P. A, Moore, J., Chakraborty, R., Jin, D. L., Goodall, J & Woodruff, D. S. (1994). Kin selection, social structure, gene flow, and the evolution of chimpanzees. *Science,* 265, 1193-1201.

[46] Noor, M. A. F. (1995). Incipient sexual isolation in *Drosophila pseudoobscura bogotana* Ayala & Dobzhansky (Diptera: Drosophilidae). *Pan-Pacific Entomologist,* 7,: 125-129.

[47] Paetkau, D., Calvert, W., Stirling, I., & Strobeck, C. (1995). Microsatellite analysis of population structure in Canadian polar bears. *Molecular Ecology,* 4, 347-354.

[48] Paetkau, D., Waits, L. P., Clarkson, P. L., Craighead, L., & Strobeck, C. (1997). An empirical evaluation of genetic distance statistics using microsatellite data from bear (Ursidae) populations. *Genetics,* 147, 1943-1957.

[49] Pilleri, G., & Gihr, M. (1977). Observations on the Bolivian (*Inia boliviensis* d´Orbigny, 1834) and the Amazonian bufeo (*Inia geoffrensis* de Blainville, 1817) with

a description of a new subspecies (*Inia geoffrensis humboldtiana*). *Investigations on Cetacea,* 8, 11-76.

[50] Pilleri, G., & Pilleri, O. (1982). *Zoologische Expedition zum Orinoco un Brazo Casiquiari 1981*. Bern, Switzerland: Hirnanatomisch Institute, University of Bern.

[51] Pope, T. (1992). The influence of dispersal patterns and mating system on genetic differentiation within and between populations of the red howler monkey (*Alouatta seniculus*). *Evolution*, 46, 1112-1128.

[52] Pope, L.C., Sharp, A., & Moritz, C. (1996). Population structure of the yellow-footed rockwallaby *Petrogale xanthopus* (Gray, 1854) inferred from mtDNA sequences and microsatellite loci. *Molecular Ecology,* 5, 629-640.

[53] Prout, T. (1981). A note on the island model with sex dependent migration. *Theoretical and Applied Genetics,* 59, 327-332.

[54] Rosel, P. E., France, S. C., Wang, J. Y., & Kocher, T. D. (1999). Genetic structure of harbour porpoise *Phocaena phocaena* populations in the northwest Atlantic based on mitochondrial and nuclear markers. *Molecular Ecology,* 8 (1), 41-54.

[55] Rothman, E. D., Sing, E. F., & Templeton, A. R. (1974). A model for analysis of population structure. *Genetics,* 78, 943-960.

[56] Ruiz-García, M. (1998). Genetic structure and evolution of different cat populations (*Felis catus*) in Spain, Italy, Argentina at microgeographical level. *Acta Theriologica,* 43, 39-66.

[57] Ruiz-García, M. (1999). Genetic structure of different cat populations in Europe and South-America at microgeographic level: Importance of the selection of a correct sampling level for accurate population genetic inferences. *Genetics and Molecular Biology,* 22, 493-505.

[58] Ruiz-García, M. (2005). The use of several microsatellite loci applied to 13 Neotropical Primates revealed a strong recent bottleneck event in the woolly monkey (*Lagothrix lagotricha*) in Colombia. *Primate Report,* 71, 27-55.

[59] Ruiz-García, M., Banguera, E. & Cárdenas, H. (2006). Morphological analysis of three *Inia* (Cetacea; Iniidae) populations from Colombian and Bolivia. *Acta Theriologica,* 51, 411-426.

[60] Ruiz-García, M., Murillo, A., Corrales, C., Romero-Aleán, N., & Alvarez-Prada, D. (2007). Genética de Poblaciones Amazónicas: La historia evolutiva del jaguar, ocelote, delfín rosado, mono lanudo y piurí reconstruida a partir de sus genes. *Animal Biodiversity and Conservation,* 30, 115-130.

[61] Ruiz-García, M., Pacheco, L., and Alvarez, D. (2009). Caracterización genética del puma andino boliviano (*Puma concolor*) en el Parque Nacional Sajama y relaciones con otras poblaciones de pumas del nor-occidente de Sudamérica. *Revista Chilena de Historia Natural,* 82, 97-117.

[62] Ruiz-García, M., Martinez-Agüero, M., Escobar-Armel, P., Gaviria, M., and Alvarez D. (2010a). Historical demographic evolution in the pink river dolphin (*Inia geoffrensis* and *Inia boliviensis*) and comparative demographic trends with other Latin American dolphins related with freshwater systems (*Sotalia fluviatilis* and *Pontoporia blainvillei*). *Evolution,* (in press).

[63] Ruiz-García, M. et al. (2010b). New morphometric data of *Inia sp.* across its distribution range: Multivariate comparison analyses. *Fieldiana Zoologica*, (to be submitted).

[64] Sambrook, J., Fritsch, E. F. & Maniatis, T. (1989). *Molecular cloning: a laboratory manual.* New York, New York: Cold Spring Harbor Laboratory Press.

[65] Say, L., Bonhomme, F., Desmarais, E., & Pontier D. (2003). Microspatial genetic heterogeneity and gene flow in stray cats (*Felis catus* L.): a comparison of coat color and microsatellite loci. *Molecular Ecology,* 12, 1669-1674.

[66] Schmidt-Lynch, C. (1994). Jungle dolphins. *Nature Conservancy*, 44(2): 16-23.

[67] Schnapp, D., and Howroyd, J. (1992). Distribution and local range of the Orinoco dolphin (*Inia geoffrensis*) in the Rio Apuré, Venezuela. *Zeitschrift für Säugertierkunde,* 57, 313-315.

[68] Schwartz, O. A., & Armitage, K. B. (1980). Genetic variation in social mammals: the marmot model. *Science,* 207, 665-667.

[69] Selander, R. K., & Kaufman, D. W. (1975). Genetic structure of populations of the brown snail (*Helix aspersa*). I. Microgeographic variation. *Evolution,* 29, 385-401.

[70] Sheridan, M., & Tamarin, R. H. (1986). Kinship in a natural meadow vole population. *Behavioral Ecology and Sociobiology,* 19, 207-211.

[71] Smith, B. D. (1993). 1990 status and conservation of the Ganges river dolphin *Platinista gangetica* in the Karnali River, Nepal. *Biological Conservation,* 66, 159-169.

[72] Smith, M. H., Manlove, M. N., & Joule, J. (1978). Spatial and temporal dynamics of the genetic organization of small mammal populations. In D. P. Snyder (Ed.), *Population of small mammals under natural conditions*. (pp. 99-113). Pittsburgh, Pennsylvania: Pymatuning Symposiums in Ecology, University of Pittsburgh.

[73] Tagliavini, F., & Pilleri G. (1984). Occasional observations on the distribution and ecology of the bufeo (*Inia geoffrensis geoffrensis*) in Ecuadorian rivers. *Investigations on Cetacea,* 16, 67-75.

[74] Taylor, A.C., Sherwin, W. B., & Wayne, R. K. (1994). Genetic variation of simple sequence loci in a bottlenecked species: the decline of the northern hairy-nosed wombat (*Lasiorhinus krefftii*). *Molecular Ecology,* 3, 277-290.

[75] Trebbau, P., and Robineau. (1974). Notes concerning the freshwater dolphin bufeo *Inia geoffrensis* (de Blainville, 1817) in Venezuela. *Zeitschrift für Säugertierkunde,* 39, 50-57.

[76] Utreras, V. M. (1996). *Estimación de la abundancia, aspectos ecológicos y etológicos del delfín amazónico Inia geoffrensis geoffrensis (Cetacea: Iniidae) en el río Lagartococha, Amazonía ecuatoriana*. Bachelor of Science Thesis. Quito, Ecuador: Pontificia Universidad Católica del Ecuador.

[77] Valsecchi, E., & Amos, W. (1996). Microsatellite markers for the study of cetacean populations. *Molecular Ecology,* 5, 151-156.

[78] Van Staaden, M. J., Chesser, R. K., & Michener, G. (1994). Genetic correlations and matrilineal structure in a population of *Spermophilus richardsonii. Journal of Mammalogy* , 75, 573-582.

[79] Vidal, O. H., Hurtado. L. A., & Ojeda, Z. V. (1993). The dolphins of the upper Amazon river: progress report. In D. Wartzok (Ed.), Is Wartzok the Editor? *Abstracts from the 10th biennial conference on the biology of marine mammals*. Galveston, Texas: Society for Marine Mammalogy.

[80] Vidal, O., Barlow, J., Hurtado, L. A., Torre, J., Cendón, P., & Ojeda, Z. (1997). Distribution and abundance of the Amazon river dolphin (*Inia geoffrensis*) and the

Tucuxi (*Sotalia fluviatilis*) in the upper Amazon river. *Marine Mammal Science,* 13, 427-445.

[81] Waser, P. M., & Strobeck, C. (1998). Genetic signatures of interpopulation dispersal. *TREE,* 11, 43-44.

[82] Waser, P. M., & Elliot, L. F. (1991). Dispersal and genetic structure in kangaroo rats. *Evolution,* 45, 935-943.

[83] Weber, J. L., & May, P. E. (1989). Abundant class of human DNA polymorphisms which can be typed using the polymerase chain reaction. *American Journal of Human Genetics*, 44, 388-396.

[84] Wilson, D. S. (1977). Structured demes and the evolution of group-advantageous traits. *The American Naturalist,* 111, 157-185.

[85] Wright, S. (1940). Breeding structure of populations in relation to speciation. *The American Naturalist* ,74, 232-248.

[86] Wright, S. (1951). The genetical structure of populations. *Annuals of Eugenetics* ,15, 323-354.

[87] Wright, S. (1982). The shifting balance theory and macroevolution. *Annual Review of Genetics,* 16, 1-19.

In: Biology, Evolution and Conservation of River Dolphins... ISBN: 978-1-60876-633-8
Editors: M. Ruiz-García, J. Shostell. pp. 161-192

Chapter 9

CHANGES IN THE DEMOGRAPHIC TRENDS OF PINK RIVER DOLPHINS (*INIA*) AT THE MICROGEOGRAPHICAL LEVEL IN PERUVIAN AND BOLIVIAN RIVERS AND WITHIN THE UPPER AMAZON: MICROSATELLITES AND MTDNA ANALYSES AND INSIGHTS INTO *INIA'S* ORIGIN

Manuel Ruiz-García*

Laboratorio de Genética de Poblaciones Molecular-Biología Evolutiva. Unidad de Genética. Departamento de Biología. Facultad de Ciencias. Pontificia Universidad Javeriana, Bogotá DC, Colombia

ABSTRACT

More than 200 pink river dolphins (*Inia geoffrensis* and *Inia boliviensis*) were sampled in diverse rivers of Colombia, Peru, Brazil and Bolivia. Ten microsatellites and 400 bp of the mitochondrial control region (D-loop) gene were analyzed with special emphasis on three Peruvian rivers (Ucayali, Marañon and Napo-Curaray) and the Bolivian Mamoré River (and tributaries). Of the different evolutionary demographic tests applied to the microsatellite and mtDNA data, the tests of Kimmel et al., (2008) and Zhivotovsky et al., (2000) provided the most insights about the demographic history of the pink river dolphin. These tests showed that initial bottlenecks occurred prior to very recent population expansions in the diverse areas studied. Two tests (Zhivotovsky and Garza & Williamson) revealed a very strong bottleneck in the origin of the Bolivian population and not during its population expansion. Together, the microsatellite and mtDNA, analyses revealed a strong population expansion for the overall upper Amazon sample and supported that the population expansion and colonization of *Inia* throughout the Amazon, Orinoco and Beni-Mamoré basins occurred in the last 200,000 years ago (and in the majority of cases between 4,000-50,000 years ago) and not several millions of years ago as was claimed by other authors. Furthermore, the original population was the

---

* mruiz@javeriana.edu.co , mruizgar@yahoo.es.

Amazon one, and not the Bolivian population as has been previously defended by several authors, such as Grabert (1984 a, b, c), Pilleri & Ghir (1977, 1980) and Pilleri et al. (1982).

Keywords: *Inia geoffrensis; Inia boliviensis*; DNA microsatellites; mtDNA; bottleneck events; population expansions; *Inia's* origins; Peru; Bolivia; upper Amazon.

INTRODUCTION

\In the last decades there have been an elevated number of threats that either potentially or directly affects marine Cetaceans as a group and especially riverine dolphins. Some local populations of river dolphins have been extirpated (Klinowska, 1991) and the Chinese river dolphin, the baiji, (*Lipotes vexillifer*) is functionally extinct (see the chapters of Turvey, 2010 and Wang & Zhao, 2010). Among these threats, that equally affect marine and river dolphins, it is very important to consider incidental catches by the fisheries. This problem can be illustrated with brief descriptions of general and particular data. For instance, the gillnet fisheries for salmon and squid in the North Pacific killed 16,000 Dall's porpoise in 1987 (Dolan, 1987) and the international fleet of tuna seiners in the eastern tropical Pacific killed about 129,000 dolphins of diverse species (*Stenella attenuata*, *S. longirostris*, *Delphinus delphis*) in 1986 alone (Hall & Boyer, 1987). In both of these cases, the animals were not utilized and were thrown back to the sea. In other cases, the catch is usually marketed locally. Alling (1985) annually reported, at least, 42,000 killed dolphins and small whales in Sri Lanka for human consumption taken in gillnets. In the case of the pink river dolphin (local names, bufeo, bugeo, boto, tonina) (*Inia geoffrensis* and *Inia boliviensis*) from the Orinoco, Amazon and Beni-Mamoré basins, the incidental capture could also have some negative effects on its demography. Da Silva & Best (1990) reported 77 dolphins incidentally caught in the central Brazilian Amazon by lampara seines, drifting gill nets, and fixed gill nets. A major fraction of the *Inia* caught (97 %) died. Besides incidental captures, other dolphins are purposefully and directly killed as sources of "love charms". The author has seen the sexual organs (especially from females) of dolphins commercialized in the "Pasaje Paquito" in the Belem market or in the "artesanal San Juan" at Iquitos (Peru) as well as observed hundreds of sexual organs (of both sexes) and eyes (especially from the other Amazon river dolphin, *Sotalia*) commercialized in the "Ver ao Peso" market at Belem do Pará in Brazil. More recently, another threat is affecting pink river dolphins. The dolphins are killed and employed like bait to attract some small catfish species like the "mota" or "mapurito" (*Calophysus macropterus*) at the Orinoco and at the Amazon (Colombia and Brazil, mainly). This fish is commercialized in important cities such as Bogotá (Colombia) or Sao Paulo (Brazil). For instance, the author observed the remains of four pink river dolphins in the Yavarí River (Amazonian frontier between Peru and Brazil) being used as bait to attract the "mota" in October 2003. But incidental or direct captures are not the unique threats affecting the Amazon River dolphins as suggested by Best & da Silva (1989), Perrin et al., (1989) and da Silva (1995). They point out other main threats such as: 1- Ecological destruction and habitat fragmentation by deforestation. The deforestation process in the Amazon causes an annual loss rate of 0.4-2.3 % (Sayer & Whitmore, 1991). This deforestation process has a disturbing impact on the Amazon hydrological dynamics, primarily affecting the raining regimes with

48 % of the local rain being recycled (Salati & Vose, 1984). Additionally, a major fraction of the deforestation is related to the destruction of the river banks by farming at the edges of the rivers and in the floodplains (Moran, 1990). The destruction of floodplain has an incontrovertible effect on the fish production (Wellcome, 1979) and the fish production is a vital process in the life of the river dolphins and their movements throughout the year (Martin et al., 2004). Mining sand and gravel from river banks alters the hydraulics and substrate composition of fish spawning and also altering the morphology of river channels (Smith & Smith, 1998). The remove of woody debris from the river channels reduces the complexity and biotic diversity of lotic systems destroying essential components for some fishes which are an important part of the dolphin's diet. 2- The use of mercury for the gold amalgamation process, where up to 10 % of the mercury is lost, has a tremendous impact in the local food chains, especially in piscivorous animals, such as river dolphins. A study in the sediments and floating plants in the Tucurui Reservoir in the Tocantins River showed the risk of mercury accumulation in the bed of non-flowing waters (Aula et al., 1995). Petts (1989) claimed that stable gravel-bed rivers might act as sinks for the progressive accumulation of heavy metals. The use of pesticides and fertilizers are other threats because they are responsible for eutrophication of the rivers. Additionally, oil spills along with gas and fuel oil explorations have an undoubtedly impacted the food chains. For instance, Denkinger et al. (2000) revealed a population density declination of *Inia* in the Cuyabeno and Lagartococha Rivers of the northeastern Ecuadorian Amazon from 1996 to 1998. This declination seemed to be due to contamination by six oil spills and waste waters of the oil fields. In fact, the other Amazon dolphin (*Sotalia fluviatilis*) disappeared after 1990 from the Lagunas Grandes in the upper Cuyabeno River, an affluent of the Aguarico River presumably because of oil spills. 3- Another extreme threat for river dolphins is the construction of hydroelectric dams. They wholly transform the hydrologic river cycles altering the flood regime and affecting the rain regimes. Da Silva (1995) claimed that dams stop fish migrations and reduce the trophic resources affecting the biomass and diversity of the fishes, primary aliments of the river dolphins. Dams also affect sediment load and water quality of rivers and suppress the seasonal flow peaks preventing the formation of adjacent floodplains which are extremely vital for the subsistence of river dolphins in certain years and seasons. But dams, also constitute absolute barriers to the potential movements of the river dolphins. Hence, they can fragment the river dolphin populations. The fragmented populations could be more susceptible to the action of endogamy and genetic drift and then more vulnerable to catastrophic environmental and demographic events, inbreeding depression or external pathogenic microorganisms. For instance, two dams in Brazil have unknown consequences on pink river dolphin populations. One of these, the Tucurui dam, is on the Tocantins River, isolating an upstream pink river dolphin subpopulation, while the second is the Balbina Dam on the Uatuma River. Projected dams could also have catastrophic consequences on pink river dolphins, such as the plan to build dams on the Atures rapids (Venezuela), in the Orinoco between Caicara and Ciudad Bolivar as well as on the Madeira River in Brazil very near to the frontier of Bolivia in the separation geographical point of *Inia geoffrensis* and *Inia boliviensis*. 4- The over-fishing could be another important problem for the river dolphins. Nylon gill nets catch more fish with the fishermen expending less effort. Small nylon-mesh gill nets are extremely damaging because they are indiscriminate in catching fishes of all size and age classes and therefore could have major and negative impacts on river dolphin prey.

Additionally, such as Leatherwood (1996) and others demonstrated, river dolphins show a significant attraction for river segments which contained confluences versus ones that didn't. Thus, the dolphin concentrations in limited and circumscribed places make them especially vulnerable to accidental entanglement in fishing nets (especially with nylon gill nets; Reeves & Leatherwood, 1994, Leatherwood, 1996), directed killing dolphins, destroying their habitat and providing local sources of pollution. Even in diverse parts of the Amazon, although illegal, explosives and electricity are used in fishing (Goulding, 1983; Smith & Smith, 1998). Best & da Silva (1989) claimed that fishermen, using explosives, commented that they sometimes attempt to kill dolphins which prey on the stunned or dead fish. The echolocation structures of the river dolphins are particularly vulnerable to the concussion effects of these explosions and acoustical contamination (noise) caused by vessel traffic could be another important risk for river dolphins.

Although Indian communities are afraid of the mythical and legendary "powers" of the dolphins, some Indians have traditionally hunted river dolphins. Such was the case of Mura in the Negro River during the nineteenth century. The destruction of their own Indian cultures by the "colonos" (settlers) and the introduction of firearms are dangerous facts which affect the "mythical invulnerability" of the river dolphins (yacurana, water spirit in quechua; Luna, 1983). Another, maybe past problem, was the capture of specimens for export to aquaria. Since 1956, over 100 pink river dolphins were caught for this task (Layne, 1958; Layne & Caldwell, 1964; Brownell, 1984). Nevertheless, imports ceased because of expensive costs, high mortality and new conservation laws.

Although the status of the pink river dolphin seems to be the most secure with regard to all the other river dolphins, if the construction of hydroelectric dams in Brazil, and in other countries of the Amazon basin, are carried out, this species may well join the list of endangered cetaceans (Klinowska, 1991). The converging threats on the South-American riverine dolphins are of sufficient magnitude for conservation action. For example, the International Union for the Conservation of Nature (IUCN) proposed an action plan for the conservation of biological diversity (1988-1992) concerning dolphins, porpoises, and whales (Perrin, 1988), where 63,000 US dollars were diverted to conservation studies of the Amazon river dolphins (the names of the projects were: 1- Monitor incidental kills of dolphins in Amazon fisheries in Brazil, 2- Promote the establishment of river dolphin conservation areas in Brazil, 3- Promote legislation to fully protect river dolphins in Peru, Ecuador, Colombia and Venezuela, 4- Promote enforcement of existing laws protecting river dolphins in South America, 5- Establish dialogue on river dolphin conservation and management among Brazil, Peru, Venezuela, Colombia, Ecuador and Bolivia).

Furthermore, the effective protection of pink river dolphins could have beneficial consequences on the protection of other species with similar conservation threats, being the case of the two Amazon otter species (*Pteronura brasiliensis* and *Lontra longicauda*), the Amazon manatee (*Trichechus inunguis*), some reptiles, such as the white and black caimans and turtles (*Podocnemys*), and large fish such as certain species of big catfish and *Arapaima gigas*.

To carry out successful conservation programs it is not enough to only undertake river dolphin censuses and to promote enforcement laws but it is a necessity to also know the evolutionary demographic history and colonization patterns of the species and if it is coming from a stable, bottlenecked or original population that has expanded. This information describes the species' evolutionary potential.

For first time, the evolutionary demographic history for three Peruvian river dolphin populations (from the Ucayali, Marañon and Napo-Curaray rivers), one Bolivian river dolphin population (Mamoré, Iruyañez, Itenez, Ipurupuru rivers) and the collective upper Amazon (these three Peruvian rivers plus the Putumayo and Caquetá river dolphin populations in Colombia) was reconstructed by means of 10 nuclear microsatellites markers and 400 bp sequences of the mitochondrial control region (D-loop) gene (mtDNA). A lot of procedures have been developed in the last 20 years, thanks to coalescence theory which captures noteworthy information from molecular markers, such as microsatellites and mtDNA, and reconstruct the evolutionary demographic history of a population or given species. The procedures for microsatellites were implemented by Cornuet & Luikart (1996), Luikart et al. (1998), Kimmel et al. (1998), Reich & Goldstein (1998), Reich et al. (1999), Zhivotovsky et al. (2000), Garza & Williamson (2001), or for mtDNA by Tajima (1989a,b), Rogers & Harpending (1992), Fu & Li (1993), Harpending et al. (1993), Harpending (1994), Rogers et al. (1996), Fu (1997), Ramos-Onsins & Rozas (2002).

In addition, some important insights about the origin of *Inia* were determined by reconstructing the evolutionary demographic history of these diverse pink river dolphin populations. Previously, from a molecular point of view, Banguera-Hinestroza et al. (2002) (mitochondrial control region and cyt-b genes) and Ruiz-García et al. (2008) (introns from autosomal and Y chromosomes) demonstrated that the Bolivian population is a different species from *Inia geoffrensis*. The current results ratify these claims, but also showed that the dispersion of *Inia* from its origin produced the split between the Bolivian form and the Amazon and Orinoco forms. Also, the colonizations of the Amazon and Orinoco river basins were different and more recent as that claimed by other authors, such as Grabert (1984a, b, c), da Silva (1994) and myself (Banguera-Hinestroza et al., 2002).

Therefore, the main objectives of the present work were as follows: 1- To determine possible bottleneck and population expansions in the pink river dolphin populations of three Peruvian and one Bolivian rivers, as well as for the upper Amazon. 2- To determine which tests were more powerful for this task and if the nuclear DNA provides the same information as mtDNA, demographically speaking. 3- To determine the geographical area of where the expansion of the current *Inia* began. 4- To determine when the demographic changes began and in which geological period, the *Inia*'s colonization began and to compare these results with the hypotheses currently in the scientific literature.

Materials and Methods

During 2002-2006, our research group carried out four expeditions in different areas of the Amazon in Colombia, Peru, Ecuador, Bolivia and Brazil. More than 7,000 km of Amazonian rivers were transected searching for pink river dolphins. A total of more than 200 dolphins were captured using special fishing nets with length of 400 meters and width of 10 meters and taking special care to ensure the physical safety of each dolphin during capture. From three (Bolivian) to six (Peru) Indian fishermen and from three (Peru) to five (Bolivia) biologists were involved in the capture of the animals (including the author). In Peru, a single wood boat of 10 meters with an attached 40 horse powered engine was utilized to capture the dolphins in low, rising and high waters. In Bolivia, the nets were of small size (80 meters of length and five meters of width) and the fishermen and the biologists captured the dolphins inside the water after of a brief body to body struggle between dolphin and humans. The individuals were brought on board (in a little wood "canoa" in the Bolivian case) and I personally biopsied the caudal fin of each dolphin captured. After the biopsy, the wound was covered with an antibiotic. Later, the animals were measured for different biometric characteristics and safety released after 5-8 minutes of manipulation. The animals were marked to avoid any recapture and the biopsies were stored in absolute alcohol until DNA extraction.

For the microsatellite analysis, a total of 180 individuals were geno-typified [69 individuals from the Mamoré, Tijamuchí, Iruyañez, Securé, Iténez (= Guaporé), and Ipurupuru Rivers in the Bolivian Amazon, 42 individuals from the Colombian Putumayo River, 33 individuals from the Napo-Curaray rivers, 13 individuals from the Marañón-Samiria rivers and 18 individuals from the Ucayali-Tapiche-Canal del Puhinauva rivers (thus, a total of 64 exemplars were analyzed in Peruvian rivers) and five individuals from the Colombian-Peruvian frontier (between Puerto Nariño in Colombia and Caballococha in Perú) in the Amazon river and tributaries]. The population sets studied for microsatellites were in the Napo-Curaray rivers, Ucayali River, Marañón River, all the collective upper Amazon (exemplars collected from the Peruvian rivers, the Colombian-Peruvian Amazon, and the Putumayo River were analyzed together) and the Bolivian rivers.

For the mtDNA analysis, 207 exemplars were analyzed [57 individuals from the Bolivian rivers, 125 individuals from the Amazon River and its tributaries (Peruvian rivers previously cited, Amazon Colombian-Peruvian frontier, Putumayo and Caquetá rivers in Colombia, and one specimen from the Negro River in the central Brazilian Amazon) and 25 individuals from the Orinoco, Bita, Inirida, Arauca and Meta rivers within the Orinoco basin in Colombia and Venezuela; the Orinoco animals were not studied in this work)]. The population sets used for mtDNA analyses were the Napo-Curaray rivers, Ucayali River, Marañón River, all the Amazon exemplars analyzed together (all the Peruvian rivers, the Colombian-Peruvian Amazon individuals and the exemplars captured in the Putumayo and Caquetá rivers) and the Bolivian rivers.

Molecular Procedures

DNA extraction from the caudal fin biopsies was performed by the phenol-chloroform method (Sambrock et al., 1989). The ten microsatellite markers studied in the *Inia* samples were Ev14, Ev37, Ev76, Ev94 and Ev96 (Valsecchi & Amos, 1996), MK5 (Krutzen et al., 2001), PPHO137 (Rosel et al., 1999), KWM2a, KWM2b and KWM12a (Hoelzel et al., 1998, 2002). The reactions were completed in 25 μl with the following conditions: 10 pmol of forward and reverse primers, 2.5 μl of reaction buffer (10X), 3.0 μl of $MgCl_2$ 3 mM, 1 μl of dNTPs 1 μM, one Taq polymerase unit, 13.5 μl of H_2O, and 2 μl between 25 and 50 ng per μl of DNA. The PCR conditions were 95 °C for 5 minutes, a number of determined cycles of 1 minute at 95 °C, 1 minute at the most accurate annealing temperature (see in the next sentence), one minute at 72 °C, and 5 minutes at 72 °C. The number of cycles and the annealing temperatures were as follows: Ev14 (35 cycles at 58° C), Ev37 (10 cycles at 46 °C and 25 cycles at 56 °C), Ev76 (10 cycles at 46 °C and 25 cycles at 56 °C), Ev94 (10 cycles at

46 °C and 25 cycles at 56 °C), Ev96 (10 cycles at 46 °C and 25 cycles at 56 °C), MK5 (35 cycles at 56° C), PPHO137 (35 cycles at 57° C), KWM2a (35 cycles at 57 ° C), KWM2b (35 cycles at 54° C) and KWM12a (35 cycles at 55° C). The PCR amplification products were run in denaturant 6 % polyacrilamide gels within a Hoefer SQ3 sequencer vertical chamber. Gels migrated for 2-3 hours depending on marker sizes, and were then stained with $AgNO_3$ (silver nitrate). Every sixth line in the gel contained molecular markers (ϕ174 cut with Hind III and Hinf I).

For the analysis of the mitochondrial control region gene (400 bp), the primers employed were H16498 and TRO (LeDuc, unpublished data). The sequences of these primers are 5' CCT GAA GTA AGA ACC AGA TG3' for H16498 and 5' CCT CCC TAA GAC TCA AGG 3' for TRO. The PCR reactions were undertaken in a final 25 µl volume with the following conditions: 0.4 µM of each primer, 2.5 µl of reaction buffer (1X), 3.0 µl of $MgCl_2$ 2.5 mM, 0,4 µM of each dNTP, one Gold Taq Polymerase (Promega) unit and 2 µl of (25 and 50 ng per µl) of DNA. The amplifications were carried out in a BioRad thermocycler with the following protocol: 95 ºC for 5 min, 30 cycles at 95 °C for 45 s, at 52 °C for 45 s and at 72°C for 45 s and at 72°C for 10 min for final extension. All DNA fragments were sequenced in both directions and in cases of mismatching, the PCR reaction was repeated and the product sequenced again. Sequence alignments and editing were performed by using Sequencer 3.0 (Gene Codes Corp.), Clustal X (Thompson et al., 1997) and MEGA 4.0.

Population Genetics Demographic Analyses

Microsatellites

Kimmel's et al. test (1998) was the first procedure to detect possible demographic changes in the evolutionary history of the pink river dolphins. This test is based on the principle that two different estimates of θ (= $4N_e\mu$) could be obtained [one is $\theta_v = V$, with V being the variance of the tandem repeat size, whose expression is $V = (2\ \Sigma_{I=1....n}(X_i - X)^2)/(n - 1)$, where *n* is the number of chromosomes analyzed, X_i is the number of repeats of each allele found and X is the average repeat number of all the alleles found in a microsatellite. The second one is $\theta_{Po} = (1/(P_o^2 - 1))/2$ (estimator of the homozygosity), where $P_o = (n\ \Sigma_{k=1.....n}\ (p_k^2 - 1))/(n - 1)$, with p_k being the allele frequency of the *kth* allele.] An imbalanced β index could be defined as:

$\beta(t) = E(\theta_v)/E(\theta_{Po}) = V(t)/[((1/P_o^2) - 1)/2$ or by the expression: $\ln \beta(t) = \ln \theta_v - \ln \theta_{Po} = \ln (V) - \ln (((1/P_o^2) - 1)/2)$.

If a population is in equilibrium, has a constant demographic size, and isn't suffering an expansion, $\beta = 1$ ($\ln \beta = 0$). On the contrary, if a population has suffered an expansion coming from a mutation-drift equilibrium situation (constant population size), $\beta < 1$ ($\ln \beta < 0$). If a population has experienced an expansion coming from a previous bottleneck , $\beta > 1$ ($\ln \beta > 0$). This last value will be present for a long time (several thousand generations) before showing the signature of a population expansion ($\beta < 1$ ($\ln \beta < 0$)). There is an exception to this general rule, when a bottleneck is so intense the population becomes monomorphic before the demographic expansion, in which case, $\beta < 1$ ($\ln \beta < 0$) all the time. All these β values are consistent in stepwise, logistic or exponential population growth and are not especially affected in diverse mutation models (Kimmel et al., 1998). Two methods were employed to determine the statistical significance of the β (ln β) obtained in diverse pink river dolphin populations. The first was to apply a jackknife procedure (Efron & Tibshirani, 1993) to obtain the variance of β and, with this variance, a Student's t test and 95 and 99 % confidence intervals were estimated. A second procedure was throughout empirical distributions of ln β from 500 coalescence simulations with a $\theta = 5$. In this case, a 95 % confidence interval was determined (- 0.23, 0.25).

A second test employed was that from Zivothovsky et al. (2000), which calculates an expansion index:

$$S_k = 1 - ((K - (R_k V/2)/5V^2),$$

where K and V are the unnormalized kurtosis (fourth central moment) and the allele size variance is estimated from a sample and corrected for sampling bias, respectively, whereas $R_k = k_m/\sigma^2_m$ (they are the kurtosis and the variance in the repeat number mutational changes). The expressions used to estimate V and K are:

$V = \Sigma_{i=1...n}\, p_i\, (X_i - X)^2$ and $K = \Sigma_{i=1...n}\, p_i\, (X_i - X)^4$, where $X = \Sigma_{i=1...k}\, p_i\, k$, and k represents the alleles in a locus given and p_i, the allele frequencies. All the other terms were defined in the previous analysis. The value of R_k employed was 6.3 because this value was obtained for dinucleotide microsatellites by Dib et al. (1996), and because dinucleotide microsatellites were used in the current study. Feldman et al. (1999), used the same data and a geometrical distribution of mutational events, and obtained an estimated σ^2_m of 2.5, which is basically the same as what was obtained by using a truncated Poisson distribution (Zhivothovsky et al., 2000) and by myself in the present work ($\sigma^2_m = 2.45$) The value of S_k is expected to be 0 in a general symmetric stepwise mutation model for a population in equilibrium and of constant size (this was derived by Zhivotovsky & Feldman, 1995). The S_k is positive if an expansion affected the population and is contrarily negative if a bottleneck affected the population. To obtain demographic conclusions of this analysis, the within-population variance and the expansion index are averaged for all the microsatellites studied within each population and their dynamics are compared. Zhivotovsky et al. (2000) showed that a significant correlation existed between V and S_k ($r = 0.58$) for a human data set, but this correlation was moderate and, in fact, both statistics could react differently to the changes in population size and have different patterns in different populations. Noteworthy consequences could be extracted from the differential behavior of both of these statistics in a given population. To measure the statistical significance of the S_k values, two methods were carried out. First, a bootstrap over loci (10,000 runs) was completed. Secondly, such as in the first test, a jackknife procedure was performed to obtain the variance of S_k and, with this variance, a Student's t test and a 95 % confidence interval were estimated.

Another procedure used to detect any possible reduction in population size was that created by Garza & Williamson (2001). This procedure is based on the ratio $M = k/r$, where k is the total number of alleles detected in a locus given and r is the spatial diversity; that is, the distance between alleles in number of repeats and the overall range in allele size. When a population is reduced in size, this ratio will be smaller than in equilibrium populations. To calculate this M value, the program will simulate an equilibrium distribution of M and give assumed values for three parameters of the two phase mutation model ($\theta = 4N_e\mu$, p_s = mean percentage of mutations that add or delete only one repeat unit, and Δ_g = mean size of larger mutations). Once M is obtained, it is ranked relative to the equilibrium distribution. Using conventional criteria, there is evidence of a significant reduction in population size if less than 5% of the replicates are below the observed value. The average θ values employed in this analysis were obtained from the MISAT program by Nielsen (1997), where p_s and Δ_g were 0.0325 (overall Amazon) and 0.1318 (Bolivia) for the first parameter and 3.5 for the second one, respectively. This analysis was carried out with the M-P-Val and Critical-M programs from Garza & Williamson (2001).

The last microsatellite demographic analysis applied to the *Inia sp.* populations reported here were the within locus kurtosis (k) test and the interlocus (g) test proposed by Reich & Goldstein (1998) and Reich et al. (1999). Both tests estimate if there is population expansion in a given species, or population.

The first test is based on the following principles. A population with constant size has gene genealogies, which tend to have a single ancient bifurcation. Therefore, the allele length distributions have multiple discrete peaks. On the contrary, in a growing population, most of the gene genealogy bifurcations tend to date back to the time expansion and as a result the allele length distribution is clearly more smoothly peaked. To measure these differences between the multipeaked allele distribution of a constant size population and the smooth single-peaked allele distribution of a population in expansion, the k statistic was calculated as follows:

$$k = 2.5\ Sig^4 + 0.28\ Var - (0.95/n) - Gam_4,$$

where Sig^4 is the unbiased variance squared of allele length, Var is the sample variance of the same concept and Gam_4 is the unbiased fourth central moment of the allele, respectively. The equations to estimate Sig^4 and Gam_4 are as follows

$$Sig^4 = \frac{(n^2-3n+3)}{n(n-1)(n-2)(n-3)}\left(\sum(Xi-\overline{X})^2\right)^2 - \frac{1}{(n-2)(n-3)}\sum(Xi-\overline{X})^4$$

$$Gam_4 = \frac{(n^2-2n+3)}{(n-1)(n-2)(n-3)}\sum(Xi-\overline{X})^4 - \frac{(6n-9)}{(n-2)(n-3)}\left(\sum(Xi-X)^2\right)^2$$

where n is the number of chromosomes analyzed, X_i is the number of repeats of each allele found and X is the average repeat number of all the alleles found in a given population for a determined microsatellite. In order to assess significance levels, a binomial distribution is used with the number of trials equal to the number of loci based on the expectation of an almost (P = 0.515) equal probability of negative and positive k-values for the set of loci analyzed ($\binom{N}{r} p^r(1-p)^{N-r} = \frac{N!}{r!(N-r)!}p^r(1-p)^{N-r}$)

When there is a smaller loci number associated with positive k values than would be expected for a constant-sized population, there is evidence of a population expansion. The significance level of the binomial distribution was measured using the program Statistics Sample written by Michael H. Kelly.

The g test focuses on the following facts. This interlocus test shows that for stable populations the allele size variance is highly variable among loci, whereas in expansion populations this variance is usually lower. Therefore, allele sizes with sufficiently low variances are taken as evidence for population expansion. The test used for statistical differences is that proposed by Reich & Goldstein (1998): $g = \frac{Var[Vj]}{\frac{4}{3}\overline{V}2 + \frac{1}{6}\overline{V}}$ where Var (V_j) is the observed variance of the allele length variances across the markers employed, and V is the average variance across loci. A low value of g is taken as a sign of population expansion. Significance levels for the interlocus test are found on Table 1 of Reich et al., (1999), which shows the fifth-percentile cutoffs for the interlocus test. For the present case, g values lower than 0.17-0.23 will indicate a population expansion. Luikart et al. (1998) noted that this last test is probably the most powerful for this task.

mtDNA

The mitochondrial gene diversity in each one of the populations studied was measured by means of the nucleotide diversity (π) and the θ per gene (which is equivalent to k, the average number of nucleotide differences).

I used two demographic methods for the mtDNA analysis. The first one corresponds to the mismatch distribution (pairwise sequence differences) calculated following the method of Rogers & Harpending (1992) and Rogers et al. (1996). The empirical observed distribution was compared to the two theoretical curves, the first one assuming a constant population size and the second one assuming a population expansion. The raggedness *rg* statistic (Harpending et al., 1993; Harpending, 1994), the Mean Absolute Error (MAE) between the observed and the theoretical mismatch distribution (Rogers et al., 1996) and the R_2 statistic (Ramos-Onsins & Rozas, 2002) were used to determine the similarities between the observed and the theoretical curves. Finally, the second method uses the Fu and Li D and F tests (Fu & Li, 1993), the Fu F_S statistic (Fu, 1997) and the Tajima D test (Tajima, 1989a). All these tests allow for the determination of possible population size changes in the pink river dolphin ensembles aforementioned (Simonsen et al., 1995; Ramos-Onsins & Rozas, 2002).

Finally, to estimate the divergence times among the mitochondrial haplotypes found in the diverse pink river dolphins studied, the median joining network (MJ; Bandelt et al., 1999) was applied by means of the software Network 4.2.0.1 (Fluxus Technology Ltd). Once the haplotype network was constructed, the ρ statistic (Morral et al., 1994) was estimated. This statistic measures the age of an ancestral node in mutational units. This value is transformed into years by multiplication with the mutation rate. In this case, I took a mutational rate of one mutation each 20,180 years, because this was the amount found for human mtDNA for the stretch from 16090 to 16365 np Additionally, the standard deviation (σ) was calculated (Saillard et al., 2000). The ρ statistic is unbiased and highly independent of past demography events. These events could have influenced the shape of a given evolutionary tree, but this only influences the error of the time estimated and does not increase or decrease this time.

RESULTS

Microsatellites

All the different procedures applied to these data sets used the 10 microsatellites of this study. Table 1 shows the imbalance indices (Kimmel et al., 1998). All the values obtained of β (and ln β) were significantly higher than 1 (β) or 0 (ln β) thus indicating later population growth from an initial bottleneck. Morever, all the rivers studied (the total upper Amazon), showed a similar degree of population expansion after an initial bottleneck. The Marañón River population had the highest population expansion or it was more recent [$\beta = 3.702$, $\ln \beta = 1.309$; $t = 6.179$, 9 df, $P < 0.01$; 99 % confidence interval (2.578 , 4.826); 500 coalescence based simulations generated an empirical distribution with the ln β values within a 95 % confidence interval (-0.23, 0.25)].

Table 1. Statistics calculated for the imbalance indices (Kimmel et al., 1998). t = Student's t test. * Significant tests at P < 0.05. IC = Interval of confidence

Statistics	Bolivia (Mamore River)	Upper Amazon	Ucayali River	Marañon River	Napo-Curaray River
$\hat{\bar{\theta}}_v$	2.03084	5.29446	5.03882	7.28930	4.35980
$\hat{\bar{P}}_o$	0.58534	0.40901	0.40051	0.45000	0.46744
$\hat{\beta}$	2.11694	2.12723	1.92536	3.70188	2.43798
$\ln \hat{\beta}$	0.74997	0.75482	0.65511	1.30884	0.89116
t	5.089*	5.645*	4.603*	6.179*	2.422*
95% IC	(1.687/2.547)	(1.736/2.519)	(1.531/2.319)	(2.845/4.559)	(1.975/2.900)
99% IC	(1.553/2.681)	(1.614/2.640)	(1.409/2.442)	(2.578/4.826)	(1.831/3.044)

The test of Zhivotovsky et al. (2000) (Table 2) showed a different picture than the previous one, indicating, that each one of the tests employed has a different power to detect diverse demographical changes at different historical moments. This test detected a strong and significant bottleneck event in the Bolivian population ($S_k = -2.253$; $t = - 5.274$, 11 df, $P < 0.01$; 99 % confidence interval (-3.351 , -1.155)). This significant negative S_k was accompanied by the lowest of the within-population variances found in this study ($V = 0.9952 \pm 0.2647$). The Napo-Curaray rivers, the Marañón River and the overall upper Amazon showed pink river dolphin populations with a constant demographic size ($S_k = -0.0001$, 0.175 and 0.208, respectively). All these S_k values were not significantly different from 0. Their respective within-population variances (V) were 2.1343 ± 0.7786, 3.4587 ± 1.7185 and 2.6254 ± 0.8257, with the V value for the Marañón River slightly higher than in the other cases. The Ucayali River yielded a demographic expansion at a significant $P < 0.05$ level ($S_k = 0.350$; $t = 2.098$, 10 df, $P < 0.05$; 95 % confidence interval (0.023 , 0.677)), but this value was not significant at a $P < 0.01$ level ($t = 2.098$, 10 df, $P > 0.01$; 99 % confidence interval (-0.078 , 0.778)) and the within-population variance was similar to that found in the other Peruvian rivers and in the overall Amazon sample ($V = 2.2580 \pm 0.7985$). Therefore, this test is more powerful than the Kimmel test when detecting stronger bottlenecks in the initial constitution of a population, or during its history, and can more powerfully detect when the populations reach a demographic stability after a rapid population growth.

Table 2. The test of Zhivotovsky et al. (2000) (S_k) to detect demographic changes applied to microsatellites. SD = Standard deviation. LCL = Lower Confidence Limit. UCL = Upper Confidence Limit. t = Student´s t test. IC = Interval of confidence. * Statistically significant value at P < 0.05

Statistics	Bolivia (Mamore River)	Upper Amazon	Ucayali River	Marañon River	Napo-Curaray River
Var ± SD	0.9951 ± 0.2647	2.6250 ± 0.8256	2.2580 ± 0.7985	3.4587 ± 1.7184	2.1342 ± 0.7786
95% LCL	0.4053	0.7575	0.4518	-0.5041	0.3388
95% UCL	1.5849	4.4932	4.0643	7.4214	3.9297
$\bar{K}$ ± SD	19.2440 ± 9.1901	35.5744 ± 16.2864	23.6875 ± 11.2147	60.2202 ± 42.5884	29.5012 ± 12.6116
95% LCL	-1.2327	-1.2680	-1.6818	-37.9888	0.4187
95% UCL	39.7208	72.4169	49.0569	158.4292	58.5836
S_k	-2.2533	0.2077	0.3498	0.1753	-0.0001
t	-5.274*	0.956	2.098*	0.703	0.0005
95% IC	(-3.090 /-1.416)	(-0.218 /0.634)	(0.023 /0.676)	(-0.313 /0.664)	(-0.455 /0.455)
99% IC	(-3.351 /-1.155)	(-0.351 /0.766)	(-0.078 /0.778)	(-0.465 /0.815)	(-0.597 /0.596)

The Garza & Williamson (2001)'s test agrees quite well with the results showed by the Zhivotovsky et al. (2000)'s test. The Bolivian population showed certain evidence of a bottleneck event, although not drastic. To carry out this test previously, I estimated the θ (= $4N_e\mu$) statistic and the percentage of multiple step mutations, which were $\theta = 2.227 \pm 1.846$ and 13.18 % for the Bolivian case, respectively. The average observed M value was 0.7846. A 10,000 simulation run with these parameters (values cited above) and sample size used resulted in an average M of 0.8143 with 95 % of the individual M values for each marker under the critical value of $M_c = 0.7094$ (indicative of a demographic constant population). Five markers showed an individual M below this critical value (Ev14, M = 0.6667; Ev76, M = 0.60; Ev37, M = 0.7072; Ev94, M = 0.3333; MK5, M = 0.6667), and therefore this fraction is statistically significant ($\chi^2 = 4.77$, 1 df, $P < 0.05$). In contrast, not a single Peruvian river or upper Amazon set showed evidence of any bottleneck. For instance, the overall Amazon set showed $\theta = 5.364 \pm 5.812$ and 3.25 % and an average M of 0.9157, whereas the 10,000 simulated M was 0.8964 with 95 % of the individual M values for each marker under the critical value of $M_c = 0.8159$ labeling it a demographic constant population. Only one marker was below this critical value (EV96, M = 0.60), but this was an insignificant quantity ($\chi^2 = 0.18$, 1 df, $P > 0.05$). Henceforth, the overall Amazonian pink river dolphin showed a historical constant demographic size, contrary to that determined for the Bolivian pink river dolphin population. The θ value of the upper Amazonian population was two times that of the

Bolivian population. A lower θ value is symptomatic of lower gene diversity and may explain the bottleneck formation.

Table 3. k and g tests to detect possible population expansions. NS = No significant; * = Significant test at P < 0.05 level

Population	Marker	k	Mean	variance	
Bolivia	EV76	-23.949	12.244	1.086	
	EV96	1.651	10.868	1.090	
	EV14	-0.639	8.130	0.246	number of loci with a negative k = 6
	KWM2b	0.061	10.717	0.207	K-test (p value) = 0.3406 NS
	KWM2a	0.499	10.800	1.200	g-test (value) = 0.6782 NS
	KWM12a	0.065	18.737	0.523	
	MK5	-9.332	28.704	1.097	
	PPHO137	-12.117	21.545	1.463	
	EV37	-81.031	28.344	3.793	
	EV94	-0.015	32.009	0.009	
Upper Amazon	EV76	-8.114	19.872	1.266	
	EV96	11.763	9.960	3.018	
	EV14	2.338	15.322	2.385	number of loci with a negative k = 5
	KWM2b	-0.025	10.112	0.099	K-test (p value) = 0.5856 NS
	KWM2a	-0.039	9.263	0.383	g-test (value) = 0.7224 NS
	KWM12a	-9.79	17.254	0.823	
	MK5	2.116	24.754	6.732	
	PPHO137	0.027	22.217	0.171	
	EV37	-19.087	25.522	7.147	
	EV94	2.846	34.967	4.236	
Ucayali River	EV76	-9.266	19.708	1.867	
	EV14	2.013	15.437	1.789	
	KWM2b	-0.038	10.156	0.136	number of loci with a negative k = 4
	KWM2a	0.039	9.292	0.302	K-test (p value) = 0.7157 NS
	KWM12a	-21.670	17.562	1.738	g-test (value) = 0.7953 NS
	MK5	60.279	25.666	8.229	
	PPHO137	0.083	22.333	0.232	
	EV37	-17.486	25.218	4.628	
	EV96	16.300	34.857	3.682	
Marañon River	EV76	-0.403	19.312	0.362	
	EV14	-3.351	14.500	0.971	
	KWM2b	-0.086	10.100	0.094	number of loci with a negative k = 5
	KWM2a	-0.156	8.997	0.196	K-test (p value) = 0.4631 NS
	KWM12a	-3.995	17.250	0.733	g-test (value) = 1.8013 NS
	MK5	63.708	25.187	8.429	
	PPHO137	0.027	22.278	0.212	
	EV37	310.811	26.687	16.629	
	EV96	27.338	34.611	4.604	
Napo-Curaray rivers	EV76	-16.721	20.074	1.806	
	EV14	0.055	14.896	1.414	
	KWM2b	-0.013	10.173	0.146	number of loci with a negative k = 7
	KWM2a	-0.136	9.130	0.294	K-test (p value) = 0.0759*
	KWM12a	-0.426	17.053	0.159	g-test (value) = 0.8736 NS
	MK5	12.037	24.625	6.189	
	PPHO137	-0.046	22.108	0.099	
	EV37	-18.237	25.280	5.022	
	EV96	-19.329	34.729	5.308	

The k and g tests of Goldstein et al. (1998) did not detect any expansion trend in all the populations analyzed (Table 3). The only case where the statistical level was borderline to the significance was for the Napo-Curaray rivers, where the k test ($p = 0.075$) indicated a possible population expansion. Thus, both of these tests did not reveal significant population changes in the population analyzed. This reveals that, in the condition of the present analysis, these tests were less powerful in detecting changes in any population trend compared to tests by Kimmel et al. (1998) and Zhivotovsky et al. (2000).

mtDNA

The overall Amazon sequence set showed levels of gene diversity considerably higher than the Bolivian sequence set ($\pi = 0.0116$ and θ per gene = 4.646 versus $\pi = 0.0014$ and θ per gene = 0.5789, respectively). This means that the Amazon set has eight times more gene diversity than the Bolivian sample. Within the Amazon sample, the Peruvian rivers showed the highest gene diversity statistics, all of them being very similar, [$\pi = 0.0164$ and θ per gene = 6.549 (Ucayali river), $\pi = 0.0150$ and θ per gene = 6.000 (Marañón river), $\pi = 0.0175$ and θ per gene = 7.003 (Napo-Curaray rivers)], whereas the dolphin populations from the two Colombian rivers yielded lower gene diversity levels, especially the Putumayo River (π = 0.0022 and θ per gene = 0.8961, Colombian Putumayo River, and $\pi = 0.0048$ and θ per gene = 1.907, Colombian Caquetá River).

The mismatch distribution (pairwise sequence differences) and their respective associated statistics (*rg*, MAE and the R_2 statistic) as well as the Fu and Li D and F tests and the Fu F_S and the Tajima D showed some demographic trend in the following pink river dolphin ensembles (Table 4): In the Ucayali River, the Fu & Li D* test revealed a significant bottleneck event (95 % confidence interval, -2.227; 1.331, with $p[D^* \leq 1.358] = 0.021$) as well as the same was detected for the Napo-Curaray rivers with the same test (95 % confidence interval, -2.449; 1.353, with $p[D^* \leq 1.306] = 0.033$). But the unique case that revealed a clear population expansion was the pink river dolphin ensemble constituted by the overall Amazon sample. The vast majority of the tests yielded striking evidence of an important population expansion (R2: 95 % confidence interval, 0.041; 0.153, with $p[R2 \leq 0.0439] = 0.038$; D*: 95 % confidence interval, -2.255; 1.563, with $p[D^* \leq -2.707] = 0.014$; F*: 95 % confidence interval, -2.208; 1.753, with $p[F^* \leq -2.906] = 0.009$; D: 95 % confidence interval, -1.609; 1.896, with $p[D \leq -2.037] = 0.002$) (Figure 1). The remaining ensembles (Marañón River, all the Peruvian rivers taken together, Bolivia) as well as the remaining tests for the Ucayali and the Napo-Curaray rivers showed stable female lineage populations.

Table 4. Tests for possible expansions or contractions in different river dolphin populations by means of the mitochondrial control region. * P < 0.05, Significant result. IC = Interval of Confidence

	Bolivia (Mamore River)		Upper Amazon		Marañon River		Ucayali River		Napo-Curaray rivers	
	95 IC	P	95 IC	P	95 IC	P	95 IC	P	95 IC	P
Raggedness r test	0.058/0.866	P[r≤0.156] = 0.219	0.013/0.187	P[r≤0.059] = 0.637	0.028/0.449	P[r≤0.141] = 0.705	0.020/0.296	P[r≤0.099] = 0.694	0.013/0.176	P[r≤0.062] = 0.696
R2 test	0.035/0.241	P[R2≤0.063] = 0.126	0.041/0.153	P[R2≤0.044] = 0.039*	0.105/0.242	P[R2≤0.144] = 0.277	0.092/0.215	P[R2≤0.146] = 0.515	0.070/0.189	P[R2≤0.111] = 0.399
Tajima D test	-1.456/1.994	P[D≤-1.099] = 0.104	1.609/1.896	P[D≤-2.038] = 0.002*	-1.765/1.677	P[D≤-0.904] = 0.197	-1.782/1.701	P[D≤-0.224] = 0.456	-1.694/1.779	P[D≤-0.477] = 0.354
Fu & Li D* test	-2.378/1.085	P[D*≤-0.881] = 0.231	-2.255/1.563	P[D*≤-2.707] = 0.014*	-2.024/1.382	P[D*≤-0.710] = 0.246	-2.227/1.331	P[D*≤1.358] = 0.021*	-2.449/1.353	P[D*≤1.306] = 0.047*
Fu & Li F* test	-2.556/1.427	P[F*≤-1.113] = 0.160	-2.208/1.754	P[F*≤-2.906] = 0.009*	-2.217/1.539	P[F*≤-0.855] = 0.235	-2.439/1.516	P[F*≤1.065] = 0.113	-2.587/1.554	P[F*≤0.869] = 0.163
Fu's Fs test	-3.215/3.760	P[Fs≤-2.621] = 0.052	-7.950/8.434	P[Fs≤-3.560] = 0.165	-3.473/4.700	P[Fs≤0.823] = 0.655	-4.119/4.870	P[Fs≤2.400] = 0.131	-5.479/5.788	P[Fs≤2.221] = 0.166

UPPER AMAZON

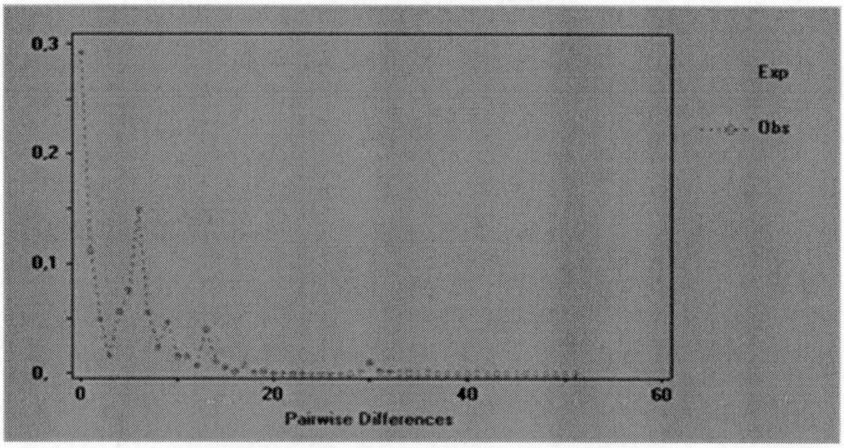

BOLIVIA (MAMORE RIVER)

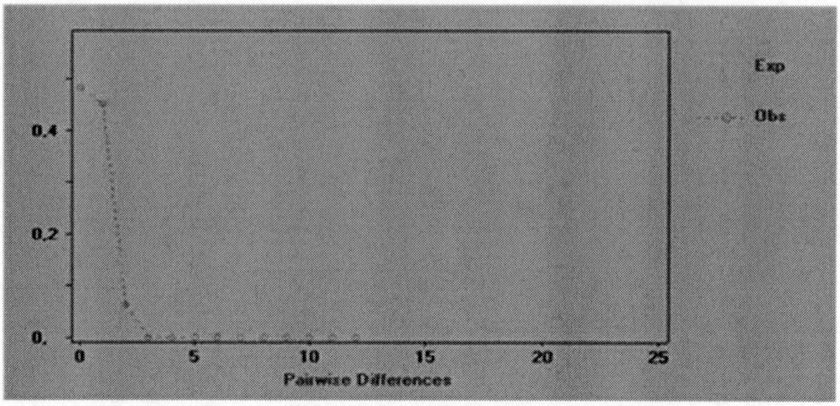

MARAÑON RIVER

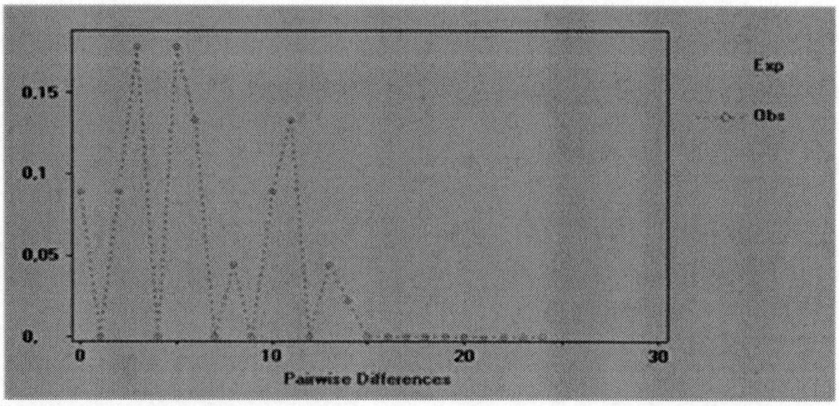

Figure 1. Continued on next page.

NAPO-CURARAY RIVERS

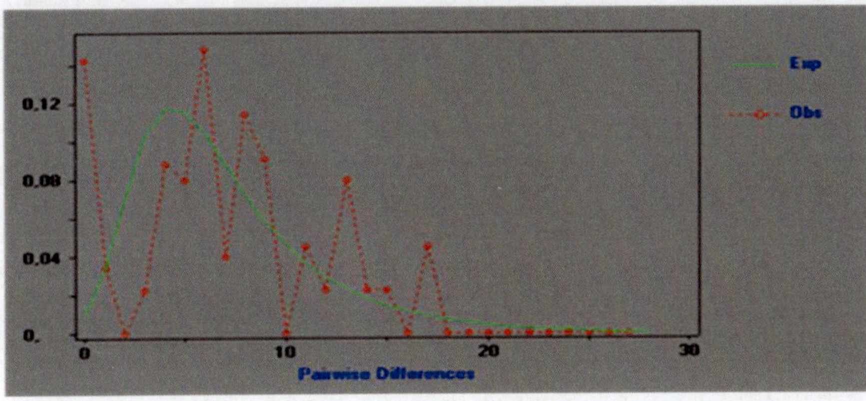

UCAYALI RIVER

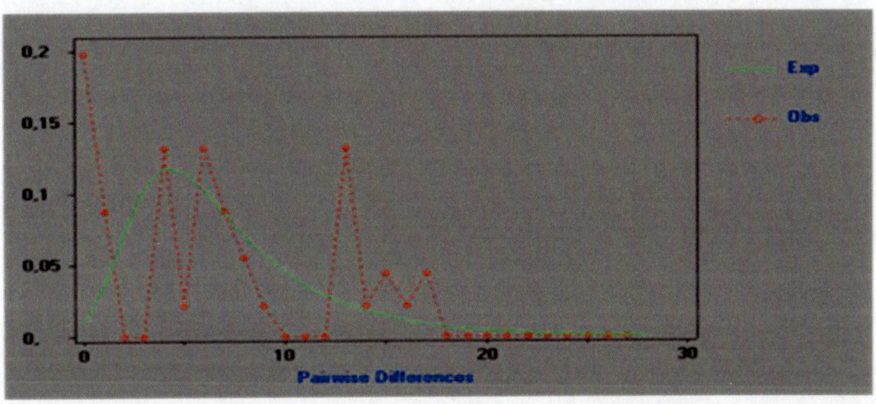

Figure 1. The mismatch distribution (pairwise sequence differences) at the mtDNA control region for the pink river dolphins of the Upper Amazon, Mamoré River (Bolivia), Marañón, Napo-Curaray and Ucayali rivers (Perú). Only the upper Amazon sample showed a clear population expansion.

The analysis of the haplotype network with the median-joining procedure showed that some Amazonian haplotypes were originals while the haplotypes, found in Bolivia and the Orinoco, were derived (Figure 2). This haplotype network has a star-like form which agrees quite well with a population expansion. Henceforth, some noteworthy different results were obtained between the microsatellites (nuclear DNA) and the control region gene (mitochondrial DNA). However, the separation time between the mitochondrial haplotypes from the Amazon, Bolivia and Orinoco, as well as inside the Amazon and the Bolivian populations, agrees quite well with some aspects obtained in the microsatellite analyses: all these expansion and colonization processes were relatively recent. The D-loop time separation between the main Amazon haplotype and the main Bolivian haplotype was about 163,800 $\pm$ 7,450 years ago ($\rho = 8.116 \pm 0.369$). Within the Amazon, the time split between the two main haplotypes was 22,700 $\pm$ 3,800 years ago ($\rho = 1.125 \pm 0.187$) and within Bolivia, the temporal separation between the two main haplotypes was 4,850 $\pm$ 4,850 years ago ($\rho = 0.24 \pm 0.24$). This last result is in agreement with tests performed by Zhivotovsky et al. (2000) and Garza & Williamson (2001) for microsatellites. This population suffered a strong bottleneck from its origin until its population expansion, which happened very recently, sometime during the last 5000 years. Although in this work, I did not analyze the Orinoco River's dolphin populations, two Amazonian lineages colonized, at different times within the Orinoco basin. The oldest Orinoco lineage diverged from the main Amazon haplotype between 7,800 ± 850 to 52,600 ± 5.100 years ago ($\rho = 0.388 \pm 0.042$ and $\rho = 2.605 \pm 0.251$, respectively), whereas the most recent lineage began between 5,400 ± 600 to 7,800 ± 600 years ago ($\rho = 0.269 \pm 0.029$ and $\rho = 0.388 \pm 0.029$, respectively). Therefore, at least, two paraphyletic molecular lineages of *Inia* are living together in the Orinoco basin. Thus a sub-specific nomenclature (*Inia geoffrensis humboldtiana*) for the Orinoco *Inia* seems to not be valid.

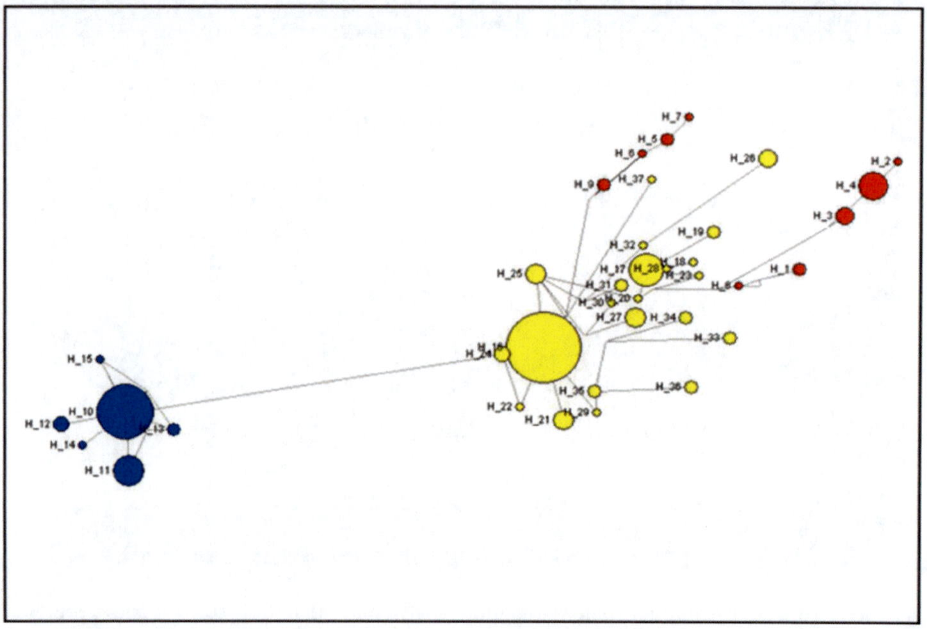

Figure 2. Median Joining network with the mtDNA control region haplotypes of pink river dolphins proceeding from the Amazon (haplotypes in yellow), Bolivian Amazon (haplotypes in blue) and Orinoco (haplotypes in red) basins. Clearly, the Amazon haplotypes were in the origin of the haplotype expansion, meanwhile the Bolivian haplotypes derived from the central Amazon haplotype and two independent Orinoquian haplotype lineages derived from the Amazon haplotypes.

Discussion

This means that in each river (and for the total upper Amazon) a small, or several small, dolphin propagules gave origin to the current populations with a rapid and/or a recent growth of these populations. The social system in small populations, detected by Ruiz-García (2010a) for this species, helps us to understand why small genetic lineages colonized the Amazonian rivers step by step instead of a single, or a few massive migration(s). The results were basically the same for all the rivers studied in Peru and Bolivia (representing the rivers of these both countries, two different and independent colonization processes) which showed that *Inia* is an efficient colonizer in small social propagules. Another very interesting result from this analysis is that if the values are (and, in fact, all they are $\beta > 1$ ($\ln \beta > 0$)) in agreement with a population expansion proceeding from a bottleneck, then these values will be present during several thousand generations (from 5,000 to 10,000 generations) before these values show the signature of a population expansion ($\beta < 1$ ($\ln \beta < 0$)). If, in the case of *Inia*, a generation is about seven years, then the population expansion process after the initial bottleneck is not older than 35,000 (5000 generations; more probable) or 70,000 (10,000 generations; less probable) years. This is one, of a series of results, which shows that the expansion and colonization of the current *Inia geoffrensis* and *Inia boliviensis* in the Amazon is a relatively recent process. The separation of both *Inia*'s forms could coincide with the second maxim peak of the Riss-Illinois glaciation (around 150,000 years ago), while the Amazon and Orinoco haplotype differentiation occurred during the last Würm-Wisconsin glaciation, which elapsed from 120,000 to 10,000 years ago with a maximum peak 18,000 years ago. Therefore, the genetics divergence process within the *Inia* genus is not an old process as a lot of researchers have previously claimed (Pilleri & Ghir, 1980; Pilleri et al., 1982; Grabert 1984 a, b, c; Cassens et al., 2000; Hamilton et al., 2001; Banguera-Hinestroza et al., 2002; Martin & da Silva, 2004).

The obtained results for S_k and V in the diverse river dolphin populations are noteworthy to be compared with the values for these statistics obtained in humans and with the theoretical simulations obtained by Zhivotovsky et al. (2000). The values of S_k and V over time weakly depend on the rate of population increase and the final population size. Only extreme differences in the growth rate could produce substantial differences between these statistics. The three Peruvian river dolphin and the overall upper Amazon populations (this last enclosing the Putumayo River's population) showed relatively similar S_k and V values with a moderate variance and little evidence of a strong population expansion. Therefore, similar growth rate conditions affected these river dolphin populations. Nevertheless, the Bolivian population presented extremely different S_k and V estimates, which demonstrated that the growth rate conditions for this population were extremely different than in the previous case and represented a totally independent colonization process. The negative S_k value and the small V value demonstrated the existence of a strong bottleneck in this population. The question is if this bottleneck occurred just before or during the population expansion. This last event may greatly influence S_k but only slightly affect V (Figure 6 from Zhivotovsky et al., 2000). In the Bolivian case, both S_k and V were extremely affected. Therefore, the striking bottleneck was in the original population formation. This result is in disagreement with Grabert (1984 a,b,c) who claimed that the original population gave origin to all other pink river dolphin populations. An extreme bottlenecked population can not generate other populations with more genetic diversity than itself. On the other hand, S_k is not affected by different mutation rates in the microsatellites studied (at least if this rate does not change over time), whereas V is greatly affected. Therefore, as all the microsatellites studied were the same in all the populations and all them were dinucleotide, different population S_k values could be not attributed to different mutation rates for the markers studied. Also, I have no evidence that the mutation rate changed, at least when the population began to grow and if the mutation rate had increased then both S_k and V would be higher than the values found.

The river dolphin population of the Ucayali River showed the higher S_k value, whereas the Marañón River's dolphin population showed the highest V estimate. Such as I previously commented, there is a significant correlation between S_k and V, but it is only moderate. Why didn't this correlation occur between the Ucayali and the Marañón River dolphin populations? A possible explanation is in Figure 4 from Zhivotovsky et al. (2000). The initial size of a population has opposite effects on the dynamics of S_k and V. The lower the initial population size, the lower the V value and the larger the S_k value. Henceforth, although the three Peruvian rivers showed a similar demographic dynamic, some subtle differences could be claimed. The original Ucayali River dolphin population was slightly smaller than the original Marañón population and the relative growth in the Ucayali River was higher than in the Marañón River. Nonetheless, the absolute population expansion was higher in the Marañón River and/or its population expansion was more recent than for the Ucayali River population which had a higher initial population size. The Napo-Curaray river dolphin population probably represented an original population slightly lower than the Ucayali and Marañón ones but without any population expansion trend or, this one, was extremely recent. But the differences in the three Peruvian river dolphin populations were quite small.

Zhivotovsky et al. (2000) showed a table (Table 5, page 761), where predicted values of expansion time for humans were based on S_k values with different initial population sizes. For an initial population of 500 humans and for S_k of 0.05, 0.20, 0.30 or 0.35, the population expansions in humans began 10,000, 32,000, 77,000 and 171,000 years ago, respectively. For an initial human population of 2,000 and for the same S_k values, these expansion population dates could be 17,000, 62,000, 141,000 and 272,000 years ago, respectively. For these estimations, the human population was assumed to be at equilibrium prior to the expansion and the growth was logistic. The generation time for humans was 25 years. The S_k values for some of our pink river dolphin populations, which showed positive values, were 0.164 (Putumayo river in Colombia), 0.175 (Marañón River), 0.208 (overall upper Amazon) and 0.350 (Ucayali River). Recall that a generation in *Inia* is about seven years. If I recalculate these time expansions for river dolphins assuming the conditions for humans from Zhivotovsky et al. (2000), although they obviously were not the same (it is only an exercise), with S_k around 0.20 (such as Putumayo, Marañón and overall upper Amazon) for initial populations of 500 or 2,000 dolphins yielded expansion times of 9,000 or 18,000 years ago, respectively. For the Ucayali River, with S_k of 0.35, it could represent a temporal expansion of 48,000-76,000 years ago. It is possible that the original population size for river dolphins was yet smaller than the initial human population sizes. It would mean that the expansion times were less than 9,000-76,000 years ago, but also that prior to expansion, the populations were coming from an initial bottleneck which reduced the S_k values during the population expansion. It means that the real population expansion times could be higher than 9,000-76,000 years ago. Even, Penny et al. (1995) and Zhivotovsky et al. (2000) claimed that more attention should be paid to the lower confidence bounds of S_k than to the estimation point. If so, the referred river dolphin populations did not reach to expand or they expanded on the last 2,000 or 3,000 years ago. Anyway, it is clear that the temporal expansion of *Inia* was relatively recent in the past (for instance, 10,000-100,000 years ago) but not several millions of years ago, such as it was previously sustained by other authors.

Really, the number of possible Iniidae fossils is scarce (but see this book, Barnes et al., 2010; Cozzuol, 2010). Some genera as *Anisodelphis*, *Ischyorhynchus* and *Saurodelphis* resemble *Inia* from the Miocene (Barnes et al., 1985). For instance, *Ischyorhynchus* is from the late Miocene of Rio Acre. Also, another species named *Plicodontina mourai* seems to be strongly related to *Saurodelphis* and therefore to *Inia* and was discovered in the Juruá River in Acre (Brazil). However, the major fraction of these fossils was very fragmented and their exact relationship with *Inia* is difficult to establish. Fittkau (1974) and Grabert (1967, 1983, 1984 a, b, c) postulated that the Iquitos-Purús arch divided the Amazon basin into two flow water directions, one to the Atlantic and other to the Pacific, during the Middle Miocene. The most extended hypothesis for the *Inia* origin is that proposed by Grabert (1984 a, b, c) and related with the Iquitos gate. This author stated some Iniidae, such as *Proinia patagonica* (although other authors like Barnes negated the relationship of this form with the Iniidae), inhabited the Pacific coastal region in the middle Miocene (15 million years ago) and penetrated to the molasse lakes which were forming when the Andean orogenesis began. With the completion of the Andean orogenesis, five million of years ago, the link between these molasse lakes (sub-Andean freshwater molasses) and the Pacific Ocean disappeared. The Iniidae forms could have penetrated to the sub-Andean freshwater molasses by way of Guayaquil Bay or through the Arica entrance, later, during the Pleistocene, when Beni lake was extinct. This is the area in the current Bolivian Amazon where *Inia boliviensis* is found. The surrounded areas of this sub-Andean freshwater lake system were savannas and arid regions and rapidly growing Andean cordilleras, where the turbidity of waters was extremely high. Following Grabert (1984a, b, c), at this moment, was the appearance of the most ancestral form of *Inia*, *Inia boliviensis*, characterized by the presence of microphthalmia. This author sustained that *Inia boliviensis* was the first form of *Inia* because it has a larger number of teeth (24 more teeth) and a smaller brain capacity (100 cc lesser) than the *Inia*'s forms from the Amazon and Orinoco (*Inia geoffrensis*). Thus, *Inia boliviensis* originated five million of years ago. Later during the late Pliocene or beginning of the Pleistocene (about 1.8 millions of years ago) throughout the Purus or the Iquitos gates, this *Inia* form migrated into the Amazon basin giving rise to *Inia geoffrensis*. Such as the Amazon basin has more biotopes and more diverse ecological conditions than the original sub-Andean freshwater lake system, the "new" *Inia* (*Inia geoffrensis geoffrensis)* form had better cerebralisation (because the species having an increased ultrasound capacity making is more efficient than all of the new biotopes that colonized) and the reduction of the dental count could be related to a better capacity to obtain a major quantity of fish species. Later, much more recently (10,000 years ago during the Holocene), with the increase of humidity, rain regimes and sea levels, the blackwater rivers (such as the Negro River) and flooded forest (várzea) formed. The aforementioned author considered that the formation of these black-water rivers between the Amazon and Orinoco basins was the main process that originated a different subspecies in the last basin, *Inia geoffrensis humboldtiana*, because black waters, with their high acidity and the lack of trophic resources, could be a barrier for *Inia*. However, the molecular genetics results herein were not in agreement with the Grabert's hypothesis. Clearly, the following insights contradicted this hypothesis:

1. The microsatellite analyses showed some historical demographic changes, and presented clear evidence that the Amazonian population expansion process occurred around 35,000 years ago (Kimmel et al.,' test) or around 9,000-76,000 years ago (Zhivotovsky et al.,' test, in the most favorable circumstances for this procedure). The mtDNA analysis showed that the two main Amazon haplotypes diverged about 23,000 years ago. Henceforth, the Amazonian expansion for *Inia* could not be 1.8 millions of years ago, such as was claimed by Grabert (1984a, b, c) or about 1.5 millions of years ago during the first Ice Age (Nebraska or Günz) as it was defended

by Pilleri et al. (1982). Furthermore, although the molecular dating estimates obtained by Cassens et al. (2000) indicated that all the river dolphin lineages diverged well before the radiation of delphinids (11-13 millions of years ago), the current *Inia* is a recent species (such as it was "prophetical" affirmed by Simpson, 1945) and not a relict species of otherwise successful Oligocene-Miocene groups, like it was sustained by Cassens et al. (2000). Also, the construction of geological or climatologic scenarios by the integration of palaeontological data with Tertiary palaeooceanography events were not useful in understanding the evolution of the river dolphins (Hamilton et al. 2001) because the actual evolution of *Inia* genera occurred more recently than assumed by other authors. Following this argumentation line, the affirmation of Martin & da Silva (2004) ("It is, nonetheless, important to recall that the entire geomorphology of the Amazon is geologically recent. Várzea Lake systems are a product of the Pleistocene and Holocene, Ayres 1993, and, consequently, appeared late in the evolutionary development of Iniid dolphins, Hamilton et al., 2001. The genus *Inia* has clearly adapted very successfully to habitat changes on a geological timescales"), is incorrect. The current *Inia* genus evolved in parallel to these Pleistocene and Holocene geological changes because it is not older than these geological changes, and probably, is a product of them.

2. For the Zhivotovsky et al.'s test as well as for the Garza & Williamson' test, there was clear evidence of a strong bottleneck affecting the Bolivian population. The microsatellite heterozygosity level (not shown here), the mtDNA sequence gene diversities (π and θ) or the gene diversity for RAPD (Polymorphism rate and expected heterozygosity in the Peruvian populations were 85.3 % and 0.211 and for the Bolivian population were 38.9 % and 0.119, respectively; Ruiz-García et al. 2007) revealed that the Bolivian population is more genetically depauperate than the other pink river dolphin populations. Therefore, *Inia boliviensis* could not be the original form of *Inia*.
3. The mitochondrial haplotypes showed that the Amazon form was original population of the current *Inia*'s forms. Thus, the origin of *Inia* was in the own Amazon River and not in the Bolivian sub-Andean freshwater lake system such as it was previously claimed. It is possible that the upper Amazon was the origin of the current *Inia* genera, because the samples analyzed in the Putumayo and Caquetá rivers (Amazon River tributary) showed a considerably lower gene diversity, at least, for mtDNA. Nevertheless, more Amazonian pink river dolphin populations should be analyzed to determine the exact geographic origin of the primary population within the Amazon.
4. The Bolivian population was derived from the Amazon population in a possible peripatric or allopatric speciation process around 160,000 years ago (mt DNA) and not five millions of year ago as established by Grabert (1984a, b, c) or two million of years ago as sustained by Pilleri & Gihr (1980) because they established this age for the historical origin of the rapids between Guayaramerin and Porto Velho. Previously Ruiz-García et al. (2008) analyzed nine nuclear and Y chromosome introns and determined a divergence of 543,000 years ago for the separation of *Inia geoffrensis* and *Inia boliviensis*. As the microsatellite analysis showed, this is a population which crossed throughout a strong bottleneck and the mtDNA analysis presented a possible diversification of haplotypes in this population very recently (around 4,000 years ago). It means that the genetic drift was extremely important in the origin of this

population. Indeed, the time splits of the Bolivian population could be overestimated, being lower than those presented in this work. Therefore, the smaller cerebral capacity and the larger teeth number of the Bolivian *Inia* could not be considered as the ancestral characters in *Inia*. In contrast, they seem to be derived characters, which could represent a violation of Willingston' law. A strong gene drift and the prolonged bottleneck effects could change specific morphological characters following a genetic revolution (Mayr, 1963) or a "flush-crash" process (Powell, 1978). If so, these characters could simply be neutral and the natural selection action invoked by Grabert (1984 a, b, c) on these morphological characters could be false. Similarly, the arguments in favor of the original Bolivian form discussed by Pilleri & Gihr (1977, 1980) and Pilleri et al. (1982) are now unsustainable from a genetics point of view. Those morphological characters considered to be the oldest and most primitive in the Bolivian population (BP), are really derived from those presented in the Amazon population (AP): rostrum longer (BP) vs. rostrum shorter and more massive (AP); premaxillares, one or one and a half times as long as wide with protuberances in front of the nares only moderately elevated (BP) vs. premaxillares, twice as long as wide with protuberances in front of the nares very prominent (AP); number of teeth by ramus 33 (BP) vs. 26 (AP); profile of supraoccipital with an indentation immediately behind the vertex (BP) vs. profile of supraoccipital with no indentation behind the vertex (AP); no condylus tertius (BP) vs. condylus tertius very frequent (AP); squamosum covers the half of the parietal (BP) vs. squamosum covering 2/3 to 4/5 of the parietal (AP); manus with fewer phalanges, variation in number of phalanges and no phalanx in the pollex (BP) vs. manus with more phalanges in all five fingers, more variation in number of phalanges and pollex often has one phalanx (AP); IV cervical with neural arch with a distinct spine (BP) vs. only rudimentary spine in the neural arch; VI cervical with upper and lower transverse processes broad and wing-like (BP) vs. upper transverse processes small and conical, lower transverse processes rod-shaped (AP); VII cervical with distinct rod-shaped lower transverse processes (BP) vs. lower transverse processes rudimentary (AP); and sternum with rostral incision wide (BP) vs. sternum with rostral incision narrow (AP). Henceforth, these morphological characters are neutral and fixed in the Bolivian population by genetic drift or they evolved by different natural selection conditions in the newly colonized Beni-Mamoré River Basin regard to the original Amazon river conditions.

5. Within the Orinoco basin, there are at least two different mtDNA lineages that derived from the Amazon population. Thus, *Inia geoffrensis humboldtiana* is a polyfiletic taxa and is possibly a non-valid one. One, of these two mtDNA lineages, was established during the Holocene (5,000-8,000 years ago), coinciding with the opinion of Grabert (1984 a, b, c) and the possible formation of black-waters and flooded forest. I quite agree with the opinion of Casinos & Ocaña (1979) that this is not a different subspecies from the Amazon form, disagreeing with the Pilleri & Gihr (1977, 1980)'s observations. However, the other mtDNA lineage could also be generated in the Holocene, but could be older than the black-water and flooded forest formation. If this last affirmation is certain, the Amazon and the Orinoco basins were connected around 50,000 years ago. Additionally, one animal sampled in the Negro River (at Nova Airao) was more related to the Orinoco haplotypes than to the

Amazon ones. This means that, at least, two migrations from the Amazon penetrated into the Orinoco basin, but, at least, one migration from the Orinoco migrated to the Amazon basin. This means that the affirmation of Grabert (1984a, b, c) is false and the black-waters are not a barrier for the dispersion of the pink river dolphins. In agreement with my affirmation, Meade & Koehnken (1991) showed that the Negro River in the Brazilian Amazon or the Atabapo River in Venezuela also contained considerable *Inia* populations. However, Pilleri & Pilleri (1982) affirmed that the density of *Inia* was 50 times less in the black-water systems such as in the upper Orinoco and the Casiquiari channel compared to a white-water river like the Apure River in Venezuela. In fact, some black water tributaries, for example, of the Napo River in Ecuador do not contain *Inia* populations (Tagliavini & Pilleri, 1984).

All these results reveal that is possible that the ancestors of *Inia* could have originated in the Atlantic Ocean (but *Inia* appeared "in situ" in the Amazon) in discordance with Grabert (1984 a, b, c), but in agreement with the opinions of Brooks et al. (1981) and Gaskin (1982). It is known, and previously mentioned, that in the late Miocene, the Iquitos-Purús arch divided the Amazon basin in two. At that time, the Paraná River basin (Cozzuol, 1996), the eastern Amazon basin (Hoorn, 1994), most of the South American Atlantic coast and a big fraction of the Orinoco basin were covered by epicontinental sea waters (Entrerriense transgression; Cozzuol, 1992; Hoorn et al., 1995; Lovejoy et al., 1998). Therefore, the Iquitos-Purús arch interconnected these three river basins (Orinoco, Amazon and Paraná-La Plata systems). The Rebeca Lagoon, the Grande Tremedal River, or the Paranense Sea, for instance, interconnected the Amazon and the Paraná basins (Klammer, 1984), at different moments, permitting the inter-change of fauna. This internal epicontinental sea is supported by sedimentological data and the fauna distribution of mollusk and foraminifera (Räsänen et al., 1995; Nuttall, 1990; Boltovsky, 1991). Recall, *Pontoporia blainvillei* is an estuarine dolphin living in the Atlantic coasts of Argentina, Uruguay and Brazil (mouth of La Plata River) and is the sister clade of *Inia* (Cassens et al., 2000; Hamilton et al., 2001). Henceforth, it is easy to imagine that *Inia* and *Pontoporia* had a common ancestor which could penetrate in that interconnected Orinoco-Amazon-Paraná system via the Atlantic Ocean. Later, when the orogeny of the Andes in the late Miocene and Pliocene, the influence of the Iquitos-Purús arch was lost and the Guayaquil Gate was closed (Pacific Amazon drainage) and the complete Amazon river drainage was towards the Atlantic ocean. In that moment, the interconnection between the three basins diminished and disappeared between the Amazon and the Paraná basins. With these changes, it is possible that *Inia* (or better an *Inia* ancestor) was isolated in the Amazon Basin, meanwhile *Pontoporia* (or a *Pontoporia* ancestor) returned, or simply stayed, at the estuarine or coastal habitats of the South America Atlantic Ocean. Some palaeontological data could be in agreement with this hypothesis. It seems that the unique and clear Iniidae fossil found in Central Florida, outside South-America, is *Goniodelphis* from the Early Pliocene (Morgan, 1994), although Muizon (1988) was not absolutely convinced that it was an Iniidae. Three clear Iniidae, such as *Ischyorhynchus*, *Saurodelphis* and *Saurocetes*, were found for the Late Miocene of the Paraná basin in Argentina. It could contribute additional proof that a marine ancestor of *Inia* could penetrate into the continental South-America via the Paraná basin and then be isolated in the Amazon Basin when the sea level declined and the Amazon drainage was only towards the Atlantic Ocean. It is easy to imagine, that *Pontoporia* is a remainder of those species of dolphins that penetrated the Paraná basin via rivers or areas situated in the current La Plata River. Or, it's also possible that *Pontoporia* (or an ancestor) returned to the coastal habitats when the geological, hydrological and climatological conditions were adverse inside the continental South America and the internal water interconnection between Paraná and Amazon basins was lost. The presence of *Pontistes*, a Late Miocene Pontoporidae, in the marine sediments of the Paraná Basin, together with the Iniidae fossils ratify the strong connection of these dolphins and their introduction to the continental South-America via the Paraná Basin. Another alternative hypothesis is that the ancestor of the current *Inia* penetrated to the Amazon Basin prior to the close of the Guayaquil gate. It could be correlated with the fact that the upper Amazon pink river dolphin populations seem to have higher gene diversity than the *Inia* populations of other Amazon tributaries such as the Putumayo and Caquetá rivers. However, a lot of other *Inia* populations should be analyzed to estimate their respective gene diversity levels in other Amazon areas (like the Juruá, Purús, Madeira, Tapajós, lower Xingú and Tocantins-Araguaia rivers, all in Brazil) to support this Pacific origin of *Inia*. Other mammals seem to have their original foci of dispersion from the upper Amazon. This could be the case of *Ateles* (Ruiz-García et al.,2006), *Cebus apella* (Ruiz-García & Castillo, 2010), the jaguar (*Panthera onca*) (Ruiz-García et al., 2010), the lowland tapir (*Tapirus terrestris*) (unpublished results) and maybe even *Saimiri* (Lavergne et al., 2009) and partially *Lagothrix* (Ruiz-García & Pinedo, 2009). Nevertheless, with the current data it's easier to understand that *Inia* has an Atlantic origin rather than a Pacific one.

Inia population sizes is another noteworthy topic. Presently our research group is using gene diversity and coalescence methods (Ruiz-García, 2010b) to address questions regarding this topic. These new results could give new insights about *Inia's* precise geographic origin. At the present, I don't know the overall size of the *Inia* population. A maximum improbable size could be derived from some data proportioned by Martin & da Silva (2004). They estimated 13,000 pink river dolphins for 11.240 km^2 of várzea at the Mamirauá Reserve (which is an extreme favorable habitat for this species) at the central Brazilian Amazon. If we assume that the entire Amazon basin is about six millions km^2, and assuming that the entire basin was in condition to maintain pink river dolphin populations (a fact that is not certain), a potential maximum total of 6.9 million of pink river dolphins could exist in the Amazon basin. Obviously, the real size must be considerably lower than this value. Futures studies on the demographic historical evolution and estimations of effective population sizes in diverse regions of the Orinoco and Amazon are needed for a complete and compressive evolutionary perspective of this species as well as to take correct measures in biological conservation.

ACKNOWLEDGMENTS

Economic resources to carry out this study were obtained from Colciencias (Grant 1203-09-11239; Geographical population structure and genetic diversity of two river dolphin species, *Inia boliviensis* and *Inia geoffrensis*, using molecular markers) and the Fondo para la Accion Ambiental (US-Aid) (120108-E0102141; Structure and Genetic Conservation of river dolphins, *Inia* and *Sotalia*, in the Amazon and Orinoco basins). Main thanks go to Pablo Escobar-Armel (Colombia), Dr. Diana Alvarez (Colombia), Ariel Rodriguez (Colombia),

Esteban Payán (Colombia), María Martínez-Agüero (Colombia), Carlos Vergara (Colombia), Maria Fernanda Gómez (Colombia), Nathalí Romero (Colombia), Mariana Escovar (La Paz, Bolivia), Juanito and Angelito (Iquitos, Perú), and, especially to Isaias and his sons (Requena, Perú) who participated in the capture of the pink river dolphins herein studied. Also, diverse Peruvian Indian communities collaborated with our pink river dolphin captures throughout the Peruvian rivers (Bora, Ocaina, Shipigo-Comibo, Capanahua, Angoteros, Orejón, Cocama, Kishuarana and Alamas) as well as diverse Bolivian communities helped to capture dolphins throughout the Mamoré river and other Bolivian Amazon tributaries (Sirionó, Canichana, Cayubaba and Chacobo). Dr. Fernando Trujillo (Omacha Fundation) gently offered some samples from the Colombian Orinoco and Amazon. The lab collaboration of María Martínez-Agüero, Magda Gaviria, Pablo Escobar-Armel and Eulalia Banguera is thanked. Additional thanks go to Hugo Gálvez (Iquitos, Perú) and Armando Castellanos (Quito, Ecuador) to collaborate in collection permits in both countries. Similarly, many thanks go to the Bolivian, Peruvian and Ecuadorian Ministry of Environment, to the Dirección General de Biodiversidad and CITES from Bolivia, PRODUCE, Dirección Nacional de Extracción and Procesamiento Pesquero and to the Instituto Nacional de Recursos Naturales (INRENA) from Perú for their role in facilitating the obtainment of the collection permits. Special thanks goes to the Colección Boliviana de Fauna (Dr. Julieta Vargas) in La Paz (Bolivia). Thanks also to the Fundación Sociedad Portuaria de Santa Marta (Colombia) for its logistical support.

REFERENCES

[1] Alling, A. (1985). Small cetacean entanglement: a case study of the incidental entrapment of cetaceans in Sri Lanka's gillnet fishery. *Thirty-seventh meeting of the IWC Scientific Committee*, 37: 8.

[2] Aula, I., Braunschweiler, H., & Malin, I. (1995). The watershed flux of mercury examined with indicators in the Tucuri reservoir in Pará, Brazil. *Science of the Total Environment,* 175, 97-107.

[3] Ayres, J. M. (1993). *As Matas de Várzea do Mamirauá. Médio Rio Solimoes*. Brasilia, Brazil: CNPq/Sociedade Civil Mamirauá.

[4] Bandelt H-J., Forster, P., Rohl, A. (1999). Median-joining networks for inferring intraspecific phylogenies. *Molecular Biology and Evolution,* 16, 37-48.

[5] Banguera-Hinestroza, E., Cárdenas H., Ruiz-García, M., Marmontel, M., Gaitán, E., Vázquez, R., & García-Vallejo F., (2002). Molecular identification of evolutionarily significant units in the Amazon River dolphin *Inia sp.* (Cetacea: Iniidae). *The Journal of Heredity,* 93, 312-322.

[6] Barnes, L. G., Domning, D. P., & Ray, C. E. (1985). Status of studies on fossil marine mammals. *Marine Mammal Sciences,* 1, 15-53.

[7] Barnes, L. G., Kimura, T., & Godfrey, S. (2010). The evolutionary history and phylogenetic relationships of the superfamily Platanistoidea. In Ruiz-García, M., and J. Shostell (Eds.), *Biology, Evolution, and Conservation of River Dolphins Within South America and Asia: Unknown Dolphins In Danger. New York, New York:* Nova Science Publishers., Inc.

[8] Best, R. C., & da Silva, V. M. F. (1989). Amazon river dolphin, Boto, *Inia geoffrensis* (de Blainville, 1817). In S. H. Ridgway, and R. J. Harrison (Eds.), *Handbook of marine mammals*. (Volume 4, pp. 1-23). London, United Kingdom: Academic Press.
[9] Boltovsky, E. (1991). Ihering's hypothesis in the light of foraminiferological data. *Lethaia,* 24, 191-198.
[10] Brooks, D. R., Thorson, T. B., & Mayer, M. A. (1981). Freshwater stingrays (Potamotrygonidae) and their evolution and coevolution.. In D. A Funk and D. R. Brooks, (Eds.), *Advances in cladistics*. (pp. 147-175). New York, New York: New York Botanic Garden.
[11] Brownell, R. L. (1984). Live-capture fisheries for small cetaceans in South America waters. *Report of the International Whaling Commission,* 34, 747.
[12] Casinos A., & Ocaña J., (1979). A craniometrical study of the genus *Inia* D'Orbigny1834, Cetacea, Platanistoidea. *Saugetierkunde Mitteilung,* 27, 194-206.
[13] Cassens, I., Vicario, S., Waddell, V. G., et al., (2000). Independent adaption to riverine habitats allowed survival of ancient Cetacean lineages. *Proceedings of the National Academy of Sciences,* 97, 11343-11347.
[14] Cornuet, J. M., & Luikart, G. (1996). Description of power analysis of two tests for detecting recent population bottlenecks from allele frequency data. *Genetics,* 144, 2001-2014.
[15] Cozzuol, M. A. (1992). Origen y evolución de los Iniidae (Cetacea, Odontoceti) Pp. 17. Quinta Reunión de especialistas en mamíferos acuáticos de América del Sur. Buenos Aires.
[16] Cozzuol, M. A. (1996). The record of aquatic mammals in southern South America. *Munchner Geowissenschaftliche Abhandlungen,* 30, 321-342.
[17] Cozzuol, M. A. (2010). Fossil record and evolutionary history of Inoidea. In M. Ruiz-García, and J. Shostell (Eds.), *Biology, Evolution, and Conservation of River Dolphins Within South America and Asia: Unknown Dolphins In Danger*. New York, New York: Nova Science Publishers., Inc.
[18] Da Silva, V. M. F. (1994). *Aspects of the Biology of the Amazonian Dolphins Genus Inia and Sotalia fluviatilis. Doctoral Thesis.* Cambridge, United England: The University of Cambridge.
[19] Da Silva, V. M. F. (1995). Conservation of the fresh water dolphin with special emphasis on *Inia geoffrensis* and *Sotalia fluviatilis*. In *Delfines y otros mamíferos Acuáticos de Venezuela.* (pp. 71-92) Caracas, Venezuela: FUDECI.
[20] Da Silva, V. M. F., & Best, R. C. (1990). Freshwater dolphin/Fisheries interaction in the central Amazon (Brazil). *IWC Symposium on mortality of cetaceans in passive fishing nets and traps*. La Jolla. CA: IWC.
[21] Denkinger, J., Campos, F., & von Fersen L. (2000). Status of the bufeo (*Inia geoffrensis*) and the tucuxi (*Sotalia fluviatilis*) in the Cuyabeno Reserve, Ecuador. (unpublished manuscript).
[22] Dib, C., Faure, S., & Fizames, C. (1996). A comprehensive genetic map of the human genome base don 5.264 microsatellites. *Nature,* 380, 152-154.
[23] Dolan, H. J. (1987). *Proposed regulations to Govern the taking of marine mammals incidental to the Japanese commercial salmon gillnet fishing operations in the United States exclusive economic zone*. Washington, D. C.: National Marine Fisheries Service.

[24] Efron, B., & Tibshirani, R. J. (1993). *An introduction to the bootstrap.* New York, NY: Chapman and Hall.

[25] Feldman, M. W., Kumm, J., & Pritchard, J. K. (1999). Mutation and migration in models of microsatellite evolution. In Goldstein, D. G., and Schlötterer, C. (Eds.), *Microsatellites: evolution and applications.* Oxford, United Kingdom: Oxford University Press.

[26] Fittkau, E. J. (1974). Zur ökologishen Gliederung Amazoniens. I. Die erdgeschichtliche entwicklung Amazoniens. *Amazoniana,* 5, 77-134.

[27] Fu, Y. X. (1997). Statistical tests of neutrality against population growth, hitchhiking and background selection. *Genetics,* 147, 915-925.

[28] Fu, Y. X., & Li, W. H., 1993. Statistical tests of neutrality of mutations. *Genetics,* 133, 693-709.

[29] Garza, J. C., & Williamson, E. G. (2001). Detection of reduction in population size using data from microsatellite loci. *Molecular Ecology,* 10, 305-318.

[30] Gaskin, D. E. (1982). *The ecology of whales and dolphins.* Oxford, United Kingdom: Heinemann Educational Books Ltd.

[31] Grabert, H. (1967). Sobre o desaguamento natural do sistema fluvial do rio Madeira desde a construcao dos Andes. *At. Simp. Biol. Amazon,* Manuel, what is the full name of this journal? 1: 209-214.

[32] Grabert, H. (1983). The Amazon shearing system. *Tectonophysics,* 95, 329-336.

[33] Grabert, H., (1984a). Migration and speciation of the South America Iniidae, Cetacea-Mammalia. *Zeitschrift Saugetierkunde,* 49, 334-341.

[34] Grabert, H., (1984b). Migracoes e especiacao dos Iniidae da América do Sul (Cetacea, Mamm). *Geografia,* 9, 145-154.

[35] Grabert, H., (1984c). Mögliche Wanderwege und Phylogenie der südamerikanischen Iniidae (Cetacea, Mammalia). *Amazoniana,* 8, 365-374.

[36] Goulding, M. (1983). Amazon fisheries. In E. F. Moran (Ed.), *The dilemma of Amazonian development.* (pp. 189-210) Boulder, CO: Westvew Press.

[37] Hall, M. A., and Boyer, S. D. (1987). *Incidental mortality of dolphins in the eastern tropical Pacific tuna fishery in 1986.* Thirty-seventh meeting of IWC Scientific Committee, Eastbourne, United Kingdom: IWC.

[38] Hamilton H., Caballero, S., Collins, A. G., & Brownell Jr, R. L., (2001). Evolution of river dolphins. *Proceedings of the Royal Society of London,* 268, 549-556.

[39] Harpending, H., (1994). Signature of ancient population growth in a low resolution mitochondrial DNA mismatch distribution. *Human Biology,* 66, 591-600.

[40] Harpending, H. C., Sherry, S. T., Rogers, A. R., & Stoneking, M. (1993). Genetic structure of ancient human populations. *Current Antrophology,* 34, 483-496.

[41] Hoelzel, A. R., Dahlheim, M & Stern, S. J. (1998). Low genetic variation among killer whales (*Orcinus orca*) in the Eastern North Pacific and genetic differentiation between foraging specialists. *The Journal of Heredity,* 89, 121-128.

[42] Hoelzel, A. R., Natoli, A., Dahlheim, M. E., Olavarria, C., Baird, R. W & Black, N. 2002. Low worldwide genetic diversity in the killer whale (*Orcinus orca*): implications for demographic history. *Proceedings of the Royal Society of London,* 269, 1467-1473.

[43] Hoorn, C. (1994). An environmental reconstruction of the palaeo-Amazon River system (Middle-Late Miocene, NW Amazonia). *Palaeogeo. Palaeclim. Palaeecol,.* 112, 187-238.

[44] Hoorn, C., Guerrero, J., Sarmiento, G., & Lorente, M. (1985). Andean tectonics as a cause for changing drainage patterns in Miocene northern South America. *Geology,* 23, 237-240.

[45] Kimmel, M., Chakraborty, R., King, J. P., Bamshad, M., Watkins, W. S., & Jorde, L. B. 1998. Signatures of population expansion in microsatellite repeat data. *Genetics,* 148, 1921-1930.

[46] Klammer, G. (1984). The relief of the extra-Andean Amazon basin. In Siolo, H (Ed.), *The Amazon. Limnology and landscape ecology of a mighty tropical river and its basins*. (pp 47-83). Dordrecht: Dr. W. Junk Publisher.

[47] Klinowska, M. (1991). *Dolphins, porpoises and whales of the world. The IUCN red data book.* Gland, Switzerland: IUCN.

[48] Krutzen, M., Valsecchi, E., Connor, R. C., & Sherwin, W. B. (2001). Characterization of microsatellite loci in *Tursiops aduncus*. *Molecular Ecology,* 1, 170-172.

[49] Lavergne, A., Ruiz-García, M., Lacaste, V., Catzeflis, F., Lacote, S & De Thoisy, B. (2009). Taxonomy and phylogeny of squirrel monkey (genus *Saimiri*) using cytochrome b genetic analysis. *American Journal of Primatology,* 71, 1-12.

[50] Layne, J. N. (1958). Observations on freshwater dolphins in the upper Amazon. *Journal of Mammalogy,* 38, 1-21.

[51] Layne, J. N., & Caldwell, D. K. (1964). Behaviour of the Amazon dolphin, *Inia geoffrensis* (Blainville) in captivity. *Zoologica,* 49, 81-108.

[52] Leatherwood, S. (1996). *Distributional ecology and conservation status of river dolphins (Inia geoffrensis and Sotalia fluviatilis) in portions of the Peruvian Amazon.* PhD dissertation. TX: Texas A&M University.

[53] Lovejoy, N. R., Bermingham, E., & Martin, A. P. (1998). Marine incursions into South America. *Nature,* 396, 421-422.

[54] Luikart, G., Allendorf, F., Sherwin, B., & Cornuet, J. M. (1998). Distortion of allele frequency distribution provides a test for recent population bottleneck. *The Journal of Heredity,* 86, 319-322.

[55] Luna, L. E. (1983). Concepto de plantas que enseñan, entre cuatro shamanes mestizos de Iquitos. *Revista Colombiana de Antropología,* 24, 43-74.

[56] Martin, A.R., & da Silva, V. M. F. (2004). Number, seasonal movements, and residency characteristics of river dolphins in an Amazonian floodplain lake system. *Canadian Journal of Zoology,* 82, 1307-1315.

[57] Martin, A.R., da Silva, V. M. F., & Salmon, D. L. (2004). Riverine habitat preferences of botos (*Inia geoffrensis*)and tucuxis (*Sotalia fluviatilis*) in the central Amazon. *Marine Mammal Science,* 20, 189-200.

[58] Mayr, E. (1963). *Animal species and Evolution*. Cambridge, Massachusetts: Harvard University Press.

[59] Meade R. H., & Koehnken, L. (1991). Distribution of the river dolphin, tonina *Inia geoffrensis*, in the Orinoco river basin of Venezuela and Colombia. *Interciencia* 16: 300-312.

[60] Moran, E. F. (1990). *A ecologia humana das populacoes da Amazonia.* Ed. Vozes. Petropolis, Rio de Janeiro.

[61] Morgan, G. S. (1994). Miocene and Pliocene marine mammal faunas from the Bone Valley formation of Central Florida. *Proceedings of the San Diego Society of Natural History*, 29, 239-268.

[62] Morral, N., Bertrantpetit, J., & Estivill, X. (1994). The origin of the major cystic fibrosis mutation (delta F508) in European populations. *Nature Genetics,* 7, 169-175.

[63] Muizon, C. (1988). Le polyphyletisme des Acrodelphidae, odontocetes longirostres du Mioceen europeen. *Bull. Mus. Natl. Hist. Nat.,* 10, 31-88.

[64] Nielsen, R. (1997). A likelihood approach to population samples of microsatellite alleles. *Genetics,* 146, 711-716.

[65] Nuttall, C. P. (1990). A review of the Tertiary non-marine molluscan faunas of the Pebasian and other island basins of northwestern South-America. *Buletin of the British Museum of Natural History, Geology,* 45, 165-371.

[66] Penny, D., Steel, M., Waddell, P. J., & Hendy, M. D. (1995). Improved analysis of human mtDNA sequences supported a recent African origin for *Homo sapiens. Molecular Biology and Evolution,* 12:, 863-882.

[67] Perrin, W. F. (1988). *Dolphin, porpoises, and whales. An action plan for the conservation of biological diversity*: 1988-1992. Gland, Switzerland: IUCN.

[68] Perrin, W. F., Brownell, R. L., Zhou, K., & Jiankang, L. (1989). Biology and conservation of the river dolphins. *Occasional Papers of the IUCN species survival commission* (SSC), 3, 173.

[69] Petts, G. E. (1989). Perspective for ecological management of regulated rivers. In J. A. Gore and G. E. Petts (Eds.), *Alternatives in regulated river management* (pp. 3-24). Boca Ratón, Florida: CRC Press.

[70] Pilleri, G., & Gihr, M. (1977). Observations on the Bolivian (*Inia boliviensis* d'Orbigny, 1834) and the Amazonian bufeo (*Inia geoffrensis* de Blainville, 1817) with a description of a new subspecies (*Inia geoffrensis humboldtiana*). *Investigations on Cetacea,* 8, 11-76.

[71] Pilleri, G., & Gihr, M. (1980). Additional considerations on the Taxonomy of the genus *Inia. Investigations on Cetacea,* 11, 15-27.

[72] Pilleri, G., & Pilleri, O. (1982). *Zoologische expedition zum Orinoco und Brazo Casiquiare 1981*. Bern, Switzerland: Hirnanatomisches Institut.

[73] Pilleri, G., Marcuzzi, G., & Pilleri, O. (1982). Speciation in the Platanistoidea. Systematic, zoogeographical and ecological observations on recent species. *Investigations on Cetacea,* 14, 15-46.

[74] Powell, J. R. (1978). The founder-flush speciation theory: An experimental approach. *Evolution,* 32, 465-474.

[75] Ramos-Onsins, S. E., & Rozas, J., (2002). Statistical properties of new neutrality tests against population growth. *Molecular Biology and Evolution,* 19, 2092-2100.

[76] Räsänen, M., Linna, A., Santos, J., & Negri, F. (1995). Late Miocene tidal deposits in the Amazonian foreland basin. *Science,* 269, 386-389.

[77] Reeves, R. R., & Leatherwood, S. (1994). *Dolphins, poprpoises and whales, 1994-1998 action plan for the conservation of cetaceans*. Gland, Switzerland: International Union for Conservation of Nature.

[78] Reich, D. E., & Goldstein, D. B. (1998). Genetic evidence for a Paleolitic human population expansion in Africa. *Proceedings of the National Academy of Sciences,* 95, 8119-8123.

[79] Reich, D. E., Feldman, M. W., & Goldstein, D. B. (1999). Statistical properties of two tests that use multilocus data sets to detect population expansions. *Molecular Biology and Evolution,* 16, 453-466.

[80] Rogers, A. R., Fraley, A. E., Bamshad, M. J., Watkins, W. S., & Jorde, L. B. (1996). Mitochondrial mismatch analysis is insensitive to the mutational process. *Molecular Biology and Evolution,* 13, 895-902.

[81] Rogers, A. R., & Harpending, H. C. (1992). Population growth makes waves in the distribution of pairwise genetic differences. *Molecular Biology and Evolution,* 9, 552-569.

[82] Rosel, P. E., France, S. C., Wang, J. Y., & Kocher, T. D. (1999). Genetic structure of harbour porpoise *Phocaena phocaena* populations in the northwest Atlantic based on mitochondrial and nuclear markers. *Molecular Ecology,* 8 (Suppl. 1), 41-54.

[83] Ruiz-García, M. (2010a). Micro-geographical genetic structure of *Inia geoffrensis* at the Napo-Curaray river basin by means of Chesser's models. In M. Ruiz-García, M. and J. Shostell (Eds.), *Biology, Evolution, and Conservation of River Dolphins Within South America and Asia: Unknown Dolphins In Danger*. Hauppauge, NY: Nova Science Publishers., Inc.

[84] Ruiz-García, M. (2010b). Estimations of historical effective numbers in the Amazon River dolphin (*Inia*) by means of gene diversity and coalescence methods in diverse Peruvian and Bolivian rivers and in the upper Amazon. *Molecular Ecology,* (submitted).

[85] Ruiz-García, M., & Castillo, M. I. (2010). Taxonomy considerations and molecular phylogeography of *Cebus apella* and related taxa with the COII mitochondrial gene. In M. Ruiz-García (Ed.). *Molecular Population Genetics, Phylogeny and Evolutionary Biology of Neotropical Primates.* Kerala, India: Transworld Research Network.

[86] Ruiz-García, M., & Pinedo, M. (2009). Molecular Systematics and Phylogeography of the genus *Lagothrix* (Atelidae, Primates) by means of mitochondrial COII gene. *Folia Primatologica,* (in press).

[87] Ruiz-García, M., Caballero, S., Martínez-Agüero, M., & Shostell, J. (2008). Molecular differentiation among *Inia geoffrensis* and *Inia boliviensis* (Iniidae, Cetacea) by means of nuclear intron sequences. In V. P. Koven, (Ed.). *Population Genetics Research Progress* (pp. 177-223). Hauppauge, NY Nova Science Publisher, Inc.

[88] Ruiz-García, M., Murillo, A., Corrales, C., Romero-Aleán, N., & Alvarez-Prada, D. (2007). Genética de Poblaciones Amazónicas: La historia evolutiva del jaguar, ocelote, delfín rosado, mono lanudo y piurí reconstruida a partir de sus genes. *Animal Biodiversity and Conservation,* 30, 115-130.

[89] Ruiz-García. M., Murillo, A., & Vásquez, C. (2010). Phylogeography and genetic differentiation among populations of jaguars (*Panthera onca*) from South and Central America by means of 12 DNA microsatellites and mitochondrial NADH5 sequences. *Molecular Ecology,* (submitted).

[90] Ruiz-García, M., Parra, A., Romero-Aleán, N., Escobar-Armel, P., & Shostell, J. M. (2006). Genetic characterization and phylogenetic relationships between the *Ateles* species (Atelidae, Primates) by means of DNA microsatellite markers and craniometric data. *Primate Report,* 73, 3-47.

[91] Saillard, J., Forster, P., Lynnerup, N., Bandelt, H-J., & Norby, S. (2000). mtDNA variation among Greenland Eskimos: the edge of the Beringian expansion. *American Journal of Human Genetics,* 67, 718-726.

[92] Salati, E., & Vose, P. B. (1984). Amazon basin: a system in equilibrium. *Science,* 225, 129-138.

[93] Sambrook, J., Fritsch, E. F. & Maniatis, T. (1989). *Molecular cloning: a laboratory manual,* New York, New York: Cold Spring Harbor Laboratory Press.

[94] Sayer, J. A. & Whitemore, T. C. (1991). Tropical moist forest: Destruction and species extinction. *Biological Conservation,* 55, 199-213.

[95] Simonsen K. L., Churchill, G. A., & Aquadro C. F. (1995). Properties of statistical tests of neutrality for DNA polymorphism data. *Genetics,* 141, 413-429.

[96] Simpson, G.G. (1945). The principles of classification and a classification of mammals. Bulletin of the American Museum of Natural History, 85, 1-350.

[97] Smith, A. M., and Smith, B. D. (1998). Review of status and threats to river cetaceans and recommendations for their conservation. *Environ*mental Reviews, 6, 189-206.

[98] Tagliavini, F., & Pilleri G. (1984). Occasional observations on the distribution and ecology of the bufeo (*Inia geoffrensis geoffrensis*) in Ecuadorian rivers. *Investigations on Cetacea,* 16, 67-75.

[99] Tajima, F. (1989a). Statistical method for testing the neutral mutation hypothesis by DNA polymorphism. *Genetics,* 123, 585-595.

[100] Tajima, F. (1989b). The effect of change in population size on DNA polymorphism. *Genetics,* 123, 597-601.

[101] Thompson, J. D., Gibson, T. J., Plewniak, F., Jeanmougin, F., & Higgins, D. G. (1997). The Clustal X windows interface: flexible strategies for multiple sequence alignment aided by quality analysis tools. *Nucleic Acids Research,* 25, 4876-4882.

[102] Turvey, S. T. (2010). Failure of the baiji recovery programme: Conservation lessons for other freshwater cetacean. In M. Ruiz-García, and J. Shostell (Eds.), *Biology, Evolution, and Conservation of River Dolphins Within South America and Asia: Unknown Dolphins In Danger*. Hauppauge, NY: Nova Science Publishers., Inc..

[103] Valsecchi, E., & Amos, W. (1996). Microsatellite markers for the study of cetacean populations. *Molecular Ecology,* 5, 151-156.

[104] Welcomme, R. L. (1979). Fisheries ecology in floodplain rivers. London, United Kingdom: Longman.

[105] Wang, D., & Zhao, X. (2010). Population status and conservation of baiji and the Yangtze finless porpoise. In M. Ruiz-García and J. Shostell (Eds.), *Biology, Evolution, and Conservation of River Dolphins Within South America and Asia: Unknown Dolphins In Danger*, Hauppauge, NY, Nova Science Publishers., Inc.

[106] Zhivotovsky, L. A., & Feldman, M. W. (1995). Microsatellite variability and genetic distances. *Proceedings of the National Academy of Sciences,* 92, 11549-11552.

[107] Zhivotovsky, L. A., Bennett, L., Bowcock, A. M., & Feldman, M. W. (2000). Human population expansion and microsatellite variation. *Molecular Biology and Evolution,* 17, 757-767.

In: Biology, Evolution and Conservation of River Dolphins… ISBN: 978-1-60876-633-8
Editors: M. Ruiz-García, J. Shostell. pp. 193-217

Chapter 10

FOSSIL RECORD AND THE EVOLUTIONARY HISTORY OF INIODEA

M. A. Cozzuol
Departamento de Biologia Geral, ICB, Universidade Federal de Minas Gerais, ICB – UFMG, Belo Horizonte, MG, Brazil.

ABSTRACT

This chapter discusses the possible phylogenetic relationships within the superfamily Inioidea (using fossil record data) and provides detailed descriptions of Brachydelphidae, Pontoporiidae and Iniidae (including *Goniodelphis, Ischyrhorhynchus, Saurocetes, Plicodontinia* and a possible new species of *Inia* that is estimated to have arisen approximately 45,000 years ago). Some previously related taxa to Iniidae are also discussed such as *Proinia patagonica.* Additionally, the chapter discusses the Lipotoidea and their relationship with Inioidea, the phylogenetic position of *Parapontoporia*, and the evolutionary process (and paths) that originated the inioid clades.

INTRODUCTION

South American river dolphins, grouped in the superfamily Inioidea (sensu Muizon, 1988a), comprise two families, Iniidae and Pontoporiidae, with only one living genus each. Genus *Inia,* strictly freshwater, has two living species, *I. geofrensis* and *I. boliviensis.* Genus *Pontoporia* is monotypic (*P. blainvilei*) and lives in shallow marine environments, with some of its distribution in proximal riverine systems.

The interrelationships of this species were obscured for a long time by their inclusion in a polyphyletic group informally called "river dolphins" with the formal name Platanistoidea (sensu Simpson, 1945). Besides the South American species, this superfamily included the Ganges and Indus dolphins (genus *Platanista*, two species) and the Baiji from China (genus *Lipotes*, one species).

Gray (1863) was the first to propose a systematic arrangement for the 'river' dolphins, at that time limited to *Platanista*, *Inia*, and *Pontoporia,* (*Lipotes* was described only in the next century). He placed *Platanista* and *Inia* into separate monotypic families, Platanistidae and Iniidae, respectively. *Pontoporia*, however, was placed by Gray in the family Delphinidae. The placement of *Pontoporia* (=*Stenodelphis*) was debated since then, and many specialist considered it as a distinct group within the true marine dolphins (Kellogg, 1928; Miller, 1923). Flower (1867), in a paper describing a complete skeleton of *Inia* and a skull of *Pontoporia*, proposed a systematic scheme for the "river dolphins" dividing the family Platanistidae into two subfamilies, the Platanistinae containing only *Platanista*, and the Iniinae, containing *Inia* and provisionally *Pontoporia*. This is the first and only reference until Muizon (1985) considered *Inia* and *Pontoporia* as closely related groups.

Lipotes, the fourth genus of extant river dolphins, was described by Miller in 1918, and referred to as Iniidae. Miller (1923) followed Gray in recognizing separate families, Platanistidae and Iniidae (now consisting of *Inia* and *Lipotes*), and in including *Pontoporia* into the Delphinidae. Since Miller's 1918 description, the taxonomic relatedness between *Inia* and *Lipotes* has been generally supported. Fraser & Purves (1960) examined multiple characters of the outer and middle ear in 37 species of cetaceans, including the four genera of river dolphins. They found the presence or absence of 32 features of the cetacean ear to be identical in *Inia* and *Lipotes*. Kasuya (1973), based on a comparative analysis of the tympano-periotic bone from 313 individual cetaceans in 30 genera, also recognized *Inia* and *Lipotes* as the two modern members of the Iniidae. While the overall resemblance (particularly in skull morphology) and shared ecological habit encouraged a grouping of *Inia* + *Lipotes*, the rarity of specimens and paucity of scientific attention left considerable room for revision.

Slijper (1936) argued at length that the three nominate river dolphin subfamilies Platanistinae, Iniinae (*Inia* + *Lipotes*) and Stenodelphininae (=*Pontoporia*) belonged together in the monophyletic Platanistidae. Simpson, in his classification of mammals (1945), followed Slijper's placement of the four genera in three subfamilies (still the Iniinae consisted of *Inia* + *Lipotes*) within the single family Platanistidae, but established the superfamily Platanistoidea to recognize the overall uniqueness of the group. However, Simpson (1945) openly questioned river dolphin monophyly, noting the association was "recognized [*sic*] as a habitus character that may have risen in sharply distinct lines of descent". Since Simpson, no other systematist has argued that *Pontoporia* belongs with the Delphinidae.

Early attempts to put them in a cladistic framework showed that not a single sinapomorphy was shared by the Platanistoidea (Zhou, 1982). The paraphyletic (or, more properly, polyphyletic) nature of the group was evidenced by Muizon (1985; 1988b), who also pointed out the sister group relationship between *Inia* and *Pontoporia* and both with the Delphinida (sensu Muizon, 1988a).

This phylogenetic scheme, with relatively minor modifications, was supported by more recent morphological and molecular studies (Heyning, 1989; Heyning & Mead 1990; Fordyce, 1994; Messenger, 1994; Messenger & McGuire 1998; Cassens et al., 2000, Hamilton et al., 2001; Nikaido et al., 2001; Verma et al., 2004; Yang et al., 2002; Yan et al., 2005).

The only recent work supporting the monophyly of the traditional river dolphins group was the one by Geisler & Sanders (2003). They recovered a monophyletic Platanistoidea, including all the living genera and some selected fossil ones in a very detailed and comprehensive osteological analysis. However, the main focus of this work was the basal, archaic and poorly known Mysticeti and Odontoceti; so, it is not surprising that some of the crown groups may appear somewhat distorted.

Zhou et al. (1979) reviewed the relationship between *Inia* and *Lipotes* with an osteological comparison that included the largest number of *Lipotes* specimens yet examined, five skeletons and four skulls. The results revealed that the differences in the skeletal characters between these two species are greater than those between the Delphinidae and the Phocoenidae. Accordingly, the authors placed *Lipotes* in the monotypic family Lipotidae. Recently, Yang et al. (2005) revised the phylogenetic position of *Lipotes* based on molecular data, and also confirmed the phylogenetic distance of *Platanista* from the other "river dolphins", and arrived at the conclusion that *Lipotes* is not close to *Pontoporia* or *Inia*.

As discussed above the morphological taxonomy of river dolphins has been quite unstable. One of the reasons is because river dolphin specimens in museum collections are relatively poorly represented, especially so for *Lipotes*. They have highly restricted distributions in widely separate areas. Their disjoint distribution does not strongly support any single biogeographic hypothesis for their evolutionary relationships. A recent view, from a molecular phylogenetic approach, interpreted the "river dolphins" as survivors due to the convergent adaptation in freshwater environments (Cassens et al., 2000). Actually, this independent adaptation to freshwater environments happens also in more "modern" lineages, like Phocoenidae (*Neophocoena phocenoides*, coastal and freshwater) and Delphinidae (*Sotalia fluviatilis*, exclusive freshwater).

From the perspective of comparative morphology, each of the four monotypic families is a relatively ancient lineage with only a single, highly modified, terminal extant species. Each species displays a unique combination of primitive and derived characters, with more autoapomorphies than the shared derived characters required to discern their relatedness (Messenger, 1994). Also, the fossil record of the "river dolphins" was quite scarce until recently and even if it goes back to the Late Oligocene (Cozzuol, 1996; Fordyce & Muizon, 2001) it lacks intermediate forms that could otherwise elucidate lineage relationships.

Fossil Record

We have known about fossilized South American "river dolphins", or more properly, Superfamily Iniodea, since the middle of the 19th century (Cozzuol, 1985). Originally they were restricted to eastern Argentina ("Mesopotamiese" bone beds), but the record dramatically increased in the second half of the 20th century, with extensive records in southern Peru (Muizon, 1983; 1984; 1988a) and in the western Amazon region of Brazil (Rancy et al., 1989; Boquentin et al., 1990; Cozzuol, 1996; Cozzuol & da Silva, 1996). Unpublished records from northern South America and the reputed North American records will be discussed.

Reputed Inioids with Dubious and No-Iniod Affinities

Proinia patagonica True, 1909

The odontocete *Proinia patagonica* True, 1909 was based on an incomplete and damaged skull found at Darwin Station, Santa Cruz Province, Argentina. The holotype comes from the "Patagonian Beds", which are now referred as the Monte León Formation by modern stratigraphic nomenclature (Leonian Stage, Late Oligocene-Earliest Miocene). Its age make it the oldest reputed iniid species.

Assignment of *P. patagonica* to the family Iniidae by True (1909) was tentatively accepted by Kellogg (1928:55), although he misquoted Cabrera (1926) in suggesting that the skull belonged to *Notocetus* (=_*Diochoticus*). Cabrera (1926) in fact, had only transferred to *Notocetus,* an isolated cervical vertebra which True (1909) had previously referred to *Proinia.* Subsequent discussions of this species have been restricted to the quoted references, and no additional material has been referred to *P. patagonica* until now.

I examined a cast of the holotype (number YPM-PU15459), now deposited in the Department of Vertebrate Paleontology at the La Plata Museum as item number PU 15459-CA. (Figure 1)

The specimen consists of a partial skull with the rostrum broken off just anterior to the nares. It has been strongly compressed dorsoventrally with a result that the supraoccipital is almost in contact with the basioccipital. All the basicranial structures beyond the basioccipital were lost, as well as the bullae, periotics, jugals, right zygomatic process and right paraoccipital process. Despite the deformation, the preserved parts of the skull permit identification of the bones and the general morphology.

The specimen is divided in two parts that separated along the unclosed sutures. In fact, all the sutures of the skull are very evident, indicating a young specimen. The smaller (anterior) part is composed of the frontals, the proximal end of the ascending processes of the maxillaries and the mesethmoid bones. Both premaxillary bones are missing, but the scars they left on the inner margin of the maxillaries permit estimation of their extent.

The cranial vertex is composed of the relatively large, high, square and flat exposure of the frontals. Both nasal bones are missing, but they were sutured to the vertical anterior border of the frontals and apparently projected onto the narial passages. The posterior part of the frontals show scars left by the suture with the supraoccipital bone. Consequently the parietals are excluded from the dorsal view of the skull as in all the modern Odontoceti.

The larger fragment comprises the parietals, part of the supraoccipital, exoccipitals, left zygomatic arch, and the basioccipital. The zygomatic process is long, massive and of triangular shape in lateral view. The parietals are largely exposed laterally, and the squamosal is very short antero-posteriorly and extended dorsally. The supraoccipital is much narrower than the exoccipitals. The paraoccipital processes are triangular, broad dorsally and with a rounded end.

In spite of the fact that the skull has been deformed by compression so that some proportions are changed, most features are visible. The sutures separating the frontals and the occipital are clearly evident, as are all other sutures. The skull is symmetrical, with a relatively small brain case. A chamber for the olfactory nerves is present in the anterior part of the brain cavity. As in several primitive odontocetes and in young specimens of modern species there are two large foramina between the mesethmoid and the ectethmoid bones. All this information indicates that the skull belongs to a young specimen.

The features mentioned by True (1909) as evidence for a close relationship with *Inia* can be regarded as plesiomorphic and are present in several odontocetes. According to True (1909) the most important feature resembling *Inia* is the strongly elevated cranial vertex. However, the morphology of this area differs from that found in all known iniids. In all species of this family the cranial vertex is massive and knob-like and does not exhibit the square shaped frontals and smooth surface found in *Proinia*.

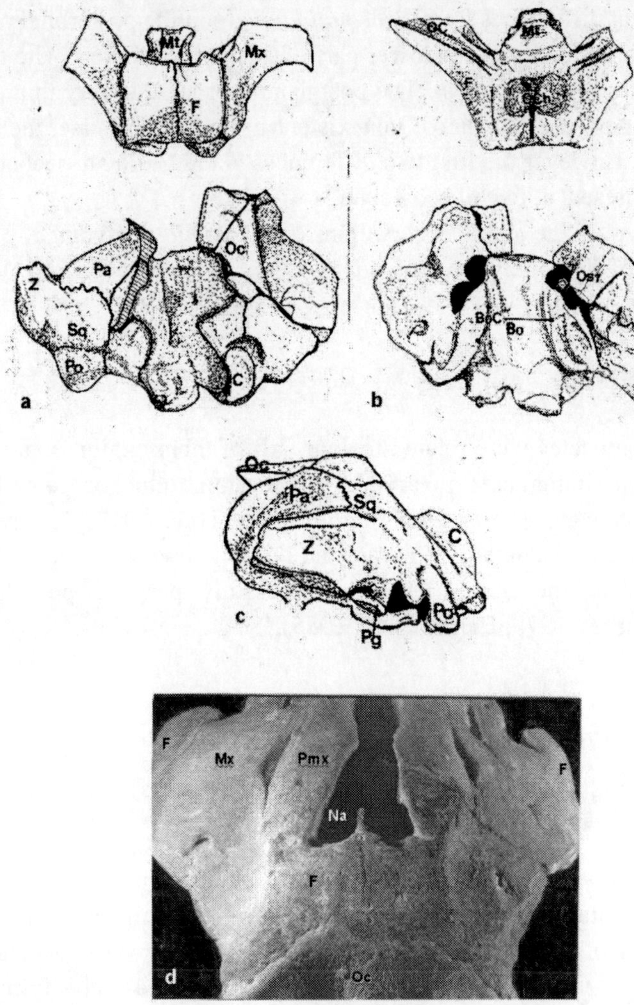

Figure 1. Holotype skull of *Pronia patagonica* (a-c) and a skull of *Prosqualodon australis* (d). a. Dorsal view, b. Ventral view, c. Lateral view, d. Dorsal view of the posterior facial surface and vertex. Bo. Basioccipital, Boc. Basioccipital crest, C. Occipital condyle, F. Frontal, G. Glenoid fossa, Mt. Mesethmoid, Mx. Maxilla, Na. Narial passage, Oc. Occipital, OC. Optic channel, Och., Olfactory chambers, Os. Orbitosphenoid, Pa. Parietal, Pg. postglenoid process, Pmx. Premaxilla, Po. Paraoccipital process, Sq. Squamosal, Z. Zygomatic process.

In fact, the holotype of *Proinia* lacks all iniid diagnostic features and most of its features are quite similar to those of *Prosqualodon australis*. One of the most important similarities with *Prosqualodon*_lies precisely in the squared and smooth frontals of the cranial vertex. Furthermore, in the skull of *Proinia* this region is relatively larger than in *Prosqualodon*; this could be due to its young age. As in *Prosqualodon* the free margin of the ascending processes of the maxilla are concave and with strong forward divergence in dorsal view. Similarly, the shape of the paraoccipital process is subtriangular, with a broad origin and a blunt end. The zygomatic process has a similar shape, a smooth to slightly convex outer-lateral face, a slight and rounded crest on the dorsal and posterior borders, broad surfaces for mandibular articulation, and no root. The squamosal is narrow and posteriorly directed. The squamosal and the posterior part of the parietal have a very convex surface without squamosal fossae.

The most important differences from *Prosqualodon* are in the postorbital processes of the frontal, the postglenoid process and the lower part of the lambdoid crest. The first are thinner and sharper that those of *Prosqualodon*. The postglenoid process is very thin and posteriorly directed. As the lambdoid crest is almost nonexistent in its lower course, the temporal fossa has no posterior limit. However the first two differences could be due to secondary distortion, and the third could represent a juvenile character.

We conclude based on the above observations, that *Proinia patagonica* is not an Iniidae, but a junior synonym of *Prosqualodon_australis*.

Other Supposed Iniodea

Several other odontocetes were related to *Inia*. All of them exhibit some characteristics actually found in *Inia,* but almost invariable are plesiomorphic or common convergent features between odontoceti. *Hesperocetus californicus* True, 1912, *Lophocetus* spp, and *Kampholophos serrulus* Rensemberg, 1969 are the better known odontoceti and all of them are presently related to the basal delphinoid (possibly para or polyphyletic) family Kentriodontiofdae (Barnes, 1978; Barnes et al., 1985).

The Position of *Parapontoporia*

Family PARAPONTOPORIIDAE Barnes, 1984

Genus Parapontoporia

The genus *Parapontoporia* Barnes, 1984 presently includes three species of exceptionally long snouted fossil odontocetes recovered from late Tertiary sediments along the west coast of California and Mexico. *Parapontoporia pacifica* Barnes, 1984 was collected from the well preserved assemblage of vertebrate fossils of the latest Miocene Almejas formation, exposed at the southeastern end of Cedros Island, off the west coast of Baja California. The holotype and only known specimen is a skull with teeth, lacking a brain case, a partial mandible that is the most complete known for the genus, and the entire rostrum, also the most complete known for this genus. *Parapontoporia wilsoni* Barnes, 1985 is a single incomplete skull recovered from sea cliffs in Santa Cruz county, California, corresponding to the lower part of the Purisima formation. This section is correlated with the Hemphillian North American land mammal age, latest Miocene, approx 6-8 Ma, and with the Almejas formation, where *P. pacifica* is found. *Parapontoporia sternbergi* (Gregory & Kellogg, 1927) is the geologically youngest parapontoporiid. Originally described as *Stenodelphis sternbergi*, the holotype section of mandibular symphysis comes from the San Diego Formation, San Diego County, California, which is correlated with the Blancan Land Mammal Age, 2-4 mya. Many additional fossils have been referred to this species, including several nearly complete skulls and mandibles, rostra lacking braincases, cranial vertices, and periotics. Barnes (1984) compared the available material and suggested the mandibular morphology of *Stenodelphis sternbergi* was sufficiently similar to assign this species to his newly described genus *Parapontoporia*, noting that *S. sternbergi* had needed a new generic allocation for some time. While each of the described parapontoporiids is known primarily by incomplete skulls that exhibit only partially overlapping morphologies, available material is sufficient to differentiate the three species.

According to the analysis of Barnes (1985), the parapontoporiids are morphologically intermediate between *Pontoporia* and *Lipotes*. He revised the taxonomy of the family Pontoporiidae accordingly. He recognized three species in the subfamily Pontoporiinae: the Late Miocene Argentine *Pontistes*, the Pliocene Peruvian *Pliopontos,* and the Recent *Pontoporia*, distinguished mainly by the apomorphy of a symmetrical cranial vertex. He erected the subfamily Parapontoporiinae for the three species of that genus, which differ from the Pontoporiinae and resemble *Lipotes* by their asymmetrical cranial vertex offset to the left side. Barnes (1985) considered several other cranial characters which diagnose this subfamily, such as braincase proportions, construction of bones around nares, and orientation of occipital shield, as intermediate between Pontoporiinae and *Lipotes*. To reflect this close relationship, he suggested a new rank and context for Lipotinae (Zhou et al., 1979) by placing it as the third subfamily of the Pontoporiidae.

Barnes' taxonomic arrangement was recognized by subsequent authors (Brownell, 1989). However, several issues cast doubt on Barnes' interpretation. The relationships among the three proposed pontoporiid subfamilies were unresolved by Barnes' analysis of cranial characters. While he stated that comparative specimens of all four extant river dolphin species, *Inia*, *Lipotes, Pontoporia,* and *Platanista* were used in formulating the descriptions and diagnoses, only the five taxa are included in his phylogeny: the three parapontoporiids, plus *Lipotes* and *Pontoporia.* A thorough methodological approach should include at least *Pontistes*, *Pliopontos* and *Inia* in any analysis of pontoporiid relationships. More recent cladistic analysis of both molecular and morphological characters are rather conclusive in establishing that, among extant cetaceans, *Inia* and *Pontoporia* are sister taxa (see below). Muizon (1985, 1988c) presents evidence that *Parapontoporia* belongs in the family Lipotidae, the sister lineage to *Inia* + *Pontoporia*. In an analysis of the auditory region of all fossil and living non-platanistid river dolphins (Muizon, 1988c), he suggested that *Parapontoporia* is allied to *Lipotes*, and not to *Inia* nor *Pontoporia.* Muizon regards characters of the auditory region of greater diagnostic value than the cranial characters that Barnes used to unite *Parapontoporia* and *Pontoporia,* some of which are subject to strong parallelisms (Muizon, 1988c). Interestingly, Barnes (1985) assigned 7 periotics found in SD formation to *Parapontoporia sternbergi* based on their resemblance to *Lipotes*. Clearly the overdue systematic analysis of the Pontoporiidae should include rigorous cladistic comparisons with the fossil parapontoporiids and *Lipotes.*

Superfamily Inioidea (Sensu Lato)

Family BRACHYDELPHIDAE Muizon, 1988c

Brachydelphis mazeasi Muizon, 1984 and Brachydelphis sp from Peru and Chile

This genus was described as the earliest member of the Pontoporiid lineage. It comes from the Pisco formation, along the southern Peruvian coast. It ranges in age from Middle Miocene (Santa Rosa level, 14-16 mya) to Late Pliocene (Sacaco level, 3.5 mya) (Muizon, 1988a). The Pisco formation is rich in both vertebrate and invertebrate marine fossils, where fossil cetaceans are represented, with over 40 taxa described.

The type species, *Brachydelphis mazeasi* was found at Cerro la Bruja level. The holotype is an almost complete cranium with both periotics and partial tympanics. Referred material includes incomplete cranial and skeletal material, including postcranial elements, and additional bones of the auditory region. The age of the Cerro La Bruja level, based on dates from volcanic tuffs and macroinvertebrates, was estimated to be approximately 12 million years.

Additional fossils referred to *Brachydelphis* were recovered from another site of the same formation, the younger level El Jahuay (calculated to be 9 Ma). Muizon (1984) described 4 periotics and a partial tympanic from this level, all which he recognized distinct enough from *B. mazeasi* to not consider them as conspecific but not enough to support the erection of a new species, so he referred to them as *Brachydelphis* sp.

Specimens referred to *B. mazeasi,* cf *Brachydelphis'* new form and cf *Brachydelphis* indeterminate were reported from the late Miocene Bahia Inglesa Formation (Gutstein et al., in press).

In the systematic discussion accompanying the type description, Muizon (1984:127) listed a suite of characters justifying the placement of *Brachydelphis* in the pontoporiid lineage. However, *Brachydelphis* possesses unique characters that Muizon used to erect the subfamily Brachydelphinae, to reflect its distinction from the other members of the Pontoporiidae - the living *Pontoporia,* and the fossil genera *Pontistes* and *Pliopontos,* described below. *Brachydelphis*, as is was originally described by Muizon (1984), has a very short rostrum, distinct among both living and fossil river dolphins, which are most often characterized by very long rostra. As it was supposed to be the earliest pontoporiid, it is significant that *Brachydelphis* possesses an asymmetrical cranial vertex. *Pontoporia* is the only living odontocete with a symmetrical cranial vertex, a character which the reconstructed skulls of *Pliopontos* and *Pontistes* also exhibited (Muizon, 1988a).

Protophocaena minima Abel, 1905

Lambert & Post (2005) reported the record of the first European Pontoporiidae, based on a restudy and reinterpretation of *Protophocaena minima* which Abel, 1905, originally identified as a Phocoenidae. They included this species in the subfamily Brachydelphinae, basically because of the asymmetrical vertex. A brief discussion of the biogeographic implications of those records is presented in their paper.

Stenasodelphis russellae Godfrey and Barnes, 2008

Godfrey and Barnes described a new genus and species of Pontoporiidae, *Stenasodelphis russellae,* from the late Miocene St. Marys Formation, in Maryland, USA. The fragmentary material resemble *Brachydelphis* in some aspects, particularly the asymmetry of the skull vertex. Despite that the authors did not refer this species to any of the subfamilies formally named, I see it as probably belonging to the Brachydelphidae.

Discussion on the phylogenetic position of Brachidelphidae

Recent phylogenetic propositions (Geisler & Sanders, 2003) suggested that *Brachydelphys* might not be a pontoporiid, but the sister group of Iniidae+Pontoporiidae (i.e. Inoidea sensu stricto).

A re-study of this genus under the light of several new specimens from Peru and Chile indicated that some of the characters used in the diagnosis of *Brachydelphis*, like the extremely short rostrum, are actually derived from the early ontogenetic age of the specimens. Also, some specimens that can be referred to this genus based on the morphology of the posterior part of the skull exhibit longer rostra compared with *B. mazeasi.* This work found additional support for the idea of excluding *Brachydelphis* from Pontoporiidae and place it as the sister group of Iniidae+Pontoporiidae and consequently obligate a reconsideration of the superfamily taxonomy (Gutstein et al, in press).

So, under the light of this observation, I do not consider Brachydelphis and allied forms as belonging to the family Pontoporiidae, but as the sister group of the remaining Iniodea. To include or not include *Brachydelphis* and allies in the superfamily Inioidea is more of a matter of choice. I think it is more economic in terms of classification to maintain it in the Inoidea and elevate the rank from subfamily to family.

Superfamily Inioidea (sensu stricto)

Family Pontoporiidae Burmeister, 1885

The family Pontoporiidae has a much richer fossil record, with greater stratigraphic and geographic range than the Iniidae. It has correspondingly received greater attention from paleocetologists. Fossil pontoporiids were found along the Pacific and Atlantic coasts of South America, in the late Miocene and Pliocene levels of the Pisco formation (Peru), the late Miocene Bahia Inglesa formation (Chile) and the late Miocene marine Paraná formation in Argentina. The isolated periotics that have been reported from several sites in North America may belong to the brachydelphid *Stenasodelphis*.

Pontistes rectifrons Burmeister, 1885

The late Miocene Paraná formation is the sedimentary source for the fossil pontoporiid *Pontistes rectifrons*. The original description was based on an almost complete skull, now badly damaged. The orbit is reduced, presumably reflecting adaptation to turbid estuarine waters. The teeth are very close set. The mandible and rostrum are strikingly depressed, uniquely so among fossil and living odontocetes. Cozzuol (1985) considered the mandibular morphology of *Pontistes* an anatomical, functional, and ecological type that had never been repeated. However, its placement as a pontoporiid has been mostly acknowledged. Cozzuol (1985) described several key synapomorphies that link *Pontistes* with *Pontoporia*, and Muizon, while remarking that *Pontistes* showed no sign of the maxillary crests characteristic of other pontoporiids (1984:61). This is a possible paedomorphic feature that places *Pontistes* in the subfamily Pontoporiinae, along with the fossil *Pliopontos* (see below) and Recent

Pontoporia (Muizon, 1988c) and emphasizes the difference with Brachydelphinae. Removing *Brachydelphis* and allied, *Parapontoporia* and *Lipotes* from Pontoporiidae makes Pontoporiinae and Pontoporiidae equivalent.

Pontistes was recorded in the Bahia Inglesa formation, Chile (Canto et al., 2002), and it is also probably present in Denmark (Pyenson & Hoch, 2007, see below).

Pliopontos littoralis Muizon, 1983

From the early Pliocene Sud-Sacaco level in the Pisco formation, dated at approximately 5 Ma, Muzion (1983) described *Pliopontos littoralis* as a large pontoporiid that he considered close to the living La Plata dolphin. The holotype is an incomplete skull, with partial rostrum and one tympanic bulla. Additional referred material includes other incomplete skulls, a sternum with almost all caudal vertebrae, other cervical and lumbar vertebrae, and other partial rostra and mandibles. The principal differences between *Pliopontos* and *Pontoporia* are the body size, orbit size, and maxillary crests.

Pontoporia sp

Remains of *Pontoporia* are relatively common in the Quaternary coastal deposits in Argentina and Brazil (Ribeiro et al., 1998; MAC, pers.obs.). Older records of this genus are reported from South America, mainly on the basis of isolated periotics (Cozzuol, 1985, 1996).

Non South American Pontoporiidae

Pyenson & Hoch (2007) reported remains of Pontoporiidae of Tortonian age from Denmark. They noted that those specimens have a symmetrical vertex, being closer to Pontoporiidea, and referred to as the best preserved specimen as cf *Pontistes*, a South American genus, and being the only pontoporids to be reported outside of South America. This is especially significant because it shows a wide distribution of pontoporids at the origin of the clade and, if the assignment to *Pontistes* is confirmed, it will also be the only South American genus recorded outside the continent.

The Interrelationships of the Pontoporiidae

The phylogenetic relationships among *Pontistes, Pliopontos* and *Pontoporia* are unresolved. Muizon (1984) does not consider that *Pliopontos* could be the direct ancestor of *Pontoporia*, due to the existence of several complex functional characters indicative of specialization for existence in the littoral environment. Rather, he suggested that the two are sister taxa, united by the apomorphic development of a crest affecting the maxillary only, and that *Pontistes* is the sister to the clade of *Pliopontos* and *Pontoporia. Pontistes,* although it has been treated as the geologically youngest of the three taxa and occurs in the same geographic area as does the living *Pontoporia.*

The late Miocene date for the Paraná formation, the provenance of *Pontistes* (at least for its upper section, the only which outcrops) is supported by the presence of several taxa, including both invertebrates and vertebrates (Zabert & Herbst, 1977; Cione et al., 2001). Del Rio (1991) pointed out that the mollusk fauna from the "Entrerriense" of northern Patagonia (formally Aonikan Stage, including, among others, the Puerto Madryn Formation) is significantly different from the "Entrerriense" of Paraná (Paranian Stage, Paraná Formation) at the level of species, indicating a probably diachronism, which agree with the evidence from vertebrates (see Cione et al., 2001; Cozzuol, 1993). More recently, Scasso et al. (2001) published radiometric dates (87Sr/86Sr) of fossilized bivalves from Patagonia ("Entrerriense") which gave an average of 10.0 +/- 0.3 My, which is the limit between middle and late Miocene.

Muizon's (1984) interpretation of the morphological characters upon which this pontoporiid phylogeny is based is questionable. For example, he lists six synapomorphies uniting *Inia* and *Lipotes* to the exclusion of the Pontoporiidae, rejecting Zhou's 1979 proposal for a monotypic Lipotidae. A year later, he reversed his position, and was the first since Flower (1869) to suggest that *Inia* and *Pontoporia* are sister taxa. However, this interpretation of river dolphin phylogeny had more taxa than characters (Muizon, 1985). If *Pontistes* and *Pliopontos* are temporally closer in age than is currently accepted, then it is plausible to argue on biogeographic grounds a sister relationship between *Pontistes* and recent *Pontoporia.*

Family INIIDAE Gray, 1863

Possibly Iniidae of uncertain affinities
Goniodelphis hudsoni Allen, 1941

This species was described on a partial skull, lacking the braincase and anterior portion of the rostrum (Allen, 1941, Figure 1). Additional referred specimens where described by Kellogg (1944). Since then, many other specimens where found, mainly rostral and mandibular fragments.

The origin of the specimens is the outcrops of the Bone Valley formation in central Florida, with a stratigraphic range from the Middle Miocene to the early Pliocene. The best known of the Bone Valley vertebrate faunas is the early Pliocene Palmetto Fauna, which has been correlated with the late Hemphillian North American Land Mammal age (Berta & Morgan, 1985). Deposition of the Palmetto Fauna sediments was during a time of sea levels 25-35 m higher than present day, in a mix of fluvial, deltaic and nearshore marine environments (Morgan, 1994).

In this original description, Allen (1941) referred *Goniodelphis* to Iniidae with some confidence, stating "The specimen proves to be of unusual interest as the first certain record of a cetacean of the family Iniidae from eastern North America". Kellogg (1944) agreed and included an extensive comparison of *Goniodelphis* with all other fossil and living Iniids (sensu Simpson) known at that time. Thus *Goniodelphis* was described by Kellogg (1944) as "somewhat similar to *Ischyrorhynchus* in the preserved portion of the type skull". However, more recently some authors consider the evidence supporting the assignment of *Goniodelphis* to the Iniidae to be inconclusive. Muizon (1988c), in an exhaustive phylogenetic analysis of both fossil and extant members of the infraorder Delphinida (defined as Lipotoidea + Iniodea + Delphinoidea; Muizon 1984), acknowledged the general similarity between *Goniodelphis* and other iniids, but stated that the available material is too incomplete to confirm this affiliation. Cozzuol (1996) followed Muizon's proposition, and considered the Iniidae as strictly endemic from the South American fresh water systems.

Morgan (1994) reviewed the marine mammal fauna of the Bone Valley formation of central Florida, from where the holotype and other specimens of *Goniodelphis* where found. Despite the doubts expressed by other authors (Muizon, 1984; Cozzuol, 1996) he continued to consider this species as an iniid. In the same paper, Morgan (1994: Figure 6A, B) described a periotic (UF 135935) he referred to as an undetermined pontoporiid. Periotics of *Pontoporia*-like species are actually present in the collection from the late Tertiary marine outcrops of the East coast of North America deposited in the National Museum of Natural History, Washington, DC (personal observation) and may well belong to *S. russellae* (see above). However, in this particular specimen, the very small posterior process, lacking the laminated posterior projection typical of Pontoporiidae and the more prominent and rounded superior process, makes this specimen very close to the periotics of other iniids. It is quite similar to one specimen from the "Mesopotamiense" of Argentina, originally considered as Pontoporiidae (Cozzuol, 1985) but lately correctly identified as Iniidae (Muizon, 1988a).

Two cranial traits can be used to identify an Iniidae. The knob-like elevated vertex and the vomer separating the palatines and proximal part of the maxillaries, being continuously exposed ventrally. Unfortunately, the vertex is not preserved in the holotype and this character as not checked in the only known skull of *G. hudsoni* (Fig. 2). In the palate region, despite the palatine bones being damaged, it is possible that the vomer is not continuously exposed ventrally, so this trait is not present. The prenarial region of *Goniodelphis* is indeed remarkably similar to other Iniidae.

The most convincing evidence of the presence of an iniid in a marine environment is based on the periotic mentioned above. If this periotic really belongs to Goniodelphis, then this genus may be included in the family, but the absence of some of the sinapomorphies shared by the other known iniids suggest it may have a basal position in the clade.

Subfamily ISCHYRORHYNCHINAE Cozzuol, 1996
Genus *Ischyrhorhynchus*

I. vanbenedeni Ameghino, 1891

The taxonomic history of the genus *Ischyrhorhynchus* and its synonyms, *Anisodelphis* and, in part, *Saurodelphis,* is quite confusing. The history is detailed by Pilleri & Gihr (1979) and further clarified by Cozzuol (1985).

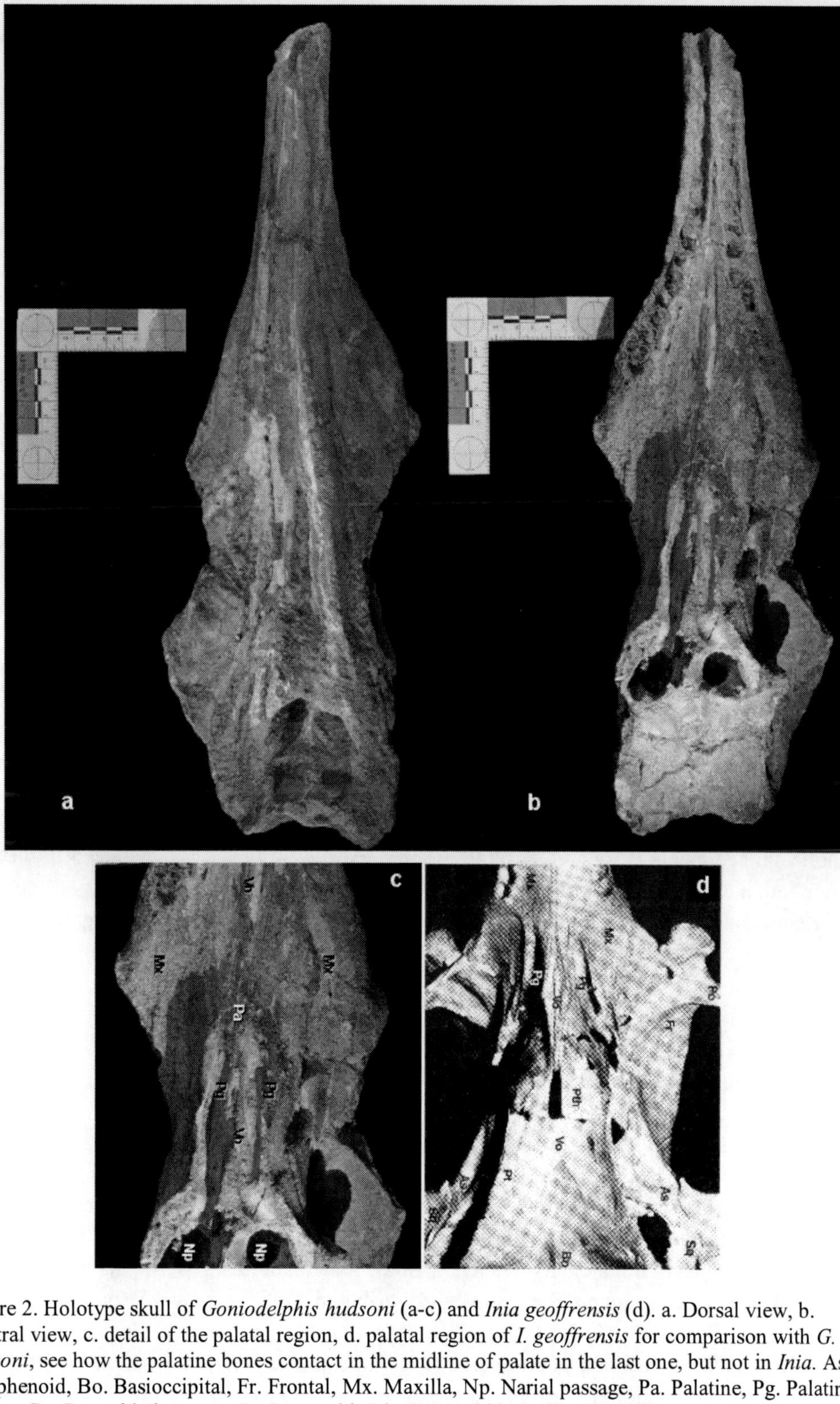

Figure 2. Holotype skull of *Goniodelphis hudsoni* (a-c) and *Inia geoffrensis* (d). a. Dorsal view, b. Ventral view, c. detail of the palatal region, d. palatal region of *I. geoffrensis* for comparison with *G. hudsoni*, see how the palatine bones contact in the midline of palate in the last one, but not in *Inia*. As. Alisphenoid, Bo. Basioccipital, Fr. Frontal, Mx. Maxilla, Np. Narial passage, Pa. Palatine, Pg. Palatine groove, Po. Postorbital process, Pt. Pterygoid, Pth. Pterygoid hamulus, Vo. Vomer.

The holotype (Ameghino, 1891) is a fragment of a rostrum with worn teeth. Additional material, including cranial and mandibular specimens, has been found and referred to this species (see Burmeister, 1891, 1892; Pilleri & Gihr, 1979; Cozzuol, 1985).

Descriptions of the skull published by Burmeister (1891, 1892) of *Ischyrhorhynchus vanbenedeni* were actually composites of fossil fragments from *I. vanvenedeni* and *Saurocetes argentinus*, then sketched in completion with the known structure of *Pontoporia.*

Ischyrhorhynchus vanbenedeni is characterized by a very long, laterally compressed beak, a pneumaticized maxillary crest, and very small orbits. These characters, all of which are convergent with the living Ganges river dolphin *Platanista gangetica*, have been interpreted as adaptations to a highly turbid fluvial environment (Cozzuol, 1985, 1988). Pilleri & Gihr (1979) reconstructed the sonar field of *Ischyrhorhynchus* and suggested that, it had a pattern of sound production and behavior similar to the living *Platanista,* that is a side-swimming cetacean with rapid scanning movements of the head. For the purposes of understanding river dolphin phylogeny and evolution, the importance of *Ischyrhorhynchus* is that it has been widely accepted as a valid member of the family Iniidae, and that this Late Miocene form was already highly specialized for existence in the freshwater riverine environment.

Fossils of *I. vanbenedeni* have been recovered originally from the fluvial Ituzaingó Formation, along the southeastern cliffs of the Paraná River in Entre Rios Province, Argentina, but it was also recorded in the Late Miocene Solimões Formation in the State of Acre, northern Brazil (Boquentin et al., 1990; Cozzuol, 2006, personal observation).

The fossil assemblage, specially vertebrates, in both localities are very similar, with several species in common. They are dated as Late Miocene and include freshwater fishes, aquatic and terrestrial turtles, Aligatoriidae, Gavialiidae and Netosuchidae crocodiles, birds and land mammals (see Cione et al., 2001; Cozzuol, 1993; Rancy et al., 1989, Cozzuol, 2006). The marine influence is stronger in the Entre Ríos (Argentina) deposits, which is clear from the underlying marine Paraná Formation and from the presence of marine fishes in the continental sediments (Cione et al., 2001). Some marine influence was detected in the Acre region too, but its importance is controversial (Cozzuol, 2006).

Remains referred to *Ischyrorhynchus* were reported originating from the contemporaneous deposit from Venezuela (Urumaco Formation; Linares, 2004).

Genus *Saurocetes*
S. argentinus Burmeister, 1871

The classification history of this species and its synonyms, has also been quite confusing, as detailed by Pilleri & Gihr (1979). The holotype of the first described species, *Saurocetes argentinus* Burmeister, 1871 is a mandibular fragment, still containing teeth. Its stratigraphic origin was described only as Tertiary outcrops of the eastern bank of the Paraná River. Subsequently, much fragmentary cranial, rostral, and mandibular fossil material has been collected from the Ituzaingó Formation and attributed to *Saurocetes argentinus* (Pilleri & Gihr 1979; Cozzuol, 1985, 1988). *S. argentinus* is principally distinguished from *Ischyrhorhynchus* by a much lesser degree of lateral compression in the mandible and rostrum and by distinct tooth morphology consisting of a very wrinkled enamel, anterior and posterior carinae, and sharp and posteriorly recurved apex.

Some authors expressed reasonable doubts about the iniid relationship of this species because the absence of more complete cranial material (Muizon, 1988a). However, despite that, the evidence from the palatal region of a preserved specimen support that *S. argentinus* has an important iniid synapomorphy because its vomer is exposed all along the palate, preventing the palatines to meet in the midline of the palate as in other odontocetes (Zhou, 1981), which support the hypothesis that it belongs to the family Iniidae (Cozzuol, 1985).

As with *I. vanbenedeni*, this species was originally found in the Late Miocene outcrops of the Paraná River in the Entre Rios Province, but subsequently remains of this species were found in the Late Miocene Solimões Formation in the Acre State, northwestern Brazil (Rancy et al., 1989; Cozzuol, 2006)

S. gigas Cozzuol, 1989

Based on a detailed review of all material identified as *Saurocetes* in the Entre Rios Museum and the La Plata Museum, a second species, *S. gigas,* was erected by Cozzuol (1989). The type specimen is a proximal fragment of the mandibular symphysis without teeth, originating from the Ituzaingó Formation. He also attributed to *S. gigas* five isolated teeth that originated from these same deposits. His justification for establishing a second species is based primarily on the large relative size of the attributed material, after rejecting the possibility of size variation in *Saurocetes argentinus.* Other minor morphological differences between the teeth of *S. argentinus* and *S. gigas* are considered as support for erecting a second species. Although Cozzuol (1989) noted that other paleocetologists discourage the establishment of new species based on isolated mandibular fragments, he considered that the results of comparing available material justified the establishment of *S. gigas.* Contrarily, the other species has yet to be found outside the Paraná outcrops nor has further material for this species been found.

The significance of this species relies is its large body size, estimated from the length of a skull (1 meter) and according the proportions for *Inia* spp, to be about 4.5 meters in total length. This is an uncommon size for a top, warm blooded predator in continental waters, reinforcing the perception that the inland aquatic environment during the Late Miocene in those areas was uncommon, extremely diverse and rich.

Subfamily INIINAE Cozzuol, 1996
Plicodontinia mourai Miranda Ribeiro, 1938

Plicodontinia mourai was described by Miranda Ribeiro (1938) based on a single tooth found in the Acre region of Brazil, probably from Pleistocene deposits. Muizon (1988c) described this specimen as "totally inadequate to define an Odontocete and should be regarded as "*incertae sedis*" (Muizon 1988c).

Although I agree with Muizon's statement, the holotype tooth has some features that deserves to be considered. It is a posterior tooth, probably an upper one, showing the postero-lingual platform characteristic of the genus *Inia*, the only known representative of the subfamily Iniinae. Consequently, it belongs to a species of the subfamily Iniinae, maybe even to the genus *Inia.*

Inia sp nov

From Pleistocene deposits of the Rondônia State, Brazil, a partial skull of *Inia* was found and is currently under study by me and Vera da Silva (INPA, Manaus, Brazil) (see Cozzuol & da Silva, 1996; Cozzuol, 1999). The specimen comes from the Rio Madeira Formation (Quadros et al., 2006) with a radiocarbon date of about 45,000 years. Today the section of the Madeira River from where this and many other fossil vertebrates comes from is characterized by a series of more than twenty rapids, some of them very strong, acting as effective barriers for *Inia* and *Sotalia* (also present down-river from the last rapid). The documented presence of *Inia* (as well as *Trichechus,* Cozzuol, 1999) in the area where today it is almost absent (personal observation), indicate a significant degree of environmental change (Rizzotto et al., 2006).

In the ongoing study of this specimen, a multivariate analysis of several cranial measures compared it with *Inia geoffrensis geoffrensis*, *I. geoffrensis humboldtiana* and *I. boliviensis* and it widely differentiated it from all the living species and subspecies, suggesting that it may represent the ancestral stock from which the living arose. As the hydrological regime and the climate in the areas changed, and the rapids became established, *Inia* populations above and below the rapids were isolated which triggered the allopatric speciation leading to the modern species.

INIIDAE: SUMMARY AND CONCUSSIONS

Cozzuol (1996) proposed a phylogenetic scheme for fossil and extant Iniids, including both species of *Saurocetes*, *Ischyrhorhynchus vanbenedeni,* and *Inia geoffrensis,* utilizing *Pontoporia blainvillei* as an outgroup. He suggested that *Saurocetes* and *Ischyrhorhynchus* were more closely related to each another than to *Inia.* He erected a new subfamily of the Iniidae, Ischyrhorhynchinae, to reflect this view of iniid phylogeny. The co-occurrence of *Saurocetes* spp. and *Ischyrhorhynchus* in the late Miocene freshwater Ituzaingó deposits suggests that these species somehow ecologically divided their fluviodeltaic environment. Cozzuol (1989, 1996) suggested that differences in size and tooth morphology may reflect alternate dietary habits that allowed the coexistence of these species of freshwater odontocetes.

The presence of two of the species found in the Argentinean deposits in sediments of the Solimões Formation in Acre, northern Brazil and, probably, in Venezuela, indicate a continuous aquatic connection between those regions at this time.

Up to now no Iniinae were recorded as being in existence prior to the Pleistocene. Cozzuol (1996) suggested that this subfamily should always be restricted to northern South America, where the records are limited. However, as more specimens are being recovered in those regions, if the absence of Iniinae persists, the hypothesis should be revised.

Up to now, with the sole, probable exception of *Goniodelphis hudsonni*, all the iniids have been found in freshwater environments. *G. hudsonni* represents a challenge of previous ideas of iniid biogeography (Cozzuol, 1996). However, this is not the only record in the Gulf region of a South American fresh water species, since remains of *Ribodon,* a Late Miocene Trichechidae were located in Argentina and the Peruvian and Brazilian Amazon as well as in

Early Pliocene deposits of North Carolina (Domning, 1982). It seems that some kind of connection existed during Late Miocene and/or Early Pliocene between the South American freshwater and Mexican Gulf faunas.

The Lipotoidea and their Relationships with Inioidea

The taxonomic status of the genus *Parapontoporia*, which has been assigned to the Pontoporiidae (Barnes 1985) or the Lipotidae (Muizon 1988c), is controversial. All the species of this genus are found in the northern hemisphere along the eastern margin of the Pacific, a location biogeographically intermediate between the present geographic distributions of *Lipotes* and *Pontoporia.* No other pontoporiids were found in this region.

The Chinese river dolphin, *Lipotes vexillifer*, historically has been linked to the South American dolphins. In the original description, Miller (1918) suggested an overall morphology of *Lipotes* similar to *Inia* and that their shared fluvial habitat was a basis for a close relationship. In an influential summary of cetacean biology, Kellogg (1928) followed Miller (1918) in grouping *Lipotes* together with *Inia* in the Iniidae. Barnes described the genus *Parapontoporia* (1978, 1985) as morphologically and biogeographically intermediate between *Pontoporia* and *Lipotes,* which he classified as sister taxa recognizing three subfamilies in Pontoporiidae: Pontoporiinae, Parapontoporiinae and Lipotinae. However, more recent morphological analyses (Muizon, 1988a; Messenger & McGuire, 1998) and multiple molecular studies (Cassens et al., 2000; Hamilton et al., 2001; Nikaido et al., 2001) do not support either of these views. While *Inia* and *Pontoporia* are now established as sister taxa, their relationship to *Lipotes* remains uncertain.

Comparative nucleic acid sequence analysis of both mitochondrial and nuclear genes failed to resolve the exact position of *Lipotes* (Cassens et al., 2000; Hamilton et al., 2001). Two SINE insertions that are shared by *Inia* and *Pontoporia* were not found in *Lipotes* (Nikaido et al., 2001). While multiple independent retroposon insertions are convincing support for the monophyly of *Inia+Pontoporia*, the current SINE data demonstrate only the monophyly of the clade (*Lipotes* (*Inia+Pontoporia*) Delphinoidea).

However, only two possible evolutionary histories are plausible for *Lipotes* and allies given the available data. Either *Lipotes* and allies are an independent early offshoot of the stem leading to (Brachydelphidae (Iniiidae+Pontoporiiidae))+Delphinoidea, or Lipotidae, Brachydelphidae, Iniidae, and Pontoporiidae share a unique common ancestry. There is some evidence supporting a monophyletic clade of (Lipotidae (Brachydelphidae (Iniidae+Pontoporiidae))).

Heyning (1989), in a detailed phylogenetic analysis of odontocete facial anatomy, grouped the three living genera of those families together in a single clade sister to the Delphinoidea. However, within this clade remained an unresolved trichotomy. As noted above, most molecular genetic studies do not statistically support a unique shared ancestry between *Lipotes* and the *Inia* + *Pontoporia* clade. One exception is the recent study by Nikaido et al. (2001). In one portion of this study, the authors analyzed almost 3 kb of nuclear sequences that flanked retroposon insertions in 14 cetaceans, including these three river dolphins. Their results find strong support for a monophyletic *Lippotes* + (*Inia* +*Pontoporia*).

The morphological and molecular data, although inconclusive, seems to favor a monophyletic grouping of (Lipotoidae (Brachydelphidae (Iniidae+Pontoporiidae))) as sister the Delphinoidea. However, a monophyletic history for the three river dolphins will remain speculative until more data (molecular and of recent and fossil morphology) will be available. Given the failure of extensive amounts of both nuclear and mitochondrial sequence data to resolve the position of *Lipotes* (Cassens et al., 2000; Hamilton et al., 2001), more robust genetic techniques will be required if a molecular approach to this issue is desired. For example, complete mitochondrial genome comparisons or SINE insertion analysis might prove to be useful. Also, more fossils and more complete specimens of each clade, particularly for Lipotidae, are needed to resolve the phylogeny of this group.

Considerations about the Evolutionary Process and Paths that Originated the Inioid Clades

After the removal of *Brachydelphis* and allied forms from Pontoporiidae, this family (Pontoporiidae) became much more homogeneous. The position of Brachydelphidae as a sister group of the other inoids help to explain some difficult points in the inioid evolution.

The time of divergence between Pontoporidae and Iniidae was estimated by molecular methods based on the age of *Brachydelphis*, considered as the oldest pontoporiid and, consequently, the first document of the divergence (Banguera et al., 2002, Hamilton et al., 2001, Nikaido et al., 2001). In the case of Iniidae, the fossil record and the age estimation for both, the origin of the family and the genus *Inia,* as well as the divergence of the living species seems to be too old. However, if Brachydelphidae are not pontoporiids, the time of divergence should be recalculated.

Also, morphologically, Brachydelphidae seem to represent an interesting "departing point" for the other two lineages. Under this interpretation it is possible to view the divergence between Iniidae and Pontoporiidae (besides a habitat shift from marine to freshwater for iniids), as a heterochronic event, with pontoporiids taking a paedomorphic pathway and iniids a peramorphic one. Pontoporiids' skulls have more rounded and relatively larger brain-cases, with less pronounced crests, non-prominent vertexes, smaller teeth and delicate zygomatic and postorbital processes. Iniids, in contrast, exhibits relatively smaller brain-cases and more pronounced crests and prominences, knob-like vertexes, larger teeth, and more robust zygomatic processes. Between pontoporiids it seems that *Pontistes* show a more advanced degree of paedomorphosis compared to the intermediate *Pontoporia* and least advanced *Pliopontos.* Looking at iniids, sufficiently complete skulls are known only for *Ischyrorhynchus* and, of course, *Inia.* The first one, seems to have a more peramorphic skull, but only marginally compared to *Inia*.

Moreover, this evolutionary path seems to not be restricted to only anatomical characters. In the living representatives, this can be observed in the life history traits too, with shorter lifespan, earlier sexual maturity and first pregnancy age, earlier weaning and shorter independence time for offspring in *Pontoporia* relative to *Inia.*

Actually, this heterochronic controlled divergent evolutionary path may also be present in other odontocete groups, like sperm whales *(Physeter* vs *Kogia)* and delphinoids (Delphinidae vs Phocoenidae) and even at the family level within the three delphinoid families.

CONCLUSION

The extant *Inia* and *Pontoporia* taxa are the only living members of two families that were more diverse during the Neogene. The fossil records for these taxa are summarized in Figure 3, and the geographic distributions of these records are in Figure 4.

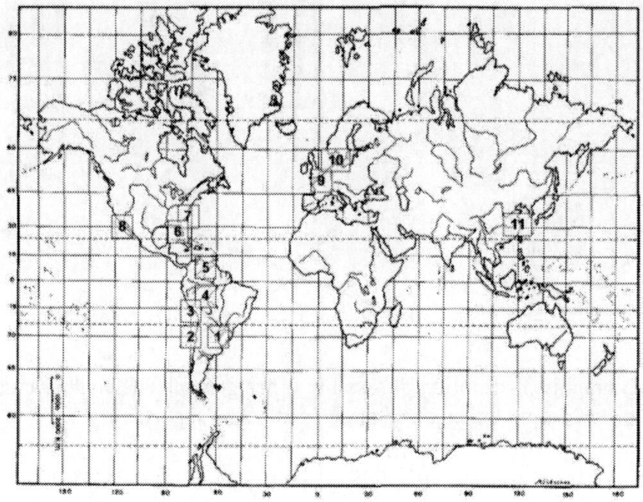

Figure 3. Map showing the distribution of the records of fossil Inioidea. 1. *Pontistes* (Late Miocene Parana Formation) and *Ischyrorhyncus, Saurocetes* and cf *Pontoporia* (late Miocene "Mesopotamian" beds), Entre Rios Province, Argentina and Uruguay. 2. *Brachydelphis, Pliopontos* and *Pontistes* (late Miocene-Pliocene Bahia Inglesa Formation), Chile. 3. *Brachydelphis* and *Pliopontos* (late middle Miocene-early Pliocene Pisco Formation), southern Peru. 4. *Ischyrorhyncus* and *Saurocetes* (late Miocene Solimões Formation) Acre state, Brazil. 5. cf *Ischyrorhyncus* and cf *Saurocetes* (late Miocene middle and upper levels of Urumaco Formation) Venezuela. 6. *Goniodelphis* (latest Miocene-early Pliocene Bone Valley Formation), Florida, USA. 7. *Stenasodelphis* (early late Miocene St. Marys Formation), Calvert County, Maryland, USA. 8. *Parapontoporia* (latest Miocene, Almejas Formation, Cedros Island, Mexico; latest Miocene Purisima Formation and late Pliocene San Diego Formation, California, USA); 9. *Protophocaena* (late Miocene Breda Formation and 'Boldérien d'Anvers'), Belgium and Netherlands. 10. cf *Pontistes* and Pontoporiidae indeterminate (Late Miocene Gram Formation), Denmark. 11. *Prolipotes* (indeterminate Miocene), China.

Fossils clearly recognized as iniids are restricted to freshwater deposits of southern South America, including areas where *Inia* do not occur today. These late Miocene fossils exhibit morphology that suggests significant evolutionary adaptation to turbid fluviodeltaic environments. The early Pliocene *Goniodelphis* of Florida may well be part of the iniid lineage and the only marine one. But it seems to be an early offshoot. All other reputed marine iniids are clearly not part of the family or even superfamily.

Pontoporiids had a wider geographic and ecologic range than the extinct iniids. The Brachydelphinae, formerly considered as pontoporiids, are removed from this family, elevated to the rank of family, and considered here as the sister group of the Iniidae+Pontoporiidae clade. Over the last few years pontoporiid and brachydelphid dolphins have been reported for the north Atlantic, including at high latitudes in the Netherlands and in Denmark.

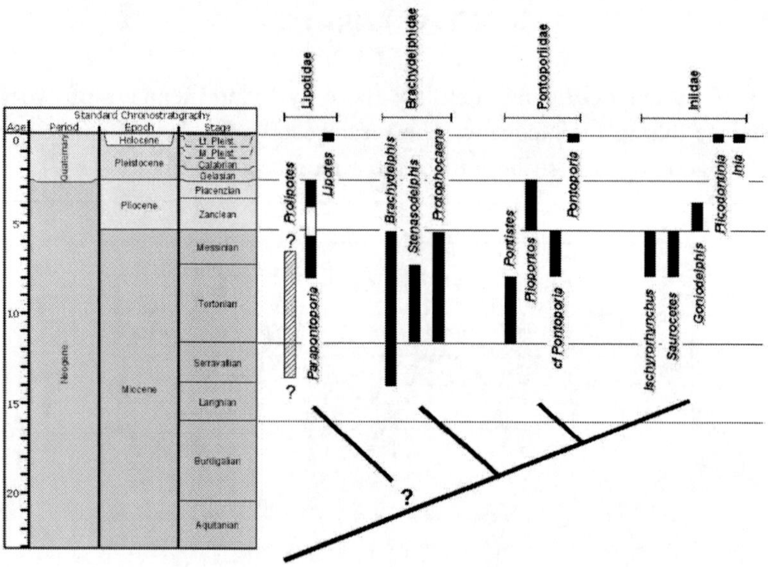

Figure 4. Chronostratigraphic distribution of fossil Inioidea genera. The cladogram at the bottom of the figure represent the proposed phylogenetic relationships of the superfamily.

The phylogenetic position of Lipotidae, including *Lipotes*, the poorly known *Prolipotes* and the pontoporiid-like *Parapontoporia*, cannot be unambiguously determined. Evidence suggests it is the sister group of the clade composed by Brachydelphidae + (Iniidae + Pontoporiiidae), and so it may be included in the superfamily Inioidea, or even as the sister group of the clade ((Brachydelphidae (Iniidae+Pontoporiiidae)) Delphinoidea), which may grant it a superfamiliar rank.

ACKNOWLEDGMENTS

I am thankful to Dr. Manuel Ruiz-García for the invitation to write this chapter. The photographs of the holotype of *Goniodelphis hudsoni* was kindly provided by the Museum of Comparative Zoology, Harvard University.

REFERENCES

[1] Allen, G. (1941). A Fossil River Dolphin From Florida. *Bulletin of the Museum of Comparative Zoology,* 89: 3-11.

[2] Ameghino, F. (1891). Mamíferos y aves fósiles Argentinas: Especies nuevas, adiciones, y correciones. *Revista Argentina de Historia Natural,* 1(3), 240-259

[3] Banguera-Hinestroza, E., Cardenas, H. Ruiz-García, M. Marmontel, M., Gaitán, E., Vázquez, R., & García Vallejo, F. (2002) Molecular identification of evolutionarily significant units in the Amazon River Dolphin *Inia* sp. (Cetacea: Iniidae). *The Journal of Heredity*, 93, 316-322.

[4] Barnes, L.G. (1978). A review of *Lophocetus* and *Liolithax* and their relationships to the delphinoid family Kentriodontidae (Cetacea: Odontoceti) *Natural History Museum of Los Angeles County, Science Bulletin*, 28, 1-35.

[5] Barnes, L. G. (1984). Fossil odontocetes (Mammalia: Cetacea) from the Almejas Formation, Isla Cedros, Mexico. *Paleobios*, 42, 1-46.

[6] Barnes, L. G. (1985). Fossil pontoporiid dolphins (Mammalia: Cetacea) from the Pacific coast of North

[7] America. *Contributions to Science, Natural History Museum of Los Angeles County*, 363, 1-34.

[8] Barnes, L.G., Domning, D. & Ray, C. (1985). Status of studies on fossil marine mammals. *Marine Mammal Science*, 1(1), 15-53.

[9] Boquentin Villanueva, J., De Muizon, C.H. & Pereira De Souza Filho, J. (1990) Novos achados de cetáceos longirrostros no Neógeno do Acre, Brasil. *Boletim do Museu Paraense Emílio Goeldi Série Ciências da Terra,* 2, 59-64.

[10] Brownell, R.L. Jr. (1989). Franciscana, *Pontoporia blainvillei.* In Ridgway, S.H. & R. Harrison (Eds.), Handbook of Marine Mammals, Vol. 4. (pp. 45-67). San Diego, CA: Academic Press.

[11] Burmeister, G. (1885). Examen crítico de los mamíferos y reptiles fósiles denominados por D. Augusto

[12] Bravard y mencionados en su obra precedente. Anales del Museo Nacional de Historia Nacional de Buenos Aires, 3, 93-174.

[13] Burmeister, G. (1891). Nuevos objetos del Museo Nacional. La Prensa, Buenos Aires 22(6690), 4-10.

[14] Burmeister, G. (1892). Continuación a las adiciones al examen crítico de los mamíferos fósiles terciarios. *Anales del Museo Nacional de Buenos Aires*, 3, 401-488.

[15] Cabrera, A. (1926). Cetáceos fósiles del Museo de la Plata. *Revista Museo de La Plata*, 29, 363-411.

[16] Canto, J., Yanez, J., & Cozzuol, M.A., (2002). Mamíferos marinos neógenos de la Formación Bahía Inglesa, Caldera, III Región, Chile. In *Primer Congreso Latinoamericano de Paleontologia de Vertebrados, Resumenes Primer Congreso Latinoamericano de Paleontologia de Vertebrados. Santiago de Chile,* 2002, 22-23

[17] Cassens, I, Saverio, W., Wadell, V. G., Balchowsky, H., Van Belle, D., Ding, W., Fan, C., Mohan, R. S. L.,Simoes-Lopes, P.C., Bastida, R., Meyere, A., Stanhope, M. J. & Milinkovitch, M. (2000) Independent adaptation to riverine habitats allowed survival of ancient lineages. *Proceedings of the National Academy of Sciences*, 97, 11343-11347.

[18] Cione, A., L, Azpelicueta, M. M., Bond, M., Carlini, A. A., Casciotta, J. R., Cozzuol, M. A., De La Fuente,M., Gasparini, Z., Goin, F. J., Noriega, J., Scillato-Yane, G. J, Soibelson, L., Tonni, E. P., Verzi, D., & Vucetich, M. G., (2000) Vertebrates in the Neogene of Paraná, Entre Rios, Argentina. In F. G. Aceñolaza & E. R. Herbst (Eds.), *Temas del Neógeno de Argentina. Serie de Correlación Geológica* 14, Buenos Aires, Argentina: *Instituto Superior de Correlación Geológica*

[19] Cozzuol, M.A. (1985) The Odontoceti of the "Mesopotamiense" of the Paraná river ravines [Argentina]:Systematic review. *Investigations on Cetacea*, 17, 39-54.

[20] Cozzuol, M.A. (1988) A new species of *Saurodelphis* Burmeister 1891 (Cetacea: Iniidae) from the Mesopotamiense (Late Miocene-Early Pliocene) of the Entre Rios Province, Argentina. *Ameghiniana,* 25, 39-45.

[21] Cozzuol, M. A. (1993). *Mamiferos aquáticos del Mioceno tardío de Argentina. Sistemática, evolucion y biogeografia. Doctoral Thesis* (unpublished). La Plata, Argentina: Universidad Nacional de La Plata, Facultad de Ciencias Naturales y Museo.

[22] Cozzuol, M.A. (1996) The record of aquatic mammals in southern South America. *MunchnerGeowissenschaftliche Abhandlungen*, 30, 321-342.

[23] Cozzuol, M.A. (1999). Mamíferos acuáticos y la antigüedad de los depósitos cuaternários del Alto Río Madeira (Rondônia, Brazil). *In Include editor names here Evolución Neotropical del Cenozoico*, Programa y Resumenes Congreso de Evolución Neotropical del Cenozoico, 18. La Paz, Bolivia. Include the Publisher name here. Also, include the page number of the chapter referenced.

[24] Cozzuol, M.A. (2006). The Acre vertebrate fauna: diversity, geography and time. *Journal of South American Earth Sciences*, 21(3), 185-203.

[25] Cozzuol, M. A., & Silva, V. M. F., (1996). Un Registro de *Inia* sp fósil del Pleistoceno tardio-Holoceno del alto Rio Madeira, Rondonia, Brasil. *7º Reunión de Trabajos de Especialistas en Mamíferos Aquáticos de América del Sur, Resúmenes VII Reunión de trabajos de especialistas en Mamíferos Acuáticos de América del Sur*. Viña del Mar, Chile: Include publisher name.

[26] Del Rio, C.J. (1991). Molúscos "entrerrienses" (Mioceno madio) de la República Argentina: Paleontología, Estratigrafía y Paleoclimatología. *XII Congreso Brasileiro de Paleontologia, San Pablo, Brazi*l,.

[27] Domning,, D.P. (1982). Evolution of Manatess: a speculative history. *Journal of Paleontology*. 56, 599-619.

[28] Flower, W. H. (1867). Description of the skeleton of *Inia geoffrensis* and the skull of *Pontoporia blainvilli*, with remarks on the systematic position of these animals in the Order Cetacea. *Transactions of the Zoological Society of London*, 6, 87-116.

[29] Fordyce, R. E. (1994). *Waipatia maerewhenua*, new genus and new species (Waipatiidae, new family), an archaic Late Oligocene dolphin (Cetacea: Odontoceti: Platanistoidea) from New Zealand. *Proceedings of the San Diego Museum of Natural History*, 29, 147-176.

[30] Fordyce, R. E. & Barnes, L. G., (1994). The evolutionary history of whales and dolphins. *Annual Review of Earth and Planetary Sciences,* 22, 419-455

[31] Fordyce, R. E. & de Muizon, C. (2001). Evolutionary history of cetaceans: a review. In J. M. Ma Mazin & V. de Buffrenil (Eds.), *Secondary Adaptation of Tetrapoda to Life in Water* (pp.169-233). Paris, France.

[32] Fraser, F. C. & Purves, P. E. (1960). Hearing in Cetaceans. *Bulletin of the British Museum (Natural History)Zoology,* 7(1), 1-140

[33] Geisler, J.H. & Sanders, A.E. (2003). Morphological evidence for the phylogeny of Cetacea. *Journal of Mammalian Evolution*, 10(1-2), 23-129

[34] Godfrey, S.J. & Barnes, L.G. (2008). A new genus and species of late Miocene pontoporiid dolphin (Cetacea: Odontoceti) from the St. Marys Formation in Maryland. *Journal of Vertebrate Paleontology,* 28(2), 520–528

[35] Gray, J.E. (1863). On the arrangement of the cetaceans. *Proceedings of the Zoological Society of London, 3,* 197-202

[36] Gregory, W. K.& Kellogg, R., (1927). A fossil porpoise from California. *American Museum Novitates,* 269, 1-7

[37] Gutstein, C.S., Cozzuol, M.A., Vargas, A.O., Suarez, M.E., & Schultz, C.L. (In press). Patterns of skull variation of *Brachydelphis* (Cetacea, Odontoceti) from the Neogene of the Southeastern Pacific. *Journal of Mammalogy*, 90(2).

[38] Hamilton, H., Caballero, S., Collins, A.G. & Brownell, R.L. Jr. (2001) Evolution of river dolphins. *Proceedings of the Royal Society of London, Series B: Biological Sciences*, 268, 549-556.

[39] Heyning, J. E. (1989). Comparative facial anatomy of beaked whales (Ziphiidae) and a systematic revision among the families of extant Odontoceti. *Contributions in science, Natural History Museum of Los Angeles County*, 405, 1-64.

[40] Heyning, J. E. & Mead, J. G., (1990). Evolution of the nasal anatomy of cetaceans. In J.Thomas & R.

[41] Kastelein (Eds.), *Sensory abilities of cetaceans (*pp. 67-79). Plenum, NY.

[42] Kasuya T. (1973). Systematic consideration of recent toothed whales based on the morphology of tympano-periotic bone. *Scientific Reports of the Whales Research Institute,* 25, 1-103.

[43] Kellogg, R. K. (1928). The history of whales – their adaptation to life in the water. *Quarterly Review of Biology,* 3, 29-76.

[44] Kellogg, R.K. (1944). Fossil cetaceans from the Florida Tertiary. Bulletin of the *Museum* of Comparative Zoology, 94(9), 433-471.

[45] Lambert, O., & Post, K. (2005). First European pontoporiid dolphins (Mammalia: Cetacea, Odontoceti) from the Miocene of Belgium and The Netherlands. *Deinsea,* 11, 7–20.

[46] Linares, O. (2004). Bioestratigrafia de la fauna de mamíferos de las formaciones Socorro, Urumaco y Codore (Mioceno-Plioceno temprano) de la región de Urumaco, Falcón, Venezuela. *Paleobiologia Neotropical*, 1, 1-26.

[47] Messenger, S. L. (1994). Phylogenetic relationships of platanistoid river dolphins (Odontoceti, Cetacea):Assessing the significance of fossil taxa. *Proceedings of the San Diego Society of Natural History,* 29, 125-133.

[48] Messenger, S. L. & McGuire, J. A., (1998). Morphology, molecules, and the phylogenetics of Cetaceans. *Systematic biology*, 47, 90-124.

[49] Miller, G. S. (1918). A new river-dolphin from China. *Smithsonian Miscellaneous Collections,* 68, 1-12

[50] Miller, G. S. Jr. (1923). The telescoping of the cetacean skul*l. Smithsonian Miscellaneous Collections,76,1-70.*

[51] Miranda-Ribeiro, A. (1938). *Plicodontinia mourai. In Livro jubilar do professor Lauro Travassos: editado para commemorar o 25* anniversario de suas actividades scientificas* (pp. 319-321). Rio de Janeiro, Brazil: Typographia do Instituto Oswaldo Cruz

[52] Morgan, G.S. (1994). Miocene and Pliocene marine mammal faunas from the Bone Valley formation of central Florida. *Proceedings of the San Diego Society of Natural History,* 29, 239-268

[53] Muizon, C. de. (1983). *Pliopontos littoralis*, un noveau Platanistidae Cetacea du Pliocène de la côte pèruvienne. *Comptes Rendus. Académie des Sciences de Paris, Série III,* 296, 625-628.

[54] Muizon, C. de. (1984). *Les vertébrés fossiles de la Formation Pisco (Pérou). Deuxieme partie: LesOdontocètes (Cetacea, Mammalia) du Pliocène inférieur de Sud-Sacaco.* Institut Français d'Études Andines, *Mémoire,* 50, 1-188.

[55] Muizon, C. de. (1985). Nouvelle données sur le diphylétisme des Dauphins de riviére (Odontoceti, Cetacea,

[56] Mammalia). *Comptes Rendus Academie Sciences de Paris*, Serie II, 301(5), 359-362

[57] Muizon, C. de. (1988a). Les relations phylogénétiques des Delphinida (Cetacea, Mammalia). *Annales dePaléontologie*, 74, 159-227.

[58] Muizon, C. de. (1988b). Le polyphylétisme des Acrodelphidae, odontocètes longirostres du Miocène européen. *Bulletin du Muséum national d'Histoire naturelle* C4 Ser, 10, 31-88

[59] Muizon, C. de. (1988c). *Les vertébrés fossiles de la Formation Pisco (Pérou). Troisième partie: lesodontocètes (Cetacea, Mammalia) du Miocène.* Institut Français d'Études Andines *Mémoire*, 78, 1-244.

[60] Nikaido, M., Matsuno, F., Hamilton, H., Brownell Jr., R. L. Ying, C., Wang, D., Zhu, Z. Y., Shedlock, A.

[61] M., Fordyce, R. E., Hasegawa, M., & Okada, N. (2001). Retroposon analysis of major cetacean lineages: the monophyly of the toothed whales and the paraphyly of river dolphins. *Proceedings of the National Academy of Sciences,* 98, 7384–7389.

[62] Pilleri, G. & Gihr, M. (1979). Skull, sonar field and swimming behavior of *Ischyrorhynchus vanbenedeni* (Ameghino 1891) and taxonomical position of the genera *Ischyrorhynchus*, *Saurodelphis*, *Anisodelphis* and *Pontoplanodes*. *Investigations on Cetacea*, 10, 17-70.

[63] Rancy, A., Boquentin Villanueva, J., Pereira De Souza Filho, J., Santos, J.C.R. & Negri, F.R. (1989). Lista preliminar da fauna do Neógeno da regiáo oriental do estado do Acre, Brasil (Material depositado em Rio Branco). *VII Jornadas Argentinas de Paleontología de Vertebrados, Buenos Aires, Ameghiniana* 26(3-4), 249.

[64] Ribeiro, A. M., Drehmer, C. J., Buchmann, F. S. DE C. & Simoes-Lopes, P.C. (1998). Pleistocene skull remains of *Pontoporia blainvillei* (Cetacea, Pontoporiidae) from the coastal plain of Rio Grande do Sul state, Brazil, and the relationships of pontoporids. *Revista da Universidade de Guarulhos, série Geociências,* 3(6), 71 - 77.

[65] Pyenson, N.D. & Hoch, E. (2007). Tortonian pontoporiid odontocetes from the Eastern North Sea. *Journal of Vertebrate Paleontology,* 27(3),757–762.

[66] Quadros, M.L., Rizzotto, G.J., Oliveira, J.G., & Castro, J.M. (2006). Depósitos Fluviais da Formação Rio Madeira, Pleistoceno Superior da Bacia do Abunã, Rondônia. *Simpósio de Geologia da Amazônia, IX, Anais, CD.*

[67] Rensberger, J.M.(1969). A new iniid cetacean from the Miocene of California. *University of California Publications in Geological Sciences*, 82, 1-33.

[68] Rizzotto, G. J., Cruz, N. M. Da Oliveira, J. G., De Quadros, M. L., Do, E.S. & Castro, J. M. (2006).

[69] Paleoambiente e o registro fossilífero pleistocênico dos sedimentos da Formação rio Madeira. In: *Simpósio de Geologia da Amazônia, IX, Anais, CD.*

[70] Scasso, R. A., McArthur, J. M., Del Rio, C. J., Martinez, S. & Thirwall, M. F. (2001). $^{87}Sr/^{86}Sr$ Late Miocene age of fossil molluscs in the "Entrerriense" of the Valdés Península (Chubut, Argentina). *Journal of South American Earth Sciences,* 14, 319-329.

[71] Simpson, G. G. (1945). The principles of classification, and a classification of mammals. *Bulletin of the American Museum of Natural History*, 85, 1-350.

[72] Slijper, E. J. (1936). Die Cetaceen: Vergleichen-anatomisch und systematisch. *Capita Zoologica*, 7/8, 1-590.

[73] True, F.W. (1909). A new genus of fossil cetacean from Santa Cruz Territory, Patagonia; and description of a mandible and vertebrae of *Prosqualodon*. *Smithsonian Miscellaneous Collections*, 52(1875), 141-156.

[74] True, F.W. (1912). A fossil toothed cetacean from California, representing a new genus and species. *Smithsonian Miscellaneous Collections*, 60(2151), 1-7.

[75] Verma, S.K., Sinha, R.K. & Singh, L. (2004). Phylogenetic position of Platanista gangetica: insights from the mitochondrial cytochrome b gene and nuclear interphotoreceptor retinoid-binding protein gene sequences. Molecular Phylogenetics and *Evolution*, 33, 280–288.

[76] Winge, H. (1921). A review of the interrelationships of the Cetacea. *Smithsonian Miscellaneous Collections*, 72, 1-97.

[77] Yan, J., Zhou, K., & Yang, G. (2005). Molecular phylogenetics of river dolphins and the baiji mitochondrial genome. *Molecular Phylogenetics and Evolution* 37, 743–750

[78] Yang, G., Zhou, K.Y., Ren, W.H., Ji, G.Q., & Liu, S. (2002). Molecular systematics of river dolphins inferred from complete mitochondrial cytochrome-b gene sequences. *Marine Mammal Science*, 18, 20–29.

[79] Zarbert, L.L. & Herbst, R. (1977). Revisión de la microfauna miocena de la Formación Paraná (entre Victoria y Villa Urquiza-Provincia de Entre Ríos-Argentina), con algunas consideraciones estratigráficas. *Facena,* 1, 131-168.

[80] Zhou, K. (1982). Classification and phylogeny of the superfamily platanistoidea, with notes on evidence of the monophyly of the cetacea. *Scientific Reports of the Whales Research Institute,* 34, 93-108.

[81] Zhou, K., Zhou, M. & Zhao, Z. (1984). First discovery of a Tertiary platanistoid fossil from Asia. *Scientific reports of the Whales Research Institute, Tokyo*, 35, 173-181.

[82] Zhou, K., Quian, W. & Yuemin, L. (1979). The osteology and systematic position of the Baiji, *Lipotesvexillifer*. *Acta Zoologica Sinica,* 25, 95-100.

SOTALIA FLUVIATILIS-SOTALIA GUIANENSIS

In: Biology, Evolution and Conservation of River Dolphins… ISBN: 978-1-60876-633-8
Editors: M. Ruiz-García, J. Shostell. pp. 221-235

Chapter 11

FISHERY ACTIVITY IMPACT ON THE *SOTALIA* POPULATIONS FROM THE AMAZON MOUTH

Sandra Beltran-Pedreros[1]* and Miguel Petrere[2]

[1]Instituto Nacional de Pesquisas da Amazônia (INPA), Coordenação de Pesquisas em Biologia Aquática, Rua Ajuricaba, Manaus, Brazil

[2]Universidade Estadual de São Paulo (UNESP), Departamento de Ecology, Rio Claro, Brazil

ABSTRACT

This chapter describes and analyzes the bycatch of *Sotalia guianensis*, in gillnets by an artisan fishing fleet within the Amazonian estuary during two time periods: 1996-1997 and 1999-2001. Number, size and gender data, as well as dolphin specimens were obtained from fishermen at Brazilian ports and analyzed. Fishing capacity and effort were determined via simple linear regression and bycatch, fishing trip and fishing effort data were analyzed between time-periods, among climatic (seasonal) periods and between strata (based on vessel length). Results indicated that the stratum two fishing fleet not only had larger vessels but longer fishing trips, used longer nets and had larger fishing crews compared to stratum one's fleet. Bycatch increased in both strata between periods but to a greater extent in stratum two. Although there was an increased percentage of fishing trips with bycatch across time, there was a reduced mean number of dolphins per bycatch. There were also differences in the bycatch by sexual maturity with an indiscriminately larger number of sexual-reproducing adults caught in stratum two. Collectively, these results in conjunction with other anthropogenic factors combined with dolphins being a k-selected species, suggest that dolphin mortality from bycatch may seriously affect *Sotalia guianensis* in the Amazonian estuary. Furthermore, the fishery-dolphin interaction was characterized and determined to be indirectly predatory.

Keywords: Amazonian estuary, *Sotalia guianensis*, bycatch, fishing-dolphin interaction.

---

* beltranpedreros@hotmail.com.

INTRODUCTION

Sotalia guianensis, one of the smallest Delphinidae, is very common in the coastal western Atlantic, where there are steady increases in human population growth rate and population density along with those activities that support this growth. One factor addressed in this chapter, fisheries, has been enhanced in the region to meet the demand of burgeoning human populations and new developmental projects along the coast. There are potential negative direct and indirect effects of fishing activities on marine life which make *Sotalia guianensis* a highly vulnerable species in this region (Trujillo, 1992; Trujillo & Beltrán 1995; Silva & Best 1996).

Although it is a species that is included in Annex I of the CITES, and classified by The International Union for Conservation of Nature and Natural Resources (IUCN) as "insufficient data", in addition to being protected by indigenous and fishermen communities' traditions, and by specific regulations in almost every country within its distribution area, *S. guianensis* is exposed to anthropogenic effects, which affects the stability of its populations. A review conducted by conservation scientists of potentially harmful anthropogenic factors suggested that two factors, new dam construction and bycatches by the coastal drift net fisheries were of principle concern. On account of these findings the continuous monitoring of those interactions became a priority, and a key element for defining the status of populations and providing for conservation mechanisms (Northridge, 1985).

Several fishing accounts recorded along the Brazilian coast stated that bycatches of this species, collected via waiting and drafting's gillnets, is often accompanied by other cetacean species such as *Pontoporia blainvillei* (Lodi & Capistrano, 1990). Fishing villages in the northeastern and southeastern Brazil, are using these types of nets where the bycatch of *S. guianensis*, has been recorded without bias to the fish harvested or fishing effort (Lodi & Capistrano, 1990; Barros & Teixeira, 1994).

In southern Brazil, where the fishing-dolphin interactions are better monitored and known, bycatch of *S. guianensis* is low and *Pontoporia blainvillei* has become the most vulnerable species (Perrin et al., 1994; Pinedo, 1994*a, b*). In the state of Pará, Siciliano (1994) mentions the villages of Algodoal, Marudá, Salinópolis, Bay of Marajó and Vigia as the places where the fisheries show the largest interaction with S. *guianensis,* with adults being the most vulnerable group. These dolphins are also harpooned and then used as shark bait (Barros, 1991; Borobia et al., 1991). It is common for fishermen to remove the genital organs and the eyes from the dolphins to be sold in the market as love charms and witchcraft devices at the Ver-o-Peso market in Belém; the teeth are sold as ornaments.

In Central American countries there are small artisan fisheries using 30 to 2,000 m long nets with mesh sizes of 4 cm up to 40 cm in which several species of cetaceans including *S. guianensis* are caught (Perrin et al,. 1994). Bycatches of S. *guianensis* in Nicaragua, Honduras, Colombia, Surinam and French Guyana have been recorded by Vidal et al. (1994).

METHODS

The sampling was carried out from 1996 to 1997 (period one) and from 1999 to 2001 (period two) by monitoring fishing vessels that used drifting gillnets in Vigia and Bragança (Pará state), Brazil (Figure 1). The fleet was characterized into two sample strata, according to the length of the fishing vessels (longer and shorter than 12 m). Additional characteristics of each vessel were noted and recorded including: ship length, hole storage capacity, engine horse power, net length, mesh size of net measured between ends, net's thread type and gauge, and number of fishermen on board.

Figure 1. The strata fishing sites for monitored fishing vessels within the Amazonian Estuary. Each vessel used drifting gillnets.

Detailed information regarding each fishing trip was recorded such as the total number of days per trip, number of fishing days, number of castings, and fishing sites. Total captures per species, abundance of each species, gender, and size of dolphins captured were also recorded. The authors interviewed the fishing crew at the ports of Vigia and Bragança after they returned from their fishing excursions. Here, at port, they recorded the number, gender, and age group of dolphins caught (released alive or dead at sea) as well as dolphin carcasses aboard the vessels. Captured dolphins (carcasses) were categorized into three age groups. Those equal to or greater than 1.6 m in total length were considered adults. Dolphins with lengths of 1.3 m to 1.6 m were recorded as young, while those less than 1.3 m were calves.

Since the data on dolphin bycatches were obtained from fishermen, it is normal to find a large variance among them. That is the sort of bias and risk one takes by working with fishermen and information collected from them. As a measure to reduce this bias, fishermen were asked multiple forms of the same question and only those answers that were consistently given were analyzed and included in this chapter.

A descriptive bycatch data analysis was completed that described the *S. guianensis* bycatch frequency subdivided by gender, age group, monthly census, and fishing fleet stratum. Significant differences in the total number of dead dolphins (in each period) between strata, were detected via a Z test for two averages. And, significant differences of the data between two periods were detected through the use of a t-test. The significance differences in the total number of dead dolphins between strata, per gender and age group were established through an analysis of variance (ANOVA Type Two). A simple linear regression analysis was carried out to determine the relationship between dolphin bycatch and fishing effort.

RESULTS

Fishing fleet activity area. Artisan fishing boats that relied on drifting gillnets to collect fish in the Amazonian estuary sailed approximately between coordinates 4° N; 1° S and 47°; 51° W (Fig. 1). This is an area that encompasses 90,000 square kilometers.

Net types used and targeted fish species. Thread gauge and mesh size varied according to depth, time of year (season) and fish species (Table 1). When targeted fish species resided in deeper depths, driftnets were designed for sinking and lowered to deeper waters. When the sea water flowed into the estuary during the summers, the most commonly captured fish species were: gillbacker sea catfish (*Arius parkeri),* weakfish (*Cynoscion acoupa*), crucifix sea catfish (*Arius proops*) and fat snook (*Centropomus parallelus*). However, during the winters an influx of freshwater from the Amazon river pushed against the sea water and the targeted species changed to piramutaba (*Brachyplatystoma vaillantii*) and dourada (*Brachyplatystoma flavicans*).

Table 1. Types of driftnet used by the artisan fishing fleet in the Amazonian estuary. Mesh size in centimeters (cm), length of net (ft) and net thread gauge (TG) in millimeters (mm) are displayed. Single-threaded (S) and multi-threaded nets are included

Net Type	Target Species	Mesh(cm)	Length(ft)	TG(mm)
Serrera	Scombridae	5-10	400-1.000	60 (S)
Douradera	*Brachyplatystoma flavicans* *Brachyplatystoma vaillantii*	14-20	400-3.500	23-35 (M)
Pescadera	*Arius parkeri* *Cynoscion acoupa*	16-22	400-3.500	36-48 (M)
Tiburonera	Sharks	40	1.500-2.000	96 (M)

Fishing Vessel Characteristics and Size of Fishing Crew

Fishing vessel characteristics and crew size varied greatly among fishing vessels both between and within strata. The average values for vessel length, engine power, hole capacity and crew size were all greater in stratum one compared to stratum two. Of course differences between strata were artificially created by the experimental design. In stratum one, vessel length varied from 6 to 11.8 m (n = 145, $\bar{x}$ = 8.9 m, ± 1.26 m), engine horse power between 9 and 160 (mode =18 HP) and, the hole capacity varied from 0.2 to 9 tons of ice ($\bar{x}$ =4.9 t ± 1.83 t). Fishing crew size varied from 2 and 7 men (mode = 4 men) and fishing nets were between 400 and 2,500 ft long ($\bar{x}$ =750 ft ±308.65 ft) in stratum one.

Length of vessel in stratum two ranged between 12 and 32 m (n = 44, $\bar{x}$ = 14.9 m, ± 3.57 m). Their engine horse power and hole capacity ranged from 18 and 230 (mode =69 HP) and from 1.5 to 35 tones ($\bar{x}$ = 12.4 t ± 6.51 t) respectively. The size of fishing crew (4 -12 men; mode = 8 men) and length of nets used (900 - 3,500 ft; $\bar{x}$ = 2027.5 ft ±508.31 ft) also varied greatly in stratum two.

Method of Fishing. Gillnets were released in the water from the bow, with the aid of a power engine and in the same direction of the wind and tide. A fisherman released the net while another fisherman set the float line to determine at which depth the net would do its work. After drifting for 5 to 6 h, during the tidal period, the net was pulled in by a fisherman, while a second fisherman pulled out the fish from the net, and a third fisherman gathered the net and coiled the line.

Fishing capacity and fishing effort. The estimated variable, number of fishermen, presented the best correlation coefficient with fish capture (n = 927; r = 0.888; r^2 = 0.789; P<0,05). Fishing effort units were defined as netting time divided by number of fishermen. There was a significant correlation between capture and fishing effort measures, however, the determination coefficient (r^2) for each stratum was low. This implies that there were other factors acting in the capture which were not being considered.

Fishing trip. During the first sampling period (oct/96 to mar/97) 440 fishing trips were registered (350 in stratum one and 90 in stratum two). In the second period of sampling (Jan/99 to sept/01) 927 fishing trips were registered (590 in stratum one and 337 in stratum two). The distribution of fishing trips across months and years is displayed in Table 2.

Table 2. Number of fishing trips with drift gillnets completed by the artisan fishing fleet across months and years in the Amazonian Estuary. The data are displayed by strata (stratum 1 = S1; stratum 2 = S2)

	1996		1997		1999		2000		2001	
	S1	S2	S1	S2	S1	S2	S1	S2	S1	S2
Jan			29	12	21	14	25	12	34	8
Feb			7	6	15	8	20	8	34	10
Mar			11	6	17	18	13	10	23	23
Apr					4	3	20	8	34	25
May							18	5	33	16
Jun							16	8	20	11
Jul							12	14	26	15
Ago							38	30	15	10
Set							18	18		
Oct	104	17					44	18		
Nov	113	30					33	17		
Dec	86	19					57	28		
Total	303	66	47	24	57	43	314	176	219	118

Bycatch. The monthly bycatch of *S. guianensis* in the Amazon estuary had great variability both within and among years, with consistently greater values in stratum two compared to stratum one. (Table 3). The total number of *S. guianensis* in bycatches per month and strata ranged from 3 to 230. In 2000, stratum two had the maximum number of *S. guianensis* caught in one year (1,245 individuals). It is very important to remember that the final bycatch value should reflect all fishing trips, including those that received no bycatch.

Table 4 shows the mean and the standard deviation of bycatch for each time period and stratum. There was a significant difference between means of bycatch/fishing excursion (df = 755; t_c = -2.41; P = 0.01) and the means of bycatch/fishing trips with bycatch (df = 179 t_c = 2.07; P = 0.04) of periods. There was a significant difference between means of bycatch/fishing trips between strata of the first (P = 0.05; Zt = 1.96; Zc = -4.68) and second (P = 0.05; Zt = 1.96; Zc = -12.73) periods.

Bycatch distribution. Bycatch distribution was only described for the first sampling period. According to fishermen, 703 (66%) of a total of 1,063 dolphins captured in bycatch were sexed. The average percent of male and females dolphins caught per bycatch were 46.37 and 53.63 respectively. Out of a total of 147 trips with bycatch, 23% only caught females and 12% only caught males. Forty-two percent of the bycatches contained both sexes and in the remaining bycatches (23%), gender was not determined (Table 5).

In relation to age, the most vulnerable group was the adults (dolphins longer than 1.6 m) with a total of 753 accidentally caught out of a total of 132 trips. A total of 202 young specimens (between 1.3 and 1.6 m long) were caught in 52 trips and 49 calves (smaller than 1.3 m long) were caught in 28 trips. It was not possible to determine the size of 59 dolphins caught in 4 trips because the fishermen did not record the data or the dolphin carcasses were returned to the water. In 147 trips presenting bycatch, 53% only caught adults; 7% only young dolphins; 0.7% caught calves; 18% caught both adults and young; 8% had adults and calves; and 10%, multiple sized (more than 2 sizes) dolphins (Table 5). Calf bycatches were only recorded from October to December; however there were larger numbers in October and November.

Table 3. Number of tucuxi, *Sotalia guianensis* caught with drift gillnets by the artisan fishing fleet across months and years in the Amazonian Estuary. The data are displayed by strata (stratum 1 = S1; stratum 2 = S2)

	1996		1997		1999		2000		2001	
	S1	**S2**	**S1**	**S2**	**S1**	**S2**	**S1**	**S2**	**S1**	**S2**
Jan			6	6	23	74	24	60	29	83
Feb			3	5	11	55	23	49	20	27
Mar			4	3	23	120	22	129	25	170
Apr					7	47	19	76	51	109
May							13	47	45	88
Jun							32	22	23	45
Jul							22	46	34	90
Ago							51	230	17	194
Set							29	153		
Oct	**26**	**13**					83	133		
Nov	**20**	**22**					45	79		
Dec	**25**	**14**					65	221		
Totals	71	49	13	14	64	296	428	1245	244	806

Table 4. Statistics of fishing trips with and without bycatch of S. *guianensis,* in the Amazonian estuary between strata (stratum 1= S1; stratum 2 = S2) and between periods. The two time-periods of when this study was conducted are labeled as 1 for 1996-1997 and 2 for 1999-2001. Mean ($\bar{x}$), standard deviation (SD) and sample size (n) statistics are included

	Fishing Excursion			Fishing Trip with bycatch		
Period	Total	S1	S2	Total	S1	S2
1 $\bar{x}$	2.4	1.3	6.8	7.2	5.3	9.7
SD	6.8	4.6	10.9	10.2	8.3	11.9
n	440	350	90	147	84	63
2 $\bar{x}$	3.3	1.2	6.9	5.4	2.6	8.1
SD	5.8	2.1	8.1	6.6	2.3	8.2
n	927	590	337	572	281	291

Table 5. Distribution according to gender and age group of S. *guianensis* in bycatch collected with drifting gillnets by the artisan fishing fleet of the Amazonian estuary

	Adult		Young		Calf	
	Female	Male	Female	Male	Female	Male
Total	284	234	71	78	22	14
$\bar{x}$	2.5	2.5	2.1	2.0	1.3	1.3
SD	2.3	2.1	2.4	1.9	0.6	0.8

Bycatch per fishing fleet stratum (Table 6) indicated that 450 dolphins were caught in stratum one and 613 in stratum two. Additionally, the smallest bycatches per age group occurred in stratum one and there was a slightly greater number of females than males caught in that same stratum. Although there were small differences in the distribution of gender and age group, adult bycatch was the only significantly different biotic variable between strata (ANOVA; $P = 0.007$). There was no significant difference in bycatch data between genders and age groups.

Table 6. Distribution per gender and age group of *S. guianensis,* bycatch in drift gillnets in each stratum of the artisan fishing fleet in the Amazonian estuary. Mean ($\bar{x}$) and standard deviation (SD) statistics are included

	Stratum 1			Stratum 2		
	Total	$\bar{x}$	SD	Total	$\bar{x}$	SD
Total specimens	450	5.36	8.29	613	9.73	11.95
Female	195	3.30	5.24	182	5.06	6.44
Male	146	3.48	4.96	180	4.86	6.32
Adult	320	4.51	5.29	433	8.17	8.26
Young	87	2.9	5.24	115	5.23	6.61
Calf	19	1.19	0.54	30	2.5	1.73
Adult female	151	2.85	3.42	123	3.73	4.02
Young female	33	2.2	3.86	38	2.8	3.36
Calf female	11	1.22	0.44	11	1.57	0.79
Adult male	98	2.72	2.90	121	4.03	3.86
Young male	42	2.33	2.89	36	3	2.76
Calf male	7	1		8	2	1.41

Bycatch and climatic period. The Amazon estuary showed four climatic periods: summer, with little rain, between August and October, a summer-winter transition between November and January; winter, with heavy rainfall, between February and April and a winter-summer transition between May and July.

For the first of two time-periods that this study was conducted, a summer (October only), summer-winter transition (November 1996 to January 1997) and winter February to March 1997) climatic periods were included. The second time-period (1999-2001) contained three winters (1999-2001), two winter-summer transitions (2000-2001), two summers (full in 2000 and partial in 2001) and one summer-winter transition was (2000).

The frequency of bycatch in fishing trips decreased, in both strata, from summer to winter-summer transition, with the number of the fishing trips with bycatch always larger in stratum one. The number of dead dolphins showed the same seasonal trend, but the number of dead dolphins was greater in stratum two (Table 7). This data trend is better observed when analyzing the 2000 data, when 679 dolphins were bycaught in summer, 522 in the summer-winter transition, 318 in winter, and 182 in the winter-summer transition.

Table 7. The number of *S. guianensis* within bycatch across climatic periods (winter = WIN, winter-summer transition = W-S, summer = SUM, and summer-winter transition = S-W) and years within the Amazonian estuary. Strata (stratum 1 – S1; stratum 2 = S2) are also displayed. % = Percentage of bycath for the climatic period

		1996		1997		1999		2000		2001	
		S1	S2	S1	S2	S1	S2	S1	S2	S1	S2
WIN	*n*			7	8	41	222	64	254	96	306
	%			46.7	53.3	15.6	84.4	20.1	79.9	23.9	76.1
W-S	*n*							67	115	102	223
	%							36.8	63.2	31.4	68.6
SUM	*n*	26	13					163	516	17	194
	%	66.7	33.3					24	76	8	92
S-W	*n*	45	36	6	6			139	383		
	%	55.5	44.5	50	50			2.7	73.3		

Bycatch and fishing effort. The determination of any fishing effort effects on dolphin bycatches has great assessment value. Although the fishing effort may not intentionally target dolphins it may be added to other factors and synergistically cause harm to dolphins trapped in nets.

The regression analysis of fishing effort versus number of dead dolphins for either stratum showed low correlation coefficients in both periods ($r = 0.18$, $n = 440$ first period; $r = 0.0639$; $n = 927$ second period). When the strata were analyzed separately, there were similar statistical findings for stratum one data ($r = 0.15$, $n = 350$ first period; $r = 0.1177$; $n = 590$ second period), even though both regressions were significant (no practical value). For stratum two, the regressions were insignificant for both periods ($r = 0.11$, $n = 90$ first period; $r = 0.008$, $n = 337$ second period).

Table 8. Monthly distribution of fishing effort (netting hours / crew number) across five years (1996-2001) using drift gillnets by the artisan fishing fleet in the Amazonian estuary. Strata (s) levels are listed and months are abbreviated

	1996		1997		1999		2000		2001	
	S1	S2	S1	S2	S1	S2	S1	S2	S1	S2
Jan			13.5	17.4	16.2	15,18	12.3	16.4	16.5	19.1
Feb			16.3	21.1	14.0	15,38	13.9	15.5	15.3	17.5
Mar			16.4	18.0	18.3	15,4	17.8	15.7	16.5	14.9
Apr					17.6	13,86	17.3	16.9	16.2	15.9
May							16.5	14.5	15,13	13.7
Jun							17.5	13.9	16.4	19.6
Jul							16.1	17.6	15.7	17.8
Aug							16.9	17.6	15.6	11.9
Sep							15.6	16.0		
Oct	11.6	13.1					13.6	19.8		
Nov	15.1	18.6					16.7	17.5		
Dec	13.8	16.0					15.0	18.1		

Fishing effort values varied little across time periods (1996-1997 vs 1999-2001), seasonal periods (winter, summer), or strata. (Table 8). Examples of mean vales were 15.35 (1999), 16.9 (2000) and 16.32 (2001).

THOUGHTS AND CONCLUSION

The lack of information on the fishing fleet effort made by the artisan fisheries is one of the greatest limiting factors in accurately assessing dolphin bycatch. There is some information regarding the interaction of *S. guianensis* with artisan fishery along its geographical distribution, although these data are incomplete because they do not include the fishing effort units. Knowledge of this information is necessary to be able to compare among fishing fleets, as has already been done for Central America (Vidal et al., 1994) and Brazil (Siciliano, 1994; Di Beneditto et al., 1998, 2001).

Every month, dolphins get caught up in nets due to the large number of artisan fishers on the West Atlantic coastal areas. The same event occurs over S. *guianensis's* distribution range along the Brazilian sea shore (Barros & Teixeira, 1994; Pinedo, 1994*b*; Siciliano, 1994).

Crew number, days per trip, and trips per year, together with the annual catch and number of vessels working in each area are the most reliable information to obtain on fishing fleets that help conservation biologists to determine the fishing fleets' direct effects on dolphins. Dolphin bycatch data presented in this chapter across years, underscored by the information on fishing effort unit for artisan vessels using drift gillnets and the number of fishermen per day of trip can be assessed in this study.

Another important fact, is that most information found in the literature on bycatch of S. *guianensis* is related to drift nets with mono-threaded nylon used to catch fish of the families Scombridae, Scianidae and sharks (Schmiegelow, 1990; Barros 1991; Pinedo, 1994*a;* Siciliano, 1994; Vidal et al*.,* 1994; Di Beneditto et al., 2001), whereas in the Amazonian estuary bycatches very rarely occur with this sort of gear, but rather with multithreaded deep fishing nets.

Our findings show that the sort of interaction between artisan fishery and dolphins is indirectly predatory, where dolphins as well as fish species are being caught by fishermen, and that both fishermen and dolphins are in pursuit of the same prey. Regarding this subject, Beverton (1985) reported on the interaction dynamics of dolphins and fishermen and concluded that their relationship can be mutualistic when there is abundant food. Feeding dolphins can be readily observed by fishermen and help the fishing fleet to pursue targeted fish, and abundant prey corralled by fishing vessels could theoretically provide a large supply of food. On the other hand, with a higher number of dolphins, the less available fishery target prey will become, lowering the resources for the fishing fleet. Such a shift in prey abundance could facilitate interspecific competition.

Bycatch of *S. guianensis,* in the Amazon estuary, presents a remarkable seasonality, with summer having the largest bycatch period and then having it decrease significantly over the winter. Change in the estuary salinity could be one of the factors generating this phenomenon, even though it doesn't directly affect the dolphins, but rather the distribution of the fishery target prey. Therefore, with the onset of winter, the stratum one fleet moves to other areas. Fishing nets are changed and the number of fishing days is lowered, decreasing the fishing effort. It is possible for dolphins to migrate in pursuit of their prey which lessens the chance of them interacting with the fishing fleet. With zero salinity, fish and dolphins move to other areas and the fishery - dolphin frequency interaction decreases. Again, there is seasonality for bycatch of *S. guianensis* in the Amazonian estuary, and summer is the season with the largest catch.

Type of net and the gauge of its thread did not have significant effects on dolphin mortality, which suggests that differences in bycatch among vessels in the Amazonian estuary may be due to additional, untested factors. The interaction between dolphins and fishing industry may be related to factors not pertaining to the fishing activity, such as ecological aspects and species behavior itself. The size of nets, depth fished, thread type and total area fished by the fleet in the Amazonian estuary, may also help to explain how bycatches in the Amazonian estuary are different from other fished areas.

Usually, artisan fishing fleets are subjected to the region's environmental and economical fluctuations. A remarkable seasonality of fish catch within the Amazonian estuary resulting from environmental factors in turn influences local economies. For example, those species with high commercial value occur during the summer and the ones caught over the winter are mainly exploited by the industrial fishing fleet. This strategy, that depends on target species, results in economic instability. Bycatches of *S. guianensis* by the fleet using this strategy should not be dangerous to this species (*S. guianensis).*

A comparison of data from the two time- periods (1996-1997 vs 1999-2001) indicated that the percentage of the fishing trips with bycatch almost doubled from 33.41% to 61.7%. while the mean number of dead animals per fishing trip with bycatch decreased significantly (7.2 to 5.4).

Similarly, the percent of fishing trips with bycatch also showed an increase in stratum one (42.33% to 47.6%) and in stratum two (57.66% to 86.3%). In contrast, the means number of dolphins caught per fishing trip (in bycatch) decreased significantly (5.3 to 2.6 in stratum one and 9.7 to 8.1 in stratum two).

S. guianensis bycatch seems quite large at 4,146 individual dolphins, for a 30 month sampling period until one considers that the size of the fleet's fishing area is nearly 90,000 km^2. Consider that this bycatch is equivalent to 4.5% of *S. guianensis's total* population size in the Amazon estuary. This is the maximum bycatch percentage estimated for *Stenella attenuata* in tuna fishing nets (Bjϕrge et al., 1994). Therefore, *S. guianensis's* estimated population size in the estuary should be approximately 92,133 dolphins, that translates into an estimated density of 1.02 ind/km^2. Densities of *S. fluviatilis* on the upper Amazon, varies from 0.9 ind/km^2 in river channels, 2-2.8 ind/km^2 in the Amazon River and 8.6 ind/km^2 in lakes (Vidal *et al.* 1997) and are the highest densities reported for small cetaceans. The closest estimated densities of other cetaceans in the Amazonian estuary are the *Phocoena phocoena* (0.56 ind/km^2) and *Delphinus delphis* (0.3 ind/km^2) (Barlow, 1988, 1995).

There are no data on the species life history and abundance, however, as a k strategist, it shows a great muscle mass, it takes a long time to reach sexual maturity (6-7 years), it has a slow growth rate (0,79), a long life span and has birth intervals of over one year apart (2 years) (Rosas, 2000). Their life history characteristics are similar to those of *Phocoena phocoena* whose annual population increase was estimated to be between 8.3 and 9.4% depending on the age (Barlow & Hanan, 1995). That is, estimated bycatch for *Sotalia guianensis* in the Amazon estuary doesn't show an impending threat for the population, since it has the potential to be able to regain 6% of its number within a year.

Still, there was a significant difference in the bycatch of adults between strata, with a significantly greater number of adults being caught in stratum 2. Specifically, there were 11 immatures, 13 pubers and 15 adults in stratum two and 6 immatures, 2 pubers and 3 adults in stratum one. For females, there were 15 immatures and 4 adults in stratum one and 22 immatures and 22 adults in stratum two.

Significant differences in the bycatch by gender and sexual maturity may affect the population because this implies an exposure of those population fractions that can undergo reproduction (adult females and males in stratum 2) or play a role in recruitment (puber males in the stratum 2 and immature females in both strata).

When the bycatch and fishing effort data were compared, we should note that there was a positive and significant correlation between the two variables for the total data and also for stratum one, but not for stratum two. An important detail here is that the value of the correlation coefficients and determinations decreased, which permits us to deduce that the increase on the number of fishing trips, doesn't pose a menacing threat to the species for a while, because the values of fishing effort do not vary much from year to year.

According to records of the geographic distribution of the species in other sites, the Amazon estuary is the region where there is higher bycatch of *S. guianensis* and where it is very common to bycatch the animals in pairs or threesomes. Bycatch happens seldomly to herds of 10 to 50 animals such as what occurred in a single night when 44 *S. guianensis* were bycaught by one vessel and three more vessels that were fishing nearby had bycatches of 20, 45 and 50. A different situation was experienced by Rosas (2000), who reported a maximum bycatch of three animals.

It's important to take into account the magnitude of the geographical area and the size of the fishing fleet analyzed. In most fisheries with a bycatch of *S. guianensis*, the fishing operation occurs near the coast, in one day, in vessels with low autonomy, with a reduced crew number (usually 4) and fishing nets smaller than 3000 m long (Siciliano, 1994; Vidal et al., 1994; Rosas, 2000; Di Benededitto et al., 2001); while in the Amazon estuary, the fishing fleet works close to the coast, but for periods of 12-15 days (stratum 1) or 20-30 days (stratum 2). The vessels have greater autonomy (especially ice to conserve the fish), nets from 1500 to 7500 m long and larger crew size (5 to 12 fishermen).

Regarding the target prey, from the fisheries described above, most are species consumed by *Sotalia guianensis* (Borobia & Barros, 1989; Beltran-Pedreros 1998), characterizing a bycatch by direct predation, that is, the fishing fleet and *S guianensis* captured the same target prey, competing between themselves. Target fish of the fisheries in the Amazon Estuary are large-sized and are not part of the *Sotalia guianensis* diet. But there are common prey items in the diet of *Sotalia guianensis* and in the diet of target fish from the fisheries, characterizing a bycatch by indirect predation and emphasizing the competition between dolphins and captured fish species rather than between dolphins and fisheries.

The International Whaling Commission, in its report about mortality of cetaceans in fishing gear (Perrin et al., 1994), considered the bycatches of *Sotalia fluviatilis* and *S. guianensis* to be non-significant relative to their whole geographical distribution which is still unknown for Caribbean and Brazil. This statement should be revised, because there are data which indicate that the bycatch events are frequently in dangerous numbers for some species populations, as shown in the present chapter, especially, when we consider synergistic effects of bycatch with habitat degradation in areas where the populations have a high level of fidelity. Therefore, it's advisable to qualify the bycatch of *S. guianensis* as potentially and possibly non-sustainable in these areas.

ACKNOWLEDGMENTS

We want to thank the National Institute of Amazonian Research (INPA) and the Ministry of Science and Technology (CNPq), for their logistical and economical support regarding the development of this research project. Special thanks go to Vera da Silva (INPA, Brazil) who provided useful comments to improve this manuscript. Also, thanks to the fishermen of Arapiranga's port in Vigia (PA), for the collaboration and information during the research.

REFERENCES

[1] Barlow, J. (1988). Harbor porpoise (*Phocoena phocoena)* abundance estimation in California, Oregon and Washington: I. Ship surveys. *Fishery Bulletin,* 86, 417-432.

[2] Barlow, J. (1995). The abundance of cetaceans in California waters. Part I: Ship surveys in summer and fall of 1991. *Fishery Bulletin*, 93, 1-14.

[3] Barlow, J. & Hanan, D. (1995). An assessment of the status of harbor porpoise in central California. In A. Bjϕrge and G. Donovan, (Eds.) *Biology of the Phocoenids. Reports of the International Whaling Commission* (Special Issue 16, pp. 123-140). Cambridge, United Kingdom: International Whaling Commission.

[4] Barros, N. (1991). Recent cetacean records for southeastern Brazil. Marine Mammal Science 7(3), 296-306.

[5] Barros, N., & Teixeira, R. L. (1994). Incidental catches of marine tucuxi, *Sotalia fluviatilis*, in Alagoas, northeastern Brazil. In W. Perrin, G. Donovan and J. Barlow (Eds.) *Gillnets and Cetaceans. Reports of the International Whaling Commission* (Special Issue 15, pp. 265 -268). Cambridge, United Kingdom: International Whaling Commission.

[6] Beltran-Pedreros, S. (1998). Captura accidental de *Sotalia fluviatilis* (Gervais, 1853) na pescaria artesanal do Estuário Amazônico. Do Amazonas (Master Thesis). Manaus, Brazil: Instituto nacional de Pesquisas da Amazônia/Universidade.

[7] Beverton, R. J. H. (1985). Analysis of marine mammal-fisheries interactions. In J. R. Beddington, R.J.H. Beverton and D.M. Lavigne, (Eds.) *Marine Mammals and Fisheries*. (pp. 3-33). London, United Kingdom: George Allen & Unwin.

[8] Bjϕrge, A., Brownell, R., Donovan, G., & Perrin, W. (1994). Significant direct and incidental catches of small cetaceans. In Perrin W., G. Donovan and J. Barlow, (Eds.) *Gillnets and cetaceans. Reports of the International Whaling Commission* (Special Issue 15 pp. 76-126). Cambridge, United Kingdom: International Whaling Commission.

[9] Borobia, M. & Barros, N. (1989). Notes on the diet of marine *Sotalia fluviatilis*. *Marine Mammal Science*, 5(4), 395-399

[10] Borobia, M., Siciliano, S., Lodi, L., & Hoek, W. (1991). Distribution of the South American dolphin *Sotalia fluviatilis*. *Canadian Journal of Zoology*, 69, 1025-1039.

[11] da Silva, V.M.F. & Best, R. (1996). Freshwater dolphin/fisheries interaction in the Central Amazon (Brazil). *Amazoniana,* XIV(1/2), 165-175.

[12] Di Beneditto, A. P., Ramos, R. M., & Lima, N. R. (1998). Fishing activity on Northern Rio de Janeiro State (Brazil) and its relation with small cetaceans. Curitiba. *Brazilian Archivo de Biological Technology*, 41(3), 296-302.

[13] Di Beneditto, A. P., Ramos, R. M., & Lima N. R. (2001). Os golfinhos. Origem, classificação, captura acidental, hábito alimentar. Cinco Continentes. Porto Alegre, Brazil.

[14] Lodi, L. & Capistrano, L. (1990). Capturas acidentais de pequenos cetáceos no litoral norte do estado do Rio de Janeiro. *Biotemas*, 3, 47-65.

[15] Northridge, S. (1985). Estudio mundial de las interacciones entre los mamíferos marinos y la pesca. *FAO Informe de Pesca,* (251), 234.

[16] Perrin, W., Donovan, G., & Barlow, J. (1994). Report of the workshop on mortality of cetaceans in passive nets and traps. In Perrin W., G.Donovan and J. Barlow, (Eds.), *Gillnets and cetaceans. Reports of the International Whaling Commission* (Special Issue 15, pp 1-72). Cambridge, United Kingdom: International Whaling Commission.

[17] Pinedo, M.C. (1994*a)*. Review of small cetacean fishery interaction in southern Brazil with special reference to the franciscana, *Pontoporia blainvillei.* In Perrin W., G. Donovan and J. Barlow, (Eds.), *Gillnets and cetaceans. Reports of the International Whaling Commission* (Special Issue 15, pp 251 -259). Cambridge, United Kingdom: International Whaling Commission.

[18] Pinedo, M.C. (1994*b*). Impact of incidental fishery mortality on the age structure of *Pontoporia blainvillei* in southern Brazil and Uruguay. In Perrin W., G. Donovan and J. Barlow, (Eds.), *Gillnets and cetaceans. Reports of the International Whaling Commission* (Special Issue 15, pp.261 -264). Cambridge, United Kingdom: International Whaling Commission.

[19] Rosas, F. (2000). *Interações com a pesca, mortalidade, idade, reprodução e crescimento de Sotalia guianensis e Pontoporia blainvillei (Cetácea: Delphinidae e Pontoporiidae) no litoral sul do Estado de São Paulo e litoral do Estado de Paraná, Brasil* (Ph.D. Thesis). Curitiba, Brazil:Universidade Federal do Paraná.

[20] Schmiegelow, J.M.M. (1990). Estudo sobre cetaceos odontocetos encontrados em praias da região entre Iguape (SP) a Baía de Paranaguá (PR) (24°42'S-25°28'S) com especial referência a *Sotalia fluviatilis* (Gervais, 1853) (Delphinidae) (Masters Thesis). São Paulo, Brazil: Universidade de São Paulo.

[21] Siciliano, S. (1994). Review of small cetaceans and fishery interactions in coastal waters of Brazil. In Perrin W., G. Donovan and J. Barlow, (Eds.), *Gillnets and cetaceans. Reports of the International Whaling Commission* (Special Issue 15, pp. 241 -259). Cambridge, United Kingdom: International Whaling Commission.

[22] Trujillo, F. (1992). Estimación poblacional de los delfines de agua dulce *Inia geoffrensis* (DeBlainville, 1817) y *Sotalia fluviatilis* (Gervais and Deville, 1853) en el sistema lacustre de Tarapoto y El Correo, amazonia colombiana. COLCIENCIAS/Universidad Jorge Tadeo Lozano. Bogotá. 49:196 p.

[23] Trujillo, F. & Beltran, S. (1995). Patrones de uso del hábitat, comportamiento y mortalidad incidental y dirigida de *Inia geoffrensis* y *Sotalia fluviatilis* en el Amazonas colombiano. COLCIENCIAS/Universidad Jorge Tadeo Lozano. Bogotá. 194 p.

[24] Vidal, O., Van Waerebeek, K., & Findley, L. (1994). Cetaceans and gillnets fisheries in Mexico,Central America and the wider Caribbean: A preliminary review. In W. Perrin, G. Donovan, & J. Barlow (Eds.), *Gillnets and cetaceans. Reports of the International Whaling Commission* (Special Issue 15, pp 221 - 233). Cambridge, United Kingdom: International Whaling Commission.

[25] Vidal, O., Barlow, J., Hurtado, L., Toree, J., Cendón, P., & Ojeda, Z.(1997). Distribution andabundance of the Amazon river dolphin (*Inia geoffrensis*) and the tucuxi (*Sotalia fluviatilis*) inthe upper Amazon River. *Marine Mammal Science*, 13(3), 427-445.

In: Biology, Evolution and Conservation of River Dolphins… ISBN: 978-1-60876-633-8
Editors: M. Ruiz-García, J. Shostell. pp. 237-246

Chapter 12

DOLPHIN-FISHERY INTERACTION: COST-BENEFIT, SOCIAL-ECONOMIC AND CULTURAL CONSIDERATIONS

Sandra Beltran-Pedreros[1]* and
Ligia Amaral Filgueiras-Henriques[2+]
[1] Projeto PIATAM. Área de Mamíferos Aquáticos. INPA (CPBA)-UFAM. Rua Ajuricaba, Aleixo. Manaus, Brazil
[2] Universidade do Estado do Pará (UEPA, Departamento de Ciências Naturais – Rua do Una, Telégrafo – Belém, PA. Brazil.

ABSTRACT

We evaluated dolphin-fishery interaction dynamics as well as the social-economic and cultural aspects of this relationship on the artisan fleet of the city of Vigia (Para) that fished with gill nets in the Amazon estuary from 1996 to 2001. Simple regression analyses were used to define the fishing-effort components (fishing-power and time) and to correlate fishing-effort and dolphin bycatch. Economical system and fish marketing dynamics analyses were used to define the interaction cost-benefit. The fishing-effort unit was FN x Hr and, the correlation between it and the dolphin bycatch was low, decreasing in the second period of this study. The trade of whole dolphins or their parts does not represent an important revenue factor for the fishermen and therefore should not be considered dangerous to the dolphin population. A new model of relationships among the variables of the fishery-dolphin system is presented.

Keywords: Dolphin-fishery relationships, bycatch, Amazon Estuary, Artisan fishing.

* beltranpedreros@hotmail.com.
+ ligiafilgueiras@yahoo.com.br.

INTRODUCTION

The Accidental Capture Problem

The accidental capture theme is important in the public's perspective for affecting species of "high visibility". However, visions that grant moral characteristics to animals are obstacles in the understanding and objective judgment of the importance of the ecological problems. According to Hall (1995) the ecology scientific concept does not accept aesthetic or moral valorizations of any species. This view affects the foundation of natural resource management and the conservation of all the ecosystem components.

Although the accidental capture of aquatic mammals has been happening for a long time, the importance of this capture was only recognized from the public alarm of dolphin mortality caused by the tuna fishery. This was the first known large-scale capture case in the world which consequently led nets to emerge as a serious problem for non target animals of the fisheries (turtles, birds, mammals and other fish species).

Hoel (1990) argues cetacean capture should be an integral part of ecosystem management; however Reeves & Leatherwood (1994) consider the idea of multispecific (ecosystem) management, valid intellectually, but in practice, difficult to implement due to the complexity of the natural systems and to the little understanding of how they work. Solving this interesting conflict is difficult and many questions need to be addressed. It is necessary to know if the aquatic mammals are affecting fishing stocks or fishing operations and vice versa, and to determine the possible expected benefits for aquatic mammals and fisheries when different management methods are implemented. Answers to those questions regarding these needs are scientifically difficult, since fish and aquatic mammal populations react differently to management.

In 1972, the FAO Fishing Committee recommended beginning and maintaining continuous evaluation of the dolphin-fishery interactions as a key element to define the population state and to supply conservation mechanisms. So far, three types of interactions were identified: operational, by parasite transmission and direct or indirect predation.

The *Sotalia* Case

The species of this genus are the most accidentally captured dolphins in the Amazonian Estuary, as well as through all its distribution range in the coastal waters of the western Atlantic Ocean, where there is a high human concentration and great development areas (Barros & Teixeira, 1994; Perrin et al., 1994; Sicilian, 1994). They are however, a protected species, being included in the CITES Appendix 1 and by specific laws in almost all the countries of its distribution area.

The Amazonian Estuary Fishery

The fishing activity is diversified, from artisan fishermen, with simple technology and small wooden vessels, to industrial fishermen with sophisticated technology and metal hulled vessels. The artisan fleet captures a wider range of fish species than the industrial fleet and its catch is unloaded mainly in the Ver-O-Peso market in Belém-PA, although there are fish landings along the State of Pará coast where the fish cargo is transferred to trucks that trade in more distant areas.

MATERIALS AND METHODS

Study Area and Samplings

The artisan fishing fleet used drift gillnet between 5° N and 1° S and between 47° to 51° W (Figure 1), to collect fish during two time periods, (October 1996 – February 1997 and January 1999 – November 2001).

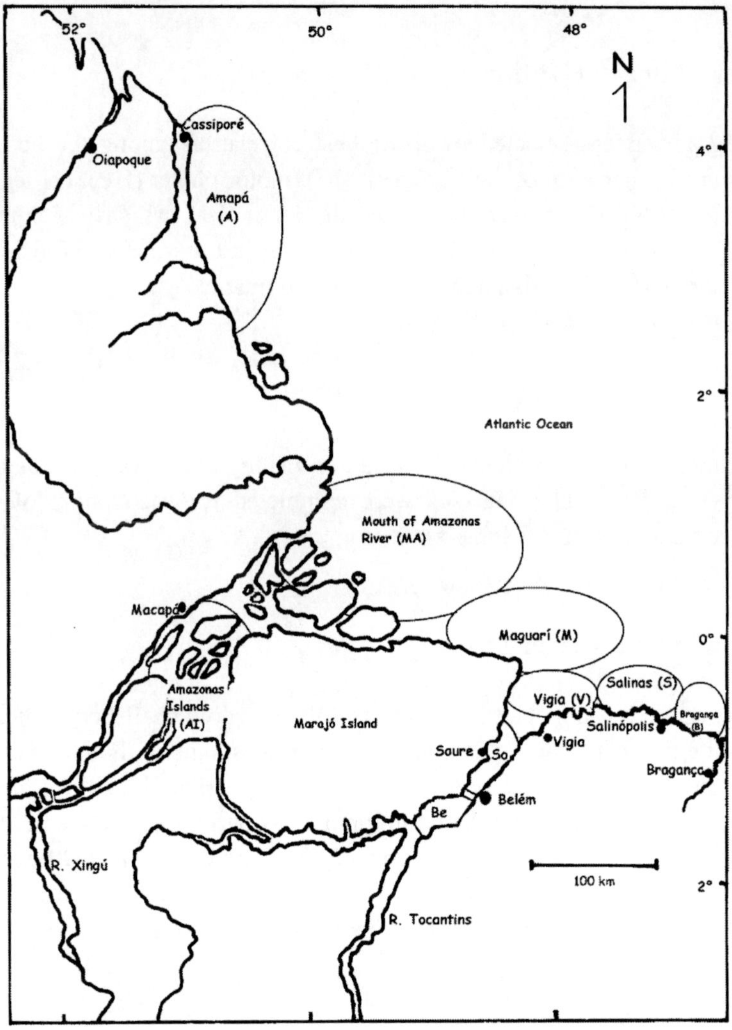

Figure 1. Artisan fishing fleet Strata, (stratum one and stratum 2) and drift gillnet sampling sites in the Amazonian estuary.

Fishing Fleet

Time and distance of the fishery divide the Vigia artisan fleet in two. One fleet of "arrives and turns", is based in the Municipality port, and consists of smaller, 12 meter long boats that use hooks, lines and monofilament nets. It supplies the fish market of the city and is operated by fishermen from Vigia or from communities of the interior of the State of Pará. And, fleet two, of "outside fishermen", is based in the Arapiranga port, and contains boats larger or smaller than 12 (this is different than the author's first chapter where the fleets were clearly divided into two groups, one with vessels larger than 12 meters and second group with vessels shorter than 12 meters long.) meters long and that depend on multifilament nets. Many of these fishermen are from cities such as Bragança, Salinas and Boa Vista. All the vessels of this port were sampled because dolphins are accidently caught here more than any other location.

Fishing Fleet Characterization

The fishing power was selected from the best correlation among the boat length (BL) with the variables: ice capacity of the basement (IC), motor power (HP), net length (NL) and fishermen number (FN). Hours were set as the time unit (Hr), which was equivalent to the throwing number (TN) times 5 (average time the net was in the water). The fishing effort was calculated as the product of the fishing power units and time.

Interaction Dolphin-Fishery

The correlation between accidental capture and fishing effort was used as an interaction dolphin-fishery parameter. The analyses were completed for each phase of the research, separated and compared among fishing strata.

Socio-Economic Considerations

The socioeconomic and cultural aspects of the dolphin-fishery interaction were investigated while the fish trade and employment contracts, among the members of a fishing unit, were verified. When there was a dolphin accidentally captured the fishermen were asked the intended destination of the animal. If a dolphin or its parts were to be sold, the fishermen were also asked, which parts and for what prices? Finally, we made a cost-benefit evaluation based on the price of the fish market, the fish number that would occupy the equivalent volume of a dolphin, the capacity of the boat basement and the prices obtained by the sale of dolphins or parts of them.

RESULTS

Fishing Fleet

There were 501 wood vessels between the municipality port and Arapiranga in Vigia. Of this complement, 24 moved by oar, 115 moved by sail and/or oar without deck and 362 deck/cabin boats moved by a combination of motor and sail. Their fishing gear consisted of hook and line, mono and multithreaded nets, cast net and puçá (a fish gear that usually catches medium and small fish). The fishing product supplies the local market, the Ver-O-Peso market in Belém (Pará), the Vigia fish processing company and the cities of São Luís (MA) and Fortaleza (CE).

The vessel registration price for the captaincy in the ports divides the fleet into two strata: smaller and larger than 12m in length. Slightly over 80% of the boats were manufactured in Vigia; all with deck, cabin and sail. Each sail was assisted by a motor to help in the navigation of Marajó Bay. Independent of the stratum, there was a strong correlation between the boat length and the variables: basement ice capacity ($r = 0.85$), motor power ($r = 0.80$) and fishermen number ($r = 0.79$), net length and fishermen number ($r = 0.74$); and a moderate correlation between boat length and net length ($r = 0.66$).

Stratum 1

"Amazônidas", fishing fleet one, comprised of smaller sized vessels, actively fished in Marajó Bay as well as in the Amazon River's mouth. They represented 76.7% ($n = 189$) of the boats between lengths of 6 m and 11.8 m ($\overline{X} = 8.9$; $S = 1.26$). These boats were powered by 9 to 160 HP engines ($\overline{X} =18$), had ice cargo holds of 0.2 to 9.0 tones ($\overline{X} = 4.9$; $S = 1.83$), crew size of 2 to 7 ($\overline{X} = 4$), and net lengths from 400 to 2,500 br ($\overline{X} = 750$; $S = 308.65$). The correlation coefficient was moderate ($r = 0.527$) between BL/IC and low between BL/HP, BL/FN and BL/NL.

Stratum 2

"Northern" outside fishermen in fleet two, fished along the coast of the State of Amapá. They comprised 23.3% ($n = 44$) of the total number of boats vessels assessed and were from 12 to 32 m in length ($\overline{X} = 14.9$; $S = 3.57$). Their boats were powered by with engines ranging from 18 to 230 HP ($\overline{X} = 69$) as well as by sails. They had 1.5 to 35.0 t potential capacity in their cargo holds ($\overline{X} = 12.4$; $S = 6.51$), crews from 8 to 12 men ($\overline{X} =8$), and used multifilament nets with lengths of 900 to 3,500 br ($\overline{X} = 2,027.5$; $S = 508.31$). The correlation coefficients were low between BL/NL; BL/FN and NL/FN, moderate between BL/IC, and strong between BL/HP ($r = 0.755$).

Fishing Effort (E)

The crew size (FN) presented the best correlation with the fish capture ($r = 0.70$), when the fishing effort unit was defined as FN x Hr. The multiple regression values were significant among capture and IC, NL and FN (at 95%) in stratum one and between NL and FN for stratum two. In both cases there was no tolerance. Besides, co-linearity between IC and FN and between IC and NL was detected; confirming FN x Hr as the unit of fishing effort.

Fishery

Stratum fleet one follows the seasonality of the estuary. From January to May, the Amazon river's discharge is at a maximum and its great force pushes the sea outside the bay. The fishermen who fish piramutaba (*Pseudoplastistoma vailantii*) stay in the Marajó Bay, while the ones who capture pescada (*Cynoscion acoupa*) go to saltier grounds. From June to December the river flow decreases and the sea approaches the coast and all the vessels fish in the Marajó Bay. The stratum two fleet always fishes in the coast of the State of Amapá (Figure 1).

When fishing, fishermen await the high tide to throw the net that drifts until the tide begins to lower (5 hours on average). Pulling the net occupies at least two fishermen; another one pulls fish out of the net, while one-two others gather the net again. Once the whole net is on the boat, the fish is gutted and stocked in the cargo-hold with ice.

Accidental Capture of Dolphins

In the two phases of the research, the analysis of linear regression between fishing effort and the accidental capture of dolphins was significant for stratum one, however the correlation coefficients were very low ($n_1 = 350$; $r_1 = 018$; e $n_2 = 590$; $r_2 = 0.12$). They were not significant for stratum two.

Socio-Economic and Cultural Aspects

In the two phases of the research the percentages of trips with accidental capture were similar and the averages did not present significant differences. Approximately, 33.4% of the trips had accidentally captured dolphins. Of the animals killed 95.77% were thrown back into the sea after eyes (14.11%), face (9.03%); penises or genital area (11.0%) parts were extracted. The animal carcasses brought to port were analyzed by researchers and later used by the fishermen as bait.

There were no significant variations in the listed sale prices of dolphins or their parts, in the port of Vigia, during the two phases of the project. Small animals were used as bait and sold for between 5.00 and 10.00 R$ (Brazilian Reais). Large animals fetched higher prices (15.00 to 20.00 R$). Genital (1.00 to 3.00 R) complete jaws (2.00 to 5.00 R), teeth (0.30 R each), and eyes (1.00 to 3.00 each) each had listed market prices. Other, un-sold body parts were used for fat and milk.

Fishermen do not normally bring dead dolphins to the port for fear of the "panema" (bad luck) and a low return for such service. Dolphins also occupy needed cargo space for at least, four medium size "pescadas amarelas" (*Cynoscion acoupa*), a medium size fish of greater commercial value (5kg/each). In the best of the situations the fisherman earns R$ 30.00 with the sale of one dolphin; while on average he earns R$ 2.00 per kg of pescada (first phase of the project) up to R$ 6.00 (second phase of the project). In other words, for four pescadas, the fisherman would earn R$ 40.00 to R$120.00. Besides the low profit for the sale of dolphins or their parts, nets are destroyed when dolphins are caught, and fishermen have to stop fishing due to the loss of their net.

The top economic positions of the artisan fishery are held by local intermediate men. They finance the trips, boat construction, repairs and reforms of the boats or the nets, and buy the fish. They facilitate the fishermen in making the transition to production, through the purchase of supply materials, fishing instruments or even providing money. When a producer lacks capital, they address the local mercantile leadership because they are more directly linked to the market. Soon afterwards it is the owner of the boat and nets that is usually in debt with the fishing boss and does not have money for the trip expenses. After payment of the trip expenses the owner of the boat receives 50% of the fish sale profit. Then the person in charge of the vessel; the trusted individual with fishery experience, receives 30% of the previous balance. The gelador (the person who is responsible for freezing), who is in charge with the disposition of the fish in the cargo-hold and is also an experienced person, receives 30% of the remaining profit monies. Finally, the fishermen divide the remaining balance in equal portions amongst themselves.

After the trip, the net has to be repaired, the motor serviced, and the boat painted or repaired. A series of unexpected situations can quickly lead to the fishermen having debts. For example they assume debt to the boat owner when they borrow monies for support of their families and houses. Such debts prevent the fishermen from having better economical status.

Historically, when dolphins were caught in times of regular or poor fishing, the fishermen would take advantage of the situation and remove parts of the animals and sell them. Nowadays the dolphin market has been decreasing because of the IBAMA surveillance, and therefore, the distribution of dolphin parts in the Ver-O-Peso market in Belém has sharply reduced. This is not true in all cases where intermediate men (one-two individuals) have increased the price of dolphin parts 200%, especially eyes and genital. The teeth are usually purchased by artisans.

Interaction Dolphin-Fishery

The causal loop in the dolphin-fishery interaction model in Figure 2 depicts how economic factors influence the fishing effort and how this effort influences the accidental capture of dolphins. Data support that this type of interaction (in this case) is indirectly predatory because dolphins and target fish of the fishery feed on the same prey and it is during this feeding behavior that these mammals are arrested in the nets.

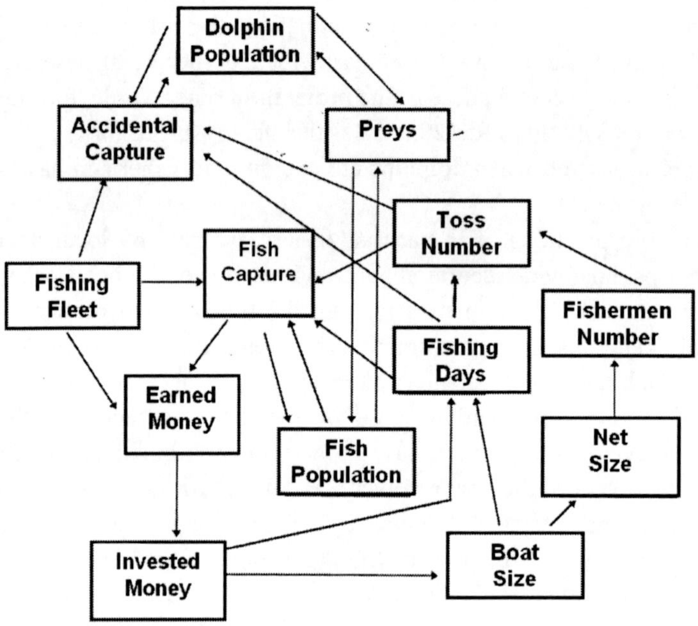

Figure 2. Causal loop Model explaining the dynamics of the interaction dolphin-fishery in the Amazonian Estuary.

THOUGHTS AND CONCLUSION

The largest problem in the characterization of the artisan fishing fleet is its structural heterogeneity, supported in the low correlation coefficients among the vessels' structural components. The variable with the largest correlation value and that best described the fleet's fishing power was cargo hold size. As noted by Refskalefsky (1985), boats with cargo holds are at an advantage compared to other vessels since they can preserve the fish in ice. This advantage explains why the number of boats with cargo holds has increased quickly. That structural adoption requests a great investment for most of the fishermen but results in the likelihood of extending the length of fishing trips to 15 days fishing, without returning to port, and also of selling the fish in Vigia or Belém reaching a better price. However, the cargo hold is not always filled in a trip. Crew size (FN) is an appropriate measure of the fishing power for both strata, preferable to the use of questionable evaluation variables which can create error. This measure was used by Petrere (1978) in the fishing effort in the Middle Amazon,

A heterogeneous fleet is vulnerable to the economic changes and challenges of the fish market. Meeting net and motor expenses can be quite challenging, and large costs may prevent the vessel's owner from adapting and utilizing more appropriate equipment. Larger nets implicate a larger crew size and expenses increase.

The low correlation coefficients between capture and fishing effort indicate that other effort factors act on capture but are not considered. Petrere (1983) highlighted biological processes, edaphic and morphologic fishing factors and fishing strategies that vary with the level of fishing experience and that influence capture frequency.

The emergence of fish-towns and the introduction of motors and nylon nets by fishermen signify the increased intensification of the fish trade in the Amazonian Estuary (Nacif, 1994). Besides those technological changes cultural mixing has occurred, creating a new hybrid culture. Estuary fishermen, even in Vigia, are from several areas, and bring with themselves their own faiths, fears and experiences. Therefore it is commonplace to listen to legends about witch crafted botos and charming entities that steal women and leave them pregnant; or, to hear of magic powers attributed to dolphin fat and milk that cure diseases (Slater, 1994).

For many years, these faiths have been incorporated into the fishermen's culture and they have been serving as a theme for speculations posted in newspapers, leaving the fisherman in an unfavorable position. The intentional capture of dolphins in the estuary to be used as shark bait has dramatically dropped (Sicilian, 1994). Now, after the break-up of this fishery, on occasion, dolphin meat is used with hook and line where one small dolphin provides enough to bait more than a thousand fishhooks.

In several areas of the geographical distribution and especially in the Amazon it is known that organs or parts of *Sotalia fluviatilis*, *Sotalia guianensis* and *Inia geoffrensis* are used as amulets and fetishes (Borobia & Rosas, 1991) and that fishermen consider dolphins as competitors in the fishery; they do not like them because they destroy or steal fish from their nets.

ACKNOWLEDGMENTS

We want to thank the National Institute of Amazonian Research (INPA) and the Ministry of Science and Technology (CNPq), for the logistical and economical support in the development of this research project. Special thanks go to Vera da Silva (INPA, Brazil). Also special thanks to the fishermen of Arapiranga port in Vigia (PA), for their collaboration and the information they provided during this research project.

REFERENCES

[1] Baros, N. & Teixeira, R. L. (1994). Incidental catches of marine tucuxi, *Sotalia fluviatilis*, in Alagoas, northeastern Brazil. In: Perrin W., G. Donovan and J. Barlow (Eds.), *Gillnets and Cetaceans*. Report of the International Whaling. Commission. (Special Issue 15, pp. 265 -268). Cambridge, United Kingdom: International Whaling Commission.

[2] Borobia, M. & Rosas, F. (1991). Tucuxi *Sotalia fluviatilis* (Gervais, 1853). In: Capozzo H.W. & M. Junin (Eds.), *Estado de conservación de los mamíferos marinos del Atlantico Sudoccidental. Informes y Estudios del Programa de Mares Regionales del PNUMA. Nairobi, PNUMA. (138):*36-41.

[3] Fabré, N.N. & Batista, V.S. (1992). Análise da frota pesqueira artesanal da comunidade

[4] Da Rapôsa, São Luis, MA. *Acta Amazonica*, 22(2), 247-259.

[5] Gulland, J.A. (1983). *Fish stock assessment. A manual of basic methods*. (Volume 1).

[6] Hoboken, New Jersey: John Wiley & Sons.

[7] Hall, M. (1995). Atunes y delfines en el océano Pacífico oriental: Situación actual y perspectivas de pesca e investigación. In: FUDECI. (Eds.), *Delfines y otros mamíferos acuaticos de Venezuela: una politica para su conservation. First Edition* (pp. 45-62). Valencia, Venezuela: FUDECI,

[8] Hoel, A. (1990). Norwegian marine policy and the International Whaling Commission. *North Atlantic Studies*, 2(1-2), 117-123.

[9] Nacif, A. M. P. (1994). Pesca artesanal, aspectos ambientais, sócio-econômicos eculturais -o cais de Marudá /PA. *Estudos do NUMA* UFPA, 5, 42.

[10] Perrin, W. Donovan, G. & Barlow, J. (1994).*Report of the workshop on mortality of cetaceans in passive fishing nets and trps. Gillnets and cetaceans. Report of the International. Whaing Commisson,* (Special Issue 15, pp. 1-72). Cambridge, United Kingdom: IWC.

[11] Petrere, M. (1978). Pesca e esforço de pesca no Estado do Amazonas. I. Esforço e Captura por unidade de esforço. *Acta Amazonica*, 8(3), 439-454.

[12] Petrere, M. (1983). Relationships among catches, fishing effort and river morphology for eight rivers in Amazonas States (Brazil), during 1976 - 1978. *Amazoniana*, VIII(2), 281-296.

[13] Refkalesfky, V. (1985) *Os parceiros do mar. Natureza e conflito social na pesca da Amazônia*. Belém, Brazil: Editora CNpq.

[14] Reeves, R. & Leatherwood, S. (1994). *Dolphins, porpoises, and whales: 1994-1998 Action Plan for the Conservation of Cetaceans*. Gland, Switzerland: IUCN/SSC.

[15] Siciliano, S. (1994). Review of small cetaceans and fishery interactions in coastal watersof Brazil. In: Perrin W., G. Donovan and J. Barlow (Eds.), *Gillnets and cetaceans*. Report of the International Whaling Commission. (Special Issue 15, pp. 241 -259). Cambridge,United Kingdom: International Whaling Commission.

[16] Slater, C (1994). *Dance of the dolphin, Transformation and disenchantment in the Amazonian imagination*. Chicago, Illinois: The University of Chicago Press.

In: Biology, Evolution and Conservation of River Dolphins... ISBN: 978-1-60876-633-8
Editors: M. Ruiz-García, J. Shostell. pp. 247-260

Chapter 13

ETHNOECOLOGY OF *SOTALIA GUIANENSIS* (GERVAIS, 1853) IN THE AMAZON ESTUARY

Sandra Beltrán-Pedreros[1]*, Miguel Petrere[2] and Ligia Amaral Filgueiras-Henriques[3]+

[1]Projeto PIATAM. Área de Mamíferos Aquáticos. INPA (CPBA)-UFAM, Rua Ajuricaba, Aleixo, Manaus, Brasil

[2]Universidade Estadual de São Paulo (UNESP), Departamento de Ecologia, Rio Claro, Brasil

[3]Universidade Estadual do Pará (UEPA), Departamento de Ciências Naturais, Rua do Uma, Belém, PA

ABSTRACT

This chapter describes a study conducted on the ecology of *Sotalia guianensis* in the Amazon estuary from 1999 to 2001, using participatory research with methodology. Interviews of 150 fishermen across 11 towns as well as surveys of the estuary by boat were completed to obtain information regarding *S. guianensis* in relation to their group size, habitat fidelity and calf-dynamics. Interactions between the ecological variables were tested using a log linear analysis of frequency tables for three factors. The results indicate that the *S. guianensis* is a gregarious species, forming groups of two or three individuals. However, groups with more than 10 individuals and herds of up to 150 were not rare. Group size was related to the behavior and kind of habitat used. In this study dolphins were commonly observed in large groups, feeding and swimming in open water habitats, however they were rarely observed in ports and near human communities. Habitats such as "igarapés", lagoons and exposed coastal beaches were visited by the dolphins in the last hours of rising tide, high tide and the beginning of receding tide, when depth facilitated the exploration of the habitat.

Keywords: *Sotalia guianensis,* ethnoecology, Amazon estuary.

---

* beltranpedreros@hotmail.com.
+ ligiafilgueiras@gmail.com.

Introduction

The taxonomy and systematic of *Sotalia* has been reassessed and changed. Until recently, it was considered mono-specific (*Sotalia fluviatilis*) having two subspecies (geographical forms or ecotypes), a marine one distributed from Nicaragua (Carr & Bonde, 2000) to Santa Catarina in Brazil (Simões-Lopes, 1988) and a second, fluvial one found in the Amazon river basin (Borobia et al., 1991). Their recognition as separate *Sotalia* species was based on cranial geometric morphometric (Monteiro-Filho et al., 2002) and molecular studies (Caballero, 2006; Caballero et al., 2007; Cunha et al., 2005)

To date, the scientific information collected about the species are related to reproduction, growth, abundance and ecology as well as feeding habits, behavior as it relates to habitat use, group size, and social structure (Best & Silva, 1984; Borobia & Barros, 1989; Silva & Best, 1994; Vidal et al., 1997; Ramos et al., 2000; Edwards & Schnell, 2001; Santos et al., 2000, 2001, 2002). But little has been accomplished in ethnobiological or ethnoecological approaches.

Inclusion of ethnoecology studies facilitate our understanding of inter-relational complexities between the human populations and the natural resources, with special attention given to perception, knowledge and use (Begossi et al., 2002). However, the inadequacy of that interface becomes clear when techniques that are adapted to the interdisciplinary objectives of the researches are sought. Although the domain of basic techniques in anthropological research, is dependent on interviews, and that for biological research, which is dependent on the collection of specimens, are primordial, the growing use of approaches and techniques of other disciplines has become even more essential (Marques, 2002).

Materials and Methods

The research was developed in the Amazonian estuary (1°S - 4°40'N and 47° - 51°W; Figure 1) from 1999 to 2001. The Amazonian estuary extends from the mouth of the Amazon River in the north and to the mouth of the Tocantins River in the south, linked through the Breves Strait. Due to the enormous discharge of the two rivers the estuary is mainly fresh water in the rainy period and slightly saline in the summer (Barthem & Schwassmann, 1994).

We utilized an ethnoecological approach experimental design, based on participatory observation technique of fishermen during their routine labor work to meet the objective of identifying, describing and quantifying the knowledge of human populations within the Amazonian Estuary about the ecology and behavior of *Sotalia guianensis*. Semi-structured interviews (n = 156) were applied, based on scientific information that contained the variables to be studied: areas (A), degree of species occurrence (OD), habitat characteristics where the species are found (H), group size (GS), frequency of calf observation (C), and behavior (B).

The variables OD and C were set up into three categories: **Common,** when fishermen observe dolphins in almost all of their fishing routine; **Seasonal,** when fishermen observe dolphins exclusively during the high tide and **Rare** when fishermen observe dolphins in other periods of the year.

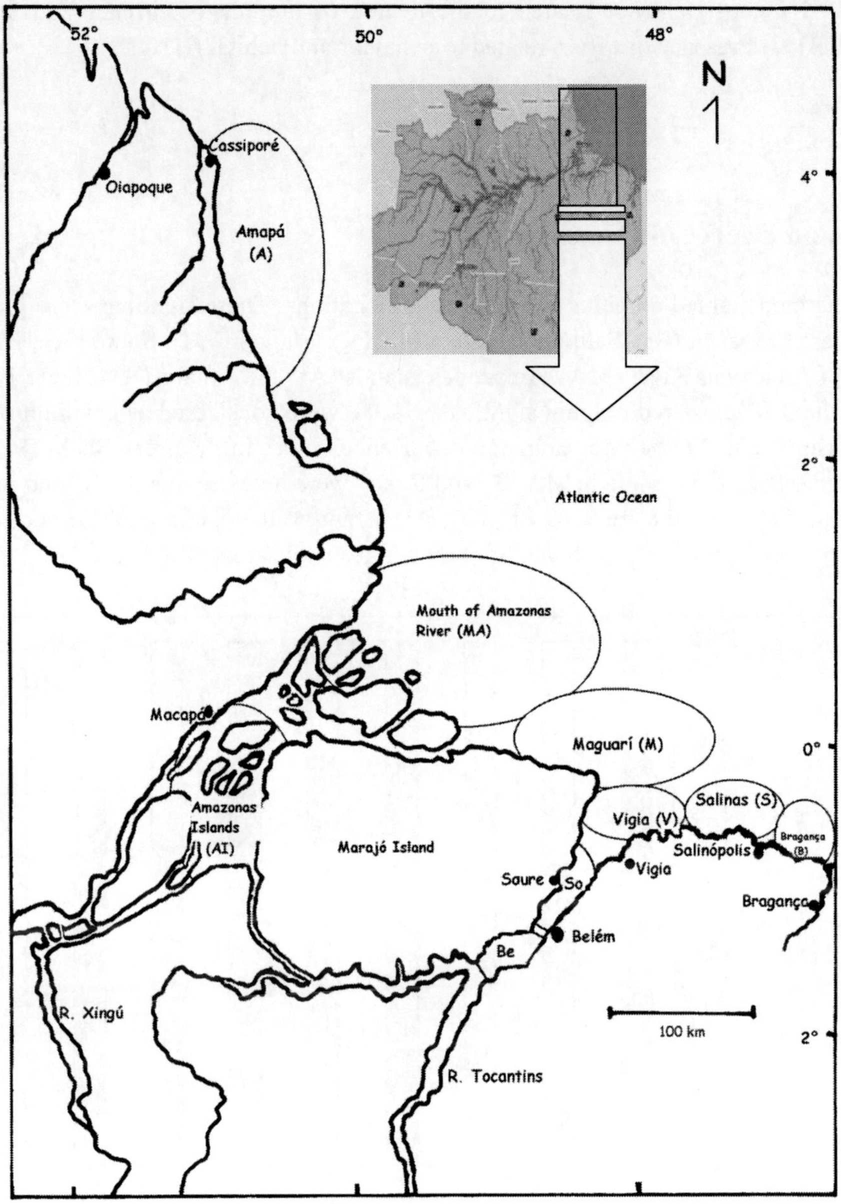

Figure 1. The Amazonian Estuary area sampled and occurrence of *Sotalia guianensis*.

For the organization and analysis of data we considered the observation by fishermen as complete, when they provided answers to all the considered variables. The data were summarized in tables of absolute frequency, used to construct three-way contingency tables and then analyzed with a log-linear model for categorical data (Sokal & Rohlf, 1995).

The analyzed interactions were: 1) Behavior related to group size and habitat (H*GS*B), 2) Presence of calves related to group size and habitat (H*GS*C), 3) Degree of Dolphin Occurrence related to group size and habitat (H*GS*OD), 4) Behavior related to the group size in each sampling area (A*GS*B), 5) Presence of calves related to habitat and sampling area (A*H*C), 6) Behavior related to the degree of dolphin occurrence and habitat type (H*OD*B), 7) Presence of calves related to behavior and habitat (H*C*B).

RESULTS

Areas and Degree of Occurrence

Fishermen sighted dolphins in a total of 69 locations. These sightings were grouped into nine areas: Bragança (Br), Salinópolis (S), Vigia (V), Maguarí (M), Soure (So), Belém (Be), Mouth of Amazonas River (MA), Amazonas Islands (AI) and Amapá (A) (Figure 1).

Of the 314 registered dolphin sightings, 64.3% were considered as common occurrence, 23% seasonal and 12.7% rare. Common occurrence values in the areas Br, M, V, A and AI were higher than 10%, while in MA, Be and S they were between 5 and 10% and in So it was 3% (Table 1). Inside the areas A, Br, MA, AI, M and V the common occurrence value was greater than 60%, while in Be, S and So it ranged from 37% to 46% (Figure 2).

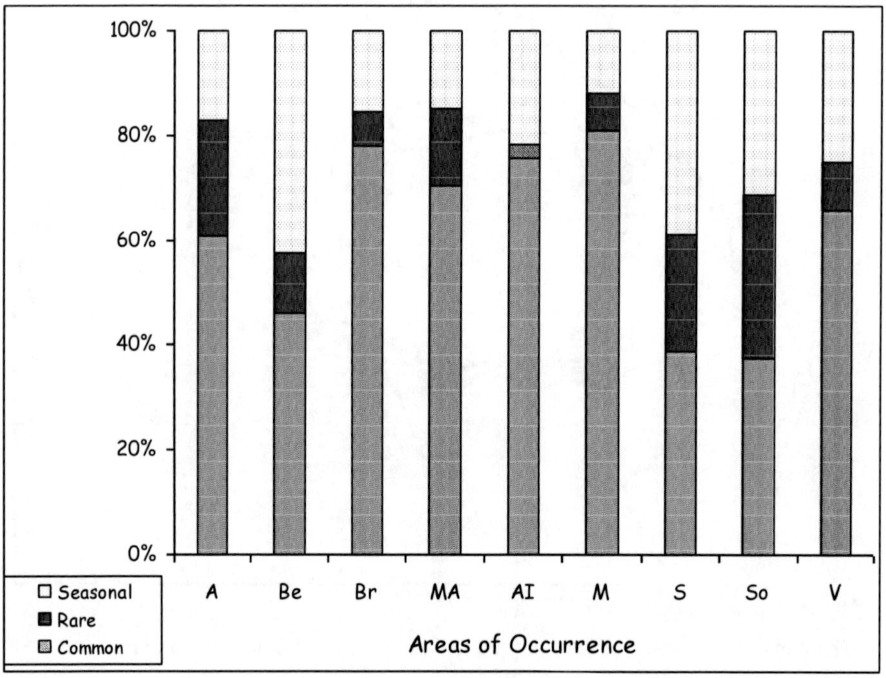

Figure 2. Percent occurrence of *Sotalia guianensis* within areas of the Amazonian Estuary from 1999 to 2001 (Br = Bragança, V = Vigia, S = Salinópolis, M = Maguari, So = Soure, Be = Belém, AI = Amazon islands, MA = Mouth of Amazonas River, A = Amapá).

Table 1. Degree of occurrence of *Sotalia guianensis* (N=314) in the Amazon Estuary between 1999 and 2001 by Area

Area	Common	Rare	Seasonal
Amapá	12.4	22.5	9.7
Belém	5.9	7.5	15.3
Bragança	17.3	7.5	9.7
Mouth of Amazon River	9.4	10	5.5
Amazon Islands	13.9	2.5	11.1
Maguarí	16.8	7.5	7
Salinas	6.9	20	19.4
Soure	3	12.5	7
Vigia	14.4	10	15.3

Habitat

Four types of habitat were identified based on morphologic characteristics provided by fishermen of their dolphin sighting locations: 1) Open waters (OW) with depths from 2 to 6 m in the low tide, 20 m in the high tide and from 10 to 50 m in the mouth of the Amazon River; 2) Beach and/or Coast (BC), beaches and mangroves that are exposed during the low tide and, in the high tide (2-20 m); 3) Igarapés and/or Bays (IB), more protected spots that dry out in the low tide (low tide 1 to 2 m, high tide 4 to 8 m) and 4) Ports and Communities (PC), populated spots or tourism beaches.

The species was more frequently present in the BC habitat (41.1%) compared to the PC habitat (9.9%). The habitats OW, IB and BC, presented common occurrence values superior to 65% and were considered favorable for the species; while PC was not (Figure 3).

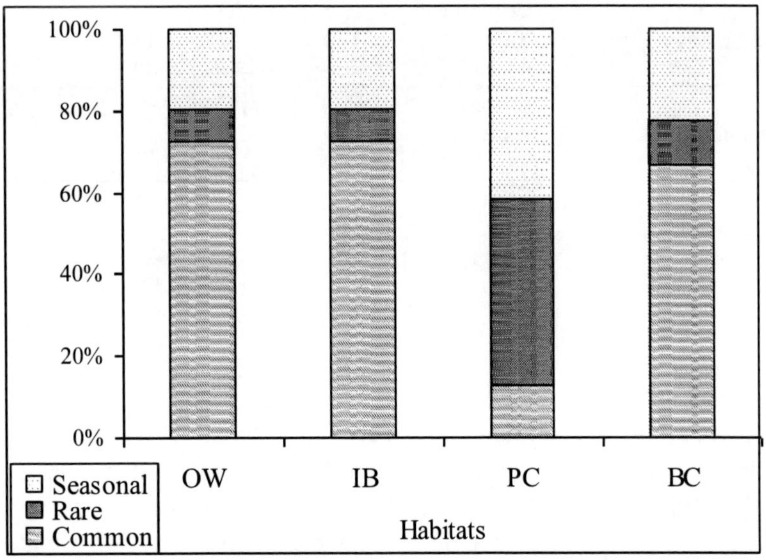

Figure 3. Percent occurrence of *Sotalia guianensis* in the Amazonian estuary between 1999 and 2001 in each of four habitats (OW = Open Waters, BC = Beach and/or Coast, IB = Igarapés and/or Bays, PC = Ports and Communities).

Group Size

Based on observations from fishermen, five group sizes of dolphins were defined: **Type I,** solitaire animals; **Type II,** 2 to 3 animals; **Type III**, 4 to 10 animals; **Type IV,** 11 to 30 animals and **Type V**, more than 30 animals.

The presence of calves among groups varied and was considered rare (48.3%), common (35.5%) or seasonal (19.4%). Their occurrence based on habitat type indicated that the calves were more common in OW (46.7%) and less common in IB (33.8%). They were more seasonal in BC (24.8%) and rare in IB (53.2%), while in BC there were no calves present in 96.8% of the cases. Within the OW habitat, the highest frequency of calf observation was in Type II groups, where they were considered common (66.7%). Calves were commonly observed in type V groups (50%) in this habitat (Table 2). Type II groups in the IB habitat were considered common and rare because both frequencies were 42.1%. But for BC, they were rarely in Type II groups (51%) and common in Type V groups (66.7%).

Calf observations inside groups of more than 10 animals were considered difficult in the OW and BC habitats on the part of the observers (the interviewees). However, calves were never observed alone in any habitat type.

Table 2. Percent occurrence of *Sotalia guianensis* calves in the Amazonian Estuary between 1999 and 2001 by size group (I: one animal, II: 2-3 animals, III: 4-10 animals, IV: 11-30 animals, V: >30 animals) in each habitat

Habitat	Occurrence	I	II	III	IV	V
	Common	-	66.7	52	28.6	50
Open waters	Rare	100	33.3	24	38.1	27.8
	Seasonal	-	-	24	33,3	22,2
	Common	-	42.1	42.9	75	66.7
Igarapés and/or Bays	Rare	100	42.1	14.2	-	33.3
	Seasonal	-	15.8	42.9	25	-
	Common	-	6.2	-	-	-
Ports and Communities	Rare	100	93.8	-	-	-
	Seasonal	-	-	-	-	-
	Common	-	32.1	38.4	26.7	66.6
Beach and/or Coast	Rare	100	50.9	30.8	33.3	16.7
	Seasonal	-	17	30.8	40	16.7

Behavior

Fishermen's narratives identified dolphin feeding behavior (FB) such as pursuit of fish, continuous immersions in the same spot with fish activity, and special formations for capturing fish in fishing spots *and close to corrals.* They also identified dolphin movement behavior (MB), like swimming in one direction without evidence of immediate return to the observation spot or following vessels.

Fishermen from the Br and V areas, who work with corralled fish mentioned fish being persecuted by groups from two to four dolphins during the ebb tide towards the curral (habitat IB). In this case, dolphins used the corral as a barrier to make their fish capture easier.

Another fishing type described was in BC, at depths between 5 and 10 m, when groups from 4 to 10 dolphins dove many times, without any specific organization, to capture yellow mouth salmon (*Macrodon ancylodon*) or White mullet (*Mugil curema*).

A more elaborate type of feeding strategy was mentioned for the open water (OW) habitat. Groups in excess of 10 animals were positioned in a line, a circle and/or a "V", to trap cutlass fish (*Trichiurus lepturus*) that swim in the stocking water during the day. The stomach contents of *S. guianensis* within the study area (N = 50) contained *Trichiurus lepturus* (Beltran-Pedreros & Araújo, 2007).

Fishermen's observations also indicated a tendency for dolphins to have similar frequencies for feeding (51.9%) and moving in BC (48.1%) and in OW (54.5% and 45.5%). In IB the feeding frequency in relation to the movement was higher (68.8% and 31.2%). There were no accounts of dolphins feeding in the PC habitat, but they did display movement behavior. When the dolphin groups were in "common occurrence" they were more often observed feeding (63.4%) than moving (36.6%), while for groups of dolphins of rare occurrence, they were often observed moving (80%).

Calf presence during the behavior observation was indicated in most cases as rare (46.8%). But the results varied when each behavior was analyzed (Table 3). The results for calf observation in different habitats for each behavior are in Table 4. Common or seasonally observed calves were best seen feeding than moving, while when they were rare they were best seen moving. Calf observation frequencies in each behavior varied according to the habitat and group size. In OW habitat they were commonly observed feeding (59.5%) especially in Type III groups (40%), while they were rarely observed moving (19.5%).

Table 3. Percent of *Sotalia guianensis* groups and occurrence of calves in the Amazon estuary between 1999 and 2001 for each behavior. I = one animal, II = from 2 to 3 animals, III = from 4 to 10 animals, IV = from 11 to 30 animals, V = > 30 animals

Behavior	Group Size					Calves		
	I	II	III	IV	V	Common	Rare	Seasonal
Feeding	29.3	59.7	59.2	50	30	75	32.7	55.9
Movement	70.7	40.3	40.8	50	70	25	67.3	44.1

In IB habitat, calves were commonly observed feeding in Type II groups (61.5%) and rarely alone and were observed moving 54.5% of the time. They were never observed feeding in port and community (PC) habitats and even their movements were considered rare when solitary (Type 1) (100%) or in small groups of two to three (Type II) (93.7 %). Calves in BC habitats were only observed commonly feeding in Type III groups (50%). For Type II groups, the occurrences of feeding behavior during common and rare categories were similar (13 and 12 %).

The most observed group sizes were Type II (37.9%) and Type III (26.7%) in BC habitat (44.5% and 61.9%). Solitary animals were more often observed in IB (53.7%) and in PC (36.6%), and only a few times in OW and BC (2.4% and 7.3%). Type V groups were frequently observed in OW (60%) while Types IV and V groups were not observed in the PC habitat. Type V groups were observed moving in 70% of the cases, as well as Type I (70.7%);

while Type II and III groups were more frequently observed feeding (59.7% and 59.5%, Table 4).

Table 4. Percent of *Sotalia guianensis* calves in the Amazon estuary between 1999 and 2001 by degree of occurrence, group size (I = one animal, II = 2-3 animals, III = 4-10 animals, IV = 11-30 animals, V= > 30 animals) and behavior for each habitat

Habitat	Behavoir	Occurrence	I	II	III	IV	V
Open waters	Feeding	Common	-	60	71.4	50	60
		Rare	100	40	14.3	16.7	-
		Seasonal	-	-	14.3	33.3	40
	Movement	Common	-	100	27.3	-	46
		Rare	-	-	36.4	66.7	38.5
		Seasonal	-	-	36.3	33.3	15.5
Igarapés and/or Bays	Feeding	Common	-	53.3	50	100	100
		Rare	100	26.7	16.7	-	-
		Seasonal	-	20	33.3	-	-
	Movement	Rare	100	100	-	-	100
		Seasonal	-	-	100	100	-
Ports and Communities	Movement	Common	-	6.3	-	-	-
		Rare	100	93.7	-	-	-
Beach and/or Coast	Feeding	Common	-	41.9	50	40	-
		Rare	100	38.7	20	20	-
		Seasonal	-	19.4	30	40	-
	Movement	Common	-	18.2	22.7	20	66.6
		Rare	100	68.2	45.5	40	16.7
		Seasonal	-	13.6	31.8	40	16.7

Observations of dolphin group size in decreasing order of frequency were: Type II (38%), III (31.2%), IV (13.4%), V (10%) and I (7.4%), although Type II and III groups also presented seasonal occurrences in 34.7% and 26.4% of the cases. Type III, IV and V groups were never seen in the PC habitat and the occurrence of Type I groups was rare (53.3%), while for Type II groups it was seasonal (43.7%). In the other habitats, the occurrences of different group sizes were similar, with the greatest percentage equal to 54% (Table 5).

Interactions (*) Among Variables

All of the interactions of three factors were not significant ($P > 0.05$, Table 6). This indicates there is no apparent association among the variables when analyzed simultaneously, but there can be when analyzed by pairs of variables:

H*GS dependent when associated to OD (degree of freedom, $df = 36$; $G = 151.7$; $P = 3.3\times10^{-16}$), B ($df = 24$; $G = 163.9$; $P = 0$) and C ($df = 36$; $G = 130.9$; $P = 1.1\times10^{-12}$).

H*OD dependent when associated to GS ($df = 30$; $G = 44.9$; $P = 0.03$) and B ($df = 12$; $G = 25.6$; $P = 0.01$).

H*B dependent when associated to GS ($df = 15$; $G = 64.9$; $P = 3.5\times10^{-8}$), OD ($df = 9$; $G = 35.3$; $P = 5.2 \times 10^{-5}$) and C ($df = 9$; $G = 24.2$; $P = 0.004$).

GS*OD independent when associated to H (df = 32; G = 29.05; P = 0.61), but dependent when associated to B (GR*C; df = 8; G = 16.4; P = 0.03).

H*C was dependent when associated to A (df = 54; G = 86.9; P = 0.003) and B (df = 12; G = 26.1; P =0.01) and independent when associated to GS (df = 30; G = 31.7; P = 0.38).

GS*B was dependent when associated to H (df = 16; G = 31.03; P = 0.01) and independent when associated to A (df = 36; G = 49.5; P = 0.06).

GS*C dependent when associated to H (df =32; G=70.3; P = 0.0001).

A*GS dependent associated to B (df = 64; G = 109.5; P = 0.0003).

A*H dependent when associated to C (df = 72; G = 192.7; P = $0.8x10^{-13}$).

C*B dependent when associated to H (df = 8; G = 42.1; P = $1.3x10^{-6}$).

A*C was independent to any kind of habitat (df = 64; G = 82.4; P = 0.06).

A*B was independent to any group size (df = 40; G = 46.4; P = 0.22).

Table 5. Percent observations of different group sizes (I = one animal, II = 2-3 animals, III = 4-10 animals, IV = 11-30 animals, V = > 30 animals) of *Sotalia guianensis* in the Amazon estuary between 1999 and 2001 by degree of occurrence in each habitat

Habitat	Occurrence	I	II	III	IV	V
Open waters	Common	100	75	80	66.7	66.7
	Rare	-	8.3	4	9.5	11.1
	Seasonal	-	16.7	16	23.8	22.2
Igarapés and/or Bays	Common	54.5	78.9	85.7	100	66.6
	Rare	18.2	2.6	-	-	16.7
	Seasonal	27.3	18.5	14.3	-	16.7
Ports and Communities	Common	6.7	18.7	-	-	-
	Rare	53.3	37.5	-	-	-
	Seasonal	40	43.8	-	-	-
Beach and/or Coast	Common	33.3	66	71.2	60	66.7
	Rare	-	17	1.9	13.3	33.3
	Seasonal	66.7	17	26.9	26.7	-

DISCUSSION

The ecology of *Sotalia guianensis* is thoroughly described in the scientific literature. Along its geographical distribution, where an ethnoecological approach is used to garner information, any collected observational data must first be verified and quantified because one of the greatest problems with obtaining information from human communities is the subjectivity and the truthfulness of their answers. In general, dolphin studies are always accompanied by the consideration of the communities' traditional information, but, these are almost always used as support elements for conservation and environmental education works.

Sotalia guianensis of the Amazonian estuary frequent coastal water habitats like bays, mangrove areas and even other estuaries, as has been documented for other points along its geographical distribution (Borobia et al., 1991; Edwards & Schnell, 2001). The Amazonian estuary offers a great variety of habitats as a consequence of the tide action and the discharge of the Amazon River causing this area to possess both marine and fluvial characteristics. The depth is determined by the tide effect and the sea-river interface which presents a width of hundreds of kilometers, according to the climatic period. For *S. guianensis*, the food availability and depth are important variables for the habitat permanence and use. But it is the depth in addition to dolphins that can restrict the distribution of some fish species.

Table 6. Statistics of the interactions among variables analyzed for *Sotalia guianensis* in the Amazon Estuary from 1999 to 2001 (A = Area; H = habitat characteristics where the species are found; GS = Group size; OD = degree of species occurrence, B = Behavior; C = frequency of calf observation; df = Degrees of freedom; G = Corrected value of the test; P = Probability). The significant probabilities are in bold

Interaction	df	G	P	Interaction	df	G	P
1. H*GS*OD	24	12.1	0.98	2. H*GS*B	12	9.36	0.67
H*GS to GO	36	151.7	**$3.3\text{x}10^{-16}$**	H*GS to B	24	163.94	**0**
H*OD to GS	30	44.9	**0.03**	H*B to GS	15	64.9	**$3.5\text{x}10^{-8}$**
GS*OD to H	32	29.05	0.61	GS*B to H	16	31.03	**0.01**
3. H*GS*C	24	13.2	0.96	4. A*H*C	48	46.9	0.51
H*GS to C	36	130.9	**$1.1\ \text{x}10^{-12}$**	A*H to C	72	192.7	**$5.8\text{x}10^{-13}$**
H*C to GS	30	31.7	0.38	A*C to H	64	82.4	0.06
GS*C to H	32	70.3	**0.0001**	H*C to A	54	86.9	**0.003**
5. A*GS*B	32	30,6	0.53	6. H*OD*B	6	1.3	0.97
A*GS to B	64	109.5	**0.0003**	H*OD to B	12	25.6	**0.01**
A*B to GS	40	46.4	0.22	H*B to OD	9	35.3	**$5.2\text{x}10^{-5}$**
GS*B to A	36	49.5	0.06	OD*B to H	8	16.4	**0.03**
7. H*C*B	6	10.4	0.10				
H*C to B	12	26.1	**0.010**				
H*B to C	9	24.2	**0.004**				
C*B to H	8	49.1	**$1.3\text{x}10^{-6}$**				

All the areas identified by the interviewees as places where species occurred were considered common. Only areas such as Belém, Salinas and Soure presented seasonal occurrence similar to the common one. Those were port areas, with high vessel traffic, as well as with leisure beaches. These characteristics in association with the tide system make the dolphins explore them more frequently in the high tide periods. It is also possible that *S. guianensis* in the Amazonian estuary avoids areas with heavy human intervention; different from other points of its geographical distribution.

Several authors indicated that lagoons, bays and estuaries as preferable habitat for this species, and to a lesser degree, the open sea (Borobia et al., 1991; Silva & Best, 1996; Carr & Bonde, 2000). While within the Amazonian estuary, *S. guianensis* will move among these habitats changing locations as they change activities. Thus, they were observed at beaches of exposed coast, as well as in igarapés, smaller estuaries and bays or mangrove lagoons. Apparently, differences in habits of this species are linked to the tide effect and the depth and distribution of its prey; relationships that have already been recognized for this species (Edwards & Schnell, 2001).

Estuaries are important areas for *S. guianensis* feeding where coast topography, water temperature, salinity and depth are decisive environmental variables for their prey distribution and abundance. According to Barthem (1985) the Amazonian estuary fishing stock distribution and abundance vary according to the depth and salinity.

In consideration of all the environmental characteristics that determine prey distribution, depth would be the most limiting factor. Prey search and hunt depend on this factor. Authors have reported the species likes to fish *at shallow depths* using the margin or beach or the fishing corral as a barrier to capture fish. This behavior is also common in the Amazonian estuary and it was also described by Monteiro-Filho (1995) in Paraná State coast. In the BC habitat, type II and III groups exhibited higher frequencies during feeding when they fished at random and/or formed attack lines on shoals and/or fish that escape from the fishing corrals.

In stream, bay and mangrove lagoon habitats (igarapés), where depth is critical during low-tide, dolphins were seen in the high tide and at the beginning of the ebb tide, when depth was sufficient for prey capture. There is an added benefit for the dolphins during the ebb tide because the current disperses the fish shoals and thus facilitating their capture.

Dolphin group size and feeding strategy were dependent on habitat characteristics as indicated by the positive interactions among group size, behavior, and habitat (GS*B*H). In igarapé, and channel habitats (IB) small groups of dolphins were observed as solitary animals (Type I), or in small groups of two to three (Type I, II), performing random fishing, without presenting special formations. In OW habitat, dolphins were most frequently found in group types III, IV and V. They demonstrated organized hunting strategies, with even line formations, in "V" and circle patterns, strategies that have already been documented for this species (Silva & Best, 1996).

Calf observations were dependent on dolphin behavior, habitat and group size, demonstrated by the significant interactions among these variables. But, Type II groups call for special attention, where group structure or group composition was linked to differences in the frequency of calf observations. In IB and BC habitats calves were more commonly observed feeding than moving. But, this frequency varied if the group was comprised of a calf and one or two adults (when they were considered common) or groups composed of a calf in the presence of another calf and/or young animal (when they were considered rare). Seasonal observations of calves for those habitats supported that feeding as well as moving were related to the periods that the tides allow for the exploration of these habitats, times when there is a reduced risk of injury.

Calf feeding was common for Type III groups in OW (40%) and BC (28.8%) habitats. But there were little evidence of groups of calves moving possibly because it is difficult to distinguish calves during successive dives and by the tendency of adults to protect them.

The most observed group sizes were of two (Type II) to ten (Type III) animals, but the observation frequency varied according to the habitat type and the dolphin activity. Therefore, habitat and behavior are the main factors that influence group size, as has been previously recognized by Edwards & Schnell (2001) and which demonstrates the significant dependence among the interaction variables.

In other studies with *S. guianensis*, group size varied from 1 to 30 animals, with groups of 2 to 10 animals being the most commonly observed (Edwards & Schnell, 2001). Those group sizes are common in internal areas of bays, while group sizes tend to be larger and more open, oceanic waters. Within the Amazonian estuary, Type I and II groups were more frequent in IB habitats. Type II and III groups were common in BC habitats and Type IV and V frequently occurred in OW habitats. There was a tendency of group size to increase as the habitat became wider and deeper (oceanic). Herds of hundreds of animals were observed in the largest habitats. That pattern is similar to that described for *Tursiops truncatus*, who observed groups from 6 to 15 individuals in coastal areas (Defran & Weller, 1999) and typically larger groups in the open sea (Shane, 1990).

A similar situation happens with *S. fluviatilis*. Their group size ranges between one and six animals, with groups of two being the most commonly observed (Silva & Best, 1996). The group formation of more than six animals are rare, but this group type becomes more common in lakes and wide rivers (ex: Solimões) during the flooding period.

The significant dependence among the variables: OD, H and B let us infer that there is a relationship between habitat use and an ecological variable that allows this use. The species invests most of its time in feeding and movement (Edwards & Schnell, 2001), but these activities are linked to food availability and depth. Depth directly affects the occurrence of dolphins because shallow depths do not allow movement without danger and because depth will also limit the distribution of dolphin food resources. Therefore, depth is the ecological variable that determines the habitat use of the species in the Amazonian estuary, which varies according to time of day and type of tide.

During the high and ebb tides the igarapés, bays and exposed coast beaches habitats, are explored by the species, for feeding as well as movement. Movement pattern and habitat use have already been described by Borobia et al. (1991), Silva & Best (1994), Edwads & Schnell (2001). Even though in the Amazonian estuary the food availability is not a limiting factor, the *S. guianensis* feeding habit in the studied area marks a preference for continuous occurrence of fish. The distribution of those prey species is linked to the salinity and depth in the system (Barthem, 1985) and, consequently, the dolphin groups would be in constant movement in the search of prey. This provides an explanation as to why movement behavior was many times associated with feeding.

ACKNOWLEDGMENTS

The present work was developed with the financial support of CNPq (Scholarship 143542/1998-2). Our gratitude to the National Institute of Researches of the Amazonia (INPA), to the members of the Aquatic Mammal Laboratory, Dr. Vera Silva and Dr. Fernando Rosas, and to the fishermen of the Amazonian estuary.

REFERENCES

[1] Barthem, R. (1985). Ocorrência, distribuição e biologia dos peixes da Baía de Marajó, Estuário Amazônico. Boletim do Museu Paraense Emilio Goeldi, *Série Zoologia*, 2(1), 49-69.

[2] Barthem, R. & Schwassmann, H. (1994). Amazon river influence on the seasonal displacement of the salt wedge in the Tocantins river estuary, Brazil, 1983-1985. Boletim do Museu Paraense Emilio Goeldi, *Série Zoologia,* 10(1), 119-130.

[3] Begossi, A., Hanazaki, N. & Silvano, R. (2002). Ecologia humana, etnoecologia e conservação. In M. Amorozo, L. Ming, S. & Silva. (Eds.), *Métodos de coleta e análise de dados em etnobiologia, etnoecologia e disciplinas correlatas* (pp. 17-32). Rio Claro, São Paulo. Editora CNPq, UNESP, SBEE.

[4] Best, R. & Silva, V. (1984). Preliminary analysis of reproductive parameters of the bouto, *Inia geoffrensis*, and the tucuxi, *Sotalia fluviatilis*, in the Amazon River system. In W. F. Perrin, R. L. Brownell Jr. & D. P. DeMaster (Eds.), *Report of the International Whaling Commission* (Special Issue 6, pp. 361-369). Cambridge, UK: International Whaling Commission.

[5] Borobia, M. & Barros, N. (1989). Notes of the diet of marine *Sotalia fluviatilis. Marine Mammal Science,* 5(4), 395-399.

[6] Borobia, M., Siciliano, S., Lodi, L. & Hoek, W. (1991). Distribution of the South American dolphin *Sotalia fluviatilis. Canadian Journal Zoology,* 69, 1025-39.

[7] Caballero, S. (2006). *Genetic characterization of South American coastal and riverine dolphins Sotalia. PhD dissertation.* Auckland, New Zealand. The University of Auckland.

[8] Caballero, S., Trujillo, F., Vianna, J. A., Barrios-Garrido, H., Montiel, M. G., Beltrán-Pedreros, S., Marmontel, M., Santos, M. C. O., Rossi-Santos, M., Santos, F. R. & Baker, C. S. (2007). Taxonomic status of the genus *Sotalia*: species level ranking for "tucuxi" (*Sotalia fluviatilis*) and "costero" dolphins (*Sotalia guianensis*). *Marine Mammal Science,* 23, 358-386.

[9] Carr, T. & Bonde, R. (2000). Tucuxi (*Sotalia fluviatilis*) occurs in Nicaragua, 800 km north of its previously known range. *Marine Mammal Science*, 16(1), 447-452.

[10] Cunha, H. A., Da Silva, V. M. F., Lailson-Brito, J. J., Santos, M. C. O., Flores, P.A.C., Martin, A. R., Azevedo, A. F., Fragoso, A. B. L., Zanelatto, R. C. & Solé-Cava, A. M. (2005). Riverine and marine ecotypes of *Sotalia fluviatilis* are different species. *Marine Biology,* 148, 449-457.

[11] Defran, R. & Weller, D. (1999). Occurrence, distribution, site fidelity, and school size of bottlenose dolphins (*Tursiops truncatus*) off San Diego, California. *Marine Mammal Science*, 15(2), 366-380.

[12] Edwards, H. & Schnell, G. (2001). Status and ecology of *Sotalia fluviatilis* in the Cayo Miskito Reserve, Nicaragua. *Marine Mammal Science,* 17(3), 445-472.

[13] Marques, J. (2002). O olhar (des)multiplicado. O papel do interdisciplinar e do qualitativo na pesquisa etnobiológica e etnoecológica. In: M. Amorozo, L. Ming, & S. Silva (Eds.), *Métodos de coleta e análise de dados em etnobiologia, etnoecologia e disciplinas correlatas* (pp. 52-71). Rio Claro, São Paulo. Editora CNPq, UNESP, SBEE.

[14] Monteiro-Filho, E. (1995). Pesca interativa entre o golfinho *Sotalia fluviatilis guianensis* e a comunidade pesqueira da região de Cananéia. *Boletim do Instituto de Pesca*, 22(2), 15-23.

[15] Monteiro-Filho, E., Monteiro, L & Reis, S. (2002). Skull shape and size divergence in dolphins of the genus *Sotalia*: A tridimensional morphometric analysis. *Journal of Mammalogy*, 83(1), 125-134

[16] Ramos, R., Di Beneditto, A. & Lima, N. (2000). Growth parameters of *Pontoporia blainvillei* and *Sotalia fluviatilis* (Cetacea) in northern Rio de Janeiro, Brazil. *Aquatic Mammals*, 26(1), 65-75.

[17] Santos, M., Rosso, S., Siciliano, S., Zerbini, A., Zampirolli, E., Vicebte, A. & Alvarenga, F. (2000). Behavioral observations of the marine tucuxi dolphin (*Sotalia fluviatilis*) in São Paulo estuarine waters, Southeastern Brazil. *Aquatic Mammals*, 26(3), 260-267.

[18] Santos, M., Acuña, L. & Rosso, S. (2001). Insights on site fidelity and calving intervals of the marine tucuxi dolphin (*Sotalia fluviatilis*) in south-eastern Brazil. *Journal of Marine Biological Association of the UK.* 81, 1049-1052.

[19] Santos, M., Rosso, S., Santos R., Lucato, S. & Bassoi, M. (2002). Insights on small cetacean feeding habits in southeastern Brazil. *Aquatic Mammals*, 28(1), 38-45.

[20] Shane, S. (1990). Behavior and ecology of the bottlenose dolphin at Sanibel Island, Florida. In: S. Leatherwood & R. Reves (Eds.), *The Bottlenose Dolphin.* (pp. 245-266), San Diego, California: Academic Press.

[21] Silva, V. & Best, R. (1994). Tucuxi *Sotalia fluviatilis* (Gervais, 1853). In: S. Ridway & R. Harrison (Eds.), Handbook *of Marine Mammals*, (pp. 43-65) London, England: Academic Press.

[22] Silva, V. & Best, R. (1996). *Sotalia fluviatilis*. *Mammalian Species*, 527, 1-7.

[23] Simões-Lopes, P. (1988). Ocorrência de uma população de *Sotalia fluviatilis* (Gervais, 1853) (Cetacea:Delphinidae), no limite sul de sua distribuição, Santa Catarina, Brasil. *Biotemas*, 1(1), 57-62.

[24] Sokal, R. & Rohlf, F. (1995). *Biometry, the principles and practice of statistics in biological research*. New York, New York: W. H. Freeman and Company.

[25] Vidal, O., Barlow, J., Hurtado, L., Torres, J., Cendon, P. & Ojeda, Z. (1997). Distribution and abundance of the Amazon river dolphin (*Inia geoffrensis*) and the tucuxi (*Sotalia fluviatilis*) in the upper Amazon river. *Marine Mammals Science*, 13(3), 427-445.

In: Biology, Evolution and Conservation of River Dolphins… ISBN: 978-1-60876-633-8
Editors: M. Ruiz-García, J. Shostell. pp. 261-283

Chapter 14

MOLECULAR ECOLOGY AND SYSTEMATICS OF *SOTALIA* DOLPHINS

H.A. Cunha[1,2*], da Silva VMF[3] and A.M. Solé-Cava[1]

[1]Laboratório de Biodiversidade Molecular, Departamento de Genética, Instituto de Biologia, Universidade Federal do Rio de Janeiro, Rio de Janeiro, RJ - Brazil

[2]Laboratório de Mamíferos Aquáticos, Faculdade de Oceanografia Universidade do Estado do Rio de Janeiro, Rio de Janeiro, RJ - Brazil

[3]Laboratório de Mamíferos Aquáticos, Coordenação de Pesquisa em Biologia Aquática Instituto Nacional de Pesquisas da Amazônia, Manaus, AM - Brazil

ABSTRACT

Molecular markers have the potential to disclose genetic variation and provide clues on macro and microevolutionary issues. The taxonomic and phylogenetic status of species lie within the realm of macroevolution while intraspecific matters, such as geographic population structure, social organization and mating system, pertain to microevolution. This chapter describes the findings on the molecular systematics and ecology of Sotalia dolphins, and is divided in two sections, each focusing on one of those topics. The first section shows how molecular markers have helped to settle the issue of species composition within the genus Sotalia – a matter of debate for over 140 years. To explain the controversy, a brief history of taxonomic changes in the genus since the first species descriptions is included. In addition, the section also makes phylogenetic considerations and discusses the timing of the speciation between the two accepted Sotalia species. The second section deals with the molecular ecology of Sotalia, presenting results and prospects of studies on population structure, phylogeography and social structure. Although many studies are still underway, some important findings have already been produced. The section also includes comments on new analytical developments that promise to widen our knowledge on those issues. The two sections close with a discussion of the relevance of results for the conservation and management of Sotalia species. At least two important results stem from molecular systematics and

* www.biologia.ufrj.br/lbdm.

ecology studies of Sotalia dolphins, both with immediate application to their conservation. At the end of the chapter there is a presentation of the prospects for new discoveries in these fields in the near future.

Keywords: *Sotalia*, population structure, phylogeography, social structure, molecular ecology.

INTRODUCTION

Sotalia dolphins are among the smallest members of the Delphinidae family. These dolphins occur along the Atlantic coast of Central and South America, as well as in the Amazon River basin (Figure 1). Marine *Sotalia* are found from Honduras to the state of Santa Catarina, in southern Brazil (Simões-Lopes, 1987; da Silva and Best, 1996), in a seemingly continuous distribution that might be limited from extending southwards by low water surface temperatures (Borobia et al., 1991). Throughout that range it has many local common names, such as "boto-cinza" (Brazil), "tonina" (Venezuela and Colombia) and "lam" (Nicaragua). The distribution of riverine *Sotalia* comprises most of the Amazon River basin from Brazil as far as Peru, Ecuador and Colombia (da Silva and Best, 1996). Locally, this dolphin is known as "tucuxi" (Brazil), "bufeo-negro" or "bufeo-gris" (Colombia and Peru). There are also records of *Sotalia* dolphins in the Orinoco River, up to 800 km inland, and some disputed reports in the Upper Orinoco (Borobia et al., 1991; Boher et al., 1995). Those sightings may be attributed to marine *Sotalia*, since it inhabits bays and estuaries and is frequently seen entering rivers along the South American coast (da Silva and Best, 1996).

Marine and riverine *Sotalia* are morphologically very alike: both are dark gray in the *dorsum*, and light gray, white or pinkish in the ventral area, with a poorly developed lateral stripe extending from the eye to the pectoral fin. The beak is moderately long and slender, and the melon small and rounded. The dorsal fin is triangular, pectoral fins are large and the body is stocky (Jefferson et al., 1993). The main morphological difference between them is size with a maximum recorded total length for marine Sotalia of 206 cm, in contrast to 152 cm for freshwater *Sotalia* (Barros, 1991; da Silva & Best, 1996). There are also meristic and morphometric differences, but those are modal rather than absolute (Fettuccia, 2006).

Marine and riverine *Sotalia* are different not only in ecology but in life history traits: they use different acoustic signals and have distinct reproductive parameters (such as gestation length and birth seasonality (da Silva & Best, 1996; Rosas & Monteiro-Filho, 2002). Some of those differences may have arisen as adaptations to the different environments they inhabit.

The infrageneric taxonomy of *Sotalia* remained uncertain for over a century, and was solved only recently, when morphological (Monteiro-Filho et al., 2002) and genetic data (Cunha et al., 2005; Caballero et al., 2007) showed that marine and riverine *Sotalia* are different species.

Figure 1. Distribution of *Sotalia fluviatilis* (yellow) and of *S. guianensis* (red), and potential area of sympatry between the two species (orange). Abbreviations correspond to localities cited in the text (PA: Pará, CE: Ceará, RN: Rio Grande do Norte, BA: Bahia, ES: Espírito Santo, South-Southeastern: includes samples from Rio de Janeiro, São Paulo, Paraná and Santa Catarina).

Genus *Sotalia* Gray 1866

Sotalia was described to accommodate a marine species from the South American continent, originally attributed to *Delphinus* (*Delphinus guianensis*). In the same year, Gray proposed the *Sotalia* sub-genus *Tucuxa* for a riverine species from the Amazon (*Steno tucuxi*): that species was later relocated to *Sotalia* by Flower (1883). Also in 1866, Gray described the sub-genus *Sousa,* using *Steno lentiginosus* (later synonymized with *Sousa chinensis*), from India, as its type species. Interestingly, many species originally described as *Delphinus* and *Steno* from the Old World were assigned to *Sotalia*, before being finally placed in *Sousa* (Iredale & Troughton, 1934; Fraser & Purves, 1960), almost 100 years after the description of that genus.

Among the South American species reclassified in *Sotalia*, three were riverine dolphins collected in Peru and Brazil (*Delphinus fluviatilis*, *D. pallidus* and *Steno tucuxi*) and the other species was estuarine, described based on three dolphins collected at the mouth of the Marowijne River, in the border between Suriname and the French Guiana (*Delphinus guianensis*). In 1875 a fifth species was added to the genus *Sotalia* (the marine *S. brasiliensis*, whose type locality was Guanabara Bay, Brazil).

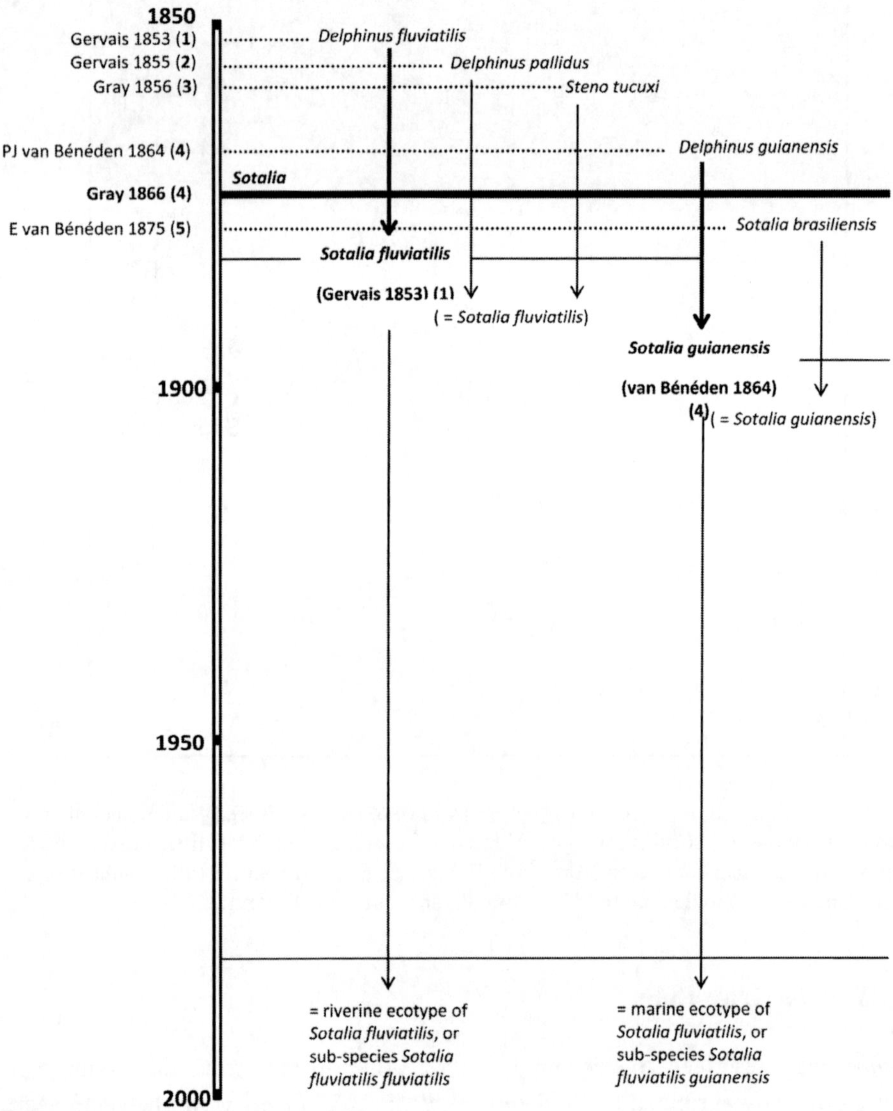

Figure 2. Time-line of descriptions of *Sotalia* species from South America and summary of subsequent nomenclature changes. (......) species description; (▬) genus description; (——) synonymization. Numbers in parentheses refer to type localities, shown in Figure 3.

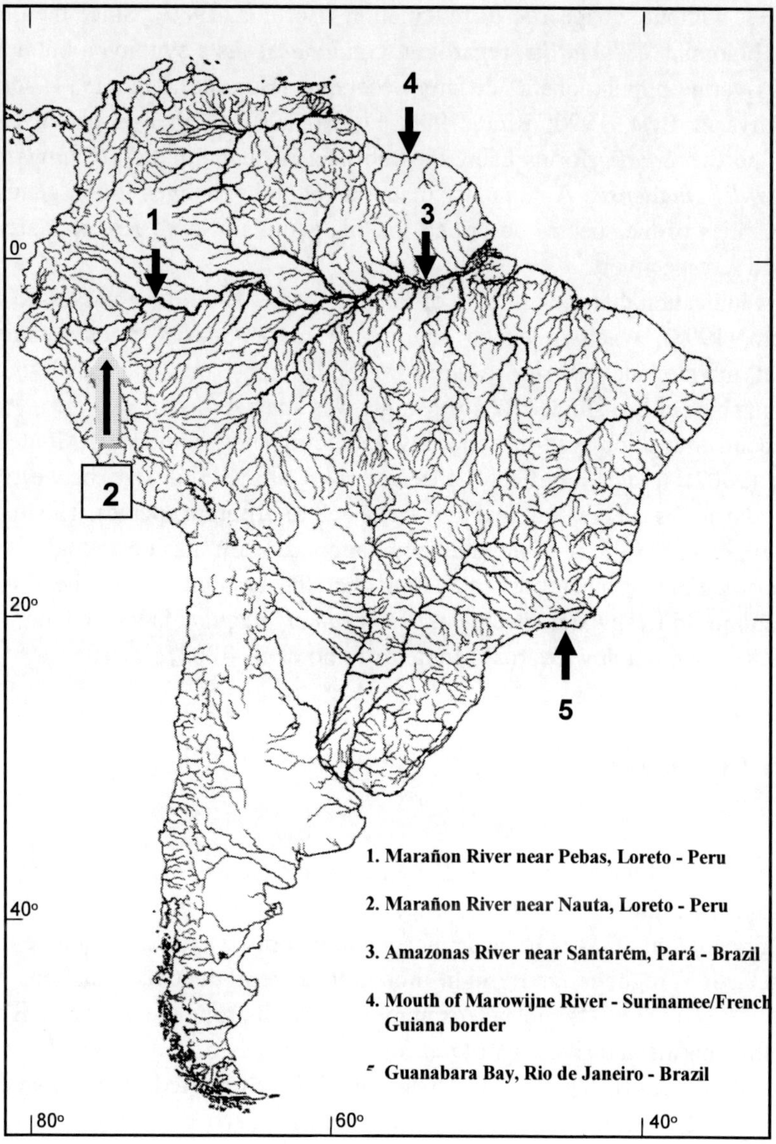

Figure 3. Type localities of *Sotalia* species from South America. Numbers correspond to those depicted between parentheses in Figure 2.

All those species were described based on few individuals from single location, at a time when barely anything was known about their ranges, so their diagnoses were incomplete and full of inconsistencies. As more specimens were examined and more data on their distribution were gathered, the three freshwater species were lumped into *Sotalia fluviatilis*, and the two marine were grouped as *Sotalia guianensis* (True, 1889; Cabrera, 1961; Carvalho, 1963).

Later, some authors argued that the differences between *S. fluviatilis* and *S. guianensis* were too subtle and attributable to phenotypic variability, and that *Sotalia* should be regarded as monotypic (Mitchell, 1975; Leatherwood & Reeves, 1983). This proposal was reinforced by a morphometric study that concluded that differences between marine and riverine *Sotalia* were mainly a consequence of size variation, and concluded that they should be considered a single species, without subspecific differentiation (Borobia, 1989). Since then, most authors adopted the binomial *S. fluviatilis*, regarding *S. guianensis* as a synonym, but acknowledging marine and riverine populations as different ecotypes (Borobia et al., 1991; Jefferson et al., 1993; da Silva & Best, 1996; Rice, 1998; Flores, 2002). Other researchers preferred to distinguish the two *Sotalia* forms using the subspecific denomination *S. fluviatilis fluviatilis* and *S. fluviatilis guianensis*. A summary of the taxonomic changes in the genus *Sotalia* and the type localities of the species described for this genus in South America are displayed in Figures 2 and 3 respectively.

The first indication that the lumping of *Sotalia* species should be reassessed was given by Furtado-Neto (1998). A phylogenetic analysis of mitochondrial cytochrome b sequences showed that marine and riverine *Sotalia* were different, but that result needed further confirmation, since only a single riverine sample was analyzed.

The second indication was provided by geometric morphometrics: Monteiro-Filho and co-workers (2002) found significant differences in shape and size between marine and riverine *Sotalia* skulls, suggesting that they belonged to different species. The main difference was in the alignment of the rostrum and occipital condyle: in marine animals, the location of the *foramen magnum* is posterior, indicating that the cranium would be in line with the vertebral column. In freshwater specimens, the *foramen magnum* is located more ventrally, so the cranium would point downwards (Monteiro-Filho et al., 2002).

Molecular Systematics

Taxonomy

As morphological analyses revealed significant differences between marine and riverine *Sotalia* (Monteiro-Filho et al., 2002), genetic analyses were essential to settle the issue of specific differentiation. This is because morphological differences might arise in response to different selection regimes and might not reflect reproductive isolation. Additionally, Monteiro-Filho et al. (2002) did not examine any skull from the Amazon Estuary, so the possibility that marine and riverine *Sotalia* formed extremes of a cline could not be ruled out. Significant differences in the skull had been previously reported by Borobia (1989), but a conservative conclusion supporting a single species was reached, among other reasons, due to the lack of samples from the Amazon Estuary, which could represent a transitional zone.

The use of molecular data in taxonomy and phylogeny has intensified over the last decades. Molecular systematics has benefits and disadvantages over traditional, morphology-based systematics (Hillis, 1987). Molecular markers are useful because they reveal a larger amount of variation, due to the large number of characters available in comparison with morphological analyses. Besides, genetic differences usually accumulate faster than phenotypic differences and, when genotypes are analyzed, environmental effects such as plasticity or convergence do not confound the analyses (Mayr, 1963; Avise, 2004). Those are invaluable features, especially in the delimitation of species. The detection of reproductive isolation and of monophyletism, which are pre-requisites of many species concepts, is also straightforward when genotypes are analyzed (Mayr, 1963; Hillis, 1987; Knowlton, 2000; Avise, 2004). On the other hand, molecular analyses demand expensive equipment and samples preserved in a way not to destroy DNA. Hence, integrating molecular and morphological approaches maximizes the likelihood of understanding true evolutionary relationships (Hillis, 1987; Knowlton, 2000; Avise, 2004).

Cunha et al. (2005) clarified the taxonomic status of *Sotalia* dolphins using sequences of the mitochondrial control region and the cytochrome b of 56 samples (12 riverine and 44 marine). This was the first study to include samples of the Amazon Estuary in analyses of differentiation between *Sotalia* ecotypes. Three phylogenetic approaches were used, and all of them recovered the same topology, displaying marine and freshwater *Sotalia* as reciprocally monophyletic groups (Figure 4). This result was corroborated by a Nested Clade Analysis (NCA; Templeton, 1998) of the same data. Notwithstanding some of its limitations in analyses of recently diverged lineages, NCA is a powerful tool that quantitatively and qualitatively investigates population structure and evolutionary history, including speciation (Templeton, 1998, 2001; Sites and Marshall, 2003). The NCA of *Sotalia* samples indicated a relatively old allopatric fragmentation event, which separated marine and riverine populations (Figure 5). Fragmentation events are evidence of speciation, especially if they: (a) are in higher level (older) clades; (b) reflect the separation of two clusters by several mutational steps and (c) coincide with independent evidence from other type of data (Templeton, 2001). The fragmentation observed between the two *Sotalia* ecotypes meets all three conditions. Interestingly, dolphins from Pará, at the mouth of the Amazon River, were genetically much closer to dolphins from Santa Catarina (4,700 km southwards, along the coast) than to the geographically closer (2,000 km) riverine dolphins (Cunha et al., 2005).

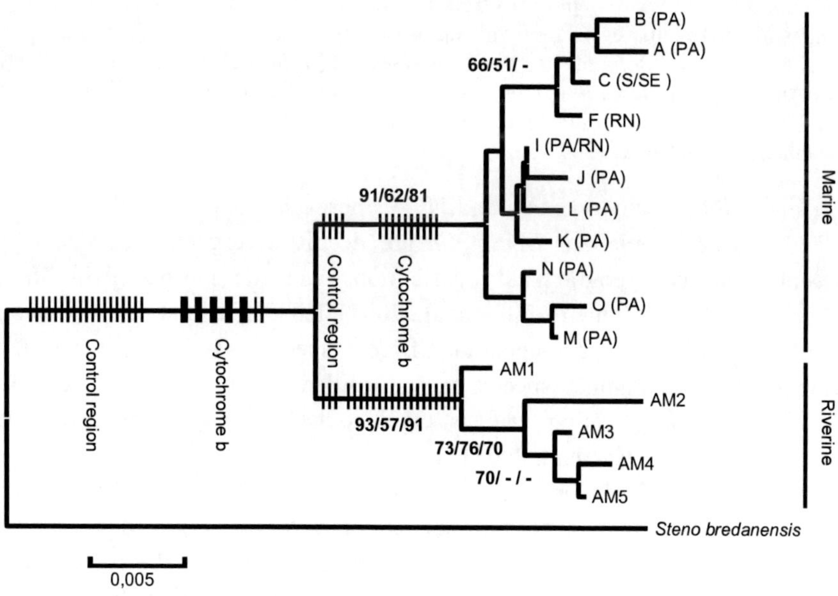

Figure 4. Neighbor-Joining (NJ) phylogenetic tree (p distances) of *Sotalia* spp. control region haplotypes. Maximum-Likelihood (ML) and Maximum Parsimony (MP) retrieved the same topology. Bootstrap values (NJ/ML/P) higher than 50% are shown. Hypothetical synapomorphies of control region and cytochrome b haplotypes from marine and riverine species are indicated by vertical bars. Thicker bar corresponds to 10 synapomorphies. Adapted from Cunha et al. (2005).

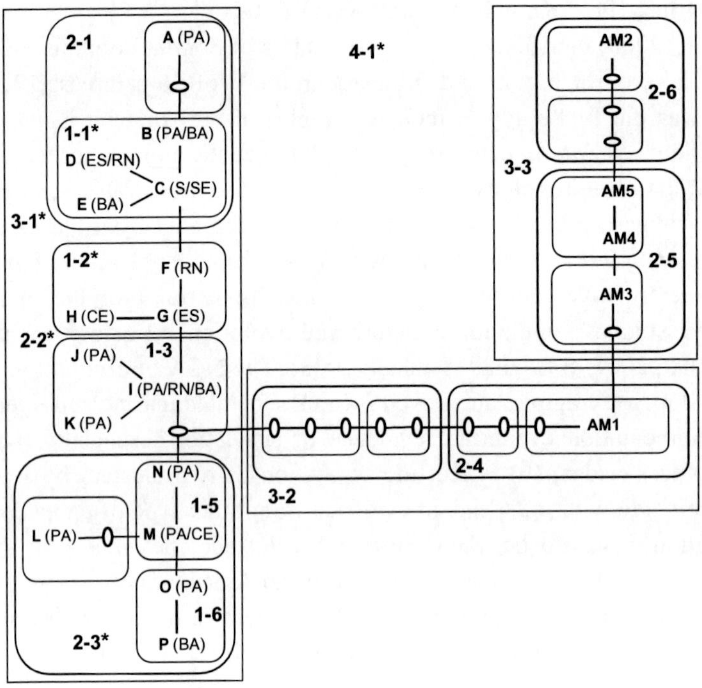

Figure 5. Parsimony network of haplotypes from the Brazilian coast, with nested clade design. Ovals represent missing intermediaries. Clades with significant associations ($P < 0.05$) are marked with an asterisk. Hierarchical level is denoted as 1-x for first level, 2-x for second, etc, where x identifies each clade. AM: Amazonas (*S. fluviatilis*). *S. guianensis* - PA: Pará; CE: Ceará; RN: Rio Grande do Norte; BA: Bahia; ES: Espírito Santo; S/SE: South-Southeastern (Rio de Janeiro, São Paulo, Paraná and Santa Catarina). Adapted from Cunha (2007).

Therefore, both the phylogenetic and NCA approaches supported the same conclusion: riverine and marine populations of *Sotalia* are deeply divergent. This result, along with distinct ecological and geographical distributions and the morphometric differentiation observed between them (Monteiro-Filho et al., 2002), led Cunha et al. (2005) to conclude that marine and riverine *Sotalia* belonged to different species. At least three criteria for the recognition of *taxa* as distinct species were fulfilled by those data (morphological and molecular population aggregation analysis, cladistic haplotype aggregation and Templeton's test of cohesion - Sites & Marshall, 2003).

In 2003, an international workshop on the molecular systematics of Cetaceans recognized that there was, in the field, a "traditional tendency to err in the direction of avoiding designating too many *taxa* rather than making sure that all potentially recognized *taxa* have been designated'' (Reeves et al., 2004). As a consequence, guidelines for the recognition of full species were established. According to the Workshop's guidelines, an argument for species status should be accepted only when there were at least two independent primary lines of evidence for its existence, such as morphology and genetics (Reeves et al., 2004). Therefore, together, the results presented by Monteiro-Filho et al. (2002) and Cunha et al. (2005) fulfilled those guidelines. Marine and riverine species of *Sotalia* could be separated not only on the basis of two primary types of evidence (morphology and genetics, respectively), but also of secondary ones (i.e., distribution and ecology). Based on priority criteria, the revalidation of *Sotalia guianensis* (van Bénéden 1864) was recommended for the marine ecotype, while the riverine form holds the binomial *Sotalia fluviatilis* (Gervais 1853, van Bree, 1974).

Another important finding was the presence of *S. guianensis* at the mouth of the Amazon River. The freshwater load of the Amazon River reaches hundreds of kilometers into the sea (Muller-Karger et al., 1988), so the animals sampled in Pará were actually living in freshwater. It would be interesting to analyze samples from intermediate locations along the Amazon River, to detect how far upriver *S. guianensis* occurs, and verify if there is sympatry in any region with *S. fluviatilis*. For that, a joint analysis of mitochondrial sequences and microsatellites would be crucial, since it would allow not only the detection of any possible hybridization in the area but also its polarity.

Recently, Caballero et al. (2007) analyzed sequences from introns of three nuclear genes (lactalbumin, actin and glucocerebrosidase) and another mitochondrial marker (ND2) including South America and Caribbean samples. Their study, based on a larger dataset both in genes analysed and geographical breadth gave support to the conclusion of Cunha et al. (2005), confirming the specific status of *S. guianensis* and *S. fluviatilis*.

Two issues related to genus *Sotalia* remain unclear: the range and species identity of *Sotalia* dolphins in the Orinoco River, and the taxonomic status of *Sotalia* dolphins from southern Maracaibo Lake. In the Orinoco River, there are frequent records of *Sotalia* dolphins at Ciudad Bolívar, some 300 km from the river's mouth, but those may correspond to *S. guianensis*, which can reach several kilometers upriver (da Silva and Best, 1996; Mead & Koehnken, 1991; Flores & da Silva, 2008). Boher et al. (1995) reported a sighting in the Middle Orinoco, 800 km inland. In addition, there are disputed reports of *Sotalia* dolphins in the Upper Orinoco, and even in the Apure River (Hershkovitz, 1963; Borobia et al., 1991; Boher et al. 1995). However, *Sotalia* dolphins were not recorded in the Upper Orinoco and Apure Rivers, nor in the lower reaches of most of the major tributaries of the Orinoco, during a long term study conducted between 1983 and 1990 (Mead & Koehnken, 1991). It is believed that *Sotalia* dolphins cannot traverse the rapids at the Casiquiare channel, which connects the Orinoco and Amazon River basins (da Silva and Best, 1996). This barrier has existed since the uplift of the Mérida Cordillera (10 mya; Lundberg et al., 1998), which predates the split between *Sotalia* species (see next section). Thus, *Sotalia* dolphins in the Middle Orinoco are likely to be an isolated population of *S. guianensis*.

Another interesting issue concerns *Sotalia* dolphins found in the southern, freshwater, portion of the Maracaibo Lake. That population is morphologically different from the marine *Sotalia* that inhabit the northern portion of the Lake, where it opens to the Gulf of Venezuela. Dolphins from southern Maracaibo are smaller than marine *Sotalia*, and about the same size as *S. fluviatilis* (Casinos et al., 1981; da Silva & Best, 1996; León, 2005). However, there is no connection between the Maracaibo Lake and the present day known range of riverine *Sotalia*, and the Maracaibo Lake has been isolated from the Amazon basin for the last 8-10 million years (Hoorn et al., 1995; Días de Gamero, 1996). The morphological distinctiveness of the southern Maracaibo Lake population could result from true phenotypic plasticity, unlike that found between *S. guianensis* and *S. fluviatilis*. However it may also indicate a lack of gene flow with the marine *Sotalia* from the mouth of the lake and the Gulf of Venezuela. Indeed, genetic differentiation between those areas was reported by Caballero et al. (2006). Those authors observed some exclusive haplotypes in samples from the lake, but did not attribute the variation to specific differentiation.

Timing of Speciation

The divergence between *S. fluviatilis* and *S. guianensis* observed by Cunha et al. (2005) was 2.5%, for both the control region and the cytochrome b. The evolutionary rates of those markers have been estimated at between 0,5% and 1% per million years (My) for the control region of cetaceans (Hoelzel et al., 1991) and 1%/My for the cytochrome b (Irwin et al., 1991). Hence, the speciation event that separated both lineages probably happened between 5 and 2.5 mya, during the Pliocene. At that time, the Amazon River was already flowing along its present course, with its outlet to the Atlantic (Hoorn et al., 1995; Lundberg et al., 1998). For the last 4 my, several sea level oscillations occurred, as a consequence of glacial and interglacial periods. During the periods of sea level rise, river discharge was prevented, and freshwater inflow into the Amazon basin increased, causing the inundation of the Amazon craton (Lundberg et al., 1998). The highest marine transgression happened around 2.5 mya (Klammer, 1984). It is possible that *Sotalia* colonized the Amazon basin during one of those transgression/inundation events. Regardless of the putative timings of speciation, dolphins that colonized the Amazon River system probably had an Atlantic origin, because the alternative explanation (entrance from the Caribbean via present day Maracaibo Lake and Paleo-Orinoco system) would require a much older divergence (>10 mya).

Caballero et al. (2007) calibrated a molecular clock for the control region using the estimated divergence between *Sotalia* and *Phocoena phocoena* based on the fossil record (10-11 my). Therefore, they arrived at a faster substitution rate, and dated the divergence between *S. fluviatilis* and *S. guianensis* at 1 to 1.2 mya, during the Pleistocene. This dating is also compatible with environmental oscillations in the Amazon basin (Caballero et al., 2007). Due to the lack of *Sotalia* fossils, it is not possible yet to decide which of the two scenarios is more likely.

Evolutionary Relationships

Sotalia is one of the several Delphinidae genera. The Delphinidae family is regarded as a taxonomic "trash basket", because its members are very diverse in shape and size, and share no exclusive characteristics. Some of the characteristics of delphinids are a marine distribution, presence of beak, presence of a falcate dorsal fin and presence of conical teeth. However, there are exceptions to each of those features (Jefferson et al., 1993).

The evolutionary relationships among delphinids are far from understood, so at present it is difficult to ascertain the phylogenetic position of *Sotalia*. Traditionally, *Sotalia* has been grouped with *Sousa* and *Steno* based on morphology. In fact, *Sousa* dolphins were originally assigned to *Sotalia*. The grouping with *Steno* might have resulted from the use of primitive morphological features in pre-cladistic analyses, but has endured to the latest classifications (reviewed in LeDuc et al., 1999). The most accepted morphological classification was proposed by Perrin (1989). This classification maintains *Sotalia*, *Sousa* and *Steno* as closely related (Subfamily Stenoninae). *Sousa* is a genus with two recognized species: *S. teuszii* from the Eastern Atlantic, and *S. chinensis* from the Indo-Pacific. A third species, *S. plumbea*, occurring in the Western Indian Ocean, is regarded by most authors as a synonym of *S. chinensis*. *Sousa* dolphins are morphologically similar to *Sotalia*. *Steno* is a monotypic genus comprised of *S. bredanensis*, a larger dolphin found around the globe in tropical and subtropical waters (Jefferson et al., 1993).

Molecular markers have also been used to investigate delphinid evolution. LeDuc et al. (1999) reassessed the phylogenetic relationships within Delphinidae using full cytochrome b sequences (about 1.2 kilobases) of 33 species. Among several interesting findings, their analysis placed *Sousa* outside Stenoninae, which comprised *Steno* and *Sotalia*. Stenoninae, however, had low bootstrap support. According to their results, *Sousa* belongs to Subfamily Delphininae.

The most recent analyses used a less complete *taxon* sampling (17 species) but a larger number of sequences (5.2 kilobases, including two mitochondrial and ten nuclear markers; Caballero et al., 2008). Differently from the work by LeDuc et al (1999), Caballero et al. showed *Sousa* and *Sotalia as* sister taxa within Delphininae, separated from *Steno*. The combined phylogeny grouped *Sousa* with the Delphininae species in the analyses, and both *Sotalia* species as a monophyletic clade branching from this grouping. *Steno* is placed with Globicephalinae, *Orcaella* and *Grampus*.

The phylogenetic position of *Sotalia* will probably remain unsettled until the taxonomy of *Steno* and *Sousa* is resolved. None of the above mentioned studies included *S. teuszii*, which is the *Sousa* species geographically closer to *Sotalia*, or *Sousa* dolphins from Australia, which may belong to a third species according to mitochondrial control region sequences (Frère et al., 2008). The existence of other species of *Steno* is also still an open issue, since very little is known about those dolphins (Jefferson, 2002).

Conservation Aspects

The uncertainty about the taxonomic situation of *Sotalia* dolphins hindered the evaluation of their conservation status, and combined with the lack of information on their biology and ecology, determined their classification as "data deficient" by the International Union for the Conservation of Nature (IUCN; 2008) and the Brazilian Institute of Environment and Renewable Natural Resources (IBAMA, 2001). The clarification of the specific status of both *Sotalia* species was an important first step toward the proper assessment of their conservation status.

One of the consequences of the recognition that the two ecotypes of *Sotalia* constitute different species is that *Sotalia fluviatilis* becomes the only exclusively freshwater delphinid in the world (Cunha et al., 2005). To date, there are only three other living species of cetaceans known to exist exclusively in freshwater, two of them belonging to the Platanistidae (*Platanista gangetica* and *P. minor*) family, and the other to the Iniidae family (*Inia geoffrensis*, which probably includes a fourth species, *Inia boliviensis*, Banguera-Hinestroza et al. 2002). The baiji (*Lipotes vexillifer*, Family Lipotidae) was another river dolphin, endemic to the Yangtze River, but is now believed extinct in the wild (Turvey et al., 2007).

At least four other dolphin species can be found both at sea and in rivers: three are delphinids (*Sousa chinensis*, *S. teuszii* and *Orcaella brevirostris*), and the other is a phocoenid (*Neophocaena phocaenoides*). However, there is no agreement about the degree of differentiation between their marine and riverine populations, except for *Orcaella brevirostris*. Beasley et al. (2005) demonstrated, using molecular analyses, that there are two *Orcaella* species (*O. brevirostris* and *O. heinsohni*), and that *Orcaella brevirostris* has both coastal and riverine populations. Therefore, *Sotalia fluviatilis* is the first delphinid living exclusively in freshwater.

S. fluviatilis is endemic of the Amazon River and its main tributaries, from Brazil to Colombia, Ecuador and Peru (da Silva & Best, 1996; Flores, 2002). The Amazon River basin has been experiencing a steep increase in human activities in the last decades, most of them potentially harmful to the Amazon river dolphins. Several anthropogenic threats have been identified (ex. direct and indirect catch, building of dams, habitat loss and degradation, heavy-metal contamination - Best and da Silva, 1989), but their effects on *S. fluviatilis* populations remain unknown (IBAMA, 2001; Reeves et al., 2003). Those potential threats, combined with the newly found endemism of *S. fluviatilis*, may jeopardize its persistence.

River dolphins are the most endangered cetaceans, because they share their endemic, restricted habitat with increasing human populations and are therefore exposed to several direct and indirect human-related threats (Reeves et al., 2003). For that reason, they have been granted special conservation status. The newly found endemism of *S. fluviatilis* implies that its conservation status should be reassessed, and it also should be included in the river dolphin category for conservation purposes.

The molecular identification of *Sotalia* species also led to an important discovery: dolphin-derived products, illegally sold in the Brazilian Amazon as love charms, do not belong to the red boto (*Inia geoffrensis*), as advertised by sellers. Instead, all samples that had actually been obtained from dolphins belonged to the marine *S. guianensis* (Cunha & Solé-Cava, 2007; Gravena et al., 2008; Sholl et al., 2008). *S. guianensis* amulets were detected not only in Belém (Pará state, at the Amazon estuary) but in Manaus and Porto Velho, despite the availability of botos and of *S. fluviatilis* in those areas. In one market (Ver-o-peso of Porto Velho, Rondônia), 90% of the eyeballs sold were in fact from pig or sheep (Gravena et al., 2008). The assessment of the impact of this illegal activity depended on the identification of the targeted species. Now that *S. guianensis* has been recognized as possibly the only species currently used, authorities can act on the sources of charms, which are likely to be the Amazon estuary and adjacent Pará and Amapá coasts. *S. guianensis* has been intentionally caught in those areas to be used as shark bait (Pinedo, 1985) - a single boat had 83 specimens on board (footage done by IBAMA and broadcasted by a Brazilian television network on 07/16/2007). Dolphin charms may originate both from by-catch from legal fisheries, and as a second commodity of the illegal bait catch.

Molecular Ecology

Molecular markers have been successfully employed to investigate other aspects of the biology of *Sotalia*, especially their population structure and social behaviour. Although studies on *Sotalia* dolphins are still in course, they promise to reveal important data for the conservation of those species.

Population Structure and Phylogeography

Phylogeography is a field of research concerned with the evolutionary and demographic processes that shaped the genealogical lineages within or between closely related species

(Avise, 2004). Phylogeographic analyses focus on the species' past, but provide important insights on its present-day population structure. Most endangered species are highly structured, because reductions in abundance contribute to the isolation of populations (O'Brien, 1994; Frankham, 1996; Avise, 2004) and small population sizes increase genetic drift, which accelerates population differentiation. As a result, endangered species are often subdivided in demographically independent units, each with a population size more affected by local birth and death rates than by migration rates. Hence, the persistence of each unit is linked to the evolutionary and demographic processes acting upon it (Moritz, 1994; Avise, 2004; Palsbøll et al., 2007). Population units that should be considered independently for evolutionary biology purposes have been named "Evolutionarily Significant Units" (ESU) (Ryder, 1986). Later, Moritz (1994) proposed the term "Management Units" (MU) to designate units for conservation purposes. MUs are different from ESUs because they are less restrictive and closer to the demographic present of species.

The population structure and phylogeography of *S. guianensis* along the Brazilian coast was investigated by Cunha (2007), using mtDNA control region sequences. Analysis of molecular variance (AMOVA; Excoffier et al. 1992), spatial analysis of molecular variance (SAMOVA; Dupanloup et al 2002) and Nested Clade Analysis (NCA; Templeton 1998, 2001) showed evidence for at least six MUs in Brazil: Pará, Ceará, Rio Grande do Norte, Bahia, Espírito Santo and the South-Southeastern area (from Rio de Janeiro to Santa Catarina states, Figure 1). Those MUs were highly differentiated (Φ_{CT} = 0,485, $P < 10^{-5}$), indicating severe restrictions to gene flow among them. An interesting finding was a lack of variation in the control region of dolphins from South-Southeastern Brazil (between parallels 22° and 25°S, extending 900 km). NCA and genetic diversity patterns suggest that this homogeneity might have been caused by a recent colonization of the Brazilian coast through a range extension from north to south, which could be linked to a warming up of the Western Atlantic during the Holocene. Thus, the observed homogeneity is probably not due to gene flow within the region, but a consequence of recent foundation (Cunha & Solé-Cava, 2006; Cunha, 2007).

Populations of *S. guianensis* from the northern part of South America and the Caribbean were analyzed by Caballero et al. (2006), who proposed two MU for that area: one for Central America, Colombia and Venezuela, and another for Guyana, Surinamee and French Guiana. The authors advised that dolphins from the Maracaibo Lake, despite being included in the first MU, had some unique haplotypes and their genetic distinctiveness should be further investigated. However, only three individuals from southern Maracaibo were analyzed: the others were from the northern portion of the lake, where it opens to the Gulf of Venezuela. Clearly, further analyses of samples from the Maracaibo must be analyzed to verify their possible genetic distinctiveness.

To date, there is no information on the population structure of *S. fluviatilis*. The only data available suggest that the species has moderate to high genetic diversity, since 12 individuals from the same location in the Central Brazilian Amazon had five different control region haplotypes (Cunha et al., 2005), and 21 dolphins from the Peruvian, Colombian and Brazilian Amazon had 13 haplotypes (combining the control region and ND2, Caballero et al., 2007). Microsatellite variation was also larger in *S. fluviatilis* (H = 0.531) than in *S. guianensis* (H = 0.364; Cunha and Watts, 2007). The reason for a higher level of gene variation in *S. fluviatilis*, in spite of its probably smaller population size, remains to be determined.

Social Structure

Undoubtedly, the newly developed microsatellite markers will be invaluable also for the investigation of the social structure of *Sotalia* dolphins. Besides being highly polymorphic, microsatellites are useful for that purpose because they are bi-parentally inherited.

During the last decade, many interesting results have been found concerning the social behaviour of *S. guianensis*, especially through long-term photo-identification studies. Three local populations in Brazil showed strong residency (North Bay, Santa Catarina - Flores 1999; Cananéia Estuary, São Paulo - Santos et al., 2001; Guanabara Bay, Rio de Janeiro - Azevedo et al., 2004), and that pattern may prove to be a feature of that species throughout its distribution.

In spite of the vast database on social associations built during the long-term monitoring of some *Sotalia guianensis* populations, studies on social structure have been hampered by the absence of easily observable sexual dimorphism. Sex determination of free-ranging *Sotalia* relies on the observation of the animal's ventral area, which is a rare event in the field. Therefore, sexing is only achieved for reproducing females, on the basis of their close, lasting and recurring association with calves. That approach demands a long-term monitoring of the population, and does not allow the detection of males and non-reproductive individuals. Fortunately, remote biopsy darting has been safely and successfully applied to *Sotalia* dolphins, providing samples that can be sexed molecularly. Two genetic systems are usually applied for sex determination in cetaceans: the ZFX/ZFY (Bérubé and Palsbøll, 1996) and the SRY (Palsbøll et al., 1992). Both systems have been tested and optimized for *Sotalia* species, and have been successfully used for sexing biopsy samples (Cunha and Solé-Cava, 2007) (Figure 6). Additionally, those molecular techniques allow the sex determination of carcasses in advanced decaying, when sexing cannot be done by the examination of the genital opening.

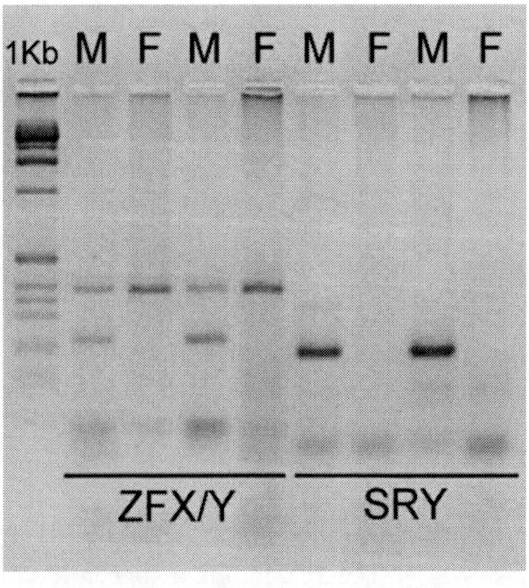

Figure 6. Sex determination patterns of *Sotalia* samples using the ZFX/ZFY and SRY systems. M: male, F: female, 1Kb: DNA size ladder.

The residency of local populations of *S. guianensis* could reflect the phylopatry of one of the sexes. In most mammals, females are the phylopatric sex while males disperse (Dobson, 1982). This pattern has been observed in almost all small cetacean species studied so far (e.g. *Tursiops truncatus* - Scott et al., 1990; Duffield Wells 1991; *Delphinapterus leucas* - O'Corry-Crowe & Lowry, 1997; *Phocoena phocoena* - Rosel et al., 1999; *Phocoenoides dalli* - Escorza-Treviño & Dizon, 2000; *Cephalorhynchus hectori* - Pichler and Baker, 2000; *Tursiops aduncus* - Möller & Beheregaray, 2004). It is possible that *S. guianensis* shares the same sex bias in dispersal, but until now that could not be evaluated due to the impossibility of visually sexing the resident animals. The hypothesis of female phylopatry can be tested with the comparison of maternally inherited mitochondrial DNA with bi-parentally transmitted markers such as microsatellites, as well as through studies of social structure coupling photo-identification and biopsy sampling.

The genetic analysis of biopsies from photo-identified dolphins will also provide a finer-scale picture of the social structure of *S. guianensis*, by seeking correlations between kinship and social affiliations, as has been done with other delphinids recently (e.g. Möller et al., 2001, 2006; Krützen et al., 2004).

The above mentioned methods can also help to unveil the social structure of *S. fluviatilis*. The only available information on the social organization of this species are from mark and recapture data, suggesting that *S. fluviatilis* in the Central Amazon is not territorial, but shows strong site fidelity (spending up to 9 years in the same area). Group structure seems to be socially organized by fusion-fission strategies, and some animals have been sighted together 8.5 years after marking (da Silva & Martin, unpublished data).

Another interesting prospect is the investigation of the mating system of *Sotalia* dolphins. Until now, the only hypothesis advanced was of polyandry of both *Sotalia* species, based on their large testis sizes (an indication of sperm competition) (da Silva & Best, 1996; Rosas & Monteiro-Filho, 2002). Mating system can be studied using microsatellites because they have the ability to ascertain paternity. That is useful when different mother-calf pairs from the same group are biopsied, and also when known siblings are sampled (for instance as calves from the same female), since the genotype of the father can be reconstructed from the calf's genotype if the mother's genotype is known. Hence, it is possible to check how many calves from the same cohort are fathered by the same male, and if calves of the same female born in different years are full siblings.

Conservation Implications

Studies on the population structure, phylogeography and social structure of *Sotalia* species will certainly help in the evaluation of their conservation status, and contribute to the design of effective measures for their conservation.

A proper evaluation of the impact of non-natural mortality on populations can only be achieved when their geographical boundaries are known. Additionally, population delimitation is fundamental for the design of effective conservation measures (O'Brien, 1994; Avise, 1997). The goal of any conservation plan should be to preserve the target species both in time and space. That means the entire range of the species should be maintained, which is an obvious challenge because there is hardly any species charismatic enough to stop human plans of growth and development in face of the low ecological responsibility of our species.

When there is enough gene flow across the species range, individuals removed by human-related factors are replaced from other areas. But when a species is split into different and isolated populations (i.e. MU), each one evolves independently, since they are not connected (and replenished) through migration. Besides, independent units harbour exclusive genetic variation (locally originated or maintained, and not spread to other units due to restricted gene flow), and it is reasonable to assume that some of that variation may encompass local adaptations. It is crucial to ensure that genetic diversity is preserved, because it constitutes the evolutionary potential of the species. Inappropriate management of units may result in the loss of adaptations, which may jeopardize the short-term viability of some populations, or even the species as a whole (Frankham, 1996; Solé-Cava, 2000; Crandall et al., 2000). Therefore, knowledge on the population structure is of paramount importance, as it enhances the probability of success of management and conservation actions (O'Brien, 1994).

Understanding the social structure of *Sotalia* dolphins may also help in their conservation. For instance, if females of either species prove phylopatric, management must be based on mitochondrial data, even if there is evidence of gene flow with nuclear markers (Avise, 1995; Dizon et al., 1997). Mitochondrial DNA is maternally inherited, so it depicts the history and structuring of female lineages. If only males disperse, populations are unlikely to be recolonized after local extinction, and the most conservative strategy would be to ensure the persistence of each population detected with mitochondrial data. In addition, if mortality rates are higher in areas between populations (which has been demonstrated for some species), that mortality would translate into a higher loss of males compared to females, causing unequal sex-ratio and reduction of the effective population size and genetic variability of the species.

The studies cited above provide the first, and most reliable, data for the establishment of MU for *S. guianensis*. Before their publication, there was no information on genetics, demography, morphology, behaviour, bioacoustics, parasites, ecology or contaminants that could argue for any delimitation of MU for the species, even provisional. That is the present situation for *S. fluviatilis*, but it will change in the near future, as the investigation of its population structure using molecular markers is currently underway.

Many threats to the persistence of both *Sotalia* species have been identified. However, the paucity of information on the taxonomy and biology of *Sotalia* dolphins hindered the evaluation of their conservation status; hence they are considered "data deficient" by the Brazilian environmental agency IBAMA (2001) and by IUCN (2008). Some countries took a precautionary approach and decided to give *Sotalia* a conservation status: in Colombia and Venezuela, both species are regarded as "vulnerable" (Rodríguez-Mahecha et al., 2006; Bolaños-Jímenez et al., 2008), and in Ecuador, *S. fluviatilis* is listed as "endangered" (Tirira, 2001). With the data now available, environmental agencies need to reassess the conservation status of both species, especially in Brazil, because that country encompasses over half of the range of *S. guianensis*, and most of the distribution of *S. fluviatilis*.

Conclusions and Prospects

This chapter reviewed the latest results on the molecular systematics and ecology of *Sotalia* dolphins. Some of the issues still require investigation, but several important results have been obtained in those fields during the last few years.

Unquestionably, the most remarkable finding to date was the elucidation that riverine and marine ecotypes of *Sotalia* are different species. Molecular markers were fundamental to settle the issue of specific differentiation between *S. fluviatilis* and *S. guianensis*. The impact of that discovery can be appreciated by considering that all articles published since the work of Cunha and co-workers (2005) accepted the revalidation of *S. guianensis* (22 articles – Web of Knowledge search on October, 2008). A major consequence of the split of *Sotalia* species is the need for reassessment of their conservation status, in recognition of the different conservation requirements of both species. The discovery of an exclusively freshwater habit for *S. fluviatilis* indicates that it should have its conservation priority raised. Secondly, the impact of non-natural mortality need to be re-evaluated for each MU of *S. guianensis* across its entire range, and conservation plans must be devised for those MU that show signs of endangerment.

References

[1] Avise JC (1995) Mitochondrial-DNA. *Conservation Biology*, 9 (3), 686-690

[2] Avise JC (1997) Conservation genetics in the marine realm. *Journal of Heredity*, 89 (5), 377- 382

[3] Avise JC (2004) *Molecular Markers, Natural History and Evolution. 2ª ed.* New York, NY: Chapman & Hall.

[4] Azevedo AF, Lailson-Brito JJr, Cunha HA, & Van Sluys M (2004) A note on site fidelity of marine tucuxis (*Sotalia fluviatilis*) in Guanabara Bay, southeastern Brazil. *Journal of Cetacean Research and Management*, 6(3), 265-268

[5] Banguera-Hinestroza E, Cárdenas H, Ruiz-García M, Marmontel M, Gaitán E, Vasquez R, & García -Vallejo (2002) Molecular identification of evolutionarily significant units in the Amazon river dolphin *Inia* sp. (Cetacea: Iniidae). *Journal of Heredity*, 93, 312-322

[6] Barros NB (1991) Recent cetacean records for southeastern Brazil. *Marine Mammal Science*, 7, 296-306.

[7] Beasley I, Robertson KM, & Arnold P (2005) Description of a new dolphin, the Australian snubfin dolphin *Orcaella heinsohni* sp. n. (Cetacea, Delphinidae). *Marine Mammal Science*, 21, 365-400

[8] Bérubé M & Palsbøll P (1996) Identification of sex in cetaceans by multiplexing with three ZFX and ZFY specific primers. *Molecular Ecology*, 5, 283-287

[9] Best RC & da Silva VMF (1989) Biology, status and conservation of *Inia geoffrensis* in the Amazon and Orinoco river basin. In W F. Perrin, R. L. Brownell, Jr., Z. Kaiya, & L. Jiankang, (Eds.), *Biology and conservation of the river dolphins,* Proceedings of the *Workshop on Biology and Conservation of the Platanistoid Dolphins* held at Wuhan, People´s Republic of China. Occasional Papers of the IUCN species survival

commissions (SSC) (Number 3, pp. 1-173). Gland, Switzerland: IUCN-The World Conservation Union.

[10] Boher SJ, Bolaños J, & Cova LJ (1995) Sobre un avistamiento del delfin estuarino o bufete (*Sotalia fluviatilis*) en el Orinoco Medio. *Acta Científica Venezolana,* 46, 217–218

[11] Bolaños-Jiménez J, Pirela D, Rojas J, &León T (2008) Bufeo negro *Sotalia guianensis* van Bénéden 1864. In: J. P. Rodríguez, & F. Rojas-Suárez F (Eds.), *Libro Rojo de la Fauna Venezolana.* (pp.112) Caracas, Venezuela: Eds. Provita y Shell Venezuela S.A.

[12] Borobia M (1989) *Distribution and morphometrics of South American dolphins of the genus Sotalia Master of Science Dissertation.* Montreal, Canada: McGill University.

[13] Borobia M, Siciliano S, Lodi L, & Hoek W (1991) Distribution of the South-American dolphin *Sotalia fluviatilis. Canadian Journal of Zoology*, 69, 1025-1039

[14] Caballero S, Trujillo F, Vianna JA, Barrios-Garrido H, Montiel MG, Beltrán-Pedreros S, Marmontel M, Santos MCO, Rossi-Santos M, Santos FR, & Baker CS (2006). Phylogeography of the South American riverine and coastal dolphins *Sotalia fluviatilis* and *Sotalia guianensis.* In *Workshop Internacional sobre Pesquisa e Conservação dos Golfinhos do Gênero Sotalia.* Búzios, Brazil:. Sociedad Latinoamericana de Especialistas en Mamíferos Acuáticos – SOLAMAC.

[15] Caballero S, Trujillo F, Vianna JA, Barrios-Garrido H, Montiel MG, Beltrán-Pedreros S, Marmontel M, Santos MC, Rossi-Santos M, Santos FR, & Baker CS (2007) Taxonomic status of the genus *Sotalia*: Species level ranking for "tucuxi" (*Sotalia fluviatilis*) and "costero" (*Sotalia guianensis*) dolphins. *Marine Mammal Science*, 23(2), 358-386

[16] Caballero S, Jackson J, Mignucci-Giannoni AA, Barrios-Garrido H, Beltrán-Pedreros S, Montiel- Villalobos MG, Robertson KM, & Baker CS (2008) Molecular systematics of South American dolphins *Sotalia*: sister taxa determination and phylogenetic relationships, with insights into a multi-locus phylogeny of the Delphinidae. *Molecular Phylogenetics and Evolution*, 46, 252-268

[17] Cabrera A (1961) Catalogo de los mamiferos de America del Sur II (Sirenia, Perissodactyla, Artiodactyla, Lagomorpha, Rodentia, Cetacea). *Rev Mus Argentino Cienc Nat Bernardino Rivadavia, Ciências Zool* 4: 309-732

[18] Carvalho CT (1963) Sobre um boto comum no litoral do Brasil (Cetacea, Delphinidae). *Revista Brasilera de Biologia,* 23, 263-276

[19] Casinos A, Bisbal F, & Boher S (1981) Sobre tres exemplares de *Sotalia fluviatilis* del Lago Maracaibo (Venezuela) (Cetacea, Delphinidae). *Publicaciones del Departamento de Zoologia, Barcelona*, 7, 93-96.

[20] Crandall KA, Bininda-Emmonds ORP, Mace GM, & Wayne RK (2000) Considering evolutionary processes in conservation biology. *TREE*, 15, 290-295

[21] Cunha HA & da Silva VMF, Lailson-Brito JJr, Santos MCO, Flores PAC, Martin AR, Azevedo AF, Fragoso ABL, Zanelatto RC, & Solé-Cava AM (2005) Riverine and marine ecotypes of *Sotalia* dolphins are different species. Marine *Biology*, 148, 449-457

[22] Cunha HA & Solé-Cava AM (2006) Phylogeography of *Sotalia guianensis* along the Brazilian coast. In: Workshop Internacional sobre Pesquisa e Conservação dos Golfinhos do Gênero *Sotalia.* Búzios, Brazil: Sociedad Latinoamericana de Especialistas en Mamíferos Acuáticos – SOLAMAC.

[23] Cunha HA (2007) *Sistemática molecular e filogeografia do gênero Sotalia Gray 1866 (Delphinidae) no Brasil. Doctoral Thesis*. Rio de Janeiro, Brazil, Universidade Federal do Rio de Janeiro.

[24] Cunha HA & Solé-Cava (2007) Molecular sexing of tucuxi dolphins (*Sotalia guianensis* and *Sotalia fluviatilis*), using samples from biopsy darting and decomposed carcasses. *Genetics and Molecular Biology,* 30 (4), 1186-1188.

[25] Cunha HA, & Watts PC (2007) Twelve microsatellite loci for marine and riverine tucuxi dolphins (*Sotalia guianensis* and *Sotalia fluviatilis*). *Molecular Ecology Notes,* 7, 1229-1231.

[26] da Silva VMF & Best RC (1996) *Sotalia fluviatilis*. *Mammalian Species,* 527, 1-7.

[27] Días de Gamero ML (1996) The changing course of the Orinoco River during the Neogene: a Review. *Palaeogeography, Palaeoclimatology, Palaeoecology,* 123, 385-402

[28] Dizon AE, Perrin WF, Amos W, Baker CS, Chivers SJ, Costa AS, Curry BE, Gaggiotti O, Hoelzel AR, Hofman R, LeDuc RG, Loughlin TR, Lux CA, O'Corry-Crowe GM, Rosel PE, Rosenberg A, Scribner KT, & Taylor BL (1997) Report of the Workshop In: A. E. Dizon, S. J. Chivers, & W. F. Perrin (Eds.), *Molecular Genetics of Marine Mammals* (Special Publication Number 3, 447 pp). Lawrence, KS: The Society for Marine Mammalogy.

[29] Dobson FS (1982) Competition for mates and predominant juvenile male dispersal in mammals. *Animal Behaviour,* 30, 1183-1192.

[30] Duffield D, Wells R (1991) The Combined Application of Chromosome, Protein and Molecular Data for The Investigation of Social Unit Structure and Dynamics in *Tursiops truncatus*. In: A. R. Hoelzel & G. P. Donovan (Eds.), *Genetic Ecology of Whales and Dolphins - Report of the International Whaling Commission* (Special Issue 13, pp. 155-176). Cambridge, United Kingdom: *International Whaling Commission.*

[31] Escorza-Treviño S., & Dizon AE (2000) Phylogeography, intraspecific structure and sex-biased dispersal of Dall's porpoise, *Phocoenoides dalli*, revealed by mitochondrial and microsatellite DNA analyses. *Molecular Ecology*, 9, 1049-1060

[32] Fettuccia DC (2006) *Comparação osteológica nas espécies do gênero Sotalia Gray, 1866 no Brazil (Cetacea, Delphinidae). Master of Science Dissertation*. Manaus, Brazil: Universidade Federal do Amazonas/Instituto de Pesquisas da Amazônia.

[33] Flores PAC (1999) Preliminary results of a photoidentification study of the marine tucuxi, *Sotalia fluviatilis*, in southern Brazil. *Marine Mammal Science,* 15, 840-847

[34] Flores PAC (2002) Tucuxi – *Sotalia fluviatilis*. In W. F. Perrin, B. Würsig, & J. G. M. Thewissen (Eds.), *Encyclopedia of Marine Mammals*. San Diego, CA: Academic Press.

[35] Flores PAC, da Silva VMF (2008) Tucuxi Dolphins – *Sotalia*. In W. F. Perrin, J. G. M. Thewissen & B. Wursig (Eds.), *Encyclopedia of Marine Mammals. 2nd edition*. San Diego, CA: Academic Press.

[36] Flower WH (1883) On the characters and divisions of the family Delphinidae. *Proceedings of the Zoological Society of London*, 1883, 466–513.

[37] Frankham R (1996) Relationship of genetic variation to population size in wildlife. *Conservation Biology,* 10 (6), 1500-1508.

[38] Fraser FC, & Purves PE (1960) Hearing in cetaceans, evolution of the accessory air sacs and the Structure of the outer and middle ears in cetaceans. *Bulletin of the British Museum of Natural History (Zoology),* 7, 1–140.

[39] Frère CH, Hale PT, Porter L, Cockcroft VG, & Dalebout ML (2008) Phylogenetic analysis of mtDNA sequences suggests revision of humpback dolphin (*Sousa* spp.) taxonomy is needed. *Marine and Freshwater Research*, 59: 259–268.

[40] Furtado Neto MAA (1998) *Molecular systematics and population genetics of marine vertebrates from Brazil. Doctoral Thesis.* St. John's, Newfoundland : Memorial University of Newfoundland.

[41] Gervais P (1853) Remarques sur les mammifères marins qui frequentent les côtes de la France et plus particulièrement sur une nouvelle espèce de Dauphin propre à la Méditerranée. *Bull Soc Centr Agricult comices agric départ l'Herault (Montpellier),* 40 année (Avril, Main, Juin), 140-156.

[42] Gervais P (1855) Description des trois especes de dauphins qui vivent dans la region du haut Amazone. In: P. Bertrand, (Ed.), *F. de Castelnau, Expedition dans les parties centrales de L'Amerique du Sud, du Rio de Janeiro a Lima et de Lima au Para, 7, Zoology* (pp. 89-94). Paris, France.

[43] Gravena W, Hrbek T, da Silva VMF, Farias IP (2008) Amazon River dolphin love fetishes: from folklore to molecular forensics. *Marine Mammal Science*, 24(4), 969–978

[44] Gray JE (1856) Description of a new species of dolphin (Steno) from the upper parts of the river Amazon. *Annals and Magazine of Natural History*, 18, 157-159.

[45] Gray JE (1866a) *Catalogue of seals and whales in the British Museum*. London, United Kingdom:British Museum of Natural History

[46] Gray JE (1866b) Notes on the skulls of dolphins or bottlenosed whales, in the British Museum. *Proceedings of the Zoological Society, London,* (1865), 211-216

[47] Hershkovitz PH (1963) Notes on South American dolphins of the genera *Inia*, *Sotalia* and *Tursiops*. *Journal of Mammalogy*, 44 (1), 98-103

[48] Hillis DM (1987) Molecular versus morphological approaches to systematics. *Annual Review ofEcology and Systematics*, 18, 23-42

[49] Hoelzel AR, Hancock JM, Dover G (1991). Evolution of the cetacean mitochondrial D-loop region.*Molecular Biology and Evolution*, 8, 475-493

[50] Hoorn C, Guerrero J, Sarmiento GA, Lorente MA (1995) Andean tectonics as a cause for changing drainage patterns in Miocene northern South America. *Geology*, 23, 237-240.

[51] Iredale T, Troughton EL (1934) Checklist of mammals recorded from Australia. *Memoirs of the Australian Museum*, 6, 1–122

[52] Irwin DM, Kocher TD, Wilson AC (1991) Evolution of the cytochrome b gene of mammals. *Journal of Molecular Evolution*, 32, 128-144

[53] International Union for Conservation of Nature and Natural Resources (IUCN). *[2008 IUCN Red list of threatened species]*. 2008 [cited year month day]. Available from:*http://www.iucn.org/themes/ssc/redlist.htm*

[54] Jefferson TA, Leatherwood S, Weber MA (1993) Marine Mammals of the World - FAO species identification guide. FAO, Roma Jefferson TA (2002) Rough-toothed dolphin - *Steno bredanensis* In W. F. Perrin, B. Würsig,.J. G. M Thewissen (Eds.), *Encyclopedia of marine mammals,* San Diego, CA: Academic Press.

[55] Klammer G (1984) The relief of the extra-Andean Amazon basin In Sioli H (Ed.), *The Amazon:limnology and landscape ecology of a mighty tropical river and its basin.* Dordrecht, Netherlands: Dr W Junk Publisher.

[56] Knowlton N (2000) Molecular genetic analyses of species boundaries in the sea. Hydrobiologia, 420, 73-90

[57] Krützen M, Barré LM, Connor RC, Mann J, Sherwin WB (2004) 'O father: where art thou?' —Paternity assessment in an open fission–fusion society of wild bottlenose dolphins (*Tursiops* sp.) in Shark Bay, Western Australia. Molecular Ecology, 13, 1975–1990

[58] Le Duc RG, Perrin WF, Dizon AE (1999) Phylogenetic relationships among delphinid cetaceans based on full cytochrome b sequences. *Marine Mammal Science*, 15, 619-648

[59] Leatherwood S, Reeves RR (1983) *The Sierra Club handbook of whales and dolphins.* San Francisco, CA: Sierra Club Books.

[60] León T (2005) *Craneometria del Genero Sotalia y Algunos Aspectos de su Ecología. BSc Monograph.* Maracaibo, Venezuela: Universidad del Zulia. Facultad Experimental de Ciencias.

[61] Lundberg JG, Marshall LG, Guerrero J, Horton B, Malabarba MCSL, Wesselingh F (1998) The stage for Neotropical fish diversification: a history of tropical South American rivers In: Malabarba LR, Reis RE, Vari RP, Lucena ZM, Lucena CAS (Eds.), *Phylogeny and classification of Neotropical fishes*. Porto Alegre, Brazil: Edipucrs.

[62] Mead RH, Koehnken L (1991) Distribution of the river dolphin, tonina Inia geoffrensis, in the Orinoco River Basin of Venezuela and Colombia. *Interciencia*, 16(6), 300-360.

[63] Mayr E (1963) *Animal Species and Evolution.* Cambridge, MA: Belknap Press.

[64] Mitchell ED (1975) Report of the meeting on smaller cetaceans. *Journal of the Fisheries Research Board of Canada,* 32, 889-983.

[65] Monteiro-Filho ELA, Monteiro LR, Reis SF (2002) Skull shape and size divergence in dolphins of the genus *Sotalia*: a tridimensional morphometric analysis. *Journal of Mammalogy*, 83, 125-134

[66] Moritz C (1994) Defining "Evolutionarily Significant Units" for conservation. TREE, 9 (10), 373-375

[67] Möller LM, Beheregaray LB, Harcourt & RG, Krützen M (2001) Alliance membership and kinship in wild male bottlenose dolphins (*Tursiops aduncus*) of southeastern Australia. *Proceedings of the Royal Society of London*, 268, 1941-1947

[68] Möller LM, Beheregaray LB (2004) Genetic evidence for sex-biased dispersal in resident bottlenose dolphins (*Tursiops aduncus*). Molecular Ecology, 13 (6), 1607-1612

[69] Möller LM, Beheregaray LB, Allen SJ, & Harcourt RG, (2006) Association patterns and kinship in female Indo-Pacific bottlenose dolphins (*Tursiops aduncus*) of southeastern Australia B*ehavioral Ecology and Sociobiology,* 61, 109–117

[70] Muller-Karger FE, McClain CR, & Richardson PL (1988) The dispersal of the Amazon's water. *Nature*, 333, 57-59

[71] O'Brien SJ (1994) A role for molecular genetics in biological conservation. *Proceedings of theNational Academy of Sciences of the United States of America*, 91, 5478-5755

[72] O'Corry-Crowe GM, & Lowry LF (1997) Genetic ecology and management concerns for the beluga whale (*Delphinapterus leucas*). In: Dizon AE, Chivers SJ, Perrin WF (Eds.), *Molecular Genetics of Marine Mammals – Special Publication Number 3*, Beaufort, NC: The Society for Marine Mammalogy.

[73] Palsbøll P, Vader A, Bakke I, & El-Gewely R (1992) Determination of gender in cetaceans by the polymerase chain reaction. *Canadian Journal of Zoology,* 70, 2166-2170.

[74] Palsbøll PJ, Berube M, & Allendorf FW (2007) Identification of management units using population genetic data. TREE, 22 (1), 11-16

[75] Perrin WF (1989) *Dolphins, popoises and whales. An action plan for the conservation of biological diversity: 1988-1992. 2nd ed.* Gland, Switzerland: International Union for Conservation of Nature and Natural Resources.

[76] Pichler FB, Baker CS (2000) Loss of genetic diversity in the endemic Hector's dolphin due to fisheries-related mortality. *Proceedings of the Royal Society of London*, 267, 97-102

[77] Reeves RR, Smith BD, Crespo EA, Notarbartolo Di & Sciara G (2003) *Dolphins, Whales and Porpoises: 2002–2010 Conservation Action Plan for the World's Cetaceans. IUCN/SSC Cetacean Specialist Group.* Gland, Switzerland and Cambridge, United Kingdom: International Union for Conservation of Nature and Natural Resources.

[78] Reeves RR, Perrin WF, Taylor BL, Baker SC, & Mesnick SL (2004) *Report of the workshop on shortcomings of cetacean taxonomy in relation to needs of conservation and management NOAA Technical Memorandum.* La Jolla, CA: NOAA-NMFS-SWFSC.

[79] Rice DW (1998) *Marine mammals of the world – Systematics and distribution. Special Publication Number 4,* Beaufort, NC: The Society for Marine Mammalogy.

[80] Rodríguez-Mahecha JV, Alberico M, Trujillo F, Jorgenson J (2006) *Libro Rojo de los Mamíferos de Colombia*. Bogotá, Colombia: Conservación Internacional.

[81] Rosas FCW, Monteiro-Filho ELA (2002) Reproduction of the estuarine dolphin (*Sotalia guianensis*) on the coast of Paraná, southern Brazil. *Journal of Mammalogy,.* 83(2), 507–515.

[82] Rosel PE, France SC, Wang JY, & Kocher TD (1999) Genetic structure of harbour porpoise *Phocoenaphocoena* populations in the northwest Atlantic based on mitochondrial and nuclear marker. *Molecular Ecolgy*, 8 (12), S41-S54

[83] Ryder OA (1986) Species conservation and systematics - the dilemma of subspecies. *TREE* 1, (1), 9-10

[84] Santos MCO, Barão-Acuña L, & Rosso S (2001) Insights on site fidelity and calving intervals of the marine tucuxi dolphin (*Sotalia fluviatilis*) in south-eastern Brazil. *Journal of the Marine Biological Association of the United Kingdom*, 81, 1049-1052

[85] Scott MD, Wells RS, & Irvine AB (1990) A Long-Term Study of Bottlenose Dolphins on the West Coast of Florida. In: S. Leatherwood, R. R. Reeves (Eds.), *The Bottlenose Dolphin.* San Diego, CA: Academic Press.

[86] Sholl TGC, Nascimento FF, Leoncini O, Bonvicino CR, & Siciliano S (2008) Taxonomic identification of dolphin love charms commercialized in the Amazonian region through the analysis of cytochrome b DNA. *Journal of the Marine Biological Association of the United Kingdom,* 88(6), 1207–1210

[87] Simões-Lopes PC (1987) Sobre a ampliação do gênero *Sotalia* Gray, 1986 (Cetacea, Delphinidae) para as águas do estado de Santa Catarina - Brazil. *Anais da 2ª Reunião de Trabalho de Especialistas em Mamíferos Aquáticos da América do Sul*, Rio de Janeiro, Brazil: Ed. FBCN.

[88] Sites JW, Marshall JC (2003) Delimiting species: a Renaissance issue in systematic biology. *TREE,* 18, 462-470

[89] Slatkin M (1985) Gene flow in natural populations. *Annual Review of Ecology and Systematics*, 16, 393-430

[90] Solé-Cava AM (2000) Biodiversidade molecular e genética da conservação. In: S. R. Matioli (Ed.), *Evolução Molecular*. Rio de Janeiro, Brazil: Editora Holos.

[91] Taylor BL, Chivers SJ, Sexton S, Dizon AE (2000) Evaluating dispersal estimates using mtDNA data: Comparing analytical and simulation approaches. *Conservation Biology*, 14 (5), 1287-1297

[92] Templeton AR (1998) Nested clade analyses of phylogeographic data: testing hypotheses about gene flow and population history. *Molecular Ecology*, 7, 381-397

[93] Templeton AR (2001) Using phylogeographic analyses of gene trees to test species status and processes. *Molecular Ecology*, 10, 779-791

[94] Tirira D (2001) *Libro Rojo de los Mamíferos de Ecuador.* Serie Libros Rojos del Ecuador, Tomo 1. Publicación Especial sobre los Mamíferos del Ecuador. Quito, Ecuador: Sociedad para la Investigación y Monitoreo de la Biodiversidad Ecuatoriana (SIMBIOE) / Ecociencias / Ministerio del Ambiente / UICN.

[95] True EW (1889) Review of the Family Delphinidae. *Bulletin of the United States Natural Museum*,36, 17-21

[96] Turvey ST, Pitman RL, Taylor BL, Barlow J, Akamatsu T, Barrett LA, Zhao XJ, Reeves RR, Stewart BS, Wang KX, Wei Z, Zhang XF, Pusser LT, Richlen M, Brandon JR, & Wang D (2007) First human-caused extinction of a cetacean species? *Biology Letters,* 3, (5): 537-540

[97] van Bénéden E (1875) Memoire sur un dolphin nouveau de la Baie de Rio de Janeiro. *Memmoires del'Academie Royale de Sciences Belgique* 41, 2-44.

[98] van Bénéden PJ (1864) Sur un dauphin nouveau et un ziphiode rare. *Memmoires de 1' Academie Royale de Sciences Belgique*, 16, 1-21.

[99] van Bree PJH (1974) On the diagnosis of the South American dolphin *Sotalia fluviatilis* and itsauthorship. Z Säugetierkd, 39, 57-58.

In: Biology, Evolution and Conservation of River Dolphins… ISBN: 978-1-60876-633-8
Editors: M. Ruiz-García, J. Shostell. pp. 285-297

Chapter 15

POPULATION STRUCTURE AND PHYLOGEOGRAPHY OF TUCUXI DOLPHINS (*SOTALIA FLUVIATILIS*)

Susana Caballero[1,2*], Fernando Trujillo[2], Manuel Ruiz-García[3], Julianna A. Vianna[4,5], Miriam Marmontel[6], Fabricio R. Santos[4] and C. Scott Baker[1, 7]

[1]Laboratory of Molecular Ecology and Evolution, School of Biological Sciences, The University of Auckland, Auckland, New Zealand.
[2]Fundación Omacha, Bogotá, Colombia
[3]Unidad de Genética (Genética de Poblaciones-Biología Evolutiva), Departamento de Biología, Facultad de Ciencias, Pontificia Universidad Javeriana, Bogotá, Colombia
[4]Laboratório de Biodiversidade e Evolução Molecular, Departamento de Biologia Geral, ICB, Universidade Federal de Minas Gerais, Av. Antonio Carlos, MG, Brazil
[5]Universidad Andrés Bello, Facultad Ecología y Recursos Naturales, Escuela de Medicina Veterinaria, Laboratorio Salud de Ecosistemas, República 252, Santiago, Chile.
[6]Instituto de Desenvolvimento Sustentável Mamirauá, Rua Augusto Correa No.1 Campus do Guamá, Setor Professional, Guamá, Brazil.
[7]Marine Mammal Institute and Department of Fisheries and Wildlife, Hatfield Marine Science Center, Oregon State University, Newport, OR, USA

ABSTRACT

Here we consider the phylogeography and population structure of the tucuxi dolphin *Sotalia fluviatilis*, based on samples (n = 26) collected across the Peruvian, Colombian and Brazilian Amazon Regions. Fourteen control region (CR) and two cytochrome *b* (Cyt-*b*) haplotypes were identified among these samples. The Amazonian population units identified showed high mitochondrial haplotype diversity and relatively high female mediated gene flow when compared to *Sotalia guianensis* and another Amazonian dolphin species, *Inia geoffrensis* throughout the sampled regions of the main river and its tributaries. A Union of Maximum Parsimonious Trees analysis generated a CR haplotype

* s.caballero@auckland.ac.nz

genealogy reflecting connectivity among sampled regions and identified divergent haplotypes found in the extremes of the species distribution. These results indicate the need to maintain connectivity between populations along the Amazon River and its tributaries as a main objective of management and conservation programs for *Sotalia fluviatilis.*

Keywords: phylogeography, *Sotalia fluviatilis*, population structure, mitochondrial DNA

INTRODUCTION

The coastal and riverine forms of the South American dolphin *Sotalia* have been recently accepted as different species (Monteiro-Filho et al., 2002; Cunha et al., 2005; Caballero et al., 2007). The riverine species, *Sotalia fluviatilis*, ranges throughout the Amazon River and most of its tributaries (Da Silva & Best, 1994). Although *Sotalia* are also reported 250 km up-river in the Orinoco, it is unclear if these animals are residents or transients from the coast (Boher et al., 1995)

Sotalia fluviatilis is considered "data deficient" by the International Union for the Conservation of Nature and Natural Resources (IUCN) (Klinowska, 1991; Reeves et al., 2003) and is listed in the Appendix I of the Convention on International Trade in Endangered Species of Wild Fauna and Flora (CITES). However, other researchers consider it endangered and in need of protection (Barros & Teixeira, 1994). The main anthropogenic threat that affects this species is gillnet entanglement in most of its Amazonian distribution (Da Silva & Best, 1996; Trujillo et al., 2000). In many places along the Amazon River, tucuxi dolphins are considered to have spiritual power while in other areas they are killed for shark bait and their eyes and genital organs sold as magical charms (Siciliano, 1994). The destruction of their habitat, oil and pesticide pollution (Trujillo et al., 2000; Monteiro-Neto et al., 2003; Yogui et al., 2003) and construction of dams for hydroelectric projects are also affecting the future of this species (Da Silva & Best, 1996).

Here we provide the first description of the phylogeography of *Sotalia fluviatilis* in the Amazonian region based on the analysis of two mitochondrial genes, control region (CR) and cytochrome *b* (Cyt-*b*).

METHODS

Sample Collection and DNA Extraction

A total of 26 samples of skin, bone or teeth were obtained from *S. fluviatilis* in eleven locations grouped into three geographic regions throughout their range (Figure 1 and Table 1). Tissue samples were obtained from dead stranded animals or animals captured in fishing nets. Bones and teeth were obtained from skeletal remains found in the field (n = 9) or from museum specimens (n = 3). Skin samples were stored in 70% ethanol at -20ºC. Bone and tooth samples were stored at room temperature in individual sealed bags. DNA extraction from tissue samples followed the protocol of Sambrook et al. (1989) modified for small samples by (Baker et al., 1994). DNA was extracted from bones following a silica-guanidinium thiocyanate based protocol described by Pichler et al. (2001)

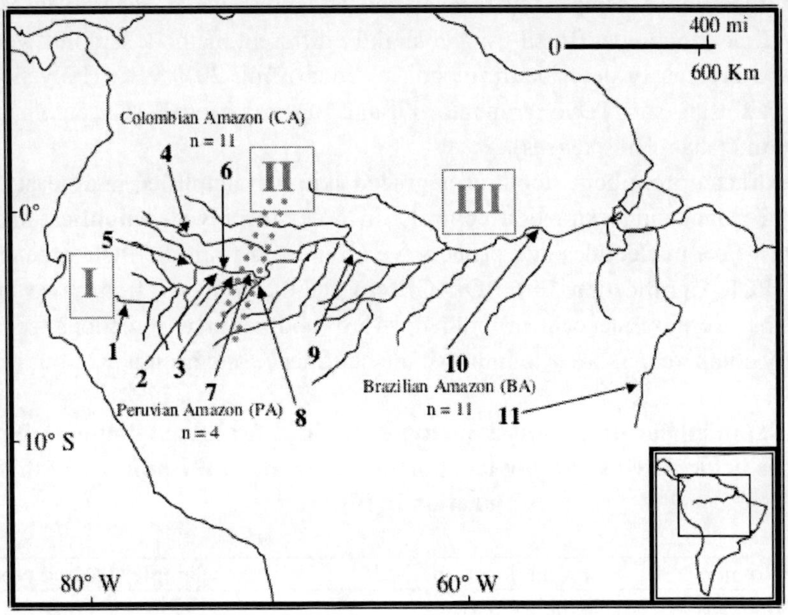

Figure 1. Map of the Amazon Region, showing the Amazon River and most of its main tributaries, as well as geographic regions, sampling locations and sizes included in this study. We also indicate the proposed genetic boundaries between *Sotalia fluviatilis* population units from the SAMOVA analysis (three groups, dotted lines, grey numbers): I = West Amazon, II = Central Amazon, III = Eastern Amazon.

PCR Amplification and Sequencing

Two mitochondrial genetic markers were analyzed; a 577 base pair (bp) portion of the mitochondrial DNA control region (CR) and a 425 bp fragment of the cytochrome *b* (Cyt-b) gene. Degradation of DNA or inhibition prevented clean amplification and sequencing of Cyt-*b* from all teeth and bone samples (n = 10). These samples are represented only by partial control region sequences. Genes were amplified via the Polymerase Chain Reaction (PCR) using standard reaction conditions (Saiki et al., 1988, Palumbi, 1996). For the CR, we used the primer combination t-Pro-whale (5'-TCACCCAAAGCTGRARTTCTA-3') and Dlp8 (5'-CCATCGWGATGTCTTATTTAAGRGGAA-3') (Baker et al., 1998). The primer combination Dlp1.5t-Pro-whale (5'-TCACCCAAAGCTGRARTTCTA-3') and Dlp4 (5'-GCGGGWTRYTGRTTTCACG-3') was used in the case of DNA extracted from bones or degraded tissue samples, since these amplify a smaller region of approximately 400 bp (M. Dalebout, pers. comm.). For Cyt-*b*, we used the primers Tglu (5'-TGACTTGAARAACCAY CGTTG-3') and CB2 (3'-ACTCCTGTTTATAGTAAGAC-5') (Palumbi, 1996). The PCR profile for all combinations of primer pairs used was as follows: an initial denaturation at 95°C for 2 min, 36 cycles of 94°C for 30 s, 55°C for 1 min and 72°C for 1.30 min, and a final extension at 72°C for 5 minutes. Free nucleotides and primers were removed from the PCR products using SAP (shrimp alkaline phosphatase) and ExoI (exonuclease I) (USB) and directly sequenced in both directions using the standard protocols of Big Dye™ terminator sequencing chemistry on an ABI 3100 automated capillary sequencer (Perkin Elmer). Alternatively, Brazilian samples were analyzed at Universidade Federal de Minas Gerais (UFMG) in Belo Horizonte, Brazil using a slightly different method: samples were amplified following the previously described protocol, cleaned using 20% PEG (Polyethyleneglycol) and sequenced using an ETDye terminator kit and run in a MegaBACE automated capillary sequencer (Amersham Biosciences).

DNA extracted from bone, tooth or degraded skin was amplified in at least two separate PCR reactions, including extraction controls, in order to prevent amplification of possible contaminants. Free nucleotides and primers were removed from the PCR products using the QIAquick PCR purification kit (QIAGEN). PCR products from two independent amplifications were sequenced in both forward and reverse directions separately and subsequently compared, in order to improve the confidence and accuracy of our results.

Table 1. Sampling locations and tissue type obtained for tucuxi dolphins. Numbers in parenthesis before each sampling location corresponds to the number of this sampling location in Figure 1

Geographic region	Sampling location	Sample size and type
Peruvian Amazon	(1) Curaray River	1 skin
(PA)	(2) Caballo Cocha	1 bone
	(Loreto province)	1 skin
	(3) Patrullero Island	1 skin
	(Loreto province)	
Colombian Amazon	(4) Caquetá River	2 bones
(CA)	(5) Puerto Alegria	1 skin
	(Amazonas province)	
	(6) Puerto Nariño	2 skins
	(Amazonas province)	4 teeth
	(7) Leticia	1 skin
	(Amazonas province)	1 tooth
Brazilian Amazon	unknown	1 DNA**
(BA)	(8) Benjamin Constant	1 skin
	(Amazonas state)	
	(9) Tefé	7 skins
	(Amazonas state)	
	(10) Santarém	1 bone
	(Pará state)	
	(11) Formoso Araguaia River	1 bone

**Sample donated as extracted DNA by the SWFSC: Southwest Fisheries Science Center (La Jolla, CA, U.S.A)

Data Analyses

Sequence quality was evaluated using the program *Phred* v.020425 (Ewing & Green, 1998, Ewing et al., 1998). Sequences with *Phred* scores ≤ 20 (a base call having a probability of more than 1/100 of being incorrectly called) were excluded from the analysis or re-sequenced. Sequences with *Phred* scores values between 20 and 40 (a probability between 1/100 and 1/10,000 of being incorrectly called) were checked by eye. All sequences were manually edited and aligned using Sequencher 4.1 software (Gene Codes Corporation). CR haplotypes were defined using MacClade (Maddison & Maddison, 2000). Haplotype sequences were submitted to GenBank as accession numbers EF027006 to EF027092.

The model of substitution for the CR was tested in Modeltest v3.06 (Posada & Crandall, 1998) and the settings for this model were used in the phylogenetic reconstructions performed in PAUP version 4.0b1 (Swofford, 2002). In order to investigate genealogical relationships among *Sotalia fluviatilis* CR haplotypes, Union of Maximum Parsimonious Trees (UPM) (Cassens et al., 2005) was used to calculate and construct a network of control region haplotypes. This method requires two consecutive steps. First, a Maximum Parsimony analysis was performed for the CR haplotype data set and the most parsimonious trees were saved with their respective branch lengths. We used the TBR branch-swapping (1000 replicates with random sequence addition) heuristic search option in PAUP* v.4b10 (Swofford, 2002). Second, all the saved MP trees were combined into a single reticulated graph, merging branches (sampled or missing) that were identical among different trees (see Cassens et al. 2005 for additional details on this analysis). The haplotype frequency was combined with the CR haplotype network, and the final network was drawn by hand.

Analyses of diversity and population structure were performed in the program Arlequin (Schneider et al., 2000) and restricted to the CR (577 bp) because of the larger sample size for this locus. To evaluate genetic boundaries between the populations studied, we performed a spatial analysis of molecular variance (SAMOVA) (Dupanloup et al., 2002). In this analysis, the sample localities (entered as geographic coordinates) are connected using an algorithm and a graphical method in order to define the genetic composition of groups or population units and to maximize the F_{CT} index, which is the proportion of total genetic variance due to differences between groups or populations (Dupanloup et al., 2002). Genetic differences among the estimated populations detected in the SAMOVA analysis were then quantified by an analysis of molecular variance (AMOVA) as implemented in Arlequin (Excoffier et al., 1992) based on conventional F_{ST} and Φ_{ST} statistics. The significance of the observed Φ_{ST} and F_{ST} statistics were tested using 10,000 random permutations. The control region haplotype and nucleotide diversity were estimated using the program Arlequin (Schneider et al., 2000). The number of female migrants per generation (N_{mf}), as a measure of gene flow among localities, was estimated based on the F_{ST} value, using the equation $N_{mf} = 1/2(1/F_{ST} - 1)$ (Takahata & Palumbi, 1985) assuming Wright's island model.

RESULTS

Phylogeography

A total of 577 bp of the CR and 425 bp of the Cyt-*b* gene were analyzed. For CR, fourteen haplotypes were defined by eleven variable sites (Table 2) (For haplotype nomenclature please refer to Caballero et al. 2007). Two haplotypes were defined for Cyt-*b*, differing by one site (for further information refer to Caballero et al., 2007). Overall, high haplotype diversity was detected in *Sotalia fluviatilis* (Table 3). Phylogenetic relationships among CR haplotypes were reconstructed by Maximum Parsimony, Maximum Likelihood (using the output model from Modeltest HKY+I+G) and Neighbor- Joining (Figure 2).

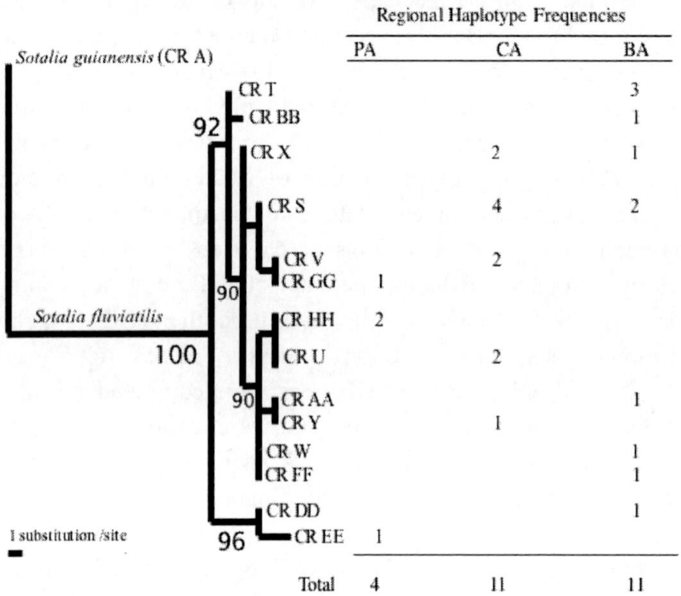

Figure 2. Neighbor-Joining phylogenetic reconstruction of Control Region haplotypes (577 bp), showing bootstrap values (1,000 replicates) and the frequency of occurrence in each geographic region. Abbreviations follow Figure 1 and Table 1. Letters on terminal branches represent haplotype codes.

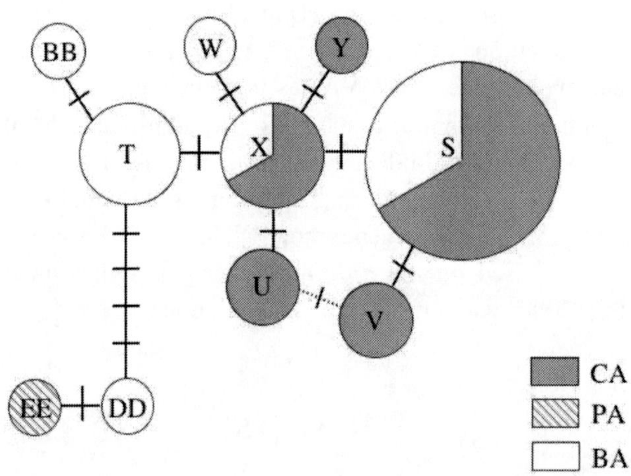

Figure 3. Haplotype genealogy obtained from the Union of Maximum Parsimonious Trees (UMP) analysis. The size of the circles reflect frequency of a particular haplotype found in the Colombian Amazon (CA), Peruvian Amazon (PA) and Brazilian Amazon (BA) geographic regions. Connections between haplotypes found in all mostly parsimonious trees are represented by a continuous line, while connections between haplotypes found in parsimonious trees are represented by a dotted line. Vertical bars represent substitutions between haplotypes.

Ten haplotypes were included in the UPM analysis. Four haplotypes were excluded since they contained too much missing data, as this can affect the performance of the algorithm used for combination of most parsimonious trees into one network or haplotype genealogy. The six most parsimonious trees were obtained and these were combined in the haplotype genealogy presented in Figure 3. Haplotypes X, S, and T were in a central position and connected with a high number of other haplotypes. The most divergent haplotypes were DD and EE. In three of the six most parsimonious trees, haplotypes U and V were connected and therefore we included this haplotype connection in the final figure.

POPULATION STRUCTURE

Very few haplotypes were shared between different geographic regions, thus indicating some degree of female phylopatry for this species. Only two haplotypes (S and X) were shared between the Colombian Amazon and Brazilian Amazon geographic regions.

For SAMOVA analysis, only sampling regions with n ≥ 2 were considered, since this method takes into account variance within each sampling location when calculating boundaries between estimated population units. Five sampling locations, Caballo Cocha (PA), Caquetá River (CA), Puerto Nariño (CA), Leticia (CA) and Tefé (BA), were considered in the SAMOVA. The largest mean F_{CT} index was found for three population units ($F_{CT} = 0.277$): (1) Western Amazon (II) Central Amazon and (III) Eastern Amazon (Figure 1).

Samples from the Central Amazon population unit had to be excluded from the AMOVA analysis due to the small sample size (n ≤ 2). For the remaining two population units (Western and Eastern Amazon), no significant differences were found at the F_{ST} level, but significant differences were detected at the Φ_{ST} level (Table 3). For these units, N_{mf} was 17 females per generation (using $F_{ST} = 0.027$).

Table 2. Eleven variable sites over 577 bp of the mtDNA control region determining fourteen haplotypes in *Sotalia fluviatilis*

	Variable sites
Haplotype	*Control Region (557 pb)*
	12222333445
	00469469777
	54483991077
CR S	GTCTCACTTTT
CR T	A..C.......
CR U	...C.T?????
CR V	.....T?????
CR W	...C......C
CR X	...C.......
CR Y	...C.....C?
CR AA	...C..?????
CR BB	A..C....A.?
CR DD	AC.CT.TC...
CR EE	ACTCT.TC.C?
CR FF	...C.....??
CR GG	.....??????
CR HH	...C.??????

Table 3. Pairwise F_{ST} (below diagonal) and Φ_{ST} (above diagonal) values for Control Region between two riverine *Sotalia* population units (population units as indicated by the SAMOVA analysis, Figure 1). Probability values based on 10,000 permutations shown in italics. Significantly different values ($P < 0.05$) in bold. Haplotype (*h*) and nucleotide (π) % ± standard deviation (SD) are shown on the diagonal for each population unit

Φ_{ST} / F_{ST}	Western Amazon	Eastern Amazon
Western Amazon (n=13)	h = 0.6026 ± 0.08 π = 0.48 ± 0.0033	0.1288 **(0.0468)**
Eastern Amazon (n=11)	0.0275 (0.1439)	h = 0.7636 ± 0.08 π = 0.39 ± 0.0026

THOUGHTS AND CONCLUSION

Population Structure

Less regional structure was found among the riverine population units compared to coastal population units (Caballero, 2006). Although the Western Amazon and Eastern Amazon population units share only two haplotypes, shorter genetic distances separate all CR lineages. This could be due to the relatively shorter evolutionary history of *Sotalia fluviatilis* when compared to the possibly longer evolutionary history of the coastal species (Caballero et al., 2007). Higher levels of gene flow could also be expected between the Amazonian population units due to the scattered distribution of small groups of individuals along the main channels and tributaries of the Amazon River. Interestingly, in our study, significant statistical differences were obtained at the Φ_{ST} level between the two population units considered in the AMOVA analysis (Table 3). This might be due to the presence of a few very distinctive haplotypes with several nucleotide differences between these population units.

The preliminary haplotype genealogy confirmed these findings, suggesting that haplotypes X, S and T may be ancestral, considering that they are geographically widespread, are connected to a higher number of other haplotypes, they have high haplotype frequencies, and are located in a central position (Castelloe & Templeton, 1994). Also, it can be observed that haplotypes EE and DD are more divergent. This is an interesting finding, since haplotypes X, S and T were determined in samples collected along the main channel of the Amazon River and also in some tributaries located centrally along the distribution of *Sotalia fluviatilis* (Tefé, Puerto Nariño, Caquetá River) while haplotypes DD and EE were determined in samples from locations located in the extremes of the distribution, for example the Cuyabeno River (EE) and Santarém (DD). This result can be reflecting patterns of connectivity among different Amazonian tributaries and channels with increasing haplotype and population differentiation in more isolated tributaries. More sampling along other Amazon River tributaries is required to describe with confidence, population units for *Sotalia fluviatilis* as well as haplotype genealogies.

Overall, haplotype and nucleotide diversities for the mitochondrial DNA control region in *Sotalia fluviatilis* are similar to those reported for species with similar distributions and habitat ranges, including the Antillean and Amazonian manatees (García-Rodríguez et al., 1998; Vianna et al., 2006) and the Amazon River dolphin *Inia geoffrensis* (Banguera-Hinestroza et al., 2002). However, although both Amazon River dolphin species present some degree of phylopatry, recent data (Vianna et al. 2010, this book) for the pink dolphin (*I. geoffrensis*), collected in the same Brazilian region as *S. fluviatilis* sampled here, indicate that the former species is highly structured, even in a microgeographic scale. As *S. fluviatilis* may have a much more recent origin as species in the Amazon, the spatial differentiation of populations is less pronounced but it is underway.

Implications for *Sotalia* Fluviatilis Conservation and Management

As can be observed in our results, gene flow appears to be higher for *S. fluviatilis* than *S. guianensis* (Caballero, 2006), at least among the Amazon regions included in our study. As can be observed in our results, gene flow appears to be high between the regions included in our study. For this reason, priority should be given to maintain this connectivity. Obstacles to connectivity could affect these population units and therefore, hydroelectric and dam constructions must be evaluated, depending on the region where they intend to be developed, taking into consideration the distribution of *S. fluviatilis* and other aquatic mammals and reptiles in the region, as well as routes in fish migration and abundance of prey items to sustain these groups (Smith & Smith, 1998). Boat traffic and fishery interactions must also be determined along the Amazon and most of its channels and tributaries, as has been done by researchers in the Colombian Amazon (Trujillo et al., 2000; Diazgranados et al., 2002). As our data have also shown, genetic differentiation is higher in the extremes of the distribution of the Amazon species, thus its conservation strategy should also take into account the relative isolation of some populations. Local takes will result in local extinction but connectivity could mask a wider decline (Taylor, 1997). This would require greater regulation and law enforcement of both commercial and artisanal fisheries. Regulation of these activities and improvement of fishing practices needs to be implemented with involvement of the local communities.

Acknowledgments

We are grateful to all the people and institutions that gave us access to samples for this study: J. G. Mead and C. Potter (United States Smithsonian Institution National Museum of Natural History), R. L. Brownell Jr., students and researchers at Fundación Omacha (Colombia), IBAMA (Brazil), the DNA Archive (NMFS Southwest Fisheries Science Center) and the Fondo para la Accion Ambiental (US-Aid) (120108-E0102141; Structure and Genetic Conservation of river dolphins, *Inia* and *Sotalia*, in the Amazon and Orinoco basins; project granted to the Pontificia Universidad Javeriana-M. Ruiz-García). All Brazilian samples were collected with the government permit IBAMA 131/2004. This research was developed according to the special authorization for access to genetic resources in Brazil # 03/2004 issued by IBAMA/CGEN. In Colombia, authorization was granted by the Ministerio del Medio Ambiente, Vivienda y Desarrollo Territorial (Contrato de Acceso a Recursos Genéticos No. 001). Thanks especially to P. Lara (UFMG, Brazil) for help with the laboratory analysis in Brazil and to P. Mardulyn (Behavioral and Evolutionary Ecology, Free University of Brussels) for his help implemetation of the UMPT analysis. Funding for fieldwork and laboratory analysis was provided by the New Zealand Marsden Fund (to C. S. Baker), a University of Auckland International PhD Scholarship (to S. Caballero), Colciencias-LASPAU (to S. Caballero), a Cetacean International Grant-In-Aid (to S. Caballero and J. A. Vianna), Universidad de los Andes (Colombia), Pontificia Universidad Javeriana (Colombia), Conselho Nacional de Pesquisas (CNPq-Brazil), The University of Auckland Graduate Research Fund and private resources.

REFERENCES

[1] Baker, C. S., Flórez-González, L., Abernethy, R. B., Rosenbaum, H. C., Slade, R. W., Capella, J. & Bannister, J. L. (1998). Mitochondrial Dna Variation And Maternal Gene Flow Among Humpback Whales Of The Southern Hemisphere. *Marine Mammal Science*, 14, 721-737.

[2] Baker, C. S., Slade, R. W., Bannister, J. L., Abernethy, R. B., Weinrich, W. T., Lien, J., Urban, J.,Corkeron, P., Calambokidis, J., Vasquez, O. & Palumbi, S. R. (1994). Hierarchical Structure Of Mitochondrial Dna Gene Flow Among Humpback Whales *Megaptera Novaeangliae*, World-Wide. *Molecular Ecology*, 3, 313-327.

[3] Banguera-Hinestroza, E., Cárdenas, H., Ruíz-García, M., Marmontel, M., Gaitán, E., Vásquez, R. & García-Vallejo, F. (2002). Molecular Identification Of Evolutionarily Significant Units In The Amazon River Dolphin *Inia* Sp. (Cetacea: Iniidae). *The Journal of Heredity*, 93, 9-19.

[4] Barros, N. & Teixeira, R. (1994). Incidental Catch Of Marine Tucuxi, *Sotalia Fluviatilis*, In Alagoas, Northeastern Brazil. In Gillnets And Cetaceans. Reports To The International Whaling Commission, (Special Issue 15, Pp. 265-268). Cambridge, United Kingdom: International Whaling Commission.

[5] Boher, S., Bolaños, J. & Cova, L. J. (1995). Sobre Un Avistamiento Del Delfín Estuarino O Bufete (*Sotalia Fluviatilis*) En El Orinoco Medio. *Acta Científica Venezolana*, 46, 217-218.

[6] Caballero, S. (2006). *Genetic Characterization Of South American Coastal And Riverine Dolphins Sotalia. Phd Dissertation.* Auckland, New Zealand. The University Of Auckland.

[7] Caballero, S., Trujillo, F., Vianna, J. A., Barrios-Garrido, H., Montiel, M. G., Beltrán-Pedreros, S., Marmontel, M., Santos, M. C. O., Rossi-Santos, M., Santos, F. R. & Baker, C. S. (2007). Taxonomic Status Of The Genus *Sotalia*: Species Level Ranking For "Tucuxi" (*Sotalia Fluviatilis*) And "Costero" Dolphins (*Sotalia Guianensis*). *Marine Mammal Science,* 23, 358-386.

[8] Cassens, I., Mardulyn, P. & Milinkovitch, M. C. (2005). Evaluating Intraspecific "Network" Construction Methods Using Simulated Sequence Data: Do Existing

Algorithms Outperform The Global Maximum Parsimony Approach? *Systematic Biology*, 54, 363-372.

[9] Castelloe, J. & Templeton, A. R. (1994). Root Probabilities For Intraspecific Gene Trees Under Neutral Coalescent Theory. Molecular Phylogenetics And Evolution, 3, 102-113.

[10] Cunha, H. A., Da Silva, V. M. F., Lailson-Brito, J. J., Santos, M. C. O., Flores, P. A. C., Martin, A. R., Azevedo, A. F., Fragoso, A. B. L., Zanelatto, R. C. & Solé-Cava, A. M. (2005). Riverine And Marine Ecotypes Of *Sotalia Fluviatilis* Are Different Species. *Marine Biology,* 148, 449-457.

[11] Da Silva, V. M. F. & Best, R. C. (1994). Tucuxi *Sotalia Fluviatilis* (Gervais, 1853). In S. H. Ridgway, S.& R. Harrison, (Eds.), *Handbook Of Marine Mammals. (*Vol. 5, Pp. 43-69). San Diego, Ca: Academic Press Inc.

[12] Da Silva, V. M. F. & Best, R. C. (1996). *Sotalia Fluviatilis*. *Mammalian Species*, 527, 1-7.

[13] Diazgranados, M. C., Mejía, P. & Acosta, A. (2002). Effect of Boat Traffic on The Vocal And Surfacing Behavior of River Dolphins: A Key For Abundance Estimation Methods. *Journal of The Acoustic Society of America*, 112, 2431.

[14] Dupanloup, I., Schneider, S. & Excoffier, L. (2002). A Simulated Annealing Approach To Define The Genetic Structure Of Populations. *Molecular Ecology*, 11, 2571-2581.

[15] Ewing, B. & Green, P. (1998). Base-Calling Of Automated Sequencer Traces Using *Phred*. Ii. Error Probabilities. *Genome Research,* 8, 186-194.

[16] Ewing, B., Hillier, L., Wendl, M. C. & Green, P. (1998). Base-Calling Of Automated Sequencer Traces Using *Phred*. I. Accuracy Assessment. *Genome Research*, 8, 175-185.

[17] Excoffier, L., Smouse, P. E. & Quattro, J. M. (1992). Analysis Of Molecular Variance Inferred FromMetric Distances Among Dna Haplotypes: Application To Human Mitochondrial Dna Restriction Data. *Genetics*, 131, 479-491.

[18] García-Rodríguez, A. I., Bowen, B. W., Domning, D., Mignucci-Giannoni, A. A., Marmontel, M.,Montoya-Ospina, R. A., Morales-Vela, B., Rudin, M., Bonde, R. K. & Mcguire, P. M. (1998). Phylogeography Of The West Indian Manatee (*Trichechus Manatus*): How Many Populations And How Many Taxa? *Molecular Ecology*, 7, 1137-1149.

[19] Klinowska, M. (1991). *The Red Data Book: Dolphins, Porpoises And Whales Of The World, InternationalUnion For The Conservation Of Nature And Natural Resources Iucn*, Gland, Switzerland: International Union For The Conservation Of Nature And Natural Resources.

[20] Maddison, D. R. & Maddison, W. P. (2000). *Macclade: Analysis Of Phylogeny And Character Evolution.*Sunderland, Ma: Sinauer.

[21] Monteiro-Filho, E. L. D. A., Rabello-Monteiro, L. & Do Reis, S. F. (2002). Skull Shape And Size Divergence In Dolphins Of The Genus *Sotalia*: A Morphometric Tridimensional Analysis. *Journal of Mammalogy*, 83, 125-134.

[22] Monteiro-Neto, C., Itavo, R. V. & Do Souza-Moraes, L. E.(2003). Concentrations Of Heavy Metals In *Sotalia Fluviatilis*. *Environmental Pollution,* 123, 319-324.

[23] Palumbi, S. R. (1996). Nucleic Acids Ii: The Polymerase Chain Reaction. In C. Moritz, C. & B. K. Mable, (Eds.), *Molecular Systematics. Hillis* (Pp: 205-247). Sunderland, Ma: Sinauer Associates.

[24] Pichler, F. B., Dalebout, M. & Baker, C. S. (2001). Nondestructive Dna Extraction From Sperm WhaleTeeth And Scrimshaw. *Molecular Ecology Notes,* 1, 106-109.

[25] Posada, D. & Crandall, K. A. (1998). Modeltest: Testing The Model Of Dna Substitution. *Bioinformatics,*14, 817-818.

[26] Reeves, R. R., Smith, B. D., Crespo, E. A. & Do Sciara, G. N. (2003). 2002-2010 *Conservation Action Plan For The World's Cetaceans: Dolphins, Whales And Porpoises,* Gland, Switzerland: International Union For The Conservation Of Nature And Natural Resources.

[27] Saiki, R. K., Gelfand, D. H., Stofell, S., Scharf, R., Higuchi, R., Horn, G. T., Mullis, K. B. & Erlich, H. A. (1988). Primer-Directed Enzymatic Amplification Of Dna With A Thermostable Dna Polymerase. *Science,* 239, 487-491.

[28] Sambrook, J., Fritsch, E. F. & Maniatis, T. (1989). *Molecular Cloning: A Laboratory Manual,* Cold New York, Ny: Spring Harbor Laboratory Press..

[29] Schneider, R., Roessli, D. & Excoffier, L. (2000). *Arlequin: A Software For Population Genetic Data Analysis*. Geneva, Switzerland: Genetic And Biometry Laboratory, University of Geneva.

[30] Siciliano, S. (1994). Review of Small Cetaceans And Fishery Interactions In The Coastal Waters of Brazil. In *Gillnets And Cetaceans. Reports To The International Whaling Commission,* (Special Issue 15, Pp. 241-250). Cambridge, United Kingdom: International Whaling Commission.

[31] Smith, A. M. & Smith, B. D. (1998). Review Of Status And Threats To River Cetaceans And Recommendations For Their Conservation. *Environmental Reviews*, 6, 189-206.

[32] Swofford, D. L. (2002). *Paup: Phylogenetic Analysis Using Parsimony, 4.0b10.* Tallahassee, Fl: Florida State University.

[33] Takahata, N. & Palumbi, S. R. (1985). Extranuclear Differentiation And Gene Flow In The Finite IslandModel. Genetics, 109, 441-457.

[34] Taylor, B. L. (1997). Defining "Population" To Meet Management Objectives For Marine Mammals. In A. Z. Dizon, S. J. Chivers, & W. F. Perrin, (Eds.), *Molecular Genetics Of Marine Mammals* (Pp. 49-65). Lawrence, Ks: The Society For Marine Mammalogy.

[35] Trujillo, F., García, C. & Avila, J. M. (2000). Status And Conservation Of The Tucuxi *Sotalia Fluviatilis* (Gervais, 1853): Marine And Fluvial Ecotypes In Colombia (Sc/52sm11/2000, Pp. 1-12). Adelaide, Australia: International Whaling Commission.

[36] Vianna, J. A., Bonde, R. K., Caballero, S., Giraldo, J. P., Lima, R. P., Clark, A. M., Marmontel, M., Morales-Vela, B., Souza, M. J., Parr, L., Rodríguez-López, M. A., Mignucci-Giannoni, A. A., Powell, J. & Santos, F. R. (2006). Phylogeography, Phylogeny And Hybridization In Trichechid Sirenians: Implications on Manatee Conservation. *Molecular Ecology*, 15, 433-447.

[37] Vianna, J.A., Hollatz, C., Marmontel, M., Redondo, R.A., & Santos, F.R. (2010). Amazon River Dolphin: High Phylopatry Due To Restricted Dispersion At Large And Short Distances. In M. Ruiz-García & J. M. Shostell (Eds.), *Biology, Evolution, And Conservation Of River Dolphins Within South America And Asia: Unknown Dolphins In Danger*. Hauppauge, Ny: Nova Science Publisher..

[38] Yogui, G. T., Santos, M. C. D. O. & Montone, R. C.(2003). Chlorinated Pesticides And Polychlorinated Biphenyls In Marine Tucuxi Dolphins (*Sotalia Fluviatilis*) From The

Cananéia Estuary, Southeastern Brazil. *The Science Of The Total Environment,* 312, 67-78.

PONTOPORIA BLAINVILLEI

In: Biology, Evolution and Conservation of River Dolphins... ISBN: 978-1-60876-633-8
Editors: M. Ruiz-García, J. Shostell. pp. 301-322

Chapter 16

LIFE HISTORY AND ECOLOGY OF FRANCISCANA, *PONTOPORIA BLAINVILLEI* (CETACEA, PONTOPORIIDAE)

Eduardo R. Secchi*

Laboratório de Tartarugas e Mamiferos Marinhos, Instituto de Oceanografia (IO-FURG)
Universidade Federal do Rio Grande/FURG Rio Grande - RS, Brazil

ABSTRACT

In the current chapter, I discuss some aspects of the life history and ecology of the Franciscana (*Pontoporia blainvillei*). This is a small dolphin which inhabits the coasts of southern Brazil, Uruguay and Argentina. Disappointedly, this species suffers an extensive loss of individuals each year due to mortality in fishing nets. Therefore, all available knowledge about the ecology of this species is useful for its conservation.

GENERAL EXTERNAL MORPHOLOGY

Pontoporia blainvillei is a small cetacean with a brownish or ochre skin on the back, which turns lighter on the flanks and ventral region. The dorsal fin is fairly tall and triangular with a slightly rounded tip while the flippers are very broad, visibly fingered, paddle-shaped with curved and irregular trailing edges. The width of the flukes is slightly over one fourth of the body length (Brownell, 1989) for both adults and juveniles. The rostrum length varies ontogenetically. It is relatively short in young individuals and extremely slender and elongated in adults, holding approximately 250 very small and sharp teeth (Bastida et al., 2007). The neck is flexible, head is small with a bulky melon and small eyes and the blowhole resembles a median transverse crescent (Figure 1).

* edu.secchi@furg.br.

Figure 1. External characteristics of a franciscana dolphin: long and narrow rostrum, broad flippers and typical ocher coloration.

ORIGIN, EVOLUTION AND PHYLOGENETIC RELATIONSHIP

Shared skull characters among *Pontoporia* and the true river dolphins such as elongated rostrum and mandibular symphysis have led many authors to consider them a monophyletic lineage and classify them as belonging to the same the family (Platanistidae) or superfamily (Platanistoidea) (Zhou, 1982; Cozzuol, 1985). These characters could, however, be ancestral and converged to adaptation to living in turbid waters which would make them uninformative for phylogenetic purposes (Cassens et al., 2000). Although this monophyly was often not contested, morphological analyses and frequent disagreement among scientists led to several re-classifications of the group's phylogenetic relationship. It was considered a group of four genera in four monotypic family: *Pontoporia* (Pontoporiidae), *Inia* (Iniidae), *Lipotes* (Lipotidae) and *Platanista* (Platanistidae) (Zhou, 1982); then later *Pontoporia*, *Inia* and *Lipotes* were placed in the same clade, while other authors clumped *Pontoporia* and *Inia* in a clade closely related to Delphinoidea and considered *Lipotes* as a sister group of this Pontoporia+Inia+Delphinoidea clade (de Muizon, 1984). To date, paleontology cannot elucidate this issue because, among other factors, fossil records of river dolphins are scanty and geographically isolated (Cassens et al., 2000). Recent multiple evidence based on both nuclear and mitochondrial DNA analyses was in accordance with recent morphological findings to demonstrate that the group is not monophyletic and most probably polyphyletic, which is also consistent with the highly disjunct geographic distribution of the group. These data also suggest that the two South American species (*Pontoporia* and *Inia*) form the sister group of the Delphinoidea (Cassens et al., 2000; Hamilton et al., 2001). In fact, fossils related to Iniidae and Pontoporiidae were found in the same layers of the La Plata region (Cozzuol, 1985). Molecular data also suggested that river dolphins lineages diverged well before the radiation of delphiniids and that they represent relict species whose adaptation to fluvial habitats might have insured their survival against environmental changes in the marine ecosystem or the emergence and dominance of delphinids (Cassens et al., 2000). In the case of franciscana, because it is mostly marine and diverged from *Inia* before the radiation of delphinids, it is unclear whether the common ancestor of the two South American species was marine or riverine. If it was riverine, it can be considered a recent ecological reversal with the reinvasion of the marine coastal habitat. If otherwise, then the franciscana's ancestors might have escaped extinction because it was ecologically specialized with adaptations to live in coastal habitat and coupe with competitive pressures after the radiation of small delphinids. The Amazon and Paraná river basins in South America were deeply invaded by marine waters during Miocene high sea levels stands. The shallow estuarine regions created by the mixing of oceanic and riverine waters probably supplied a diversity and abundance of food resources, particularly for the species able to tolerate osmotic stress. The draining of epicontinental seas occurred with sea level regression which took place during the Late Miocene and Pliocene. During the highest global sea level, the Amazon and Paraná basins may have been connected, originating an interior sea known as Paranaese Sea (sensu Von Ihering, 1927), dividing the continent. It has been hypothesized that dolphins entered the Paranaese Sea from the north, diversified within its complex fluvial-estuarine-marine system and colonized as far as the western South Atlantic Ocean (Hamilton et al., 2001). Evidences include: isolated periotics bones of *Pontoporia* from the late Miocene-Pliocene found in the eastern ravines of the Paraná River (Cozzuol, 1985); pleistocene specimens recovered in “Piso Querandino”, near La Plata, Argentina (Ameghino 1918) and; incomplete, Late Pleistocene, skulls of *P. blainvillei* found on Rio Grande do Sul State coast, Brazil (Buchmann and Rincón, 1997; Ribeiro et al., 1998). Furthermore, closely related early Pliocene Miocene (*Pliopontos littoralis,* de Muizon, 1983) and middle Miocene (*Brachydelphis mazeasi,* de Muizon, 1988) pontoporiids fossils were collected as north as the Pisco Formation in Peru and Pontistes rectifrons (Bravard, 1885) was recovered from the late Miocene in the Paraná Formation, Argentina (Cozzuol, 1985). The lowering of the global sea level the inland sea was drained separating the northern and southern river basins and isolating river dolphins. While Inia remained isolated to fluvial system in the Amazon and Orinoco, *Pontoporia* followed the marine waters, the recede of the Paraná basin, to colonize the coastal zone north and south of the La Plata estuary (Hamilton et al., 2001).

DISTRIBUTION AND HABITAT

The franciscana is endemic to the western South Atlantic Ocean, ranging from Itaúnas (18°25′S), Espírito Santo State, Brazil (Siciliano, 1994) to Golfo San Matias (~42°10S’), Rio Negro Province, Argentina (Crespo et al., 1998) (Figure 2). Although it has been considered by many to be a member of the so-called river dolphins (superfamily Platanistoidea – currently thought to be a polyphyletic taxon - *e.g.* Cassens et al., 2000), franciscanas are found mainly in coastal marine waters with occasional occurrences in estuaries (*e.g.* Santos et al., 2007; 2009). It is, however, relatively common in the Uruguayan part of the La Plata River (Praderi, 1986) and Babitonga Bay estuaries (Cremer & Simões-Lopes, 2005). There is evidence that the species is not continuously distributed throughout its range. Siciliano et al.,

(2002) reported that there are two areas in its northern range where franciscana are extremely rare or absent. One of these is situated between Macaé (circa 22°25'S) (southern Rio de Janeiro State) and Ubatuba (*ca* 23°18'S) (northern São Paulo State); the other occurs between southern Espírito Santo (circa 19°40'S) and northern Rio de Janeiro States (*ca* 21°37'S) (Figure 2). Although unusual records of stranded franciscanas have been documented by Azevedo et al., (2002) within one of these gaps (in southern Rio de Janeiro State), the hypothesis that the species is rare in this area remains valid, due to a marked genetic difference between samples collected from northern Rio de Janeiro and from São Paulo State southward (Secchi et al., 1998; Ott, 2002). This suggests the existence of two isolated populations, one small in northern Espírito Santo and, possibly, another in northern Rio de Janeiro. The reason for these hiatuses is still unclear, but, due to the species' preference for turbid waters less than 30-35 m deep (Pinedo et al., 1989; Secchi & Ott, 2000; Danilewicz et al., 2009), water transparency and depth may be among the factors (Siciliano et al., 2002).

The species is restricted to coastal waters and two criteria have been suggested as offshore limits to its distribution: a) the area between the shoreline and the 30 m isobaths and b) the area between the shoreline and 30 nautical miles (NM, 1.853 kilometers) from the coast (Pinedo et al., 1989). Nevertheless, based on the depth distribution of incidentally caught dolphins (Moreno et al., 1997; Secchi et al., 1997), it was considered that the 30 m isobath best fits as the outer distribution limit of the species in southern Brazil (Secchi & Ott, 2000), though a few animals have been incidentally caught or sighted in deeper waters (*e.g.* Secchi et al., 1997; Crespo et al., 2009; Danilewicz et al., 2009). In its northern range, the species seems to occur in relatively deeper water (Di Beneditto & Ramos, 2001). No size, age or sex-related difference in habitat use patterns in relation to depth have been observed in southern Brazil (Denilewicz et al., 2009).

Figure 2. The franciscana distribution is restricted to coastal waters of the western South Atlantic, from Itaúnas, southeastern Brazil to Golfo San Matias, Argentina.

POPULATION STRUCTURE

Morphological and molecular data strongly support the existence of different franciscana populations. Evidence of population structuring was first demonstrated through multivariate analysis of morphometric data, which revealed the existence of two geographical forms: a smaller form in the northern part of the species' range (north of 27°S) and a larger form in the coastal waters of southern Brazil, Uruguay and Argentina (south of 32°S) (Pinedo, 1991, 1995). Body length of individuals from São Paulo (ca. 23°30'S-25°30'S) and Rio de Janeiro (21°35'S-22°25'S) states were significantly different with animals from Rio de Janeiro being larger (Ramos et al., 2002). Similar growth patterns were observed in cranial dimensions (Ramos et al., 2002) as well as other body metrics (Barbato et al., 2008). These results indicated that metric differences in body and skull variables were not clinal. Analyses of a highly variable region of mitochondrial DNA (mtDNA) also supported these two geographic forms (Secchi et al., 1998). Ott (2002) and Lázaro et al., (2004) compared the mtDNA of franciscanas from Uruguay and Argentina with those published by Secchi et al., (1998). These studies found support for the existence of a large southern population (composed of animals from Rio Grande do Sul State in Brazil, Uruguay and Argentina) that is clearly differentiated from animals in the waters of Rio de Janeiro. In addition, they revealed fixed genetic differences between the populations that suggest essentially no effective genetic exchange (Secchi et al., 1998; Ott, 2002; Lázaro et al., 2004). Ott's results also showed that individuals inhabiting waters of the Paraná and São Paulo states belong to a genetically distinct population. A pairwise analysis of haplotype distances between different geographic locations showed increasing differentiation in the haplotype frequencies with increasing geographic distance, following an isolation-by-distance pattern (Lázaro et al., 2004). These authors and Ott (2002) also indicated that haplotypic frequencies of samples from Claromecó (in Argentina) were significantly different from the rest of the southern population. Recent results by Mendez et al., (2007), however, do not fit this model. Rather, they suggest that ecological forces can be more relevant than geographic distance in determining population structuring by regulating gene flow. Individuals from Claromecó are most similar to those from coastal oceanic areas, including those from Uruguay, than with those from the estuary-influenced Samborombon Bay. A similar pattern was observed in the Uruguayan coast (Costa et al., 2008). The authors noticed that, among several sampled areas including oceanic and estuarine coasts, the most genetically distinct individuals were those collected in the La Plata estuary, regardless of the geographic distance. This suggests that a fine-scale structuring occurs in areas with higher influence of the La Plata River estuary in both Argentina and Uruguay. These areas are shallow, relatively enclosed and have high abundances of estuary-dependent prey, which make them a suitable habitat for calving. In fact, telemetry studies have demonstrated that individuals from Samborombon Bay are residents and have a very restricted movement range of about 20 km (Bordino & Wells, 2005). Furthermore, Rodriguez et al., (2002) found that individuals from this area prey upon different species when compared to those from the coastal oceanic environments. If part of the population has evolved and is mostly adapted to occupy an estuarial niche, intra-specific competition for resources is minimized, representing an advantage for the species as a whole.

Abundance

There is no current abundance estimate for the species as a whole. The only estimates available are for the animals inhabiting the Rio Grande do Sul State, southern Brazil, and Argentina. In 1996, thirty-four franciscanas (in 29 groups) were recorded during aerial surveys in the area, giving a mean density of 0.657 individuals/km^2 (95%CI: 0.516 to 0.836) for the 435 km^2 study area after correcting for the probability of missing submerged dolphins. Extrapolating this density to the entire stock's range, *i.e.* Rio Grande do Sul State and Uruguay (Secchi et al., 2003a) would result in an estimate of 42,078 franciscanas (95% CI: 33,047-53,542). This extrapolated result, however, should be interpreted cautiously as it is based on a density estimate for a small fraction of the coastline, representing only 0.7% of the possible range of the stock (*ca* 64,045 km^2). If only the Rio Grande do Sul coast, from shoreline up to the 30 m isobath (ca. 24,315 km^2) is taken into account, the extrapolated abundance would be around 15,975 animals. More recently, in 2004, another aerial survey was conducted covering a much larger area off the coast of Rio Grande do Sul State (Danilewicz et al., 2007). Thirty-one animals were seen in 25 groups, from which the authors estimated a density of 0.51 ind/km^2. If extrapolated to the same area (i.e. 24,315 km^2), this density would result in an abundance of 12,400 individuals. This difference should not be viewed as a population decline because the area covered in both studies differed greatly in magnitude, mainly because of the constraints imposed by the flight autonomy of the single-engine aircraft utilized in 1996. The first study was covered an area about 30 times smaller than the second study and was concentrated in shallower waters close the Patos Lagoon estuary, a potentially more productive area for both fish and franciscanas. The aircrafts and the observers also differed among surveys. The main difference was that the aircraft had only flat windows in the first survey while in the second survey bubble windows were present in the rear allowing the observers to see right below the plane.

Crespo et al., (2009), carried out aerial surveys in 2003 and 2004 for estimating franciscana abundance in Argentina. The area was divided into two sections, northern sector, from Lavalle to Mar del Plata and from Mar del Plata to Claromecó, Buenos Aires Province; and a southern sector, from Bahía Blanca to the mouth of Río Negro River and along the northern coast of Golfo San Matías, Rio Negro Province. One hundred and one franciscanas were observed in 71 sightings. For the northern sector, the density was estimated to be 0.106 ind/km^2. Density declined with depth (0.05 ind/km^2 between the 30 m and 50 m isobaths) and was lower in the southern sector (0.056 ind/km^2). After correcting for submerged dolphins and extrapolation for unsurveyed areas, the abundance for the Argentine coast was estimated to be between 15,062 to 16,335 individuals.

Although these numbers cannot be considered absolute abundance estimates because the surveyed area does not cover the entire population (or stock, Secchi et al., 2003a) range, they can be viewed as reasonable approximations for assessing the potential effects of non-natural removals (*e.g.* Secchi, 1999, 2006; Kinas, 2002; Secchi et al., 2003b; Crespo et al., 2009). Densities, on the other hand, provide good insights about habitat use in both latitudinal and depth gradients. The values presented above suggest that although franciscana might occur up to the 50 m isobath or even deeper, their density is higher along the coast up to a depth of 30 m. Furthermore, the density decreases from southern Brazil towards the austral distribution limit of the species.

Although data on abundance exist only for the southern range of the species (Secchi et al., 2001; Crespo et al., 2009), empirical evidence suggests that the southern population is larger than the northern one. Relative abundance of franciscanas is suspected, based on bycatch per unit of effort (CPUE) data, to be much higher to the south (*e.g.* Crespo et al., 1986; Corcuera, 1994; Praderi, 1997; Secchi et al., 1997; Ott, 1998; Secchi & Ott, 2000), than to the north of Santa Catarina (*e.g.* Di Beneditto et al., 1998; Di Beneditto & Ramos, 2001; Bertozzi & Zerbini, 2002). Furthermore, as previously hypothesized by Secchi et al., (2003a) the franciscana abundance might be limited in the north by the presence of abundant sympatric species such as the Guiana dolphin (*Sotalia guianensis*), which has its southern limit at Santa Catarina (Borobia et al., 1991) and other less abundant or occasionally sympatric species such as the Atlantic spotted (*Stenella frontalis*), rough-toothed (*Steno bredanensis*) and bottlenose (*Tursiops truncatus*) dolphin (Moreno et al., 2005; Bastida et al., 2007). Because these sympatric species may compete for the same resources (habitat or food), it may be reasonable to expect that the northern population is less abundant and a more opportunistic feeder than the southern one. In northern Rio de Janeiro, franciscana feed upon about 25 species and a low degree of competition for the same resources with Guina dolphin has been reported (Di Beneditto & Ramos, 2001). To the south, franciscana is probably widely distributed from northern Argentina to Santa Catarina (as suggested by the aerial surveys in southern Brazil and northern Argentina). The only cetaceans that are sympatric with the southern population year round are the highly coastal or estuary-dependent small populations of bottlenose dolphins, mainly in southern Brazil and Uruguay (Bastida et al., 2007), and Burmeister's porpoises, *Phocoena spinipinnis*, in northern Argentina (Brownell and Praderi, 1984; Corcuera et al., 1994; Molina et al., 2005). Moreover, genetic diversity was greater within samples collected from animals from the southern populations than within the samples of franciscanas from the northern populations (Secchi et al., 1998; Ott, 2002; Lázaro et al., 2004; Mendez et al., 2007).

PREY AND PREDATORS

Likewise the true river dolphins, franciscana has an elongated, narrow rostrum filled with many small and sharp teeth (up to 65 in each jaw) well adapted to its trophic niche, characterized by small and soft prey. Franciscanas feed upon several small-sized shallow-water fish, cephalopods, and crustaceans (Brownell, 1989; Di Beneditto & Ramos, 2001; Rodriguez et al., 2002; Danilewicz et al., 2002). Prey are typically smaller than 80 mm in length and 5 g (fish) or 10 g (squids) in weight, regardless of the geographic location (*e.g.* Bassoi, 1997; Di Beneditto & Ramos, 2001; Rodriguez et al., 2002). The diet of adults consists of at least 76 food items (see Danilewicz et al., 2002 for a revision) that includes fishes (83%) with the teleost Sciaenidae family as the predominant prey (mainly *Cynoscion guatucupa),* crustaceans (9%), and molluscs (8%) (specially, the small squid species such as Loligo *sanpaulensis)*, although geographic variation exists (*e.g.* Fitch & Brownell, 1971; Ott, 1994, Bassoi, 1997, 2005; Di Beneditto & Ramos, 2001; Rodríguez et al., 2002). Shrimps are very important in the diet of these young dolphins (Rodríguez et al., 2002; Bassoi, 2005). Three diet categories were defined during the first year of life of franciscanas: lactating, mixed diet, and solid diet (Rodríguez et al., 2002). The first solid food is composed of very small fish and shrimps. This ontogenetic variation might be related to a learning process with young franciscana preying upon slower prey. Franciscana's feeding strategy seems to be opportunistic, eating the most abundant prey in the area (Danilewicz et al., 2002). Seasonal fluctuations in the franciscana's diet match with patterns of variation in the abundance of the prey species throughout the year (Di Beneditto & Ramos, 2001; Rodriguez et al., 2002; Bassoi, 2005). Furthermore, decadal changes in franciscana's diet seems also to match with decline in fish stock abundance (*e.g.* Secchi et al., 2003b; Secchi, unpubl. data), though further studies are necessary for certainty. To date there is no information about daily food consumption in both juvenile and adult specimens, although feeding studies carried out in captivity indicate that a medium sized franciscana might eat approximately 10% of its body weight daily with a diet composed by a variety of fishes and invertebrates of low, medium and high caloric values (Loureiro et al., 2000). Determining franciscana's nutritional requirements and preferred prey is crucial for assessing the potential competition with coastal fisheries and, most importantly, to understand its role in the ecosystem functioning.

Predation is a natural cause of mortality for *P. blainvillei.* The top most predators in the marine ecosystem, the killer whales (*Orcinus orca*) and several shark species are known to prey upon franciscana throughout most of its distribution range (Di Beneditto, 2004; Monzón et al., 1994; Ott & Danilewicz, 1996; Praderi, 1985; Santos & Netto, 2005). In Uruguayan waters, the broadnose seven-gill (*Notorhynchus cepedianus*), hammerhead (*Sphyrna* spp), the sand tiger (*Eugomphodus taurus*), tiger (*Galeocerdo cuvieri*), and requiem (*Carcharhinus* sp.) sharks are known predators of franciscana (Brownell, 1975; Pilleri, 1971; Praderi, 1985). Other shark species are also potential predators. Fresh wounds and scars caused by shark bites found on individuals incidentally caught in fishing nets suggest that predation might also be an important source of natural mortality of franciscana in Argentina (Monzón et al., 1994) as well as in Brazil. A male killer whale was observed preying on a free-swimming franciscana in coastal waters off Parana State, southern Brazil (Santos & Neto 2005). Ott and Danilewicz (1996) found a female killer whale washed ashore in Rio Grande do Sul State, southern Brazil, with remains of three franciscana in its stomach.

PARASITES

A few species of ectoparasite or epizoit crustaceans have been reported for *P. blainvillei.* The barnacle, *Xenobalanus globicipitis*, was found attached to the trailing edge of the flukes and fin of franciscana (Brownell, 1975; Di Beneditto & Ramos, 2000). Specimens of an unidentified necked-barnacle were also observed in the teeth row of the lower jaw of a large and skinny, possibly old and sick franciscana. These epizoits are known to settle during their larval stage on slow moving animals or drifting objects. The isopods *Cirolana* sp. and *Nerocila* sp. which normally infect the gill of some fishes and sharks were found on several occasions, respectively, in the blowholes or stomachs and on the skin of franciscanas from Uruguay and Rio Grande do Sul State, Brazil (Brownell, 1975). Likewise, the isopod *Riggia* sp. was found in the vagina of one franciscana in Santos, São Paulo State, Brazil (Ferreira et al., 1998). The association of isopods and franciscana is not yet well understood. Films of diatoms *Navicula* sp. and mainly *Cocconeis ceticola* have been documented to partially covering the skin of franciscana (Nemoto et al., 1977; Santos et al., 2009).

Franciscana is the final host of several gastrointestinal parasites. The frequency of occurrence and infection levels of helminths varies according to the host's geographical distribution (Andrade et al., 1997; Aznar et al., 1994, 1995; Marigo et al., 2002). Adult acanthocephalans *Corynosoma cetaceum* (= *Polymorphus arctocephali* and *P. cetaceum* - Aznar et al., 1999) typically found in the intestine of vertebrates, have been reported in the pyloric stomach and, to a lesser degree, in the duodenal ampulla and the main stomach of franciscana (Aznar et al., 2001). Franciscanas from southern São Paulo State showed very low prevalence and intensity of this infection (Marigo et al., 2002) as do those from Rio Grande do Sul State, Brazil, and Uruguay, while franciscanas from Argentina exhibit higher infection levels (Andrade et al., 1997; Aznar et al., 1994; Brownell, 1975). Other acanthocephalans, such as *Corynosoma australe*, commonly found free in the pyloric stomach, and *Bolbosoma* turbinella, often firmly attached to the walls of the large intestine, showed moderate to high prevalence with low abundance in franciscanas from Rio Grande do Sul State (Andrade et al., 1997). The nematode *Contracaecum* sp. was found in the main stomach, with a low prevalence, in dolphins from Argentina (Aznar et al., 1995), and Uruguay (Brownell, 1975), while *Anisakis typica* was found in the stomachs of specimens from Rio Grande do Sul State (Andrade et al., 1997) and Uruguay (Kagei et al., 1976), and *A. simplex* in stomachs of franciscana from Argentina (Aznar et al., 2003). Although considered rare, both the trematode *Pholeter gastrophilus* and the nematode *Procamallanus* sp. were found in the stomachs of franciscanas from Argentina (Aznar et al., 1994) and Uruguay (Kagei et al., 1976), respectively. The trematode *Hadwenius pontoporiae*, only known to infect *P. blainvillei*, was found in the small intestine of specimens from São Paulo and Paraná States (Marigo et al., 2002), Rio Grande do Sul State (Andrade et al., 1997), and Argentina (Aznar et al., 1994; 1997; Raga et al., 1994). No parasites were found in the intestines of individuals from Uruguay (Brownell, 1975), nor in the internal organs of franciscana from northern Rio de Janeiro State, Brazil (Di Beneditto & Ramos, 2001; Santos et al., 1996). No parasites were found in the lungs of specimens from São Paulo and Paraná States (Marigo et al., 2002).

AGE AND GROWTH

The franciscana's life span seems shorter than 20 years. The oldest aged franciscana was a 21 year old female (Pinedo, 1991). The maximum observed age for males was 16 years (Kasuya & Brownell, 1979). The age frequency distribution of incidentally caught and washed ashore animals suggests that only a percentage of the population lives more than 12 years, regardless of their geographic location (Kasuya & Brownell, 1979; Pinedo, 1994; Pinedo & Hohn, 2000; Di Beneditto & Ramos, 2001; Secchi et al., 2003b). Although this age-at-death frequency distribution is negatively biased, the very small fraction of individuals older than 12 years indicates that franciscana has one of the shortest life spans among all cetaceans.

Mean asymptotic length for *P. blainvillei* from Rio de Janeiro State, Brazil, is 117.1 cm for males (n = 43) and 144.7 cm for females (n = 43 - Ramos et al., 2000). For specimens from São Paulo and Paraná states, Brazil, this measure is 113.3 cm for males (n = 23) and 128.9 cm for females (n = 18 - Barreto & Rosas, 2006). For individuals from Rio Grande do

Sul State, Brazil, it is from 129.8 – 130.6 cm for males (*n* = 59; *n* = 96) and 146.3 – 152.6 cm for females (*n* = 48; *n* = 69 - Barreto & Rosas, 2006; Botta et al., 2007). For specimens from Uruguay, the asymptotic length is 133.3 cm for males (*n* = 137) and 153.0 cm for females (*n* = 123 - Kasuya & Brownell, 1979) and for franciscanas from Argentina it is 135.8 cm for males (*n*=14) and 150.5 cm for females (*n* = 12 - Botta et al. 2007). Body weight also follows the same geographic pattern (*e.g.* Rosas, 2000; Botta et al., 2006). It is worthwhile to note that franciscana variation in size is not clinal as one would expect. Franciscanas are larger in their southern range (Rio Grande do Sul State, Brazil; Uruguay and Argentina) followed by animals from the northern range (Rio de Janeiro and Espírito Santo States, Brazil) and are much smaller in the middle of its range (São Paulo and Paraná States, Brazil). Larger sizes in the southern range might be explained as an evolved characteristic to cope with colder waters as well as increased abundance of potential predators. Both sharks and killer whales are predators of franciscanas (see below) and are more abundant in cold water (Heyning & Dahlheim, 1988; Compagno, 1984). The larger size of franciscana in the north than in mid-range latitudes could be due to some influence of the upwelling cold waters near Cabo Frio (Rio de Janeiro State) northwards. The possibility, however, that franciscanas in the northern range originate from larger individuals from the south (founder effect) should not be overlooked.

Reversed sexual size dimorphism (RSSD) is observed in franciscana (Higa et al., 2002; Kasuya & Brownell, 1979; Pinedo, 1995; Ramos et al., 2002 – see also references on growth above). Females are larger than males in both the total body length as well as weight (Brownell, 1989; Rosas, 2000; Botta et al., 2006). Larger females might represent an evolutionary advantage for small species by allowing gestation and birth of larger, even slightly, calves. Larger calves potentially have a higher likelihood of survival due to several factors including a relatively lower metabolic rate and a lower surface/volume ratio which reduces heat loss. Predation of larger calves might also be less likely. In fact, most of the odontocetes presenting RSSD are the smallest species in the group. In general these species produce relatively larger calves at birth than the other species within the taxonomic group (*e.g.* Ralls, 1976).

Reproduction and Survival

Franciscana has one of the fastest reproductive cycles among all cetaceans (Danilewicz et al., 2000). Sexual maturity of franciscanas from the coast of Uruguay is attained at about 131 cm in males and at 140 cm in females, with an estimated age at sexual maturity between two and three years for both sexes (Kasuya & Brownell, 1979). The mean age at sexual maturity of specimens from the coast of Rio Grande do Sul, southern Brazil, was estimated to be around 3.5 years for both sexes (Danilewicz et al., 2000; 2004; Danilewicz, 2003). Mean length and weight at sexual maturity in the same area was estimated at 138.9 cm and 32.8 kg for females and 127.4 cm and 26.6 kg for males (Danilewicz et al., 2000; 2004; Danilewicz, 2003). In Rio Grande do Sul, the annual pregnancy rate was estimated to be 0.66, which is equivalent to an average birth interval of 1.5 years. This suggests that within the population, half of the females reproduce annually and the other half biannually (Danilewicz et al., 2000; Danilewicz, 2003). Ramos et al., (2000) and Di Beneditto and Ramos (2001), estimated that franciscana males from the northern coast of Rio de Janeiro State reach sexual maturity at two years, and a total length of 115 cm, and the females at three years and a total length of 130 cm. Franciscanas from the southern coast of São Paulo and coast of Paraná states are sexually mature between 112 and 116 cm for males, and between 122 and 126 cm for females, with age at sexual maturity for both between four and five years (Rosas & Monteiro-Filho, 2002). These results suggest that the populations from Paraná and southern São Paulo, particularly females, attain sexual maturity at smaller sizes, but at an older age, than those inhabiting areas to the north and to the south of these two neighboring states. The small relative size of the testes (testes/body weight ratio of 0.12%) suggests a single-male breeding system without sperm competition (Danilewicz et al., 2004; Rosas & Monteiro-Filho, 2002). Most of the females from Rio Grande do Sul State and Uruguay have a larger and heavier left ovary, a weight difference correlated to a higher number of corpora (Harrison et al., 1981 Brownell, 1984; Danilewicz, 2003). However, no ovulation polarity is observed in franciscanas from São Paulo and Paraná states, with both ovaries being functional (Rosas & Monteiro-Filho, 2002).

The gestation period does not appear to vary much according to geographic location, with estimates of between 10.2 and 11.2 months (Kasuya & Brownell, 1979; Harrison et al., 1981; Di Beneditto & Ramos, 2001; Rosas & Monteiro-Filho, 2002; Danilewicz, 2003). Reproduction is markedly seasonal in the southern range, with births occurring "in a pulse" from October to February (Brownell, 1984; Danilewicz, 2003). In the north, births "flow" year round (Di Beneditto & Ramos, 2001). Therefore, they can be classified as birth-pulse and birth-flow populations (definition of the terms, in either biological or mathematical language, can be found in ecology text books – *e.g.* Caughley, 1977; Krebs, 1994; Caswell, 2001). Lactation periods last approximately. 8.4 months in northern Rio de Janeiro State (Di Beneditto & Ramos, 2001), around 7.4 months in São Paulo and Paraná states (Rosas & Monteiro-Filho, 2002), 6 to 7 months in northern Argentina (Rodríguez et al., 2002), and 8-9 months for individuals in Uruguay (Harrison et al., 1981; Kasuya & Brownell, 1979). Weaning is gradual with early predation on shrimps and small fish (Pinedo et al., 1989; Rodriguez et al., 2002). Litter size is limited to one in utero and at birth. The length and weight at birth of *P. blainvillei* are 70 to 80 cm and about 5 to 6 kg in the southern range (Danilewicz, 2003; Kasuya & Brownell, 1979; Harrison et al., 1981). In the northern range the newborn calves are smaller (Ramos et al., 2000; Rosas & Monteiro-Filho, 2002). In Argentina, weaned calves exceed 97 cm in length and weigh 13 to 17 kg (Rodríguez et al., 2002). Females do not show any evidence of reproductive senescence (Kasuya & Brownell, 1979; Danilewicz, 2003). Taking the age of first reproduction, life span and calving interval into account, it is suggested that a female franciscana might produce four to eight offspring in her lifetime (Danilewicz, 2003), though investment in reproduction seems to vary geographically.

Survival probably varies accordingly as there seem to be a competing relationship between reproduction and survivorship. There is empirical evidence for some bird and mammal species that lower reproductive effort is compensated by higher survival. For example Stearns (1976) and Millar and Zammuto (1983) stated that there is positive correlation between age at maturity and life expectancy. It is therefore expected that survival rates are lower in areas where franciscanas attain sexual maturity at younger ages and have shorter calving intervals. Unfortunately very little is known about survivorship for franciscana due to the lack of unbiased age-at-death data to construct its life table and due to unsuitability to apply mark-recapture studies to this species. Secchi (2006) used two approaches for estimating survival rates of franciscana from Rio Grande do Sul and Uruguay: by fitting the Siler model (Siler, 1979) to age-at-death data of bycatch and beach cast animals and by using life tables from similar model species. Despite their intrinsic limitations, the two methods, after careful treatment to reduce bias effects, resulted in identical mean survival rates for calf (0.67 to 0.74), juvenile (0.88 to 0.90) and adult franciscana (0.86 to 0.87). Life-tables from other species may be the only option to model the survivorship of many cetacean species due to a lack of data, although caution is needed when selecting the model species. This approach relies on the assumption that the chosen species have similar life histories.

The use of distributions of age-at-death data to estimate mortality relies on the assumption that the population has a stable age distribution. For age distribution to be stable, age-specific differences in both death rates and birth rates across age classes must be constant, and need to have been long enough for the age structure to equilibrate. Removal of franciscana from the population through bycatch can lead to deviations from the stable age distribution if the age structure of the bycatch fluctuates through time. This fluctuation leads to unstable age distribution biasing survival estimates. If biases are small, it can be concluded that survival rates of franciscana are lower than other small cetaceans (*e.g.* bottlenose and Hector's dolphins – Wells & Scott, 1990; Cameron et al., 1999; Du Fresne, 2004) and much lower than medium size and large cetaceans (*e.g.* killer whales – Olesiuk et al., 1990; bowhead and humpback whales - Zeh et al., 1995; Givens et al., 1995; Barlow & Clapham, 1997). This seems biologically reasonable as there is strong evidence that body mass is positively correlated with survival in mammals (Millar & Zammuto, 1983). Hector's dolphins however, have similar body mass to franciscanas. The explanation for the higher survival rate in Hector's dolphin is related, to some extent, to its much lower reproductive potential. A female Hector's dolphin first reproduces at approximately 8 years of age and produces one offspring every two to three years (Slooten, 1991; Slooten & Dawson, 1994). A female franciscana, on the other hand, has its first calf at an age of approximately 4 years and reproduces every one or two years.

POPULATION GROWTH RATE

Based on matrix population models using reproduction and survival rates data as input parameters, growth rates for different franciscana populations were estimated at 0.8% to 3.8% (Secchi, 2006). The estimates of population growth rates have to be interpreted with caution due to the fact that survival rates used as input parameters were estimated based on limited data and on the model life tables of similar species. These survival rates were also assumed to be the same for all franciscana populations. Growth rate estimates can be sensitive to estimates of survival rates. If the assumption that all franciscana populations have the same survival rate is valid, differences in population growth rate will be determined by reproduction. The higher estimated growth rate for the franciscanas from Rio Grande do Sul State/Uruguay and Rio de Janeiro State is due to a higher reproductive potential of the females from those areas. Females are approximately one year older when they attain sexual maturity in the two populations adjacent to Rio Grande do Sul/Uruguay (*i.e.* from Santa Catarina/Paraná/São Paulo states and Argentina, to the north and to the south, respectively).

The higher reproductive potential of females from Rio Grande do Sul/Uruguay area might be due to a density-dependent response to several years of incidental mortality in fisheries. Franciscanas from this area have experienced high levels of bycatch for a long period (Praderi, 1997; Secchi et al., 2003b). Although density-dependent response to bycatch has yet to be documented in franciscana, the possibility that animals from Rio Grande do Sul/Uruguay have been responding to high levels of bycatch by reducing the age at first reproduction and/or by increasing reproductive rates cannot be ruled out.

The lower estimated growth rate for the population from Santa Catarina/Paraná/São Paulo States is possibly due to a combination of intrinsically low reproductive potential and poor parameter estimation. Some parameters for this area were obtained from small sample sizes. For example, Rosas and Monteiro Filho (2002) suggest that franciscanas are sexually mature between 4 and 5 years of age and reproduce biannually based on a sample of only six mature animals. Franciscanas from adjacent areas have much higher reproductive potential (Di Beneditto & Ramos, 2001; Danilewicz et al., 2000; Danilewicz, 2003). If the much lower reproductive potential is not an artifact of poor estimation, then a compensatory higher survival could be expected. If this is the case, then the population growth rate for this area and for Argentina as well could be underestimated as the survival rates estimated with data from the Rio Grande do Sul/Uruguay were assumed to be the same for these populations. Given this uncertainty, it is recommended that more precise estimates on reproductive parameters are obtained for those areas. Despite geographic variation and parameter uncertainties, the estimated growth rate for the species is within the range of likely growth rate for small cetaceans (*e.g.* Reilly & Barlow, 1986; Slooten & Lad, 1991; Stolen & Barlow, 2003).

REFERENCES

[1] Ameghino, F. (1918). Los mamíferos fósiles de la República Argentina. Parte III. Homalodontes. Pp. 383-392. In Torcelli, A.J., (Ed.) Obras completas y correspondencia científica de Florentino Ameghino. Impresiones Oficiales, La Plata, Argentina.

[2] Andrade, A. L. V., Pinedo, M. C. & Pereira Jr, J. (1997). The gastrointestinal helminths of Franciscana, *Pontoporia blainvillei*, in Southern Brazil. *Reports of the International Whaling Commission,* 47, 669-73.

[3] Azevedo, A. F., Fragoso, A. B. L., Lailson-Brito Jr., J., & Cunha, H. A. (2002). Records of the franciscana (*Pontoporia blainvillei*) in southwestern Rio de Janeiro and northernmost São Paulo State coasts – Brazil. *The Latin American Journal of Aquatic Mammals* (special issue), 1, 191-192.

[4] Aznar, F. J., Balbuena, J. A. & Raga, J. A. (1994). Helminth communities of *Pontoporia blainvillei* (Cetacea: Pontoporiidae) in Argentinean waters. *Canadian Journal of Zoology,* (72), 1-5

[5] Aznar, F. J., Raga, J. A., Corcuera, J. & Monzón, F. (1995) Helminths as biological tags for franciscana (*Pontoporia blainvillei*) (Cetacea, Pontoporiidae) in Argentinean and Uruguayan waters. *Mammalia* 59 (3): 427-435.

[6] Aznar, F. J., Balbuena, J. A., Bush, A. O. & Raga, J. A. (1997). Ontogenetic habitat selection by *Hadwenius pontoporiae* (Digenea: Campulidae) in the intestine of franciscanas (Cetacea). *Journal of Parasitology,* 83(1), 13-18.

[7] Aznar, F. J., Bush, A. O., Fernandez, M. & Raga, J. A. (1999). Constructional morphology and mode of attachment of the trunk of *Corynossoma cetaceum* (Acantocephala: Polymorphidae). *Journal of Morphology,* 241, 237-249.

[8] Aznar, F. J., Bush, A. O., Balbuena, J. A. & Raga, J. A. (2001). *Corynosoma cetaceum* in the stomach of franciscanas, *Pontoporia blainvillei* (Cetacea): an exceptional case of habitat selection by an Acanthocephalan. *Journal of Parasitology*, 87, 536-541.

[9] Barbato, B. H. A., Secchi, E. R., Kinas, P. G., DiBeneditto, A. P. M., Ramos, R. M. A., Bertozzi, C., Marigo, J., & Bordino, P. (2008). In *Variação geográfica em caracteres morfológicos externos de Pontoporia blainvillei (Cetacea:Pontoporidae). XIII Workshop of Specialists in Aquatic Marine Mammals of South America* held in Montevideo, Uruguay (p. 107): Rio de Janiero, Brazil: Status of Aquatic Mammals in Latin America (SOLAMAC).

[10] Barlow, J. and Clapham, P. J. (1997). A new birth-interval approach to estimating demographic parameters of humpback whales. *Ecology*, 78, 535-546.

[11] Barreto, A. S., & F. C. W. Rosas. (2006). Comparative growth analysis of two populations of *Pontoporia blainvillei* on the Brazilian coast. *Marine Mammal Science,* 22, 644-653.

[12] Bassoi, M. (1997). *Avaliação da dieta alimentar de toninha, Pontoporia blainvillei (Gervais and D' Orbigny, 1844), capturadas acidentalmente na pesca costeira de emalhe no sul do Rio Grande do Sul. Bachelor Dissertation.* Rio Grande, Brazil: Fundação Universidade do Rio Grande.

[13] Bassoi, M. (2005). *Feeding ecology of franciscana dolphin, Pontoporia blainvillei (Cetacea: Pontoporiidae), and oceanographic processes on the Southern Brazilian coast. Doctorate dissertation.* Southampton, United Kingdom: University of Southampton School of Ocean and Earth Sciences.

[14] Bastida, R., Rodriguez, D., Secchi, E. R., & da Silva, V. M. F. (2007). *Mamíferos Acuáticos de Sudamérica y Antártida.* Buenos Aires, Brazil: Vazquez Mazzini Editores (publisher).

[15] Bertozzi, C. & Zerbini, A. N. (2002) Incidental mortality of franciscana, *Pontoporia blainvillei*, in the artisanal fishery of Praia Grande, São Paulo State, Brazil. *The Latin American Journal of Aquatic Mammals* (special issue), 1, 153-160.

[16] Bordino, P. & Wells, R.S. (2005). Radiotracking of franciscana dolphins (*Pontoporia blainvillei*) in Bahia Sanborombom, Buenos Aires, Argentina. In *16th Biennial Conference on the Biology of Marine Mammals*. Society for Marine Mammalogy.

[17] Borobia. M., Siciliano, S., Lodi, L. & Hoek, W. (1991). Distribution of the South American dolphin *Sotalia fluviatilis*. *Canadian Journal of Zoology,* 69, 1025-1039.

[18] Botta, S., Muelbert, M. & Secchi, E. R. (2006). Morphometric relationships of franciscana dolphin, *Pontoporia blainvillei* (Cetacea), off Rio Grande do Sul coast, southern Brazil. The Latin American Journal of Aquatic Mammals, 2, 117-123.

[19] Botta, S., Muelbert, M., Secchi, E. R., Danilewicz, D., Negri, M. F., Cappozzo, L.H., & Hohn, A. A. (2007). Age and growth of franciscana dolphin, *Pontoporia blainvillei* (Cetacea), incidentally caught from Rio Grande do Sul, southern Brazil, and Buenos Aires, Argentina. In *17th Biennial Conference on the Biology of Marine Mammals, Cape Town, South Africa*: Society for Marine Mammalogy

[20] Bravard, A. (1858). Monografía de los terrenos terciarios de las cercanias del Paraná. (Reimpresión facsimilar con prólogo de F. G. Aceñolaza, Imprenta del Congreso de la Nación, Buenos Aires, Argentina, 1995, V-XII + 1-107 pp.).

[21] Brownell, R. L. Jr. (1975). Progress report on the biology of the franciscana dolphin, *Pontoporia blainvillei*, in Uruguayan waters. *Journal of the Fisheries Research Board of Canada* 32, 1073-1078.

[22] Brownell, R. L. Jr. (1984). Review of reproduction in platanistid dolphins. *Reports of the International Whaling Commission*, (special issue), 6, 149-158.

[23] Brownell, R. L. Jr. (1989) Franciscana, *Pontoporia blainvillei* (Gervais and d'Orbigny 1844). In S. H. Ridgway & R. J. Harrison, (Eds.), *Handbook of Marine Mammals* (volume 4, pp 45-67). London, United Kingdom: Academic Press.

[24] Brownell, R. L. & Praderi, R. (1984) *Phocoena spinipinnis*. *Mammalian Species,* 217, 1-5.

[25] Buchmann, F. S. de C. & Rincón, G. F. (1997). Fósseis de vertebrados marinhos do Pleistoceno Superior na porção sul da planície costeira do Rio Grande do Sul, Brasil. CECO/IG/UFRGS - Notas técnicas 10:7-16.

[26] Cameron, C., Baker R., Fletcher, D., Dawson, S, & Slooten, E. (1999). Modeling survival of Hector's dolphins around Banks Peninsula, New Zealand. *Journal of Agricultural, Biological and Environmental Statistics,* 4, 126-135.

[27] Cassens, I., Vicario, S., Waddell, V. G., Balchowsky, H., Van Belle, D., Ding, W., Fan, C., Lal Mohan, R. S., Simões-Lopes, P. C., Bastida, R., Meyer, A., Stanhope, M. J., & Milinkovitch, M. C. (2000). Independent adaptation to riverine habitats allowed survival of ancient cetacean lineages. *Proceedings of the National Academy of Science. USA,* 97, 11343-11347.

[28] Caswell, H. (2001). *Matrix Population Models: Construction, Analysis and Interpretation.* 2nd Edition. Sinauer Associates, Inc.: Sunderland, Massachusetts.

[29] Caughley, G. (1977). *Analysis of Vertebrate Population*. Hoboken, N. J.: John Wiley and Sons.

[30] Compagno, L. J. V. (1984) *FAO species catalog. Sharks of the world. An annotated and illustrated catalogue of shark species known to date. Part 1. Hexanchiformes to Lamniformes*. Rome, Italy: FAO

[31] Corcuera, J. (1994). Mortality of *Pontoporia blainvillei* in Northern Buenos Aires Province: the threat of small fishing camps. *Reports of the International Whaling Commission*, (special issue), 15, 291-294.

[32] Corcuera, J., Monzón, F., Crespo, E.A., Aguilar, A., & Raga, J.A. (1994). Interactions between marine mammals and the coastal fisheries of Necochea and Claromecó (Buenos Aires, Argentina). *Reports of the International Whaling Commission*, (special issue), 15, 283-290.

[33] Costa, P., Lessa, E. P., & Secchi, E. R. (2008) Microestrutura poblacional del delfin franciscana, *Pontoporia blainvillei.* In *XIII Reunión de Trabajo de Especialistas en Mamíferos Acuáticos de América del Sur held in Montevideo, Uruguay* (p. 126). Rio de Janiero, Brazil: Status of Aquatic Mammals in Latin America (SOLAMAC).

[34] Cozzuol, M. A. (1985). The Odontoceti of the "Mesopotamiense" of the Parana River ravines. Systematic review. *Investigations on Cetacea* 17:39-51.

[35] Cremer, M. J. & Simões-Lopes, P. C. (2005). The occurrence of *Pontoporia blainvillei* (Gervais & d'Orbigny) (Cetacea, Pontoporiidae) in an estuarine area in southern Brazil. *Revista Brasileira de Zoologia,* 22, 717-723.

[36] Crespo, E.A., Perez-Macri, G., & Praderi, R. (1986). Estado atual de la población de franciscana (*Pontoporia blainvillei*) en las costas uruguayas. In Actas I Reunión de Trabajo de Especialistas en Mamíferos Acuáticos de America del Sur held in Buenos Aires, Argentina (pp. 92-105). Rio de Janiero, Brazil: Status of Aquatic Mammals in Latin America (SOLAMAC).

[37] Crespo, E. A., Harris, G. & González, R. (1998). Group size and distributional range of the franciscana, *Pontoporia blainvillei. Marine Mammal Science,* 14, 845-849.

[38] Crespo, E. A., Pedraza, S. N., Grandi, M. F. Dans, S. L. & Graffo, G.. V. (In press) Abundance and distribution of endangered Franciscana dolphins (*Pontoporia blainvillei*) in Argentine waters and conservation implications. *Marine Mammal Science*

[39] Danilewicz, D. S. (2003). Reproduction of female franciscana (*Pontoporia blainvillei*) in Rio Grande do Sul, southern Brazil. *The Latin American Journal of Aquatic Mammals,* 2, 67-78.

[40] Danilewicz, D. S., Secchi, E. R., Ott, P. H., & Moreno, I. M. (2000). Analysis of the age at sexual maturity and reproductive rates of franciscana (*Pontoporia blainvillei*) from Rio Grande do Sul, southern Brazil. *Comunicações do Museu de Ciências e Tecnologia, Pontifícia Universidade do Rio Grande do Sul,*13, 89-98.

[41] Danilewicz, D., Rosas, F., Bastida, R., Marigo, J., Muelbert, M., Rodriguez, D., Lailson-Brito Jr.,J., Ruoppolo, V., Ramos, R., Bassoi, M., Ott, P. H., Caon, G., Rocha, A. M., Catão-Dias, J. L. & Secchi, E. R. (2002). Report of the Working Group on Biology and Ecology. In E. R. Secchi (Ed.), *Special Issue on the Biology and Conservation of Franciscana. The Latin American Journal of Aquatic Mammals* 1 (special issue 1, pp. 25-42). Rio de Janeiro, Brazil: Sociedad Latinoamericana de Especialistas en Mamiferos Acuaticos (SOLAMAC).

[42] Danilewicz, D., J. A. Claver, A. L. P. Carrera, E. R. Secchi, & N. F. Fontoura. (2004). Reproductive biology of male franciscanas (*Pontoporia blainvillei*) (Mammalia: Cetacea) from Rio Grande do Sul, southern Brazil. Fishery Bulletin, 102, 581-592.

[43] Danilewicz, D., Moreno, I. B., Ott, P. H., Tavares, M., Azevedo, A., Secchi, E. R.& Andriolo, A. (2006). Abundance estimation for a threatened population of franciscana dolphins in southern Brazil. Paper SC/58/SM6 presented at the International Whaling Commission's 58th Meeting in St. Kitts in June 2006.

[44] Danilewicz, D., Secchi, E. R., Ott, P.H., Moreno, I. B., Bassoi, M. & Borges-Martins, M. (In press). Habitat use patterns of franciscana dolphins (*Pontoporia blainvillei*) off southern Brazil in relation to water depth. *Journal of the Marine Biological Association of the United Kingdom.*

[45] Di Beneditto, A. P. M. (2004). Presence of franciscana dolphin (*Pontoporia blainvillei*) remains in the stomach of a tiger shark (*Galeocerdo cuvieri*) captured in Southeastern Brazil. *Aquatic Mammals,* 30, 311-314.

[46] Di Beneditto, A. P. M., Ramos, R. M. A. & Lima, N. R. W. (1998). Fishing activity in northern Rio de Janeiro State (Brazil) and its relation with small cetaceans. *Brazilian Archives of Biology and Technology,* 41, 296-302.

[47] Di Beneditto, A. P. & Ramos, R. (2000). Records of the barnacle *Xenoblanus globicipitis* (Steenstrup, 1851) on small cetaceans of Brazil. *Biotemas,* 13(2), 159-165
[48] Di Beneditto, A. P. M. & Ramos, R. M. A. (2001). Biology and conservation of the franciscana (*Pontoporia blainvillei*) in the north of Rio de Janeiro State, Brazil. *Journal of Cetacean Research and Management,* 3, 185-192.
[49] DuFresne, S. (2004). *Conservation biology of Hector's dolphins*. Doctorate Dissertation. Dunedin, New Zealand: University of Otago.
[50] Ferreira, H. S., Cunha, C. M. da and Gonzalez, M. M. B. (1998) Endoparasitismo de franciscana *Pontoporia blainvillei* (Cetacea, Platanistidae) por *Riggia* sp. (Isopoda, Cymothoidae). In *VIII Reunião de Trabalhos de Especialistas em Mamíferos Aquáticos da América do Sul.held in Olinda, Brazil* (p. 80). Rio de Janiero, Brazil: Status of Aquatic Mammals in Latin America (SOLAMAC).
[51] Fitch, J. E. & Brownell, R. L. Jr. (1971) Food habits of the franciscana *Pontoporia blainvillei* (Cetacea: Platanistidae) from South America. *Bulletin of Marine Science,* 21 (2), 626-636.
[52] Givens, G. H., Zeh, J. E. & Raftery, A. E. (1995). Assessment of the Bering-Chukchi-Beaufort Seas stock of bowhead whales using the BALEEN II model in a Bayesian synthesis framework. *Reports of the International Whaling Commission,* 45, 345-364.
[53] Hamilton, H., Caballero, S., Collins, A. G. & Brownell, Jr., R. L. (2001). Evolution of river dolphins. *Proceedings of the Royal Society of London Series B*, 268, 549-556.
[54] Harrison, R. J, Bryden, M. M., Mcbrearty, D. A., and Brownell, R. L. (1981). The ovaries and reproduction in *Pontoporia blainvillei* (Cetacea:Platanistidae). *Journal of Zoology, London,* 193, 563-580.
[55] Heyning, J. E. & Dahlheim, M. E. (1988). Orcinus orca. *Mammalian Species,* 304, 1-9.
[56] Higa, A., E. Hingst-Zaher, & M. De Vivo. (2002). Size and shape variability in the skull of *Pontoporia blainvillei* (Cetacea: Pontoporiidae) from the Brazilian coast. The Latin American Journal of Aquatic Mammals, 1, 145--152.
[57] Kagei, N. Tobayama, T. & Nagasaki, Y. (1976). On the helminthum of franciscana *Pontoporia blainvillei. Scientific Reports of Whales Research Institute,* 28, 161-166.
[58] Kasuya, T. & Brownell, R. L. (1979). Age determination, reproduction and growth of franciscana dolphin, *Pontoporia blainvillei. Scientific Reports of the Whales Research Institute,* 31, 45-67.
[59] Kinas, P. G. (2002). The impact of incidental kills by gillnets on the franciscana dolphin (*Pontoporia blainvillei*) in southern Brazil. *Bulletin Marine Science,* 70, 409-421
[60] Krebs, C.J. (1994). *Ecology: The Experimental Analysis of Distribution and Abundance*. 4th Edition. Menlo Park, CA: Harper Collins College Publishers
[61] Lázaro, M., Lessa, E. & Hamilton, H. (2004). Geographic genetic structure in the franciscana dolphin (*Pontoporia blainvillei*). *Marine Mammal Science*, 20, 201-214.
[62] Loureiro, J., Quse, V., Bastida, R., Rodríguez, D. & Rodríguez, S. (2000) Sobre un caso de rehabilitación exitosa de *Pontoporia blainvillei* en la Fundación Mundo Marino (San Clemente del Tuyú, Argentina). In UNEP/CMS (Eds.), *Report of the Third Workshop for Coordinated Research and Conservation of the Franciscana Dolphin (Pontoporia blainvillei) in the Southwestern Atlantic (*pp. 89-92). Bonn, Germany: *UN* Environment Programme's Convention on Migratory Species.

[63] Marigo, J., Rosas, F. C. W., Andrade, A. L. V., Oliveira, M. R. Dias, R. A., & Catão-Dias, J. L. (2002). Parasites of franciscana (*Pontoporia blainvillei*) from São Paulo and Paraná States, Brazil. The Latin *American Journal of Aquatic Mammals*, 1, 115-122.

[64] Mendez, M., Rosembaum, H. C. and Bordino, P. (2008). Conservation genetics of the franciscana dolphin in Northern Argentina: population structure, by-catch impacts and management implications. *Conservation Genetics*, 9, 419-435.

[65] Millar, J.S. and Zammuto, R.M. (1983). Life histories of mammals: an analysis of life tables. *Ecology*, 64, 631-635.

[66] Molina-Schiller, D., Rosales, S. A. & Freitas, T. R. O. (2005). Oceanographic conditions off coastal South America in relation to the distribution of Burmeister's porpoise, *Phocoena spinipinnis. The Latin American Journal of Aquatic Mammals,* 4, 141-156.

[67] Monzón, F., Perez, J., Varela, G., & Corcuera, J. (1994) *Indices de predacion de tiburones sobre delfines franciscana en la costa Argentina.* In *VI Reunião de Trabalhos de Especialistas em Mamíferos Aquáticos da América do Sul held in Florianópolis, Brazil* (p. 101). Rio de Janiero, Brazil: Status of Aquatic Mammals in Latin America (SOLAMAC).

[68] Moreno, I. B., Ott, P. H., & Danilewicz, D. S. (1997). Análise preliminar do impacto da pesca artesanal costeira sobre *Pontoporia blainvillei* no litoral norte do Rio Grande do Sul, sul do Brasil. In M. C. Pinedo & A. S. Barreto (Eds.), '*Anais do 2º Encontro sobre Coordenação de Pesquisa e Manejo da Franciscana'* (pp 31 – 41). Rio Grande, Brazil: Federal University of *Rio Grande* Foundation.

[69] Moreno, I. B., Zerbini, A. N., Danilewicz, D. Santos. M. C. O., Simões-Lopes, P. C. Lailson-Brito Jr., J., Azevedo, A. F. (2005). Distribution and habitat characteristics of dolphins of the genus *Stenella* (Cetacea: Delphinidae) in the southwest Atlantic Ocean. *Marine Ecology Progress Series,* 300, 229–240.

[70] Muizon, C., de (1983). *Pliopontos litoralis* un nouveau Platanistidae Cetacea du Pliocène de la côte péruvienne. *Compte Rendus de l'Académie des Sciences, Paris, Série II,* 296, 1101-1104.

[71] Muizon, C., de (1984). Les vertébrés fossiles de la Formation Pisco (Pérou). Deuxième partie: Les Odontocètes (Cetacea, Mammalia) du Pliocène infèrieur de Sud-Sacaco. *Travaux de l'Institut Fraçais d'Études Andines,* 25, 1-188.

[72] Muizon, C., de (1988). Les relations phylogénétiques des Delphinida (Cetacea, Mammalia). *Annales de Paléontologie (Vertébrés et Invertébrés),* 74, 159-227.

[73] Nemoto, T., R. L. Brownell, Jr., & T. Ishimaru. (1977). *Cocconeis* diatom on the skin of franciscana. *Scientific Reports of the Whales Research Institute,* 29, 101-105.

[74] Olesiuk, P. K., Bigg, M. A. and Ellis, G.. M. (1990). Life history and population dynamic of resident killer whales (Orcinus orca) in the coast of British Columbia and Washington State. *Reports of the International Whaling Commission* (special issue 12), 209-213.

[75] Ott, P. H. (1998). *Análise das capturas acidentais de Pontoporia blainvillei (Gervais & D'Orbigny, 1844) (Cetacea, Pontoporiidae) nas comunidades pesqueiras do litoral Norte do Rio Grande do Sul, Sul do Brazil.* Master Thesis. Porto Alegre, Brazil: Pontifícia Universidade Católica do Rio Grande do Sul.

[76] Ott, P. H. (2002). *Diversidade genética e estrutura populacional de duas espécies de cetáceos do Atlântico Sul Ocidental: Pontoporia blainvillei e Eubalaena australis.* Ph.D. Thesis. Porto Alegre, Brazil. Universidade Federal do Rio Grande do Sul.

[77] Ott, P. H. & Danilewicz, D. S. (1996). Presence of franciscana dolphins (*Pontoporia blainvillei*) in the stomach of a killer whale (*Orcinus orca*) stranded in southern Brazil. *Mammalia*, 62 (4), 605-609.

[78] Pilleri, G. (1971). On the La Plata dolphin *Pontoporia blainvillei* off the Uruguayan coast. *Investigation Cetacea,* 3, 69-73.

[79] Pinedo, M.C. (1991). *Development and variation of the franciscana Pontoporia blainvillei.* Ph.D. Thesis, Santa Cruz, CA: University of California.

[80] Pinedo, M. C. (1994). Impact of incidental fishery mortality on the age structure of *Pontoporia blainvillei* in southern Brazil and Uruguay. *Reports of the International Whaling Commission,* (special issue), 15, 261-264.

[81] Pinedo, M. C. (1995). Development and variation in external morphology of the franciscana, *Pontoporia blainvillei. Revista Brasileira de Biologia,* 55, 85-96.

[82] Pinedo, M. C. & Hohn, A. (2000). Growth layer patterns in teet from the franciscana, *Pontoporia balinvillei*: developing a model for precision in age estimation. *Marine Mammal Science,* 16, 1-27.

[83] Pinedo, M. C., Praderi, R., & Brownell, R. Jr. (1989). Review of the biology and status of the franciscana *Pontoporia blainvillei.* In W. F. Perrin, R. L. Brownell, Z. Kaiya & L. Jiankang.(Eds.), Biology *and Conservation of the River Dolphins,* Occasional Papers. IUCN SSC 3. (pp. 46-51). Gland, Switzerland: *International Union for Conservation of Nature*

[84] Praderi, R. (1985). Relaciones entre *Pontoporia blainvillei* (Mammalia: Cetacea) y tiburones (Selachii) de aguas Uruguayas. *Comunicaciones Zoologicas del Museo de Historia Natural de Montevideo,* 11, 1-19.

[85] Praderi, R. (1986). Comentarios sobre la distribución de *Pontoporia blainvillei* en aguas del Rio de La Plata. *Actas I Reunión de Trabajo de Especialistas en Mamíferos Acuáticos de America del Sur held in* (pp. 206-214). Buenos Aires, Argentina. Rio de Janiero, Brazil: Status of Aquatic Mammals in Latin America (SOLAMAC).

[86] Praderi, R. (1997). Análisis comparativo de estadísticas de captura y mortalidad incidental de *Pontoporia blainvillei* en Uruguay durante 20 años. In M. C. Pinedo & A. S. Barreto (Eds.), *Anais do 2° Encontro sobre Coordenação de Pesquisa e Manejo da Franciscana* (pp 42 – 53). Rio Grande, Brazil: Federal University of *Rio Grande* Foundation

[87] Praderi, R. (2000). Estado actual de la mortalidad de franciscana en las pesquerías artesanales de Uruguay. In UNEP/CMS (Eds.), *Report of the Third Workshop for Coordinated Research and Conservation of the Franciscana Dolphin (Pontoporia blainvillei) in the Soutwestern Atlantic* (pp. 13-15). Bonn, Germany: *UN* Environment Programme's Convention on Migratory Species.

[88] Raga, J. A., Aznar, J., Balbuena, J. A. & Dailey, M. D. (1994). *Hadwenius pontoporiae* sp.n. (Digenea: Campulidae) from the intestine of Franciscana (Cetacea: Pontoporidae) in Argentinian Waters. *Journal of the Helminthological Society of Washington*, 61(1), 45-49.

[89] Ramos, R. M. A., Di Beneditto, A. P. M. and Lima, N. R. W. (2000). Growth parameters of *Pontoporia blainvillei* in northern Rio de Janeiro, Brazil. *Aquatic Mammals,* 26, 65-75.

[90] Ramos, R. M. A., Di Beneditto, A. P. M. Siciliano, S., Santos, M. C. O., Zerbini, A. N., Bertozzi, C. , Vicente, A. F. C., Zampirolli, E., Alvarenga, F. S. and Lima, N. R. W. (2002). Morphology of the franciscana (*Pontoporia blainvillei*) off southeastern Brazil: sexual dimorphism, growth and geographic variation. *The Latin American Journal of Aquatic Mammals,* 1, 129 - 144.

[91] Reilly, S. B. & Barlow, J. (1986). Rates of increase in dolphin population size. *Fishery Bulletin,* 84, 527-533.

[92] Ribeiro, A. M., Drehmer, C. J., Buchmann, F. S. C. & Simões-Lopes, P. C. (1998). Pleistocene skull remains of *Pontoporia blainvillei* (Cetacea, Pontoporiidae) from the coast plain of Rio Grande do Sul State, Brazil, and the relationship of Pontoporiids. *Geociências*, 3, 71-77.

[93] Rodríguez, D. H., Rivero, L. & Bastida, R. O. (2002). Feeding ecology of the franciscana (*Pontoporia blainvillei*) in marine and estuarine waters of Argentina. T*he Latin American Journal of Aquatic Mammals,* 1, 77--94.

[94] Rosas, F. C. W. & Monteiro-Filho, E. L. A. (2002). Reproductive parameters of *Pontoporia blainvillei* (Cetacea, Pontoporiidae), on the coast of São Paulo and Paraná States, Brazil. *Mammalia,* 66, 231-245.

[95] Santos, C. P., Rohde, K., Ramos, R. & Di Beneditto, A. P. (1996). Helminths of cetaceans on the southeastern coast of Brazil. *Journal of the Helminthological Society of Washington,* 63(1), 149-152.

[96] Santos, M. C. O., & Netto, D. F. (2005). Killer whale (*Orcinus orca*) predation on a franciscana dolphin (*Pontoporia blainvillei*) in Brazilian waters. *The Latin American Journal of Aquatic Mammals,* 4, 69-72.

[97] Santos, M. C. O., Pacífico, E. S. & Gonçalves, M. F. (2007). Unusual record of franciscana dolphins (*Pontoporia blainvillei*) in inner waters of the Cananéia estuary, s

[98] Santos, M. C. O., Oshima, J. E. F. & Silva. E. (2009). Sightings of franciscana dolphins (*Pontoporia blainvillei*): the discovery of a population in the Paranaguá estuarine complex, southern Brazil. *Brazilian Journal of Oceanography*, 57(1), 57-63.

[99] Secchi, E. R. (1999). *Taxa de crescimento potencial intrínseco de um estoque de franciscanas, Pontoporia blainvillei (Gervais & D'Orbigny, 1844) (Cetacea, Pontoporiidae) sob o impacto da pesca costeira de emalhe. Master Thesis*. Rio Grande, Brazil: Fundação Universidade Federal do Rio Grande.

[100] Secchi, E. R. (2006). *Modeling the population dynamics and viability analysis of franciscana (Pontoporia blainvillei) and Hector's dolphins (Cephalorhynchus hectori) under the effects of bycatch in fisheries, parameter uncertainty and stochasticity. Ph.D. Thesis.* Dunedin, New Zealand: University of Otago.

[101] Secchi, E. R. & Ott, P. H. (2000). A profundidade como um fator determinante da distribuição de toninhas, *Pontoporia blainvillei*, conforme indicado pelos índices de CPUE. In UNEP/CMS (Eds.), *Report of the Third Workshop for Coordinated Research and Conservation of the Franciscana Dolphin (Pontoporia blainvillei) in the Southwestern Atlantic* (pp. 55-57). Bonn, Germany: *UN* Environment Programme's Convention on Migratory Species

[102] Secchi, E. R., Zerbini, A. N., Bassoi, M., Dalla Rosa, L., Möller, L. M. & Rocha-Campos, C. C. (1997). Mortality of franciscanas, *Pontoporia blainvillei*, in coastal gillnetting in southern Brazil: 1994-1995. *Reports of the International Whaling Commission,* 47, 653-658.

[103] Secchi, E. R., Wang, J. Y., Murray, B., Rocha-Campos, C. C. & White, B. N. (1998). Populational differences between franciscanas, *Pontoporia blainvillei*, from two geographical locations as indicated by sequences of mtDNA control region. *Canadian Journal of Zoology*, 76, 1622-1627.

[104] Secchi, E. R., Ott, P. H., Crespo, E. A., Kinas, P. G., Pedraza, S. N. & Bordino, P. (2001). A first estimate of franciscana (*Pontoporia blainvillei*) abundance off southern Brazil. *Journal of Cetacean Research and Management*, 3, 95-100.

[105] Secchi, E. R., Danilewicz, D. & Ott, P. H. (2003a). Applying the phylogeographic concept to identify franciscana dolphin stocks: implications to meet management objectives. *Journal of Cetacean Research and Management*, 5, 61-68.

[106] Secchi, E. R., Ott, P. H. & Danilewicz, D. S. (2003b). Effects of fishing by-catch and conservation status of the franciscana dolphin, *Pontoporia blainvillei.* In N. Gales, M. Hindell, & R. Kirkwood (Eds.), *Marine Mammals: Fisheries, Tourism and Management Issues* (pp. 174-191). Collingwood, Australia:.CSIRO Publishing.

[107] Siciliano, S. (1994). Review of small cetaceans and fishery interactions in coastal waters of Brazil. *Reports of the International Whaling Commission,* (special issue), 15, 241-250.

[108] Siciliano, S., Di Beneditto, A. P., & Ramos, R. M. A. (2002). A toninha, *Pontoporia blainvillei* (Gervais & d'Orbigny, 1844) (Mammalia, Cetacea, Pontoporiidae), nos estados do Rio de Janeiro e Espírito Santo, costa sudeste do Brasil: caracterizações dos habitats e fatores de isolamento das populações. *Boletim do Museu Nacional, Zoologia*, 476, 1-15.

[109] Siler, W. (1979). A competing-risk model for animal mortality. *Ecology,* 60, 750-757.

[110] Slooten, E. (1991). Age, growth and reproduction in Hector's dolphins. *Canadian Journal of Zoology*, 69, 1689-1700.

[111] Slooten, E. & Dawson S. M. (1994). Hector's Dolphin *Cephalorhynchus hectori* (van Beneden, 1881). In S. H. Ridgway & R. Harrison, *Handbook of Marine Mammals. The First Book of Dolphins* (Volume 5, pp. 311-333). Burlington, MA: Academic Press.

[112] Slooten, E. & Lad, F. (1991). Population biology and conservation of Hector's dolphin. *Canadian Journal of Zoology,* 69, 1701-1707.

[113] Stearns, S.C. (1976). Life-history tactics: a review of the ideas. *Quarterly Review of Biology*, 51, 3-47.

[114] Stolen, M. K. & Barlow, J. (2003). A model life table for bottlenose dolphins (*Tursiops truncatus*) from the Indian River Lagoon System, Florida, USA. *Marine Mammal Science,* 19, 630-649.

[115] Von Ihering, H. (1927). *Die Geschichle des Atlantischen Ozeans*. Jena, Germany: Gustav Fisher.

[116] Wells, R. S. & Scott, M. D. (1990). Estimating bottlenose dolphin population parameters from individual identification and capture-release techniques. *Reports of the International Whaling Commission,* (special issue), 12, 407-415.

[117] Zhou, K. (1982). Classification and phylogeny of the superfamily platanistoidea, with notes on evidence of the monophyly of the cetacea. *Scientific Reports of the Whales Research Institute,* 34, 93-108.

[118] Zeh, J. E., George, J. C. & Suydam, R. (1995). Population size and rate of increase, 1978-1993, of bowhead whales, *Balaena mysticetus*. *Reports of the International Whaling Commission*, 45, 339-344.

In: Biology, Evolution and Conservation of River Dolphins… ISBN: 978-1-60876-633-8
Editors: M. Ruiz-García, J. Shostell. pp. 323-339

Chapter 17

REVIEW ON THE THREATS AND CONSERVATION STATUS OF FRANCISCANA, *PONTOPORIA BLAINVILLEI* (CETACEA, PONTOPORIIDAE)

Eduardo R. Secchi*

Laboratório de Tartarugas e Mamiferos Marinhos, Instituto de Oceanografia (IO-FURG)
Universidade Federal do Rio Grande/FURG, Rio Grande - RS, Brazil

ABSTRACT

Franciscana's restriction to shallow coastal waters makes it highly vulnerable to anthropogenic threats. Habitat degradation (noise pollution, chemical pollution and overfishing) and loss affect many coastal cetacean species around the world. Nonetheless, incidental catches in fishing gear are believed to be the main threat to franciscana conservation. This chapter aims at providing a review about the main conservation issues for franciscana with emphasis on bycatch in fishing gear. It also discusses the species conservation status, the potential alternatives for minimizing incidental mortality in fisheries and the constraints for the effective establishment and implementation of conservation measures.

CONSERVATION THREATS

Habitat Degradation

Plastic pollution and ingestion of debris: Ingestion of plastic debris by cetaceans has been a worldwide concern (*e.g.* Laist, 1997). In the western South Atlantic, both coastal and oceanic species are vulnerable to ingesting debris accidentally (*e.g.* Secchi & Zarzur, 1999; Bastida et al., 2000). Analysis of stomach contents of franciscanas have shown that this species is also vulnerable to ingesting many kinds of debris including discarded fishing gear

* edu.secchi@furg.br.

(Bassoi, 1997; Bastida et al., 2000 - Danilewicz et al., 2002). For instance, stomach contents of franciscanas from Rio Grande do Sul, southern Brazil contained discarded fishing gear such as pieces of nylon net (17% of 36 stomachs), cellophane, and plastic fragments (6%) (Bassoi, 1997). The rate of debris ingestion by franciscanas varies spatially with higher values observed in northern Argentina, where cellophane, fishing debris, and plastic were found in 45%, 32% and 16% of the stomachs. Fishing-related debris were more often found in stomachs of franciscanas collected in estuarial waters while, in contrast, cellophane debris were more abundant (100% greater) in marine animals (Bastida et al., 2000). The frequency of marine debris seems lower in animals found to the north of Rio Grande do Sul State, though no assessment has been made so far (Danilewicz et al., 2002). Regardless if the ingestion occurs directly or indirectly through the prey, the effects of such debris ingestion on the health status of individual franciscanas have not been determined, and the population-level implications are uncertain but should not be ignored.

Chemical pollution: Coastal oil spills have affected other marine species (*e.g.* penguins and pinnipeds) but there is no evidence that they also affect franciscanas. On the other hand, trace elements and organochlorines have been documented in the tissues of franciscanas along its distribution range. O'Shea et al. (1980) were the first to analyze the concentration of organochlorine in franciscanas incidentally killed in fisheries off Uruguay. Only a decade later, the presence of these pollutants in tissues of franciscanas were investigated in Argentina (*e.g.* Borrell et al., 1995, 1997). More recently levels of organochlorine contamination were compared between samples of franciscana from southern Brazil and Buenos Aires Province, with higher concentration found in the latter (Castello et al., 2000). Concentrations of DDTs and PCBs in the three countries were considered low and not regarded as a threat to the species. Comparatively, the concentrations of DDTs in Brazil and Argentina were lower than in Uruguay. A wide range of organochlorine residues have been discovered in the blubber of franciscanas from the Brazilian coastal waters (Kajiwara et al., 2004). In contrast to the lower residue levels of CHLs, HCB, HCHs, and heptachlor epoxide, the concentrations of DDT's and PCB's are surprisingly, the highest compared to those of north Atlantic dolphins, possibly reflecting high levels of industrialization or poor ecological enforcement in Brazil (Kajiwara et al., 2004).

With regard to trace elements, Seixas et al. (2008) found that the concentrations of selenium (Se), total mercury (Hg) and organic mercury (OrgHg) were higher in the livers and kidneys of franciscanas from Rio Grande do Sul than Rio de Janeiro, Brazil. For both areas the values were of the same order of magnitude as those reported in earlier studies with the same species from Brazil (Lailson-Brito et al., 2002; Kunito et al., 2004; Seixas et al., 2007) and Argentina (Marcovecchio et al., 1994; Gerpe et al., 2002). Franciscana livers showed higher concentrations of mercury, zinc, and copper relative to concentrations in other organs, whereas their highest cadmium concentrations were mostly found in kidneys (Marcovecchio et al., 1990; Gerpe et al., 2002; Lailson-Brito et al., 2002; Kajiwara et al., 2004, Seixas et al., 2008). Hepatic cadmium concentrations were low in franciscanas from both Rio de Janeiro and Rio Grande do Sul states (Lailson-Brito et al., 2002; Dornelles et al., 2007*b*). Hepatic cadmium accumulates in franciscanas through their feeding upon loliginid squids because this squid family is known to contain relatively low levels of cadmium (Dornelles et al., 2007*a*). The concentrations of these heavy metals seem to be positively correlated with age (*e.g.* Seixas et al., 2008). For example, mercury is an exogenous and harmful metal (no benefit at any concentration), which accumulates in the tissues of higher food web organisms (such as marine mammals) as they grow (Caurant et al., 1994; Feroci et al., 2005). Conversely, selenium is recognized as an essential element for metabolic activity of aquatic mammals, acting as a protective agent against the toxicity of exogenous metals such as mercury (Feroci et al., 2005). Due to the relatively low level that these exogenous elements were found in the tissues of franciscanas, there probably is no reason for concern about their potential effect on the populations' viability. However, the additive effect of other pollutants, habitat degradation and bycatch should be taken into account.

Depletion of fish stocks and temporal changes in franciscana's diet: Historical catch records of commercial fishes have demonstrated a decline in yearly landings of the sciaenids *Micropogonias furnieri* and *Macrodon ancylodon* in southern Brazil (Haimovici et al., 1997; Haimovici, 1998). This is consistent with a reduction in the occurrence of these two species in franciscana's diet (Bassoi & Secchi, 2000; Secchi et al., 2003*b*). *M. furnieri* has been heavily exploited by gillnet and trawl fisheries for more than three decades (Reis, 1992; Haimovici, 1998) and a drastic decrease in the density of juveniles in coastal waters has been observed (Ruffino & Castello, 1992). During that same period, *M. ancylodon* and *M. furnieri* decreased drastically from 41% to 7% and 27.5% to 4% in frequency of occurrence, respectively, in the diet of franciscana (Bassoi & Secchi, 2000). Haimovici (1998) showed that stocks of these sciaenid species have been extensively exploited and are currently at very low levels in the region. On the other hand, frequency of occurrence of cutlassfish *Trichiurus lepturus* and sciaenid *Umbrina canosai* in the diet of the franciscana increased from about 5% and 3% in the late 1970s to about 39% and 20% in the mid 1990s, respectively. *T. lepturus* together with *Cynoscion guatucupa* represents 47% of the total estimated bony fish biomass in this region (Haimovici et al., 1996). Both species have only experienced moderate fishing pressure (Haimovici et al., 1997; Haimovici, 1998). While *C. guatucupa* has always been the most important prey for franciscana, *T. lepturus* has had only a little importance in the franciscana's diet in the past. However, now it is the second most important prey for the species in this region. These values suggest that changes in the franciscana diet parallels reduced availability of certain prey species due to their over-exploitation. Although the effects of this major dietary change on the franciscana are unknown, the energetic implications might be of some concern.

Incidental Mortality in Fishing Nets

Mortality due to incidental entanglement in gillnets seems to be by far the greatest threat to the franciscana (Figure 1). There is no indication of direct exploitation of the species. Reports of bycatch in the shark gillnet fisheries of Punta del Diablo, Uruguay, date back to the early 1940s (Van Erp, 1969). Although gillnetting in southern Brazil began around this time (Haimovici et al., 1997), gillnet fisheries for bottom-dwelling fish were only documented as a major threat to the franciscana in the 1980s. Bycatch has since been reported from the main fishing villages along most of the species' distribution (*e.g.* Corcuera, 1994; Moreno et al., 1997; Praderi, 1997; Secchi et al., 1997, 2003*b*; Di Beneditto et al., 1998; Bertozzi & Zerbini, 2002; Rosas et al., 2002; Ott et al., 2002). In Uruguay, Praderi (1997) estimated that between 1974 and 1994 at least 3,683 dolphins were killed. The highest and lowest annual estimates were 418 and 66 dolphins caught in 1974 and 1994, respectively. The bycatch was even higher prior to that period. In the late 1960s the annual bycatch was estimated to be as high as 1,500 to 2,000 animals (Brownell & Ness, 1970; Pilleri, 1971). Large mesh-size nets targeting sharks were responsible for about 70 to 90% of the captures (*e.g.* Praderi, 1997; 2000). The depletion of the target shark species led to a drop of 25% in the fishing effort using these nets from the 1960s and 1970s to the early 1980s. Since the mid 1990s, only 20% of the fishery targets sharks (Praderi, 1997). Changes in the Uruguayan coastal fishery may be beneficial to the recovery of the franciscana from the intense bycatch pressure of the past (Praderi, 1997). However, an uncontrolled increase in fishing effort closer to shore using small mesh-sized nets to catch bony fishes, with an associated high rate of franscicana bycatch, in adjacent areas of southern Brazil is likely to have offset or nullified any recovery.

The gillnet coastal fishery in southern Brazil emerged in the 1940s and increased greatly during the 1980s. Throughout this time, vessels increased in size and engines became more powerful, which allowed for longer trips and the use of larger nets (Haimovici et al., 1997). The first information regarding franciscana bycatch, however, was published only in the 1980s (*e.g.* Pinedo, 1986; Praderi et al., 1989). This information was based on the number of animals found washed ashore in southern Brazil. Strandings of franciscanas were also published later for the coast of São Paulo State, southeastern Brazil (*e.g.* Pinedo, 1994; Santos et al., 2002). It is well known that stranding data greatly underestimate the magnitude of the bycatch (Secchi et al., 1997) as only a small fraction of the bycatch is washed ashore (Prado et al., 2007). Incidental mortality of franciscanas based on the monitoring of fishing operations started in the late 1980s (Lodi & Capistrano, 1990) and have been conducted systematically in several places along the Brazilian coast. Incidental catches occur mostly in coastal gillnets set at the bottom for sciaenid fish and at the surface for sharks and other fish (Ott et al., 2002; Secchi et al., 2003*b*). Estimated annual mortality of franciscanas in coastal gillnet fisheries off Brazil range from several hundred up to more than one thousand in the Rio Grande do Sul State coast (Moreno et al., 1997; Secchi et al., 1997; 2004; Kinas & Secchi 1998; Ferreira, 2009); tens to hundreds along the coast of Santa Catarina, Paraná and São Paulo states altogether (*e.g.* Cremer et al., 1995; Bertozzi & Zerbini 2002; Rosas et al., 2002) as well as for the Rio de Janeiro and Espirito Santo state coasts together (Di Beneditto & Ramos, 2001; Di Beneditto, 2003; Freitas-Neto & Barbosa, 2003).

In Argentina, most of the franciscana bycatch occurred in inshore gillnets targeting croakers (Sciaenidae fish) and sharks (*Galeorhinus galeus*, *Mustelus* spp., *Eugomphodus taurus*, *Squatina argentina*) in waters less than 20 m deep (Corcuera et al., 1994, Crespo et al., 1994). In the mid 1980s, based on information provided by fishermen, annual mortality of franciscana was estimated to be at least 340-350 animals (Perez-Macri & Crespo, 1989). Data collected over several years, from the mid 1980s and early 1990s, resulted in estimated annual mortality of around 230 and 240 franciscanas in the northern and southern Buenos Aires Province, respectively (Corcuera, 1994; Corcuera et al., 1994, 2000). In the late 1990s, the estimates of overall mortality of the species in the entire Buenos Aires Province was around 450 to 500 dolphins/year based on interviews (Cappozzo et al., 2000; Corcuera et al., 2000). Research carried out onboard artisanal fishing boats off Cabo San Antonio, however, has shown that annual by-catch was much higher than previous estimates obtained from interviews (Bordino & Albareda, 2005). Since most of the available data on by-catch estimates (*e.g.* Corcuera, 1994; Corcuera et al., 1994; Cappozzo et al., 2000) were obtained from interviews, it is likely that the total annual by-catch for Argentina is considerably underestimated. Gillnet fishing effort has decreased in some important fishing ports due to the decline of some shark stocks (see Chiaramonte, 1998) or due to depredation and damage of caught fish and nets by southern sea lions (*Otaria flavescens*) (Cappozzo et al., 2000). Although this decline of gillneting activities might have positive outcomes for franciscanas, recent trawling for shrimps has been responsible for high by-catch around Ingeniero White and Puerto Rosales, southern Buenos Aires Province and is an additional reason for concern (Cappozzo et al., 2000).

Figure 1. Franciscanas incidentally caught in coastal gillnetting in southern Brazil.

Assessing the Species Conservation Status

The franciscana is possibly the cetacean species most seriously and immediately affected by human activities in the western South Atlantic, especially incidental mortality in fisheries. Mortality due to incidental entanglement in coastal gillnets is suspected to be by far the greatest threat to the franciscana. Regrettably, despite the collapse of some fish stocks (*e.g.* Haimovici, 1998), fishing effort is increasing and the number of franciscanas annually caught remains high in some areas (*e.g.* Secchi et al. 2004, Ferreira, 2009). Nevertheless, the actual effect of bycatch on the chances of population persistence is likely to vary geographically. This is because both the level of bycatch (Ott et al., 2002; Secchi et al., 2003*b*) and the life history strategies differs along the species range (*e.g.* Secchi et al., 2003*a*; Secchi, this volume). Some life history parameters such as age at the attainment of sexual maturity, fecundity and survival rates influence the population's potential to respond to non-natural removals and on its likelihood for long term persistence (Caughley, 1977; Caswell, 2001; Morris & Doak, 2003). Therefore, for a proper assessment on the species conservation status to be possible, management units (*i.e.* discrete populations) need to be identified. Once these units are identified, abundance, removal rates due to bycatch (or other causes) and population-specific biological rates (*e.g.* reproductive and survival) can be estimated and changes of long-term persistence, with or without non-natural removal may be assessed.

By applying the hierarchical classification scheme (Dizon et al., 1992) for defining stocks for management purposes, Secchi et al., (2003*a*) comprehensively reviewed the genotypic, phenotypic, population response and distributional data and proposed that the franciscana distribution range be divided into four areas. The range limits for each area were defined as provisional Franciscana Management Areas (FMA)(Figure 2), as follows: FMA I - coastal waters of Espírito Santo and Rio de Janeiro states (note: confirmation of the hiatus in the Espírito Santo State with increased survey effort will require further division of this FMA); FMA II - São Paulo, Paraná and Santa Catarina states; FMA III - coastal waters of Rio Grande do Sul State and Uruguay; and FMA IV - coastal waters of Argentina, including the provinces of Buenos Aires, Rio Negro, and Chubut. It is not meant to apply new management dogmas to these stocks, rather, it is strongly recommended that limits and sub-structuring of these FMAs be constantly reassessed as new data become available.

Combining all information on bycatch from fleet monitoring programs and interviews along the species' range resulted in an annual bycatch estimate of about 110 (min: 44; max:176) franciscanas for FMA I; 279 (min: 63; max: 497) for FMA II; 1,245 (min: 562; max: 1,778) for FMA III and; 405 (min: 241; max: 567) for FMA IV (Ott et al., 2002; Secchi et al., 2003*b*; Di Beneditto, 2003). These results might represent an underestimate of the actual bycatch for several reasons. For example, there are captures in other non-monitored types of fisheries such as gillnetting (Secchi et al., 1997) and shrimp trawling (Cappozzo et al., 2000) and fishermen generally tend to under report bycatch (Lien et al., 1994, Hall, 1999). Also, bycaught dolphins may fall from the net before or during the hauling-in process (Bravington & Bisack, 1996) and some small fishing villages may not have been monitored, especially in the central and northern portions of the species' range (*e.g.* Bertozzi & Zerbini, 2002). Nevertheless, these values together with information on population dynamics can be useful as baselines for modeling the potential effects of fishing bycatch on the viability of the species on a local basis.

Bearing in mind the uncertainty on abundance and the between area variation in the quality data on bycatch rates and population dynamics (see Secchi's chapter), Secchi (2006) projected the four management units 25 years into the future based on a stage-structured matrix model (*e.g.* Caswell, 2001) using a variety of scenarios of fishing effort. Because there were estimates of franciscana density and abundance only for FMA III and IV, Secchi (2006) used the density estimated for FMA III and applied a correction factor based on the ratio of capture per unit of effort (CPUE) between the other areas and FMA III. This was assumed to represent a valid index of abundance because the unit of fishing effort is the same and the fishing gears are similar among management units. The corrected densities were multiplied by the entire area of both FMA I and II to obtain the estimate of total abundance. Uncertainty in the parameter estimates was incorporated through appropriate probability distributions. The scenarios considered most realistic (*i.e.* those that aimed to compensate for underestimation of the bycatch and that modeled environmental stochasticity) resulted in relatively high probabilities that each management unit would decline by at least 30% below its initial size with the exception of FMA I. However, it should be noted that estimates of bycatch in FMA I come from only one fishing village and it is known that bycatch occurs in other parts of this FMA (*e.g.* Freitas-Neto & Barbosa, 2003). The modeling exercise described above is considered to underestimate the risk of franciscana decline. The most recent data on bycatch (*e.g.* Rosas et al., 2002; Bordino & Albareda, 2005; Secchi et al., 2004; IWC, 2005) indicate that the numbers caught annually in FMAs II and IV are roughly twice as high as the values used by Secchi (2006) in his projections. In addition, other sources of potential threat as described above were not considered in Secchi's study.

Since the projections suggest a population decline of more than 30% over three generations considering actual and potential levels of fishing-related mortality as well as potential effects of environmental stochasticity (Secchi, 2006), the species qualified to be classified as VU under criterion A3d of the World Conservation Union's Red List of Endangered Species of Fauna and Flora (IUCN, 2008). The rate of decline is probably underestimated because a period of only 25 years was considered and other sources of non-natural mortality were not incorporated into the analysis. The causes of the inferred population decline have not ceased and are likely increasing because of fishery expansion (causing higher bycatch and also potentially reducing prey base) and lack of mitigation actions. In southern Brazil, for example, both gillnet and trawl fishing effort have been increasing since the early 1980s (Haimovici et al., 1997). The very high fishing effort has led to the stock depletion of many bottom-dwelling fish species of the family *Sciaenidae* because they are targeted by gillnet fisheries (Reis, 1992; Haimovici, 1998). The natural reaction of fishermen is to further increase fishing effort to compensate for lower catches per unit of effort until a profitable level is reached. Since the mid 1990s, the mean net length of most of the coastal gillnet fleet has increased fourfold in this area (Secchi et al., 1997; 2004, Ferreira, 2009). Unfortunately, perspectives for action to mitigate bycatch on a short-term basis are minimal (see below).

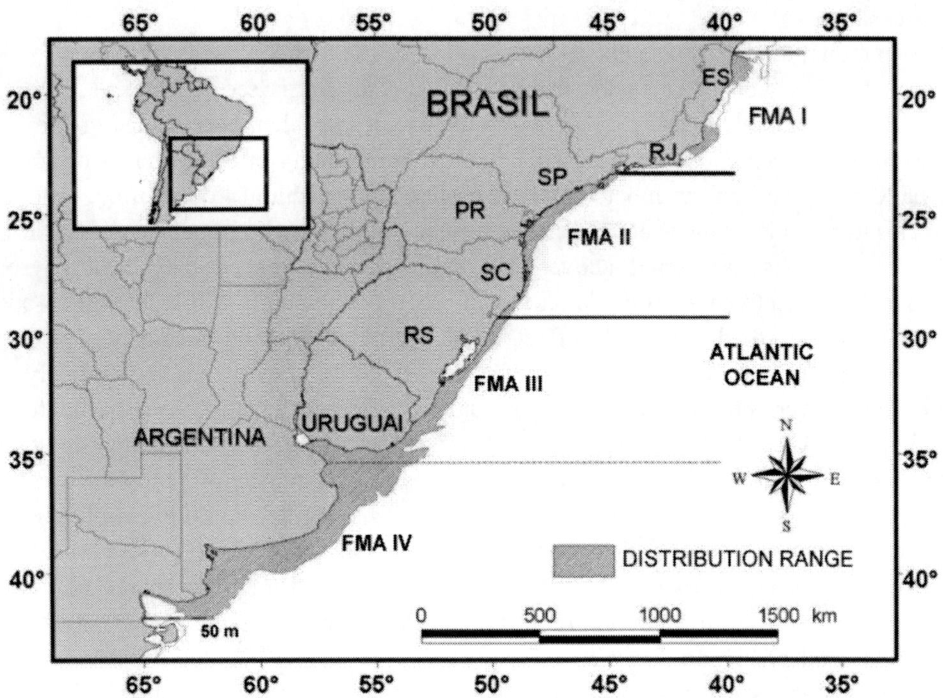

Figure 2. The species distribution range showing the four Franciscana Management Areas (FMA – sensu Secchi et al., 2003*a*).

Franciscana vs. Fisheries: Conservation from an Ecological and Socio-Economical Perspective

Over the second half of the twentieth century, many developing nations have sought to improve the efficiency of their fisheries and have received assistance from various development agencies and banks to make this possible. In many cases, such assistance has indeed resulted in more productive fisheries. Just as often, it has led to a rapid over-exploitation of many fish stocks along with the decline of several marine mammal species around the globe due to unsustainable levels of bycatch (*e.g.* Vidal, 1993; Perrin et al., 1994) or possibly due to competition for the same resource. Despite some controversies regarding the actual level of decline, it is suspected that large predatory fish populations might be only at small percentages of their original size (Myers & Worm, 2003; Hampton et al., 2005). Mitigating these problems is not an easy task, regardless of the economic situation of the nation where the problem occurs (*e.g.* Ritcher, 1998). In many developing countries, however, high foreign debt along with other socio-economic priorities have played a major role in constraining the ability of governments to allocate resources to, and properly respond to, environmental concerns (Vidal, 1993). Although fishing yield per fisherman has decreased in the last decades in many areas, fast demographic growth and high unemployment has led to a steady increase in the number of fishermen in many Latin American countries (Morrissey, 1989). Thus, a lack of options in some areas is perhaps the major cause for the continued increase in fishing effort and for the unsustainable level of fishery-related mortality of some franciscana stocks.

Even though franciscana is legally protected in Brazil, Uruguay and Argentina (for details on Federal and Regional Legislation that can benefit franciscana see Arias et al., 2002), law enforcement is unlikely to offer a solution to the bycatch problem because the greatest threat to the species is incidental capture. Legislation to limit fishing effort, in terms of maximum allowable net length and number of boats, or restricting fishing grounds (*e.g.* time and/or local closures) could be more effective. The former could easily be inspected at port, however the three countries mentioned above have few resources for policing fishing grounds. Therefore, the effectiveness of the latter would rely on fishermen's willingness to co-operate. Since all options could negatively affect their income or even be unsafe (*e.g.* if they are forced to go fishing in deeper waters further offshore), they would be unlikely to be implemented over the short term. Other potential alternatives are likely to be found in experiments related to fishing practices. Corcuera et al., (1994) suggested in vain the replacement of gear type, from gillnets to longlines, as a means of reducing bycatch off Argentina. In my experience, fishermen are usually conservative and skeptical of new fishing practices. They would not try other gear types if they were suspected to be less profitable. As stated in the article 'the tragedy of the commons' (Hardin, 1968), a resource user will not reduce his/her profit if others do not reduce their's first. Since the fisheries have also affected fish stocks, and other non-target species, a wider management strategy that considers the marine ecosystem as a whole is needed. Moreover, cultural and social needs of the fishing communities have to be taken into account to avoid adding yet another social problem to the already difficult socio-economic situation of Latin American countries.

A similar dilemma of high mortality in fisheries is faced by many other dolphin species inhabiting coastal waters of both developing and developed nations. The main difference is that, with few exceptions (*e.g.* Taiwan), in developed countries, a high average standard of living and comprehensive social welfare allow decision-makers to take a much more pro-conservation stance. A fisherman from any of these countries, prevented from fishing in a particular place with a particular gear type, will not starve or fail to provide his or her family education, medical assistance food and other basic needs. This is not necessarily so in Latin America. Resources, already limited in Brazil, Uruguay and Argentina, are minimally allocated to conservation because of other priorities. Comparatively, in many developed nations the magnitude of problem is much less and resources for conservation much greater and more readily available. In New Zealand for example, strict regulations such as the North Island fishing area closures and the establishment of the Banks Peninsula Marine Mammals Sanctuary to protect Hector's dolphins, have been possible without compensation (DOC & MAF, 2007). Generally, in developed countries, if management options have to be tested or implemented, socio-economical "side-effects" will probably be minimized through governmental compensation. However, bycatch mitigation or ecosystem restoration is not an immediate concern for governments of many developing countries. A lack of options has led some of these governments to promote fishing in times when fishing should be halted or, at least, reduced.

Conservation priorities are not necessarily priorities in the agendas of nations' governments with important health, education and other basic social needs. The Brazilian Government, for example, with its plan for accelerating development named PAC (*Plano para Aceleração do Crescimento*), has increased investment in development and cut allocation of its resources to conservation. Thus, alternative and inexpensive approaches are urgently needed in the meantime. Bordino et al., (2002), for the first time, designed and implemented an experiment to reduce franciscana bycatch. They used acoustic pingers in the nets set off Cabo San Antonio, Argentina. Although the experiment showed reduced bycatch of franciscana, the rate of depredation of the catch by southern sea lions increased. Therefore, implementation of this kind of acoustic device seems to be inappropriate as a long-term management option for the region. Another problem for this kind of device is its high cost, making it most suitable for valuable fisheries in developed countries (Dawson et al., 1998). Nevertheless, further studies should be encouraged, particularly in areas where sea lions do not occur (*e.g.* many small fishing villages along FMA I and II), with a view to quantifying long-term effectiveness. Additionally, robust trials of other gillnet modifications (*e.g.* stiffnets), and alternative fishing gear should be encouraged along with the promotion of alternative livelihoods (*e.g.* fishermen could engage the dolphin watching industry, proved economically viable and socially beneficial in many developing nations – Hoyt, 2000). International development agencies could play a role in supporting such trials – after all, via their aggressive promotion of gillnetting they played an important role in creating the problem.

If ecosystem-level management is desired, the complexity increases. A comprehensive understanding of trophic relations is needed for the framework of ecosystem management. The ecosystem-level impacts should be reduced to levels that result in stability of the system involved, but this is difficult to define or assess (Hall, 1996). True ecosystem management would combine and balance the needs of humans, marine mammals, and the fish stocks upon which they both depend (Manning, 1989). Perhaps, this could be achievable in the medium to long-term with the collection of good ecological data, co-operation of fishermen, and through education of fishing communities, in order to increase their awareness and participation in conservation. Some conservationists argue that seeking for ecosystem management can be disastrous as it is often implemented by adaptive management. The underlying mechanism driving an ecosystem and the moment when a new management approach is needed are difficult to determine. As a consequence, continued environmental destruction and species extirpation would be allowed in the name of modern resources management (Simberloff, 1998). The simplest option to achieve some level of ecosystem conservation would be the protection of umbrella species, *i.e.* a species inhabiting an extensive habitat and that protecting it would conserve many other species (Simberloff, 1998). The process could be facilitated if the umbrella species is also a flagship (a charismatic large vertebrate) that would anchor conservation campaigns. Franciscana is a good example. Regulating fishing effort to minimize bycatch mortality would also benefit some already collapsed bottom-dwelling fish stocks. Conservation of flagship species is often expensive and may take too much time until implementation. Delay may be such that some populations/stocks might decline to such low levels that recovery would be difficult.

References

[1] Arias, A., da Rocha, J., Weiskel, H. W., Fidelix, L., De Haro, C., Cremer, M., Lázaro, M., & Siciliano, S. (2002). Report of the Working Group on Legislation and Education. In E. R. Secchi (Ed.), Special Issue on the Biology and Conservation of Franciscana. *The Latin American Journal of Aquatic Mammals,* 1 (special issue 1, pp. 67-70). Rio de Janeiro, Brazil: Socieiedad Latinoamericana de Especialistas en Mamiferos Acuaticos

[2] Bassoi, M. (1997). *Avaliação da dieta alimentar de toninha, Pontoporia blainvillei (Gervais and D' Orbigny, 1844), capturadas acidentalmente na pesca costeira de emalhe no sul do Rio Grande do Sul. Bachelor Thesis,* Rio Grande, Brazil: Fundação Universidade do Rio Grande.

[3] Bassoi, M. & Secchi, E.R. (2000). Temporal variation in the diet of franciscana *Pontoporia blainvillei* (Cetacea, Pontoporiidae) as a consequence of fish stocks depletion off southern Brazil. IV Encontro para a Coordenação da Pesquisa e Conservação da Franciscana, *Pontoporia blainvillei*, no Atlântico Sul Ocidental, Working paper No. 9.

[4] Bastida, R., Rivero, L., & Rodríguez, D. (2000). Presencia inusual de elementos de origen antrópico en los contenidos estomacales de la franciscana (*Pontoporia blainvillei*). IV Encontro para a Coordenação da Pesquisa e Conservação da Franciscana, *Pontoporia blainvillei*, no Atlântico Sul Ocidental, Working paper No. 26.

[5] Bertozzi, C. & Zerbini, A. N. (2002) Incidental mortality of franciscana, *Pontoporia blainvillei*, in the artisanal fishery of Praia Grande, São Paulo State, Brazil. *The Latin American Journal of Aquatic Mammals* (special issue), 1, 153-160.

[6] Bordino, P & Albareda, D. (2005). Incidental mortality of Franciscana dolphin, *Pontoporia blainvillei*, in coastal gillnet fisheries in Buenos Aires, Argentina. DT Nº 12 (in Abstracts), V Taller para la Coordinación de la Investigación y Conservación del delfín Franciscana (*Pontoporia blainvillei*) en el Atlántico Sudoccidental, 28-30 November, Mar del Plata, Argentina.

[7] Bordino, P., Kraus, S., Albareda, D., Fazio, A., Palmiero, A., Mendez, M. & Botta, S. (2002). Reducing incidental mortality of franciscana dolphin (*Pontoporia blainvillei*)

with acoustic warning devices attached to fishing nets. *Marine Mammal Science,* 18, 833-842.

[8] Borrell, A., Pastor, T., Aguilar, A., Corcuera, J. & Monzón, F. (1995). DDTs and PCBs in La Plata dolphins (*Pontoporia blainvillei*) from Argentina: age and sex trends. *European Research on Cetaceans,* 8, 273-276.

[9] Borrell, A., Pastor, T., Aguilar, A., Corcuera, J. & Monzón, F. (1997). Contaminación por DDTs y PCBs en *Pontoporia blainvillei* de aguas argentinas: variación con la edad y el sexo. Pages 62-69. In M. C. Pinedo & A. S. Barreto (Eds.), *Anais do 2°. Encontro sobre a Coordenação de Pesquisa e Manejamento da Franciscana.* Rio Grande, Brazil: Fundação Universidade Federal do Rio Grande (FURG,).

[10] Bravington, M. V. & Bisack, K.D. (1996). Estimates of harbor porpoise bycatch in the Gulf of Maine sink gillnet fishery, 1990-1993. *Reports of the International Whaling Commission,* 46, 567-574.

[11] Brownell, R. L. Jr. & Ness, R. (1970). Preliminary notes on the biology of the franciscana, *Pontoporia blainvillei*, (Cetacea: Platanistidae). *Proceedings of the Sixth Annual Conference on Biological Sonar Diving Mammals* – 1969, 23-28.

[12] Cappozzo, H. L., Monzón, F., Perez, J. E., Albareda, D., & Corcuera, J. (2000). Mortalidad del delfín franciscana (*Pontoporia blainvillei*) en la Provincia de Buenos Aires. IV Encontro para a Coordenação da Pesquisa e Conservação da Franciscana, *Pontoporia blainvillei*, no Atlântico Sul Ocidental, Working Paper No. 24.

[13] Castello, H. P., Junin, M., Rotman, F. & Sartí, G. C. (2000) Análisis de Contaminantes Organoclorados y metales Pesados en Franciscana, *Pontoporia blainvillei,* de Argentina y Brasil. Pages 46-50 In UNEP/CMS (Eds.), *Report of the Third Workshop for Coordinated Research and Conservation of the Franciscana Dolphin* (Pontoporia blainvillei) *in the Southwestern Atlantic*. United Nations Environmental Program/The Convention on the Conservation of Migratory Species of Wild Animals (UNEP/CMS), Bonn, Germany.

[14] Caswell, H. (2001). *Matrix Population Models: Construction, Analysis and Interpretation.* 2nd Edition. Sunderland, United Kingdom: Sinauer Associates, Inc.

[15] Caughley, G. (1977). *Analysis of vertebrate population*. New York, USA: Wiley.

[16] Caurant, F., Amiard, J. C., Amiard-Triquet, C., & Sariau, P., (1994). Ecological and biological factors controlling the concentrations of trace elements (As, Cd, Cu, Hg, Se, Zn) in delphinids *Globicephalas melas* from the North Atlantic Ocean. *Marine Ecology Progress Series* 103, 207-219.

Chiaramonte, G. E. (1998). Shark fisheries in Argentina. *Marine and Freshwater Research,* 49, 601-610.

[17] Corcuera, J. (1994). Mortality of *Pontoporia blainvillei* in Northern Buenos Aires Province: the threat of small fishing camps. *Reports of the International Whaling Commission*, (special issue) 15, 291-294.

[18] Corcuera, J., Monzón, F., Crespo, E. A., Aguilar, A., & Raga, J. A. (1994). Interactions between marine mammals and the coastal fisheries of Necochea and Claromecó (Buenos Aires, Argentina). *Reports of the International Whaling Commission*, (special issue), 15, 283-290.

[19] Corcuera, J., Monzón, F., Cornejo, I., Perez, J., Beilis, A., Gingarelli, M., Albareda, D. & Arias, A. (2000) Mortalidad de *Pontoporia blainvillei* en el norte de la Provincia de Buenos Aires. In UNEP/CMS (Eds.), *Report of the Third Workshop for Coordinated*

*Research and Conservation of the Franciscana Dolphin (*Pontoporia blainvillei*) in the Southwestern Atlantic* (pp. 75-80). Bonn, Germany: UNEP/CMS.

[20] Cremer, M. J., Brutto, L. F., Simões-Lopes, P. C. & Diefenbach, C. O. (1995). The incidental catch of *Pontoporia blainvillei* in Santa Catarina state, southern Brazil. In Eleventh Biennial Conference on the Biology of Marine Mammals (pp. 26). Orlando, FL: Society for Marine Mammalogy.

[21] Crespo, E. A., Corcuera, J. F. & Cazorla, A. L. (1994). Interactions between marine mammals and fisheries in some coastal fishing areas of Argentina. In W. F. Perrin, G. P. Donovan & J. Barlow (Eds). *Gillnets and cetaceans*. International Whaling Commission (special issue 15, pp. 269-282), Cambridge, United Kingdom: International Whaling Commission.

[22] Danilewicz, D., Rosas, F., Bastida, R., Marigo, J., Muelbert, M., Rodriguez, D., Lailson-Brito Jr., J., Ruoppolo, V., Ramos, R., Bassoi, M., Ott, P. H., Caon, G., Rocha, A. M., Catão-Dias, J. L. & Secchi, E. R. (2002). Report of the Working Group on Biology and Ecology. In E. R. Secchi (Ed.), Special Issue on the Biology and Conservation of Franciscana. *The Latin American Journal of Aquatic Mammals* (special issue 1, pp. 25-42).

[23] Dawson, S. M., Read A., & Slooten, E. (1998). Pingers, porpoises and power: Uncertainties with using pingers to reduce bycatch of small cetaceans. *Biological Conservation,* 84, 141-146.

[24] Di Beneditto, A. P. M. (2003). Interactions between gillnet fisheries and small cetaceans in northern Rio de Janeiro, Brazil: 2001-2002. *The Latin American Journal of Aquatic Mammals*, 2, 79-86.

[25] Di Beneditto, A. P. M., Ramos, R. M. A. & Lima, N. R. W. (1998). Fishing activity in northern Rio de Janeiro State (Brazil) and its relation with small cetaceans. *Brazilian Archives of Biology and Technology,* 41, 296-302.

[26] Di Beneditto, A. P. M. & Ramos, R. M. A. (2001). Biology and conservation of the franciscana (*Pontoporia blainvillei*) in the north of Rio de Janeiro State, Brazil. *Journal of Cetacean Research and Management,* 3, 185-192.

[27] Dizon, A.E., Lockyer, C., Perrin, W.F., Demaster D.P., and Seisson, J. (1992). Rethinking the stock concept: a phylogeographic approach. *Conservation Biology,* 6, 24-36.

[28] DOC and MAF (Department of Conservation, and Ministry of Agriculture and Fisheries). Hector's dolphin threat management discussion document [online]. 2007 [cited 2009 July 21]. Available from: www.fish.govt.nz/en-nz/Environmental/default.htm

[29] Dorneles, P. R. Lailson-Brito, J. Jr., Santos, R. A., Costa, P. A. S., Malm, Azevedo, A. F. & Torres, J. P. M. (2007*a*). Cephalopods and cetaceans as indicators of offshore bioavailability of cadmium off Central South Brazil Bight. *Environmental Pollution,* 148, 352-359.

[30] Dorneles, P. R., Lailson-Brito, J. Jr, Secchi, E. R., Bassoi, M., Lozinsky, C. P. C., Torres, J. P. M. & Malm, O. (2007*b*). Cadmium concentrations in franciscana dolphin *(Pontoporia blainvillei*) from south Brazilian coast. *Brazilian Journal of Oceanography,* 55, 179-186.

[31] Feroci, G., Badiello, R., & Fini, A., (2005). Interactions between different selenium compounds and zinc, cadmium and mercury. *Journal of Trace Elements in Medicine and Biology,* 18, 227-234.

[32] Ferreira, E.C (2009). A dinâmica da pesca costeira de emalhe e o efeito nas taxas de captura acidental de toninhas, *Pontoporia blainvillei* (Cetacea, Pontoporiidae), na costa sul do Rio Grande do Sul. Masters Dissertation. Rio Grande, Brazil: Universidade Federal do Rio Grande.

[33] Freitas Neto, R. & Barbosa, L. A. (2003). Cetaceans and fishery interactions along the Espírito Santo state, southeastern Brazil during 1994-2001. *The Latin American Journal of Aquatic Mammals,* 2, 57-60.

[34] Gerpe, M., Rodríguez, D., Moreno, V., Bastida, R. & de Moreno, J.E. (2002). Accumulation of heavy metals in the franciscanas (*Pontoporia blainvillei*) from Buenos Aires Province, Argentina. *The Latin American Journal of Aquatic Mammals* (special issue), 1, 95-106.

[35] Haimovici, M. (1998). Present state and perspectives for the southern Brazil shelf demersal fisheries. *Fishery Management and Ecology,* 5, 277-289.

[36] Haimovici, M., Martins, A. S. & Vieira, P. C. (1996). Distribuição e abundância de peixes teleósteos sobre a plataforma continental do Sul do Brasil. *Revista Brasileira de Biologia,* 56, 27-50.

[37] Haimovici, M., Castello, J. P., & Vooren, C. M. (1997). Fisheries. In U. Seeliger, C. Odebrecht & J. P. Castello (Eds.), Subtropical Convergence Environments – the coasts and sea in the southwestern Atlantic (pp. 184-96). Berlin, Germany: Springer-Verlag.

[38] Hall, M. A. (1996). On bycatches. *Reviews in Fish Biology and Fisheries*, 6, 319-352.

[39] Hall, M. A. (1999). Estimating the ecological impacts of fisheries: what data are needed to estimate bycatches? In C. P. Nolan, C.P. (Ed.), *Proceedings of the International Conference on Integrated Fisheries Monitoring* (pp. 175-184). Sydney, Australia: Food and Agriculture Organization *of the United Nations* (FAO).

[40] Hampton, J., Sibert, J. R., Kleibert, P., Maunders, M. N. & Harley, S. J. (2005). Decline of Pacific tuna populations exaggerated? *Nature*, 434, E1-E2.

[41] Hardin, G. (1968). The tragedy of the commons. *Science,* 162, 1243-1248.

[42] Hoyt, E. Whale watching 2000: worldwide tourism numbers, expenditures, and expanding socioeconomic benefits [online]. 2000 [include the year, month, and day cited here]. Available from: The International Fund for Animal Welfare (IFAW) http://www.ifaw.org.

[43] International Union for Conservation of Nature (IUCN), (2008). Red Data List [online]. 2008 [cited 2009 July 21]. Available from the World Conservation Union: www.redlist.org.

[44] International Whaling Commission (IWC) (2005). Annex L. Report of the Sub-Committee on Small Cetaceans. J*ournal of Cetacean Research Management* (Supplement), 7, 307-326.

[45] Kajiwara, N., Matsuoka, S., Iwata, H., Rosas, F. C. W., Fillmann, G. & Readman, J. W. (2004). Contamination by persistent organochlorines in cetaceans incidentally caught along Brazilian coastal waters. *Archives of Environmental Contamination and Toxicology,* 46, 124-134.

[46] Kinas, P. G. & Secchi, E. R. (1998). Modeling truncated data to estimate incidental kills of franciscana, *Pontoporia blainvillei*, by gillnets. *Reports of the International Whaling Commission,* 48, 533-536.

[47] Kunito, T., Nakamura, S., Ikemoto, T., Anan, Y., Kubota, R., Tanabe, S., Rosas, F. C. W., Fillmann, G., Readman, J. W. (2004). Concentration and sub-celular distribution of

trace elements in liver of small cetaceans incidentally caught along the Brazilian coast. *Marine Pollution Bulletin,* 49, 574-587.

[48] Lailson Brito Jr, J., Azevedo, M. A. A., Malm, O., Ramos, R., Di Beneditto. A. P. M. & Saldanha, M. F. C. (2002) Trace metals in liver and kidney of the franciscana, *Pontoporia blainvillei*, from the northern coast of Rio de Janeiro State, Brazil. *The Latin American Journal of Aquatic Mammals* (special issue), 1, 107-114.

[49] Laist, D.W. (1997). Impacts of marine debris: entanglement of marine life in marine debris including a comprehensive list of species with entanglement and ingestion records. In J. M. Coe & D. B. Rogers (Eds.), Marine Debris Sources, Impacts and Solutions (pp. 99-139). Berlin, Germany: Springer-Verlag.

[50] Lien, J., Stenson, G. B., Carver, S., & Chardine, J. (1994). How many did you catch? The effect of methodology on bycatch reports obtained from fishermen. *Reports of the International Whaling Commission,* (special issue), 15, 535-540.

[51] Lodi, L. & Capistrano, L. (1990). Capturas acidentais de pequenos cetáceos no litoral norte do Rio de Janeiro. *Biotemas,* 3, 47-65.

[52] Manning, L. L. (1989). Marine mammals and fisheries conflicts: a philosophical dispute. *Ocean and Shoreline Management*, 12, 217-232.

[53] Marcovecchio, J., Moreno, V. J., Bastida, R. O., Gerpe, M. S., & Rodríguez, D. H. (1990). Tissue distribution of heavy metals in small cetaceans from the Southwestern Atlantic Ocean. *Marine Pollution Bulletin,* 21(6), 299-304.

[54] Marcovecchio, J., Gerpe, M. S., Bastida, R. O., Rodríguez, D. H. & Morón, S. G. (1994). Environmental contamination and marine mammals in coastal waters from Argentina: an overview. *The Science of the Total Environment,* 154, 141-151.

[55] Moreno, I. B., Ott, P. H., & Danilewicz, D. S. (1997). Análise preliminar do impacto da pesca artesanal costeira sobre *Pontoporia blainvillei* no litoral norte do Rio Grande do Sul, sul do Brasil. In M. C. Pinedo & A. S. Barreto (Eds.), Anais do 2° Encontro sobre Coordenação de Pesquisa e Manejo da Franciscana (pp. 31 – 41). Rio Grande, Brazil: Fundação Universidade Federal do Rio Grande (FURG).

[56] Morris, W. F., & Doak, D. F. (2003). *Quantitative Conservation Biology: Theory and Practice of Population Viability Analysis*. Sunderland, United Kingdom: Sinauer Associates.

[57] Morrissey, M. T. (1989). Commercial alternatives for small-scale fisheries in Latin-America. In R. B. Pollnac and M. T. Morrissey (Eds.), *Aspects of small-scale fisheries development* (pp. 99-114). Kingston, RI: International Center for Marine Resource Development, University of Rhode Island.

[58] Myers, R. A. & Worm, B. (2003). Rapid worldwide depletion of predatory fish communities. Nature, 423, 280-283.

[59] O'Shea, T. J., Brownell Jr., R. L., Clark, D. R., Walker, W. A., Cay, M. L. & Lamont, T. G. (1980). Organochlorine pollutants in small cetaceans from the Pacific and South Atlantic Ocean. *Pesticide Monitoring Journal,* 2, 35-46.

[60] Ott, P. H., Secchi, E. R., Moreno, I. B., Danilewicz, D., Crespo, E. A., Bordino, P., Ramos, R., Di Beneditto, A. P., Bertozzi, C., Bastida, R., Zanelatto, R., Perez, J. E. & Kinas, P. G.. (2002). Report of the Working Group on Fishery Interactions. Pages 55-64 In E. R. Secchi, (Ed.), *Special Issue on the Biology and Conservation of Franciscana The Latin American Journal of Aquatic Mammals,* (special issue 1): Rio de Janeiro, Brazil: Socieiedad Latinoamericana de Especialistas en Mamiferos Acuaticos

[61] Perez-Macri, G. & Crespo, E. A. (1989) Survey of the franciscana, *Pontoporia blainvillei*, along the Argentine coast, with a preliminary evaluation of mortality in coastal fisheries. Pages 57-63 In W. F. Perrin, R. L. Brownell, Z. Kaiya & L. Jiankang (Eds.), *Biology and Conservation of the River Dolphins* (Occasional papers IUCN SSC 3, pp. 57-63). Gland, Switzerland: IUCN.

[62] Perrin, W. F., Donavan, G. P. & Barlow, J. (Eds.), (1994). *Gillnets and cetaceans. Reports of the International Whaling Commission* (special issue 15). Cambridge, United Kingdom: International Whaling Commission.

[63] Pilleri, G. (1971). On the La Plata dolphin *Pontoporia blainvillei* off the Uruguayan coast. *Investigation Cetacea,* 3, 69-73.

[64] Pinedo, M.C. (1986). Mortalidade de *Pontoporia blainvillei*, *Tursiops gephyreus*, *Otaria flavescens* e *Arctocephalus australis* na costa do Rio Grande do Sul, Brasil, 1976-1983. In H. P. Castello, (Ed.), *Actas I Reunión de Trabajo de Especialistas en Mamíferos Acuáticos de América del Sur*, (pp. 187-199) held in Buenos Aires, Argentina (pp. 187-199). Rio de Janiero, Brazil: Status of Aquatic Mammals in Latin America (SOLAMAC).

[65] Pinedo, M. C. (1994). Impact of incidental fishery mortality on the age structure of *Pontoporia blainvillei* in southern Brazil and Uruguay. *Reports of the International Whaling Commission,* (special issue) 15, 261-264.

[66] Praderi, R. (1997). Análisis comparativo de estadísticas de captura y mortalidad incidental de *Pontoporia blainvillei* en Uruguay durante 20 años. In M. C. Pinedo & A. S. Barreto (Eds.), Anais do 2° Encontro sobre Coordenação de Pesquisa e Manejo da Franciscana (pp. 42 – 53). Rio Grande, Brazil: FURG.

[67] Praderi, R. (2000). Estado actual de la mortalidad de franciscana en las pesquerías artesanales de Uruguay. In UNEP/CMS (Eds.), *Report of the Third Workshop for Coordinated Research and Conservation of the Franciscana Dolphin (Pontoporia blainvillei) in the Soutwestern Atlantic* (pp. 13-15). Bonn, Germany: UNEP/CMS.

[68] Praderi, R., Pinedo, M. C. & Crespo, E. A. (1989). Conservation and management of *Pontoporia blainvillei* in Uruguay, Brazil and Argentina. In W. F. Perrin, R. L. Brownell, Z. Kaiya & L. Jiankang (Eds.), *Biology and Conservation of the River Dolphins*. (Occasional Papers IUCN SSC 3, pp. 52-56): Gland, Switzerland: IUCN.

[69] Prado, J. H. F., Secchi, E. R., Kinas, P. G. & Ferreira, E. (2007). Mark-recapture of incidentally killed franciscana to estimate past bycatch from time series of stranded carcasses. In *XVII Biennial Conference on the Biology of Marine Mammals, 2007, Cape Town*. (in abstracts). Society for Marine Mammalogy.

[70] Reis, E. G. (1992). An assessment of the exploitation of the white croaker *Micropogonias furnieri* (Pisces, Sciaenidae) by the artisanal and industrial fisheries in coastal waters of southern Brazil. Ph.D. Thesis. Norwich, UK: University of East Anglia.

[71] Richter, C. (1998). Harbour porpoise and people: strategies for bycatch reduction in the Bay of Fundy. M.Sc. Thesis, Memorial University of Newfoundland, St. John's.

[72] Rosas, F. C. W., Monteiro-Filho, E. L. A. & Oliveira, M. R. (2002). Incidental catches of franciscana (*Pontoporia blainvillei*) on the southern coast of São Paulo State and the coast of Paraná State, Brazil. *The Latin American Journal of Aquatic Mammals* (special issue), 1, 161-168.

[73] Ruffino, M. L. & Castello, J. P. (1992). Alterações na fauna acompanhante da pesca do camarão barba-ruça (*Artemesia longinaris*) nas imediações da barra de Rio Grande, RS. *Nerítica*, 7, 43-55.

[74] Santos, M. C. O., Vicente, A. F. C., Zampirolli, E., Alvarenga, F. & Souza, S. P. (2002). Records of franciscana (*Pontoporia blainvillei*) from the coastal waters of São Paulo State, southeastern Brazil. *The Latin American Journal of Aquatic Mammals* (special issue), 1, 169-174.

[75] Secchi, E.R. (2006). *Modelling the population dynamics and viability analysis of franciscana (Pontoporia blainvillei) and Hector's dolphins (Cephalorhynchus hectori) under the effects of bycatch in fisheries, parameter uncertainty and stochasticity Ph.D. Dissertation*. Dunedin, New Zealand: University of Otago.

[76] Secchi, E. R. (2009) *Life History and Ecology of Franciscana,* Pontoporia blainvillei *(Cetacea, Pontoporiidae).* In M. Ruiz-García & J. Shostell (Eds.), *Biology, Evolution, and Conservation of River Dolphins Within South America and Asia: Unknown Dolphins In Danger*. New York, New York: Nova Science Publishers, Inc.

[77] Secchi, E. R. & Zarzur, S. (1999). Plastic debris ingested by a Blainville's beaked whale, *Mesoplodon densirostris*, washed ashore in Brazil. *Aquatic Mammals,* 25, 21-24.

[78] Secchi, E. R., Zerbini, A. N., Bassoi, M., Dalla Rosa, L., Möller, L. M. & Rocha-Campos, C. C. (1997). Mortality of franciscanas, *Pontoporia blainvillei*, in coastal gillneting in southern Brazil: 1994-1995. *Reports of the International Whaling Commission,* 47, 653-658.

[79] Secchi, E. R., Danilewicz, D. & Ott, P. H. (2003*a*). Applying the phylogeographic concept to identify franciscana dolphin stocks: implications to meet management objectives. *Journal of Cetacean Research and Management*, 5, 61-68.

[80] Secchi, E. R., Ott, P. H. & Danilewicz, D. S. (2003*b*). Effects of fishing by-catch and conservation status of the franciscana dolphin, *Pontoporia blainvillei*. Pages 174-191 In N. Gales, M. Hindell, & R. Kirkwood (Eds.), *Marine Mammals: Fisheries, Tourism and Management Issues*. Collingwood, Australia: CSIRO Publishing.

[81] Secchi, E. R., Kinas, P. G. & Muelbert, M. (2004). Incidental catches of franciscana in coastal gillnet fisheries in the Franciscana Management Area III: period 1999-2000. *The Latin American Journal of Aquatic Mammals,* 3, 61-68.

[82] Seixas. T. G., Kehrig, H. A., Fillmann, G., Di Beneditto, A. P. M., Souza, C. M. M., Secchi, E. R., Moreira, I. & Malm, O. (2007) Ecological and biological determinants of trace elements accumulation in liver and kidney of *Pontoporia blainvillei. Science of the Total Environment*, 385, 208-220.

[83] Seixas. T. G., Kehrig, H. A., Costa, M. Fillmann, G., Di Beneditto, A. P. M., Secchi, E. R., Malm, O., Souza, C. M. M., & Moreira, I. (2008). Total mercury, organic mercury and selenium in liver and kidney of a South American coastal dolphin. *Environmental Pollution* (London), 154, 98-106.

[84] Siciliano, S. (1994). Review of small cetaceans and fishery interactions in coastal waters of Brazil. *Reports of the International Whaling Commission,* (special issue), 15, 241-250.

[85] Siciliano, S., Di Beneditto, A.P., and Ramos, R.M.A. (2002) A toninha, *Pontoporia blainvillei* (Gervais & d'Orbigny, 1844) (Mammalia, Cetacea, Pontoporiidae), nos estados do Rio de Janeiro e Espírito Santo, costa sudeste do Brasil: caracterizações dos

habitats e fatores de isolamento das populações. *Boletim do Museu Nacional, Zoologia*, 476, 1-15.

[86] Simberloff, D. (1998). Flagships, umbrellas and keystones: is single-species management passé in the landscape era? *Biological Conservation,* 83, 247-257.

[87] Van Erp, I. (1969). In quest of the La Plata dolphin. *Pacific Discovery*, 22, 18-24.

[88] Vidal, O. (1993). Aquatic mammal conservation in Latin America: problems and perspectives. *Conservation Biology*, 7, 788-795.

ASIAN RIVER DOLPHINS

In: Biology, Evolution and Conservation of River Dolphins… ISBN: 978-1-60876-633-8
Editors: M. Ruiz-García, J. Shostell. pp. 343-355

Chapter 18

DETECTION OF YANGTZE FINLESS PORPOISES IN THE POYANG LAKE MOUTH AREA VIA PASSIVE ACOUSTIC DATA-LOGGERS

Songhai Li[1], Shouyue Dong[1,2] Satoko Kimura[3], Tomonari Akamatsu[4], Kexiong Wang[1], and Ding Wang[1]

[1]Institute of Hydrobiology, Chinese Academy of Sciences, People's Republic of China

[2]Graduate School of Chinese Academy of Sciences, Beijing, China

[3]Graduate School of Informatics, Kyoto University, Kyoto, Japan

[4]National Research Institute of Fisheries Engineering, Hasaki, Kashima, Ibaraki, Japan

ABSTRACT

This chapter presents preliminary results on the distribution pattern of Yangtze finless porpoises (*Neophocaena phocaenoides asiaeorientalis*) in the Poyang Lake mouth area by using passive acoustic data-loggers at four different stations. Porpoise sounds were detected at all stations but their abundance decreased as the distance from the Yantze River increased. Porpoises were detected swimming both upstream to the Poyang Lake and downstream to the Yangtze River as well as between railway and highway bridges at the end of the lake. They were detected 13.9% of the total time monitored, and detected less frequently between 05:00 and 10:00 and between 15:00 and 18:00 during heavier shipping traffic. Also, there were relatively vacant periods between July 12 and July 28, 2007, and between August 5 and August 22, 2007, when virtually no porpoises were detected while there was a reversal of water current or increased water turbulence in the mouth area. These results suggest that movement and genetic communication between porpoise groups in the Yangtze River section and Poyang Lake might still remain, and therefore, the groups should be considered collectively, as a uniform unit for conservation. Bridge construction, shipping traffic, and current (turbulence and direction), might have affected the presence or movement pattern of porpoises in the study area and should be included in future conservation plans.

Keywords: Yangtze finless porpoise, Movement, Acoustic, Poyang Lake, Yangtze River.

INTRODUCTION

The Yangtze finless porpoise (*Neophocaena phocaenoides asiaeorientalis*), the only freshwater subspecies of finless porpoise (*Neophocaena phocaenoides*), shares the same habitat with the Yangtze River dolphin, locally called baiji (*Lipotes vexillifer*), which has been declared to be functionally extinct (Turvey et al., 2007). Yangtze finless porpoises historically distribute in the middle and lower reaches of the Yangtze River from Yichang to Shanghai and its conjoint lakes, such as Poyang and Dongting (Figure 1). Due to human activities such as fishing, transportation, pollution, and dam construction etc, the Yangtze finless porpoise has been declining in population size and reducing its distribution range sharply in the past thirty years (Wang et al., 2006), and has been listed as an endangered species under the Red Data List criteria (C2b) by the International Union for the Conservation of Nature and Natural Resources (IUCN) since 1996.

In November and December 2006, an intensive six-week visual and acoustic survey (Yangtze Freshwater Dolphin Expedition 2006, YFDE 2006) to find baiji and to document the status of the Yangtze finless porpoise was carried out by the Institute of Hydrobiology, Chinese Academy of Sciences (IHB) and Baiji.org Foundation (Swiss organization) with international collaborators (Turvey et al., 2007; Akamatsu et al., 2008b; Zhao et al., 2008). The survey covered the entire historical distribution range of the porpoise in the main channel of the Yangtze River and mouth area of Poyang Lake. The results indicated that the porpoise population in the main channel of the Yangtze River was approximately 1,200 (Zhao et al., 2008), which was less than half of its population size in the early 1990s (Zhang et al., 1993). Fragmentation of habitat and apparent long-distance (over 100 km) gap, where no animals were detected, was also observed (Zhao et al., 2008). Also, mtDNA haplotype analysis indicated that differences in genetic structure were present among populations of the Yangtze finless porpoise (Zheng et al., 2005).

The mouth area of Poyang Lake, a channel connecting the lake and the main stem of the Yangtze River, is a traditional "hot spot" area for porpoises. Historically, large groups of porpoises could be frequently observed in this area moving back and forth between the Yangtze River and Poyang Lake (Zhang et al., 1993; Wei et al., 2002). Unfortunately, the mouth area is also a geographical "bottleneck" between the Poyang Lake and Yangtze River. It is not only a heavy shipping traffic channel, but also an appropriate site to construct bridges for terrestrial traffic. In the two recent decades, along with economic development, human activities, such as fishery, transportation, and bridge constructions etc, have been remarkably expanding in the mouth area. Since sand-digging activity was initiated in Poyang Lake after 1998, there have been hundreds of additional sand- transporting ships passing through the mouth area day and night. In addition, two bridges, one for a highway and the second for a railway cross this channel, approximately 3 km from each other (Figure 1), were recently constructed (2000 and 2008).

Human activities have caused a serious threat on porpoise survival in this area. By visual observation, Wei et al. (2002) found that the group size of animals in this area had been decreasing continually from 1989 to 1999, and the "back and forth" movement of animals was tending to disappear in this channel. Recent visual observation also showed that porpoises who appeared there were often single and rare. It was also difficult to encounter porpoises in the section between two bridges (unpublished data). As separation or fragmentation of groups will cause genetic isolation and negatively affect the sustainable survival of wild animals, we are very much concerned that "back and forth" movements of the porpoises between the river section and the lake no longer occur in this channel. Furthermore, habitat isolation should be avoided for the *in situ* conservation of animals, and identification of isolated groups is crucial for the conservation and management of wildlife.

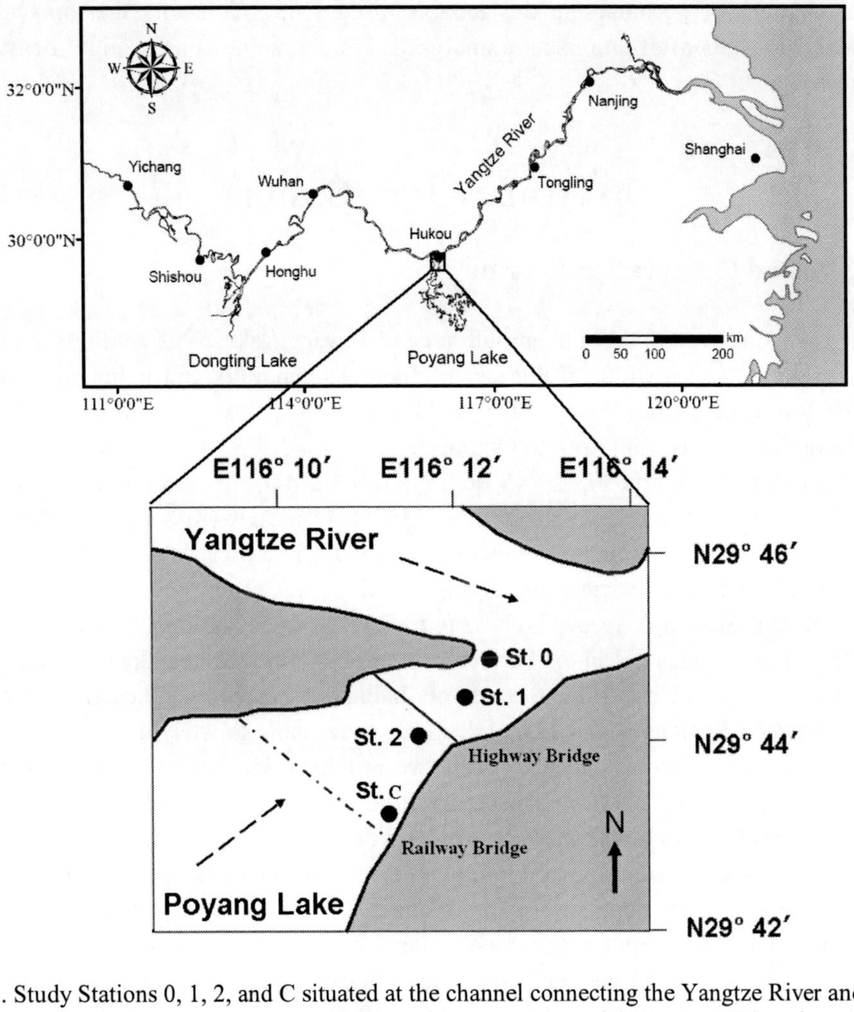

Figure 1. Study Stations 0, 1, 2, and C situated at the channel connecting the Yangtze River and Poyang Lake, China. The upper panel shows the middle and lower reaches of the Yangtze River from Yichang to Shanghai. The dashed arrows in the lower panel indicate the directions of the water current at the Yangtze River and Poyang Lake.

Since the finless porpoise is one of the smallest odontocetes and has no dorsal fin, it is difficult to detect them visually, especially when their group size is small (Akamatsu et al., 2008b). However, the porpoises emit high-frequency click trains frequently (Akamatsu et al., 2005) and passive acoustic observation which receives high-frequency sound from the animals has proved to be effective both on moving (Akamatsu et al., 2001; Akamatsu et al.,

2008b) and stationary platforms (Wang et al., 2005; Akamatsu et al., 2008a). An automatic acoustic data-logger (A-tag, Marine Micro Technology, Saitama, Japan) system, which only records the high-frequency sound events but not waveform (see below for details), has proven powerful and convenient for acoustic observation of finless porpoises (Wang et al., 2005; Akamatsu et al., 2008a, b).

In the present study, we applied the A-tag to observe the finless porpoises acoustically in a boat-based stationary platform at Stations 0, 1, and 2, to document their presence pattern among these stations, and a buoy-based stationary platform at Station C to document long-term presence patterns (Figure 1). Thereafter, fragmented or isolated status, and potential movement pattern of porpoises in this area is discussed, and factors that may affect the presence and movement of animals are analyzed. Finally, some conservation measurements are suggested.

MATERIALS AND METHODS

Study Site and Observation System

This study was performed in the mouth area of Poyang Lake, at its confluence point with the river (Figure 1). The width of this water changes temporally and is hundreds of meters during the low-water season (November to March of next year) and is several kilometers during the high-water season (April to October).

Three stations, (station 0, 29°45'06" N, 116°12'41" E; station 1, 29°44'34" N, 116°12'10" E, and station 2, 29°44'02" N, 116°11'47" E), were selected for boat-based stationary acoustic observation to document the spatial presence pattern of porpoises. The acoustic observation was carried out during April 27–29, 2006 and May 9–10, 2007 at Stations 1, 2 and Station 0, respectively (Figure 1). All three stations were along the north shore of the shipping channel in the "bottleneck" mouth area, with sandbank, shallow water, and aquatic grass, which constitute the favorite habitat of porpoises (Chen et al., 1997). The distances between Stations 0 and 1, and 1 and 2 were approximately 1300 m and 1200 m, respectively. Station 2 was situated between two bridges (highway and railway bridges, see Figure 1), approximately 300 m upstream from the highway bridge. During acoustic observations, boats at each station were fixed by double anchors to minimize drifting. The directions of the boats were relatively immovable and boat engines were completely stopped during observation. Water depths were approximately 3 m at the three stations.

To document the temporal presence pattern of porpoises, Station C (29°42'43" N, 116°11'26" E), which was based on a buoy and approximately 300 m downstream from the railway bridge, and 2,400 m upstream to Station 2 (Figure 1), was selected for long-term acoustic observation since June 27, 2007. Station C is situated beside the deep channel along the south shore and near the railway bridge. The water depth of this station is over 13 m during high-water seasons, and over 1.5 m even during the low-water seasons. These depth values justified station C as a suitable station for year-round underwater acoustic observation. The buoy is held by only one anchor, and its direction was variable, changed by water current, wind, water waves made by passing ships, etc.

Underwater acoustic observations were performed by stereo acoustic data loggers (A-tags; Little Leonardo Ltd., Tokyo, Japan, in 2006; Marine Micro Technology, MMT, Saitama, Japan, in 2007). Each A-tag consisted of two hydrophones, approximately 170 mm apart, which were used to identify the sound source direction, a preamplifier with band-pass filter between 70–300 kHz (in 2006) or 55–235 kHz (in 2007) to eliminate noise outside the frequency bands, a PIC18F6620 CPU, a 128 MB flash memory, and a waterproof tube to encase a CR2 lithium battery cell for the boat-based observation at Stations 0, 1, and 2 or two alkaline UM-1 battery cells for the buoy-based long-term observation at Station C. The lifetime of A-tag with CR2 lithium battery cell is over 30 hours, and the lifetime with two alkaline UM-1 battery cells is about one month. To ensure getting unabridged data in the long-term observation, each deployment period would be approximately month. The hydrophone sensitivity is -201 dB re 1 V/μPa at 120 kHz (100–160 kHz within 5 dB), which is close to the dominant frequency of sonar signal of finless porpoises (Li et al., 2005). Each A-tag is an event data logger that records sound pressure and the travel time difference (Td) between two hydrophones every 0.5 ms (2 kHz event sampling frequency). It does not record the waveforms of received sound. The active range of the A-tags for porpoise observation is approximately 300 m, according to the source levels of porpoise signals (Li et al., 2006).

For the boat-based acoustic observation at Stations 0, 1, and 2, a bamboo rod was used to fix the A-tag to be an underwater depth of 1-m at the side of each double-anchored boat. The two hydrophones of each A-tag were roughly set parallel to the current direction to monitor the moving direction of porpoises. The primary hydrophone of the A-tag was directed upstream towards Poyang Lake, and the secondary hydrophone was directed downstream towards the Yangtze River. This would mean that if the travel time difference of porpoise signal between the two hydrophones of each A-tag was changing from positive to negative, the phonating porpoise was moving from the Poyang Lake direction to the Yangtze River direction. For the buoy-based long-term acoustic observation at Station C, an iron bar was used to tightly fix the A-tag approximately at a depth of 1 m. As the direction of buoy is not fixed, the relative direction of the two hydrophones in each A-tag to the current direction is uncertain, and the moving direction of phonating porpoise upstream to the Poyang Lake or downstream to the Yangtze River would not be identified in this case.

Data Analysis

The acoustic data were analyzed by using a custom-made program developed on Igor Pro 5.03 (WaveMetrics, Lake Oswego, OR, USA). The high-frequency click trains produced by porpoises present regular or gradual changes in sound pressure and interclick interval (ICI). The interclick intervals are typically between 10–80 ms (Akamatsu et al., 2005; 2007). These characteristics can distinguish porpoises click trains from the noise of background or cargo ships passing nearby, which have randomly changing sound pressures and interclick intervals. Figures 2 and 3 illustrate sound pressure (SP), travel time difference (Td) between two hydrophones of each A-tag, and interclick interval (ICI) of typical porpoise click trains and cargo ship noise, respectively. Since the interclick intervals of porpoise click trains are usually shorter than 130 ms (Li et al, 2007; Akamatsu et al., 2007), a click train was defined as a series of over 5 clicks with ICIs shorter than 130 ms.

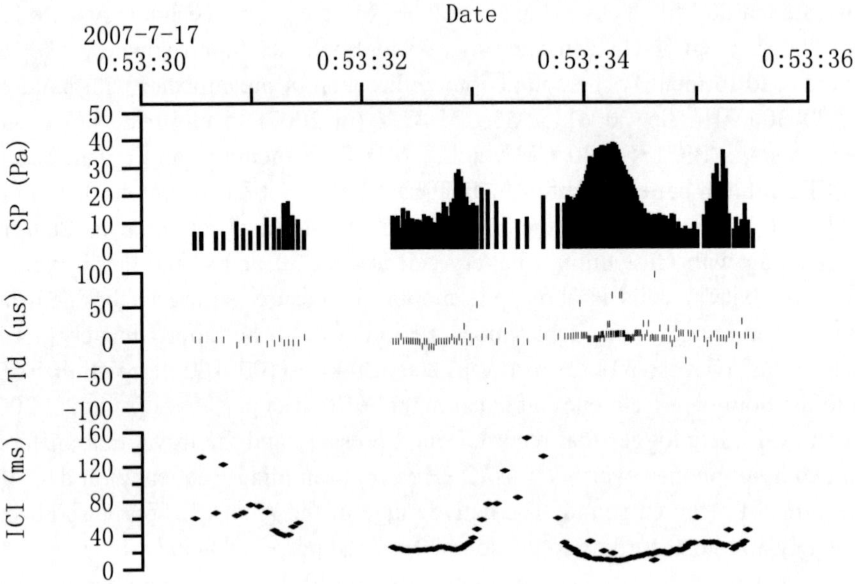

Figure 2. Sound pressure (SP), travel time difference (Td) between two hydrophones of each A-tag, and interclick interval (ICI) of typical porpoise click trains recorded by the stereo acoustic data logger (A-tag). Note that the sound pressures and ICIs of porpoise click trains were changing smoothly.

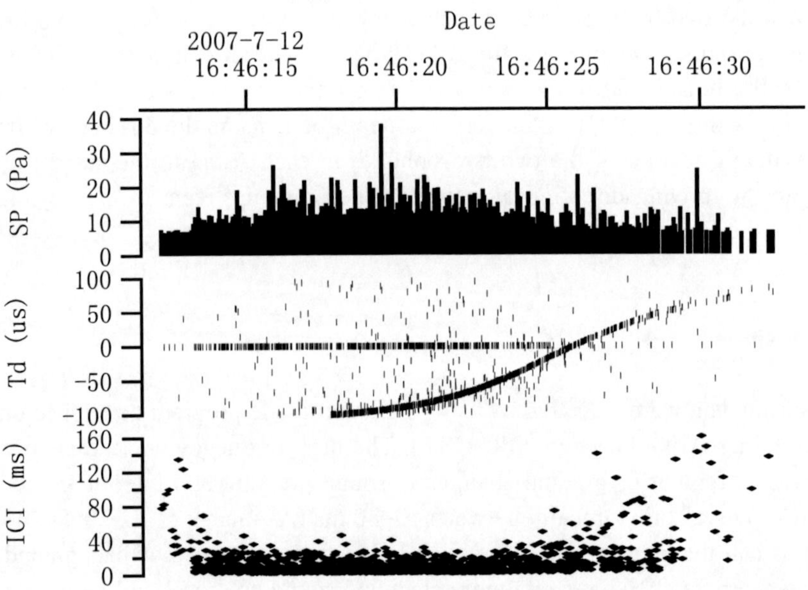

Figure 3. Sound pressure (SP), travel time difference (Td) between two hydrophones of each A-tag, and interclick interval (ICI) of typical cargo ship noise recorded by the stereo acoustic data logger (A-tag). Note that the sound pressures and ICIs of porpoise click trains were changing randomly.

For the boat-based acoustic observations at Stations 0, 1, and 2, owing to the immobility of the relative direction between the two hydrophones of each A-tag and the current direction, it was possible to count the number and determine the swimming direction of the phonating animals by analysis of their click trains. During off-line analysis, click trains of 10 s or less (apart from each other) and having smoothly changed traces of travel time difference were considered produced by one individual; these trains were defined as a single track. The number of independent tracks in each 1-min time bin was defined as the observed number of animals (or group size) in a unit of time (1 minute). The swimming direction of an animal was determined by the gradual change of travel time difference traces (see above).

For the buoy-based long-term acoustic observation at Station C, on account of the uncertainty of the buoy direction, the relative direction between the two hydrophones of each A-tag and uncertain current direction, and it was difficult to determine the number and swimming directions of phonating animals. Instead, only detection (or group) of porpoise was documented. A detection (or group) was determined when porpoise click trains were identified and the click trains were within 5 minutes of each other. Porpoise presence time was defined as the start time of the first click train in a detection; and detection duration was the time period between the start time and the end time of a detection. Once a detection was identified, the presence time and detection duration were documented. To describe the temporal presence pattern of porpoises in Station C, the presence ratio of animals was analyzed. The presence ratio of animals was the ratio of the accumulated detection duration over the unit duration such as one day or one hour.

Number of passing cargo ships and hydrology data were also collected to compare with the presence of porpoises in the long-term acoustic observation station, Station C. These cargo ships could be identified and counted acoustically by their changing travel time difference (Td) (Figure 3). The hydrology data of the study area during deployment of the long-term acoustic observation, including flux and direction of water current, were acquired from the Hydrological Bureau of the Yangtze River Water Resources Commission.

RESULTS

In the boat-based acoustic observation, we obtained 1,216 and 504 minutes of effective observation time at Stations 1 and 2, respectively, during April 27–29 of 2006, and 464 minutes at Station 0 during May 9–10 of 2007. At Stations 0, 1, and 2, animals were detected acoustically in 92.9, 76.2, and 76.0% of the effective observation time, respectively. In unit time (1 minute), Station 0 counted the most porpoises, which was on average 1.85 individuals/min; Station 1 counted an average of 1.41 individuals/min; and Station 2 counted the least porpoises, at only 0.83 individuals/min (Figure 4b). Swimming direction could only be determined for a few porpoises. At Stations 0, 1, and 2, averages of only 0.35, 0.30, and 0.01 individuals/min were determined with swimming directions, respectively (Figure 4a). Porpoises were observed swimming both upstream to the Poyang Lake and downstream to the Yangtze River at all three stations (Figure 4a).

The buoy-based long-term acoustic observation has been deployed since June 27, 2007, at Station C (Figure 1). The data shown in this chapter were obtained between June 27 and September 28, 2007, with a total observation time of 80 days (Table 1). In total, 578 porpoise detections were identified, and the total detection duration was 15,411 minutes, which occupied 13.9% of the effective observation time (Table 1). The presence pattern of porpoises at Station C was displayed by both the diurnal presence ratio (per hour) in Figure 5a and by the day presence ratio (per day) in Figure 6b. The number of ships detected by the A-tag in (per hour) over the entire observation period is shown in Figure 5B. Figure 6A shows the flux and direction of water current measured daily over the entire observation period.

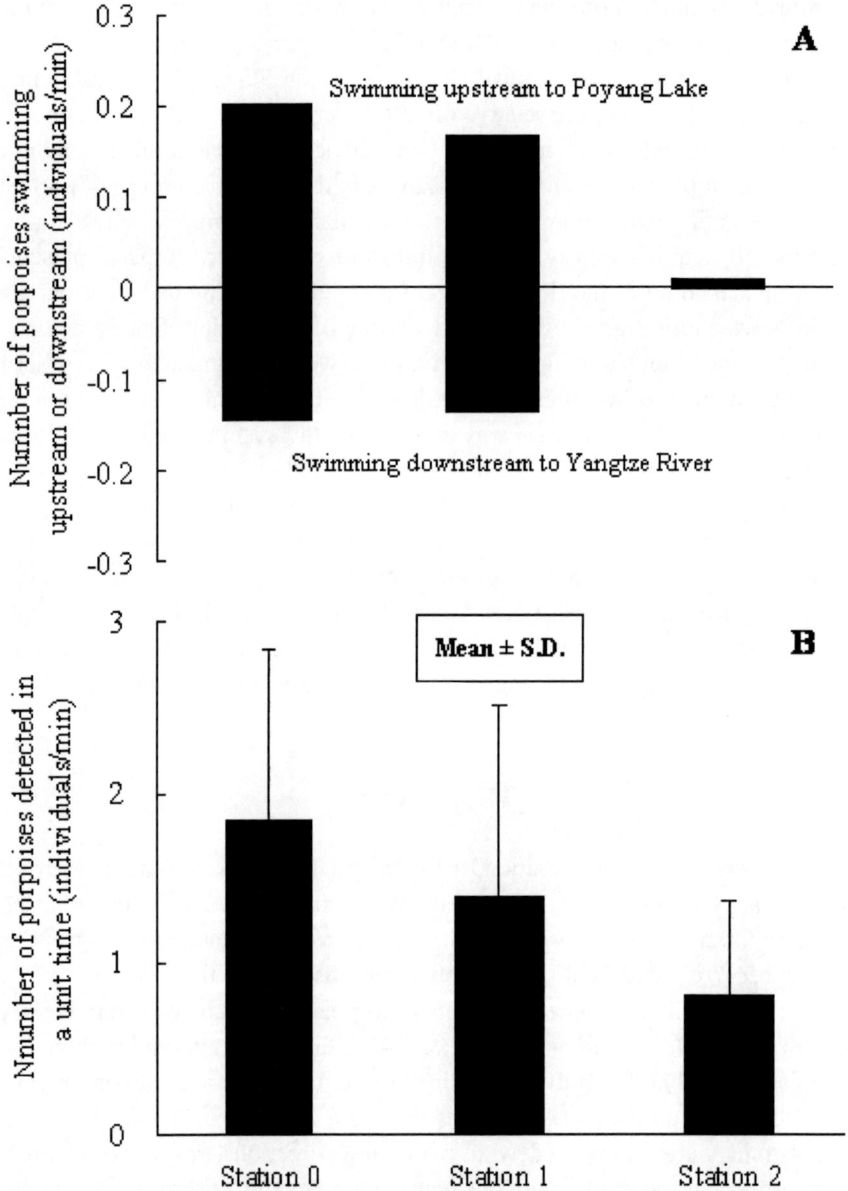

Figure 4. (a) Average number of porpoises per unit time (1 minute) at Stations 0, 1, and 2, to which swimming direction could be determined. Positive bars indicate animals swimming upstream to the Poyang Lake, negative bars indicate swimming downstream to the Yangtze River. (b) Average number of porpoises detected per unit time (1 minute) at the three stations. Standard deviations (S.D.) are also included.

Table 1. Details of buoy-based long-term acoustic observation deployed at Station C (see Figure 1), and total porpoise detection duration (min) and presence ratio (%) during each deploying period

Start day	End day	Days	Porpoise detection duration (min)	Presence ratio (%)
Jun. 27	Jul. 25	29	7,727	19.4
Jul. 26	Aug. 28	34	3,148	6.6
Sep. 12	Sep. 28	17	4,536	19.5
Total		80	15,411	13.9

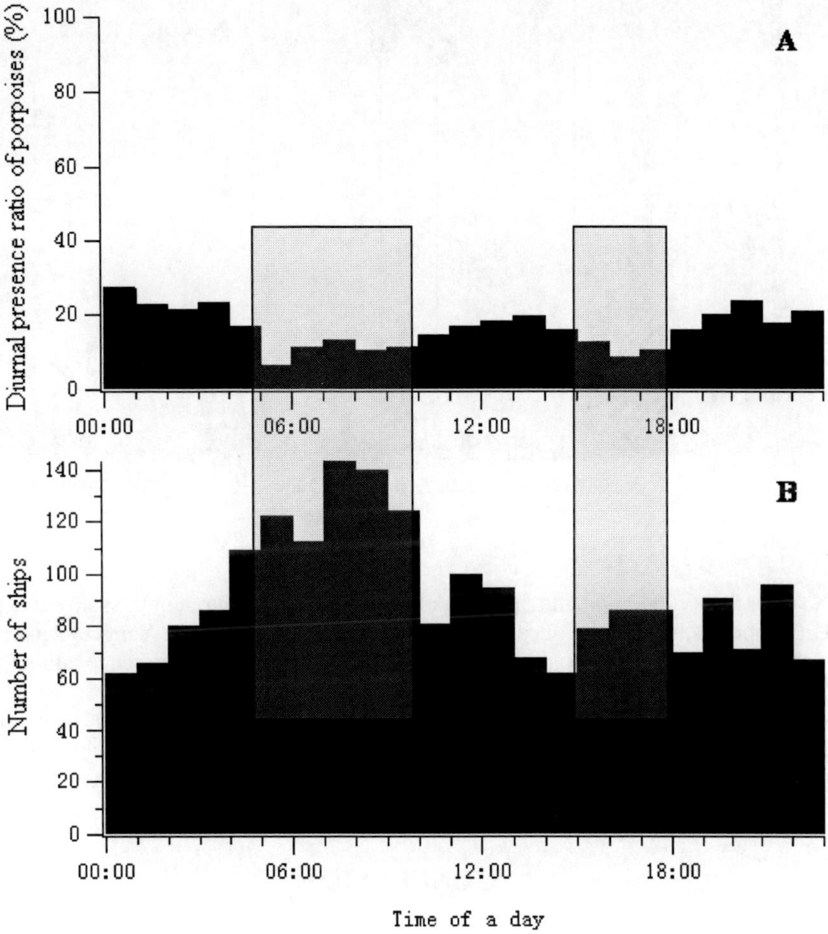

Figure 5. (a) The diurnal presence ratio of porpoises detected by the A-tag in each 1-hour time unit over the entire observation period at Station C. (b) The number of ships detected by the A-tag in each 1-hour time unit over the entire observation period at Station C. Matches between hollows of porpoise presence ratio and ridges of ship number, are indicated by grey transparent panes.

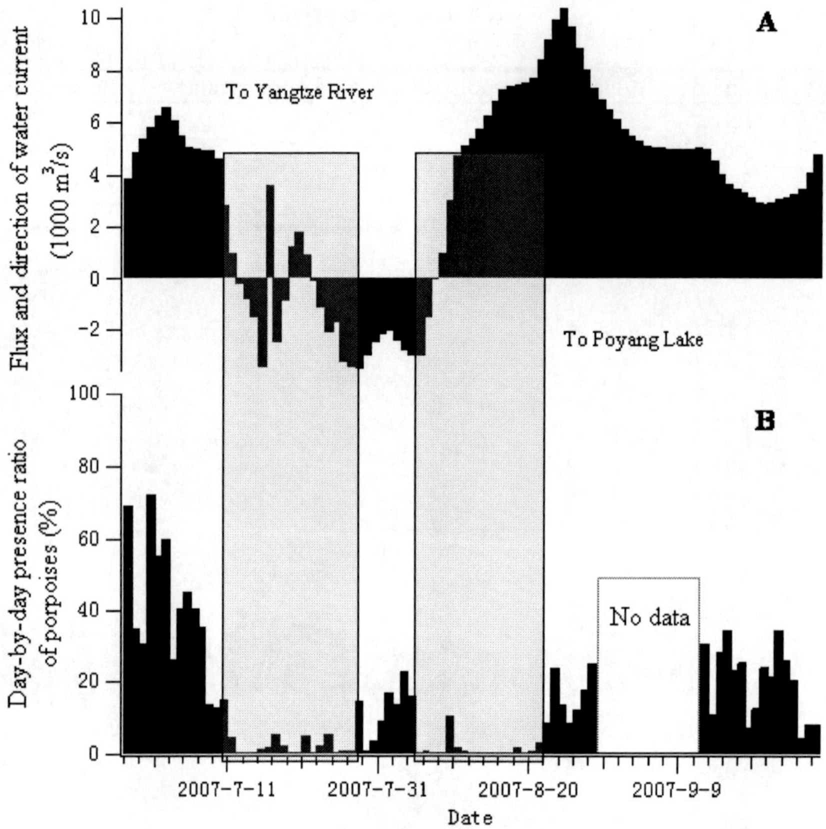

Figure 6. (a) The flux (1000 m^3/s) and direction of water current measured in the study area per day over the entire observational period. Positive values indicate direction to the Yangtze River, and negative values indicate direction to the Poyang Lake; (b) The day-by-day presence ratio of porpoises detected by the A-tag in each 1-day time unit over the entire observation period at Station C. A white pane indicates that there were no porpoise data between Aug. 29 and Sep. 11, 2007. Gray transparent panes match indicate decreased porpoise presence Note that the matches between hollows of porpoise presence ratio and turbulence or reversing of current direction are indicated by grey transparent panes.

CONCLUSION

Effect of Bridges

Porpoise presence was successfully detected acoustically at all four stations by using the stationary stereo A-tags (Figures 4, 5a, and 6b). Among stations 0, 1, and 2, leaving out the account that the data between Station 0 and Stations 1, 2 were obtained from different years (Station 0 in 2007, and Stations 1, 2 in 2006), the maximum average animal density (1.85 individuals/min) occurred at Station 0, and the minimum density (0.83 individuals/min) occurred at Station 2. The animal density gradually decreased along Stations 0, 1, and 2 (Figure 4b). These results were consistent with recent visual observations (unpublished data).

As described above, both Stations 2 and C are situated in the section between the two bridges (highway and railway bridges). The distance between Station 2 and C is only approximately 2,400 m, without any other visible physical barriers between them. Porpoises could be acoustically detected at stations 2 (Figure 4) and C (Figures 5a and 6b), and detected swimming both upstream to Poyang Lake and downstream to the Yangtze River at stations 0, 1, and 2 (Figure 4a), despite being difficult to visually observe the animals in the section between the two bridges (unpublished data). It is supposed that the "back and forth" movement behaviors, which were often observed visually before (Zhang et al., 1993), might still occur in the mouth area of Poyang Lake. There could still be a chance of genetic communication among the groups in the Yangtze River main stem and Poyang Lake. However, the proportion or degree of the "back and forth" movement and genetic communication might be limited, since the animal density at Stations 2 was relatively low (0.83 individuals/min on average), which is about half of the presence at Stations 0 and 1 (1.85 and 1.41 individuals/min, respectively). The presence ratio of porpoises in Station C is only 13.9% of the effective observation time. These data suggest that the lowest density area exists between the two bridges, a busy construction area. We need to carefully monitor finless porpoises in these areas to hopefully prevent population fragmentation. The data presented in this chapter verify the usefulness of stereo A-tags for stationary acoustic monitoring and the detection of porpoise presence, especially in the cases, where the animal density is relatively low and finding animals by visual observations is hard.

The low animal density detected in Stations 2 and C, situated in the section between the two bridges (Figure 1), might be a bridge effect. The bridges might have blocked the movement of porpoises through them by both changing local bottom topography and environments of hydrology and underwater noise. All vibration transferred by piers, engine noise produced by vehicular traffic and the construction activity at the upstream railway bridge (the railway bridge was still under construction during the observation period) could have produced extra underwater noise.

Effects of Shipping Traffic and Water Current

At Station C, porpoise presence was observed in both day and night over the entire observation period between June 27 and September 28, 2007. There were two distinct time-periods when fewer porpoises were detected; between 05:00 and 10:00, and between 15:00 and 18:00 o'clock, respectively (Figure 5a). It seems that the shipping traffic did have a negative effect on the porpoise presence (Figure 5). When the shipping traffic was high, the presence ratio of porpoises was low, and vice versa (Figure 5). The shipping traffic might affect the porpoise presence by strong engine noise and generated water waves.

Porpoises were almost observed everyday during the acoustic observation period at Station C, except for the period when the system did not work, and between July 12 and July 28, 2007, and between August 5 and August 22, 2007, when the presence ratio data was reduced (Figure 6b). Figure 6 shows that there were some matches between hollows of day-by-day presence ratio of porpoises in each 1-day time unit and turbulences or reversing of current direction. When the turbulence or reversing of current direction appeared, the presence ratio of porpoises dropped dramatically (see the grey transparent panes in Figure 6). The turbulence or reversing of current direction might have indirectly affected the presence or distribution of porpoises by disordering the distribution of fishes, which are the prey of porpoises.

Conservation Suggestions

Since the "back and forth" movement and genetic communication among the porpoise groups in the Poyang Lake and Yangtze River main stem may still occur, the groups should be considered collectively as a uniform unit for conservation.

Bridges and their construction, shipping traffic, and current direction along with its turbulence, might have affected the presence or movement pattern of porpoises in the Poyang Lake mouth area. To mitigate these effects we suggest: (1) the local bottom topography under the bridges should be recovered back to its arenaceous, quagmiry, and adlittoral state; characteristics of the finless porpoise's favorite habitat (Chen et al., 1997); (2) to reduce the water vibration and underwater noise the speed of road and water traffic should be restricted, and whistling (boat and car horns) should be avoided when vehicles cross these bridges and ships pass along the water channel,; (3) reservoir discharge events (such as from Three-Gorges Dam), should be tightly regulated to reduce artificially-high river turbulence and changes in current direction within Poyand Lake's mouth.

Clearly, there are limitations to these data as they were collected over a relatively short period of time (buoy based-3 months; boat based-two days) and there was a detection range limit of only 300 m for each probe. We recommend that future studies incorporate year-round monitoring and a series of probes.

Acknowledgments

This work was supported by the National Basic Research Program of China (2007CB411600), the Chinese National Natural Science Foundation (30730018), the Ocean Park Conservation Foundation of Hong Kong (OPCFHK), the President's Fund of the Chinese Academy of Sciences, Special Funds for Presidential Scholarships of the Chinese Academy of Sciences (082Z01), Research and Development Program for New Bio-industry Initiatives of Japan, and Grant-in-Aid for Scientific Research (B) of Japan (19405005).

References

[1] Akamatsu, T., Teilmann, J., Miller, L.A., Tougaard, J., Dietz, R., Wang, D., Wang, K., Siebert, U., & Naito, Y., (2007). Comparison of echolocation behaviour between coastal and riverine porpoises. *Deep-Sea Research Part II*, 54, 290–297.

[2] Akamatsu, T., Nakazawa, I., Tsuchiyama, T., & Kimura, N., (2008a). Evidence of nighttime movement of finless porpoises through Kanmon Strait monitored using a stationary acoustic recording device. *Fisheries Science*, 74, 970–975.

[3] Akamatsu, T., Wang, D., Wang, K., Li, S., Dong, S., Zhao, X., Barlow, J., Stewart, B. S., & Richlen, M., (2008b). Estimation of the detection probability for Yangtze finless

porpoises (*Neophocaena phocaenoides asiaeorientalis*) with a passive acoustic method. *Journal of the Acoustical Society of America*, 123, 4403–4411.

[4] Akamatsu, T., Wang, D., Wang, K., & Naito, Y., (2005). Biosonar behaviour of free-ranging porpoises. *Proceedings of the Royal Society B-Biological Sciences*, 272, 797–801.

[5] Akamatsu, T., Wang, D., Wang, K., & Wei, Z., (2001). Comparison between visual and passive acoustic detection of finless porpoises in the Yangtze River, China. *Journal of the Acoustical Society of America*, 109, 1723–1727.

[6] Chen, P., Liu, R., Wang, D., & Zhang, X., (1997). *Biology, Rearing and Conservation of Baiji* Beijing, China: Science Publisher.

[7] Kimura, S., Akamatsu, T., Wang, K., Wang, D., Li, S., Dong, S., & Arai, N., (2009). Comparison of stationary acoustic monitoring and visual observation of finless porpoises. *Journal of the Acoustical Society of America* (in press).

[8] Li, S., Wang, K., Wang, D., & Akamatsu, T., (2005). Echolocation signals of the free-ranging Yangtze finless porpoise (*Neophocaena phocaenoides asiaeorientialis*). *Journal of the Acoustical Society of America*, 117, 3288–3296.

[9] Li, S., Wang, D., Wang, K., & Akamatsu, T., (2005). Sonar gain control in echolocating finless porpoises (*Neophocaena phocaenoides*) in an open water. *Journal of the Acoustical Society of America,* 120, 1803–1806.

[10] Li, S., Wang, D., Wang, K., Xiao, J., & Akamatsu, T., (2007). The ontogeny of echolocation in the Yangtze Finless porpoise (*Neophocaena phocaenoides asiaeorientalis*). *Journal of the Acoustical Society of America,* 122, 715–718.

[11] Turvey, S. T., Pitman, R. L., Taylor, B. L., Barlow, J., Akamatsu, T., Barrett, L. A., Zhao, X., Reeves, R. R., Stewart, B. S., Wang, K., Wei, Z., Zhang, X., Pusser, L. T., Richlen, M., Brandon, J. R., & Wang D., (2007). First human-caused extinction of a cetacean species? *Biology Letters,* 3, 537–540.

[12] Wang, D., Zhang, X., Wang, K., Wei, Z., Wursig, B., Braulik, G. T., & Ellis, S., (2006). Conservation of the baiji: No simple solution. *Conservation Biology,* 20, 623–625.

[13] Wang, K., Wang, D., Akamatsu, T., Li, S., & Xiao, J., (2005). A passive acoustic monitoring method applied to observation and group size estimation of finless porpoises. *Journal of the Acoustical Society of America*, 118, 1180–1185.

[14] Wei, Z., Wang, D., Zhang, X., Zhao, Q., Wang, K., & Kuang, X., (2002). Population size, behavior, movement pattern and protection of Yangtze finless porpoise at Balijiang section of the Yangtze River. *Resources and Environment in the Yangtze Basin,* 11, 427–432.

[15] Zhang, X., Liu, R., Zhao, Q., Zhang, G., Wei, Z., Wang, X., & Yang, J., (1993). The population of finless porpoise in the middle and lower reaches of Yangtze River. *Acta Theriol Sinica*, 16, 490–496.

[16] Zhao, X., Barlow, J., Taylor, B. L., Pitman, R. L., Wang, K., Wei, Z., Stewart, B. S., Turvey, S. T., Akamatsu, T., Reeves, R. R., & Wang, D., (2008). Abundance and conservation status of the Yangtze finless porpoise in the Yangtze River, China. *Biological Conservation,* 141, 3006–3018.

[17] Zheng, J., Xia, J., He, S., & Wang, D., (2005). Population genetic structure of the Yangtze finless porpoise (*Neophocaena phocaenoides asiaeorientalis*): implications for management and conservation. *Biochemical Genetics,* 43, 307–320.

In: Biology, Evolution and Conservation of River Dolphins… ISBN: 978-1-60876-633-8
Editors: M. Ruiz-García, J. Shostell. pp. 357-375

Chapter 19

POPULATION STATUS AND CONSERVATION OF BAIJI AND THE YANGTZE FINLESS PORPOISE

***Ding Wang*[1]* and *Xiujiang Zhao*[2]**
[1]Institute of Hydrobiology, Chinese Academy of Sciences, Wuhan, China
[2]Graduate School of Chinese Academy of Sciences, Beijing, China

ABSTRACT

The Yangtze River is home to two endemic cetaceans, the baiji or Yangtze River dolphin (*Lipotes vexillifer*) and Yangtze finless porpoise (*Neophocaena phocaenoides asiaeorientalis*). Both cetaceans suffered great abundance reduction and range contraction during the last three decades. Baiji had at one point been abundant in the river, but in 2006 was declared likely extinct because an extensive survey conducted by a team of international scientists throughout baiji's geographical range failed to observe a single baiji. The latest abundance estimate of the Yangtze finless porpoise, based on data collected in the same survey is approximately 1,800 which indicates that one half of the population has vanished since 1991. It is because the baiji and the Yangtze finless porpoise share the same river and almost the same habitat, they also have been facing the same kind of threats, i.e. over- and illegal fishing, heavy boat traffic, water constructions and water pollution. We provide an analysis of the effectiveness of our conservation methods over the last three decades regarding three measures (in situ, ex situ and captive breeding). We also provide suggestions for the future protection of the baiji and Yangtze finless porpoise including, forbidding fishing in the river or at least in the current reserves, expansion of the current Tian-e-Zhou Oxbow Reserve and establishing new similar ex situ reserves, and intensifying the captive breeding program.

Keywords: baiji, *Lipotes vexillifer*, Yangtze finless porpoise, *Neophocaena phocaenoides*, population size, abundance, conservation, Yangtze River.

---

* wangd@ihb.ac.cn.

Introduction

There are two cetaceans endemic to the Yangtze River, the baiji or Yangtze River dolphin (*Lipotes vexillifer*) and Yangtze finless porpoise (*Neophocaena phocaenoides asiaeorientalis*). Both cetaceans occur in the middle and lower reaches of the river and two appended lakes (Poyang and Dongting), China (Figure 1). As a mammal species at the top of food chain, their survival heavily depends on habitat stability and food resource availability. However, the Yangtze River, as the third largest river in the world and so called "golden channel of the country" in China, has been heavily used and explored by all kinds of human activities which have led to the likely extinction of the baiji (Turvey et al., 2007). Additionally, the Yangtze finless porpoise is now listed in the Second Order of Protected Animals in China and has also been listed as an endangered population in the IUCN Red Data Book since 1996 (Baillie & Groombridge, 1996).

Abundance and Distribution

Baiji

Baiji once occurred in the Qiantang River but disappeared in the 1950s (Zhou et al., 1977) (Figure 1). As a member of the true river dolphins, a particularly rare group on this planet, baiji was considered to be the most threatened cetacean (Reeves et al., 2003), and probably the rarest animal within the category of large mammals (Dudgeon, 2005). This species, as the sole representative of the Lipotidae family lineage diverging from other cetacean more than 20 million years ago (mya) (Nikaido et al., 2001), has long been listed as "Critically Endangered" by IUCN (Reeves et al., 2003) until very recently when it was announced to be possibly extinct after an intensive range-wide survey concluded without a single sighting in 2006 (Turvey et al., 2007). This would mean, although a few individuals might still survive somewhere in the wild outside of detection limits, presumably, there is only a slim chance of reversing its upcoming extinction. This will be the first aquatic mammal species to be extinct since the demise of the Japanese Sea Lion (*Zalophus japonicus*) and the West Indian Monk Seal (*Monachus tropicalis*) in the 1950s, as well the first cetacean species to be extinguished as a result of human activity (Turvey et al., 2007).

There are occasional records on baiji in the historical Chinese literature dating back to 200 B.C. (~2,200 years ago, Guo, 200 B.C.). The baiji was well observed by the ancient Chinese people and they could discriminate the precise differences between the baiji and Yangtze finless porpoise that co-inhabited the same river. However, the international scientific community didn't know this species until its scientific nomination by Miller in 1918 (Miller, 1918). No data was available on the abundance of baiji before the late 1970s, but we speculate that baiji had at one time been quite abundant in the Yangtze River evidenced by its description in ancient books, e.g., Er-Ya (Guo, 200 B.C.) and Ru-Fan (Li, 1874). The first systematic modern surveys of baiji were carried out during the late 1970s and early 1980s and provided the first population abundance estimate. Approximately 300 individuals were observed across their whole range (Lin et al., 1985; Chen & Hua, 1987, 1989) with about 100 individuals in the downstream section (Zhou & Li, 1989) in 1980s. Then the subsequent landmark surveys described a consistent rapid decline: ~200 individuals in 1990 (Chen et al., 1993), less than 100 individuals in 1995 (Liu et al., 1996) and zero individuals in 2006 and thus likely to be extinct (Turvey et al., 2007). Additional surveys (more regular) were conducted to monitor their abundance and look into their major threats (Akamatsu et al., 1998; Wang et al., 1998, 2000, 2006; Zhou et al., 1998a; Zhang et al., 2003; Wang et al., 2006).

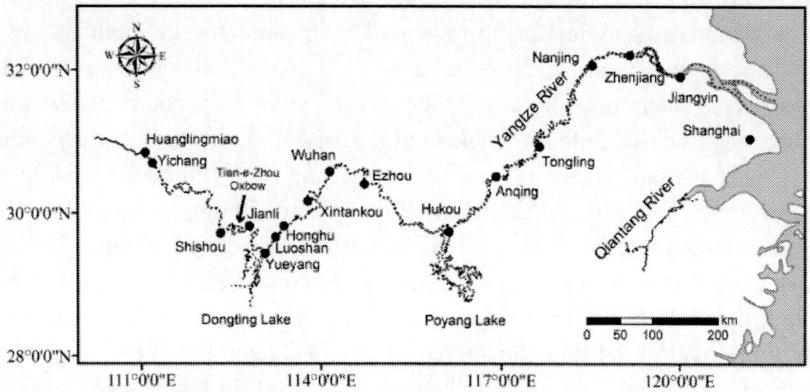

Figure 1 Historical distribution map of the baiji (dashed line and area in Yangtze and Qiantang Rivers and two lakes) and Yangtze finless porpoise (dashed line and area only in Yangtze River and two lakes). The positions of extant reserves related to Yangtze cetaceans are marked in the map.

The baiji's habitat continuously shrank and became fragmentary after its earlier abundance record in the 1950s. Baiji had widely inhabited the middle-lower Yangtze River drainage in 1970s (Figure 1, i.e. river section from Yichang to Shanghai and two appended lakes, the Poyang and Dongting, Zhou et al., 1977). It also once occurred in the Qiantang River, based on dozens of observations conducted in the river in 1955 (Zhou et al., 1977) (Figure 1). The most upriver sighting ever recorded in the Yangtze occurred in 1940s at Huanglingmiao, a small town ~30 km upriver Yichang (Zhou et al., 1977) (Figure 1). The density was very sparse and its distribution range was adversely reduced to a restricted section from Shishou to Zhenjiang in 1990s (Hua et al., 1995). Surveys in 1997-1999 suggested that the baiji has been extirpated from Poyang and Dongting lakes (Yang et al., 2000) and survived exclusively in a few sections (Honghu, Hukou - Tongling, Nanjing - Zhenjiang) with tiny population size (Zhang et al., 2003).

Yangtze Finless Porpoise

There was no systematic research on the Yangtze finless porpoise until late 1978 when the Baiji Research Collaboration Group was organized by the Chinese Academy of Science (CAS) whose members included the Institute of Hydrobiology, CAS, Nanjing Normal College (now re-named as Nanjing Normal University), the Institute of Acoustics, CAS, and the Institute of Biophysics, CAS. While this group mainly focused on the study of baiji at the beginning, information of the Yangtze finless porpoise was also collected throughout the study. Since the formation of this collaboration, the Institute of Hydrobiology, CAS and Nanjing Normal College began to survey in the river (Zhang et al., 1993; Zhou et al., 1998b; Wang et al., 1998, 2000; Xiao & Zhang, 2000, 2002; Yang et al., 2000; Wei et al., 2002). Most of the earlier surveys were conducted by a single survey ship, or a single survey ship with several small fishing boats, and no standard methods were applied. The first range-wide estimate of finless porpoise abundance in the Yangtze River system (~2700 porpoises) was based on many such kinds of small-scale surveys conducted between 1984 and 1991 (Zhang et al., 1993). Thereafter, fragmentary surveys in different sections of the Yangtze River were carried out by various researchers using essentially the same survey methods (Wang et al., 1998, 2000; Zhou et al., 1998b; Yang et al., 2000; Yu et al., 2001). During 1997 to 1999, a series of so-called "Synchronous Surveys", one in each year, were conducted by the Ministry of Agriculture and the Institute of Hydrobiology, CAS. For those surveys, the historic distribution ranges of baiji and the Yangtze finless porpoise within the middle and lower Yangtze River from Yichang to Shanghai, Poyang and Dongting Lake, and their main tributaries were divided into 21 sections (lengths varied from 50 - 200 km). Two large boats (~30 m long) simultaneously searched each section for one week during November of each year. Preliminarily analyses on the data collected showed that there were approximately 2,000 animals left in the river at the time the surveys were conducted (D. Wang, unpublished data; for the design of the surveys, please see Zhang et al., 2003). In November and December of 2006, a systematic survey was conducted in the entire current range of the population in the main stem of the Yangtze River (except for lakes Poyang and Dongting) by using a modified standard Line-transect Survey method which was pre-designed based on the results of a pilot survey between Wuhan and Yueyang (Figure 1) (Zhao et al., 2008). Both visual and acoustic methods were utilized in the survey (Akamatsu et al., 2008; Zhao et al., 2008) and experts and researchers from the United States, United Kingdom, Germany, Japan, Switzerland, Canada and China participated as part of an international collaborative effort. The findings of this extensive survey indicated that the finless porpoise population within the Yangtze's main stem is approximately 1,000 to 1,200 individuals. If the two lakes are included, the overall estimate of the population increases to approximately 1,800 (Zhao et al., 2008). This means that the current population size of the porpoise in the main stem of the river is less than half of the estimate (2,550) from surveys completed between 1984 and 1991 (Zhang et al., 1993), and it implies an annual rate of decline of at least 5% for the whole population in the main stem of the river (Zhao et al., 2008).

The Yangtze finless porpoise is now primarily restricted to the main river channel and its two largest appended lakes (Poyang and Dongting). It had occasionally occurred in some large adjacent tributaries of the river and lakes, but now has been extirpated from most of these areas (Zhang et al., 1993; Yang et al., 2000; Xiao & Zhang, 2002). Of the six extant species of porpoise (Phocoenidae), this is the only population found in fresh water (Gao & Zhou, 1995). The amount of river and lake habitat available to this subspecies is relatively small compared to that available to marine populations of finless porpoises, which occur in coastal waters from Japan to the Arabian Sea (Kasuya, 1999). Based on the finding of a range-wide survey in 2006, most porpoises are concentrated in the middle and lower reaches from Ezhou to Jiangyin, with the lowest densities in the upper region and the estuaries of the Yangtze River (Zhao et al., 2008). The current distribution pattern is almost the same as what Zhang et al. (1993) reported and the porpoises in the upper region from Yichang to Ezhou (~130 porpoises in 716.4 km) appear to be at the highest risk of local extirpation. The observed density in this region decreased from 0.11 porpoises/km in 1991 (Zhang et al., 1993) to 0.02 porpoises/km in 2006 (Zhao et al., 2008). Moreover, there appeared to be significant gaps in the distribution in this part of the river, since no porpoises were detected during either the upstream or the downstream passes by the two survey-boats in the 150 km subsection between Yueyang and Shishou in 2006 (Zhao et al., 2008). Despite the possibility of false negatives in determining the presence of finless porpoises in that study, the number of porpoises in this region must be extremely low or nil. The ~90 km subsection upstream of this gap included the most-upriver population (roughly 60 porpoises, Zhao et al., 2008). If the porpoises in this subsection were to become extirpated, the linear extent of the recent historical range of this subspecies on the river would shrink by ~400 km, or by about 24% of the whole range in the main stem of the river (Zhao et al., 2008). It may be noteworthy that this was also the river section where the baiji were first eliminated (Zhou et al., 1977; Chen, et al., 1997; Zhang et al., 2003). Although limited photo-identification studies suggested that baiji traveled over hundreds of kilometers up and down the river (Zhou et al., 1998a), the significantly different patterns of mtDNA haplotypes among finless porpoises in different sections of the Yangtze River implies that these animals do not move far (Zheng et al., 2005). This means that even if all threats were eliminated and habitat conditions improved, there is little chance that porpoises from other areas would repopulate the upper region of the Yangtze River below Yichang and above Yueyang. Therefore, unless the current trend is reversed, there seems to be a good chance that finless porpoises will soon disappear permanently from that area. In the middle and lower regions between Wuhan and Jiangyin, the porpoise distribution appeared continuous but their abundance decreased from the (presumably underestimated) level of 1,652 (surveys of 1984 to 1991, Zhang et al., 1993) or 1,481 (surveys of 1989 to 1992, Zhou et al., 1998b) to the current level of ~800 (Zhao et al., 2008).

THREATS AND CONSERVATION

A number of anthropogenic factors are known or suspected to be responsible for the population decline and range contraction of the Yangtze cetaceans. For example, Chen et al. (1997) reported that among 64 baiji specimens collected (33 were collected from 1973 to 1983 in middle reaches from Yichang to Hukou, and 31 were collected from 1978 to 1985 in lower reaches from Hukou to Shanghai) (Figure 1), 53 were the result of different kinds of human activities, use of harmful fishing gears, boat collisions, and explosives used to widen and deepen the shipping channel. Since baiji and the Yangtze finless porpoise share the same river and almost the same habitat, the porpoise must have been facing the same kind of threats as that of the baiji.

Turvey et al. (2007) concluded that entanglement in gear used in unregulated and unselective fishing (rolling hooks, electrofishing gear and gillnets) was the main factor responsible for the probable extinction of the baiji. This same factor likely explains much of the ongoing decline of the Yangtze finless porpoise (Wang et al., 1998, 2000, 2005; Wang et al., 2006). Illegal fishing is widespread in the Yangtze River (Reeves et al., 2000b; IWC, 2001; Smith et al., 2007) and was observed daily during a rang-wide survey in 2006 (Turvey et al., 2007). Zhou & Wang (1994) reported that 'most' of the 80 finless porpoise specimens collected by Nanjing Normal University since 1974 had been killed by rolling hooks or gillnets. Other studies indicate that bycatch in gillnets is adversely impacting marine populations of finless porpoises (Jefferson & Curry, 1994; Zhou et al., 1995; Reeves et al., 1997; Yang et al., 1999). Because the preferred habitat of the Yangtze finless porpoise overlaps extensively with gillnetting areas in the river (Yu et al., 2005), the impact of gillnet mortality may be much more serious than has been generally assumed based on the infrequency of actual reports.

Boat traffic, which is increasing rapidly in the Yangtze River and lakes, also likely causes mortality of cetaceans (from propeller strikes) and boat noise may mask their social communication and ability to forage efficiently (Wang et al., 1998, 2000; Wang et al., 2006). Widespread mining of the river bed, lake beds and banks (much of it is illegal) has been destroying important habitat of the porpoise's prey and adversely affecting primary productivity. This problem is especially serious in Poyang Lake, currently with a population of around 400 finless porpoises (Xiao & Zhang, 2000; Wang et al., 2006; Zhao et al., 2008). Compared with cetaceans that live in marine habitats, riverine forms may be at a higher risk from pollution. Indeed, cetaceans in rivers generally occur in the world's most densely populated human environments (Reeves et al., 2000a). Four hundred million people live in the Yangtze River basin and thousands of factories along the river bank discharge tremendous quantities of domestic sewage and industrial effluents. Furthermore, because rivers are relatively small water bodies, their water quality can be degraded much more easily than larger water bodies, such as the oceans are. However, relatively few data exists which assesses the impacts of pollutants on Yangtze finless porpoise health, fertility or population status. In April 2004, five porpoises died in Dongting Lake within one week, apparently due to the combination of a short-term exposure to the pesticide hostathion and a long-term exposure to mercury and chromium (D. Wang, unpublished data). Finally, water development projects, especially dams, have major effects on river ecology. In the Yangtze River system, structures can block porpoise movements between the river and adjoining lakes or tributaries (Liu et al., 2000; Smith & Reeves, 2000), as well as the movements of their prey (Xie & Chen, 1996). The Three Gorges Dam in particular has altered and will continue to alter downstream hydrologic conditions in the Yangtze River (Tong et al., 2008), and consequently, may adversely affect the habitat of the baiji and finless porpoises, in the river.

Although the relative importance of each of the above threats has not been quantified, all have contributed to the decline of the Yangtze finless porpoise. And despite the fact that for many years these same factors were also known to be pushing the baiji towards likely extinction, none has been aggressively or seriously addressed and most of them have escalated dramatically over recent decades. Consequently, we must reiterate that immediate action is urgently needed to reduce the threats, with highest priority given to areas with greatest abundance in all regions (see above).

PROGRESS OF CONSERVATION

On the first Workshop on Biology and Conservation of the Platanistoid Dolphins held at the Institute of Hydrobiology of CAS, Wuhan on October, 1986, Chen & Hua (1989) proposed three measures for protecting baiji: 1. in situ conservation by establishing natural refuges in the river; 2. ex situ conservation by establishing semi-natural reserves in some oxbows or other places; and 3. Intensifying captive breeding studies and establishing captive colonies. At the same meeting, Zhou & Li (1989) also suggested that it was of urgent need to set-up protection measures for the development of breeding colonies in semi-natural reserves. Since the Yangtze finless porpoise has been facing the same kind of threats as the baiji has, and is also very much endangered, the applicability of these three measures on the conservation of the porpoises were discussed at Workshop to Develop a Conservation Action Plan for the Yangtze River Finless Porpoise, held at Hong Kong in 1997 (Reeves et al., 2000a). On the Second Meeting of the Asian River Dolphin Committee which was held in Bangladesh on February, 1997, Wang et al. (2000) made three further recommendations: 1. establish a breeding group of the Yangtze finless porpoise in the Shishou Baiji Semi-natural Reserve; 2. Establish more natural reserves, such as in the mouth areas of Poyang and Dongting Lakes and adjacent waters in the Yangtze River; and 3. Carry out breeding programs in captivity. Then, a Conservation Action Plan for Cetaceans in the Yangtze River was developed by scientists of the Institute of Hydrobiology of CAS, and was approved by the Chinese government in 2001 (MOA-China, 2001). This plan emphasized the importance of protecting the Yangtze finless porpoise, and proposed that the three measures identified at the 1986 workshop should also be carried out in the protection of the Yangtze finless porpoise. The Chinese government and scientists have been pushing forward to carry out these three measures since then. Here, follows an updated summary of the work completed, encountered difficulties, and overall progress.

Progress and Difficulty of In Situ Protection

In 1992, the first two national baiji reserves were established. One is called Honghu Xin-Luo Baiji National Natural Reserve, which is a 135 km section of the Yangtze River between Xintankou and Luoshan located in Honghu City of Hubei Province (Figure 1). The second is Shishou Tian-e-Zhou Baiji Natural Reserve, which includes an 89 km section of the Yangtze River in Shishou and a 21 km long Tian-e-Zhou Oxbow connected with this section (Figure 1). Baiji and the Yangtze finless porpoise are two main protected target animal species for these two reserves. In 1996, the Ministry of Agriculture of China organized a workshop on the conservation measures that targeted the baiji and the Yangtze finless porpoise. Another five so called protecting stations were set up in Jianli, Chenglingji (a small town nearby Yueyang), Hukou, Anqing, and Zhenjiang (Figure 1). Yueyang City set up a local reserve in east Dongting Lake in 1996 which covers a 66,700 ha area (Figure 1). A provincial Yangtze freshwater cetacean natural reserve located in Tongling section of Anhui Province was established in 2000, and upgraded to a national reserve in 2006. It covers a 58 km river section in Tongling City (Figure 1). Zhenjiang Protecting Station was upgraded as a provincial reserve in 2003, which covers approximately a 15 km river section located in a side channel of the river in Zhenjiang (Figure 1). A provincial Poyang Lake Yangtze Finless Porpoise Reserve was established in 2004, which covers an area of 8,600 ha. in the lake (Figure 1). Anqing Protecting Station was upgraded as a local (city) reserve in 2007, which covers the total 243 km section of the river in Anqing.

By now, most of the areas or sections of the Yangtze River and two lakes with relatively high density of the baiji and Yangtze finless porpoise are covered by these reserves. But, the Yangtze River basin is also the most densely populated area for humans, approximately 40%

of Chinese people living in the basin – home to approximately 10% of the world's human population. The Yangtze River is also called "golden channel" of the country, which means it plays very important role for development of the country. Although reserve management staff may try very hard to lessen harmful human activity impacts on the baiji and the Yangtze finless porpoise they are overwhelmed because many of these human activities are still on-going, and worse, expanding on a great scale. For example, transportation through the Three Gorges Dam was 147,500,000 tons (t) in 2003, and it reached 439,300,000 t in 2005 (Yi et al., 2007), the number tripled in three years. The number of boats in the river has increased approximately five-fold since the late 1980s (Wang et al., 2006). Futhermore, during a survey between Yichang and Shanghai in 2006, a minimum of 19,830 large shipping vessels were counted, which translates into more than one ship per hundred meters of river (Turvey et al., 2007). Because of over fishing and habitat loss, fish production of the Yangtze River has been decreasing remarkably (Wei et al., 2007). In contrast, the influx of sewage into the Yangtze River has significantly increased from 9,500,000,000 t/a at the end of the 1970s, to 15,000,000,000 t/a at the end of the 1980s, and it reached 29,640,000,000 t/a in 2005 (Wu & Tu, 2007). While protective regulations for the baiji and Yangtze finless porpoise and their habitats are in place, effective enforcement is an immense problem in such a huge river, in a densely populated area of a developing country. For example, even though some harmful fishing gears are listed as illegal, they have never-the-less been used frequently (Turvey et al., 2007). Therefore, these reserves may help to slow down the process of extinction of both the baiji and the Yangtze finless porpoise, yet they cannot prevent the occurrence of harmful human activities. Unfortunately, the success of in situ conservation is highly limited (Wang et al., 2006; Turvey et al., 2007).

Establishment of Semi-Natural Protected Populations

The conditions in the Yangtze River are considered highly unlikely to be improved in the foreseeable future, which make the outlook for the barely surviving baiji and finless porpoise populations in the river bleak. We have to seek some other ways to help the porpoise before they become extinct. As early as in middle 1980s, our research group started to search for a place to set-up semi-natural reserve to establish an ex situ protected population of the porpoise. Tian-e-Zhou Oxbow (Figure 1), an old course of the Yangtze River, lies in the north bank of the river in Shishou County, Hubei Province of China. This oxbow used to be a section of the Yangtze River, and was cut off from the main stem of the river naturally in 1972. It is approximately 21 km long and 1 - 2 km wide. Zhang et al. (1995) made a systematic investigation on its water quality, biological productivity, and fish production etc., and concluded that the oxbow is ideal as a semi-natural habitat for the finless porpoise. The first group of 5 finless porpoise, 3 females and 2 males, were captured in the Yangtze River, and released in the oxbow in 1990 (Table 1). Since then, several more groups of Yangtze finless porpoises have been captured or rescued from the river and also transplanted into the oxbow. The animals have been left to live in the oxbow freely without the intervention of any factitious variable. For example, no artificial feeding is needed. The result confirms that these animals can not only survive, but can also reproduce naturally and successfully in this reserve. Approximately two calves are born each year, with at least 29 babies born in the reserve by the end of 2007 (Table 1). Accounting that some animals have moved, died naturally or died accidentally, there were approximately 30 individuals living in the reserve at the end of 2007 (Table 1). In early spring of 2008, a huge, long time lasting snow storm swept through southern China, causing the Tian-e-Zhou oxbow to be almost completely covered by ice which had never previously happened. Five porpoises were confirmed died because of wounds caused by ice when they were trying to break it to breathe. By a rescue operation for treating wounded animals in April 2008, five of eight matured females were confirmed pregnant. Currently if adding these five new born calves, presumably more than 27 individuals are living in the area (Table 1). "Thus, a viable population capable of breeding and expanding has been established. This effort represents the world's first attempt and a successful example of ex situ preservation of a cetacean species" (Wang et al., 2005). As Braulik et al. (2006) pointed out "China's successful program of capture, translocation and maintenance of finless porpoise in the Shishou oxbow has demonstrated the oxbow's adequacy as an ex situ environment for cetaceans". The successful story of Shishou Tian-e-Zhou Reserve has shed light on the protection of the Yangtze finless porpoise.

One other smaller scale semi-natural reserve was set up in Tongling of Anhui Province in 1994 (Figure 1). This reserve is located in a small channel (1.6 km long, and 80 - 220 m wide) between two sandbars of the Yangtze River. A small group of 5 porpoises were introduced into the channel in 2001, and one calf was born there in 2003, 2005, 2006, 2007 and 2008 respectively (Wenhua Jiang, personal communication).

Table 1. Establishment of the Yangtze finless porpoise breeding colony in the Tian-e-Zhou National Natural Reserve. Pregnant females in the Yangtze (+) and reserve (++) are noted as well as the least confirmed population size of the current colony (*). This number reflects 13 males and 9 females (8 matured females), five of which were confirmed pregnant in April 2008

Dates	Source Location	No. of porpoises introduced		No. of porpoises born in the reserve	No. of Loss	Loss Reasons	Remained
		F.	M.				
Mar, 1990	Chenglingji	3	2		--	--	5
Spring, 1990	--	--	--	2+	--	--	7
Apr, 1990	--	--	--	--	2	2 deaths/One infant was killed accidentally by rolling hooks, one male died on April 25, 1990 from injuries during capture.	5
Spring, 1992	--	--	--	1++	--	--	6
May 28, 1992	--	--	--	--	1	1 death/ One male was killed accidentally by rolling hooks.	5
Apr, 1993	Chenglingji	3	2	3+	--	--	13
Apr, 1993	--	--	--	--	1	1 death/One infant was found dead on April 26, 1993, born prematurely due to capture.	12
Oct 18,1993	--	--	--	--	7	7 deaths/seven killed accidentally.	5
May, 1995	Chenglingji	1	2	1+	--	--	9
Dec 6, 1995	Chenglingji	2	2	--	--	--	13
Apr 20, 1996	Jianli	3	2	2+	--	--	20
Jun - Aug, 1996	--	--	--	--	14	14 escaped into the Yangtze river	6
Dec, 1996	Chenglingji & Shishou	5	9	--	--	--	20
Spring, 1997	--	--	--	--	15	15 released into the Yangtze river	5

Table 1. Continued

Dates	Source Location	No. of porpoises introduced		No. of porpoises born in the reserve	No. of Loss	Loss Reasons	Remained
		F.	M.				
Autumn, 1997	--	--	--	2++	--	--	7
Spring, 1998	--	--	--	1++	--	--	8
Dec, 1998	Shishou	2	1	--	--	--	11
Apr, 1999	--	--	--	1++	--	--	12
Dec, 1999	Shishou & Jiayu	2	4	--	--	--	18
Dec, 1999	--	--	--	--	1	1 translocated to Wuhan Baiji Dolphinarium at Institute of Hydrobiology, CAS	17
Spring, 2000	--	--	--	2++	--	--	19
Spring, 2001	--	--	--	1++	--	--	20
Spring, 2002	--	--	--	2++	--	--	22
Jun, 2002	--	--	--	--	5	5 death/ one found died naturally and four killed accidentally by capture operation.	17
Spring, 2003	--	--	--	1++	--	--	18
Nov, 2003	Shishou	3	--	--	--	--	21
Jan, 2004	Honghu	1	--	--	--	--	22
Spring, 2004	--	--	--	1+2++	--	--	25
Oct, 2004	--	--	--	--	1	1 translocated to Wuhan Baiji Dolphinarium at Institute of Hydrobiology, CAS	24
Spring, 2005	--	--	--	2++	--	--	26
Oct, 2005	--	--	--	--	2	2 deaths/two killed accidentally by capture operation.	24
Spring, 2006	--	--		2++	--	--	26
Spring, 2007	--	--	--	3++	--	--	29
Apr, 2008	--	--	--		6	6 deaths/five killed by ice, and one died naturally	> 22*(the number was confirmed the least one by capture operation for physical check)
End of 2008	--	--	--	5++	--	--	> 27 (Presumably, the five pregnant females gave births successfully, and the calves survive)
Total	--	25	24	34	55	--	> 27

Progress of Captive Breeding Program

As one of the protection measures, captive breeding plays an important role for understanding the animals, particularly with regards to their breeding biology, to help conservation work in the wild.

As early as in 1980s, the Institute of Hydrobiology, CAS started carrying out the baiji captive breeding program. One male baiji, named QiQi, was rescued when it was stranded on Jan 11th, 1980 at the mouth of Dongting Lake and was relocated to the Institute's aquarium where it lived for almost 23 years. QiQi was badly injured by fishing hooks when a fisherman tried to capture him. He became stranded and developed an infection, but miraculously, QiQi recovered gradually after the creative and careful treatment by the Institute's staff using both Chinese and Western medicines (Liu & Lin, 1982). To carry out a breeding program in captivity, scientist tried to find a female dolphin to couple with QiQi, but this was constantly a problem owing to the extremely low density of wild population. It wasn't until March of 1986, that a young female baiji, named ZhenZhen and another adult male LianLian, presumedly ZhenZhen's father, were captured from the river and then translocated immediately into aquarium. LianLian was very weak when it was captured because of sickness presumably, and died 203 days later, and sadly ZhenZhen also died in September of 1988 because of pneumonia before reaching sexual maturity (Chen et al., 1994; Chen et al., 1997). In March, 1981, a stranded female baiji was found in Taicang, Jiangsu Province and then translocated to Nanjing Normal University (Braulik et al., 2006), but died 17 days later because of multiple organ failure. In April of the same year, a stranded male baiji RongRong was rescued in the middle reaches of the river and then reared in the Institute of Hydrobiology, CAS for 228 days (Chen et al., 1986). One other male dolphin was accidentally captured by rolling hooks in Zhenjiang, Jiangsu Province in this year, and then moved to the Nanjing Fishery Institute where it survived for 129 days. Through rearing and researching on captive baiji, scientists obtained valuable experiences and knowledge including animal behavior, acoustic, physiology, diagnosis and treatment of common diseases, hematology, and breeding biology. All of these should have greatly promoted the preservation of baiji if more individuals could have been recruited into either ex situ or captive breeding populations.

Yangtze finless porpoises were first reared in captivity in China back to the mid-1960s, but most of the animals only survived for very short times in pools, usually less than one year (Liu et al., 2002). The Baiji Dolphinarium, a new facility for rearing baiji and the porpoise, was established in 1992 at the Institute of Hydrobiology, CAS in Wuhan. The first two Yangtze finless porpoises, one 1.5 years old male and one 1.5 years old female, were captured from the Yangtze River, and introduced into the Dolphinarium's in-door pools at the end of 1996. One other 1.5 year old female and one other adult male were introduced into it from the Tian-e-Zhou oxbow in 1999 and 2004, respectively. Since then, a good deal of research has been conducted in the Dolphinarium on their rearing, behavior, acoustics, physiology and breeding biology (e.g., D. Chen et al., 1997, 2005; P. Chen et al., 1997; Akamatsu et al., 1998; Wei et al., 2004; Popov et al., 2005, 2006; Li et al., 2007, 2008). All of the individuals, except for one female introduced in 1999 and who died accidentally in 2007, are now in good health within the facility. Both individuals introduced in 1996 have survived in captivity for almost 13 years. This success marks great progress of rearing the Yangtze finless porpoise in captivity as no such achievement has been reached previously (Liu et al., 2002; Wang et al., 2005).

For having the porpoises successfully bred in captivity, their physiology status, such as cycles of serum reproductive hormones have to be fully understood. We used to take blood samples for physical status examination of the animals monthly, but this sampling rate is insufficient for appropriately monitoring hormone levels. Therefore, we undertook veterinary training and became proficient in the collection of feces, mouth saliva and blowhole secretion everyday or even during every feeding time. We also established a laboratory protocol that used feces to evaluate cycles of serum reproductive hormones. Meanwhile, their growths were monitored monthly and behavior was observed daily. All of these results indicated that the two animals that arrived in 1996 reached sexual maturity in 1999, and the animal that arrived in 1999 reached sexual maturity in 2002. They started to mate at a very early age but no confirmed copulation was recorded (Wei et al., 2004). Beginning in 2004, we physically separated the females and the male for a short period in different two pools that are connected by a water channel with a fence between them so they could communicate to each other through this channel prior to ovulation without physical touching, and cancelled routine physical examinations during ovulation in order to avoid disturbing the animals during this sensitive period. We also petted the female's genital regions to stimulate sexual behaviors in the females, making the females more accessible to the males during the breeding season (Wang et al., 2005). After the males and females were reintroduced into the same pool, the younger female became pregnant (introduced in 1999) and later gave birth to a male on July 5th, 2005. This represented the first freshwater cetacean ever born in captivity in the world (Wang et al., 2005). This baby porpoise is still alive in captivity and in good health. On June 2, 2007, the same female gave birth to another male. Unfortunately, this female ate some cast from the pool wall in which she lived, and consequently died 39 days later. Her second baby also died 11 days later, even when we tried to feed him mixed milk. On July 5, 2008, the elder female who was introduced in our pools in 1996, gave a birth to another male baby. But for some unknown reason, she did not excrete milk for nursing, and the baby died 5 days later.

FUTURE PROSPECT OF CONSERVATION

Since China is still on its route of fast economic development, we cannot expect that the Yangtze River's environment is going to improve in the near future, and it may get worse. Under severe impacts caused by human activities in the Yangtze River, baiji is likely extinct (Turvey et al., 2007). What we should do to prevent the Yangtze finless porpoise to become the second baiji? Are there any ways to prevent this tragedy from happening again?

In general, in situ protection always comes first as a choice of conservation measures. Even though the habitat of the Yangtze River has been degrading, we should first explore every possibility to protect the Yangtze finless porpoise in its natural habitat. Fortunately, we still have a relatively good number of the animals in the river and lakes, so it may provide us a good base for us to carry out some measures to protect them.

Overfishing and illegal fishing of baiji and the Yangtze finless porpoise's prey are blamed as some of the main reasons for causing the decline of both species (e.g.,D Wang et al., 1998, 2000, 2005, 2006; K Wang et al., 2006). Meanwhile, fish production of the Yangtze River has been decreasing from its high of 427,000,000 kg in 1954 to approximately 100,000,000 kg in recent years, even with a much more intense fishing effort (Wei et al., 2007). This warns that fish resources of the river are almost dried up. But, annual freshwater fish aquaculture production of the whole country is quite high at approximately 21,000,000,000 kg (Wei et al., 2007) in recent years. This means that fish production of the Yangtze River plays a relatively minor role for the fishery economic development. On the other hand, fish germplasm of the Yangtze River is the best overall in the country (Wei et al., 2007). For protecting fish resources in the river, the Chinese government has been prohibiting any fishing activity in the middle and lower reaches of the Yangtze River from April 1 to the end of June each year since 2003. Even though this measure may have improved the status of fish resources in the river (Wei et al., 2007), it is still far away from solving the problem, as fisherman may just simply spend more time and much effort on fishing right after the period to compensate their loss during fishing ban period and because there may be more fish for fishing, and any improvement of fish resources could be destroyed right away. For protecting fish resources in the river to benefit aquiculture development and Yangtze cetacean protection, we suggest that fishing should be forbidden year-round in the whole river. In the least, fishing should be forbidden in each reserve. Furthermore, because the disconnection between river and lakes within the middle and lower reaches of the Yangtze River has directly resulted in decreasing of fish recourses (Wei et al., 2007), re-establishing linkage between the Yangtze River and its appended lake clusters could greatly improve the habitat status of fish resources of the river, which could greatly help the conservation of the Yangtze finless porpoise.

We already established some natural reserves in the river and lakes that cover almost every hot spot of the animal distribution (Figure 1). But, most of the reserves are in many challenging areas since the river is being used by many kinds of human activities, and they can do little for managing most of them. For example, we can’t expect to stop transportation in the river that is blamed to be very harmful for the baiji and the porpoise (Chen et al., 1997; D. Wang et al., 1998, 2000). In this case, some regulations have to be worked out and put into practice to at least control navigation. We suggest that the speed of every ship passing the reserve should be limited, possibly below 10 km/h and that blasting cannot be used to deepen and widen the shipping channel in the reserve.

The demonstration of Tian-e-Zhou reserve proves that ex situ is a possible way to establish a sustainable population of the Yangtze finless porpoise. It provides a possibility that we could establish additional off site protected populations of the porpoise for assuring long term survival of it in nearly natural habitats, such as in other similar oxbows of the Yangtze River. A systematic survey should be done soon to investigate these sites to select some as other semi-natural reserves for the porpoise. Meanwhile, after the Three-Gorges Dam was finished, the water current above the dam is much slower than it was before which was the main restriction to effectively block the porpoise from the upper reaches of the Yangtze River. We suggest that the huge reservoir above the dam should be explored for the possibility of establishing a population of the Yangtze finless porpoise. Should this occur, it could provide another reliable solution for saving the Yangtze finless porpoise.

Some progresses have been made on captive breeding. Even we cannot expect that captive breeding can solve all of the problems (Wang et al., 2005), we should consider expanding the captive colony to establish a possible sustainable group. While doing so, much more research in the areas of rearing biology, breeding biology, physiology, behavior and acoustics can be carried out in captive animals to help conservation in the wild. After a sustainable group is established in captivity, we may establish an exchange program of individuals of the porpoise between captive and off site protected populations, even wild ones.

For effectively protecting the baiji and the porpoise, a network that is composed by governmental administrations, reserves, and research institutions was just organized by the Ministry of Agriculture of China and Institute of Hydrobiology, CAS. It is our hope that this network promotes conservation efforts concerning the baiji and porpoise by serving as a platform to exchange information, train staff, organize surveys, and educate the public.

We have to point out that most of the measures we proposed above have been repeated for many times in workshops, published papers and reports to the government, but they have received little attention and little progress has been made for carrying them out. Most of the threats are still present and at least some of them are getting worse. Under the pressure of rapid economic development, perhaps the best thing for the government to do could be to seek a balance between development and conservation. But, development almost always comes as a priority when there is conflict between them in a developing country like China. In this type of situation, no matter what research-based conservation suggestions are put forward, conservation results will likely be limited and most likely will be nothing more than "conservation on paper" (for example, see Bearzi, 2007). Will of government agencies and care and support of public are the two keys for any possible success of any conservation program. Eventually, we have to ask ourselves if we are prepared to lose one more mammal species in the Yangtze River. Are we? The Yangtze finless porpoise may be the only one left in the river since we may have already lost the baiji. Can we really afford the cost of losing them and eventually the whole biodiversity of the river? Our hope is that the international community has learned a lesson from the baiji tragedy and will react accordingly (in post-haste) to remediate the Yangtze River, save and improve its biodiversity, and protect the finless porpoise.

ACKNOWLEDGMENTS

The writing of this paper is supported by National Basic Research Program of China (2007CB411600), National Natural Science Foundation of China (30730018), and the President's Fund of the Chinese Academy of Sciences.

REFERENCES

[1] Akamatsu T., Wang D., Nakamura K. & Wang K. (1998). Echolocation range of captive and free-ranging baiji (*Lipotes vexillifer*), finless porpoise (*Neophocaena phocaenoides*), and bottlenose dolphin (*Tursiops truncatus*). *Journal of the Acoustical Society of America,* 104, 2511-2516.

[2] Akamatsu T., Wang D., Wang K., Li S., Dong S., Zhao X., Barlow J., Stewart B. & Richlen M., (2008). Estimation of the detection probability for Yangtze finless

porpoises (*Neophocaena phocaenoides asiaeorientalis*) with a passive acoustic method. *Journal of the Acoustical Society of America,* 123, 4403-4411.

[3] Baillie J. & Groombridge B. (1996). 1996 International Union for Conservation of Nature Red List of threatened animals, Gland, Switzerland: *International Union for Conservation of Nature,* .

[4] Bearzi G., (2007). Marine Conservation on paper. *Conservation Biology,* 21, 1-3.

[5] Braulik G. T., Reeves R. R., Wang D., Ellis S., Wells R. S. & Dudgeon D. (2006). Report of the workshop on conservation of the baiji and Yangtze finless porpoise. Zurich, Switzerland Baiji Foundation.

[6] Chen D., Zhao Q. & Liu R., (1994). Preldiminary study on some serum hormones of female and male baiji dolphin, *Lipotes vexillifer* in captivity. *Acta Hydrobiologica Sinica,* 18, 292-295.

[7] Chen D., Zhao Q. & Liu R., (1997). Preliminary study on some hormones of *Neophocaena phocaenoides* in the Yangtze River. *Acta Theriologica Sinica,* 17, 43-47.

[8] Chen D., Hao Y., Zhao Q. & Wang D., (2005). Reproductive seasonality and maturity of male *Neophocaena phocaenoides asiaeorientalis* in captivity: A case study based on the hormone evidence. *Marine Freshwater Behavior and Physiology,* 39, 163-173.

[9] Chen P., Liu P., Liu R. & Lin K., (1986). Studies on the rearing of *Lipotes. Acta Hydrobiologica Sinica,* 10, 128-135.

[10] Chen P. & Hua Y., (1987). Projected impacts of the Three Gorges Dam on the baiji, *Lipotes vexillifer*, and needs for conservation of the species. In Anonymous (Ed.), *A collection of articles on the impacts of the Three-Gorges Dam project on aquatic ecosystem along the Changjiang and research on their countermeasures* (pp. 30-41). Beijing, China: China Science Press.

[11] Chen P. & Hua Y., (1989). Distribution, population size and protection of *Lipotes vexillifer*. In K. Zhou & J. Liu (Eds.), *Biology and conservation of the river dolphins.* International Union for Conservation of Nature *SSC* (Occasional Paper No. 3, pp. 81-85). Gland, Switzerland: *International Union for Conservation of Nature.*

[12] Chen P., Zhang X., Wei Z., Zhao Q., Wang X., Zhang G. & Yang J., (1993). Appraisal of the influence upon baiji, *Lipotes vexillifer* by the Three-Gorge Project and conservation strategy. *Acta Hydrobiologica Sinica,* 11, 101-111.

[13] Chen P., Liu R., Wang D. & Zhang X., (1997). *Biology, rearing and conservation of Baiji.* Beijing, China: Science Press, Beijing.

[14] Dudgeon D., (2005). Last chance to see . . .: ex situ conservation and the fate of the baiji. *Aquatic Conservation: Marine and Freshwater Ecosystems,* 15, 105-108.

[15] Gao A. & Zhou K., (1995). Geographical variation of external measurements and three subspecies of *Neophocaena Phocaenoides* in Chinese waters. *Acta Theriologica Sinica,* 15, 81-92.

[16] Guo P., (200 B.C.) Er-Ya.

[17] Hua Y., Zhang J., Zhang X. & Xu X., (1995). Research on the status quo and endangering factors of baiji, *Lipotes vexillifer*, and the strategy for its protection. *Resources and Environment in the Yangtze Valley,* 4, 45-51.

[18] IWC, (2001). Annex K Report of the standing sub-committee on small cetaceans. *Journal of Cetacean Research and Management (supplement),* 3, 263-291.

[19] Jefferson T. A. & Curry B. E., (1994). A global review of porpoise (Cetacea: Phocoenidae) mortality in gillnets *Biological Conservation,* 67, 167-183.

[20] Kasuya T., (1999). Finless porpoise *Neophocaena phocaenoides* (G. Cuvier, 1829). In S. H. Ridgway & R. Harrison (Eds.), *Handbook of Marine Mammals Volume 6: The Second Book of Dolphins and the Porpoises* (pp. 411-442). San Diego, CA: Academic Press.

[21] Li S. H., Wang D., Wang K. X., Akamatsu T., Ma Z. Q. & Han J. B., (2007). Echolocation click sounds from wild inshore finless porpoise (*Neophocaena phocaenoides sunameri*) with comparisons to the sonar of riverine *N. p. asiaeorientalis*. *Journal of the Acoustical Society of America,* 121, 3938-3946.

[22] Li S. H., Wang K. X., Wang D., Dong S. Y. & Akamatsu T., (2008). Simultaneous production of low- and high-frequency sounds by neonatal finless porpoises. *Journal of the Acoustical Society of America,* 124, 716-718.

[23] Li Y., (1874). Ru-Fan.

[24] Lin K., Chen P. & Hua Y., (1985). Population size and conservation of *Lipotes vexillifer. Acta Ecologica Sinica,* 5, 77-85.

[25] Liu R. & Lin K., (1982). Treatment of injury of Chinese River dolphin, *Lipotes Vexillifer. Oceanologia Et Limnologia Sinica,* 13, 548-552.

[26] Liu R., Zhang X., Wang D. & Yang J., (1996). Once again studies on the conservation of *Lipotes vexillifer* and *Neophocaena phocaenoides*. *Resources and Environment in the Yangtze Valley,* 5, 220-225.

[27] Liu R., Wang D. & Zhou K., (2000). Effects of water development on river cetaceans in China. In R. R. Reeves, S. Brian & K. Tasuya (Eds.), *Biology and conservation of freshwater cetaceans in Asia*. International Union for Conservation of Nature *SSC* (Occasional Paper No. 23, pp. 40-42). Gland, Switzerland and Cambridge, United Kingdom: *International Union for Conservation of Nature*.

[28] Liu R., Wang K. & Zhao Q., (2002). Rearing of cetaceans in captivity in China. *Acta Theriologica Sinica,* 22, 130-135.

[29] Miller C. M. J., (1918). A new river-dolphin from China. *Smithsonian Miscellaneous Collections,* 68, 1-12.

[30] MOA-China, (2001). Conservation *Action Plan for Cetaceans of the Yangtze River*. Ministry of Agriculture, People's Republic of China.

[31] Nikaido M., Matsuno F., Hamilton H., Robert L. Brownell J., Cao Y., Wang D., Zhu Z., Shedlock A. M., R. Ewan Fordyce, Hasegawa M. & Okada N., (2001). Retroposon analysis of major cetacean lineages: the monophyly of toothed whales and the paraphyly of river dolphins. *Proceedings of the National Academy of Sciences,* 98, 7384-7389.

[32] Popov V., Supin A., Wang D., Wang K., Xiao J. & Li S., (2005). Evoked-potential audiogram of the Yangtze finless porpoise *Neophocaena phocaenoides asiaeorientalis*. *Journal of the Acoustical Society of America,* 117, 2728-2731.

[33] Popov V., Supin A., Wang D. & Wang K., (2006). Nonconstant quality of auditory filters in the porpoises, *Phocoena phocoena* and *Neophocaena phocaenoides* (Cetacea, Phocoenidae). *Journal of the Acoustical Society of America*, 119, 3173-3180.

[34] Reeves R. R., Wang J. Y. & Leatherwood S., (1997). The finless porpoise, *Neophocaena phocaenoides* (G. Cuvier 1829): a summary of current knowledge and recommendations for conservation action. *Asian Marine Biology,* 14, 111-143.

[35] Reeves R. R., Jefferson T. A., Kasuya T., Smith B. D., Wang D., Wang P., Wells R. S., Würsig B. & Zhou K., (2000a). Report of the workshop to develop a conservation

action plan for the Yangtze River finless porpoise. In R. R. Reeves, B. D. Smith & T. Kasuya (Eds.), *Biology and conservation of freshwater cetaceans in Asia. Occasional Paper of the* International Union for Conservation of Nature *Species Survival Commission* (No. 23, pp. 67-80). Gland, Switzerland and Cambridge, United Kingdom: *International Union for Conservation of Nature.*

[36] Reeves R. R., Smith B. D. & Kasuya T., (2000b). Biology and conservation of Asian river cetaceans. In R. R. Reeves, B. D. Smith & T. Kasuya (Eds.), *Biology and conservation of freshwater cetaceans in Asia. Occasional Paper of the* International Union for Conservation of Nature *Species Survival Commission* (No. 23, IUCN, pp. 86-107). Gland, Switzerland and Cambridge, United Kingdom: *International Union for Conservation of Nature.*

[37] Reeves R. R., Smith B., E.A.Crespo & Notarbartolo di Sciara G., (2003). *Dolphins, whales and porpoises: 2002-2010 conservation action plan for the world's cetaceans.* Gland, Switzerland and Cambridge, United Kingdom: *International Union for Conservation of Nature.*

[38] Smith B. & Reeves R. R., (2000). Report of the workshop on the effects of water development on river cetaceans, 26-28 February 1997, Rajendrapur, Bangladesh. In R. R. Reeves, B. D. Smith & T. Kasuya (Eds.), *Biology and conservation of freshwater cetaceans in Asia. Occasional Paper of the.* International Union for Conservation of Nature *SSC* (No. 23, pp. 15-22). Gland, Switzerland and Cambridge, United Kingdom: *International Union for Conservation of Nature.*

[39] Smith B. D., Zhou K., Wang D., Reeves R. R., Barlow J., Taylor B. & Pitman R., (2007). *Lipotes vexillifer*. In International Union for Conservation of Nature (Ed.), *2007 IUCN Red List of Threatened Species*. Gland, Switzerland and Cambridge, United Kingdom *International Union for Conservation of Nature*.

[40] Tong H., Zheng Y. & Xu Q., (2008). Preliminary analysis of flow and sediment variation of the Yangtze River from Yichang to Wuhan. *Yangtze River,* 39, 37-41.

[41] Turvey S., Pitman R. L., L.Taylor B., Barlow J., Akamatsu T., Barrett L. A., Zhao X., Reeves R. R., Stewart B. S., Wang K., Wei Z., Zhang X., Pusser L. T., Richlen M., Brandon J. R. & Wang D., (2007). First human-caused extinction of a cetacean species? *Biology Letters,* 3, 537-540.

[42] Wang D., Zhang X. & Liu R., (1998). Conservation status and its future of baiji and Yangtze finless porpoise in China. In Z. L. Hua, B. Fu and Y. Yang (Eds.), *Ecology and Environmental Protection of Large Irrigation Projects in Yangtze River in 21st Century* (pp. 218-226). Beijing, China: Environmental Science Press.

[43] Wang D., Liu R., Zhang X., Yang J., Wei Z., Zhao Q. & Wang X., (2000). Status and conservation of the Yangtze finless porpoise. In R. R. Reeves, B. D. Smith & T. Kasuya (Eds.), *Biology and conservation of freshwater cetaceans in Asia.* International Union for Conservation of Nature *SSC Occasional Paper* (No. 23, pp. 81-85). Gland, Switzerland and Cambridge, United Kingdom: *International Union for Conservation of Nature.*

[44] Wang D., Hao Y., Wang K., Zhao Q., Chen D., Wei Z. & Zhang X., (2005). The first Yangtze finless porpoise successfully born in captivity. *Environmental Science and Pollution Research,* 12, 247-250.

[45] Wang D., Zhang X., Wang K., Wei Z., Wursig B., Braulik G. T. & Ellis S., (2006). Conservation of the baiji: No simple solution. *Conservation Biology,* 20, 623-625.

[46] Wang K., Wang D., Zhang X., Pfluger A. & Barrett L., (2006). Range-wide Yangtze freshwater dolphin expedition: The last chance to see Baiji? *Environmental Science and Pollution Research International,* 13, 418-424.

[47] Wei Q., Wang D. & Wang L., (2007). Aquatic Biodiversity Conservation. In G. S. Yang, L. D. Weng & L. F. Li (Ed.). *Yangtze Conservation and Development Report* (pp. 90-113). Wuhan, China: Changjiang Press.

[48] Wei Z., Wang D., Zhang X., Zhao Q., Wang K. & Kuang X., (2002). Population size, behavior, movement pattern and protection of Yangtze finless porpoise at Balijiang section of the Yangtze River. *Resources and Environment in the Yangtze Basin,* 11, 427-432.

[49] Wei Z., Wang D., Zhang X., Wang K., Chen D., Zhao Q., Kuang X., Gong W. & Wang X., 2004. Observation on some sexual behavior of the Yangtze finless porpoise (*Neophocaena phocaenoides asiaeorientalis*) in captivity. *Acta Theriologica Sinica,* 24, 98-102.

[50] Wu G. & Tu J., (2007). Water Pollution. In G. S. Yang, L. D. Weng & L. F. Li (Eds.), *Yangtze Conservation and Development Report.* (pp. 59-70). Wuhan, China: Changjiang Press.

[51] Xiao W. & Zhang X., (2000). A preliminary study on the population size of Yangtze finless porpoise in Poyang Lake, Jiangxi. *Chinese Biodiversity,* 8, 106-111.

[52] Xiao W. & Zhang X., (2002). Distribution and population size of Yangtze finless porpoise in Poyang Lake and its branches. *Acta Theriologica Sinica,* 22, 7-14.

[53] Xie P. & Chen Y., (1996). Biodiversity problems in freshwater ecosystems in China: Impact of human activities and loss of biodiversity. In J. MacKinnon & S. Wang (Eds.). *Conserving China's biodiversity* (pp. 160-168). Wuhan, China: China Environmental Science Press.

[54] Yang G., Zhou K., Xu X. & Leatherwood S., (1999). A survey on the incidental catches of small cetaceans in coastal waters of China. *Chinese Journal of Applied Ecology,* 10, 713-716.

[55] Yang J., Xiao W., Kuang X., Wei Z. & Liu R., (2000). Studies on the distribution, population size and the activity of *Lipotes vexillifer* and *Neophocaena phocaenoides* in Dongting Lake and Boyang Lake. *Resources and Environment in the Yangtze Basin,* 9, 444-450.

[56] Yi W., Dai C. & Qian J., (2007). Water Utilization and Development. In G. S. Yang, L. D. Weng & L. F. Li (Eds.), Yangtze Conservation and Development Report (pp. 114-123).Wuhan, China: Changjiang Press.

[57] Yu D., Dong M., Wang J. & Zhang X., (2001). Population status of Yangtze finless porpoise in the Yangtze River section from Hukou to Nanjing. *Acta Theriologica Sinica,* 21, 174-179.

[58] Yu D., Wang J., Yang G. & Zhang X., (2005). Primary analysis on habitat selection of Yangtze finless porpoise in spring in the section between Hukou and Digang. *Acta Theriologica Sinica,* 25, 302-306.

[59] Zhang X., Liu R., Zhao Q., Zhang G., Wei Z., Wang X. & Yang J., (1993). The population of finless porpoise in the middle and lower reaches of Yangtze River. *Acta Theriologica Sinica,* 13, 260-270.

[60] Zhang X., Wei Z., Wang X., Yang J. & Chen P., (1995). Studies on the feasibility of establishment of a semi-natural reserve at Tian-e-Zhou (Swan) Oxbow for baiji, *Lipotes vexillifer. Acta Hydrobiologica Sinica,* 19, 110-123.

[61] Zhang X., Wang D., Liu R., Wei Z., Hua Y., Wang Y., Chen Z. & Wang L., (2003). The

[62] Yangtze River dolphin or baiji (*Lipotes vexillifer*): population status and conservation issues in the Yangtze River, China. *Aquatic Conservation: Marine and Freshwater Ecosystems,* 13, 51-64.

[63] Zhao X., Barlow J., Taylor B. L., Pitman R. L., Wang K., Wei Z., Stewart B. S., Turvey S. T., Akamatsu T., Reeves R. R. & Wang D., 2008. Abundance and conservation status of the Yangtze finless porpoise in the Yangtze River, China. *Biological Conservation,* 141, 3006-3018.

[64] Zheng J., Xia J., He S. & Wang D., (2005). Population genetic structure of the Yangtze finless porpoise (*Neophocaena phocaenoides asiaeorientalis*): implications for management and conservation. *Biochemical Genetics,* 43, 307-320.

[65] Zhou K., Qian W. & Li Y., (1977). Studies on the distribution of baiji, *Lipotes vexillifer Miller. Acta Zoologica Sinica,* 23, 72-79.

[66] Zhou K. & Li Y., (1989). Status and aspects of the ecology and behavior of the baiji, *Lipotes vexillifer,* in the lower Yangtze River. In F. Perrin, R. L. Brownell Jr., K. Zhou & J. Liu (Eds.). *Biology and conservation of the river dolphins. Occasional papers of* the International Union for Conservation of Nature */SSC* (pp. 86-91). Gland, Switzerland and Cambridge, United Kingdom: *International Union for Conservation of Nature.*

[67] Zhou K. & Wang X., (1994). Brief review of passive fishing gear and incidental catches of small cetaceans in Chinese Waters. In W. F. Perrin, G. P. Donovan & J. Barlow (Eds.), *Gillnets and Cetaceans. Report of International Whaling Commission* (special issue 15, pp. 347-354). Cambridge, United Kingdom: International Whaling Commission.

[68] Zhou K., Leatherwood S. & Jefferson T. A., (1995). Records of small cetaceans in Chinese waters: a review. *Asian Marine Biology,* 12, 119-139.

[69] Zhou K., Sun J., Gao A. & Wursig B., (1998a). Baiji (*lipotes vexillifer*) in the lower Yangtze River: Movements, numbers threats and conservation needs. *Aquatic Mammals,* 24, 123-132.

[70] Zhou K., Yang G., Gao A., Sun J. & Xu X., (1998b). Population abundance and distribution characteristics of finless porpoise in the River Section from Nanjing to Hukou of the Yangtze River. *Journal of Nanjing Normal University (Natural Science),* 21, 91-98.

In: Biology, Evolution and Conservation of River Dolphins… ISBN: 978-1-60876-633-8
Editors: M. Ruiz-García, J. Shostell. pp. 377-394

Chapter 20

FAILURE OF THE BAIJI RECOVERY PROGRAM: CONSERVATION LESSONS FOR OTHER FRESHWATER CETACEANS

Samuel T. Turvey
Institute of Zoology, Zoological Society of London, Regent's Park, London,UK

ABSTRACT

The Yangtze River dolphin or baiji, a freshwater cetacean found in the mid-lower Yangtze River and neighboring lake and river systems, experienced a precipitous population decline throughout the late twentieth century driven by unsustainable by-catch in local fisheries and habitat degradation. An intensive survey in 2006 failed to find any evidence that the baiji still survives, and the species is now highly likely to be extinct. Although considerable protective legislation was put in place from the late 1970s onwards in China, notably laws banning harmful fishing practices and the establishment of a series of reserve sections in the main Yangtze channel, regulations were difficult or impossible to enforce and *in situ* reserves proved unable to provide adequate protection for baiji. More intensive species-specific recovery strategies also received considerable national and international attention, with extensive deliberation for over twenty years about an *ex situ* recovery program that aimed to establish a translocated breeding population of baiji under semi-natural conditions. However, minimal financial or logistical support for this active baiji conservation strategy was ever provided by the international conservation community. A more dynamic international response is required if other threatened river dolphin species are to be conserved in the future.

Keywords: baiji, *ex situ* conservation, extinct, recovery program, translocation, Yangtze River dolphin

The story of the baiji's decline ... needs to be told and re-told.
Reeves et al. (2000: vi)

INTRODUCTION

The Yangtze River dolphin or baiji (*Lipotes vexillifer*), an obligate river dolphin (*sensu* Leatherwood & Reeves, 1994) endemic to the mid-lower Yangtze River [Changjiang] drainage and the neighboring Qiantang River in eastern China, has long been recognized as one of the world's rarest and most threatened mammal species (Figures 1-2). No meaningful baiji population estimates are available before the late twentieth century, but using densities of other river dolphins in areas uncompromised by development as a model, it has been suggested that the baiji population may have formerly consisted of a few thousand animals (Zhou et al., 1994). However, the mid-lower Yangtze region has experienced intensive development driven by extremely high human population densities since the advent of rice cultivation in the region approximately 7000 years ago (Scott, 1989), and the lowlands of eastern China lost most of their Holocene large mammal fauna centuries or millennia ago (Gu, 1989; Elvin, 2004; Wen, 2006). Historical writings spanning at least 2000 years indicate that baiji were widely hunted, primarily to provide oil for lamps, caulking for boats, and for the supposed medicinal properties of their blubber and meat, and it has been suggested that this long history of exploitation had already greatly reduced their numbers before the twentieth century (Pilleri, 1979; see also Hoy, 1923).

Figure 1. Yangtze River dolphin or baiji (*Lipotes vexillifer*). This animal was shot in February 1914 in the channel connecting Dongting Lake to the main Yangtze; its head and cervical vertebrae were sent to the United States National Museum of Natural History, and represent the holotype of the species. From Hoy (1923).

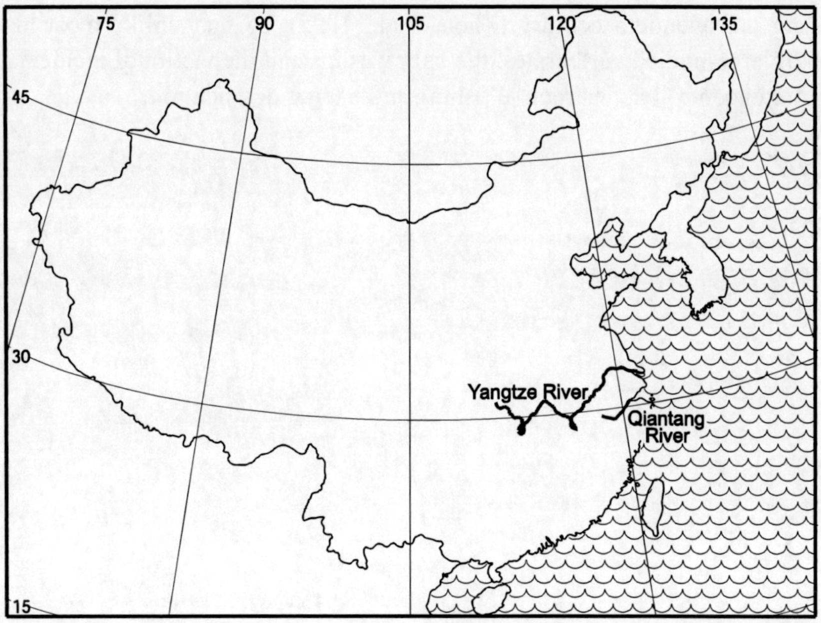

Figure 2. Historical distribution of the baiji until the mid-twentieth century. The baiji formerly occurred in the main Yangtze channel as far upstream as Huanglingmiao and Liantuo (downstream section of the Three Gorges), approximately 1900 kilometers from the estuary (see Zhou et al., 1977). It also occurred in Dongting and Poyang Lakes, two large lake systems appended to the main Yangtze channel, and the neighboring Qiantang River.

Continued wide-scale anthropogenic impacts caused by increasing human population density, aggressive environmental exploitation and industrialization over the past 50 or so years (Shapiro, 2001) were responsible for a further precipitous decline in the remaining baiji population. However, the range of different potential extinction drivers operating in the Yangtze region, and the limited available data on baiji mortality, complicates our understanding of the relative importance of different threat processes in this population collapse. The primary factor was probably unsustainable by-catch in local fisheries; at least half of all known baiji deaths observed by Chinese researchers from the 1950s to the 1980s were caused by rolling hook long-lines and other fishing gear (Figure 3), and electro-fishing accounted for 40% of the limited number of known baiji deaths recorded during the 1990s. Further mortality is also known to have been caused by collisions with boat hulls and propellers, explosives used in channel clearance, and chemical spills (Lin et al., 1985; Chen & Hua, 1989; Zhou & Li, 1989; Zhou & Zhang, 1991; Zhou & Wang, 1994; Zhou et al., 1994, 1998; Sheng, 1998b; Zhang et al., 2003). Agricultural and industrial intensification and water development projects have led to escalating habitat degradation in the main Yangtze and Qiantang river channels and their tributaries and appended lakes, including increased siltation, elimination of optimal baiji counter-current habitat, and decreases or extirpation of many fish species, all of which are likely to have had substantial further impacts on baiji populations (Liu et al., 2000; Smith et al., 2000; Xie, 2003; Fang et al., 2006). In particular, industrial and agricultural pollutants may have severely impacted baiji health and fertility, but data to assess the significance of this likely extinction driver remain very limited (Yang & Liu, 2005; Shao et al., 2006). Although dolphins stranded on sandbars were sometimes beaten to death by local residents (Perrin & Brownell, 1989), direct exploitation of baiji largely ceased during the twentieth century (Zhou et al., 1995), so that unlike most historical-era extinctions of large-bodied vertebrates, the baiji was instead the victim of incidental mortality resulting from uncontrolled destructive fishing and habitat degradation.

Figure 3. Rolling hook long-lines used by fishermen at Hexiao harbour, Nanjing, photographed by the author in March 2008.

Baiji have not been seen in the Qiantang River since the 1950s, following construction of the Xinanjiang Dam (Zhou et al., 1977; Zhou & Zhang, 1991; Smith et al., 2000), and have apparently not been seen in either Dongting or Poyang Lake since the late 1970s (Yang et al., 2000; Fang et al., 2006). Chinese researchers reported a steady, rapid decline of the baiji population in the main Yangtze channel through the 1980s and 1990s from an estimated 400 individuals in 1979 (Zhou, 1982; Chen & Hua, 1989; Zhou & Li, 1989; Zhou et al., 1994, 1998), with an apparent range contraction of several hundred kilometers from the former upstream limit of its distribution during this period (Zhou et al., 1977; Chen et al., 1997). Surveys during 1997-1999 provided a minimum estimate of only 13 surviving animals (Zhang et al., 2003). The last verified baiji reports are of a pregnant female found stranded at Zhenjiang in November 2001, and an individual photographed in the Tongling river section in May 2002. Subsequent unverified sighting reports suggested that a remnant baiji population continued to persist in the river (Braulik et al., 2005). However, an intensive six-week multi-vessel visual and acoustic survey in 2006 that covered the entire historical range of the baiji in the main Yangtze channel failed to find any evidence that the species survives (Barrett et al., 2006; K. Wang et al., 2006; Turvey et al., 2007), and the baiji is now highly likely to be extinct. This represents not only the first documented global extinction of a 'megafaunal' vertebrate for over 50 years, but also the disappearance of an entire mammal family (Lipotidae), only the fourth such event in the past 500 years (MacPhee & Flemming, 1999; Isaac et al., 2007). Furthermore, this is the first probable extinction of a large-bodied vertebrate species since the emergence of an international network of conservation organizations that have tended to prioritize conservation efforts on such charismatic animals (cf. Leader-Williams & Dublin, 2000; Entwistle & Stephenson, 2000). How then was it possible for a species of river dolphin to become extinct when it should have been the focus of intensive international conservation attention and activity?

Most of the remaining obligate and facultative dolphins and porpoises are also highly threatened by intensive anthropogenic pressures, and these species are regarded as among the world's most threatened mammals (Perrin et al., 1989; Reeves et al., 2000, 2003; Jefferson & Smith, 2002). In particular, species and populations of freshwater cetaceans found in other Asian river systems (Ganges River dolphin *Platanista [gangetica] gangetica*; Indus River dolphin *Platanista [gangetica] minor*; Yangtze finless porpoise *Neophocaena phocaenoides asiaeorientalis*; Irrawaddy dolphin *Orcaella brevirostris*) are all classified as Endangered or Critically Endangered by the International Union for Conservation of Nature (IUCN, 2008). It is therefore imperative to identify the key lessons that can be learned from the history of Chinese and international attempts to conserve the baiji, and the ultimate failure of these attempts to prevent the extinction of this species. In particular, it is necessary to consider whether conservation efforts for the baiji were hindered by unique and insurmountable theoretical and/or practical challenges associated with the specific ecology of river dolphins, river systems, and associated threat processes, or whether the lack of successful conservation action resulted instead from institutional failure to implement a feasible recovery program.

CONSERVATION CONCERN, LEGISLATION AND ACTION

Although scientific research was effectively halted in China during the 1960s and 1970s, investigations into the ecology, distribution and status of the baiji commenced shortly after the end of the Cultural Revolution (Zhou et al., 1977). By the end of the 1970s, researchers and officials at a number of Chinese institutions had already become aware of the threatened status of the baiji, and were planning active measures to conserve the species (Pilleri, 1979). In addition to ongoing scientific research from 1978 into baiji biology directed by the Coordination Group on Lipotes Research (Chen 1981; see e.g. Perrin et al., 1989; Chen et al., 1997; Chen, 2007), a series of surveys were conducted in the main Yangtze channel during the late 1970s, 1980s and 1990s to monitor the remaining baiji population, although the wide variation in methodology employed between different surveys (e.g. distance surveyed, number of boats and observers, height of observers above water, boat speed, correction factors) made it difficult to identify meaningful population trends over time before the baiji population was critically low (Zhou et al., 1994; Zhang et al., 2003; Braulik et al., 2005; Turvey et al., 2007). The baiji was listed in the Key Protected List of the Aquatic Resources Regulation in 1979, and on the First Category of the List of National Protected Wild Animals (State Key Protected Wildlife List) in 1989, for which hunting is strictly prohibited (Zhou et al., 1994; Sheng, 1998b). Additional protective legislation is also officially in place, such as the 'Baiji and Yangtze Finless Porpoise Protection Act' drafted by the Institute of Hydrobiology, Chinese Academy of Sciences, and approved by the Chinese Ministry of Agriculture in 2001 (Dudgeon, 2005). Rolling hook long-lines, dynamite fishing, poison fishing, electro-fishing, and fixed fyke nets were all banned in the main Yangtze channel due to recognition of the threats that these methods posed to baiji through incidental mortality, and also because they were harmful to fisheries resources (Zhou & Wang, 1994; Zhou et al., 1998). This protective legislation led to prosecution and imprisonment of a small number of fishermen found guilty of killing baiji (Zhou & Zhang, 1991). Awareness-raising efforts addressing the importance of baiji conservation were carried out, notably through the production of considerable baiji 'souvenirs' or merchandise (e.g. stamps, key rings, badges, clothing, beer), distribution of educational materials (brochures, posters) among riverside communities by research institutes working with the Fisheries Management Bureau in Hunan, Hubei, Jiangxi, Anhui and Jiangsu provinces, and national newspaper and television reports on the status of the species (Adams & Carwardine, 1990; Chen et al., 1997; Zhou et al., 1998). Since 1986, a series of river sections at Shishou, Honghu (Xin-Luo), Tongling and Zhenjiang were officially designated as National or Provincial Baiji Reserves, where more stringent regulations on fishing, pollution and vessel traffic were proposed; the longest of these protected sections, the Xin-Luo National Baiji Reserve between Xintankou and Luoshan, was 135 km long (Zhou et al., 1994, 1998; Figure 4). A further series of protection stations were set up along the river at Jianli, Chenglingji, Hukou, Anqing and Zhenjiang, where it was intended that reserve staff would make daily patrols to monitor baiji populations, control fishing restrictions, rescue injured, sick or stranded animals, and provide further conservation education for riverside communities (Zhou et al., 1994; Figure 4).

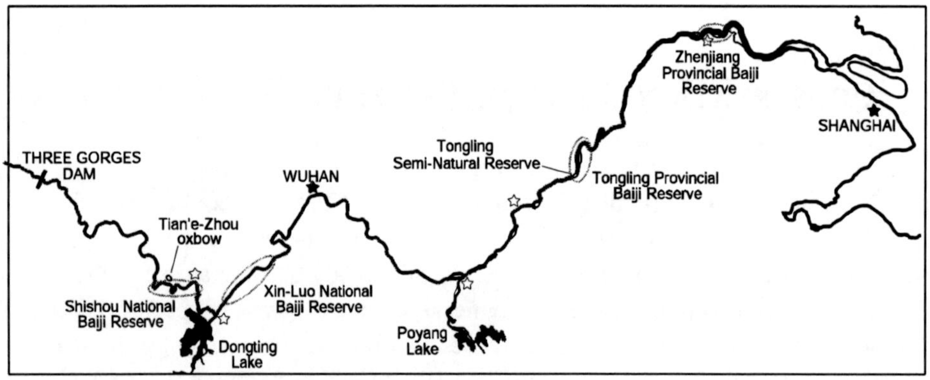

Figure 4. Locations of National and Provincial Baiji Reserves (dashed circles), protection stations (open stars), and semi-natural reserves in the main Yangtze channel and associated water bodies.

However, these measures proved to be inadequate in preventing the continued decline of baiji in the main Yangtze channel. Although it has been suggested that the establishment of *in situ* reserves helped slow the decline of Yangtze cetaceans through the effective banning of harmful and illegal fishing methods (K. Wang et al., 2006), these regulations were considered difficult or impossible to enforce (Zhou et al., 1998), despite recommendations to strengthen enforcement in combination with further public education (Sheng, 1998b). In reality it is difficult to assess the extent to which practical enforcement was ever attempted by regional fisheries authorities and reserve staff, as rolling hook long-lining remains one of the commonest fishing methods in the Yangtze in both protected and unprotected river sections today, with fishermen prepared to discuss illegal fishing practices openly with foreign researchers and officials (pers. obs.; Figure 3). Administrative agencies in charge of *in situ* reserves were recognized to lack the resources either to reduce the rapid ongoing increase in vessel traffic in protected river sections or to patrol these sections frequently (Zhou et al., 1998); indeed, it is unlikely that vessel traffic would ever have been controlled within China on the basis of environmental concerns alone, given the Yangtze's importance as the 'Golden Channel' in supporting large-scale national economic development in recent decades (D. Wang et al., 2006). Similar problems also continue to surround effective management of pollutant release into the river channel, with negligible control of point and non-point pollution sources (Dudgeon, 2005). Furthermore, wider questions over whether protection of limited river sections could ever provide adequate conservation for baiji were raised by ecological observations and photo-identification studies of wild baiji, which indicated that although individual animals may stay in the same restricted geographical area for up to a month, they could also make migrations of more than 200 kilometers up and down the Yangtze channel, with anecdotal information provided by fishermen supporting the idea of large-scale seasonal movements (Zhou et al., 1994, 1998; Zhang et al., 2003). Survey data interpreted by Zhou et al. (1998) suggested that even the Xin-Luo Reserve section would only be inhabited by six baiji at any one time, and at least some of these animals would move between protected and adjacent non-protected areas, making *in situ* conservation efforts of limited usefulness.

Because of these potentially insurmountable obstacles to effective conservation of baiji in their natural habitat, more intensive species-specific recovery strategies also received considerable attention from both Chinese and international conservation practitioners. As early as the late 1970s, Pilleri (1979) noted the potential for conserving baiji through captive breeding, which he considered would represent 'a splendidly original achievement'. This approach was soon also widely supported within China. Between 1980 and 1986, six baiji were brought into captivity at the Institute of Hydrobiology, Chinese Academy of Sciences (four individuals), Nanjing Normal University (one individual) and the Jiangsu Aquatic Institute (one individual) (Zhou & Zhang, 1991; Chen et al., 1997). However, only two of these animals survived for more than a few months, and reproductively viable male and female individuals were never maintained in captivity together (the only captive female baiji died before reaching sexual maturity; Zhou & Zhang, 1991). It therefore remains impossible to assess whether successful reproduction could have eventually been achieved under these conditions (contra Yang et al., 2006), especially because Yangtze finless porpoises have now bred successfully in the modern well-equipped dolphinarium at the Institute of Hydrobiology (Wang et al., 2005). Although the official view within China appears to have increasingly supported placing baiji into this dolphinarium in recent years (Dudgeon, 2005), *ex situ* baiji conservation under strict captive conditions received little support from the international conservation community (Braulik et al., 2005) other than Japan (Chen & Liu, 1992), even given the marked international advances in captive cetacean maintenance, welfare and husbandry that have been achieved in recent decades. 'Qi Qi', a male baiji that survived in captivity at the Institute of Hydrobiology for over 22 years, displayed stereotypical behaviour (Dudgeon, 2005), and it is unlikely that animals bred and maintained for long periods under such circumstances could have been successfully reintroduced into the wild, or even whether Chinese authorities and research staff would have permitted such a move given the national importance attached to any institution possessing captive individuals. Attempts to cryopreserve sperm from captive baiji also proved unsuccessful.

The alternative *ex situ* conservation strategy that was widely promoted by both Chinese and international conservationists was the establishment of a translocated breeding population of baiji under semi-natural conditions, in a protected environment away from the main river where human impacts could be closely managed and minimized compared to the degraded Yangtze channel. This approach was therefore more similar to translocation of insular species to predator-free offshore islands than to traditional *ex situ* propagation in artificial facilities. The semi-natural reserve strategy first received serious consideration in the published literature by Lin et al. (1985), and was pursued independently by the two Chinese research groups actively working on baiji in the 1980s. The research group at Nanjing Normal University initiated a project to create a semi-natural reserve in a 1550-meter channel between Heyuezhou Island and Tiebanzhou Island near Tongling in 1985 (Zhou, 1986, 1989). The Institute of Hydrobiology's Baiji Research Group had already begun surveying the mid-lower Yangtze to identify suitable semi-natural reserve sites in 1984 (Chen & Liu, 1992), and proposed the establishment of a reserve at Tian'e-Zhou (a 21 kilometer oxbow near Shishou which formed part of the main Yangtze channel until 1972) at the Workshop on Biology and Conservation of the Platanistoid Dolphins in 1986 (Baiji Research Group, 1989; Zhang et al., 1995). Both sites were developed into potential baiji reserves, but attention both within China and from the international community soon focused on the Tian'e-Zhou oxbow as the more suitable prospective site for a semi-natural baiji population, and this was designated as a National Natural Reserve for baiji conservation by the Chinese Ministry of Agriculture in 1992 (Zhou et al., 1994; Braulik et al., 2005).

A translocated population of Yangtze finless porpoises introduced to the oxbow from 1990 onwards, as a surrogate to test the suitability of the reserve for baiji, began to breed successfully in 1992, suggesting that conditions were also favorable for introduced baiji to survive and breed (Wang et al., 2000). However, despite assurances by regional authorities that fishing and associated human impacts would be strictly controlled at Tian'e-Zhou, 30% of the reserve budget is still met by income from fishing in the reserve, and only 200 of the 500 fishermen have been moved away from the area and provided with alternate livelihoods, leading to continued serious problems of competition for fish resources and the dangers of accidental by-catch of translocated cetaceans (Dudgeon, 2005; pers. obs.). Indeed, two translocated porpoises have been killed by rolling hook long-lines in the reserve, two other animals died from injuries associated with their capture and translocation, a further seven animals were killed accidentally in the reserve by inexperienced researchers, and fourteen escaped during the flood season of 1996; despite this high level of mortality, fifteen more animals were released back into the Yangtze to reduce competition for fish with local fishermen operating in the reserve (Zhou et al., 1994; Wang et al., 2000). The continued presence of porpoises in the reserve also led to ongoing concerns from international conservationists about possible risks to any translocated baiji from agonistic interactions and competition for limited food resources between the two species (Zhou et al., 1994; Braulik et al., 2005; Dudgeon, 2005; Yang et al., 2006). Recommended on-site infrastructural improvements, e.g. cetacean holding pens to allow effective post-translocation health monitoring and veterinary care before soft-release into the reserve, were also never adequately adopted (Figure 5). Only one baiji was ever translocated to the Tian'e-Zhou oxbow, in December 1995, and was found dead of unknown causes a few months later, emaciated and entangled in fence nets in a region of strong current (Liu et al., 1998). No further attempts were made to capture baiji for translocation to Tian'e-Zhou, although subsequent unsuccessful capture efforts to establish a baiji population at the Tongling semi-natural reserve continued until 2001 (Tongling Provincial Baiji Reserve staff, pers. comm.,

2008). A new initiative to generate momentum and increased international support for a more carefully managed semi-natural recovery program at Tian'e-Zhou from 2004 onwards, involving development of a detailed budget and implementation plan and extensive fund-raising efforts, ultimately proved unsuccessful (Turvey et al., 2006; Turvey, 2008).

Figure 5. Incomplete cetacean holding pens at Tian'e-Zhou. Construction on these holding pens finally commenced shortly before the November-December 2006 survey that documented the probable extinction of the baiji, even though they had been repeatedly recommended at international workshops as an essential infrastructural improvement needed for the baiji recovery program. Note also the extensive amount of fishing gear in the two boats in the foreground. Photograph by the author.

Would it have been Possible to Save the Baiji?

The extensive range of *in situ* and *ex situ* conservation approaches outlined above was deliberated by both Chinese and international researchers for three decades. Substantial conservation recommendations for the baiji were developed during this period, notably in four major baiji-focused workshop reports (Perrin et al., 1989; Zhou et al., 1994; Ministry of Agriculture, 2001; Braulik et al., 2005), two further IUCN Species Survival Commission documents (Reeves et al., 2000, 2003), at considerable further small-scale or more general workshops and meetings (e.g. Reeves & Leatherwood, 1995; Turvey et al., 2006), and in numerous scientific publications (e.g. Zhang et al., 1995; Smith & Smith, 1998; Dudgeon, 2005). However, all of these efforts still failed to prevent the probable extinction of the baiji by the first decade of the twenty-first century. Indeed, progressive international meeting reports increasingly acknowledged that participants were reaffirming recommendations made at previous meetings, were surprised by the extreme similarity of views given more than a decade earlier on key issues, and expressed disappointment and concern that so little progress had been made towards implementing conservation actions (e.g. Smith & Reeves, 2000; Braulik et al., 2005). Why, then, was so little actually achieved in the struggle to save the baiji?

In situ conservation measures must always be addressed as a primary recovery strategy for any endangered species. However, the escalating anthropogenic impacts on the Yangtze ecosystem throughout the late twentieth century that drove unsustainable levels of incidental dolphin mortality were probably irreversible in the time period required to prevent the disappearance of the baiji from its natural habitat, due to human-wildlife conflicts not only with local communities across the baiji's range but also with ongoing national-scale economic and industrial development that relied heavily on the river's resources. The series of existing *in situ* National and Provincial Baiji Reserves in the main Yangtze channel could certainly have been more adequately managed and policed, with greater control of illegal fishing practices in particular; but it is highly unlikely that protection in a series of discrete, unconnected river sections, exposed to uncontrollable through-flow of pollutants, vessel traffic and other threat factors, would have been able to delay the decline of a wide-ranging dolphin species with unknown site fidelity in any substantial way. Wider-scale Yangtze regeneration projects were also certainly necessary for long-term baiji persistence, as well as for the conservation of the river system's many other highly threatened endemics, but they remained impossible to effect in time, and were sadly insufficient for continued short-term survival of the species. This tragic situation was increasingly recognized by international conservationists such as Dudgeon (2005), who concluded that 'the baiji is certain to become extinct if left to languish in the Yangtze'.

It is the lack of success in developing a viable *ex situ* recovery program that raises more fundamental concerns about the efficacy of both national and international conservation efforts to save the baiji. Intensive species-level manipulations have long been recognized as crucial for conserving species with tiny population sizes and rapid rates of decline and where major cause(s) of decline cannot be determined or quickly corrected, and such approaches have been widely credited for effecting successful species recoveries impossible by other methods. The establishment of a closely managed baiji breeding program at Tian'e-Zhou was first proposed over twenty years ago, when the wild baiji population was estimated at 300 individuals (Chen & Hua, 1989), and it has since been repeatedly recommended in workshop reports and the scientific literature as an urgent recovery strategy. The semi-natural conservation approach was also widely publicized in popular international accounts of Chinese conservation (e.g. Adams & Carwardine, 1990; Schaller, 1993; Laidler & Laidler, 1996). However, *ex situ* propagation remains controversial, as it inevitably involves higher-risk intensive contact activity compared to ecosystem-scale programs, and there is widespread caution amongst policy-makers towards such interventionist techniques (see Clark, 1997; Snyder & Snyder, 2000; Groombridge et al., 2004; Flueck & Smith-Flueck, 2006; VanderWerf et al., 2006). Although the international conservation community eventually concluded that removal and translocation of baiji to a safer environment was the only feasible option to save the species from extinction, and identified this action as the key short-term goal in a longer-term recovery strategy for the species (Braulik et al., 2005), earlier more equivocal attitudes outside China about the potential success or necessity of such a strategy are reflected in the minimal support for active baiji conservation that was ever provided by the international conservation community.

Despite the apparent commitment expressed to the baiji by several major conservation organizations (e.g. Ellis, 2005), and frequent insistences that it was a 'popular' species dominating conservation resources (see Turvey, 2008), nearly all of the international funding that was ever made available for baiji conservation was restricted to supporting passive survey work, meetings and workshops, awareness-raising campaigns, and infrastructural improvements at the dolphinarium of the Institute of Hydrobiology (Chen & Liu, 1992) rather than active implementation of the proposed *ex situ* recovery program. However, while surveys have provided invaluable data on baiji abundance, ecology and decline, in themselves they were unable to save the species from extinction; and even Chinese researchers recognized that whereas community education may well have helped save some baiji, the overall trend of small and decreasing numbers was not likely to be reversed by public awareness alone (Zhou et al., 1998). Although establishing a viable baiji breeding population at Tian'e-Zhou represented a major conservation challenge, the lack of success in capturing sufficient numbers of baiji from the main Yangtze channel and the human-wildlife conflicts and problems with maintaining cetaceans at the reserve could undoubtedly have been substantially addressed by increased international assistance. For example, the series of six two-three month baiji capture attempts conducted between 1993 and 1995, which eventually led to the translocation of a single female baiji to Tian'e-Zhou, were funded and conducted entirely using in-country finances, equipment and methods; the lack of greater success in these operations was attributed to the limited number of available boats and personnel (Zhou et al., 1994), and Chinese researchers reported that they saw other dolphins which they were unable to catch as they had 'primitive equipment and not enough manpower' (Liu et al., 1998). Regularly revised budgets for necessary infrastructural improvements and running costs for the baiji recovery program (Zhou et al., 1994; Ministry of Agriculture, 2001; Turvey et al., 2006), whilst far from low, were comparable to those employed in attempts to save other species of extreme rarity (e.g. Rabinowitz, 1995; Clark, 1997; Groombridge et al., 2004), and lower than other marine mammal conservation projects which have conversely been readily funded (Morell, 2008). Whereas financial and logistical support for implementing the baiji recovery program should also certainly have been more forthcoming from within China, in the absence of concerted national-level actions for baiji conservation, the unfortunate unwillingness on the part of western organizations to provide direct financial assistance, applied skills transfer, capacity-building and associated project support, and/or international pressure constituted one of the most significant barriers to effective protection of the species. This factor was even increasingly appreciated by international conservationists themselves before recognition of the baiji's probable extinction (e.g. Reeves et al., 2000; Reeves & Gales, 2006).

Whether or not a viable breeding population of baiji could have been established in time at Tian'e-Zhou, it is crucial to recognize that there were no fundamental obstacles preventing the implementation of the *ex situ* baiji recovery program from being considerably further advanced before the probable extinction of the species was discovered in 2006. However, international interest in the baiji's plight at the beginning of the twenty-first century was instead maintained largely through scientific debate over both the possibility and the value of attempting to preserve this Critically Endangered species rather than concerted efforts to support active conservation measures (Kleiman, 2006; Reeves & Gales, 2006; Wang et al.,

2006; Yang et al., 2006). Some authors even chose to advocate the supposed inevitability of the baiji's demise (Yang et al., 2006), contrasting markedly with the extreme caution typically displayed by conservationists when ruling other species to be beyond help (Butchart et al., 2006; Roberts & Kitchener, 2006). However, whilst open discussion and debate remain invaluable in refining and establishing effective conservation strategies for threatened species, a more dynamic international response was also ultimately required to prevent the baiji's extinction.

Wider Conservation Lessons

Although the baiji is now likely to be extinct, the mid-lower Yangtze drainage still contains many other increasingly threatened endemic species (Qiu & Chen, 1988; Zhong & Power, 1997; Fu, 2003), several of which are also 'charismatic' megafaunal taxa (e.g. Yangtze paddlefish *Psephurus gladius*; Yangtze giant soft-shell turtle *Rafetus swinhoei*; Chinese alligator *Alligator sinensis*; see Wei et al., 1997; Thorbjarnarson et al., 2002; Xie, 2003; Stone, 2007). Data collected during the 2006 range-wide baiji survey indicate that the Yangtze finless porpoise, the world's only freshwater porpoise, has experienced a population decline of over 50% since the early 1990s (Zhao et al., 2008), and population viability analysis (PVA) conducted a decade ago suggested that this cetacean was likely to become extinct within 24-94 years (Zhang & Wang, 1999), although the relative importance of different anthropogenic threat factors and the dynamics of this population decline again remain poorly understood. Given the continuing massive-scale intensification of human impacts on this freshwater system, it is difficult to suggest optimal recovery strategies to conserve its remaining unique biodiversity, especially since most of these other threatened species have been the focus of far less national and international conservation attention than was received by the baiji in recent decades. It is likely that *ex situ* conservation will also be required to prevent the further extinctions of many Yangtze species. In particular, the failure of existing *in situ* conservation measures to prevent the disappearance of the baiji suggests that it is also highly unlikely that the river's finless porpoise population will be able to persist without an intensive and well-managed *ex situ* recovery program. Semi-natural breeding groups of porpoises have already been established at both Tian'e-Zhou and Tongling (despite the ongoing problems with on-site cetacean mortality and human-wildlife conflict described above), and further introductions have also been proposed for the Hei-Wa-Wu oxbow (Hubei Province) and the Three Gorges Dam reservoir. However, more unified efforts – and greater international support – are once again required if the conservation of the porpoise is to stand a strong chance of success, and unlike the baiji, this species is still only listed in the Second Category of the List of National Protected Wild Animals (Sheng, 1998a).

The implications of the baiji's probable extinction for the conservation of other threatened freshwater cetaceans, and indeed other globally threatened species, are more general. Accidental by-catch in fishing gear, the likely main extinction driver for the baiji, remains the principal cause of mortality in many populations of small cetaceans worldwide (Reeves et al., 2003). However, river dolphins in other geographical regions, notably other Asian river systems also experiencing large-scale and escalating anthropogenic impacts, are declining due to a range of extinction drivers which may vary in relative severity compared to the threats faced by cetaceans in the Yangtze. For example, in addition to accidental by-catch and bioaccumulation of industrial, agricultural and domestic pollutants, Indian river dolphins (*Platanista* spp.) are known to be severely threatened by construction of irrigation barrages, which have fragmented dolphin subpopulations and greatly decreased water flow in major river channels, and also by deliberate killing for meat and oil (Mohan & Kunhi, 1996; Bairagi, 1999; Smith et al., 2000). The specific actions advocated for the baiji recovery program are therefore not necessarily the most appropriate solutions for preserving these threatened species, and expert consideration is required to identify optimal conservation strategies. However, given the ongoing declines of each of these species, it is imperative that such strategies are identified in the immediate future, and that robust, dynamic efforts are made to implement all recommended conservation actions.

Snyder & Snyder (2000) reflected that one searches in despair for signs that lessons learned in conservation efforts with one species have been applied to conservation efforts for any others. This applies to management, bureaucracy and implementation of recovery plans as much as utilization of specific techniques (Clark, 1997). In addition to the particular anthropogenic extinction drivers operating in the Yangtze River, the scientific and conservation communities must acknowledge that it was the slow pace of decision-making, widespread international conservatism about subjectively unfavorable conservation actions, and a concomitant lack of adequate global support for such ultimately essential actions which are responsible for the tragic extinction of the baiji. This is a mistake that we cannot permit to happen again.

REFERENCES

[1] Adams, D., Carwardine, M., (1990). *Last chance to see*. New York, NY: Ballentine Books.

[2] Baiji Research Group, (1989). A proposal for establishment of a semi-natural reserve at Shishou for conservation and management of the baiji (*Lipotes vexillifer*). *Occasional Papers of the IUCN Species Survival Commission*, 3, 21-22.

[3] Bairagi, S. P., (1999). Oil bait fishery of catfishes in Brahmaputra River affecting river dolphin populations in Assam, India. *Journal of the Bombay Natural History Society*, 96, 424-426.

[4] Barrett, L. A., Pfluger, A., & Wang, D., (2006). Successful pilot Yangtze freshwater dolphin expedition. *Oryx*, 40, 259-260.

[5] Braulik, G. T., Reeves, R. R., Wang, D., Ellis, S., Wells, R. S., & Dudgeon, D., (2005). *Report of the workshop on conservation of the baiji and Yangtze finless porpoise*. Gland, Switzerland: World Conservation Union.

[6] Butchart, S. H. M., Stattersfield, A. J., & Brooks, T. M., (2006). Going or gone: defining 'Possibly Extinct' to give a truer picture of recent extinctions. *Bulletin of the British Ornithologists' Club*, 126A, 7-24.

[7] Chen P., (1981). *Lipotes* research in China. *Report of the International Whaling Commission*, 31, 575-578.

[8] Chen, P., (2007). *Chen Peixun and study on Lipotes vexillifer*. Beijing, China: Ocean Press.

[9] Chen P., & Hua, Y., (1989). Distribution, population size and protection of *Lipotes vexillifer*. *Occasional Papers of the IUCN Species Survival Commission*, 3, 81-85.

[10] Chen, P., & Liu, R., (1992). *Baiji: a rare treasure*. Tokyo and Wuhan: Enoshima Aquarium and Institute of Hydrobiology.

[11] Chen, P., Liu, R., Wang, D., & Zhang, X., (1997). *Biology of baiji, and its rearing and conservation*. Beijing: Science Press.

[12] Clark, T. W., (1997). *Averting extinction*. New Haven and London: Yale University Press.

[13] Dudgeon, D., (2005). Last chance to see...: *ex situ* conservation and the fate of the baiji. *Aquatic Conservation of Marine and Freshwater Ecosystems*, 15, 105-108.

[14] Ellis, S., (2005). Goodbye to the baiji? Not if these scientists can help it. *Conservation Frontlines*, Spring 2005, 3.

[15] Elvin, M., (2004). *The retreat of the elephants: an environmental history of China*. New Haven and London: Yale University Press.

[16] Entwistle, A. C., & Stephenson, P. J., (2000). Small mammals and the conservation agenda. In Entwistle, A., & Dunstone, N., (Eds.), *Priorities for the conservation of mammalian diversity: has the panda had its day?* (pp. 119-139). Cambridge, United Kingdom: Cambridge University Press.

[17] Fang, J., Wang, Z., Zhao, S., Li, Y., Tang, Z., Yu, D., Ni, L., Liu, H., Xie, P., Da, L., Li, & Z., Zheng, C., (2006). Biodiversity changes in the lakes of the central Yangtze. *Frontiers in Ecology and the Environment*, 4, 369-377.

[18] Flueck, W. T., & Smith-Flueck, J. M., (2006). Predicaments of endangered huemul deer, *Hippocamelus bisulcus*, in Argentina: a review. *European Journal of Wildlife Research*, 52, 69-80.

[19] Fu, C., Wu, J., Chen, J., Wu, Q., & Lei, G., (2003). Freshwater fish biodiversity in the Yangtze River basin of China: patterns, threats and conservation. *Biodiversity and Conservation*, 12, 1649-1685.

[20] Groombridge, J. J., Massey, J. G., Bruch, J. C., Malcolm, T., Brosius, C. N., Okada, M. M., Sparklin, B., Fretz, J. S., & VanderWerf, E. A., (2004). An attempt to recover the po'ouli by translocation and an appraisal of recovery strategy for bird species of extreme rarity. *Biological Conservation,* 118, 365-375.

[21] Gu, Y., (1989). Preliminary research on the fossil gibbons of the Chinese Pleistocene and recent. *Human Evolution*, 4, 509-514.

[22] Hoy, C. M., (1923). The "white-flag" dolphin of the Tung Ting Lake. *The China Journal of Science & Arts*, 1, 154-157.

[23] Isaac, N. J. B., Turvey, S. T., Collen, B., Waterman, C., & Baillie, J. E. M., (2007). Mammals on the EDGE: conservation priorities based on threat and phylogeny. *PLoS One*, 2, e296.

[24] IUCN (International Union for Conservation of Nature), (2008). IUCN Red List of threatened species: a global species assessment [online]. [cited 2008-5-05]. Available from: http://www.iucnredlist.org

[25] Jefferson, T. A., & Smith, B. D., (2002). Facultative freshwater cetaceans of Asia: their ecology and conservation. *The Raffles Bulletin of Zoology, Supplement*, 10, 1-187.

[26] Kleiman, D. G., (2006). Conservation focus: baiji, freshwater dolphin. Introduction. *Conservation Biology*, 20, 620.

[27] Laidler, L., & Laidler, K., (1996). *China's threatened wildlife*. London, United Kingdom: Blandford.

[28] Leader-Williams, N., & Dublin, H. T., (2000). Charismatic megafauna as 'flagship species'. In Entwistle, A., Dunstone, N., (Eds.), *Priorities for the conservation of mammalian diversity: has the panda had its day?* (pp. 53-81). Cambridge, United Kingdom: Cambridge University Press.

[29] Leatherwood, S., & Reeves, R. R., (1994). River dolphins: a review of activities and plans of the Cetacean Specialist Group. *Aquatic Mammals*, 20, 137-154.

[30] Lin, K., Chen, & P., Hua, Y., (1985). Population size and conservation of *Lipotes vexillifer*. *Acta Zoologica Sinica*, 5, 77-85.

[31] Liu, R., Wang, D., & Zhou, K., (2000). Effects of water development on river cetaceans in China. *Occasional Papers of the IUCN Species Survival Commission*, 23, 40-42.

[32] Liu, R., Yang, J., Wang, D., Zhao, Q., Wei, Z., & Wang, X., (1998). Analysis on the capture, behavior monitoring and death of the baiji (*Lipotes vexillifer*) in the Shishou semi-nature reserve at the Yangtze River, China. *IBI Reports*, 8, 11-21.

[33] MacPhee, R. D. E., & Flemming, C., (1999). Requiem aeteman: last five hundred years of mammalian species extinctions. In MacPhee, R. D. E., (Ed.), *Extinctions in near time: causes, contexts, and consequences* (pp. 333-371). New York, NY: Kluwer Academic/Plenum.

[34] Ministry of Agriculture, (2001). *Conservation action plan for Chinese river dolphins*. Beijing, China: Ministry of Agriculture.

[35] Mohan, R. S. L., & Kunhi, K. V. M., (1996). Fish oils as alternative to river dolphin, *Platanista gangetica* (Lebeck) oil for fishing catfish *Clupisoma garua* in the River Ganges, India. *Journal of the Bombay Natural History Society*, 93, 86-89.

[36] Morell, V., (2008). Puzzling over a Steller whodunit. *Science*, 320, 44-45.

[37] Perrin, W. F., & Brownell, R. L. Jr, (1989). Report of the workshop. *Occasional Papers of the IUCN Species Survival Commission*, 3, 1-22.

[38] Perrin, W. F., Brownell, R. L. Jr, Zhou, K., & Liu, J. (Eds.), 1989. Biology and conservation of the river dolphins. *Occasional Papers of the IUCN Species Survival Commission*, 3, 1-173.

[39] Pilleri, G., (1979). The Chinese river dolphin (*Lipotes vexillifer*) in poetry, literature and legend. *Investigations on Cetacea*, 10, 336-349.

[40] Qiu, S., & Chen, D., (1988). Analyses on generations of Chinese shad in the Yangtze River and preliminary assessment of its abundance. *Freshwater Fisheries*, 6, 3-5.

[41] Rabinowitz, A., (1995). Helping a species go extinct: the Sumatran rhino in Borneo. *Conservation Biology*, 9, 482-488.

[42] Reeves, R. R., & Gales, N. J., (2006). Realities of baiji conservation. *Conservation Biology*, 20, 626-628.

[43] Reeves, R. R., & Leatherwood, S. (Eds.), 1995. *Report of the first meeting of the Asian River Dolphin Committee, Ocean Park, Hong Kong, 5-7 December 1994*. Hong Kong: Ocean Park Conservation Foundation.

[44] Reeves, R. R., Smith, B. D., Crespo, E. A., & Notarbartalo di Sciara, G., (2003). *Dolphins, whales and porpoises. 2002-2010 Conservation Action Plan for the world's cetaceans*. Cambridge, United Kingdom: International Union for Conservation of Nature.

[45] Reeves, R. R., Smith, B. D., & Kasuya, T. (Eds.), (2000). Biology and conservation of freshwater cetaceans in Asia. *Occasional Papers of the IUCN Species Survival Commission* 23: 1-152.

[46] Roberts, D. L., & Kitchener, A. C., (2006). Inferring extinction from biological records: were we too quick to write off Miss Waldron's red colobus monkey (*Piliocolobus badius waldronae*)? *Biological Conservation*, 128, 285-287.

[47] Schaller, G. B., (1993). *The last panda*. Chicago, IL: University of Chicago Press.

[48] Scott, D. A., (1993). *A directory of Asian wetlands*. Gland, Switzerland: International Union for Conservation of Nature.

[49] Shao, M., Tang, X., Zhang, Y., & Li, W., (2006). City clusters in China: air and surface water pollution. *Frontiers in Ecology and the Environment*, 4, 353-361.

[50] Shapiro, J., (2001). *Mao's war against nature: politics and the environment in revolutionary China*. Cambridge, United Kingdom: Cambridge University Press.

[51] Sheng, H., (1998a). *Neophocaena phocaenoides*. In Wang, S., (Ed.), *China Red Data book of endangered animals: Mammalia* (pp. 199-202). Beijing, China: Science Press.

[52] Sheng, H., (1998b). *Lipotes vexillifer*. In Wang, S., (Ed.), *China Red Data book of endangered animals: Mammalia* (pp. 203-207). Beijing, China: Science Press.

[53] Smith, A. M., & Smith, B. D., (1998). Review of status and threats to river cetaceans and recommendations for their conservation. *Environmental Reviews*, 6, 189-206.

[54] Smith, B. D., & Reeves, R. R. (Eds.), (2000). Introduction: Report of the Second Meeting of the Asian River Dolphin Committee, 22-24 February 1997, Rajendrapur, Bangladesh. *Occasional Papers of the IUCN Species Survival Commission*, 23, 1-14.

[55] Smith, B. D., Sinha, R. K., Zhou, K., Chaudhry, A. A., Liu, R., Wang, D., Ahmed, B., Aminul Haque, A. K. M., Mohan, R. S. L., & Sapkota, K., (2000). Register of water development projects affecting river cetaceans in Asia. *Occasional Papers of the IUCN Species Survival Commission*, 23, 22-39.

[56] Snyder, N. F. R., & Snyder, H., (2000). *The California condor: a saga of natural history and conservation*. Princeton: Princeton University Press.

[57] Stone, R., (2007). The last of the leviathans. *Science*, 316, 1684-1688.

[58] Thorbjarnarson, J., Wang, X., Ming, S., He, L., Ding, Y., Wu, Y., & McMurry, S. T., (2002). Wild populations of the Chinese alligator approach extinction. *Biological Conservation*, 103, 93-102.

[59] Turvey, S. T., (2008). *Witness to extinction: how we failed to save the Yangtze River dolphin*. Oxford, United Kingdom: Oxford University Press.

[60] Turvey, S. T., Barrett, L. A., Braulik, G. T., & Wang, D., (2006). Last chance to save? Implementing the recovery programme for the Critically Endangered Yangtze River dolphin. *Oryx*, 40, 258-259.

[61] Turvey, S. T., Pitman, R. L., Taylor, B. L., Barlow, J., Akamatsu, T., Barrett, L. A., Zhao, X., Reeves, R. R., Stewart, B. S., Pusser, L. T., Wang, K., Wei, Z., Zhang, X., Richlen, M., Brandon, J. R., & Wang, D., (2007). First human-caused extinction of a cetacean species? *Biology Letters*, 3, 537-540.

[62] VanderWerf, E. A., Groombridge, J. J., Fretz, J. S., & Swinnerton, K. J., (2006). Decision analysis to guide recovery of the po'ouli, a critically endangered Hawaiian honeycreeper. *Biological Conservation*, 129, 383-392.

[63] Wang, D., Hao, Y., Wang, K., Zhao, Q., Chen, D., Wei, Z., & Zhang, X., (2005). The first Yangtze finless porpoise successfully born in captivity. *Environmental Science and Pollution Research*, 5, 247-250.

[64] Wang, D., Liu, R., Zhang, X., Yang, J., Wei, Z., Zhao, Q., & Wang, X., (2000). Status and conservation of the Yangtze finless porpoise. *Occasional Papers of the IUCN Species Survival Commission*, 23, 81-85.

[65] Wang, D., Zhang, X., Wang, K., Wei, Z., Würsig, B., Braulik, G. T., & Ellis, S., (2006). Conservation of the baiji: no simple solution. *Conservation Biology*, 20, 623-625.

[66] Wang, K., Wang, D., Zhang, X., Pfluger, A., & Barrett, L., (2006). Range-wide Yangtze freshwater dolphin expedition: the last chance to see baiji? *Environmental Science and Pollution Research*, 13, 418-424.

[67] Wei, Q., Ke, F., Zhang, J., Zhuang, P., Luo, J., Zhou, R., & Yang, W., (1997). Biology, fisheries, and conservation of the sturgeons and paddlefish in China. *Environmental Biology of Fishes*, 48, 241-255.

[68] Wen, R., (2006). *The changes of plants and animals in China during different historical periods*. Chongqing, China: Chongqing Publishing Company.

[69] Xie, P., (2003). Three-Gorges Dam: risk to ancient fish. *Science*, 302, 1149.

[70] Yang, G., Bruford, M. W., Wei, F., & Zhou, K., (2006). Conservation options for the baiji: time for realism? *Conservation Biology*, 20, 620-622.

[71] Yang, J., & Liu, H., (2005). Bioaccumulation of trace elements in a captive Chinese river dolphin (*Lipotes vexillifer*). *Acta Theriologica Sinica*, 25, 307-309.

[72] Yang, J., Xiao, W., Kuang, X., Wei, Z., & Liu, R., (2000). Studies on the distribution, population size and the active regularity of *Lipotes vexillifer* and *Neophocaena phocaenoides* in Dongting Lake and Boyang Lake. *Resources and Environment in the Yangtze Basin*, 9, 444-450.

[73] Zhang, X., Wang, D., Liu, R., Wei, Z., Hua, Y., Wang, Y., Chen, Z., & Wang, L., (2003). The Yangtze River dolphin or baiji (*Lipotes vexillifer*): population status and conservation issues in the Yangtze River, China. *Aquatic Conservation of Marine and Freshwater Ecosystems*, 13, 51-64.

[74] Zhang, X., & Wang, K., (1999). Population viability analysis for the Yangtze finless porpoise. *Acta Ecologica Sinica,* 19, 529-533.

[75] Zhang, X., Wei, Z., Wang, X., Yang, X., & Chen, P., (1995). Studies on the feasibility of establishment of a semi-natural reserve at Tian-e-Zhou (Swan) oxbow for baiji, *Lipotes vexillifer*. *Acta Hydrobiologica Sinica*, 19, 110-113.

[76] Zhao, X., Barlow, J., Taylor, B. L., Pitman, R. L., Wang, K., Wei, Z., Stewart, B. S., Turvey, S. T., Akamatsu, T., Reeves, R. R., & Wang, D., (2008). Abundance and conservation status of the Yangtze finless porpoise in the Yangtze River, China. *Biological Conservation*, 141, 3006-3018.

[77] Zhou, K., (1982). On the conservation of the baiji, *Lipotes vexillifer*. *Journal of Nanjing Normal College (Natural Science)*, 1982(4), 71-74.

[78] Zhou, K., (1986). A project to translocate the baiji, *Lipotes vexillifer*, from the main stream of the Yangtze River to Tongling Baiji Semi-nature Reserve. *Aquatic Mammals*, 12.1, 21-24.

[79] Zhou, K., (1989). Progress report on the semi-natural reserve at Tongling. *Occasional Papers of the IUCN Species Survival Commission*, 3, 19-21.

[80] Zhou K., Ellis, S., Leatherwood, S., Bruford, M., Seal, U. S. (Eds.), (1994). *Baiji population and habitat viability assessment report*. Apple Valley, MN: IUCN/SSC Conservation Breeding Specialist Group.

[81] Zhou, K., Leatherwood, S., & Jefferson, T. A., (1995). Records of small cetaceans in Chinese waters: a review. *Asian Marine Biology*, 12, 119-139.

[82] Zhou, K., & Li, Y., (1989). Status and aspects of the ecology and behavior of the baiji, *Lipotes vexillifer*, in the lower Yangtze River. *Occasional Papers of the IUCN Species Survival Commission*, 3, 86-91.

[83] Zhou, K., Qian, W., & Li, Y., (1977). Studies on the distribution of the baiji, *Lipotes vexillifer* Miller. *Acta Zoologica Sinica*, 23, 73-79.

[84] Zhou, K., Sun, J., Gao, A., & Würsig, B., (1998). Baiji (*Lipotes vexillifer*) in the lower Yangtze River: movements, numbers threats and conservation needs. *Aquatic Mammals,* 24, 123-132.

[85] Zhou, K., & Wang, X., (1994). Brief review of passing fishing gear and incidental catches of small cetaceans in Chinese waters. *Report of the International Whaling Commission, Special Issue*, 15, 347-354.

[86] Zhou K., & Zhang, X., (1991). *Baiji: the Yangtze River dolphin and other endangered animals of China*. Washington, D.C.: Stone Wall Press.

[87] Zhong, Y., & Power, G., (1997). Fisheries in China: progress, problems, and prospects. *Canadian Journal of Fisheries and Aquatic Sciences*, 54, 224-238.

In: Biology, Evolution and Conservation of River Dolphins… ISBN: 978-1-60876-633-8
Editors: M. Ruiz-García, J. Shostell. pp. 395-417

Chapter 21

HIGH LEVEL OF MHC POLYMORPHISM IN THE BAIJI AND FINLESS PORPOISE, WITH SPECIAL REFERENCE TO POSSIBLE CONVERGENT ADAPTATION TO THE FRESHWATER YANGTZE RIVER

Shixia Xu, Wenhua Ren, Kaiya Zhou and Guang Yang[1]
Jiangsu Key Laboratory for Biodiversity and Biotechnology, College of Life Sciences, Nanjing Normal University, Nanjing, China

We surveyed the sequence variability at exon 2 of the MHC class I and class II (*DRA* and *DQB*) genes in the baiji (*Lipotes vexillifer*) and finless porpoise (*Neophocaena phocaenoides*). Little sequence variation was detected at the *DRA* locus whereas considerable variation was found at *DQB* and MHC-I. Three exon 2 MHC loci of the baiji revealed striking similarity with those of the finless porpoise. Some identical alleles shared by both species at the MHC-I and *DQB* loci suggest that convergent evolution as a consequence of common adaptive solutions to similar environmental pressures in the Yangtze River. As for the *DRA* locus, the identical alleles were shared not only by baiji and finless porpoise but also by some other cetacean species of the families Phocoenidae and Delphinidae, suggesting trans-species evolution of this gene.

Keywords: *Lipotes vexillifer*; *Neophocaena phocaenoides*; MHC; trans-species evolution; convergent evolution

INTRODUCTION

The major histocompatibility complex (MHC) consists of a group of closely linked genes that constitute the most important genetic component of the mammalian immune system (Klein, 1986). Two major groups of MHC genes can be distinguished, i.e., Class I and II. The fundamental role of class I genes is to recognize antigens from intracellular proteins, including those from viruses. The primary role of class II genes is to recognize antigens from extracellular proteins, including those from bacteria and other pathogens and parasites (Klein & Horejsi, 1997; Dengjel et al., 2005). The ability of both class I and II genes to face various pathogens is believed to be mainly associated with sequence variation among MHC alleles in the functionally important peptide-binding region or PBR which is responsible for antigen recognition (Ohta, 1998; Hughes & Yeager, 1998). Variation within the PBR suggests that there has been evolutionary pressure for organisms to combat a wide range of immunological challenges (Abbott et al., 2006). MHC variability reflects evolutionary relevant and adaptive processes within and between populations and is very suitable to investigate a wide range of open questions in evolutionary ecology and conservation. Certain MHC loci exhibit an extensive genetic polymorphism in most vertebrate species studied so far (Parham & Ohta, 1996; Babik et al., 2005; Sachdev et al., 2005). Despite the extensive polymorphism within species, a remarkable sharing of polymorphic sequence motifs even identical alleles have been observed between different mammalian species (Gustafsson & Andersson, 1994; Kriener et al., 2000; Otting et al., 2002; Suárez et al., 2006; Huchard et al., 2006). Three possible mechanisms have been put forward to explain this phenomenon. First, the similarity of alleles between related species can be explained by their common ancestry — the persistence of allelic lineages through speciation and their passage from species to species (Klein, 1987). This "trans-species polymorphism" has been documented to occur in primate (Otting et al., 2002; Huchard et al., 2006), artiodactyl (Sena et al., 2003), and cetacea (e.g., Hayashi et al., 2003). The second mechanism for convergent evolution, the occurrence of convergent evolution at the amino acid sequence level, has been controversial (Doolittle, 1994). Convergent evolution is the emergence of biological structures or species that exhibit similar function and appearance but that evolved through widely divergent evolutionary pathways (Gustafsson & Andersson, 1994; Hughes, 1999). The similarities that are shared in the case of convergent evolution are not the result of evolution from a common ancestor sharing those similarities. Instead, the similarities are typically explained as the result of common adaptive solutions to similar environmental pressures (Kriener et al., 2000). However, evidence for molecular convergence is either lacking or disputed (Doolittle, 1994). A third possibility is that the similarity has arisen by chance (Kriener enmjt al., 2000).

The baiji or Yangtze River dolphin (*Lipotes vexillifer*) is endemic to the Yangtze River of China, and is probably the most threatened cetacean in the world (Reeves et al., 2003). It has become a flagship species for the conservation of endangered aquatic animals and the entire aquatic ecosystem. The baiji is a relict species and the only living representative of the family Lipotidae (Rice, 1998). This species was listed as critically endangered in the International Union for Conservation of Nature (IUCN) Red List of Threatened Species due to its very low abundance and projected continuing decline (Reeves et al., 2003). The finless porpoise (*Neophocaena phocaenoides*) is a small cetacean widely distributed along the coast waters of Indo-Pacific Oceans and the Yangtze River (Reeves et al., 1997). The Yangtze finless porpoise, a sole freshwater population, is sympatric with the baiji in the middle and lower reaches of the Yangtze River. Due to its unique and limited distribution in freshwater, its small and rapidly declining population size and highly endangered status, and its special adaptation to the freshwater environment, the Yangtze population has been categorized as endangered in the IUCN Red List (Reeves et al., 2003). In addition, a systematic survey recently conducted by a team of scientists from China, USA, and four other countries could not find a single baiji during a 6-week search, which suggested that this species might have gone extinct in the wild. Meanwhile, the abundance of the Yangtze finless porpoise was estimated to be much less than before (Hutzler, 2006). More and more conservation biologists in China have proposed to increase the conservation grade of the finless porpoise from II to I in the List of Key Nationally Protected Animals.

The baiji and finless porpoise both have low levels of genetic variability at neutral markers such as at the mitochondrial control region and microsatellites (Yoshida et al., 2001; Yang et al., 2003 2008; Xia et al., 2005; Zheng et al., 2005). However, no systematic information on sequence variation in adaptive markers is currently available for the baiji and finless porpoise. In the present study, sequences of exon 2 of the MHC class I gene and class II (*DRA* and *DQB*) gene were determined in both species. It is expected to have an in-depth understanding on the behavior of these molecules, esp. sequence variability possibly caused by selection pressure. Findings from this study will provide basic information for studying the MHC immunogenetics at a population level, and especially to identify a genetic basis for its adaptation to freshwater by the Yangtze finless porpoise. Moreover, the MHC data sets reveal a striking interspecific identity and similarity, which suggests that convergent evolution is a response to the common freshwater environment they have inhabited.

MATERIALS AND METHODS

Samples

Fifteen baiji and 195 finless porpoise samples were available for this study. The baiji samples were collected from the lower reaches of the Yangtze River, whereas the finless porpoise samples were collected over a period of more than 20 years from 20 locations along the coast of China, as well as from the middle and lower reaches of the Yangtze River. The finless porpoise samples were assigned to different populations *a priori*, i.e. the Yangtze River population, the Yellow Sea population, and the South China Sea population, according to the discriminant features suggested by Gao and Zhou (1995). All these samples were taken from stranded or incidentally captured/killed individuals. Voucher specimens are preserved in the Jiangsu Key Laboratory for Biodiversity and Biotechnology, College of Life Sciences, Nanjing Normal University (NNU), China.

DNA Isolation and PCR

Myologic and skeletal samples were extracted using the DNeasy Tissue Kit (QIAGEN) and Geneclean for Ancient DNA kit (Q. Biogene), respectively, following the manufacturer's protocol. The exon 2 fragments for MHC-I, *DRA* and *DQB* were amplified using three primer sets as shown in Table 1. The primers used to amplify the *DRA* gene were designed against a conserved region among sheep (*Ovis aries*, GenBank Accession, M73983), cattle (*Bos taurus*, M30120), horses (*Equus caballus*, L47174) and humans (*Homo sapiens*, M60334) (Sena et al. 2003). Polymerase chain reactions (PCR) were carried out in a total volume of 50 μl containing 2.5 mM $MgCl_2$, 10 mM Tris- HCl (pH 8.4), 50 mM KCl, 0.2 mM each dNTP, 0.4 μM each primer, 1.0 unit Ex-Taq DNA polymerase (Takara, Japan) and 10-100 ng DNA template. The PCR cycling scheme included an initial denaturation of 5 min at 94°C, followed by 35 cycles of 94 °C for 30 s, 55 °C for 30 s, 72 °C for 30 s, and a final extension at 72°C for 10 min. The PCR products were purified using Wizard PCR Preps DNA Purification Kit (Promega, USA) according to the manufacturer's instruction.

Table 1. PCR primers used to amplify three MHC loci in the present study

Locus	Size of amplifications (bp)	Primer sequences	Reference
DRA	189	5'-AATCATGTGATCATCCAAGCTGAGTTC-3' 5'-TGTTTGGGGTGTTGTTGGAGCG-3'	This study
DQB	172	5'-CTGGTAGTTGTGTCTGCACAC-3' 5'-CATGTGCTACTTCACCAACGG-3'	Murray et al. (1995)
MHC-I	147	5'-TACGTGGMCGACACGSAGTTC-3' 5'-CTCGCTCTGGTTGTAGTAGCS-3'	Flores-Ramirez et al. (2000)

SSCP, Cloning and Sequencing

All the finless porpoise samples were first screened for consistent polymorphism at exon 2 of the MHC-I, *DRA* and *DQB* loci using single-strand conformation polymorphism (SSCP). The selected samples were then characterized at the genetic level by DNA sequence comparison. For SSCP analysis, 1 μl of the purified PCR product was mixed with 9 μl of loading dye (95% v/v formamide, 20 mM EDTA, 0.05% w/v Bromophenol Blue, 0.05% w/v xylene cyanol). After denaturing at 95°C for 10 min and cooling on ice for 5 min, 5 ul of the mixture was loaded into a 10% polyacrylamide gel (38:1, acrylamide/bisacrylamide). Electrophoresis was performed in 1×TBE buffer at 150 V for 16~20 h at room temperature. After completion of the run, SSCP bands were visualized by silver staining procedures. To avoid categorizing PCR artifacts as a new allele based on the SSCP bands, the PCR products were rearranged and separated again on the gel according to assessed similarities. In this study, each sample was analyzed at least twice following the same procedure. For new samples, all known alleles were run as references on each SSCP gel.

PCR products of the finless porpoise showing the same SSCP pattern in the replicates were cloned into the pMD-18T vectors using the TA cloning kit (Takara, Japan). For each locus, five to six randomly chosen PCR products were cloned from each SSCP genotype. While for the baiji, all the PCR products were cloned into a pMD-18T vector. Four to six clones were picked for each cloned PCR product and sequenced in the forward and/or reverse directions. The sequence reaction was using the BigDye Terminator Cycle Sequencing Ready Reaction Kit (ABI). Automated DNA sequence analysis was performed on an ABI 3730 automated genetic analyzer.

Data Analysis

Statistical analysis of nucleotide and amino acid sequences were computed in MEGA version 4 (Tamura et al., 2007). The average rate of nonsynonymous (d_N) and synonymous (d_S) substitutions in the overall domain, PBR, and non-PBR were calculated according to the Nei–Gojobori method (Nei & Gojobori, 1986) with the Jukes–Cantor correction for multiple substitutions. The standard errors were obtained by 1000 bootstrap replicates. To test whether positive selection was operating at each locus we compared the relative abundance of synonymous and nonsynonymous substitutions using a Z-test at the 5% level (Tamura et al., 2007). PBR and non-PBR were identified assuming homology with predictions made for human MHC molecules (Brown et al., 1993; Bjorkman et al., 1987).

Mismatch distribution analyses (Figueroa et al., 2000; Go et al., 2002; Suárez et al., 2006) were used to detect convergence in each locus and demographic history of the baiji and finless porpoise. The sudden expansion model (Rogers & Harpending, 1992) and goodness-of-fit tests (sum of squared deviations, SSD; Harpending's raggedness index, R; Schneider & Excoffier, 1999) of the observed to the estimated mismatch distributions were computed in ARLEQUIN version 3.0 (Excoffier et al., 2005). The program GENEVONV version 1.81 (Sawyer, 1999) was employed to find the most likely candidate alleles for intragenic recombination/gene conversion events in the baiji and finless porpoise. This method uses pairwise comparison of sequences in the alignment to find blocks of sequence pairs that are more similar than would be expected by chance. GENEVONV finds and ranks the highest-scoring fragments globally for the entire alignment. Global permutation test P-values of <0.05 (derived from BLAST-like global scores using 10,000 replicates) were considered as evidence of intragenic recombination. Phylogenetic trees employing the neighbor-joining (NJ) method were constructed according to Kimura 2-parameter (or K2P) nucleotide distances in MEGA version 4 (Tamura et al., 2007) in order to reveal relationships between alleles of the baiji and finless porpoise, and alleles from some other cetacean species. Bootstrap confidence intervals were obtained from 1000 replicates.

RESULTS

Allelic Diversity of the Finless Porpoise at Three MHC Genes

DRA

A total of six unique sequences were detected for the *DRA* gene from all the samples. All individuals tested had no more than two sequences, suggesting that the *DRA* primers used in this study amplified a single locus. Allele *Neph-DRA*06* had a stop codon in the middle part of the sequence, representing most likely a pseudogene, and it was excluded from further analyses. The remaining allele sequences have been submitted to GenBank with accession numbers DQ843609–DQ843613. Sequence polymorphism analysis showed that the overall variation was very low, with 2.1% variability at the nucleotide level (four polymorphic sites of 189 base pairs sequenced) and 4.8% at the amino acid level (three polymorphic sites of 63 amino acid residues) (Figure 1a). On the other hand, the similarity between *DRA* alleles was extremely high, with each pair containing only a maximum of two different nucleotides. For the deduced amino acid sequences, three of the four polymorphic sites represented non-synonymous substitutions (Figure 1a) but none were found in the $\alpha 1$ domain which encodes the putative MHC class II PBR (Brown et al. 1993; Stern et al. 1994). Of all the alleles identified, *Neph-DRA*01* had the highest frequency while the other alleles only appeared once or twice.

DQB

Fourteen unique sequences were detected in this study, and they were labeled from *Neph-DQB*01* to *Neph-DQB*14* (GenBank accession nos. DQ843614–DQ843623 and EF056477–EF056480). Five of these sequences, i.e. *Neph-DQB*06*, 07, 08, 09 and *10*, have previously been reported from porpoises in Japanese waters (Hayashi et al. 2003). In the 14 sequences, five were found to be population-specific, i.e. *Neph-DQB*02, Neph-DQB*13* and *Neph-DQB*11, 12, 14* appeared in the South China Sea population, Yellow Sea population, and Yangtze River population, respectively, whereas the remaining nine sequences were shared by population pairs. None of the sequences contained insertions/deletions (or indels), or stop codons, suggesting that all sequences might come from functional molecules in the genome. More than two *DQB* sequences were detected in 8 out the 195 individuals examined in this study. Five of the eight porpoises had three unique sequences, two of them had four, and the last one had five, suggesting at least three copies of the *DQB* gene existed in finless porpoises. Compared with *Neph-DRA*, *DQB* had a relatively higher variability of 12.8% (22/172) at the nucleotide level and 24.6% (14/57) at the amino acid level (Figure 1b). The number of pairwise nucleotide differences between pairs of sequences ranged from 1 (*Neph-DQB*01* vs. *Neph-DQB*02*) to 16 (*Neph-DQB*10* vs. *Neph-DQB*14*), and the number for amino acid varied from 0 (*Neph-DQB*06* vs. *Neph-DQB*11*) to 11 (*Neph-DQB*10* vs. *Neph-DQB*14*). These values indicate the divergence both within and between loci as we were not able to distinguish unique sequences of particular loci according to gene trees. But in further analyses, we considered all sequences as if they would be alleles of one locus. In addition, the rate of nonsynonymous substitutions was more than four times higher than that of synonymous substitutions ($P = 0.071$, *Z*-test of positive selection) in the putative PBR, while the rate decreased to two ($P = 0.148$, *Z*-test of positive selection) in the Non-PBR (Table 2).

MHC-I

Thirty-four unique sequences were identified for the MHC-I gene (GenBank accession nos. DQ843624–DQ843657). No indels or stop codons were detected. However, these sequences are possibly active members of three loci rather than a single specific locus because three to five distinct sequences were detected from most of the individuals. In addition, it is very difficult to divide loci on a genetic tree and we considered all sequences as if they would be alleles of one locus in the following analyses. The variability was 47 of 147 (32.0%) in nucleotide sequences and 25 of 49 (51.0%) in amino acid residues (Figure 1c). Nucleotide sequence variation between all pairwise comparisons of *Neph-I* sequences ranged from 1 (*Neph-I*02* vs. *Neph-I *26*) to 25 nucleotides (*Neph-I *14* vs. *Neph-I *26*), whereas amino acid substitutions ranged from 1 (*Neph-I *02* vs. *Neph-I *26*) to 19 (*Neph-I *14* vs. *Neph-I *22*). The detected number of unique sequences was 13 for 134 clones from 22 Yangtze samples, 22 for 100 clones from 20 Yellow Sea samples, and 26 for 133 clones from 26 South China Sea samples, respectively. The five common sequences of *Neph-I*01, 02, 03, 04* and *I*07* were the most frequent and widespread in the three populations, and 17 sequences were shared only by some population pairs. In addition, 12 population-specific sequences were identified, four (*I*06, I*14, I*23, I*32*) in the Yangtze River population, five (*I*21, I*24, I*25, I*31, I*34*) in the South China Sea population, and three (*I*12, I*26, I*27*) in the Yellow Sea population, respectively. The *Z*-test showed that d_N was significantly greater than d_S in the PBR (d_N/d_S= 5.8, $P = 0.002$), but could not reach the significant level in the non-PBR (d_N/d_S=1.6, $P = 0.188$) (Table 2).

Allelic Diversity of the Baiji at Three MHC Genes

DRA

Three *DRA* unique sequences were identified from 75 clones of 15 baiji (Figure 1a). All individuals have no more than two sequences, which suggested that only one *DRA* locus was amplified in each sample. Sequences are available in GenBank under Accession Nos. DQ851844–DQ851846. The allele *Live-DRA*03* most likely represented a pseudogene, as evidenced by the absence of a base in the middle part of the sequence. This allele, therefore, was not included in the following analyses. The allele frequency of *Live-DRA*01* and *Live-DRA*02* was 0.633 and 0.367, respectively. In the 189 bp *DRA* sequence, only two nucleotide sites (1.1%) were variable in the baiji. An alignment of the 63 unique inferred amino acid sequences revealed two variable sites (3.2%), both of which represented nonsynonymous substitutions in the non-PBR (Figure 1a & Table 2).

DQB

Eight *DQB* alleles of *L. vexillifer* were identified (Figure 1b). Sequences are available in GenBank under Accession Nos.: *Live-DQB*4*: AY177153; *Live-DQB*5*: AY177283; *Live-DQB*8*: AY177286; *Live-DQB*11*: AY177289; *Live-DQB*13*: AY177291; *Live-DQB*16*: AY333383; *Live-DQB*28*: AY333395; *Live-DQB*29*: AY333396). The variability was 16 of 172 (9.30%) in nucleotide sequences and 9 of 57 (15.79%) in amino acid residues (Figure 1b). The number of pairwise nucleotide differences between pairs of sequences ranged from 1 (*Live-DQB*8* vs. *Live-DQB*11*) to 16 (*Live-DQB*28* vs. *Live-DQB*11*), and the number for amino acid varied from 0 (*Live-DQB*8* vs. *Live-DQB*11*) to 9 (*Live-DQB*28* vs. *Live-DQB*11*). Considerable sequence variation was also evident because the rate of nonsynonymous substitutions was more than eight times higher than that of synonymous substitutions ($P = 0.033$, *Z*-test of positive selection) in the putative PBR, while the rate decreased to 0.2 ($P = 0.136$, *Z*-test of positive selection) in the Non-PBR (Table 2).

MHC-I

A 147 bp fragment of the MHC-I exon 2 was amplified from each of 15 baiji. A total of 111 clones were sequenced and six unique sequences, *Live-I*01* to *Live-I*06* (GenBank Accession Nos. DQ851847–DQ851852), were identified (Figure 1c). We named these sequences as MHC alleles although we were aware that they stemmed from at least three different loci rather than a single specific locus. Ten out of 15 individuals had three to five discrepant sequences. But in further analyses, we considered all sequences as if they would be alleles of one locus. None of the sequences contained insertions/deletions (or indels) or stop codons, suggesting that all sequences come from functional alleles in the genome. Thirty-five out of 147 (23.8%) nucleotide positions were variable. The number of pairwise nucleotide differences between pairs of alleles ranged from six (*Live-I*04* vs. *Live-I*06*) to 30 (*Live-I*05* vs. *Live-I*06*). In the deduced amino acid sequences, 21 out of 49 (42.9%) were variable

(Figure 1c). Amino acid replacements between alleles ranged from 4 (*Live-I*03* vs. *Live-I*05*) to 18 (*Live-I*03* vs. *Live-I*04*).

(a) MHC-DRA

```
                   10         20         30         40         50         60
              *  *  *   **        *      *  *** *   *   *  ** *
Neph-DRA*01 SLSPDQSNEF MFDFDGDEIF HVDMEKRETV WRLKEFGNFA SFQAQGALAN MAVGKANLDI LIK
Neph-DRA*02 .......... .......... .........A .......... .......... .......... ...
Neph-DRA*03 .....L.... .......... .......... .......... .......... .......... ...
Neph-DRA*04 .......... .......... ..G....... .......... .......... .......... ...
Neph-DRA*05 .......... .......... .......... .......... .......... .......... ...
Live-DRA*01 .......... .......... .......... .......... .......... .......... ...
Live-DRA*02 .......... .......... ......K... .......... .......... .......... M..
Live-DRA*03 .......... .......... .......... .......... .......... .........- ...
```

(b) MHC-DQB

```
                   10         20         30         40         50
              *  *  *  **       *      *  * *  * * * * *
Neph-DQB*01 TERVRLVERH IYNREEYVRF DSDVGEYRAV TELGRRTAEY WNGQKDILEQ KRAELDT
Neph-DQB*02 .......... .......... .......... .......... .......... E......
Neph-DQB*03 .....F...Y .......... .......... .......... ..S...L... ...VV..
Neph-DQB*04 .........Y ......F... .......... .....PD... ......L... .....G.
Neph-DQB*05 .....F...Y ......F... .......... .....PD.K. .......... N......
Neph-DQB*06 .....F.... .......... .......... .....PD.K. .......... .......
Neph-DQB*07 .........Y ......F... .......... .....PD.K. .......... .......
Neph-DQB*08 .........Y .......... .......... .....PD.K. .......... .......
Neph-DQB*09 .....F.... ......FL.. .......... .....QI..N .......... .......
Neph-DQB*10 .......... .......... .......... .....PD.K. .......... .......
Neph-DQB*11 .....F.... .......... .......... .....PD.K. .......... .......
Neph-DQB*12 .....F...Y ......F... .......... .....PD.K. .......... .......
Neph-DQB*13 .....F...Y .......... .......... .......... .......... .......
Neph-DQB*14 .....F...Y ......FT.. .......... .......... ..S...L... ...VV..
Live-DQB*4 .....F...Y ......FT.. .......... .......... ..S...L... ...VV..
Live-DQB*5 .....F...Y ......FT.. .......... .......... ..S...L... ...VV..
Live-DQB*8 .....YMT.. .......... .......... .......... ..S...L... R...V..
Live-DQB*11 .....YMT.. .......... .......... .......... ..S...L... R...V..
```

Figure 1 (Continued).

```
Live-DQB*13 .....YMT.. .......... .......... .......... ..S...L... ...VV..
Live-DQB*16 .....F...Y .......... .......... .......... ..S...L... ...VV..
Live-DQB*28 .....F...Y ......FT.. .......W.. .......... ..S...L... ...VV..
Live-DQB*29 .....F...Y ......FA.. .......... .......... ..S...L... ...VV..
```

(c) MHC-I

```
               10         20         30         40
                *         ** **      ** * * * * **
Neph-I*01 VRFDSDAPNP RGEPRAPWVE QVGPEYWDRN TRIYKDAAQF YRESLNNLR
Neph-I*02 .......... .K......M. .E.......E .Q.S...... ..VN.....
Neph-I*03 .......... .M..W..... .E........ P....HH..T F.VN..T.C
Neph-I*04 .......... ........M. .E.......E ..NF.EN... ...D..T..
Neph-I*05 .......... .E......M. .E.......E .Q.S..T..L .........
Neph-I*06 .......... .M......M. .E.......E .Q.S.....V ...D..T..
Neph-I*07 .......... .......... .........E .Q.S...... .........
Neph-I*08 .......... ........M. .E........ .........L ...D..I..
Neph-I*09 .......... .E......M. .E.S...... ...C..T..I ...D..T..
Neph-I*10 .......... .E......I. .E.S...... ...C..T..I ...D..T..
Neph-I*11 .......... ........I. .E.......E ..NF.G...I ...D..T..
Neph-I*12 .......... .K......M. .E.......E .Q.S.EN..I ...D..I..
Neph-I*13 .......... .K......M. .E.......E ...S.EN..I ...D..T..
Neph-I*14 L...G..... .M.LW..... .K........ P....HH..T F.VN..T.C
Neph-I*15 .......... .E......I. .E........ .........I ...N..I..
Neph-I*16 .......... .K......M. .E.......E ...S.EN..I ...N..T..
Neph-I*17 .......... ........I. .E........ ..NF.G...I ...D..T..
Neph-I*18 .......... .K......M. .........E ...S.EN..I .........
Neph-I*19 .......... .......... .E........ ...F.....I ..VN.ST..
Neph-I*20 .......... .M......I. .E.......E ...C..T..L ...D..T..
Neph-I*21 .......... .E......I. ...S...... ...C..T..I ...D..T..
Neph-I*22 .......... .K......M. .E.......E .Q.S.EN... .........
Neph-I*23 .......... .K......M. .E.......E .Q.S..T..L ...D..T..
Neph-I*24 .......... .E......I. .E.......E .Q.S..T..L ...D..T..
Neph-I*25 .......... .E......I. .E.......E .Q.S..T... ..VN..T..
Neph-I*26 A......... .K......M. .E.......E .Q.S...... ..VN.....
```

Figure 1 (Continued).

```
Neph-I*27 .......... .K......M. .E.....V.E .Q.S...... ..VN.....
Neph-I*28 ....G..... .K......M. .E.......E .Q.S...... ..VN.....
Neph-I*29 .......... .E......I. .E........ ...S..T..L .........
Neph-I*30 .......... .......... .........E .Q.S...... ..VN.....
Neph-I*31 .......... .......... .......... .........I ...D..T..
Neph-I*32 .......... .M........ .E.....EEQ ..GC.....I ..VD..T..
Neph-I*33 .......... .K........ .E.....EEE ...S.....I ..VN..I..
Neph-I*34 .......... .M......I. .E.......E ...C..T..L ...N..I..
Live-I*01 .......... .......... .E........ ...F.....I ..VN.ST..
Live-I*02 .......... .K........ .E.....EEE ...S.....I ..VN..I..
Live-I*03 .......... .......... .ER....EEE ...L.....I ..VN.ST..
Live-I*04 L...G..... .M.LW..... .K........ P...HH...T F.VN..T.C
Live-I*05 L......... .......... .E.....EEE ..KL.G...I ..VN.ST..
Live-I*06 ....G..... .M..W..... .Q........ P...HH...T F.VN..T.C
```

Figure 1. Alignment of predicted amino acid sequences of MHC-DRA (a), *DQB* (b) and MHC-I (c) exon 2 from the baiji. Dots (.) indicate residues identical to the reference sequences. Putative peptide binding sites (Brown et al. 1993, Bjorkman et al., 1987) are marked with asterisks (*).

Table 2. The estimated rates (± standard error) of nonsynonymous (d_N) and synonymous (d_S) substitutions for overall domain, peptide (PBR) and non-peptide (non-PBR) binding region and the results of the *Z*-test of positive selection for exon 2 of MHC-I, *DQB*, and *DRA* sequences in the baiji and finless porpoise. d_S and d_N values are given as percentages per site

Loci		Sites	d_N	d_S	d_N/d_S	*P*
Finless porpoise	*DRA*	PBR	0 (0)	3.8 (4.5)	0	1.000
		Non-PBR	1.2 (0.7)	0 (0)	-	0.038
		All	0.9 (0.5)	0.8 (0.9)	1.1	0.476
	DQB	PBR	10.0 (3.5)	2.3 (3.0)	4.3	0.071
		Non-PBR	3.7 (1.7)	1.8 (1.0)	2.1	0.148
		All	5.2 (1.5)	2.0 (0.9)	2.6	0.037
	MHC-I	PBR	26.9 (8.0)	1.7 (1.3)	15.8	0.002
		Non-PBR	4.8 (1.5)	3.0 (1.4)	1.6	0.188
		All	10.7 (2.5)	3.5 (1.4)	3.1	0.003
Baiji	*DRA*	PBR	0.0 (0.0)	0.0 (0.0)	-	-
		Non-PBR	2.0 (1.4)	0.0 (0.0)	-	0.075
		All	1.4 (1.0)	0.0 (0.0)	-	0.073
	DQB	PBR	17.7 (8.7)	2.2 (2.4)	8.0	0.033
		Non-PBR	1.5 (0.9)	6.5 (3.3)	0.2	0.136
		All	5.0 (1.9)	5.4 (2.7)	0.9	0.886
	MHC-I	PBR	42.8 (17.3)	16.9 (8.9)	2.5	0.105
		Non-PBR	8.5 (2.3)	4.8 (2.9)	1.8	0.178
		All	15.7 (3.6)	7.8 (3.1)	2.0	0.065

The baiji amino acid sequences correspond to sites 34–82 of the human leukocyte antigen (HLA) and owl monkeys (*Aotus nancymaae*) MHC class I sequences (Cardenas et al., 2005). The rates of synonymous (d_S) and nonsynonymous (d_N) substitutions separately for the PBR, non-PBR, as well as the overall domain are shown in Table 2. Of the 13 sites predicted to be involved in the PBR (Brown et al., 1993), nine (69.2%) were variable whereas 12 out of 36 (33.3%) in the non-PBR were polymorphic. The relative frequency of d_N was higher than that of d_S especially in the PBR (Table 2).

Phylogenetic Reconstruction

The phylogenetic reconstruction based on the NJ algorithm showed that sequences of the baiji at three loci did not form a monophyletic clade but grouped with those of the finless porpoise (Figure 2). For example, in the NJ tree of MHC-I gene, three alleles of the baiji (*Live-I*01*, *02*, *04*), respectively, grouped with three alleles of the finless porpoise (*Neph-I*19*, *33*, *14*) (Figure. 2c). In addition, some alleles from different species of the family Phocoenidae were within a same lineage. For instance, *Neph-I*04* clustered with five alleles of the vaquita (*Phocoena sinus*) with 94% bootstrap support (Figure 2c), whereas two alleles, respectively, from the vaquita (*Phsi-DQB*01*) and harbor porpoise *P. phocoena* (*Phph-a*) were included in a lineage constituted mainly by the finless porpoise alleles (Figure 2b).

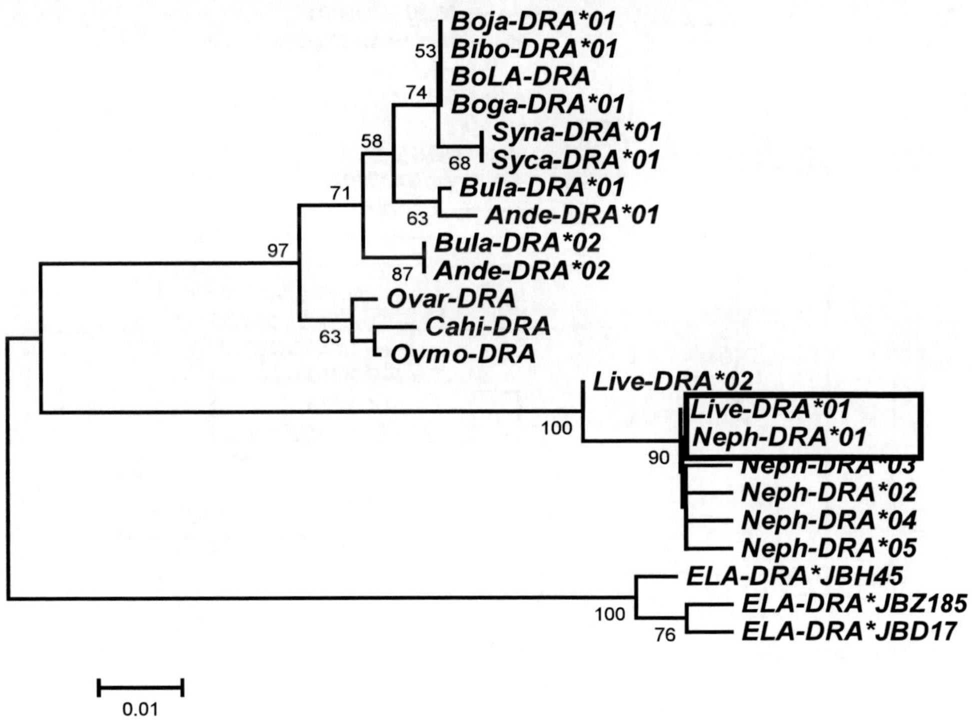

Figure 2. Neighbor-joining phylogenetic trees reconstructed based on nucleotide sequences of the *DRA* (a), *DQB* (b) and MHC-I (c), from a matrix of Kimura 2-parameter nucleotide distances. Numbers at branch-points represent bootstrap support values and only bootstrap values P ≥ 50% (1000 replicates) were shown. Identical alleles between the baiji and finless porpoise were marked with frame. Other allelic sequences downloaded from GenBank were included in the analyses, they are: (a)DRA: *B. javanicus* (Boja-DRA*01: AF385487), *Bison bonasus* (Bibo-DRA*01: AF385485), *B. gaurus* (Boga-DRA*01: AF385486), *Syncerus caffer nanus* (Syna-DRA*01: AF385491), *S. c. caffer* (Syca-DRA*01:

AF385490), *Bubalus bubalis* (Bula-DRA*01-02: AF385488–AF385489), *Anoa depressicornis* (Ande-DRA*01-02: AF385483–AF385484), *Ovibos moschatus* (Ovmo-DRA: AF227193), *O. aries* (Ovar-DRA: M73983), *Capra hircus* (Cahi-DRA: AB008755), *E. caballus* (ELA-DRA*JBZ185: AJ575299; ELADRA*JBD17: AJ575297; ELA-DRA*JBH45: AJ575298). (b)DQB: three alleles *of N. phocaenoides* in Japan waters (Neph-a: AB164212; Neph-g: AB164218; Neph-h: AB164219), *P. phocoena* (Phph-a: AB164211), *P. sinus* (Phsi-DQB*01: AY170897), *Physeter macrocephalus* (Phma-a: AB164208), *Balaenoptera bonaerensis* (Babo-a: AB164202), *B. acutorostrata* (Baac-a: AB164201), *B. physalus* (Baph-a: AB164199), *Delphinus delphis* (Dede-a: AB164220), *Lagenorhynchus obliquidens* (Laob-a: AB164224), *Globicephala macrorhynchus* (Glma-a: AB164226), *Monodon monoceros* (Momo-DQB*0201: U16991), *Delphinapterus leucas* (Dele-DQB*0201: U16989; Dele-DQB*0101: U16986), *Sus scrofa* (SLA-DQB*P06: AF272715), and *B. taurus* (BoLA-DQB*1001: U62318).(c) MHC-I: *P. sinus* (*Phsi*01-Phsi*06*: AY170890–AY170895).

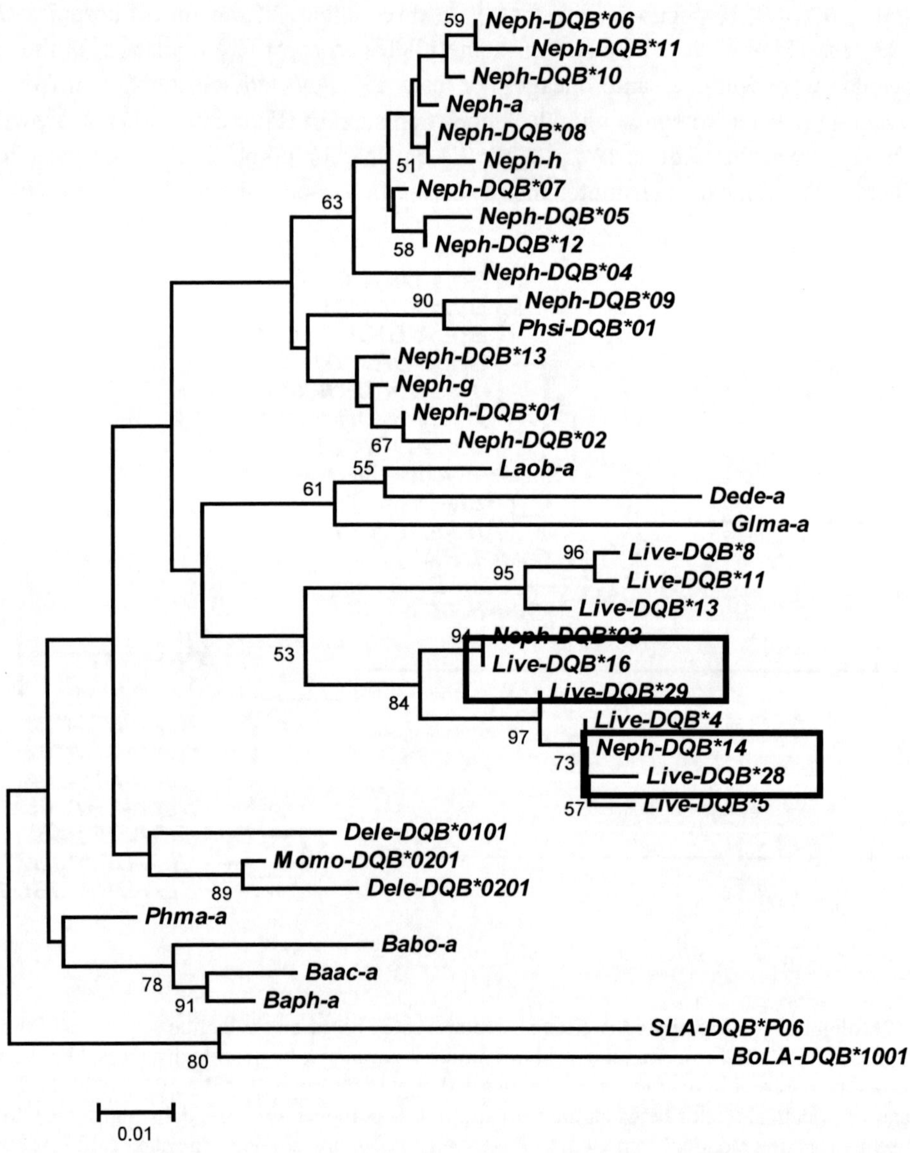

Figure 2b.

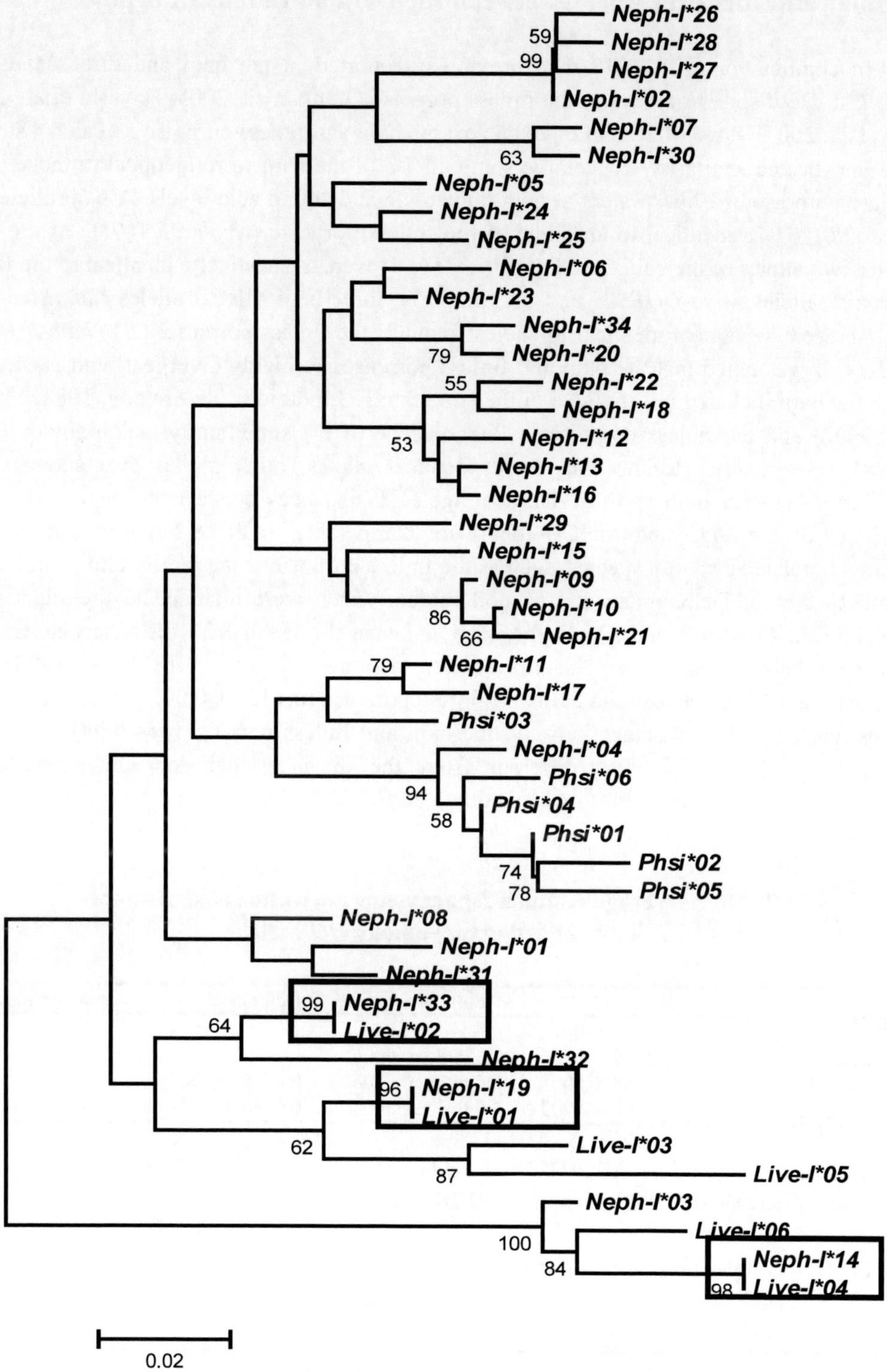
Neph-I*26
Neph-I*28
Neph-I*27
Neph-I*02
Neph-I*07
Neph-I*30
Neph-I*05
Neph-I*24
Neph-I*25
Neph-I*06
Neph-I*23
Neph-I*34
Neph-I*20
Neph-I*22
Neph-I*18
Neph-I*12
Neph-I*13
Neph-I*16
Neph-I*29
Neph-I*15
Neph-I*09
Neph-I*10
Neph-I*21
Neph-I*11
Neph-I*17
Phsi*03
Neph-I*04
Phsi*06
Phsi*04
Phsi*01
Phsi*02
Phsi*05
Neph-I*08
Neph-I*01
Neph-I*31
Neph-I*33
Live-I*02
Neph-I*32
Neph-I*19
Live-I*01
Live-I*03
Live-I*05
Neph-I*03
Live-I*06
Neph-I*14
Live-I*04
59
99
63
79
55
53
86
66
79
94
58
74
78
99
64
96
62
87
100
84
98
0.02

Figure 2c.

Similar and Identical Alleles between the Baiji and Finless Porpoise

In combination with *DQB* data previously reported in the baiji and three MHC loci (MHC-I, *DQB* and *DRA*) data in the finless porpoise (Yang et al., 2005; Hayashi et al., 2006; Xu et al., 2007), three MHC loci on exon 2 of the baiji and finless porpoise revealed a striking and unexpected similarity. Especially, some alleles of the baiji were identical to those of the finless porpoise of China waters at both nucleotide and amino acid levels. A baiji allele, i.e., *Live-DRA*01*, is identical to an allele of the finless porpoise (*Neph-DRA*01*). At the *DQB* locus, two alleles of the baiji, i.e., *Live-DQB*16*, *04* were, respectively, identical to the finless porpoise alleles *Neph-DQB*03* and *14*. Similarly, three baiji MHC-I alleles (i.e., *Live-I*01*, *02*, *04*) were separately identical to alleles found in the finless porpoise, i.e., *Neph-I*19*, *33*, and *14*. It was noted that the baiji and finless porpoise are highly divergent with each other, with the baiji included in Lipotidae of the superfamily Lipotoidea (de Muizon, 1988; Yang et al., 2002) and the finless porpoise in Phocoenidae of the superfamily Delphinoidea (Rice, 1998), respectively. Besides the above identical alleles, other alleles also showed high similarity between both species. The average K2P distances between the baiji and finless porpoise at the *DQB* and MHC-I loci were comparable to those between each pair of relatively related cetacean species, such as the finless porpoise in the family Phocoenidae, and some species of Delphinidae and Monodonitdae, which were included in the superfamily Delphinidea (Tables 3 and 4). At the MHC-I locus, the mean K2P distance between two closely related species, i.e., the finless porpoise and vaquita (0.1014 ± 0.0011) was comparable to that between the baiji and finless porpoise (0.1321 ± 0.0021). And at the *DQB* locus, the mean K2P distance between the baiji and finless porpoise was 0.0815 ± 0.0006, which was not significantly different from the distances between either species of Delphinidae (0.0972 ± 0.0003) or Monodontidae (0.0731 ± 0.0003) and the finless porpoise (P > 0.05) (Table 3).

Table 3. Average Kimura 2-parameter nucleotide acid distances (± standard error) among *DQB* alleles

	Baiji	Finless porpoise	Delphinidae	Monodontidae
Baiji	0.0511±0.0012			
Finless porpoise	0.0815±0.0006	0.0386±0.0006		
Delphinidae[a]	0.1088±0.0001	0.0972±0.0003	0.0736±0.0008	
Monodontidae[b]	0.0919±0.0003	0.0731±0.0003	0.0750±0.0141	0.0360±0.0003

[a] Including *D. delphis* (Dede-a: AB164220), *L. obliquidens* (Laob-a: AB164224), and *G. macrorhynchus* (Glma-a: AB164226), [b] Including *M. monoceros* (Momo-DQB*0201: U16991), *D. leucas* (Dele-DQB*0201: U16989; Dele-DQB*0101: U16986)

Table 4. Average Kimura 2-parameter nucleotide acid distances (± standard error) among MHC-I alleles

	Baiji	Finless porpoise	Vaquita
Baiji	0.1335±0.0047		
Finless porpoise	0.1321±0.0021	0.0863±0.0015	
Vaquita[c]	0.1470±0.0019	0.1014±0.0011	0.0322±0.0087

[c] Six alleles, i.e. *Phsi*01-Phsi*06* (AY170890-AY170895), were used

Mismatch Distributions and Intragenic Recombination Analyses

Mismatch distribution analyses supported a pattern of demographic expansion or high mutation rate leading to sequence convergence in the finless porpoise at three MHC loci. The goodness-of-fit tests were not significant ($P > 0.05$) (Figure 3), while the analyses for the baiji differed significantly from expectations under the sudden-expansion model ($P < 0.05$) (Figure 3). This suggested that the baiji did not undergo a historical population expansion, or total recombination at sequences of three loci.

The GENECONV analysis showed that intragenic recombination (or homologous gene conversion) events in the baiji and finless porpoise have occurred at the *DQB* locus, but not at the MHC-I and *DRA* loci (Table 5). Intragenic recombination events were not only detected within segmental variants of the baiji but also between alleles of the two species. As a whole, three *Neph-DQB* and nine *Live-DQB* alleles were found to be involved in intragenic recombination events ($P < 0.05$, Table 5). Further, some sequence blocks (i.e. DNA block 30–152 and DNA block 30–172, see Table 5) were repeatedly involved in recombination events and may have served as recombination hot spots.

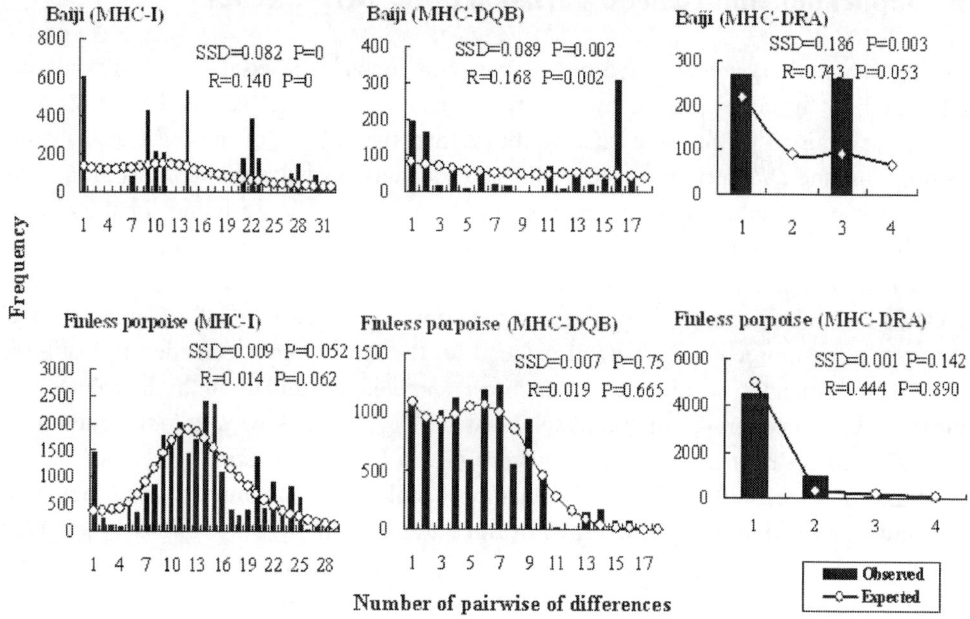

Figure 3. The observed pairwise difference (bars), and the expected mismatch distributions under the sudden expansion model (solid line) of the MHC-I, *DQB*, and *DRA* alleles in the baiji and finless porpoise.

Table 5. Gene conversion events between MHC *DQB* sequences from the baiji and finless porpoise, identified using GENECONV. Sim P = Simulated P values based on 10,000 permutations; Begin = first nucleotide of the converted region; End = last nucleotide of the converted region; Length = length of the converted region; MisM Pen indicates the mismatch penalty

Sequence 1	Sequence 2	Sim P	Begin	End	Length	MisM Pen
*Neph-DQB*03*	*Live-DQB*8*	0.0406	30	152	123	None
*Neph-DQB*03*	*Live-DQB*11*	0.0406	30	152	123	None
*Neph-DQB*03*	*Live-DQB*13*	0.0406	30	172	143	None
*Live-DQB*16*	*Live-DQB*8*	0.0406	30	152	123	None
*Live-DQB*16*	*Live-DQB*11*	0.0406	30	152	123	None
*Live-DQB*16*	*Live-DQB*13*	0.0406	30	172	143	None

CONCLUSION

Gene Duplication and Genetic Variation of the MHC Genes

Many mammalian species have one single *DRA* locus (Chu et al., 1994; Takada et al., 1998), which is further approved by the present chapter. All the baiji and finless porpoise individuals examined in this study had no more than two alleles of the *DRA* gene, strongly suggesting that the *DRA* primers used in this study had amplified a single locus in this study. However, some examples of strong gene duplication evidence were found for the *DQB* and MHC-I loci. For example, as shown in the Results section, three, four, or five distinct sequences were detected separately in eight individuals of the finless porpoise and 10 individuals of the baiji, suggesting at least three copies of *DQB* gene existed in both species. In contrast, at least three copies were also found for the MHC-I gene considering that three to five distinct sequences were detected in most samples examined in both species. Gene duplication was corroborated in humpback whales (*Megaptera novaeangliae*), southern right whales (*Eubalaena australis*) and grey whales (*Eschrichtius robustus*) (Baker et al., 2006; Flores-Ramirez et al., 2000). These duplications could be analogous or homologous with cattle, which are also known to have two or three transcribed *DQB* and MHC-I loci (Ellis et al., 1999a, b). However, there were no significant groupings of sequences that would indicate divergence of the duplicate genes, and so it was unable to attribute each sequence to specific loci. Further, although Baker et al. (2006) suggested that *DQB* duplication in the baleen whale (suborder Mysticeti) and baiji (suborder Odonotoceti), an early divergence of the toothed whales (suborder Odonotoceti; Cassens et al., 2000), is consistent with retention of an ancestral condition shared with the ruminants and loss in the more derived cetaceans such as the beluga, the narwhal (family Monodontidae) and the true dolphins (*Delphinus delphis* included in family Delphinidae), this was not supported by the finless porpoise (family Phocoenidae) examined in this chapter.

Although populations have dramatically declined in numbers, the baiji (which may now be extinct) and the finless porpoise still retain considerable MHC genetic diversity, which is supported by the large number of unique sequences examined in this chapter. For the baiji, a total of two *DRA*, eight *DQB*, six MHC-I alleles were identified in 15 samples. In contrast, a total of 5 *DRA*, 14 *DQB* and 34 MHC-I unique sequences were identified in 195 finless porpoises. In addition, a high level of sequence variation between sequences also indicates that the MHC genes have significant genetic diversity.

Similar to other mammalian species, the *DRA* gene showed very low sequence divergence in the baiji and the finless porpoise, with 3.17% sequence variation in the baiji and 4.8% in the finless porpoise at the amino acid level. In contrast, a considerable sequence variation was detected at the *DQB* locus. This was evidenced by the high level of nucleotide sequence variation among pairwise comparisons, which ranged from 0.58% to 9.30% in both the baiji and the finless porpoise. However, of the three genes investigated in this study, MHC-I showed the most extensive variability. For example, nucleotide sequence variation among all pairwise comparisons of Live-I sequences, corrected for multiple substitutions, ranged from 12.24% to 61.22%. The high divergence between alleles of MHC-I was also supported by the relatively higher ratio of d_N/d_S than those of *DQB* and *DRA* as shown in Table 2.

Variability was higher in the functionally important antigen recognition and binding sites of the MHC-I and *DQB* genes, as supported by more nonsynonymous than synonymous substitution rates. This is a clear indication of balancing selection (positive selection) maintaining new variants and increasing allelic polymorphism in the baiji and finless porpoise. However, for the *DRA* gene, no such phenomenon was found in the PBR, which was contrary to the normal pattern of substitution in the PBR for MHC class II genes (Hughes & Nei, 1989), and may suggest no balancing selection on this gene.

Population Expansion and Intragenic Recombination

Evolution by random bifurcation, without population expansion, recombination and/or convergence, is expected to yield a multimodal histogram with many peaks and ragged appearance as a result of differentiation of sequences into allelic lineages and extinctions of intermediates (Go et al., 2002; Figueroa et al., 2000; Suárez et al. et al., 2006). In this chapter, the mismatch distribution analyses for the baiji were clearly multimodal, suggesting that this species did not undergo population expansion or total recombination at the MHC-I, *DRA* and *DQB* loci. In contrast, the mismatch distribution analyses supported that the finless porpoise underwent a historical population expansion, which was congruent with the analysis by the mitochondrial control region sequences (Yang et al., 2008).

Additional analysis following the method implemented in GENECONV software revealed that three *Neph-DQB* and eight *Live-DQB* alleles were found to be involved in intragenic recombination events in some sequence blocks (i.e., DNA block 30–152 and DNA block 30–172, see Table 5). However, the intragenic recombination was not detected at the *DRA* and MHC-I loci. As suggested by Yeager and Hughes (1999), intragenic recombination was related to the search for MHC genes' diversity as an adaptive response. Thus, the present high polymorphism at the *DQB* locus might be an outcome of intragenic recombination.

Highly Similarity between the Baiji and Finless Porpoise: Convergent Evolution?

It was interestingly noted that certain identical alleles were shared by the baiji and finless porpoise in China waters at all three MHC loci. Of all the MHC alleles identified in this chapter and those reported by Yang et al. (2005), Hayashi et al. (2006) and Xu et al. (2007), six pairs of alleles, i.e., one at *DRA*, two at *DQB*, and three at MHC-I, were identical between the two species. Each of these identical alleles was identified from at least two individuals or independent clones. For example, *Live-DRA*01*and *Neph-DRA*01* are two identical alleles, the former of which was identified from 30 clones of 10 baiji samples, whereas the latter of which was detected in 94 clones of 35 finless porpoises. At the MHC-I locus, three pairs of identical alleles in both species were from 54 clones in 13 baiji individuals and 18 clones in five finless porpoises, respectively. Two pairs of identical *DQB* alleles were shared by 16 baiji individuals and four finless porpoises, respectively. In addition to the identical alleles, other alleles of the baiji and finless porpoise had highly interspecific similarity as shown in Tables 3 and 4. Actually, mitochondrial control region sequences were determined from the same DNA extractions, and all these sequences correctly correspond to either baiji or finless porpoise, without any haplotype shared by both species (Yang et al., 2003, 2008). For this reason, the possibility that the identity or similarity between the baiji and finless porpoise may be due to sample contamination, "PCR artifacts", or chance, should be excluded.

Up to now, cases of total identity amongst MHC alleles from different species have been reported (Leuchte et al., 2004; Otting et al., 2002; Suárez et al., 2006; Huchard et al., 2006), but most of them are restricted to congeneric species and rarely from above genus level (Suárez et al., 2006; Otting et al., 2002; Huchard et al., 2006). The identity or high similarity between different but closely related species, as a result of long-term effect of selection on MHC, was usually explained by trans-species mode of evolution (Sena et al., 2003; Huchard et al., 2006). Trans-species evolution refers to polymorphism that predates speciation events, whereby allelic lineages are passed from species to species and persist over long periods of evolutionary time (Klein, 1987). This was evidenced by this chapter that some alleles from harbor porpoises (*Phph-a*) and vaquita (*Phsi-DQB*01*) clustered with those of finless porpoises (Figure 2b). Other evidence came from the *DRA* data. Forty-five individuals of other cetacean species from Pontoporiidae, Phocoenidae, and Delphinidae, were also examined at the *DRA* locus for comparison, which revealed that the alleles shared between the baiji and finless porpoise (i.e., *Live-DRA*01* and *Neph-DRA*01*) were also found in species of Phocoenidae and Delphinidae (data not shown).

However, it is difficult to explain the identity and high similarity between distantly related species, e.g., the baiji and finless porpoise, using the trans-species mode of evolution. The two species are highly divergent with each other, with the baiji included in Lipotidae of the superfamily Lipotoidea (de Muizon, 1988; Yang et al., 2002) and the finless porpoise in Phocoenidae of the superfamily Delphinoidea (Rice, 1998), respectively. As suggested by some other authors (Andersson et al., 1991; Gustafsson & Andersson, 1994; Kriener et al., 2000), identity and similarity between distantly related species can be explained by convergent evolution. Unlike trans-species evolution, the identity and similarity that are shared in the case of convergent evolution are not the result of evolution from a common ancestor, but typically explained as the result of common adaptive solutions to similarly environmental pressures. As for the baiji and finless porpoise, it is well known that they are sympatric in the middle and lower reaches of the Yangtze River and face similar selection pressures such as pathogens. As a consequence, shaping the same motifs or alleles in both species in order to adapt to the similar environmental pressures may be inferred. Furthermore, identical alleles at the *DQB* and MHC-I loci were only shared by the baiji and finless porpoise. Although we sequenced some individuals from eight species in three families (i.e., Pontoporiidae, Phocoenidae, and Delphinidae) at the MHC-I and *DQB* loci, no identical allele was detected between the baiji and these species. Also, Hayashi et al. (2003) sequenced the *DQB* gene of 16 cetacean species but did not find any allele shared by different species. The other evidence to support convergent evolution between the baiji and finless porpoise came from the sequence divergence between both species which was comparable to those between each pair of relatively related cetacean species, as shown in Tables 3 and 4. Further studies, however, are needed to clarify the convergent evolution between the baiji and finless porpoise with more MHC loci or other molecular data.

ACKNOWLEDGMENTS

We thank Mr Anli Gao, Xinrong Xu, Hua Chen, and Qing Chang for collecting samples for many years, and members of the Institute of Genetic Resources, Nanjing Normal University, for their contributions to this paper. This study was supported by the National Natural Science Foundation of China grant numbers 30830016, 30670294 and 30470253, the Program for New Century Excellent Talents in Universities (NCET-07-0445), the Ministry of Education of China, the Specialized Research Fund for the Doctoral Program of Higher Education (SRFDP 20060319002), the Ministry of Education of China, and the Major Project for Basic Researches of Jiangsu Province Universities (07KJA18016).

REFERENCES

[1] Abbott, K.M., Wickings, E.J., & Knapp, L.A. (2006). High levels of diversity characterize mandrill (Mandrillus sphinx) Mhc-DRB sequences. *Immunogenetics,* 58, 628–640.

[2] Andersson, L., Sigurdardóttir S, S., Borsch, C., & Gustafsson, K., (1991). Evolution of MHC polymorphism: extensive sharing of polymorphic sequence motifs between human and bovine DRB alleles. *Immunogenetics* 33: 188–193.

[3] Babik, W., Durka, W. & Radwan, J. (2005). Sequence diversity of the MHC DRB gene in the Eurasian beaver (Castor fiber). *Molecular Ecology,* 14, 4249–4257.

[4] Baker, C.S., Vant, M.D., Dalebout, M.L,, Lento, G.M,, O'Brien, S.J., & Yuhki, N. (2006). Diversity and duplication of *DQB* and *DRB*-like genes of the MHC in baleen whales (suborder: Mysticeti). *Immunogenetics,* 58, 283 – 296.

[5] Bjorkman, P.J., Saper, M.A., Samraoui, B., Bennett, W.S., & Strominger, J.L., Wiley, D.C. (1987). Structure of the human class-I histocompatibility antigen, Hla-A2. *Nature,* 329: 506–512.

[6] Brown, J.H., Jardetzky, T.S., Gorga, J.C., Stern, L.J., Urban, R.G., Strominger, J.L., & Wiley, D.C. (1993). 3-Dimensional structure of the human class-II histocompatibility antigen HLADR1. *Nature* 364: 33–39.

[7] Cardenas, P.P., Suarez, C.F., Martinez, P., Patarroyo, M.E. & Patarroyo, M.A. (2005). MHC class I genes in the owl monkey: mosaic organization, convergence and loci diversity. *Immunogenetics,* 56, 818–832.

[8] Chu, Z.T., Carswell-Crumpton, C., Cole, B.C., & Jones, P.P. (1994). The minimal polymorphism of class II E alpha chains is not due to the functional neutrality of mutations. *Immunogenetics,* 40, 9 – 20.

[9] De Muizon, C. (1988). Les relations phyloge'ne'tigues des Delphinida (Cetacea, Mammalia). *Annales de pale'ontologie* 74: 159–227.

[10] Dengjel, J., Schoor, O., Fischer, R., Reich, M., Kraus, M., Muller, M., Kreymborg, K., Altenberend, F., Brandenburg, J., Kalbacher, H., Brock, R., Driessen, C., Rammensee, H.G., & Stevanovic, S., (2005). Autophagy promotes MHC class II presentation of peptides from intracellular source proteins. *Proceedings of the National Academy of Sciences, USA* 102, 7922–7927.

[11] Doolittle, R.F. (1994). Convergent evolution: the need to be explicit. *Trends in Biochemical Sciences,* 19, 15–18.

[12] Ellis, S.A., & Ballingall, K.T. (1999a). Cattle MHC: evolution in action? *Immunological Reviews,* 167, 159 – 168.

[13] Excoffier, L., Laval, G., & Schneider, S. (2005). ARLEQUIN (version 3.0): an integrated software package for population genetics data analysis. *Evolutionary Bioinformatics Online,* 1, 47–50.

[14] Figueroa, F., Mayer, W.E., Sultmann, H., O'hUigin, C., Tichy, H., Satta, Y., Takezaki, N., Takahata, N., & Klein, J. (2000). Mhc class II B gene evolution in East African Cichlid fishes. *Immunogenetics,* 51, 556–575.

[15] Flores-Ramirez, S., Urban-Ramirez, J. & Miller, R.D. (2000). Major histocompatibility complex class I loci from the gray whale (*Eschrichtius robustus*). *Journal of Heredity* 91: 279–282.

[16] Go, Y., Satta, Y., Kawamoto, Y., Rakotoarisoa, G., Randrianjafy, A., Koyama, N., & Hirai, H. (2002). Mhc-DRB genes evolution in lemurs. *Immunogenetics,* 54, 403–417.

[17] Gustafsson, K., & Andersson, L., (1994). Structure and polymorphism of horse MHC class II DRB genes: convergent evolution in the antigen binding site. *Immunogenetics,* 39, 355–358.

[18] Hayashi, K., Nishida, S., Yoshida, H., Goto, M., Pastene, L.A., & Koike, H. (2003). Sequence variation of the DQB allele in the cetacean MHC. *Mammal Study,* 28, 89–96.

[19] Hayashi, K., Yoshida, H., Nishida, S., Goto, M., Pastene, L.A., Kanda, N., Baba, Y., Koike, H. (2006). Genetic variation of the MHC DQB locus in the finless porpoise (*Neophocaena phocaenoides*). *Zoological Science,* 23, 147–153.

[20] Huchard, E., Cowlishaw, G., Raymond,M., Weill, M., & Knapp, L.A., (2006). Molecular study of Mhc-DRB in wild chacma baboons reveals high variability and evidence for trans-species inheritance. *Immunogenetics,* 58, 805–816.

[21] Hughes, A.L. (1999). Adaptive Evolution of Genes and Genomes. Oxford University Press, New York.

[22] Hughes, A.L.& Nei, M. (1989). Nucleotide substitution at major histocompatibility complex class II loci: evidence for overdominant selection. *Proceedings of the National Academy of Sciences, USA* 86, 958–962.

[23] Hughes, A.L., Yeager, M. (1998). Natural selection at major histocompatibility complex loci of vertebrates. *Annual Review of Genetics,* 32, 415–435.

[24] Hutzler C (2006) Rare Chinese dolphin gone, scientists say. Downloaded from http://www.chicagotribune.com/services/newspaper/premium/printedition/Thursday/chi-0612140201dec14,2,2565873.story?coll=chi-printnewsthursday-hed.

[25] Klein, J. (1986). Natural History of the Major Histocompatibility Complex. Wiley and Sons, New York.

[26] Klein, J., (1987). Origin of major histocompatibility complex polymorphism: the trans-species hypothesis. *Human Immunology* 19: 155–162.

[27] Klein, J., Horejsi, V. (1997). Immunology, 2nd ed. Blackwell Science, Oxford.

[28] Kriener, K., O'hUigin, C., Tichy, H., & Klein, J. (2000). Convergent evolution of major histocompatibility complex molecules in humans and New World monkeys. *Immunogenetics*, 51, 169–178.

[29] Kumar, S., Tamura, K., & Nei, M. (2004). MEGA3: integrated software for molecular evolutionary genetics analysis and sequence alignment. *Briefings in Bioinformatics* 5: 150–163.

[30] Leuchte, N., Berry, N., Köhler, B., Almond, N., LeGrand, R., Thorstensson, R., Titti, F., Sauermann, U. (2004). MhcDRB-sequences from cynomolgus macaques (Macaca fascicularis) of different origin. *Tissue Antigens,* 63, 529–537.

[31] Murray, B.W., Malik, S. & White, B.N. (1995). Sequence variation at the major histocompatibility complex locus DQB in beluga whales (*Delphinapterus leucas*). *Molecular Biology and Evolution,* 12, 582–593.

[32] Nei, M., & Gojobori, T. (1986). Simple methods for estimating the numbers of Synonymous and nonsynonymous nucleotide substitutions. *Molecular Biology and Evolution,* 3, 418–426.

[33] Nei, M. & Gojobori, T. (1986). Simple methods for estimating the numbers of Synonymous and nonsynonymous nucleotide substitutions. *Molecular Biology and Evolution,* 3, 418–426

[34] Ohta, T. (1998). On the pattern of polymorphisms at major histocompatibility complex loci. *Journal of Molecular Evolution,* 46, 633–638.

[35] Otting, N., de Groot, N.G., Doxiadis, G.G. & Bontrop, R.E. (2002). Extensive Mhc-DQB variation in humans and non-human primate species. *Immunogenetics,* 54, 230–239.

[36] Parham, P. & Ohta, T., (1996). Population biology of antigen presentation by MHC class I molecules. *Science*, **272,** 67–74.

[37] Pimtanothai, N., Hurley, C.K., Leke, R., Klitz, W. & Johnson, A.H. (2001). HLA-DR and -DQ polymorphism in Cameroon. *Tissue Antigens,* 58, 1–8.

[38] Reeves, R.R., Smith, B.D., Crespo, E.A. & Notarbartolo di Sciara, G. (2003).*Dolphins, Whales and Porpoises: 2002–2010 Conservation Action Plan for the World's Cetaceans. IUCN/SSC Cetacean Specialist Group*. Gland, Switzerland and Cambridge, United Kingdom: The International Union for Conservation of Nature,

[39] Reeves, R.R., Wang, J.Y., & Leatherwood, S. (1997). The finless porpoise,*Neophocaena phocaenoides* (G. Cuvier 1829): a summary of current

knowledge and recommendations for conservation action. *Asian Marine Biology,* 14, 111–143.

[40] Rice, D.W. (1998). *Marine Mammals of the World: Systematics and Distribution. Marine Mammal Society, Special Publication No. 4.* Lawrence, KS: The Society for Marine Mammalogy.

[41] Rogers, A.R. & Harpending, H. (1992). Population growth makes waves in the distribution of pairwise genetic differences. *Molecular Biology and Evolution,* 9, 552–569.

[42] Sachdev, M., Sankaranarayanan, R., Reddanna, P., Thangaraj, K., & Singh, L. (2005). Major histocompatibility complex class I polymorphism in Asiatic lions. *Tissue Antigens,* 66, 9–18.

[43] Sawyer, S.A. (1999) GENECONV: A computer package for the statistical detection of gene conversion. Distributed by the author, Department of Mathematics, Washington University in St. Louis (available at http://www.math.wustl.edu/~sawyer/geneconv/).

[44] Schneider, S. & Excoffier, L. (1999). Estimation of demographic parameters from the distribution of pairwise differences when mutation rates vary among sites: applications to human mitochondrial DNA. *Genetics,* 152, 1079–1089.

[45] Sena, L., Schneider, M.P.C., Brenig, B., Honeycutt, R.L., Womack, J.E., & Skow, L.C. (2003). Polymorphisms in MHC-DRA and -DRB alleles of water buffalo (*Bubalus bubalis*) reveal different features from cattle DR alleles. *Animal Genetics,* 34, 1 – 10.

[46] Stern, L.J., Brown, J.H., Jardetzky, T.S., Gorga, J.C., Urban, R.G., Strominger, J.L., & Wiley, D.C. (1994). Crystal structure of the human class II MHC protein HLA-DR1 complexed with an influenza virus peptide. *Nature,* 368, 215 – 221.

[47] Suárez, C. F., Patarroyo, M. E., Trujillo, E., Estupiñán, M., Baquero, J. E., Parra, C. & Rodriguez R. (2006) Owl monkey MHC-DRB exon 2 reveals high similarity with several HLA-DRB lineages. *Immunogenetics,* 58, 542–558.

[48] Takahata, N., & Satta, Y. (1998). Selection, convergence, and intragenic recombination in HLA diversity. *Genetica,* (102/103), 157–169.

[49] Tamura, K., Dudley, J., Nei, M., & Kumar, S. (2007). MEGA4: Molecular Evolutionary Genetics Analysis (MEGA) software version 4.0. *Molecular Biology and Evolution,* 24, 1596–1599.

[50] Xia, J. H., Zheng, J. H., & Wang, D. (2005). Individual identification of the Yangtze finless porpoises Neophocaena phocaenoides asiaeorientalis inhabiting the Tian-e-Zhou Natural Reserve based on microsatellite fingerprints. *Acta Zoologica Sinica,* 51, 142–148.

[51] Xu, S.X., Sun, P., Zhou, K.Y., & Yang, G. (2007). Sequence variability at three MHC loci of finless porpoises (*Neophocaena phocaenoides*). *Immunogenetics,* 59, 581–592.

[52] Yang, G., Guo, L., Bruford, M., Wei, F.W. & Zhou, K.Y. (2008). Mitochondrial phylogeography and population history of finless porpoises in Sino-Japanese waters. *Biological Journal of the Linnean Society*, 95, 193–204.

[53] Yang, G., Liu, S., Ren, W.H., Zhou, K.Y. & Wei, F.W. (2003). Mitochondrial control region variability of baiji and the Yangtze finless porpoises, two sympatric small cetaceans in the Yangtze River. *Acta Theriogica,* 48, 469–483.

[54] Yang, G., Yan, J., Zhou, K.Y. & Wei, F.U. (2005). Sequence variation and gene duplication at MHC DQB loci of Baiji (*Lipotes vexillifer*), a Chinese river dolphin. *Journal of Heredity,* 96, 310–317.

[55] Yang, G., Zhou, K.Y., Ren, W.H., Ji, G.Q., Liu, S., Bastida, R., & Rivero, L. (2002). Molecular systematics of river dolphins inferred from complete mitochondrial cytochrome b gene sequences. *Marine Mammal Science,* 18, 20–29.

[56] Yeager, M. & Hughes, A.L. (1999). Evolution of the mammalian MHC: natural selection, recombination, and convergent evolution. *Immunological Reviews,* 167, 45–58.

[57] Yoshida, H., Yoshioka, M., Shirakihara, M., & Chow, S. (2001). Population structure of finless porpoises (Neophocaena phocaenoides) in coastal waters of Japan based on mitochondrial DNA sequences. *Journal of Mammalogy,* 82, 123–130.

[58] Zheng, J.S., Xia, J.H., He, S.P. & Wang, D. (2005). Population genetic structure of the Yangtze finless porpoise (*Neophocaena phocaenoides asiaeorientalis*): implications for management and conservation. *Biochemical Genetics,* 43, 307–320.

In: Biology, Evolution and Conservation of River Dolphins… ISBN: 978-1-60876-633-8
Editors: M. Ruiz-García, J. Shostell. pp. 419-443

Chapter 22

POPULATION STATUS AND CONSERVATION OF THE GANGES RIVER DOLPHIN (*PLATANISTA GANGETICA GANGETICA*) IN THE INDIAN SUBCONTINENT

R. K. Sinha[1]*, Sunil Kumar Verma[2] and Lalji Singh[2+]

[1]Environmental Biology Laboratory, Department of Zoology, Patna University, Patna, INDIA

[2]Centre for Cellular and Molecular Biology, Uppal Road, Hyderabad, INDIA

ABSTRACT

Herein we discuss the Ganges River dolphin (*Platanista gangetica gangetica* or susu) which inhabits the Ganges-Brahmaputra-Meghna and Sangu-Karnaphuli river systems of India, Nepal and Bangladesh. The chapter begins with a discussion of the origin, evolution, and phylogeny of the Ganges River dolphin as well as river dolphins in general. Also included are descriptions of past and present distribution patterns of the Ganges River Dolphin along with its anatomical structure, including primitive characters and morphological characters of interest. In the second section of the chapter we elaborate on Ganges River dolphin population surveys we conducted within a 500 km section of the Ganges River in the state of Bihar during 2005 to 2007. Both upstream and downstream surveys were performed three times per year. A significantly greater number of Ganges dolphins were observed per kilometer upstream compared to downstream surveys (1.28 versus 1.0 respectively) and the mean number of dolphins observed per upstream survey ranged from 559 to 808. Our results also support spatial and temporal variation of the Ganges dolphin population with for example a greater number of animals in confluence areas. These survey results are similar to those obtained from other Ganges River surveys that used similar methods. The chapter concludes with a discussion on the Ganges River dolphin's conservation status and major threats to its existence. Direct catch, incidental catch, pollution, and habitat degradation are all serious threats.

Keywords: *Platanista*, susu, Ganges River, Phylogeny, India.

---

* rksinha.pu@gmail.com; rksinha_54@sancharnet.in
\+ lalji@ccmb.res.in.

INTRODUCTION

The Ganges River dolphin, commonly known as 'susu', is discontinuously distributed in the Ganges-Brahmaputra-Meghna and Karnaphuli-Sangu river systems of India, Nepal and Bangladesh between tidal zones and as far up as the rivers are navigable from the foothill of the Himalayas to the Bay of Bengal (Smith et al., 1994, 1998; Sinha, 1997; Sinha et al., 2000). It belongs to the Order Cetacea, suborder Odontoceti (toothed whales), family Platanistidae (South Asian river dolphin), genus *Platanista*, species *P. gangetica* and sub-species *P. g. gangetica*. Rivers and associated freshwater ecosystems in the Indian subcontinent are under threat due to a wide range of intensive human use and developmental activities. The Ganga-Brahmaputra-Meghna river basins cover only 0.12% of the world's land mass where about 10% of the world's population live. Increased population and development pressures have led to depletion of fish stocks, severe pollution from point and non-point sources, degradation of habitats, sediment load changes and hydrological alterations (Mohan, 1989; Ansari et al., 1999; Dudgeon, 2000; Sinha 2006). These in turn have had detrimental effects on the flora and fauna of the river ecosystems, including the Ganges River dolphin *Platanista gangetica gangetica*, an endemic species of the Ganges-Brahmaputra-Meghna river systems in India, Nepal and Bangladesh (Sinha et al., 2000; Sinha, 2006). The total estimated population of the dolphin in its entire distribution range is about 2000-2500. River dolphin conservation has become a very critical issue owing to the recently reported extinction of the Baiji or Chinese River dolphin *Lipotes vexillifer* Miller 1918 (Turvey et al., 2007). The Ganges River dolphin *Platanista gangetica gangetica* Roxburgh 1801 has been declared endangered by the IUCN (IUCN Red List 2007) and the species has been listed in Schedule-I in the Indian Wildlife (Protection) Act, 1972.

Different studies have proposed that water depth, channel width, direction and velocity of flow, geomorphologic complexities, and substrate type affect dolphin habitat use (Smith et al., 1998; Sinha et al., 2000, 2006; Choudhary et al., 2006). Along with these, prey availability is another factor that can affect population size and habitat selection. River dolphins in the Ganges have been recorded to feed on small fish, and occasionally on crustaceans and snails (Sinha, 2006). The species is known to be mostly solitary, except mother-calf pairs. They are also known to congregate sometimes in shallow water zones for feeding on small fish groups in such areas (Sinha, 2006). For the Ganges River dolphins in India, Nepal, and Bangladesh, population and threat assessment surveys have been carried out throughout their distribution range in order to assess the conservation status and obtain information on threats (Smith et al., 1994, 1998; Sinha et al., 2000; Wakid; 2005, Biswas & Boruah, 2006; Choudhary et al., 2006; WWF, 2006).

The main threat to the Ganges dolphins, especially in Bihar and Assam states, were reported to be direct killing by fishermen for extraction of oil from the blubber, which was used as a fish-bait (Mohan, 1989; Sinha, 2002). One novel approach to prevent further hunting has been the use of fish scrap oil instead of oil from hunted dolphins (Sinha, 2002). Such approaches have contributed to an increase of awareness and conservation efforts leading to the reduction of directly killing dolphins (Sinha, 2006; Choudhary et al., 2006). Threats to dolphins and their habitat such as construction of dams and barrages have been suspected to cause genetic isolation of dolphin populations (Reeves et al., 2000; Smith et al., 2000; Sinha, 2006). However, accidental by-catch through the entanglement in gillnets continues to be a threat (Sinha, 2006). Depletion of big size and major carp fishery in the rivers over the years has resulted in greater exploitation of smaller fishes, which are considered main prey of dolphins (Sinha, 2006; Choudhary et al., 2006). Conservation efforts across India, Nepal and Bangladesh have mainly focused on sensitizing fishermen to stop the killing of dolphins, seeking people's co-operation for prevention of illegal hunting and creating awareness about the adverse effects of dams and barrages on river flows and catchments (Smith et al., 1998; Sinha, 2006; WWF, 2006; Choudhary et al., 2006). Apart from population surveys and threat mitigation measures, the detailed, scientific ecological knowledge about the species is still bereft of empirical information. Considering the high human impacts on the river systems the species inhabits, information regarding population size, space use, and habitat preferences that influence distribution and survival is required to systematically plan conservation strategies.

Most of the Ganges' major tributaries originate in the Himalayas and merge with the Ganges in the state of Bihar and almost one-half of the total dolphin populations are expected to survive in the Bihar stretch of the Ganges and its tributaries. The present study is based on the surveys conducted during post-monsoon (November-December), winter (February-March) and summer (May-June) of 2005-2007. It also includes information on biology including origin, evolution and phylogenetic position of the species, as well as habitats (different rivers / stretch of the rivers), threats and conservation efforts made to save the animal from extinction.

Origin of Cetacea

Phylogenetic analyses of molecular data on extant animals strongly support the notion that hippopotamids are the closest relatives of cetaceans (Millinkovitch et al., 1998; Nikaido et al., 1999; Gatesy & O'Leary, 2001). In spite of this, it is unlikely that the two groups are closely related when extant and extinct artiodactyls are analyzed, for the simple reason that cetaceans originated about 50 million years ago in south Asia, whereas the family Hippopotamidae is only 15 million year old, and the first hippopotamids to be recorded in Asia are only 6 million year old (Boisserie et al., 2005). The middle Eocene artiodactyl family Raoellidae is broadly coeval with the earliest cetaceans, and both are endemic to south Asia. Thewissen et al. (2007) studied new dental, cranial and postcranial material for *Indohyus*, a middle Eocene raoellid artiodactyl from Kashmir, India. Their analysis identifies raoellid as the sister group to cetaceans and bridges the morphological divide that separated early cetaceans from artiodactyls. Bajpai et al. (2009) reviewed the first steps of whale evolution, i.e. the transition from a land mammal to obligate marine predators, documented by the Eocene cetacean families of the Indian subcontinent: Pakicetidae, Ambulocetidae, Remingtonocetidae, Protocetidae, and Basilosauridae, as well as their artiodactyl sister group, the Raoellidae and concluded that the Eocene origin and evolution of whales is one of the best documented examples of macroevolutionary change.

Indohyus was a small, stocky artiodactyl, roughly the size of the racoon *Procyon lotor*. It was not an adept swimmer; instead it waded in shallow water and may have fed on land, although a specialized aquatic diet is also possible. It probably spent a considerably greater amount of time in the water either for protection or when feeding. As indicated by the evidence from stable isotopes, *Indohyus* spent most of its time in the water and either came onshore to feed on vegetation (as the modern *Hippopotamus* does) or foraged on invertebrates or aquatic vegetation. Raoellids are the sister group to cetaceans, and this implies that aquatic habitats originated before the Order Cetacea. The great evolutionary change that occurred at the origin of cetaceans is thus not the adoption of an aquatic lifestyle but dietary change was the event that defined cetacean origins. Cetaceans originated from an *Indohyus*-like ancestor and switched to a diet of aquatic prey. Significant changes in the morphology of the teeth, the oral skeleton and the sense organs made cetaceans different from their ancestors and unique among mammals (Thewissen et al., 2007).

The River Dolphins

The four genera of toothed cetaceans, i.e., the baiji, *Lipotes vexillifer*; the susu, *Platanista gangetica*, the boto, *Inia geoffrensis*; and the franciscana, *Pontoporia blainvillei*; comprise the peculiar and poorly known 'river dolphins'. The modern river dolphins occur only in two continents: Asia and South America. The baijii *Lipotes vexillifer* (the Yangtze River Dolphin), is endemic to China but was declared functionally/effectively extinct in December 2006, an event which the news media broadcasted worldwide. Either with a glimmer of hope, or more likely, not-updated, the IUCN Red List of Threatened Species lists the Baiji, Yangtze River dolphin (*Lipotes vexillifer*) as 'critically endangered (possibly extinct)' (Hopkin, 2007). The two populations of *Platanista gangetica* have been isolated for a considerable time with *P. gangetica minor* being confined to the Indus drainage in Pakistan. However, a couple of this species were sighted in the Sutlej River, a tributary of the Indus, in Punjab state in India in 2008. *P. g. gangetica* occurs in the Ganges, Brahmaputra, Meghna, Karnaphuli, and Sangu drainage systems of India, Bangladesh and Nepal. The boto, *Inia geoffrensis* has an extraordinarily wide distribution. It can be found along the entire Amazon River and its tributaries, small rivers and lakes, throughout the Orinoco river basin. It occurs in six countries of South America: Bolivia, Brazil, Colombia, Ecuador, Peru, and Venezuela. The franciscana, *Pontoporia blainvillei* is the only one of the four river dolphin species living in the marine environment. It lives in coastal marine waters of eastern South America between Argentina and Uruguay.

Freshwater dolphins in Asia are among the world's most endangered mammals and there is an urgent need to establish conservation priorities based on scientifically credible abundance estimates (Perrin & Brownell, 1989; Smith & Reeves, 2000a; IWC, 2001; Smith & Jefferson, 2002). The complex geomorphology of freshwater and estuarine systems tends to concentrate the distribution of cetaceans in counter-current associated with confluences, meanders and mid-channel islands (Hua et al., 1989; Smith, 1993; Smith et al., 1997, 1998). The murky river water makes it totally invisible under water. Many times surfacing is very quiescent and whenever they come up it is usually for only a fraction of a second. All these limitations make the credible abundance estimates of these dolphins a great challenge. Extensive population fragmentation has resulted from the widespread construction of barrages (Smith & Reeves, 2000a; Smith et al., 2000).

Evolution of River Dolphins

The Middle Miocene was a time of globally high sea levels, with three significant marine transgressive-regressive cycles recorded worldwide (Haq et al., 1987). With the resulting large-scale marine transgressions on to low lying regions of the continents, shallow epicontinental seas became prominent marine ecosystems. The Indo-Gangetic plain of the Indian subcontinent, the Amazon and Parana River basins of South America, and the Yangtze River basin of China are vast geomorphic systems whose fluvio-deltaic regions were deeply penetrated by marine waters during high sea-level stands. The shallow estuarine regions created by the mixing of riverine and marine waters probably supported diverse food resources, particularly for aquatic animals able to tolerate osmotic differences between fresh and saltwater systems. Hamilton et al. (2001) proposed that the ancestors of the four extant river dolphin taxa were inhabitants of Miocene epicontinental seas. Draining of epicontinental seas and reduction of the near shore marine ecosystem occurred with a late Miocene trend of sea level regression, which continued throughout the Pliocene, interrupted by only moderate and relatively brief events of sea-level rise (Hallam, 1992). As sea levels fell, these archaic odontocetes survived in river systems, while their marine relatives were superseded by the radiation of Delphinoidea. Cassens et al. (2000) also noted the persistence of river dolphins during the radiation of delphinoids. They suggest that extant river dolphin lineages 'escaped extinction' by adaptation to their current riverine habitats. By integrating phylogenetic, palaeoceanographic and fossil data, an explicit hypothesis for the evolution and modern distribution of river dolphins has been provided by Hamilton et al. (2001).

The Indo-Gangetic foreland basin is a broad, flat plain of sediment delivered throughout the Cenozoic by an intricate network of migrating rivers descending from the tectonically dynamic Himalayan Mountain (Burbank et al., 1996). The increased sea levels of the middle Miocene would have inundated large areas of the foreland basin, creating a shallow marine habitat. Fossils have not yet been recovered from these regions, but platanistids are known to have inhabited Miocene epicontinental seas in North America (Morgan, 1994; Gottfried et al., 1994).

Ganges River Dolphin

Platanista is the only surviving descendant of an archaic odontocete that ventured into the epicontinental seas of the Indo-Gangetic basin, and remained through its transition to an extensive freshwater ecosystem during the Late Neogene trend of sea-level regression. Although the paleogeography of the two river systems would suggest a history of isolation, the genetic distance in the sample of *P. gangetica* (Ganges population) and *P. minor* (Indus population) is surprisingly low. The Indus and Ganges populations were long regarded as identical until Pilleri & Gihr (1971) divided them into two species based on differences in skull structure, but Kasuya (1972) reduced the two taxa to subspecies of a single species. This is supported by the results of Yang & Zhou (1999), who found that the difference between cytochrome-b sequences of Ganges and Indus river dolphins was very small. Even until historical times there was probably sporadic faunal exchange between the Indus and Ganges drainages by way of head-stream capture on the low Indo-Gangetic plains between the Sutlej (Indus) and Yamuna (Ganges) rivers (Rice, 1998 and refs. therein). Rice (1998), in his taxonomic classification of cetaceans that has become standard in the field, found that there were insufficient morphological differences to warrant distinction at the species level. Thus one species is recognized in the genus *Platanista* and currently the Ganges River dolphins are *Platanista gangetica gangetica* and the Indus River dolphins of Pakistan are *Platanista gangetica minor.* The species has many primitive characters and is one of the charismatic megafauna of the rivers of the Indian subcontinent. The animal is facing threats of extinction in its entire distribution range due to overexploitation and habitat degradation caused by various anthropogenic pressures. In most of the rivers, its distribution range has shrunk in the last couple of decades.

Phylogenetic Position of the Ganges River Dolphin

The four genera of classical river dolphins are associated with six separate great river systems on three subcontinents and have been lumped into a single taxon (family Platanistidae or super family Platanistoidea) based on their similarity in some morphological characters. This grouping has been proven unreasonable by all modern phylogenetic analyses (Yan et al., 2005). Muizon (1984, 1988) regarded *Platanista* as only distantly related to the other three, and Heyning (1989) also restricted Platanistoidea to *Platanista.* In cladistic analysis, the family Platanistidae fell into a clade with the extinct families Squalodelphinidae. Delpiazinidae, Waipatiidae, and Squalodontidae (Fordyce & Barnes, 1994; Fordyce & Muizon, 2001). Studies on cetacean phylogeny using DNA sequence data have become very prominent since the 1990s. Arnason & Gullberg (1996) first supplied molecular evidence (Cyt b) with the view that *Platanista* had no affinity to *Inia* and *Pontoporia*. Yang & Zhou (1999) first included all of the four classical river dolphins in their studies. Subsequently, Cassens et al. (2000), Hamilton et al. (2001), Nikaido et al. (2001), and Yang et al. (2002) analyzed phylogeny of river dolphins using different DNA markers respectively. In all analyses mentioned above, *Platanista* was identified as an independent lineage of odontocetes, and had no affinity to the nonplatanistoid river dolphins. The overview that the classical river dolphins including Platanistidae and three other families are an unnatural group has been widely accepted. Yan et al. (2005), in their analyses, split classical river dolphins into two distinct lineages, *Platanista* and *Lipotes* + (*Inia* + *Pontoporia*), having no sister relationship with each other and opined that such a phylogenetic pattern strongly supports the paraphyletic relationship of the classical river dolphins.

Numerous arrangements have been proposed for the phylogenetic relationships of the world's river dolphins to one another and to other odontocete cetaceans. Based on phylogenetic analysis of three mitochondrial genes for 29 cetacean species, Hamilton et al. (2001) concluded that the four genera of freshwater dolphins represent three separate, ancient branches in odontocete evolution. Further, they suggested that ancestors of the four extant river dolphin lineages colonized the shallow epicontinental seas that inundated the Amazon, Parana, Yangtze and Indo-Gangetic river basins, subsequently remaining in these extensive waterways during their transition to freshwater within the Late Neogene trend of sea-level lowering

While studying the molecular phylogeny of river dolphins, Guang & Kaiya (1999) observed that the difference between *cyt b* sequence of Ganges River dolphin and Indus River dolphin was very small, which supported that Ganges River dolphin (*P. g. gangetica*) and

Indus River dolphin (*P. g. minor*) were probably two subspecies of a single species. They also suggested that among the four river dolphin families, Platanistidae was the earliest divergent clade, the Lipotidae was the next, and then the Iniidae and Pontoporiidae. Further they suggested that the river dolphins were paraphyletic, and it was reasonable to place *Platanista* at a superfamily level as no affinity was revealed between the Platanistidae and other river dolphin families. The long-suspected polyphyly of river dolphins is supported by the mitochondrial sequence data. In both trees, *Platanista gangetica* (and *Platanista mino*r, representing Platanistidae), is sister to the remaining odontocetes, although bootstrap support for this node is low (Hamilton et al., 2001).

Extinct taxa assigned to the Platanistidae are well documented, particularly *Zarhachis* and *Pomatodelphis*, long beaked Middle to Late Miocene cetaceans recovered primarily from shallow epicontinental sea deposits of the Atlantic coast of North America (Kellog, 1995; Gottfried et al., 1994; Morgan 1994). Possible platanistid relatives are Squalodelphinidae and at least some members of Squalodontidae (Muizon, 1994; Fordyce, 1994), two well known, extinct families of archaic, medium sized heterodonts. Other fossil relatives of the Platanistidae include members of the Delpiaziniidae (Muizon, 1994) and Waipatiidae (Fordyce, 1994). If these lineages are monophyletic, then Platanista is the sole extant member of a once-abundant and diverse clade of archaic odontocetes. The side-swimming, blind and highly endangered Indian River dolphin has long been recognized as 'the genus presenting the greatest total of modifications known in any cetaceans' (Miller, 1923). However, both fossil and extant platanistids warrant further investigation for potential insights into cetacean evolution. Verma et al. (2004) established the evolutionary relationship of the Ganges River dolphin with extinct and extant cetaceans based on comprehensive analyses of the mitochondrial cytochrome b and nuclear interphotoreceptor retinoid-binding protein gene sequences, obtained from 15 specimens of Ganges dolphin from India and Bangladesh. The study suggested that *P. g. gangetica,* a toothed cetacean, is significantly closer to Mysteceti (Toothless whales) than to any other group of toothed whales. However, Yan et al. (2005) observed that the *Platanista* lineage is always within the odontocete clade instead of having a closer affinity to Mysticeti. Nevertheless, they opined that the position of the *Platanista* is more basal, suggesting separate divergence of this lineage well before the other one. And they agree that they could not resolve with high significance the exact phylogenetic position of Platanista. The more basal position of Platanistidae is also supported by the records of platanistoid fossils in the late Oligocene (Fordyce & Barnes, 1994). Muizon (1991), Heyning (1989) as well as Messenger & McGuire (1998) proposed that Platanistidae branched after the divergence of sperm and beaked whales. However, others placed Platanistidae and beaked whales in a clade between the sperm whale and more crown-ward odontocetes (Cassens et al., 2000) or placed Platanistidae between sperm whale and beaked whales (Hamilton et al., 2001; Nikaido et al., 2001). The position of Platanistidae is not very clear. This may be, at least in part, because the susu, sperm, beaked, and baleen whales lineages seem to have been produced through a very rapid succession of splitting events in the Eocene (Nikaido et al., 2001). Meanwhile, additional evidence is needed to resolve this issue. There is a consensus that these four river dolphins belong to two different groups of dolphins: Platanistoidea, which is an early divergent superfamily of odontocetes, and nonplatanistoid river dolphins, a monophyletic clade closely related to superfamily Delphinoidea.

Distribution of the Susu: Historical and Current

Historical Distribution

The water-hog (P. *khuk-abi*, *Platanista gangetica*, the porpoise) is in all Hindustan rivers (quoted in the 'Babur Nama', a book written by Babur in the 15th Century, the first ruler of the Mughal Dynasty in India). The book contains a miniature painting depicting the Ganges River dolphin. By stating "all Hindustan rivers" Babur probably meant all the rivers of North India as he had widely traveled mainly in the Indo-Gangetic plains of northern India. Anderson (1879) recorded its distribution in the Ganges over an area comprised between 77^{0}E and 89^{0}E Longitude; in the Brahmaputra it occurred throughout the entire main river, as far eastwards as longitude 95^0 by latitude 27^{0}30' north. He also reported that even in the month of May, when the Ganges was very low, it extended up the Yamuna as far as Delhi. Anderson emphasized that the upstream range of this dolphin was apparently only limited by insufficiency of water and by rocky barriers. The last record of susu in Yamuna at Delhi was in 1967, when a dead dolphin caught in a fishing net was brought to the Delhi Zoo (personal communication Dr. K. S. Sankhla, the then Director, Delhi Zoo).

Current Distribution

The Ganges River dolphins live mainly in the rivers originating from the Himalayas and some tributaries of the Ganges originating in the central India below an elevation of about 250 m. In the Ganges valley it ranges into most of the large tributaries: the Yamuna, Son, Sind, Chambal, Ramganga, Gomti, Ghaghara, Rapti, Gandak, Kosi, etc besides the main channel of the Ganges. In the Brahmaputra valley it also ranges into many of the major tributaries: the Tista, Adadhar, Champamat, Manas, Bhareli, Subhansiri, Dihang, Dibang, Lohit, Disang, Dikho and Kulsi rivers. Downstream it ranges through most of the rivers in Bangladesh, as far as the tidal limits at the mouth of the Ganges. They are also reported to be within the Fenny, Karnaphuli, and Sangu rivers to the southeast of the mouths of the Ganges (Rice, 1998). The uppermost distribution is said to be restricted only by the lack of water and rocky barriers (Reyes, 1991). Relatively high population densities (approx.125) have been observed in the 60 km stretch of the Ganga, the Vikramshila Gangetic Dolphin Sanctuary, in the state of Bihar in India. There is a small (perhaps 20 individuals) but potentially viable population in the Karnali River, the largest river system in Nepal, now isolated by the Girija Barrage located about 25 km downstream of the Nepal-India border (Smith et al., 1994). During a continuous survey in the Ganga from Haridwar downward, in the month of December 1996 when water was low, we could not find susu in the 100 km stretch of the river between Haridwar and Middle Ganga Barrage at Bijnor. However, in September 1994 one susu was sighted in the Ganges at Nangal about 30-40 km downstream Haridwar (Pers. Comm. Raju Kumar). During a status survey conducted in 1978, the susu were found most abundant from Munger to Sahibganj in Bihar; common up to the Farakka Barrage towards the east and up to Varanasi or slightly more westwards (Gupta, 1986). Gangetic dolphins were fairly common in tidal waters but never entered the sea (Agrawal, 1991).

Table 1. Summary of Dolphin sightings in the River Ganges between Buxar and Manihari ghat during 2005 – 2007

Date (Survey direction)	Survey distance (km)	Boat's speed (km/hr.)	Sightings by Primary observer	Sightings by Secondary observer	Sum of group size estimates from Primary & Secondary sightings			Percentage of error in sightings missed by Primary observer	Sightings/ km.	Dolphin/ km.	No. of calves (percentage)
					Best	High	Low				
November, 05(US)	507.4	6.32	321	97	664	750	578	23.21	0.84	1.31	77 (11.60)
November, 05(DS)	494.8	11.19	262	81	517	576	465	23.62	0.71	1.04	62 (11.99)
March, 06 (US)	517.8	6.05	274	91	576	692	514	24.93	0.72	1.11	47 (8.16)
March, 06 (DS)	499.2	9.72	200	78	418	496	370	28.06	0.58	0.84	39 (9.33)
May - June, 06(US)	506	6.24	284	103	559	644	502	26.61	0.78	1.1	88 (15.74)
May - June, 06(DS)	499.5	10.38	223	75	439	507	392	25.17	0.61	0.93	68 (15.49)
December, 06 (US)	517.1	6.34	437	110	808	931	729	20.11	1.08	1.56	67 (8.29)
December, 06 (DS)	501	10.22	324	72	559	668	486	18.18	0.81	1.12	58 (9.68)
March, 07 (US)	513.1	6.46	398	79	706	813	635	16.56	0.94	1.38	50 (7.08)
March, 07 (DS)	506.9	10.63	257	63	482	584	411	19.69	0.64	0.95	37 (7.68)
May - June, 07(US)	514.8	6.05	372	89	631	755	517	19.31	0.91	1.23	72 (11.41)
May - June, 07(DS)	508.5	10.19	299	78	577	700	470	20.69	0.76	1.13	69 (11.96)
Mean ± S.D. (US)	**512.7**± 4.96	**6.24**± 0.17	**348**± 65.26	**95**± 10.96	**657**± 91.75	**764**± 100.15	**579**± 89.03	**21.79**± 3.78	**0.88**± 0.13	**1.28**± 0.17	**67 (10.38)** ± 15.84 (3.21)
Mean ± S.D. (DS)	**501.7**± 5.15	**10.39**± 0.49	**261**± 46.03	**75**± 6.41	**499**± 63.96	**589**± 82.62	**432**± 47.61	**22.57**± 3.72	**0.69**± 0.09	**1.00**± 0.11	**56 (11.02)** ± 14.15 (2.74)

Table 2. Population (sighting records) of the Ganges River Dolphin in different rivers/stretch of the rivers of the Ganga Basin

Name of the river	Length of the river surveyed	Number of susu	Source
In India			
The Ganga Mainstem			
The Ganga (Haridwar to Bijnor Barrage)	100 km	Nil	Sinha et al (2000)
The Ganga (Bijnor Barrage to Narora Barrage	169 km	36 (d/s survey)	Sinha et al (2000)
The Ganga (Narora to Allahabad)	600 km	10 (discrete segment survey)	Sinha et al (2000)
The Ganga (Allahabad to Buxar)	425 km	172 (d/s survey)	Sinha et al (2000)
The Ganga (Buxar to Maniharighat)	500 km	808 (u/s survey)	unpublished data of 2007-08 (Sinha)
The Ganga (Maniharighat to Farakka)	100 km	24 (d/s survey)	unpublished data of Dec. 2004 (Sinha)
The Farakka Feeder canal	38 km	21 (d/s survey)	Sinha et al (2000)
The Bhagirathi (Jangipur Barrage to Triveni)	320 km	119 (d/s survey)	Sinha et al (2000)
The Hooghli (Triveni to Ganga Sagar)	190 km	97 (d/s survey)	pers. comm. Gopal Sharma (2007)
Tributaries of the Ganga)			
The Yamuna (from confluence of Chambal to Hamirpur)	250 km	25-40 (d/s survey)	Sinha et al (2000)
The Yamuna (Kausambi to Allahabad)	90 km	18 (d/s survey)	Sinha et al (2000)
The Kosi (Kosi Barrage to Kursela)	200 km	85 (discrete survey)	Sinha and Sharma (2003)
The Gandak (Confluence with Ganga at Patna to	101 km	106 (u/s survey)	unpublished data of 2007-08 (Sinha)
The Gherua (India-Nepal border to Girijapuri Barrage)	20 km	23 (d/s survey)	Smith et al (1994)
The Sarda (Sarda Barrage to Palya)	100 km	Nil	Sinha and Sharma (2003)
The Chambal (Pali to Barhi)	370 km	29 (d/s survey)	Sinha et al (2000)
The Ken(from confluence of Yamuna at Chilla to Sindhan Kala village)	30 km	08 (d/s survey)	Sinha et al (2000)
The Kumari (from confluence of Sind River)	100 km	Nil	Sinha et al (2000)

Table 2. Continued

The Betwa (from confluence of the Yamuna at Hamirpur to Orai	84 km	06 (d/s survey)	Sinha et al (2000)
The Sind (from confluence with the Yamuna)	110 km	05 (d/s survey)	Sinha et al (2000)
The Son	130 km	10 (d/s survey)	Sinha et al (2000)
The Brahmaputra	600 km 856 km	400 (1996) 197 (2004-05)	Mohan (1997) pers. comm. A. Wakid (2006)
The Barak River	17 km	12 (Nov. 1999) 8 (2004) 6 (2006)	Pers. comm. Paulan Singh
The Subhansiri River	99 km	26	pers. comm. A. Wakid(2006)
The River Kulsi	76 km	27	pers. comm. A. Wakid(2006)
In Bangladesh			
The Jamuna	189 km	38-50	Smith et al (1998)
The Kushiyara	113 km	34-43	Smith et al (1998)
The Burhi Ganga		03	Smith et al (1998)
The Karnaphuli-Sangu	222 km	131	Smith et al (2005)
The Sundarbans	1488 km	225	Smith et al (2005)
In Nepal			
The Karnali (from Kachali to Kotiaghat)	60 km	06	Smith et al (1994)
The Saptakosi (from confluence of Arun and Sun Kosi to Kosi Barrage)	60 km	03	Smith et al (1994)
The Narayani (Devghat to Triveni Barrage)		1-2	Smith et al (1994)
The Mahakali		Nil	Smith et al (1994)

In the Brahmaputra River system the susu are present as far north-east as the Dihang, Buri Dihing and Lohit rivers in eastern Assam, and as far north as the Teesta River and its tributaries, which extend into Sikkim and Bhutan (Mohan, 1989). The population status in different rivers in its distribution range, based on surveys conducted by different workers in the last two decades has been depicted in Table 2. The survey methods adopted by different workers were not consistent and therefore there is a lack of scientifically credible population estimates for this species. Nevertheless, the estimated total dolphin population in its fairly extensive distribution range is about 2500.

During the dry season, when the water levels are low in the rivers, the dolphins stay in the main river channels, however, they stay back in the deep pools in the tributaries also, where they face threats of being caught in the fishing nets as such pools attract intensive fishing. During the monsoon season, they spread out and move into even smaller tributaries and creeks. There are more dolphins at confluences, meanderings, and behind sand bars, where counter currents and complex hydro-geo-morphological formations exist. Such complexities provide habitats for diversified biota in the river ecosystems.

Physical Description of the Ganges Dolphin

The Ganges River dolphins have long, pointed snout characteristic of all river dolphins. Both the upper and lower jaw sets of long sharp teeth are visible even when the mouth is closed. The snout is long and widens at the tip. In females, the snout is generally longer and may curve upwards and to one side. The eyes are extremely small resembling pinhole openings slightly above the mouth. The species does not have crystalline eye lenses, rendering it effectively blind, although it may still be able to detect the intensity and direction of light. The river water where they live is so murky that good eyesight would most-likely not be advantageous. Navigation and hunting are carried out using echo-location. The body is subtle and robust, attenuating posteriorly from the dorsal fin to a narrow tail stalk. The body is deep and has a brownish color and is stocky at the middle. They have round bellies. The dorsal fin is very low triangular hump located two-thirds body length from the anterior end. The broad flippers have a crenellated margin, with visible hand and arm bones. Its flukes are broad and these along with their flippers are thin and large in relation to body size, which normally ranges from 2-2.2 m in adult males and 2.4-2.6 m in adult females. At the time of birth they measure 70-90 cm and weigh 4 – 7.5 kg. Adults (2 – 2.6 meter) weigh between 70 and 90 kg, however, an adult pregnant female (2.5 m) caught at Araria in Northeastern Bihar, near Indo-Nepal border, weighed 114 kg (caught in February 1993 and brought to Patna Zoo). We also recorded a 70 cm male fetus weighing 4 kg, 77.5 cm male 6.6 kg and a 91 cm female weighing 11 kg, all collected from the Ganges in and around Patna.

Primitive Characters

Platanista g. gangetica bears some of the very primitive characters not known in other cetaceans, videlicet the presence of the ceacum at the junction of the small and large intestines, a testis position that is much more dorsal compared to other marine cetaceans (testes are extra-peritoneal in terrestrial mammals), and subcutaneous muscle between two layers of blubber. The following informations were recorded while working on a carcass of a male dolphin at Patna:

Body Length: 171cm, Body Weight: 55 kg, Blubber thickness (cm): Dorsal Skin (0.1) + Blubber (2.5), Lateral Skin (0.1) + Outer blubber (1.3) + Panniculus (0.2) + Inner blubber (2.7), Ventral Skin (0.1) + Outer blubber (1.9) + Panniculus (0.2) + Inner blubber (0.7), Intestine length (cm): Small intestine (620) + Caecum (8) + Large intestine (80).

Morphological Characters of Interest

1. Texture of subcutaneous tissue: Cetacean fatty tissues are accumulated in blubber, whereas in terrestrial mammals fatty tissues can be found here and there in the subcutaneous connective tissues. In *Platanista*, they have certain thickness of blubber, but at the same time texture of the deeper connective tissues is somewhat more similar to those of the terrestrial mammals. We are not sure if this has anything to do with primitiveness or similarity to ancestral terrestrial mammals (Personal communication Tadasu Yamada).

2. Disposition of subcutaneous tissue layer especially deeper to the Panniculus carnosus: Panniculus carnosus is subcutaneous muscle, which usually is a thin sheet within the subcutaneous tissue (or superficial fascia) layer. In cetaceans the panniculus usually originates from the proximal portion of the humerus. Healthy wild oceanic dolphins have a scanty amount of connective tissues deeper to the panniculus. This has much to do with the first character above.
3. Existence of the Caecum.
4. Simple air sacs around nasal passage: Accessory air sacs around the nasal passage might indicate that the *Platanista* are more primitive than other oceanic dolphins.
5. Ventrally situated testes compared to marine dolphins: In terrestrial mammals, descended testes are standard which allow them to have a cooler temperature compared to the rest of the body. Male oceanic dolphins have the testes dorsally and much less descended than *Platanista*. This might have something to do with more terrestrial characteristic of *Platanista*.
6. Specific structure of stomach: The stomach of *Platanista* consisted of three chambers and a connecting channel. The basic arrangement of the chambers is similar to that of the dolphins of delphinidae, however, the connecting chamber is short and straight whereas in delphinid dolphins it is longer and takes a hairpin bend.
7. Peculiar muscular arrangement around the shoulder: more studies are needed to arrive at any conclusion regarding 6 and 7.

METHODS OF SIGHTING RECORDS

From November 2005 to June 2007 we assessed the abundance and distribution of Ganges River dolphins in the 500 km stretch of the Ganges between Buxar (25°33'32"N and 83°56'23"E) and Maniharighat (25°19'56"N and 87°36'44"E) in Bihar state (Figure 1). A map of the Ganges Basin in the Indian subcontinent has been depicted in Figure 2. Six continuous vessel-based visual surveys for dolphins were conducted in the entire length of the river in both upstream and downstream directions using motorized wooden boats. Two primary observers, one each on the right and left sides of the vessel searched dolphins by eye in a 90-degree cone in front of the vessel. A third observer served as data recorder and also searched for dolphins when not filling out the data forms. Two independent observers positioned behind the primary observers recorded any sightings missed by the primary team. Sightings made by the primary and secondary teams were pooled for calculating encounter rates and the best minimum abundance estimate. A Global Positioning System was used to record the distance traveled and the geographical coordinates of dolphin sightings. River depth was recorded at every 2 km interval using an automatic depth finder.

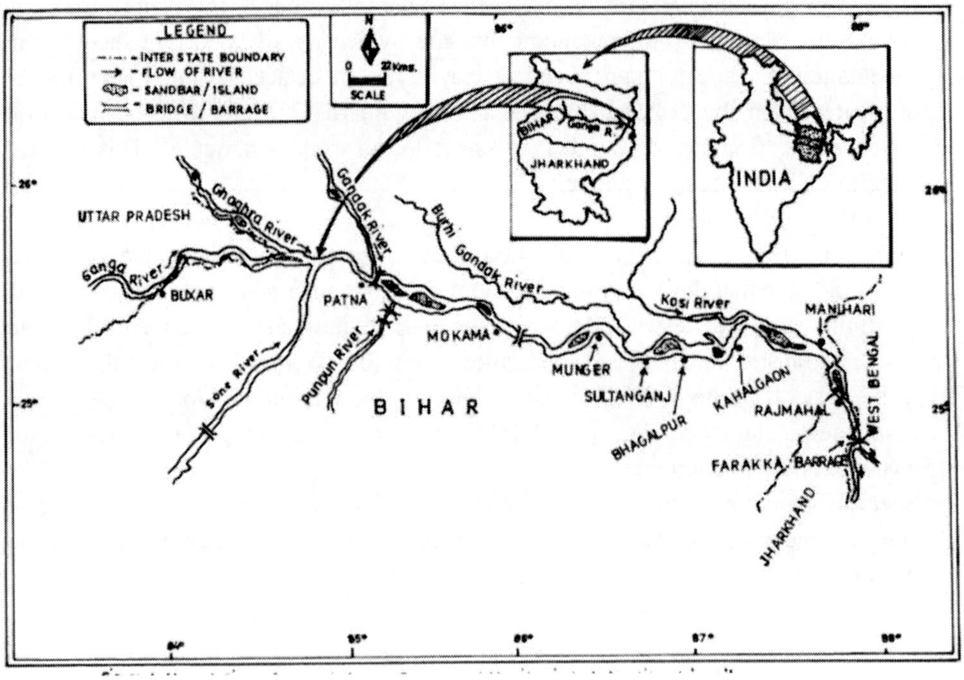

Figure 1. Map of the River Ganga between Buxar and Maniharighat (Katihar) in Bihar.

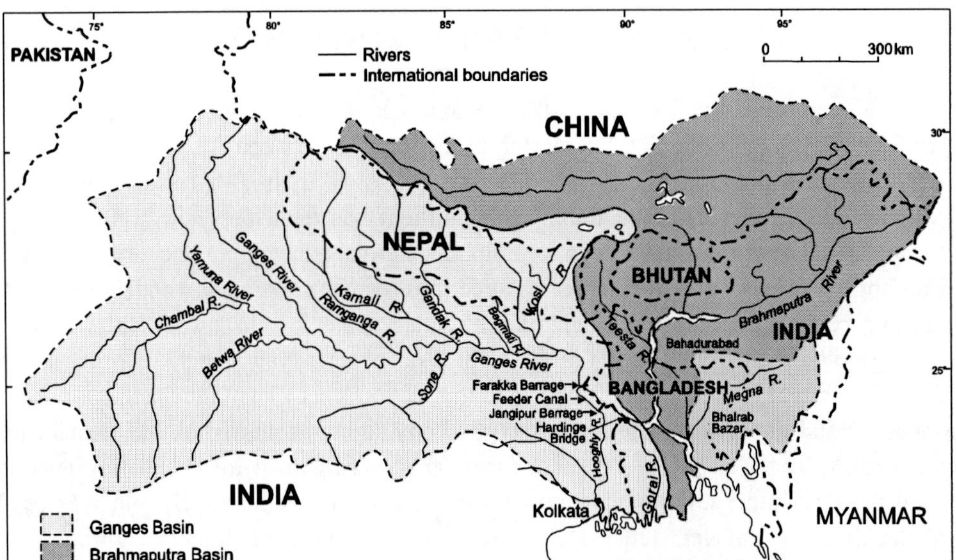

Figure 2. Map of the Ganges Basin in Indian Subcontinent.

There was an impact of water current on the speed of the vessel which had direct bearing on the sighting records. During downstream surveys the average speed of the boat was 10.73 km/hr (range 9.72-12.43; S.D. 0.66), whereas during upstream surveys the average boat speed was 6.46 km/hr (range 6.05 – 6.99; S.D. 0.31).

Group sizes were recorded according to best, high, and low estimates which allowed us to evaluate sightings in terms of a range of abundance estimates, rather than an absolute count, which would not reflect the inherent uncertainty about the actual number of animals present in a certain area (Smith & Reeves, 2000). High and low estimates were used to reflect the confidence of observers in the accuracy of the best estimate. The low estimate was considered a minimum count and the high estimate a maximum count. Identical best, high, and low estimates indicated a high level of confidence in the best estimate. Sightings that could not be substantiated by subsequent surfacing or confirmation by a second member of the survey team were given a best and low estimate of zero and a high estimate of one. Distinctive physical characteristics of individual animals (e.g. scarring, pigmentation patterns, length of the rostrum relative to height of melon, and body size) and the location of surfacing relative to shoreline features and other animals was used to assist observers in making group-size estimates. Estimates were agreed upon by a consensus of the research team. If observers did not agree, the lowest estimate by any team member was used for the low, the highest estimate for the high, and the best estimate by either the observer with the most experience or the observer who first sighted the animal(s) for the best. Double counts were avoided by maintaining close communication among observers and, for some sightings; we used a zero for our low and occasionally best group size estimates, if there was a possibility that the animals had already been counted (Smith et al., 1994). The number of calves, defined as <1.0 m long (Brownell, 1984), was also recorded. Age-class of dolphins was recorded as adult, sub-adult and calf based on Smith & Reeves (2000). This relatively simple direct count survey technique was selected because of the need for a standardized methodology that could be economically and consistently applied by a small team to monitor long-term population trends.

The river condition varied in different seasons. It was quiet and calm but wide (in some segments >2 km) after flooding in November-December. February-March is the leanest season, whereas in May-June water level rises due to melting of snow in the Himalayas. But the surface becomes choppy on many occasions due to high winds. July – September is the period of monsoon and flood. Thus during the November-December period, the possibility of missing dolphins is quite low because the river's surface is very quiet and calm. In contrast, wide river widths and choppy waters make it difficult to locate and observe dolphins.

RESULTS AND DISCUSSION

Mean encounter rates calculated from the best estimates of group size were 1.28 dolphins km^{-1} (range 1.10-1.56; S.D. 0.17) and 1.0 dolphins km^{-1} (range 0.84-1.13; S.D. 0.11) for the six upstream and six downstream surveys, respectively (Table 1). Upstream counts were significantly different from downstream counts (Chi Square $p<0.001$, df = 1), with upstream counts averaging 31.8% greater than downstream. The overall mean survey speed was 10.39 km h^{-1} (range 9.72 – 11.19; S.D. 0.49) for downstream and 6.24 km h^{-1} (range 6.05 – 6.46; S.D. 0.17) for upstream surveys. The mean count for all upstream surveys was 657 ± SD 91.75 dolphins (range 559-808) based on the sum of best estimates of group size for both primary and secondary observers. Counts based on the sum of the low and high estimates of group size were on average 0.88% lower and 1.16% higher respectively. The mean percentage of sightings by the primary observer was 348 ± SD 65.26 whereas by the secondary observer was 95 ± SD 10.96. Thus the mean percentage of sightings missed by the primary observers but detected by the secondary observers was 27.3%. The percentage of observed neonates was relatively high during the surveys conducted in May-June 2006 (15.74 and 15.49% for upstream and downstream surveys, respectively) and low for those conducted in March 2007 (7.08 and 7.68% for upstream and downstream surveys, respectively).The difference in the proportion of neonates observed during May-June 2006 versus March 2007 may be due to the fact that maximum breeding takes place during April-May in the Ganges. Choudhary et al. (2006) recorded 21% and 18.4% calves in May and December 2001, respectively in the Ganges and opined that the difference in the proportion of neonates may be accounted for by the apparent preference of cow-calf pairs to congregate in large counter-current pools and the fact that the availability of these features varies in different river segments from year to year.

The Ganges River dolphins were sighted all along the stretch of the Ganges under study, however, the dolphin population showed spatial and temporal variation. The climatic and weather conditions, viz. choppy water surface due to high wind, glaring due to sunlight, foggy weather, rain etc also had impacts on sighting records of the dolphins. Usually more dolphins were sighted near confluences, meanderings, behind sand bars, and in the river stretches having deep pools, shallow waters and sand bars.

Our dolphin encounter rates in the Ganges compare favorably to other rivers and in the Ganges where the species has been surveyed using similar techniques. For example, the encounter rate for upstream and downstream surveys in 65 km of the Vikramshila Gangetic Dolphin sanctuary between Sultanganj and Kahalgaon in Bihar within the Ganges was 1.8 dolphin km-1 and 1.2 dolphins km-1, respectively during 2001-2003 (Choudhary et al., 2006). This stretch falls within our area of survey. Similarly, the encounter rate for a downstream survey in the middle Brahamaputra River between Guwahati and Goalpara was 0.03 dolphin km^{-1} during April 1999, and 0.76 dolphins km^{-1} for downstream surveys in the Karnaphuli-Sangu river of Bangladesh from January to April 1999 (Smith et al., 2001). 197 dolphins were recorded in the entire 856 km (i.e. 0.23 dolphin km^{-1}) of the River Brahmaputra stretch from Assam-Arunachal Pradesh border to the India-Bangladesh border (Wakid, 2009). Overall the pattern of dolphin occurrence was consistent with a preference for reaches characterized by complex morphological features that induce hydraulic heterogeneity and bottom scouring (Smith, 1993; Smith et al., 1998, 2001).

In areas with large seasonal fluctuations in water level, dolphins are most abundant in the long stretches of deep water that remain during the dry season (Singh & Sharma, 1985). Occasionally susu enter shallow water (2 m) while chasing prey, but they remain primarily in, or near (with ready access to) deeper river channels (Reeves et al., 1993). Habitat preferences of the susu have been studied in the River Karnali in Nepal by Smith (1993). Two categories of habitats were identified: primary habitats characterized by an eddy-counter system in the main river flow caused by a point bar formed from sediment deposits of a convergent stream branch or a tributary, and marginal habitats having a smaller eddy counter-current system caused by an upstream meander. The stretch of the Ganges under study has very low gradient (1:13000) and receives the three major tributaries, the Ghaghara (Mahakali+ Karnali rivers in Nepal), Gandak (Narayani in Nepal), and Kosi (saptakosi in Nepal) originating in the Himalayas and smaller tributaries like the Sone and the Punpun from the south, originating in central India. The primary habitats were identified at the confluence of the tributaries. Bottom substrates were fine silt and clay deposited by the tributaries creating a large number of relatively shallow sloping sand bars especially during low water seasons (February to June). The sand bars deflect the river flow and create eddy counter-current systems. Another type of primary habitat was deep pools, each pool with an eddy counter-current system due to hillocks in the mid channel or hard rocks jutting out. This creates deep zone and meandering of the river at Munger, Sultanganj and Kahalgaon. The river flow in a single channel with meanderings formed marginal habitats downstream. Dolphins were also frequently sighted in large groups in counter-current pools induced by bridge pilings at Patna and Bhagalpur, and the monadnocks (rock islands) at Munger and Kahalgaon, and below the confluence of Ghaghara, Gandak and Kosi at Doriganj, Patna and downstream of Kahalgaon, respectively. Large counter current pools were also the primary sites for fishing, ferry crossings, and religious and domestic bathing. The same conditions that make these areas suitable for dolphins (i.e. hydraulic refuge and abundant fish), also make them desirable sites for human use.

The higher counts for the upstream surveys can be explained by the greater number of opportunities to detect surfacing dolphins at a slower surveying speed. The relatively large number of sightings missed by the primary observers but detected by secondary observers indicates that the actual number of dolphins occurring in the stretch of the Ganges under study is probably greater than our minimum abundance estimates (Marsh & Sinclair, 1989), even though we included the secondary observer sightings.

Conservation Status of the Ganges River Dolphins

The Ganges dolphin is listed in CITES Appendix 1. All of the Asian river dolphins have suffered dramatic declines in both their range and numbers over the last three decades to the point at which they rank as some of the most endangered of all mammals. In addition to CITES listings, the IUCN-World Conservation Union has classified the *susu* as an endangered species in 1996. Furthermore, the Ganges River dolphin has been included as a Schedule-1 animal under the Indian Wildlife (Protection) Act 1972.

The future of Ganges dolphins appears to be bleak unless action is taken immediately to reduce pressure on their freshwater ecosystems. They face unprecedented threats through the exploitation of freshwater river systems by people, causing high incidental mortality. In order to prosper, the dolphin ironically requires the same conditions as people who use the river, namely, a healthy, flowing, living river.

Threats

The Ganges dolphins face many threats –incidental capture in fisheries (by catch), depletion of food resources, chemical and noise pollution, habitat destruction, over-fishing and climate change, and in some cases, directed killing for oil and meat.

Direct Catch

Deliberate killing of *susus* is believed to have declined in most areas but some individuals are still taken each year and their oil and meat used as liniment, as an aphrodisiac and as bait for catfish. A few cases are recorded in the middle Ganges in Bihar (Sinha et al 2000), in the Kalni-Kushiyara River of Bangladesh, and in upper reaches of the Brahmaputra River in Assam, India (Mohan et al., 1997). The magnitude of direct take in recent years is unknown, but probably not high (IWC, 2000). Based on a report of dolphin poaching in the Ganges in 2001, the Patna High Court (C.J.W.C. No. 5628) intervened and directed both the state government of Bihar and central government of India to allocate funds for supporting dolphin conservation efforts in Bihar and to stop poaching of the susu. It created awareness among the common masses and at the same time forced the government officials to act which resulted in an apparent decline in directed killings of the dolphins. However much more needs to be accomplished. Cooperation among regulatory authorities, NGOs and local fishermen will be essential for advancing science and community based management and conservation of this species.

Incidental Catch

Entanglement in fishing gillnets causes significant damage to local population numbers. Accidental killing is a severe problem for Ganges River dolphins throughout their range. Gillnets are problematic because from an economic perspective their use is of primary importance to impoverished fishing communities. The primary cause is believed to be entanglement in fishing gear, most often in nylon gillnets, mainly because the dolphin's preferred habitat is often in the same location as primary fishing grounds. It is not easy to get actual estimates of dolphin mortality due to unorganized fishing in the very vast areas of the Ganges and its tributaries but the problem of accidental killing is expected to worsen with increasing fishing intensity and use of monofilament fishing gillnets. Regulations of gillnets should include strictly limiting their numbers and configurations (e.g. five per family with a mesh size of no less than 24 mm and a length and width of no more than 150 m and 7 m, respectively, and these numbers may be adjusted if additional families enter the gillnet fishery) and requiring fishermen to monitor their nets and release dolphin bycatch if entangled (Choudhary et al., 2006). Incidental killing via boating also occurs, evidenced by a pregnant female hit and killed by a boat propeller at Patna in September of 2005. Thus, motorized river traffic is also a threat to Ganges dolphins.

Habitat Degradation

Construction of dams and barrages for hydroelectric development and irrigation in the Ganges system has fragmented the dolphin population and prevented migrations, thus leading to the segregation of populations and a narrowing of the gene pool in which dolphins can breed. These same barriers have also reduced food availability and also drastically altered the dolphin's habits and habitat. The population above the Kaptai dam in the Karnaphuli River in Bangladesh disappeared over a period of 6 or 7 years after dam construction. Similarly, the dolphin population disappeared from the main stem of the Ganges above the Middle Ganga Barrage at Bijnor (about 100 km downstream Haridwar) after 12 years of its construction (Sinha, 1999). Dolphins in Nepal are almost extinct in Mahakali, Narayani and Sapta Kosi due to construction of barrages at their heads at the India-Nepal border (Smith et al., 1994).

The construction of embankments in the dolphin's distribution range within India and Bangladesh has drastically affected the ecosystem where overflow and flooding are important cycles in the movements and spawning of flood plain dependent fishes (Smith et al., 1998).

In addition to fragmenting dolphin populations, dams and barrages degrade downstream habitat and create reservoirs/pondage with high sedimentation and altered assemblages of fish and invertebrates (IWC, 2000). Luxuriant growth of macrophytes and excessive siltation have eliminated suitable habitat immediately above Farakka Barrage (Sinha, 2000). A detail register of water development projects and their effects on the river dolphins in Asia was published by Smith et al. (2000).

Excess abstraction of the river water for irrigation has lowered water levels throughout the species' range and has threatened suitable habitat especially in the Ganges where the mean dry-season water depth has been dramatically reduced in the recent years. The long-term implications of the reduction of dry-season flows in the Ganges are catastrophic for the survival of *susus*.

Dredging and development of the river environment has altered its nature and eliminated counter currents, one of the most preferred habitats of the dolphin and where the dolphins spend much of their time. Heavy river traffic is drastically increasing in the Ganges and Brahmaputra. This may result in habitat restriction, noise pollution, depletion of prey base and changes in feeding behavior of the susu in the rivers.

Other sources of habitat degradation include dredging (Smith et al., 1998) and the removal of stones, sand (Mohan et al., 1997) and woody debris (Smith, 1993). These activities threaten the ecological integrity of the river environments, especially in small tributaries where suitable habitat is more confined and therefore more vulnerable to local sources of degradation.

Pollution

Pollution by fertilizers, pesticides, and industrial and domestic effluents is dramatic in the Ganges River; about 1.5 million metric tons of chemical fertilizers and about 21,000 tons of technical grade pesticides were dumped annually to the Ganges-Brahamaputra river system in India during 2002-2003 (Source: www.ncipm.org.in/asps/pesticides.asp). Senthilkumar et al. (1999) determined concentrations of polychlorinated biphenyls (PCBs), hexachlorocyclohexane (HCH), chlordane compounds, and hexachlorobenzene (HCB) in the Ganges River dolphin blubber, muscle, kidney, liver and prey collected from stomach of the dolphins collected during 1993 through 1996 from the River Ganges in and around Patna, India. Comparison of organochlorine concentrations with values reported for samples analyzed during 1988 through 1992 suggested that the contamination by these compounds has increased in the river. Kannan et al. (1997) determined concentrations of butyltin compounds in dolphins, fish, invertebrates and sediments collected from the Ganges in and around Patna. Total level in dolphin tissues was up to 2000 ng g^{-1} wet wt, which was about 5-10 times higher than in their diet. The biomagnification factor for butyltins in river dolphin from its food was in the range of 0.2-7.5. Butyltin concentrations in Ganges River organisms were higher than those reported for several persistent organochlorine compounds.

Recently perfluorinated compounds (PFCs) were analyzed in the biological samples including the Ganges dolphins (n = 15) from the Ganges at Patna. The arithmetic mean PFOS (perfluorooctanesulfonate) concentration found in the liver of the Ganges River dolphin was 27.9 ng g^{-1} ww. Biomagnification factors (BMF) of PFcs were estimated for the Ganges River dolphins, based on concentrations in water from the Ganges River at Patna, shrimps, fish livers and Ganges River dolphins. The estimated BMFs of PFOS from fishes to Ganges River dolphin were similar to the BMFs reported for narwhals (*Monodon monoceros*) and beluga whales (*Delphinapterus leucas*) in an eastern Arctic food web (4.0-8.4), and in bottlenose dolphin food web from costal Florida (1.5-35) USA (Yeung et al., 2009). Poisoning of the water from industrial and agricultural chemicals may have also contributed to dolphin population decline.

The UN Convention on the Conservation of Migratory Species of Wild Animals (CMS), known as the Bonn Convention, which focuses on wild animals crossing the national boundaries declared 2007 as the "Year of the Dolphin". The UNEP and UNESCO, governments and non-governmental organizations are building a strong alliance to achieve a common objective: to protect dolphins. Crucial factors in achieving this objective include the education to create awareness of dolphin species and the threats facing them, informing decision makers and involving local communities. The Year of the Dolphin is also part of the UN Decade on Education for Sustainable Development. The campaign is a tangible contribution towards meeting the targets of significantly reducing the loss of biodiversity by 2010, which all governments have agreed upon at several global UN meetings.

The Ganges dolphin as National aquatic Animal of India

On the initiatives of the author (RKS), the Indian Prime Minister on 5th October 2009 declared the Ganges River dolphin as National Aquatic Animal. Hopefully, this will ensure long term survival of this species by joint efforts of policy makers, planners, executives, scientists, conservationists and common mass. The Government of India has initiated a mega plan to maintain ecological flow and pollution abatement in the River Ganges and its tributaries in 2009 and gradual increase in the number of the Ganges dolphin will be an indicator of success of these plans.

ACKNOWLEDGMENTS

We gratefully acknowledge financial assistance from the Ministry of Environment and Forests, Government of India.

REFERENCES

[1] Agrawal, V. C. (1991). Mammalia. In *Faunal resources of Ganga, part – 1* (pp. 13-20). Calcutta, India: Zoological Survey of India.

[2] Anderson, J. (1879). "*Anatomical and Zoological researches: Comprising an account of zoological results of the two expeditions to western Yunnan in 1868 and 1875; and a monograph of the two cetacean genera Platanista and Orcella, Two Volumes*" London, United Kingdom: Bernard Quaritich.

[3] Ansari, A. A., Singh, I. B. & Tobschall, H. J. (1999). Status of anthropogenically induced metal pollution in Kanpur-Unnao industrial region of the Ganga plain, India. *Environmental Geology*, **38**, 25-33.

[4] Aranson, U. & Gullberg, A.(1996). Cytochrome-b nucleotide sequences and the identification of five primary lineages of extinct cetaceans. *Molecular Biology and Evolution*, **13**, 407-417.

[5] Bajpai, S., Thewissen, J. G. M. & Sahni, A. (2009). The Origin and early evolution of whales: macroevolution documented on the Indian Subcontinent. *J. Bioscience*, **34** 673-686.

[6] Biswas, S. P. & Baruah, S. (2006). Ecology of the River Dolphin *Pltanista gangetica gangetica* in the Upper Brahmaputra. *Hydrobiologia*, **430**, 97-111.

[7] Boisserie, J. R., Lihoreau, F. & Brunet, M. (2005). Origins of Hippopotamidae (Cetartiodactyla): towards resolution. *Zoologica Scripta*, **34**, 119-143.

[8] Brownell, R. L. Jr.1984. Review of reproduction in Platanistid dolphins. *Rep. International Whaling Commission (Special issue)* **6:**149-158.

[9] Burbank, D. W., Beck, R. A. & Mulder, T. (1996). The Himalayan Foreland basin. In A. Yin, & M. Harrison (Eds.), *The tectonic evolution of Asia* (pp. 149-188). Cambridge, United Kingdom: Cambridge University Press.

[10] Cassens, I., Vicario, S., Waddell, V. G., Balchowsky, H., Belle, D. V., Ding, W., Fan, C., Mohan, R. S. L., Simões-Lopes, P. C., Bastida, R., Meyer, A., Stanhope, M. J. & Milinkovitch, M. C. (2000). Independent adaptation to riverine habitats allowed survival of ancient lineages. *Proceedings of the National Academy of Sciences,* **97**, 11343-11347.

[11] Choudhary, S. K., Smith, B. D., Dey, S. & Satya Prakash (2006). Conservation and biomonitoring in the Vikramshila Gangetic Dolphin Sanctuary, Bihar, India. *Oryx,* **40**, 1-9.

[12] Dudgeon, D. (2000). The ecology of tropical Asian rivers and streams in relation to biodiversity conservation. *Annual Review of Ecology and Systematics,* **31**, 239-263.

[13] Fordyce, R. E. (1994). *Waipatia maerewhenua*, new genus and new species, an archaic late Oligocene dolphin. *Proceedings of the San Diego Society of Natural History,* **29**, 147-176.

[14] Fordyce,R. E. & Barnes, L. G. (1994). The evolutionary history of whales and dolphins. *Annual Review of Earth and Planetary Sciences*, **22**, 419-455.

[15] Fordyce,R. E. & Muizon, C. de. 2001. Evolutionary history of Cetaceans: A Review In V. de Buffrenil, & J. M. Mazin (Eds.), *Secondary adaptations to the life in water* (pp. 169 -233), Munich, Germany: Pfeil Verlag.

[16] Gatesy, J. & O'Leary, M. A. (2001). Deciphering whales origins with molecules and fossils. *Trends Ecology and Evolution*, **16**, 562-570.

[17] Gottfried, M. D., Bohaska, D. J., Whitmore, Jr. F. C. 1994. Miocene cetaceans of the Chesapeake of the group. *Proc. San Diego Soc. Nat. Hist.***29**: 229-238.

[18] Guang, Y. & Kaiya, Z. (1999). A study on the molecular phylogeny of river dolphins. *Acta* Theriologica *Sinica,*. **19**(1), 1-19.

[19] Gupta, P. D. (1986). The Gangetic dolphin, *Platanista gangetica* (Lebeck, 1801). In T. C. Majupuria (Ed.), *Wildlife Wealth of India: Resources and managements*, Bangkok, Tailand: Teepress Service, L. P.

[20] Hallam A. (1992). *Phanerozoic sea-level changes*. New York, NY: Columbia University Press.

[21] Hamilton, H., Cabellero, S., Collins, A. G. & Brownell, R. L. Jr. (2001). Evolution of river dolphins. *Proceedings of the Royal Society of London B,* **268**, 549-556.

[22] Haq, B. H., Hardenbol, J. & Vail, P. R. (1987). Chronology of fluctuating sea levels since the Triassic. *Science*, **235**, 1156-1162.

[23] Hopkin, M. (2007). Gorillas on the list. *Nature*, **449**(13), 127.

[24] Hua, Y., Zhao, Q., & Zhang G. (1989). The habitat and behavior of *Lipotes vexillifer*. In W. F. Perrin, R. L. Jr. Brownell, K. Zhou & J. Liu (Eds.), *Biology and conservation of the river dolphins* Occasional *Paper of the IUCN Species Survival Commission* (No.3., pp. 92-98). Gland, Switzerland: International Union for Conservation of Nature.

[25] Heyning, J. E. 1989. Comparative facial anatomy of the beaked whales (Ziphiidae) and a systematic revision among the families of extant Odontoceti. *Contributions in Science (Las Angeles) of the Natural History Museum,* **405**, 1-64.

[26] International Union for Conservation of Nature (IUCN). IUCN Red List of threatened species [online]. 2007 [cited 2007, August, 11]. Available from: URL: http://www.iucnredlist.org/info/stats.

[27] International Whaling Commission (IWC). (2001). Report of the standing sub-committee on small cetaceans. *Journal of Cetacean Research and Management,* Annex- K. **3**(Suppl.), 263-291.

[28] Kannan, K., Senthilkumar, K. & Sinha, R. K.(1997). Sources and accumulation of butyltin compounds in Ganges River dolphin, Platanista gangetica. *Applied Organometallic,***11**, 223-230.

[29] Kasuya, T. (1972). Some information on the growth of the Ganges dolphin with a comment on the Indus dolphin. *The Scientific Reports of the Whales Research Institute,* **24**, 87-108.

[30] Kellog, A. R. (1995). Description of the skull of *Pomatodelphis inaequalis. Bulletin of the Museum of Comparative Zoology,***121**, 1-26.

[31] Marsh, H. & Sinclair, D. F. (1989). Correcting for visibility bias in strip transect aerial surveys of aquatic fauna. *The Journal of Wildlife Management*, **53**, 1017-1024.

[32] Messenger, S.L & McGuire, J.A. (1998). Morphology, molecules and the phylogenetics of cetaceans. *Systematic Biology*, **47**, 90-124.

[33] Miller GS. (1923) The telescoping of the cetacean skull. *Smithsonian Miscellaneous Collections,***76**, 1-70.

[34] Millinkovitch, M.C., Berube, M. & Palsbol, P.J. (1998). Cetaceans are highly specialized artiodactyls. In J. G. M. Thewissen (Ed.), *The Emergence of Whales: Evolutionary Patterns in the Origin of Cetacea* (pp. 113-131). New York, NY: Plenum Press.

[35] Mohan, R.S.L. (1989). Conservation and management of the Ganges River dolphin, Platanista gangetica, in India. In W.F. Perrin, R. L. Brownell, Jr. Zhou Kaiya & Liu Jiankang (Eds.), *Biology and conservation of river dolphins* Occasional paper of the IUCN (Species Survival Commission No. 3). Gland, Switzerland: International Union for Conservation of Nature.

[36] Mohan RSL, Dey SC, Bairagi, SP, & Roy, S. (1997). On a survey of the Ganges River dolphin *Platanista gangetica* of Brahmaputra River, Assam. *Journal of Bombay Natural History Society,* **94**:483-495.

[37] Morgan GS. (1994). Miocene and Pliocene marine mammal faunas from the Bone Valley formation of central Florida. *Proceedings of the San Diego Society of Natural History,* **29**, 239-268.

[38] Muizon, C de. (1984). Les Vertebres fossiles de la Formation Pisco (Perou). Deuxieme partie: Les odontoctes (Cetacea, Mammalia). du Pliocene inferieur de Sud-Sacaco. *Tracaux Instit. Francais. de Etudes Andines*, **25**, 1-188.

[39] Muizon, C de. (1988). Les relations phylogenetique des Delphinida (cetacea, Mammalia). *Annales de paleontogie*, **74**, 159-227

[40] Muizon, C de. (1991). A new Ziphiidae from the early Miocene of Washington state and a phylogenetic analysis of the major group of odontocetes. *Bull. Mus. Natl. Hist. Nat.* Paris. **3-4**: 279-326.

[41] Muizon, C de. (1994). Are the squalodontids related to the platanistoids? *Proceedings of the San Diego Society of Natural History*, 29, 135-146.

[42] Nikaido, M., Rooney, A.P., & Okada, N. (1999). Phylogenetic relationship among cetartiodactyls based on insertion of short and long interspersed elements: hippopotamuses are the closest extant relatives of whales. *Proceedings of the National Academy of Sciences,* **96**, 10261-10266.

[43] Nikaido, M., Matsuno, F., Hamilton, H., Brownell, R.L. Jr., Cao, Y., Ding, W., Zuoyan, Z., Shedlock, A.M., Fordyce, R.E., Hasegawa, M., & Okada, N. 2001. Retroposon analysis of major cetacean lineages: The monophyly of toothed whales and the paraphyly of river dolphins. *Proceedings of the National Academy of Sciences,* **98**, 7384-7389.

[44] Perrin WF, & Brownell Jr. RL Report of the workshop. (1989) In W. F. Perrin, R. L. Brownell Jr., K. Zhou, J. Liu (Eds.), *Biology and conservation of the river dolphins. Occasional Paper of the IUCN Species Survival Commission* (No.3, pp. 1-21). Gland, Switzerland: IUCN.

[45] Pilleri G, & Gihr M. (1971) Zür systematic der gattung *Platanista* (Cetacea). *Revue* Suisse *de Zoologie,* **78**(3), 746-759.

[46] Reeves, R.R. Leatherwood, S. & Mohan, R.S.L. (1993). *Report from a Seminar on the conservation of river dolphins of the Indian subcontinent. 18-19 Aug.1992, New Delhi, India.* Bath Avon, United Kingdom: Whale and Dolphin Conservation Society.

[47] Reeves, R.R., Smith, B.D. & Kasuya, T. (2000). *Biology and Conservation of Freshwater Cetaceans in Asia.* Occasional Paper No.23 of IUCN Species Survival Commission. Gland, Switzerland and Cambridge, United Kingdom: IUCN.

[48] Reyes JC. (1991). The conservation of small cetaceans: A review. Report prepared for the Secretariat of the Convention on the Conservation of Migratory Species of Wild Animals. Bonn, Germany: UNEP / CMS Secretariat.

[49] Rice DW. (1998). *Marine mammals of the world.* Special Publication no.4. Lawrence, KS: Marine Mammal Society.

[50] Senthilkumar K, Kannan K, Sinha RK, Tanabe S, & Giesy JP. (1999). Bioaccumulation profiles of polychlorinated biphenyl congeners and organochlorine pesticides in Ganges River dolphins. *Environmental Toxicology and Chemistry*, **18**, 1511-1520.

[51] Singh, L.A.K. & Sharma R.K. (1985). Gangetic dolphin *Platanista gangetica*: observations on habits and distribution pattern in National Chambal Sanctuary. *Journal of the Bombay Natural History Society*, **82**(3), 648-653.

[52] Sinha RK. Status and conservation of Ganges River dolphin in Bhagirathi – Hooghly River systems in India. *International Journal of Ecology and Environmental Sciences,* 1997, **23**:343-355.

[53] Sinha, R. K. (1999). *The Ganges River dolphin- a tool for baseline assessment of biological diversity in River Ganges, India. Final Technical Report No. 1/99*. Patna University, Patna, India.

[54] Sinha RK. Status of the Ganges River dolphin (*Platanista gangetica*) in the vicinity of Farakka Barrage, India. (2000) In R. R. Reeves, B. D. Smith, T. Kasuya (Eds.), *Biology and conservation of freshwater cetaceans in Asia. Occasional Paper of the IUCN Species Survival Commission* (No. 23, pp. 42-48). Gland, Switzerland and Cambridge, United Kingdom: IUCN.

[55] Sinha RK, Smith BD, Sharma G, Prasad K, Choudhary BC, Sapkota K, Sharma RK, & Behera SK (2000). Status and distribution of the Ganges susu (*Platanista gangetica*) in Ganges River system of India and Nepal. In R. R. Reeves, B. D. Smith, T. Kasuya, (Eds.), *Biology and conservation of freshwater cetaceans in Asia. Occasional Paper of the IUCN Species Survival Commission* (No. 23, pp. 42-48) Gland, Switzerland and Cambridge, United Kingdom: IUCN.

[56] Sinha, R. K. (2002). An alternative to dolphin oil as a fish attractant in the Ganges River system: conservation of the Ganges River dolphin. *Biological Conservation*, **107**, 253-257.

[57] Sinha, R.K. (2006). The Ganges River dolphin *Platanista gangetica gangetica. Journal of the Bombay Natural History Society*, **103**, 254-263.

[58] Smith B. D. (1993). Status and conservation of the Ganges River dolphin *Platanista gangetica* in the Karnali River, Nepal. *Biological Conservation*, **66**, 159-169.

[59] Smith B. D, & Jefferson TA. (2002). Status and conservation of facultative freshwater cetaceans in Asia. *The Raffles Bulletin of Zoology*, **10** (Suppl.), 173-187.

[60] Smith B. D & Reeves R. R. (2000). Report of the second meeting of the Asian River Dolphin Committee, Rajendrapur, Bangladesh, 22-24 February 1997. In R. R. Reeves, B. D. Smith, T. Kasuya (Eds.). *Biology and conservation of freshwater cetaceans in Asia. Occasional Paper of the IUCN Species Survival Commission* (No. 23, pp. 1-14). Gland, Switzerland and Cambridge, United Kingdom: IUCN.

[61] Smith B. D & Reeves R. R. (2000). Survey methods for population assessment Asian river dolphins. In R. R. Reeves, B. D. Smith, T. Kasuya (Eds.), *Biology and conservation of freshwater cetaceans in Asia. Occasional Paper of the IUCN Species Survival Commission* (No. 23, pp. 97-115). Gland, Switzerland and Cambridge, UK: IUCN, pp. 97-115.

[62] Smith B. D, Sinha R. K, Regmi U, & Sapkota K. (1994). Status of Ganges River dolphins (*Platanista gangetica*) in the Karnali, Narayani and Saptakosi Rivers of Nepal and India in 1993. *Marine Mammal Science,* **10**, 68-75.

[63] Smith B. D, Thant U. H, Lwin J. M, & Shaw, C. D. (1997). Investigations of cetaceans in the Ayeyarwadi River and northern coastal waters of Myanmar. *Asian Marine Biology*, **14**, 173-194.

[64] Smith B. D, Aminul Haque A. K. M, Hossain M. S, & Khan A. (1998). River dolphins in Bangladesh: Conservation and the effects of water developments. *Environmental Management,* **22**, 323-335.

[65] Smith B. D, Sinha R. K, Zhou K, Chaudhry A. A, Renjun L, Wang D, Ahmed B, Aminul Haque AKM, Sapkota K, & Mohan RSL.(2000). Register of water development projects affecting Asian river cetaceans. In R. R. Reeves, B. D. Smith, T. Kasuya (Eds.), *Biology and conservation of freshwater cetaceans in Asia. Occasional*

Paper of the IUCN Species Survival Commission (No. 23 pp. 22-39). Gland, Switzerland and Cambridge, United Kingdom: IUCN,

[66] Smith BD, Ahmed B, Ali ME, & Braulic G. (2001) Status of the Ganges River dolphin or shushuk *Platanista gangetica* in Kaptai lake and the southern rivers of Bangladesh. *Oryx*, **35**, 61-72.

[67] Turvey, S. T., Pitman, R.L., Taylor, B.L., Barlow, J. Akamatsu, T., Barrett, L.A., Xiujiang, Z., Reeves, R.R., Stewart, B.S., Wang, K., Zhou, W., Zhang, X., Pusser, L.T., Richlen, M., Brandon, J. & Ding, W. (2007). First human caused extinction of a cetacean species? Biology Letters, **3**, 537-540.

[68] Thewissen, J.G.M., Cooper, L.N., Clementz, M.T., Bajpai, S. & Tiwari, B.N. (2007). Whales originated from aquatic artiodactyls in the Eocene epoch of India. *Nature*, **450**, 1190-1194.

[69] Verma S. K, Sinha R. K, & Singh L. (2004). Phylogenetic position of *Platanista gangetica*: insights from the mitochondrial cytochrome *b* and nuclear interphotoreceptor retinoid-binding protein gene sequences. *Molecular Phylogenetics and Evolution*, **33**, 280-288.

[70] Wakid, A. (2005). Status and distribution of newly documented residential Gangetic Dolphin (*Platanista gangetica gangetica*) population in eastern Assam. *Journal of the Bombay Natural History Society*, **102**, 158-161.

[71] Wakid, A. (2009). Status and distribution of the endangered Gangetic dolphin (Platanista gangetica gangetica) in the Brahmaputra River within India in 2005. *Current Science*, **97** (8), 1143-1151.

[72] World Wide Fund for Nature (WWF). (2006). Conservation and Management of river dolphins in Asia. Nepal, Kathmandu: WWF.

[73] Yan, J., Zhou, K. & Yang, G. 2005. Molecular phylogenetics of 'river dolphins' and the baiji mitochondrial genome. *Molecular Phylogenetics and Evolution*, 37, 743-750.Yang, G. & Zhou, K. (1999). A study on the molecular phylogeny of river dolphins. Acta *Theriologica* Sinica, **19**, 1-9.

[74] Yang, G., Zhou, K., Ren, W.H., Ji, G.Q. & Liu, S. (2002). Molecular systematics of river dolphins inferred from complete cytochrome-b gene sequences. *Marine Mammal Scence,* **18**, 20-29.

[75] Yeung, L. W. Y., Yamashita, N., Taniyasu, S., Lam, K. S., Sinha, R. K., Borole, D. V. & Kannan, K. (2009). A survey of perfluorinated compounds in surface water and biota including dolphins from the Ganges River and in other water bodies in India. *Chemosphere*, **76** (1): 55-62.

In: Biology, Evolution and Conservation of River Dolphins... ISBN: 978-1-60876-633-8
Editors: M. Ruiz-García, J. Shostell. pp. 445-488

Chapter 23

THE EVOLUTIONARY HISTORY AND PHYLOGENETIC RELATIONSHIPS OF THE SUPERFAMILY PLATANISTOIDEA

Lawrence G. Barnes[1]*, Toshiyuki Kimura[2]+, and Stephen J. Godfrey[3]±

[1]Curator of Vertebrate Paleontology, Department of Vertebrate Paleontology, Natural History Museum of Los Angeles County, Los Angeles, California, USA
[2]Assistant Curator of Paleontology, Gunma Museum of Natural History, Kamikuroiwa, Tomioka, Gunma, JAPAN
[3]Curator of Paleontology, Department of Paleontology, Calvert Marine Museum, Solomons, Maryland, USA

ABSTRACT

The fossil record demonstrates that in the past the "river dolphin" superfamily, Platanistoidea, was much more widespread geographically, and more diverse ecologically and taxonomically than it is now, and that most of its early members lived in salt water, not fresh water. Families in the Platanistoidea comprise a significant initial radiation of dolphin-like toothed whales (suborder Odontoceti). Platanistoids were predominant odontocetes in some late Oligocene and early Miocene age fossil assemblages, from approximately 25 to 15 million years ago. However, the Platanistoidea gradually declined in abundance and diversity approximately 15 million years ago, and they were gradually replaced, largely by another rapidly diversifying odontocete superfamily, the Delphinoidea. During their evolutionary histories, these two superfamilies have had an inverse relationship of diversity and abundance. Among the archaic groups of Platanistoidea, the essentially cosmopolitan Oligocene and Miocene family Squalodontidae is the most primitive dentally, having heterodonty (teeth still recognizable as incisors, canines, premolars, and molars), large and projecting anterior

---

* e-mail lbarnes@nhm.org.
+ kimura@gmnh.pref.gunma.jp.
± e-mail godfresj@co.cal.md.us.

teeth, and serrated and broad-crowned cheek teeth, but well-telescoped crania with their nares moved posteriorly, and relatively primitive body skeletons showing them to have been medium-size whales compared to living species. The Miocene family Allodelphinidae comprises strictly marine North Pacific odontocetes that had primitive braincases, with relatively small nares, and very long and dorsoventrally flattened rostra and symphyseal portions of their mandibles, which contained numerous small teeth. Late Oligocene and Early Miocene marine members of the family Waipatiidae from the South Pacific and northern hemisphere were smaller than squalodontids, and had smaller teeth with less recognizable heterodonty. Platanistoids in the more derived clade that ultimately culminated in the recent family Platanistidae have a modified zygomatic process of the squamosal that is compressed from side to side. Within this clade of Platanistoidea, the Atlantic and Southern Ocean family Squalodelphinidae includes the more primitive, small to medium-size species that have tuberosities superior to the orbits that are not invaded by the pterygoid sinuses, and teeth that still retained remnants of heterodonty. The more highly derived Miocene to Recent family Platanistidae includes two named subfamilies, the Miocene Pomatodelphininae, and the Miocene to Recent Platanistinae. Species of the North Atlantic subfamily Pomatodelphininae are relatively large, long-snouted dolphins that had many small teeth, rostra and symphyseal parts of the mandibles that are compressed dorsoventrally, and many species in this subfamily, but not all of them, have enlarged bony tuberosities over the orbits that are invaded by extensions of the pterygoid air sinuses. These dolphins have been found in near shore marine, estuarine, and fresh water deposits, and these are the first indications of any fresh water-dwelling Platanistoidea. The more derived species of Platanistidae, those in the subfamily Platanistinae, have fenestrations within the supraorbital crest caused by invasion of the pterygoid sinus, and have a transversely flattened rostrum and symphyseal part of the mandible. Miocene members of the subfamily Platanistinae are known from North Pacific marine deposits, but the living members of the genus *Platanista* live only in rivers of south Asia. A cladistic analysis provides a framework for a classification of the Platanistoidea that is presented here.

INTRODUCTION

Among the cetacean suborder Odontoceti, or echolocating toothed whales, species of the superfamily Platanistoidea constitute an early evolutionary radiation of small to medium-size dolphin-like animals that diversified during the Oligocene and early part of the Miocene epochs. The Platanistoidea reached the apex of their diversity during Late Oligocene and Early Miocene time, approximately 28 to 16 million years ago. Subsequently platanistoids gradually declined in both abundance and diversity. They were gradually replaced ecologically by the now very diverse and numerous members of the superfamily Delphinoidea, which includes, in addition to such extinct families as the generalized Kentriodontidae, the Albireonidae, and the very strange and large-tusked Odobenocetopsidae, and the fossil and Recent pelagic dolphins, killer whales, true porpoises, belugas, and narwhals (families Delphinidae, Phocoenidae, and Monodontidae) (Barnes, 2002a).

The taxon Platanistoidea, first recognized at the superfamily level by Simpson (1945), has had a convoluted history in cetacean taxonomic studies (for example see Kellogg, 1926:2; Messenger & McGuire, 1998; Hamilton et al., 2001). In some previous classifications (including that of Simpson, 1945) this superfamily included other so-called "river dolphins" of the families Iniidae, Lipotidae, and Pontoporiidae, but these groups are now usually classified in other odontocete superfamilies (Muizon, 1988b, 1991; Fordyce & Barnes, 1994; Fordyce et al., 1995; Barnes, 2002b, 2006; Cassens et al., 2000; Nikaido, 2001; McKenna & Bell, 1997:386-387). Geisler & Sanders (2003), however, supported a more inclusive Platanistoidea, as it was used by Simpson (1945).

The current concept of the superfamily Platanistoidea (e.g., Muizon, 1987, 1991, 1994; Heyning, 1989; Fordyce, 1994, 2006; Barnes, 2002b, 2006; Godfrey et al., 2006) differs greatly from its original concept, with the inclusion of Oligocene to Middle Miocene "shark-toothed" whales of the family Squalodontidae (see Muizon, 1994; Fordyce, 1994; Dooley, 2005), archaic Late Oligocene dolphin-like members of the family Waipatiidae (Fordyce, 1994), long-snouted Miocene species of the family Allodelphinidae (Barnes, 2006), Miocene dolphin-like species of the family Squalodelphinidae (see True, 1910; Muizon, 1987; Fordyce, 1994; Godfrey et al., 2006), and with a more restrictive definition of the extant family Platanistidae (see Fordyce, 1994; Barnes, 2002a, 2002b, 2006; Godfrey et al., 2006; Bohaska et al., 2007). Among these various family-level platanistoid clades, only members of the family Platanistidae survive today, and these are fresh water dolphins of the genus *Platanista* Wagler, 1830, that have relict distributions in the Ganges and Bramaputra River systems of south Asia.

The purpose of this study is to provide a historic perspective of the evolution of the Platanistoidea, the group that includes the extant genus *Platanista* Wagler, 1830, to provide illustrations of skulls (some with their parts labeled) of some of the better-known fossil members, to provide a new genus name for a long-problematic fossil platanistoid species, to prepare a phylogenetic analysis of the group, and to provide a classification of the superfamily Platanistoidea based on that analysis.

Materials and Methods

Institutional Abbreviations

CAS Division of Birds and Mammals, California Academy of Sciences, San Francisco, California, U.S.A.

LACM Department of Vertebrate Paleontology, Natural History Museum of Los Angeles County, Los Angeles, California, U.S.A.

USNM Department of Paleobiology, United States National Museum of Natural History, Smithsonian Institution, Washington, D.C., U. S. A.

YPM Peabody Museum of Natural History, Yale University, New Haven, Connecticut, U.S.A.

The cranium, CAS 16340, of *Platanista gangetica* (Roxburgh, 1801) that appears in Figure 11 is part of a skeleton of an adult female individual, collected by T. A. Khan, in October 1969, at Mian Sahib Jo Got, approximately 48 km north of Lloyd Barrage, Sukkar, Sind, Pakistan. It has not previously been documented in publication.

Millions of years before present is abbreviated as Ma. Geologic ages and terminology generally follow Berggren et al. (1995). Terminology for cranial anatomy follows Kellogg

(1927), Fraser & Purves (1960), and Barnes (1978, 1985), and authorships and nomenclature of family groups and of modern taxa follow Hershkovitz (1966) and Rice (1998).

A Synopsis of the Superfamily Platanistoidea

Not included in this review is the problematic family Dalpiaziniidae. It was originally proposed with the name Dalpiazinidae by Muizon (1988a), its spelling was subsequently emended by Muizon (1994: Figure 1), and includes only *Dalpiazina ombonii* (Longhi, 1898). Muizon (1988a, 1991) suggested that *Dalpiazina* has close relationships with Squalodontidae, but he later (Muizon 1994:141) stated it has "…. none of the platanistoid synapomorphies."

Also not included in this analysis is the problematic and poorly characterized family Acrodelphinidae (originally named Acrodelphidae by Abel, 1905; and its spelling emended by Rice, 1998, to Acrodelphinidae). This family has sometimes been discussed in relationship to odontocetes that are now included in the Platanistoidea, but Muizon (1988a) relegated the family to Odontoceti, *incertae sedis,* and restricted it to the type species of the type genus, because it was based on non-diagnostic material. Abel (1905) based the family Acrodelphidae on the genus *Acrodelphis* Abel, 1900. The type species of *Acrodelphis* is *Champsodelphis denticulatus* Probst, 1886 (see Probst, 1886:124), and the type material of *C. denticulatus* is four isolated teeth (see Probst, 1886:pl. III, figs. 18-21; and Pilleri, 1986:36, Table 35, pl. 21, Figures 5-7). These four teeth do not necessarily belong to the same individual cetacean, and they definitely do not belong to the same species, and such isolated odontocete teeth are not considered to be diagnostic.

Therefore, on the basis of the above observations and facts, we do not include the families Dalpiaziniidae or Acrodelphinidae in our review of the superfamily Platanistoidea.

General

No phylogenetic study has definitively determined the origin of the Platanistoidea, but the group must have originated from archaic odontocetes that were at least as primitive as *Agorophius pygmaeus* (see Fordyce, 1981). The time of origin of the Platanistoidea must have been at least by the early part of the Oligocene, because platanistoids were diverse at the family levels by late Oligocene time.

All species of the Platanistoidea share the presence of a pointed process, or spine, on the anterior end of the tympanic bulla. This spine in life partly surrounds the ventral side of the Eustachian tube. This may seem like an inconsequential structure, (and if a fossil is not properly cleaned from the rock, or not properly handled in collections, this small spine can be easily destroyed), however, platanistoids have this structure to the exclusion of all other Cetacea, and it is a shared derived (synapomorphic) character that is diagnostic for the superfamily (see Fordyce, 1994: 1776; Barnes, 2006; and see Character number 32 in the following text).

Platanistoids have in their skull a thick, vertically-oriented, bony lamina that extends from the posterior part of the palate, across the medial side of the orbit, to the anterior part of the ear region (see Character number 31 in following text). This bony lamina is composed, in varying degrees, by the palatine bone and the lateral lamina of the pterygoid bone (see Figure 1C, where it is labeled on the holotype skull of *Allodelphis pratti* Wilson, 1935), which extend posteriorly and connect with the alisphenoid bone at a location anteromedial to the glenoid fossa. Muizon (1994) discussed this structure, noting that it exists in all platanistoids except for species of *Prosqualodon.* He described the lamina as being comprised of the palatine bone, covered in its middle by maxilla and pterygoids, and divided into a small anteroventromedial part, and a larger posterodorsolateral area. This laminar plate of bone was called the "reduplicated pterygoid" by Miller (1923). This structure is in part a modification of the smaller pterygoid of generalized mammals. Among the Odontoceti the pterygoid hamulus is invaded by the pterygoid sinus (see Fraser and Purves, 1960). The participation of the lateral lamina of the pterygoid in this prominent bony wall within the orbit of the skull is noteworthy, because in most other Odontoceti except members of the Platanistoidea, it is an atrophied and thin sheet of bone, or it is replaced by a membranous sheath.

Several platanistoids have been shown to lack a coracoid process on the anterior surface of the scapula (Fordyce, 1994; 2006), and because this process is present in more primitive Cetacea, and in most other more derived cetacean groups, its loss from the scapula in the platanistoids is interpreted as being a synapomorphic character. This character, the lack of a coracoid process, exists in the primitive platanistoids, the family Squalodontidae, and in the Recent *Platanista,* but because scapulae are not yet described for many other platanistoids, we did not include it in our phylogenetic analysis.

Muizon (1994) also suggested that the superfamily Platanistoidea is characterized by having the acromion process located on the anterior edge of scapula rather than the lateral surface, the latter location being where it is in generalized mammals, and in primitive Cetacea. The apomorphic condition, however, might characterize some but not all Platanistoidea, because on the scapula of an as yet un-named late Early Miocene allodelphinid platanistoid from Japan (Kimura & Ozawa, 2001) the acromion is not located on the anterior edge of the scapula, but is on its lateral side, as is typical of primitive Cetacea. If it should be demonstrated that this primitive position of the acromion, on the lateral side of the scapula, is characteristic of species of the family Allodelphinidae, then this might re-enforce other indications from the phylogenetic analysis presented herein that the Allodelphinidae are basal members of the superfamily Platanistoidea.

Platanistoids retain relatively large dorsal vertebrae, a primitive character state for Cetacea. Commensurate with their large dorsal vertebrae, the ribs of platanistoids are also relatively large for their body size. Furthermore, through time, members of the family Allodelpinidae evolved progressively larger cervical vertebrae. This phenomenon is the opposite trend that is exemplified by most other groups of Cetecea, in which the cervical vertebrae become progressively very much shortened anteroposteriorly, and in some taxa becoming fused to one another to varying degrees.

A primitive condition that is exemplified by squalodontids is the persistence of very large and anteriorly procumbent incisors and canines. In life these anterior teeth certainly must have protruded from the front of the mouth, probably appearing in a fan-like array of small tusks. In members of the later and more highly derived platanistoid families Waipatiidae and Squalodelphinidae, these anterior teeth became more reduced in size, but they remained still relatively prominent. In members of the families Allodelphinidae and Platanistidae, the anterior teeth are no longer tusk-like, but they did remain larger in comparison with the posterior teeth.

Family Allodelphinidae

Species in the extinct platanistoid family Allodelphinidae are known only from marine sedimentary deposits around the margin of the North Pacific Ocean. Their fossils have been found in California, Oregon, Washington, and Japan. The presently known geochronologic range of the Allodelphinidae, currently represented by two named species and several undescribed species, is from the earliest Miocene, approximately 23 million years ago, to the Late Miocene, approximately 10 to 12 million years ago.

Their fossils indicate that in life allodelphinids attained adult body lengths ranging from slightly less than four meters to possibly as much as five meters, and had skull lengths of approximately one meter or larger, very long and slender rostra with the lower jaw reaching the same length as the rostrum, many small and single-rooted teeth, and unusually long cervical vertebrae. The more derived and later-occurring species developed wide braincases, cranial asymmetry, cranial tuberosities and crests (see Figure 2), and even longer and larger neck vertebrae. This latter evolutionary trend is the opposite of virtually all other cetacean groups, in which the cervical vertebrae became shorter, smaller, and in many cases fused together. The brain cases of these animals are in several characters more primitive than squalodontids (anteriorly positioned nasal openings, anteroposteriorly elongate nasal bones), and their primitively constructed zygomatic arches distinguish them from some of the similarly long-snouted species in the family Platanistidae.

The family Allodelphinidae was named by Barnes (2006), having as its type and only included genus *Allodelphis* Wilson, 1935. The type and only included species of *Allodelphis* is *Allodelphis pratti* Wilson, 1935 (Figure 1), which is known only from earliest Miocene sediments in the San Joaquin Valley of California, U.S.A. After it was named in 1935, this species was rarely cited in the scientific literature, although it is a valid taxon, and as its name implies, it is a very strange dolphin. Wilson (1935) originally classified it in the modern oceanic dolphin family Delphinidae, but it was reassigned by Barnes (1977) to the Platanistidae.

Deering et al. (2003) and Barnes et al. (2003) reported the discovery of an *Allodelphis*-like platanistoid from the Early Miocene age Vaqueros Formation in Orange County, southern California, U. S. A. Barnes & Reynolds (2007) reported another species of early allodelphinid from elsewhere in the same formation in southern California. Kimura and Ozawa (2001) reported a partial skeleton of a long-snouted platanistoid of late Early Miocene age from Japan, and we now identify this animal as an unnamed taxon of the family Allodelphinidae.

The Middle Miocene age *Squalodon errabundus* Kellogg, 1931, from the Sharktooth Hill Local Fauna in central California, was originally named on the basis of isolated petrosals, and was assigned by Barnes (1977) to the family Platanistidae. This species is now known by a complete cranium (Figures 2 and 3) from the same deposit, associated with a mandible, petrosal, and tympanic bulla of the same individual, which demonstrate that this species is a member of the family Allodelphinidae. Because it is not a species of *Squalodon,* in the following text we propose a new genus name for it. Barnes (1977:320) reported an isolated petrosal of a closely related species from the Late Miocene Santa Margarita Formation near Santa Cruz in coastal central California, and this is the geochronologically youngest known occurrence of any allodelphinid.

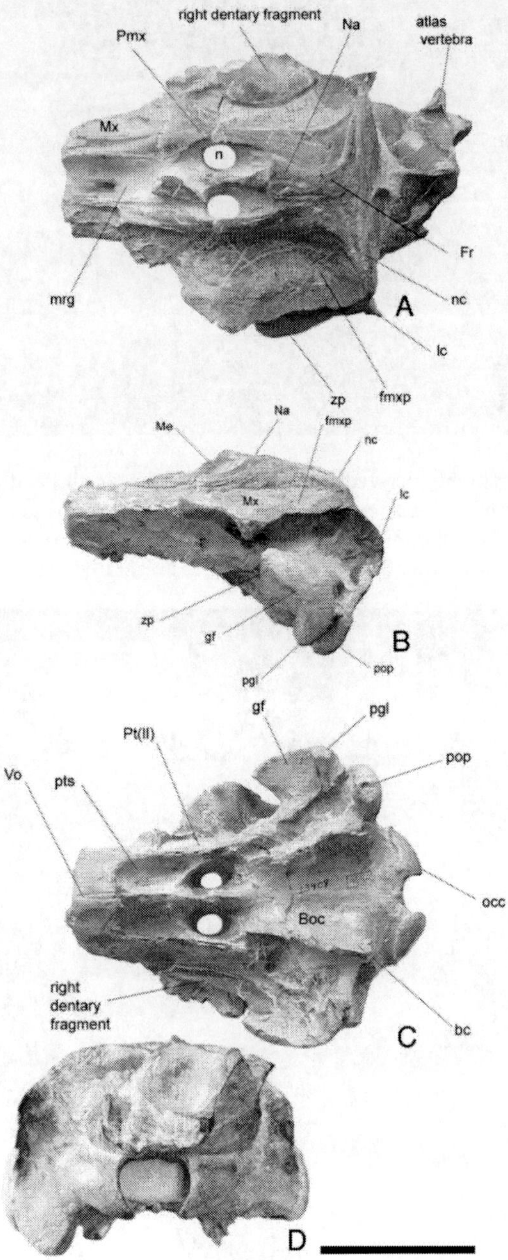

Figure 1. Cranium of the holotype of the first recognized species of the North Pacific platanistoid family Allodelphinidae, *Allodelphis pratti* Wilson, 1935, of earliest Miocene age, central California, U.S.A.; A, dorsal view; B, left lateral view; C, ventral view; D, posterior view. *Allodelphis pratti* is the type species of the genus *Allodelphis* Wilson, 1935, which is the type genus of the family Allodelphinidae; scale bar equals 10 cm; anatomical abbreviations are: bc-basioccipital crest, Boc-basioccipital bone, fmxp-posterior maxillary foramen, Fr-frontal bone, gf-glenoid fossa, lc-lambdoidal crest, Me-mesethmoid bone, mrg-mesorostral groove, Mx-maxillary bone, Na-nasal bone, n-narial passage (or naris), nc-nuchal crest, occ-occipital condyle, pgl-postglenoid process, Pmx-premaxillary bone, pop-paroccipital process, Pt(ll)-lateral lamina of the pterygoid bone, pts-fossa for pterygoid sinus, Vo-vomer bone, zp-zygomatic process of the squamosal bone.

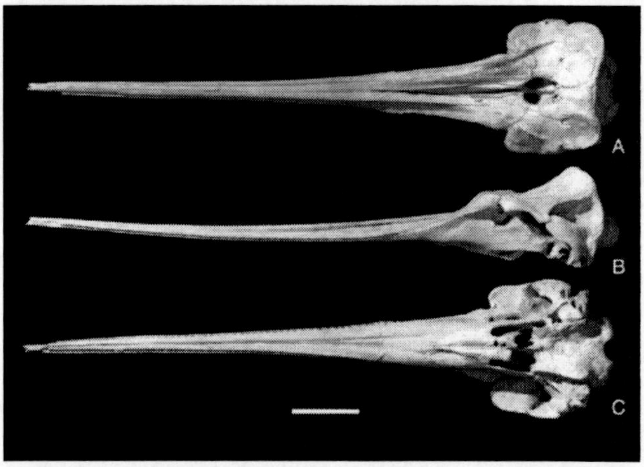

Figure 2. Skull of the most highly derived named species of the North Pacific long-snouted platanistoid family Allodelphinidae, *Zarhinocetus errabundus* (Kellogg, 1931), new generic allocation, referred cranium, LACM 149588, from the middle Middle Miocene age Sharktooth Hill Bonebed, Kern County, California, U. S. A.; A, dorsal view; B, left lateral view; C, ventral view; scale bar equals 10 cm.

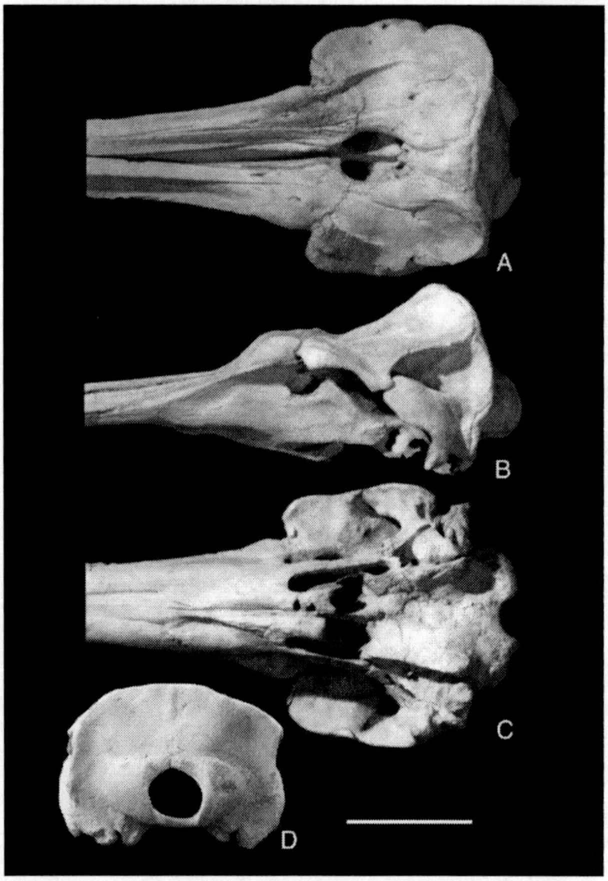

Figure 3. Enlarged views of cranium of the North Pacific long-snouted platanistoid allodelphinid, *Zarhinocetus errabundus* (Kellogg, 1931), new generic allocation, referred specimen, LACM 149588, from the middle Middle Miocene age Sharktooth Hill Bonebed, Kern County, California, U.S.A.; A, dorsal view; B, left lateral view; C, ventral view; D, posterior view; scale bar equals 10 cm.

Family Squalodontidae

The extinct family Squalodontidae includes species of medium size to large size archaic platanistoids that have among their primitive characters a symmetrical cranial vertex, zygomatic process of squamosal not covered in dorsal view by the margin of the facial surface, large cranial crests, stout rostra, unfused mandibular symphyses, heterodont dentitions, large teeth, elongated incisors and canines, and multiple-rooted cheek teeth with accessory cusps on their crowns. Their derived characters include posteriorly-positioned nares and concomitantly anteroposteriorly short and blocky nasal bones, wide posterior ends of the premaxilla, and a very deeply keeled posterior part of the rostrum (Dooley, 2005). Our classification includes those genera and species that are the most universally accepted as being squalodontids.

Squalodontids are known from all major ocean basins and their chronostratigraphic range is from the later part of the Oligocene to the middle Miocene. Fordyce (2006:766) pointed out that most reputed squalodontids from Australia and New Zealand do not belong to this family, and this is reflected in our classification. Rothausen (1968) and Dubrovo and Sanders (2000) regarded the Patriocetinae as a primitive subfamily of the Squalodontidae, with which we concur. This group in the past has been interpreted as having other affinities, and as being a separate family of Odontoceti.

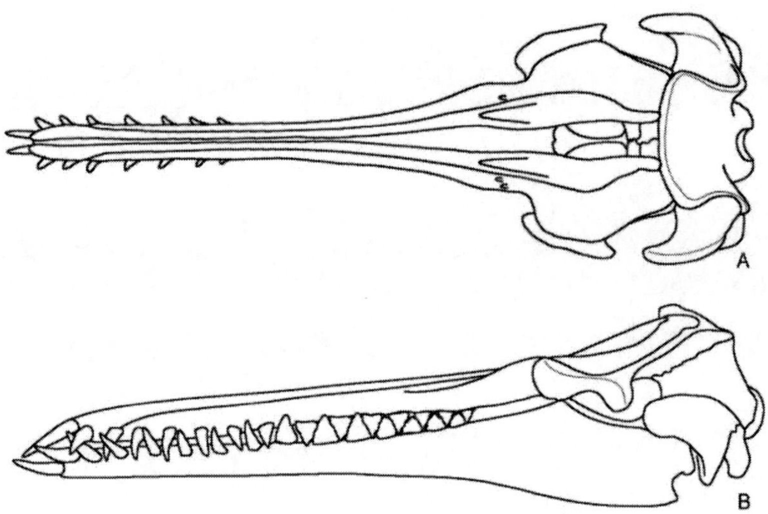

Figure 4. Skulls of typical North Atlantic Platanistoidea of the family Squalodontidae, which are sometimes called "shark-toothed whales"; A, *Squalodon calvertensis* Kellogg, 1923, Middle Miocene age, Maryland, U.S.A., dorsal view; B, *Squalodon bariensis* (Jourdan, 1861), Middle Miocene age, France, left lateral view. *Squalodon* is the type genus of the family Squalodontidae; images modified from Kellogg (1928: Figures 5 & 6); images are not to the same scale.

Among the more highly derived squalodontids, those members of the subfamily Squalodontinae, are *Squalodon calvertensis* Kellogg, 1923 (Figure 4), of Middle Miocene age from the Calvert Formation in Maryland, U. S. A. (and see Dooley, 2005). Squalodontids have been reported from the South Atlantic (Cabrera, 1926) and from the North Pacific

(Mitchell & Tedford, 1973). Some members of the subfamily Squalodontinae have very large and procumbent incisors and canines, and in *Squalodon whitmorei* Dooley, 2005, the anterior end of the rostrum is prominently expanded laterally (Dooley, 2005). The relationships of southern hemisphere species of *Prosqualodon* are uncertain (Fordyce, 2006), but here we included them in the family Squalodontidae until work has been done to resolve their precise relationships (fide Cozzuol, 1996).

Family Waipatiidae

The extinct family Waipatiidae was originally based on *Waipatia maerewhenua* Fordyce, 1994. The holotype and only known specimen of this species is of Late Oligocene age from New Zealand, and its characters were illustrated and described by Fordyce (1994, 2006). Its braincase is telescoped to the same extent as in members of the Squalodontidae, resulting in the nares being at the level of the posterior parts of the orbits, and the nasal bones being short, wide, and blocky. The rostrum is not elongated as it is in members of the Allodelphinidae, and the mandible is relatively stout. The dentition retains traces of heterodonty, but its teeth are smaller than are those of the squalodontids. The anterior teeth have long and tapered crowns, and the two-rooted cheek teeth have crowns that are expanded anteroposteriorly and have accessory cusps.

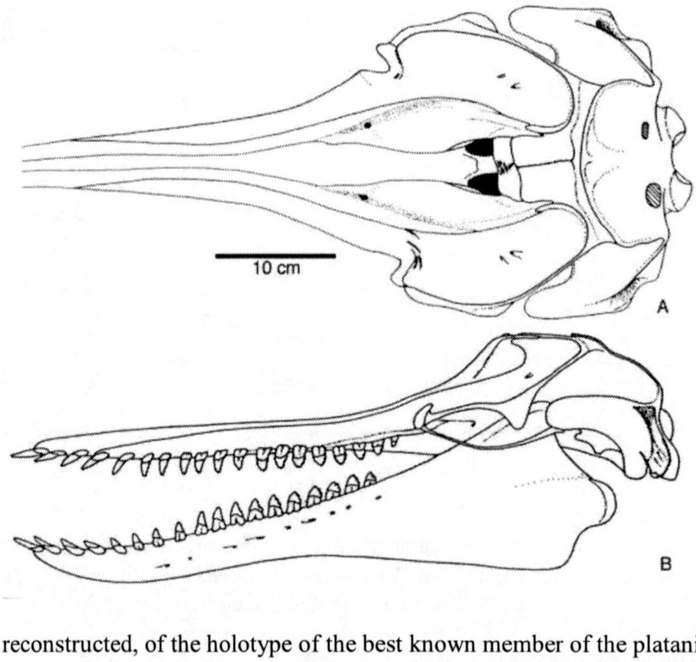

Figure 5. Skull, reconstructed, of the holotype of the best known member of the platanistoid family Waipatiidae, the South Pacific *Waipatia maerewhenua* Fordyce, 1994, Late Oligocene age, New Zealand; A, dorsal view; B, left lateral view. *Waipatia maerewhenua* is the type species of the genus *Waipatia,* which is the type genus of the family Waipatiidae; modified from Fordyce (1994:figs. 5A, 6A); scale bar equals 10 cm.

Fordyce (1994) noted similarities between *Waipatia maerewhenua* and the relatively poorly understood Northern Hemisphere fossil odontocetes *Sulakocetus dagestanicus*

Mchedlidze, 1976, of Late Oligocene age from Caucasus, and *Sachalinocetus cholmicus* Dubrovo, 1971, of Miocene age from Sakhalin Island, Russia, and he suggested that these cetaceans might also be waipatiids. We have classified them therefore in the family Waipatiidae (Figure 5).

Family Squalodelphinidae

The type genus of Squalodelphinidae is *Squalodelphis* Dal Piaz, 1916, whose type and only included species is *Squalodelphis fabianii* Dal Piaz, 1916, from Early Miocene deposits in Italy. Members of the Squalodelphinidae have relatively stout rostra that are not very elongated, anterior teeth that are reduced in size so that they are not so procumbent as they are in species of Squalodontidae or Waipatiidae, and cheek teeth with single roots, their primitively double roots having become coalesced into one root. Squalodelphinids have dorsoventrally thickened surpaorbital processes, but unlike the species of Platanistidae, they do not have supraorbital eminences. They do share with species of Platanistidae very asymmetrical cranial vertices, on which the posterior ends of the premaxillae and the midline sutures between the nasal and frontal bones are skewed to the anatomical left side of the skull. They also share with the Platanistidae, to the exclusion of all other platanistoids, transverse compression of the zygomatic process of the squamosal, so that its medial side is very concave.

Lydekker (1894) placed the Early Miocene *Notocetus vanbenedeni* Moreno, 1892, from Patagonia, in the family Platanistidae, and various authors have placed it in other families. Following Muizon (1987) and Fordyce (2006), we classify it in the family Squalodelphinidae, which in our phylogenetic analysis is indicated as being the sister family of the Platanistidae. *Notocetus vanbenedeni* is known by at least two well preserved and relatively complete specimens, the holotype in the La Plata Museum, Argentina (see Lydekker, 1894), and a referred specimen, described by True (1910; and see Muizon, 1987; and Fordyce, 1994) (Figure 6).

Fordyce (2006) indicated that Late Oligocene members of this family from New Zealand include *Notocetus marplesi* (Dickson, 1964), which is in neither the family nor the genus to which it was originally assigned, and "*Microcetus" hectori* (Benham, 1935), which is in need of a new generic assignment.

An apparent western North Atlantic member of this family is *Phocageneus venustus* Leidy, 1869, a Middle Miocene odontocete known from the Chesapeake Group in Virginia and Maryland, which was originally named on the basis of a tooth. It is represented by additional material from the Middle Miocene Pungo River Formation in North Carolina (Whitmore & Kaltenbach, 2007), and by unpublished material in the USNM from the Middle Miocene Calvert Formation, that suggest that it is referable to the Squalodelphinidae.

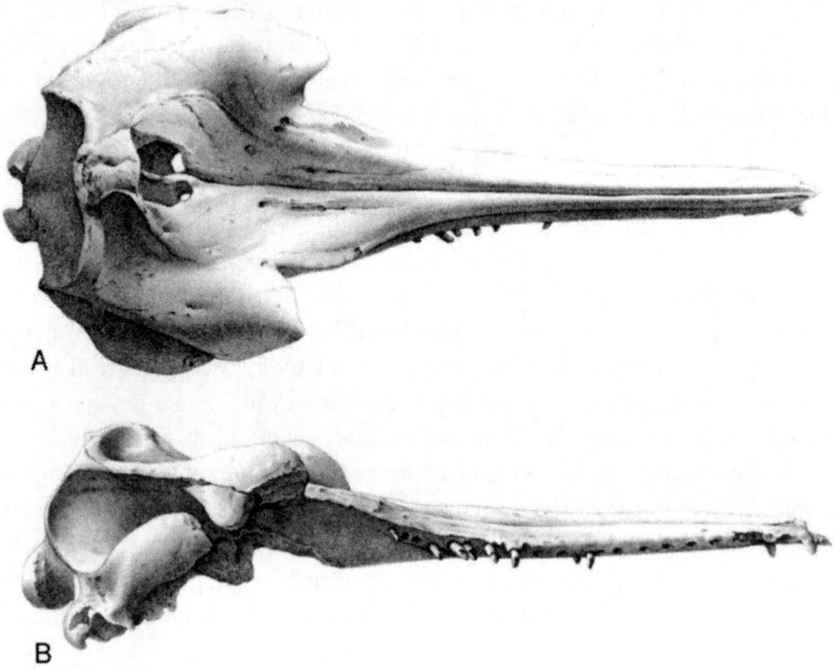

Figure 6. Cranium of a member of the platanistoid family Squalodelphinidae, *Notocetus vanbenedeni* (Moreno, 1892), of earliest Miocene age from the southwestern Atlantic coast of South America; **A,** dorsal view; B**,** right lateral view; modified from Lydekker (1894:pl. 6, figs. 1, 1b).

Platanistidae

The monophyly of the platanistoid crown group, the family Platanistidae, is supported by the phylogenetic analysis herein. The family is represented by fossils from Miocene marine, brackish, and fresh water deposits of the Northern Hemisphere, but it is today represented only by the living dolphins called the susus, members of the genus *Platanista* Wagler, 1830, which live in fresh water environments of south Asia.

Members of the Platanistidae have a cranium with an asymmetrical cranial vertex that is skewed to left side, posterior ends of premaxillae much expanded posterolaterally, frontals exposed between the mesethmoid and the nasal bones on the posterior wall of dorsal naris, a sheath of bone present in the medial part of the orbit formed by combination of the lateral lamina of the pterygoid (formed as an outer lamina or bony plate of the pterygoid within the orbit that extends posteriorly from the palate and contacts with the styliform process of the squamosal) and a posterior extension of the palatine bone. They also have a highly modified zygomatic process of the squamosal that is compressed transversely (and that has a concave medial surface and a broad connection to the postorbital process of the frontal), and an elongated and narrow rostrum and symphyseal portion of the mandible, especially elongated in some of the geochronologically earlier Miocene species. The anterior teeth, when preserved in the fossil species, are larger than the posterior teeth, as is the case also with the living species, *Platanista gangetica.* In most odontocetes, including members of all of the other families of Platanistoidea, the posterior maxillary foramen is located posterolateral to the dorsal narial opening, and it is separated by a considerable distance from the premaxillary bone. This is the primitive condition for the Odontoceti. Members of the family Platanistidae, in contrast, have a derived character state, in which the posterior maxillary foramen is located very close to the posterolateral corner of the posterior end of the premaxilla, and in some individuals and in some species, this foramen is under the edge of the premaxilla.

The most primitive member of the Platanistidae is *Araeodelphis natator* Kellogg, 1957, which occurs stratigraphically low within the Miocene Calvert Formation in Maryland (Godfrey et al., 2006; Bohaska et al., 2007), and also in the Middle Miocene age Pungo River Formation in North Carolina (Whitmore & Kaltenbach, 2007). The exact relationships of this species have yet to be determined, so it cannot now be assigned to either of the following two platanistid subfamilies. When Kellogg (1957) described *Araeodelphis natator,* he did not recognize the platanistid affinities of the holotype skull because it included only the rostrum and part of the mandible. A more complete cranium (USNM 526604) is now known from the lower part of the Calvert Formation (Early Miocene in age, approximately 17 MA), and it has a rostrum that is similar in morphology but slightly smaller than that of the holotype of *Araeodelphis natator*. This skull, USNM 526604 (Figure 7), we now identify as *Araeodelphis,* cf. *A. natator*, and it allows *Araeodelphis* to be identified as a member of the family Platanistidae. As in all Platanistidae, the posterior maxillary foramen is located at the posterior-most end of the premaxilla, behind the nasals, but unlike other Platanistidae, the bone over the large, dorsally convex supraorbital processes is not particularly thickened (Figure 7B), and there is no development of pneumaticity above these processes. The large size of its orbits suggests that sight was still an important sense in this fully marine platanistid, in contrast to the living *Platanista gangetica.*

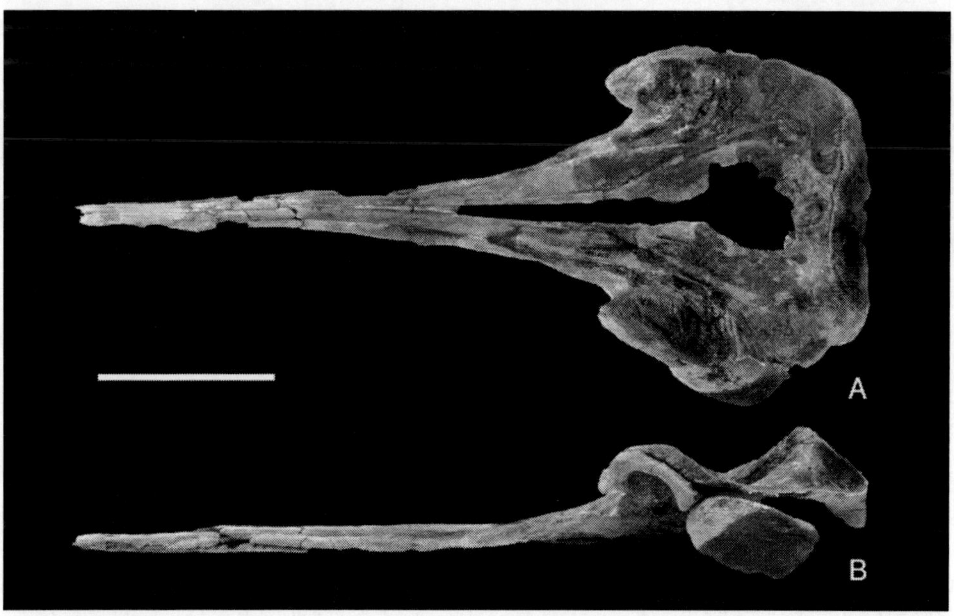

Figure 7. Cranium of a very primitive member of the platanistoid family Platanistidae, *Araeodelphis,* cf. *A. natator* Kellogg, 1957, USNM 526604, early Miocene in age, approximately 17 Ma, from the Calvert Formation, western North Atlantic Ocean; A, dorsal view; B, left lateral view; scale bar equals 10 cm.

Two subfamilies have been recognized (Barnes, 2002b) within the family Platanistidae; the Pomatodelphininae and the Platanistinae. *Araeodelphis,* discussed above, cannot be assigned to either of these. The subfamily Pomatodelphininae, named by Barnes (2002b), includes extinct species of the North Atlantic realm belonging to the genera *Prepomatodelphis* Barnes, 2002b, *Zarhachis* Cope, 1868, and *Pomatodelphis* Allen, 1921. Members of the subfamily Pomatodelphininae have dorsoventrally flattened rostra and symphyseal portions of their mandibles (Figure 8).

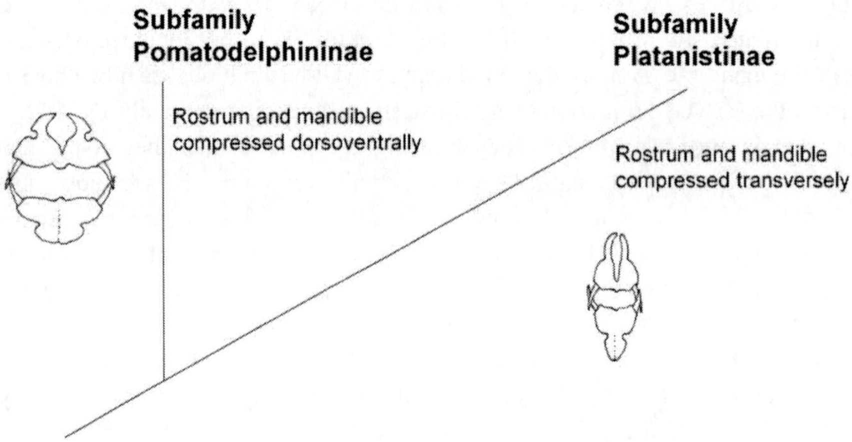

Figure 8. Schematic cross sections of the rostra and mandibles, at approximately mid-length, of representatives of the two different subfamilies of the family Platanistidae. The example for the subfamily Pomatodelphininae is based upon USNM 10485, a specimen that was referred by Kellogg (1924) to Middle Miocene *Zarhachis flagellator* Cope, 1868, from the Calvert Formation in Maryland. The example for the subfamily Platanistinae is based upon CAS 16340, an adult female individual of Recent *Platanista gangetica* (Roxburgh, 1801) from Pakistan; after Barnes (2006: Figure 8). Images are diagrammatic and are not to scale.

The geochronologically earliest named pomatodelphinine is the late Early Miocene *Prepomatodelphis korneuburgensis* Barnes, 2002b, from the Karpatian age (late Burdigalian correlative) Korneuburg Formation in the Korneuburg Basin, Austria, and is between approximately 16.5 Ma and 16.7 Ma in age. The Korneuburg Basin was a small, sometimes brackish-water estuary that was connected in Miocene time to the larger Vienna Basin. P*repomatodelphis korneuburgensis*, and *Araeodelphis,* demonstrate that a supraorbital maxillary crest is not a diagnostic character of the family Platanistidae. A maxillary crest does occur in some of the more derived taxa, and in some of the taxa that have the crest, it is fenestrated by extensions of the supraorbital lobe of the pterygoid sinus (Fraser and Purves, 1960; Bohaska et al., 2007).

Zarhachis flagellator Cope, 1868, was originally named on the basis of a vertebra. Kellogg (1924, 1926) referred more complete specimens to this species (see Figure 9) from the Calvert Formation in Maryland, U. S. A., and these are of middle Middle Miocene age (Gottfried et al., 1994:233), approximately 13 to 15 Ma. *Zarhachis flagellator* differs from *P.*

korneuburgensis by being larger, having a thickened supraorbital process, an enlarged supraorbital maxillary tuberosity, and a more enlarged and more elevated nuchal crest.

In species of the genus *Pomatodelphis* the lateral margin of the maxilla immediately anterior to the antorbital notch is thickened and laterally expanded, and the supraorbital maxillary crest is greatly enlarged and knob-like. *Pomatodelphis* is represented in the eastern North Atlantic by the poorly known species, *Pomatodelphis stenorhynchus* (Holl, 1829), which was based on a fragment of a rostrum from France. The holotype of this species, No. 2228, Laboratoire de Paléontologie, Muséum National d'Histoire Naturelle in Paris, is of Middle Miocene age from Maine-et-Loire, Department de l'Orne, France, (see Kellogg 1959:19-20), and no other specimens can now be confidently referred to this species.

The type species of *Pomatodelphis* is *Pomatodelphis inaequalis* Allen, 1921, which is known from the late Middle Miocene Agricola Fauna from central Florida, and which is derived from near shore marine deposits of the Lower Bone Valley Formation (Morgan, 1994), and is approximately 10.5 Ma to 11.5 Ma in age (Morgan, 1994:251-252). Hulbert & Whitmore (2006) reported this species from a Middle Miocene age river-laid deposit in Alabama, and this is the geochronologically earliest indication of a fresh water occurrence of a platanistoid. *Schizodelphis depressus* Allen, 1921, also named on the basis of fossils from the marine Bone Valley Formation in Florida, has been synonymized with *P. inaequalis* (Morgan, 1994).

Another Florida pomatodelphinine species is *Pomatodelphis bobengi* (Case, 1934), which was originally described as a species of *Schizodelphis* (see Morgan, 1994). It is also from the marine Lower Bone Valley Formation, of late Middle Miocene age, and approximately 10.5 Ma to 11.5 Ma in age (Morgan, 1994:251-252; and see also Kellogg, 1944). Whitmore & Kaltenbach (2007) have reported *Pomatodelphis* sp. from the marine, Middle Miocene age, Pungo River Formation in North Carolina.

The prominent eminences over the orbits on skulls of derived species of pomatodelphinine platanistids are formed by the frontal bones. These eminences are not solid bony structures. A CT scan (Figure 9C) through an adult cranium (USNM 10911) that was referred to *Zarhachis flagellator* by Kellogg (1926) shows that these structures are excavated in their medial parts and have within their lateral parts layers of bone. We conclude that the layers within these eminences formed during ontogeny, and that newer layers were added in the lateral parts of the enlarging frontal bone as the skull increased in size. Much of the medial part of each supraorbital eminence is seen to be secondarily excavated, and these excavations cut across the different layers of interior bone. We suggest that this excavation of the medial sides of these eminences is the result of invasion by the supraorbital lobe of the pterygoid air sinus. The supraorbital lobe of the pterygoid air sinus, as it does in Recent *Platanista gangetica,* (see Fraser & Purves, 1960), appears to have extended dorsally from the orbital region, via the anterior maxillary foramina, and invaded the medial sides of these tuberosities. The invasion by the sinus tissue is interpreted to have resorbed the medial side of the eminence, as evidenced by the way that the vacuities cut across the early ontogenetic layers of bone. It is significant that the enlarged supraorbital eminence of *Zarhachis flagellator* is formed by the frontal bone, and the maxillary bone is not exceptionally thickened (Figure 9C). In contrast, in Recent *Platanista gangetica,* the frontal bone is not exceptionally thickened, but the maxillary bone is greatly enlarged (Figure 11B), and it is the maxilla that is invaded by the pterygoid sinus (Figure 9C). These two different manifestations of bone invasion by the dorsal expansion of the pterygoid sinus may be diagnostic differences between the two named subfamilies of Platanistidae: invasion of the enlarged supraorbital process of the frontal bone in the subfamily Pomatodelphininae, and invasion of the enlarged supraorbital process of the maxillary bone in the subfamily Platanistinae. If this is true, then *Araeodelphis natator* is indeed a basal platanistid, in which neither the maxilla nor the frontal are extremely thickened dorsal to the orbit, and neither bone is invaded by the pterygoid sinus dorsally.

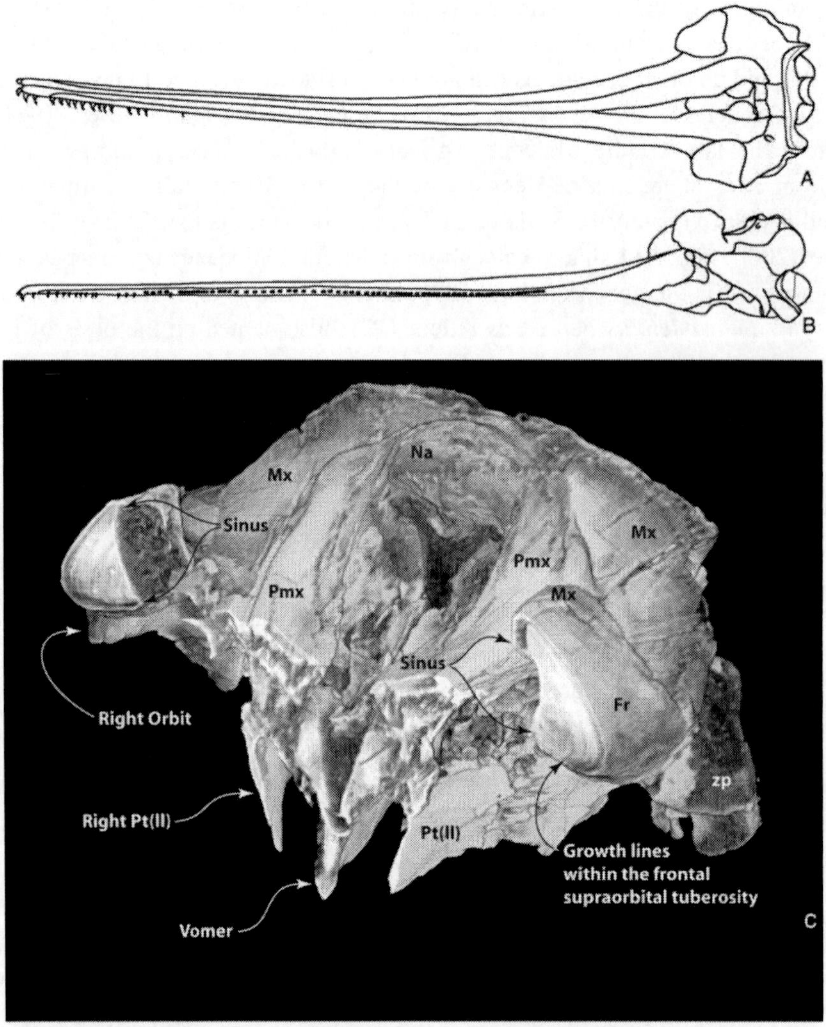

Figure 9. Cranium of a fossil western North Atlantic marine-adapted member of the platanistoid family Platanistidae, subfamily Pomatodelphininae, *Zarhachis flagellator* Cope, 1868, of Middle Miocene age from the Calvert Formation in Maryland, U.S.A.; A, dorsal view; and B, left lateral view; C, high resolution CT scan made transversely through a cranium, USNM 10911, referred to *Zarhachis flagellator* by Kellogg (1926) transecting the orbital area, showing vacuities within the supraorbital tuberosities created as the supraorbital lobe of the pterygoid sinuses invaded them; figures not to scale; the images in A and B are composites of two referred specimens, USNM 10485 and USNM 10911, and are modified from Kellogg (1924:pls. 1, 2; 1926:pl. 1, 4); anatomical abbreviations in C are: Fr-frontal bone, Mx-maxillary bone, Na-nasal bone, Pmx-premaxillary bone, Pt(ll)-lateral lamina of the pterygoid bone, zp-zygomatic process of the squamosal bone.

The platanistid subfamily Platanistinae, in the context used by Barnes (2002b, 2006), includes Platanistidae that have an anteroposteriorly lengthened zygomatic process of the squamosal, transversely flattened rostrum and symphyseal portion of the mandible (Figure 8), and a large pneumaticized supraorbital crest (Figure 11).

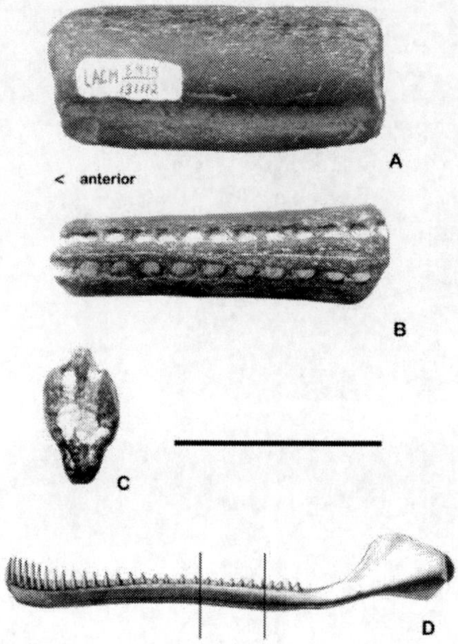

Figure 10. Part of the symphyseal region of a mandible, LACM 131112, of an unidentified genus and species, an Early Miocene marine member of the subfamily Platanistinae of the family Platanistidae, from the Nye Formation in Lincoln County, coastal Oregon, U. S. A; A, left lateral view; B, occlusal view; C; anterior view of broken cross section; D, lateral view of a mandible of Recent *Platanista gangetica* (Roxburgh, 1801), with lines bracketing the part of the specimen that is represented by the fossil; A-C modified from Barnes (2006: Figure 7); D modified and reversed from Van Beneden and Gervais (1868-1880: pl. XXXI, Figure 2); arrow indicates the anterior direction of the specimen as shown in A and B; scale bar for A-C equals 5 cm; D is not to scale.

Around the eastern margin of the North Pacific, fossils have been found that appear to belong to the subfamily Platanistinae (Crowley et al., 1999), but none of these have been named. One of these apparent platanistine fossils, from the marine, Early Miocene age, Nye Formation on the coast of Oregon, U. S. A., is part of the symphysis of a mandible (Figure 10, and see Barnes, 2006: Figure 7) that is approximately three times the size of the corresponding part of the mandible of *Platanista gangetica.* It includes the posterior parts of both fused dentaries, extending posteriorly as far as the anterior-most part of the divergence of the two horizontal rami. It shares with *P. gangetica* transverse compression of the symphysis of the mandible, close approximation of the right and left tooth rows, a longitudinal groove at approximately one-third of the height on the lateral side of the dentary, nutrient foramina positioned dorsal to this groove, and a very slight posterior divergence of the horizontal rami. It differs from *P. gangetica* by being larger, by having a greater separation between the alveolar rows, and by having dental alveoli which at their alveolar rims are round rather than being transversely flattened. Its presence in a marine deposit indicates that Platanistinae were probably previously more widely distributed in oceans.

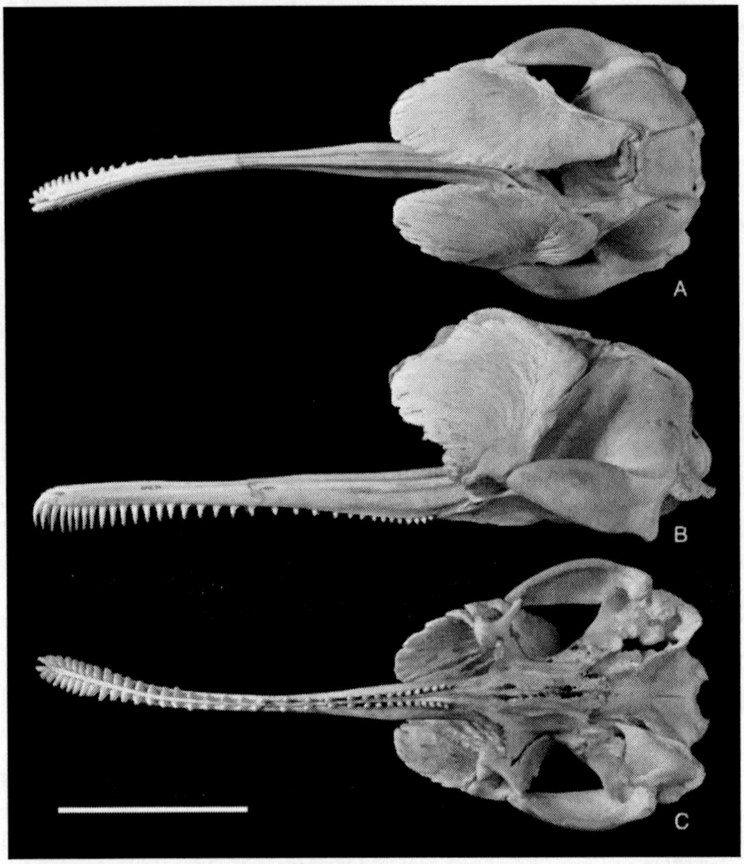

Figure 11. Cranium of the only living member of the superfamily Platanistoidea, a highly derived member of the family Platanistidae, subfamily Platanistinae, *Platanista gangetica* (Roxburgh, 1801), an adult female individual, CAS 16340, from Sind, Pakistan; A, dorsal view; B, left lateral view; C, ventral view; scale bar equals 10 cm. The rostral curvature is not natural, and is the result of unequal drying of the bone. *Platanista* is the type genus of the subfamily Platanistinae, the family Platanistidae, and the superfamily Platanistoidea.

Platanista Wagler, 1830, is the only named genus that is now included in the subfamily Platanistinae. Its recent species include the freshwater susus of south Asia. Usually one species is recognized, *Platanista gangetica* (Roxburgh, 1801), and the subspecies name, *Platanista gangetica gangetica* (Roxburgh, 1801), is applied to the Ganges River Dolphin of the rivers of India, Bangladesh, Nepal, and the name *Platanista gangetica minor* Owen, 1853, is given to the Indus Dolphin of Pakistan. These are among the most highly derived odontocetes ever to have lived. Among their derived characters are: small size, enlarged and anteriorly extended zygomatic process of the squamosal, atrophied eye, extreme left-skew asymmetry of the cranial vertex, reduced nasal bones, greatly enlarged supraorbital crests (formed by maxillary bones and pneumaticized by extensions from the middle ear air sinus system), extremely narrow supraoccipital-nuchal crest region, reduced lambdoidal crests, secondarily inflated zygomatic process of the jugal, transversely flattened rostrum and symphyseal part of the mandible, secondary heterodonty (crowns of anterior teeth greatly elongated, crowns of posterior teeth widened), and what we suggest is paedomorphism (which may account for some of the derived characters of this dolphin).

A New Genus of Fossil Platanistoid Dolphin

Zarhinocetus, New Genus

Synonymy

Squalodon (part). Kellogg, 1931:373, for *Squalodon errabundus* Kellogg, 1931.

"Squalodon". Barnes, 1977: 327, Table 3, for *Squalodon errabundus* Kellogg, 1931, as a taxon of the family Platanistidae; Barnes, 2006:31-34, [part], for S*qualodon errabundus* Kellogg, 1931, as a taxon of the platanistoid family Allodelphinidae.

Diagnosis of genus: A genus of the platanistoid family Allodelphinidae, differing from *Allodelphis* Wilson, 1935, by having cranium with relatively wider facial region, dorsoventrally higher occipital shield, transversely narrower mesorostral gutter in posterior part of rostrum anterior to nares, ventrally depressed medial part of dorsal surface of proximal part of rostrum, anteroposteriorly-oriented crest of maxilla on dorsal surface of supraorbital process of maxilla, nasal bones narrower anteriorly and wider posteriorly, posterior ends of premaxillae atrophied and retracted anteriorly, and not extending posteriorly as far as nares, area where posterior end of premaxilla formerly existed (forming premaxillary sac fossa (or spiracular plate)) formed of smooth, convex, and dense maxillary bone, asymmetrical cranial vertex with mid-line between nasal and frontal bones located to left of sagittal plane of cranium, and swollen tubercle of bone on lateral part of rostral maxilla anterior to antorbital notch.

Type species: Squalodon errabundus Kellogg, 1931.

*Included species: Zarhinocetus errabund*us (Kellogg, 1931).

Etymology: The genus name is derived from *za-*, Greek, an intensive particle, meaning very or exceedingly; combined with *rhinos,* Greek for nose, snout, beak, or bill; plus *ketos,* Greek for a large sea animal or whale, here using the Latin form, *cetus;* and alludes to the extremely long rostrum of this type of platanistoid cetacean. We use the suffix *cetus,* rather than *delphis,* to intentionally emphasize that platanistoids are not in the strict sense dolphins, as so many of the living members of the Delphinoidea are commonly called, but are in the more broad sense toothed whales.

Geographic and geochronologic ranges: Middle and Late Miocene, eastern North Pacific coast in the area of central and southern California, U.S.A.

*Zarhinocetus Errabund*us (Kellogg, 1931), New Combination

Synonymy

Squalodon errabundus Kellogg, 1931:373; as a species of Squalodontidae.

"Squalodon" errabundus Kellogg, 1931. Barnes, 1977:327, Table 3; as a species of Platanistidae.

"Squalodon" errabundus Kellogg, 1931. Barnes, 2006:31-34; as a species of Allodelphinidae.

Emended diagnosis of species: Until further species are described in the genus, the diagnosis of the species shall remain identical to that of the genus.

Holotype: USNM 11573, left petrosal, collected by Charles Morrice in 1924.

Type locality: CAS locality 905, in the Sharktooth Hill Bonebed "on a hill west of Round Mountain, locally known as one of the Shark Tooth Hills. Near latitude 35° 28' North, and longitude 119° 27' West, about 6.5 miles northeast of Bakersfield, 4 miles east of the Kern River Oil Field and 0.5 miles north of Kern River, Kern County, California. Section 25, Township 28 South, Range 28 East, Caliente Quadrangle, United States Geological Survey." From CAS locality catalog.

Referred specimens: USNM 11574, left petrosal from CAS locality 905 collected by Charles Morrice in 1924; LACM 149588, cranium and mandible with associated right petrosal and left tympanic bulla from LACM locality 4314 (field number L.G. Barnes 1818), collected by Gregory Art in 1976.

Formation and age: All known specimens of *Zarhinocetus errabundus* are from the middle Middle Miocene age Sharktooth Hill Bonebed, which is in the upper part of the marine Round Mountain Silt. The species is part of the Sharktooth Hill Local Fauna (Mitchell, 1965:iii; and see Barnes, 1977:326-327; 2006), which is derived from the Sharktooth Hill Bonebed, a horizon that bears densely packed fossils, and is widely exposed in outcrops of the upper part of the Round Mountain Silt in the area northeast of Bakersfield in Kern County, California. The age and stratigraphic relationships of the upper part of the Round Mountain Silt, which includes the Sharktooth Hill Bonebed, are well known, and the horizon is correlated with the Temblor Provincial mega-invertebrate stage of Addicott (1972), the Relizian and/or Luisian foraminiferal stages, and the later part of the Barstovian North American Land Mammal Age, and is approximately 15.3 million years old (Barnes, 1977; Barnes and Mitchell, 1984; Tedford et al., 1987:156, 201, fig. 6.2 (chart in pocket); Tedford et al., 2004:172, Figure 6.2; Prothero et al., 2008).

Geographic and geochronologic ranges: Middle Middle Miocene, eastern North Pacific coast in the area of central California.

Comments: The odontocete, *Squalodon errabundus* Kellogg, 1931, was originally based by Kellogg (1931) on a holotypic isolated left petrosal, USNM 11573, and a referred left petrosal, USNM 11574, both of which were collected from the same locality in the Sharktooth Hill Bonebed. Kellogg (1931:373) explained his referral of this new species to the genus *Squalodon* by writing: "The two ear bones hereinafter described are referred to the genus *Squalodon,* for reasons which are almost indefinable, and yet all known squalodonts have similar peculiarly shaped periotics. The subtle characters that distinguish the periotics of squalodonts from those of other porpoises are apparent to anyone who has studied these bones, although it is difficult to point out any tangible feature which will invariably identify them."

The characters resembling *Squalodon* that Kellogg observed in the periotics that he called *Squalodon errabundus* are primitive characters that are shared by various primitive Platanistoidea, the superfamily that is now interpreted as including the family Squalodontidae. Kellogg (1931) stated that *"Squalodon" errabundus* was the first documented fossil occurrence of a squalodontid from the Pacific coast of North America. Unquestioned squalodontids are known from the North Pacific. Squalodontid teeth have been reported by Mitchell and Tedford (1973) from a lower stratigraphic unit in the area where *"Squalodon" errabundus* occurs, and they are from the earliest Miocene Pyramid Hill Sand in Kern County, California.

Subsequent to Kellogg's (1931) publication, a complete cranium and mandible (LACM 149588) were collected from the Sharktooth Hill Bonebed, and these are directly associated with it a petrosal and tympanic bulla. The petrosal of this referred specimen is identifiable as *Squalodon errabundus,* but the skull is unquestionably that of an allodelphinid platanistoid. Therefore, *Squalodon errabundus* Kellogg, 1931, clearly does not belong in the genus *Squalodon* (see specimens of Squalodontidae in Kellogg, 1923; Whitmore & Sanders, 1977; Dooley, 2005; and Figure 4 herein), and this is why the generic name *Squalodon* has been used in quotation marks when the species *Squalodon errabundus* has been referred to in some previous publications (e.g. Barnes, 1977, 2002b, 2006). No generic name has been proposed that is appropriate for this species, and for this reason the new genus *Zarhinocetus* is proposed for it here, yielding the new combination *Zarhinocetus errabundus* (Kellogg, 1931).

ANALYSIS OF PHYLOGENETIC RELATIONSHIPS

Our phylogenetic analysis presented here is nearly the same as was presented by Barnes (2006). The analysis includes 64 characters that were scored for nine taxa (Table 1). Fordyce (1994) prepared a character list and phylogenetic analysis for the superfamily Platanistoidea when he described the primitive platanistoid *Waipatia maerewhenua.* Barnes (2006) and we here have used many of the characters from that study, some of which are modified, and in the following list of characters, these are identified by Fordyce's (1994) page and character numbers. Geisler and Sanders (2003) presented a phylogenetic analysis of all Cetacea, and several of their characters are used here, also cited in the list of characters by their page and character numbers.

Only cranial and mandibular characters were used because of the lack of uniform information about postcranial bones for many of the taxa. Similarly, although characters of the petrosal and the tympanic bulla are pivotal in recognizing and defining the superfamily Platanistoidea *(sensu* Muizon, 1994; Fordyce, 1994) and the family Platanistidae *(sensu* Barnes, 2002a, 2002b, 2006), because these bones are not known for all species, their characters were with one exception (the anterior bullar spine) omitted from this analysis.

Definitions of morphologic characters used in the phylogenetic analysis: The following characters are those that we found to be useful among the Platanistoidea. The numbers of the characters are the same that appear in the matrix (Table 1). Character states are: [0] indicates a plesiomorphic character state, and [1] indicates an apomorphic character state.

1. **Mesorostral groove** absent [0]; or present [1]. The mesorostral groove is formed dorsal to the vomer bone, and is flanked on either side by the premaxillae (see Rommel, 1990: Figure 2). In life the mesorostral groove holds the anteroposteriorly elongated mesethmoid cartilage. In some derived species of Odontoceti, notably in some species of the family Ziphiidae, the groove is occupied by a mesorostral ossification. The mesorostral groove occurs in no species of the Archaeoceti, and this

structure is characteristic of the Mysticeti and Odontoceti (Barnes, 2006:38, character 1).

2. **Lacrimal foramen (or groove)** present [0]; or absent [1]. (See Geisler & Sanders, 2003:102, character 52; and Barnes, 2006:38, character 2.)
3. **Zygomatic portion of jugal bone** thick both dorsoventrally and transversely, as in terrestrial mammals and as in the Archaeoceti [0]; or narrower, thin, and rod-like [1]. (See Miller, 1923; Geisler and Sanders, 2003:102, character 56; and Barnes, 2006:38, character 3.) The zygomatic arch of the jugal may be secondarily modified, as explained for character 61, described below.
4. **Dorsal infraorbital foramen (anterior maxillary foramina for cranial nerve V_2)** having a single dorsal aperture [0]; or having multiple dorsal apertures [1]. (See Geisler and Sanders, 2003:103, character 64; illustrated by Rommel, 1990: Figure 2; Barnes, 2006:38, character 4.)
5. **Lacrimal and jugal bones** distinguishable as separate bones [0]; or fused [1]. (See Miller, 1923; Heyning, 1989; Barnes, 1990:21, node 10; 2006:38, character 5; and Geisler and Sanders, 2003:102, character 53.)
6. **Antorbital notch** absent [0]; or present [1]. The antorbital notch is formed between the base of the rostrum and the enlarged anterolateral corner of the supraorbital process, and the opening of this notch faces either anteriorly or anterolaterally. This is a character that is present in all Odontoceti (see Barnes, 1990:21, node 10; 2006:38, character 6; and Rommel, 1990: Figure 2; modified from Geisler and Sanders, 2003:102, character 49).
7. **Ascending (or frontal) process of the maxilla** abuts the anterior edge of the supraorbital process of the frontal [0]; or the ascending process of the maxilla to some degree covers the dorsal surface of the supraorbital process of the frontal [1]. The derived state is a character present in all Odontoceti, and any posterior extension of the maxilla over the dorsal surface of the frontal bone is considered to represent the derived character state. (See Miller, 1923; Barnes, 1990:21, node 10; 2006:38, character 7; modified from Fordyce, 1994:175, character 3; Rommel, 1990: Figures 1-2; and Geisler and Sanders, 2003:104, character 76).
8. **Premaxillary foramen** absent [0]; or present [1]. The derived state is a character of all Odontoceti, and there is usually a single foramen in each premaxilla (Barnes, 1990:21, node 10; 2006:38, character 8; modified from Geisler and Sanders, 2003:103, character 69). The foramen, however, may have more than one aperture, and in some members of the superfamily Physeteroidea (sperm whales and pygmy sperm whales) there may be two or three apertures in the right premaxilla, and none in the left premaxilla, the latter being the result of secondary loss.
9. **Posterior maxillary foramen** absent [0]; or present [1]. This is the foramen that carries the posterior branches of the internal maxillary artery and the maxillary division of the infraorbital nerve. This is a character of all Odontoceti (Barnes, 1990; 2006:38, character 9; modified from Geisler & Sanders, 2003:104, character 75).
10. **Premaxillary sac fossa** absent [0]; or present [1]. The premaxillary sac fossa (see Rommel, 1990: Figure 2) is formed on the dorsal surface of the posterior part of the premaxilla that is anterolateral to, or adjacent to, each dorsal narial opening, and posterior to the premaxillary foramen. The ventral wall of the premaxillary sac, which is a diverticulum of the narial passage, lies directly upon this smooth and

usually flat, and usually oval-shaped part of the premaxilla. This is a character of all Odontoceti (Barnes, 1990:21; 2006:38, character 10; and see Au, 2002: Figure 1).

11. **Posterolateral sulcus** absent [0]; or present [1]. This sulcus originates from the premaxillary foramen, traverses posterolaterally to the lateral side of the premaxilla, and usually demarcates the anterolateral and lateral sides of the premaxillary sac fossa (Barnes, 1978:13, 2006:38, character 11; modified from Muizon, 1988b; and Geisler & Sanders, 2003:104, character 72).
12. **Posteromedial sulcus** of premaxilla absent [0]; or present [1]. This sulcus originates from the premaxillary foramen, traverses posteromedially toward the medial side of the premaxilla, and usually demarcates the anteromedial edge of the premaxillary sac fossa (Barnes, 1978:13; 2006:38, character 12).
13. **Anteromedial sulcus** of premaxilla absent [0]; or present [1]. This sulcus originates from the premaxillary foramen and traverses anteromedially toward the medial side of the premaxilla (Barnes, 1978:13, 2006:38, character 13).
14. **Lateral surface of maxilla** dorsal to alveolar row and immediately anterior to antorbital notch thickened and expanded laterally to form a flange [0]; or lateral margin of maxilla immediately anterior to antorbital notch relatively thin and anteroposteriorly nearly straight [1]. This character is present in Odontoceti only. Most of the earliest-occurring Odontoceti (as exemplified by *Agorophius pygmaeus,* see Fordyce, 1981) have this lateral expansion of the lateral margin of the maxilla, and it is considered here to be a character that is shared among stem Odontoceti. The loss of this lateral flare of the edge of the maxilla among more derived clades of Odontoceti is a derived character state (Barnes, 2006:38, character 14).
15. **Maxilla** covers anterior part of supraorbital process of frontal and does not contact nuchal crest [0]; or extends posteriorly so far as to make contact with the nuchal crest [1]. For this analysis, we scored simply whether or not the maxilla reaches the nuchal crest (Barnes, 2006:38, character 15; modified from Geisler & Sanders, 2003:104, character 77).
16. **Mesorostral groove** open dorsally [0], or roofed over at least in part by the medial margins of the premaxillae [1]. The structure is labeled by Rommel (1990: Figure 2), and was explained by Barnes (2006:38, character 16). The plesiomorphic character state is present in *Agorophius pygmaeus* (see Fordyce, 1981) and in *Allodelphis pratti* (see Fig. 1A herein), and the derived character state is present in *Zarhachis flagellator* (see Kellogg, 1924: pl. 1; and see Figure 9A herein) and in *Platanista gangetica* (see Van Beneden & Gervais, 1868-1880:pl. XXXI, Figures 2a, 9a; and see Figure 11A herein).
17. **Right and left parietal bones** make contact along their medial margins on the dorsal surface of the brain case [0]; or right and left parietals are separated in dorsal exposure of the cranial surface by posterior extension of maxillae to contact the nuchal crest [1]; (see Barnes, 1990:21, node 17; 2006:38, character 17).
18. **Ascending process of each premaxilla** contacts only the lateral side of its respective nasal bone [0]; or ascending process of premaxilla extends posteriorly to a point that is posterior to the posterior end of its respective nasal bone [1]. The derived character state is shown by the right premaxilla of the specimens of *Zarhachis flagellator* (see Kellogg, 1926:pl. 2; and Figure 9A herein). (Character explained by Barnes, 2006:38-39, character 18.)

19. **Anterior ends of the nasal bones** extend anteriorly, to overhang at least to some extent, the posterior side of the dorsal narial openings [0]; or nasal bones retracted posteriorly so as not to hang over the posterior side of the dorsal narial openings [1]. In the derived character state (for example as exists in *Zarhachis flagellator,* see Kellogg, 1926:pl. 2; and see Figure 9A herein), the nasal bones do not prevent viewing the entire diameters of the dorsal narial passages in a standard dorsal view of the cranium. (Character explained by Barnes, 2006:39, character 19.)
20. **Nasal bones elongate,** having a greater anteroposterior dimension than a transverse dimension [0]; or nasal bones shortened and broadened, having lesser anteroposterior dimension than transverse dimension [1]. The derived character state exists in *Zarhachis flagellator* (example illustrated by Kellogg, 1926:pl. 2; and see Figure 9A herein). (Character explained by Barnes, 2006:39, character 20).
21. **Premaxillary sac fossae** relatively narrow, being approximately the same width as the more anterior parts of the premaxilla anterior to the narial region [0] (as in *Allodelphis pratti,* see Figure 1A herein); or premaxilla wider in the area of the premaxillary sac fossa on either side of the dorsal narial passages [1]. The derived character state exists in *Prepomatodelphis korneuburgensis* (see Barnes, 2006: Figure 4), in *Araeodelphis,* cf. *A. natator* (see Figure 7A herein), and in *Zarhachis flagellator* (see Figure 9A herein). (See Barnes, 2006:39, character 21; modified from Geisler & Sanders, 2003:98, character 8.)
22. **Rostrum** narrows in width anteriorly or anterior half of rostrum approximately the same width as the posterior half [0]; or anterior end of rostrum widened transversely [1], as in *Squalodon* (character explained by Muizon, 1991, 1994; Geisler & Sanders, 2003:98, character 2; and Barnes, 2006:39, character 22).
23. **Anterior end of zygomatic process of squamosal** not contacting the postorbital process of the frontal [0]; or anterodorsal part of zygomatic process of squamosal having a broad contact with ventral extremity of the postorbital process of frontal [1], (Character defined by Barnes, 2006:39, character 23). The derived character state is present in *Araeodelphis,* cf. *A. natator* (see Figure 7B herein), in *Zarhachis flagellator* (see Kellogg, 1926:pl. 4; and Figure 9B herein), and in *Platanista gangetica* (see Van Beneden & Gervais, 1868-1880:pl. XXXI, Figures. 2, 9; Fraser & Purves, 1960:pls. 17-18; and Figure 11 herein).
24. **Posterior ends of right and left maxillae** on posterior part of facial region bilaterally symmetrical [0]; or posterior end of right maxilla, compared to posterior end of left maxilla, curving farther medially toward the mid-line of the cranium [1]. The derived character state is present in *Notocetus vanbenedeni* (see Fig. 6A herein), in *Araeodelphis,* cf. *A. natator* (see Fig. 7A herein), in *Zarhachis flagellator* (see Kellogg, 1926; and Fig. 9A herein), and in *Platanista gangetica* (see Van Beneden & Gervais, 1868-1880:pl. XXXI, figs. 2a, 4, 9a; and Figure 11A herein). (Character explained by Barnes, 2006:39, character 24.)
25. **Posterior ends of right and left maxillae** on posterior part of facial region both of the same height [0]; or posterior end of right maxilla, compared to posterior end of left maxilla, having a more concave dorsal surface in the area of the bone that is medial to the temporal fossa [1]. The derived character state in some taxa of Odontoceti is associated with the derived state of Character 24 (described above), but it is not associated in all taxa, and it is thus regarded here as a separate character. The

character is visible in illustrations of *Platanista gangetica* by Van Beneden & Gervais (1868-1880:pl. XXXI, Figures 2a, 9a), and Kellogg (1926: pl. 2), and see Figure 11A herein. (Character defined by Barnes (2006:39, character 25.)

26. **Right and left halves of nuchal crest** bilaterally symmetrical [0]; or left half of nuchal crest, compared to right half of nuchal crest, curving anteriorly farther than right half of nuchal crest [1]. The location of the nuchal crest is shown by Rommel (1990:fig.2). The derived character state exists in *Notocetus vanbenedeni* (see Figure 6A herein), in *Araeodelphis,* cf. *A. natator* (see Figure 7A herein), and in *Platanista gangetica* (see Van Beneden & Gervais (1868-1880:pl. XXXI, Figures. 2a, 9a; and Figure 11A herein), and was defined by Barnes (2006:39, character 26).
27. **Rostrum** not remarkably elongated [0]; or rostrum elongated [1]. The rostrum is considered to be elongated when it is 2 ½ times the anteroposterior length of the braincase or more, the latter as measured from the antorbital notches to the occipital condyles. The derived character state is present in *Prepomatodelphis korneuburgensis* (see Barnes 2002b: Figure 1), in *Zarhinocetus errabundus* (see Figure 2 herein), in *Zarhachis flagellator* (see Kellogg, 1924:pls. 1, 2; and Figure 9 herein), and *Platanista gangetica,* as illustrated by Van Beneden & Gervais (1868-1880:pl. XXXI, Figures 1, 2, 2a, 2b, 9, 9a; and Figure 11 herein).
28. **Premaxillae** alone forming rostral extremity [0]; or premaxillae and maxillae both reaching the anterior rostral extremity [1]. The plesiomorphic state exists in Squalodontidae (Figure 4) and Waipatiidae (Figure 5). (Character was defined by Barnes (2006:39, character 28.)
29. **Premaxillae and maxillae** not fused at distal end of rostrum [0]; or fused [1]. The derived character state is present in *Platanista gangetica,* seen in images of that taxon published by Van Beneden & Gervais (1868-1880: pl. XXXI, Figurel; and Figure 11 herein). The character was defined by Barnes (2006:39, character 29), and is modified from Fordyce (1994) and Messenger & McGuire (1998). Following Barnes (2006:39, character 29), in this study we interpret the opposite polarity for the character as was indicated by Geisler & Sanders (2003:99, character 10).
30. **Groove on the lateral side of the rostrum** approximately following the maxilla/premaxilla suture: absent [0]; or groove present [1]. The derived character state is exemplified by the holotype cranium of *Prepomatodelphis korneuburgensis* (see Barnes, 2002b: Figures 1a-b), a referred specimen of *Zarhachis flagellator* (see Kellogg, 1924:pl. 1; and Figure 9 herein), and referred specimens of *Platanista gangetica* (illustrated by Van Beneden & Gervais, 1868-1880:pl. XXXI, Figure 1; and Figure 11 herein), and was defined by Fordyce (1994:176, character 36), and by Barnes (2006:39, character 30).
31. **Lateral lamina of the pterygoid bone** extending posteriorly sufficiently far to contact the alisphenoid bone and/or the falciform process of the squamosal, thus forming an extensive ossified lateral lamina of the pterygoid bone in the ventromedial part of the orbit [0]; or lateral lamina of the pterygoid absent (vestigial), not formed as an outer lamina or bony plate of the pterygoid within the orbit, and not extending posteriorly from the palate to contact the alisphenoid and squamosal [1]. The primitive character state is present in *Allodelphis pratti* (see Figure 1 herein), in *Zarhinocetus errabundus* (see Figs. 2 and 3 herein), in *Prepomatodelphis korneuburgensis* (see Barnes, 2006: Figure 6), in *Zarhachis flagellator* (see Kellogg,

1926:pl. 6; and Figure 9 herein), and in *Platanista gangetica* (see Fraser & Purves, 1960:pl.18; and Figure 11 herein), and was described by Fordyce (1994:75, character 9) and by Barnes (2006:39, character 31). The derived character state is clearly shown in an illustration of a skull of *Steno bredanensis* that was published by Fraser & Purves (1960: pl. 24).

32. **Anterior end of tympanic bulla** rounded [0]; or having an elongate and pointed anterior process or spine [1]. The derived character state is shown in illustrations of *Platanista gangetica* (see Van Beneden & Gervais, 1868-1880:pl. XXXI, Figures 7, 7a, 7b; Kellogg, 1924:pl. 7, Figures. 1-4; Fraser & Purves, 1960:pl. 18; and Figure 11C herein), and was explained by Fordyce (1994:176, character 45), and by Barnes (2002b: 409; 2006:39, character 32). Although the petrosal and tympanic bulla are not known for several fossil species of platanistoids, this very distinctive platanistoid character is included in this analysis because an anterior bullar spine is present in all known species of Platanistoidea for which a bulla is known.
33. **Symphyseal portion of mandible** not greatly elongated anteroposteriorly [0]; or symphyseal portion of mandible anteroposteriorly elongated [1]. The symphyseal portion of the mandible is considered to be elongated in Cetacea when it is more than one-half of the total length of the mandible (for example as it is in *Platanista gangetica* [see illustrations by Van Beneden & Gervais, 1868-1880:pl. XXXI, Figure 3], and in *Zarhachis flagellator* [Kellogg, 1924:pl. 3]), and was defined by Barnes (2006:39-40, character 33).
34. **Mandibular symphysis:** unfused [0]; or firmly ankylosed [1], (as shown on a referred specimen of *Zarhachis flagellator* [see Kellogg, 1924: pl. 3]; and described by Fordyce [1994:175, character 5] and Barnes [2006:40, character 34].
35. **Cheek teeth** posterior to the first premolar having at least two roots [0], or all premolars and molars single-rooted [1]. The homologies of the teeth in heterodont dentitions of primitive Cetacea are clearly demonstrated by the Archaeoceti (see Kellogg, 1936: Figures. 30, 31a), and the canine tooth, as is typical for Mammalia, is the anterior-most tooth that is rooted in the maxilla. The tooth following the P1 is in the Archaeoceti a single-rooted tooth, or is in some taxa a double-rooted tooth. We consider that a single-rooted P1 (and also the p1) is the plesiomorphic character state for Cetacea. In all species of Archaeoceti, the P2 (and p2) are two-rooted teeth (see Kellogg, 1936: Figure 31a), and the presence of single-rooted second premolars, and subsequent premolars and molars, in other Cetacea is the derived character state, and this is a step in the development of homodonty (Barnes, 2006:40, character 35).
36. **Facial surface of cranium** not arched transversely in the area of the dorsal narial openings; may be nearly flat transversely or may be ascending posteriorly [0]; or arched transversely across the area of the narial openings and the cranial vertex, and sloping laterally onto the supraorbital process [1]. The derived character state is present in *Allodelphis pratti* and *Zarhinocetus errabundus* (see Figures 1-3 herein; character defined by Barnes, 2006:40, character 36).
37. **Fossa in the anterior side of the paroccipital process** marking the presence in life of a posterior sinus of the middle ear sinus system; absent [0] (as in *Allodelphis pratti* [Figure 1C]); or present [1] (as in *Prepomatodelphis korneuburgensis* [see Barnes, 2006: Figure 6]). The fossa for this sinus is also present in various genera of Odontoceti as shown by Fraser & Purves (1960:pls. 13 *[Monodon],* 15

[Delphinapterus], 17 *[Platanista]*, 19-20 *[Pontoporia*, labeled as *Stenodelphis]*, 21-22 *[Inia]*, 23 *[Lipotes]*, 25 *[Sousa]*, 27 *[Phocoena]*, 28 *[Neophocaena*, labeled as *Neomeris]*, 31 *[Orcinus]*, 32 *[Orcaella]*, 34 *[Globicephala]*, 35 *[Feresa]*, 36 *[Cephalorhynchus]*, 38-40 *[Lagenorhynchus]*, 42 *[Grampus]*, 44 *[Tursiops]*, 45 *[Stenella]*, and 47 *[Delphinus]);* and see Barnes, 2006:40, character 37).

38. **Posterior-most termination of premaxilla** with an entire, or rounded margin [0]; or posterior termination of premaxilla bifurcated, with the bifurcation containing an exposed wedge of the maxilla [1]. The derived character state (defined by Barnes, 2006:40, character 38) exists on a referred specimen of *Zarhachis flagellator,* as shown by Kellogg (1926:pl. 2).
39. **Zygomatic process of squamosal** with glenoid fossa facing anteroventrally, and not dorsoventrally expanded on its lateral side [0] (as in *Allodelphis pratti,* and *Zarhinocetus errabundus,* see Figures 1C, 2C, and 3C; and Barnes, 2006: Figures 2 and 3); or zygomatic process much expanded dorsoventrally, so that it is flattened transversely, the glenoid fossa is concave medially and closed on its lateral side, and the area of the zygomatic process that is anterior to the glenoid fossa has a medially-directed concavity [1]. The derived character state is present on the holotype of *Prepomatodelphis korneuburgensis* (see Barnes, 2002b:415, Figure 2b; 2006:40, character 39, and shown on Figures 4, 5, and 6), on a referred specimen of *Zarhachis flagellator* (see Kellogg, 1926:pl. 5; and Figure 9 herein), and in Recent *Platanista gangetica* (see Van Beneden & Gervais, 1868-1880: pl. XXXI, Figure 2b; Kellogg, 1924:pl. 6; Fraser & Purves, 1960: pls. 17-18; and Figure 11 herein).
40. **Zygomatic process of squamosal** in lateral view not markedly deeper at the posterior end, having a nearly equal dorsoventral thickness for most of its full length [0]; or posterior part of zygomatic process of squamosal much deepened dorsoventrally in its posterior part [1]. The derived character state is present on the holotype of *Prepomatodelphis korneuburgensis* (see Barnes, 2002b:415, Figure 2b; 2006:40, character 40, Figure 5).
41. **Posterolateral sulcus on premaxilla** that emanates from the premaxillary foramen; uniform in depth or shallow throughout its length [0]; or sulcus very deep, particularly in its posterior part, where its lateral margin may overhang the sulcus [1] (see specimens illustrated by Van Beneden & Gervais, 1868-1880:pl. XXX, Figure 19; by Kellogg, 1926: pl. 2; and by Barnes, 2002: Figure 2a; 2006:40, character 41, Figure 4).
42. **Surface of premaxillary sac fossa** on dorsal surface of posterior part of premaxilla; nearly smooth or only slightly convex [0]; or undulating, having in its mid-part a sulcus that is bounded both medially and laterally by ridges of bone, and abruptly sloping ventrally at both its medial and lateral margins [1]. The derived character state is present on the holotype of *Prepomatodelphis korneuburgensis* (see Barnes, 2002b: Figure 2a; 2006:40, character 42, Figure 4).
43. **Ventrolateral-most part of lambdoidal crest** narrow and only slightly projecting from lateral surface of braincase [0]; or much thickened anteroposteriorly and having a prominent rounded edge, that is at least 10 mm thick around the posterior margin of the temporal fossa and spanning from the dorsal surface of the zygomatic process of the squamosal to the lateral wall of the braincase [1]. The derived character state is

present on the holotype of *Prepomatodelphis korneuburgensis* (see Barnes, 2002b: Figure 2b; 2006:40, character 43, and Figure 5).

44. **Nuchal crest** relatively low, and not elevated significantly above the adjacent bones of the posterior part of the facial region [0]; or thickened dorsoventrally and elevated above the adjacent maxillary and frontal bones [1]. The derived character state is present on a referred specimen of *Zarhachis flagellator* (see Kellogg, 1926:pls. 1-4; Barnes, 2006:40, character 44; and Figure 9 herein).
45. **Crowns of teeth** relatively broad and wide anteroposteriorly at the base, the width at the base of the crown being at least one-half of the crown height [0]; or tooth crowns dorsoventrally elongate, slender, with pointed apices, the width at the base of the crown being less than one-third of the crown height [1]. The derived character state is exemplified by a referred specimen of *Zarhachis flagellator* (see Kellogg, 1924: pls, 1-3; and Figure 9B herein) (character defined by Barnes, 2006:40, character 45). An exception to this is the acquisition of secondary heterodonty by Recent *Platanista gangetica* (see explanation of character 62 following; see specimens illustrated by Van Beneden & Gervais, 1868-1880:pl. XXXI, Figure 1; Figures. 10D and 11 herein).
46. **Roots of teeth** conical, elongate, and implanted nearly straight into the rostrum and the dentary [0]; or apices of the roots of teeth expanded anteroposteriorly, and in some taxa curved posteriorly [1]. The derived character state was described by Barnes (2006:40, character 46) and is present in Recent *Platanista gangetica* (see Van Beneden & Gervais, 1868-1880:pl. XXXI, Figure 1; and see Figure 11 herein).
47. **Posterior end of premaxilla** relatively narrow and not significantly expanded transversely [0]; or posterior end of premaxilla much expanded transversely [1]. The derived character state was described by Barnes (2006:40, character 47), and is exemplified by the holotype of *Prepomatodelphis korneuburgensis* (see Barnes, 2006: Figure 4) and by a referred specimen of *Zarhachis flagellator* (see Kellogg, 1926:pl. 2; and see Figure 9 herein).
48. **Nasal bones,** whether elongate or shortened, of typical width [0]; or nasal bones much narrowed transversely [1]. The derived character state was described by Barnes, 2006:40-41, character 48), and is present on the holotype of *Allodelphis pratti* (specimen illustrated as Figure 1A herein; by Wilson, 1935: Figure 1; and by Barnes, 2006: Figure 1).
49. **Supraorbital maxillary crest** not developed on dorsal surface of supraorbital process [0]; or supraorbital crest (approximately anteroposteriorly aligned) present on dorsal surface of the maxilla on the supraorbital process [1]. The derived character state was described by Barnes, 2006:40, character 49, and is illustrated by Van Beneden & Gervais, 1868-1880:pl. XXX, Figure 1; pl. XXXI, Figures 2, 2a, 9, 9a; by Kellogg, 1926: pl. 2; and in Figures 9 and 11 herein).
50. **Rostrum cross section** not dorsoventrally flattened [0]; or rostrum dorsoventrally flattened so that for at least the proximal two-thirds of its length its transverse width is greater than its dorsoventral height [1]. The derived character state is present in such pomatodelphinine platanistids as the species of *Prepomatodelphis, Zarhachis,* and *Pomatodelphis* (see Barnes, 2006: 41, character 50, Figure 8; Figure 8 herein).
51. **Mandible cross section** not dorsoventrally flattened [0]; or symphyseal portion of the mandible flattened dorsoventrally so that so that in at least its middle one-half the

transverse width is greater than its dorsoventral height [1]. This derived character state is present in the pomatodelphinine platanistids, for example the species of *Prepomatodelphis, Zarhachis,* and *Pomatodelphis* (see Barnes, 2006:41, character 51, Figure 8; Figure 8 herein). This derived character state could possibly be combined with the preceding character, because in the Platanistidae the two appear to be linked. However, the acquisition of the two characters may have occurred at different times, so they are here listed separately.

52. **Supraorbital process** not significantly thickened, for example in *Prepomatodelphis korneuburgensis* [0] (see Barnes, 2006: Figure 5); or supraorbital process dorsoventrally thickened, involving parts of both the maxilla and frontal that are dorsal to the orbit [1]. (See Van Beneden & Gervais, 1868-1880:pl. XXXI, Figures 2, 9; Kellogg, 1926:pl. 4; Figures 6, 7, 9, and 11 herein; character explained by Barnes, 2006:41, as character 52.)
53. **Anterior end of zygomatic process of the squamosal** not remarkably extended anteriorly [0]; or anterior end of zygomatic process of the squamosal extended anteriorly [1]. The derived character state is present in *Platanista gangetica* (see Van Beneden & Gervais, 1868-1880: pl. XXX, Figure 1; pl. XXXI, Figures 2, 9; Kellogg, 1924:pl. 6; Fraser and Purves, 1960:pls. 17-18; and Geisler and Sanders, 2003:character 188, Figure 13b; Barnes, 2006:41, character 53; and Figure 11 herein.)
54. **Cranial vertex** symmetrical, with mid-line sutures between the right and left nasal and frontal bones at the cranial vertex aligned with the mid-line sagittal plane of the cranium [0]; or asymmetrical, with mid-line sutures between the nasal and frontal bones and dorsal narial openings skewed asymmetrically to the left side of the mid-line sagittal plane of the cranium [1]. (For the derived character state see Van Beneden & Gervais, 1868-1880:pl. XXXI, Figures 2a, 4, 9a; and listed by Barnes, 1990:21, at node 13; Barnes, 2006:41, character 54; Figures 6A, 7A, 9A, and 11A herein.)
55. **Rostrum** not transversely compressed [0]; or rostrum transversely compressed so that for at least the proximal two-thirds of its length its transverse width is less than its dorsoventral height [1]. The derived character state is present in *Platanista gangetica* (see Van Beneden & Gervais, 1868-1880:pl. XXXI; and Figure 8 and 11 herein; and was described by Barnes (2006:41, character 55)).
56. **Symphyseal portion of mandible** not transversely compressed [0]; or symphyseal portion of mandible transversely compressed so that in at least its middle one-half the transverse width is less than its dorsoventral height [1]. The derived character state is present in *Platanista gangetica* (see Van Beneden & Gervais, 1868-1880:pl. XXXI; and Figure 8 herein), and in a probable platanistine of Early Miocene age from Oregon (see Barnes, 2006: Figure 7c; and Figure 10D herein), and was described by Barnes (2006:41, character 56).
57. **Size of nasal bones:** right and left nasal bones of normal size for an odontocete, the transverse width of each bone being approximately equal to the width of the corresponding dorsal narial opening [0]; or nasal bones greatly reduced in size [1]. The derived character state is present in *Platanista gangetica,* in which the nasal bones are reduced to small tubercles on the anterodorsal surfaces of the frontal bones posterior to the dorsal narial openings (see Van Beneden & Gervais, 1868-1880:pl.

XXXI, Figure 4; and Figure 11A herein), and was described by Barnes (2006:41, character 57).

58. **Relationship of falciform process of squamosal bone to petrosal:** anterior process of petrosal has little contact with the falciform process [0]; or falciform process of squamosal bone is enlarged and extends ventrally to have a wide contact with the dorsal side of the anterior process of the petrosal [1]. The falciform process descends from the ventral surface of the squamosal (Rommel, 1990: Figure 3; see Barnes, 2006:41, character 58). The plesiomorphic character state is present in many species of the superfamily Delphinoidea, for example in the phocoenid *Neophocaena phocaenoides* (Fraser & Purves, 1960: pl. 28). The derived character state is present in *Platanista gangetica* (Fraser & Purves, 1960:pl. 17; and Figure 11C herein).
59. **Supraorbital process of the frontal** bone not fenestrated by extensions of the pterygoid sinus [0]; or fenestration exists to some extent, spreading from the ventral surface of the supraorbital process of the frontal via the area of the infraorbital foramen system [1]. This occurs because of expansion of a lobe of the pterygoid air sinus from the postorbital area dorsally into the ventral surface of the supraorbital process of the frontal. In the derived character state, which exists in *Platanista gangetica,* multiple branches of the pterygoid sinus extend dorsally toward the infraorbital foramina and spread via the anterior maxillary foramina onto the dorsal surface of the supraorbital process and into the medial side of the maxillary crest (see Van Beneden & Gervais, 1868-1880:pl. XXX, Figures 19a, 19b; Kellogg, 1924: pl. 6; Fraser & Purves, 1960:91, Figure 17a, pls. 17-18; and Figure 11 herein). (Character described by Barnes, 2006:41, character 59.)
60. **Size of the orbit** of normal proportions [0]; or eye much atrophied [1]. Defined by Barnes (2006:41, character 60), the derived character state is present in *Platanista gangetica* (see illustrations in Van Beneden & Gervais, 1868-1880: pl. XXXI, Figures 2, 9; Kellogg, 1924: pl. 5; and Fraser & Purves, 1960:pls. 17-18; and Figure 11 herein). (Character defined by Barnes, 2006:41, character 60.)
61. **Zygomatic process** of the jugal narrow, thin, and rod-like in shape, as is the condition in most odontocetes [0]; or zygomatic process of jugal secondarily thickened and shortened anteroposteriorly [1]. This modification, defined by Barnes (2006:41, character 61) is correlated with reduction of the size of the orbit, and the derived character state is present in *Platanista gangetica* (see illustration of this species by Van Beneden & Gervais, 1868-1880:pl. XXXI, figs. 2, 6, 9; and Fraser & Purves, 1960:pl. 18, and Figures 11B and 11C herein).
62. **Secondary heterodonty:** crowns of teeth are similar in shape throughout the tooth row, and are simple and conical [0]; or crowns of the posterior teeth are short and widened transversely, in contrast to the crowns of the anterior teeth, which are greatly elongated apically [1]. Defined by Barnes (2006:41-42, character 62), the derived character state is present in *Platanista gangetica* (see Van Beneden & Gervais, 1868-1880: pl. XXX, Figure 1; and pl. XXXI, Figures 1, 2, 2b, 9; and Figures 10D and 11 herein).
63. **Lambdoidal crest on the lateral side of the braincase** present along the squamosal-exoccipital suture on the surface of the lateral side of the braincase, dorsal to the zygomatic process of the squamosal, the lambdoidal crest is elevated and discernible [0]; or the lambdoidal crest in this area is reduced in size and is barely protruding

laterally from the side of the cranium [1]. Defined by Barnes (2006:42, character 63), the derived character state is present in *Platanista gangetica,* for which see illustrations published by Van Beneden & Gervais (1868-1880: pl. XXXI, Figures 2, 2a, 9, 9a); by Kellogg (1924:pl. 5); and Figures 11A and 11B herein.

64. **Tympanosquamosal recess, or fossa for middle sinus:** no fossa developed for middle sinus [0]; or fossa is present [1]. This fossa, called the tympano-squamosal recess (Fraser and Purves, 1960; and see Fordyce, 1994), in the living animal is a recess in the bone that marks the location of the middle sinus (see Barnes, 1990:21, node 10; 2006:42, character 64), which is a branch of the middle ear sinus system. The fossa is on the ventral surface of the squamosal bone between the ear region and the medial margin of the glenoid fossa on the ventral surface of the squamosal (see for example Fraser and Purves, 1960:pls. 17-18; and Figure 11C herein, where it is shown on specimens of *Platanista gangetica).*

RESULTS

No Mysticeti were included in the analysis. *Zygorhiza* is represented in this analysis by the Late Eocene North American relatively derived archaeocete, *Zygorhiza kochii,* and this member of the archaeocete family Basilosauridae serves in the present analysis to root the phylogenetic tree. *Agorophius* is represented by the Late Oligocene *Agorophius pygmaeus* (see images and characters in Fordyce, 1981) and this stem odontocete, which was not included in the (1994) analysis by Fordyce, serves as the outgroup, and helps to demonstrate the polarity of the characters of Platanistoidea.

Morphological characters that we list for *Squalodon,* a taxon that was included in the (1994) analysis by Fordyce, were based on *Squalodon* as reported by Muizon (1988a, 1988b, & 1994), on *S. calvertensis* as reported by Kellogg (1923), and on *S. calvertensis* and *S. whitmorei* as reported by Dooley (2005). *Notocetus* is based on *Notocetus vanbenedeni,* and it also represents the characters of *Squalodelphis,* a closely related genus in the family Squalodelphinidae (Muizon, 1987; Fordyce, 2006). *Zarhachis flagellator* serves to represent the characters of the derived pomatodelphinine platanistid genus *Pomatodelphis.*

All characters in the analysis are binary; none are multistate. The plesiomorphic character states were determined by outgroup comparisons with the archaeocete Zygorhiza kochii and with the stem odontocete Agorophius pygmaeus. The numbers of the characters in the above list are the same as those that appear in the Matrix of Character Codings (Table 1). Some characters that are not preserved on available specimens, and thus not known for a particular taxon, were scored on the matrix as ?. The entries with a ? are treated the same way in PAUP (Swofford, 2002) as are characters that are known to be absent in a taxon, which in Table 1 are indicated by a dash (-). Examples of the latter are character states relating to the premaxillary sac fossa and premaxillary foramen for Zygorhiza, because these are structures that never evolved in the Archaeoceti.

The phylogenetic analysis presented here includes the 64 characters that are listed and defined above, and that were scored for nine taxa. Figure 12 shows the only tree that was generated by PAUP from a manipulation of the data in Table 1 using MacClade (Version 3.1.1; Maddison and Maddison, 1992).

Table 1. Matrix of coding of characters that were used in the analysis of relationships of nine taxa of Cetacea, of which seven are Platanistoidea, and which when analyzed yielded the tree shown in Figure 12. The 64 characters are listed and explained in previous text. Character codings are: [0] postulated plesiomorphic condition of a character; [1] postulated apomorphic condition of a character; [-] character not present in the taxon, and [?] condition of the character is not preserved on available specimen(s)

	Characters			
Taxon	1 - 10	11 - 20	21 - 30	31 - 40
Zygorhiza	0000000000	000-0-0000	-000000000	0000000000
Agorophius	11?1111111	1110000000	0?00000??0	??0?000100
Squalodon	1111111111	1111111001	0100000000	0100010000
Waipatia	1111111111	1110101101	1000000000	1100010100
Allodelphis	1111111111	1110101100	0000001111	0111100000
Notocetus	1111111111	1111111101	1001110000	0101111010
Prepomatodelphis	1111111111	111?1111??	100???1111	0?11111011
Zarhachis	1111111111	1111111111	1011111111	0111111010
Platanista	1111111111	1111111111	1011111111	0111111010

	Characters		
Taxon	41 - 50	51 - 60	61-64
Zygorhiza	--01000000	0000000000	-000
Agorophius	0001000000	0000000000	?000
Squalodon	0000000000	0000000000	0000
Waipatia	1000000000	0000000000	0001
Allodelphis	0001100101	1000000100	0000
Notocetus	1000001000	0101000110	0001
Prepomatodelphis	1111?11001	100000?110	0001
Zarhachis	1001111011	1101000110	0001
Platanista	1000110010	0111111111	1111

Barnes (2006) presented a cladistic analysis of the Platanistoidea, and the same year, Fordyce (2006: Figure 3, and see p. 764) provided a tree that was based on several other previous analyses. These two authors independently concurred in their recognition of the monophyly of the superfamily Platanistoidea, the monophyly of each of its included families, the basal placement of the family Squalodontidae, the more derived placement of the family Waipatiidae, the sister relationship of the family Squalodelphinidae to the family Platanistidae, and the placement as a crown group the extant family Platanistidae.

Fordyce's (2006) study was prepared prior to description by Barnes (2006) of the family Allodelphinidae. In our present analysis, and that of Barnes (2006), the Allodelpinidae appear at a basal position in the tree because of their possession of a relatively primitive braincase, on which the relatively small nares are positioned relatively anteriorly, and the nasal bones are relatively elongate anteroposteriorly and transversely narrow. The members of the family Squalodontidae, while being dentally the most primitive platanistoids, share with the more crown-ward families Waipatiidea, Squalodelphinidae, and Platanistidae, a more telescoped (sensu Miller, 1923) cranium on which the nares are enlarged and the nasal bones have become anteroposteriorly shortened and transversely widened. The families Squalodontidae, Waipatiidae, and Squalodelphinidae demonstrate progressive simplification of the dentition, reduction in size of the anterior teeth, and coalescing of the roots of the premolars and molars. The families Squalodelphinidae and Platanistidae share such derived characters as a tuberosity or crest on the maxilla over the orbit, the location of the posterior maxillary foramen very close to the posterolateral corner of the ascending process of the maxilla, and the medial excavation and transverse compression of the zygomatic process of the squamosal. The clade that is represented by the subfamily Pomatodelphininae of the family Platanistidae is notable by the dorsoventral compression of the rostrum and the symphyseal part of the mandible, and the clade that is represented by the subfamily Platanistinae of the family Platanistidae is characterized by the transverse compression of the rostrum and the symphyseal part of the mandible.

Classification of the Platanistoidea

The following classification of the superfamily Platanistoidea is derived in part from the classifications that have been proposed by Fordyce (1994), Fordyce & Barnes (1994), Fordyce et al. (1995), and Barnes (2002b, 2006), and on some taxonomic comments by Fordyce (2006), and has been modified on the basis of the morphological observations and phylogenetic analysis in this study.

We place the family Allodelphinidae first, despite their long rostra and single-rooted teeth, because they are basal in the phylogenetic tree (Figure 12), and in recognition of their primitive cranial structures, including the anterior location of the nares on the facial part of the cranium.

Bold text indicates new taxonomic determinations in this study. Parentheses indicate names that originally were proposed by authors at different ranks than they are used here, followed by the name of the reviser, and the date of publication of that revision. Names that are in quotations ("   ") indicate the need for additional study and probable re-assignment.

Order Cetacea Brisson, 1762
Suborder Odontoceti Flower, 1864

Superfamily Platanistoidea (Gray, 1846) Simpson, 1945 (including Squalodontoidea (Brandt, 1873) Simpson, 1945

Family Allodelphinidae Barnes, 2006

Allodelphis Wilson, 1935

cf. *Allodelphis,* species undescribed (Early Miocene, Orange County, California)

Allodelphis pratti Wilson, 1935 (Earliest Miocene, Woody, Kern County, California)

Allodelphis, species undescribed (Early Miocene, Cajon Pass, San Bernardino County, California)

Genus and species undescribed (late Early Miocene, Japan)

***Zarhinocetus* Barnes, Kimura, and Godfrey, new genus**

***Zarhinocetus errabundus* (Kellogg, 1931), new combination** (middle Middle Miocene, Kern County, California)

aff. *Zarhinocetus errabundus* (Kellogg, 1931) (Late Miocene, Santa Cruz County, California)

Family Squalodontidae Brandt, 1872

Subfamily Patriocetinae (Abel, 1913) Rothausen, 1968

Patriocetus Abel, 1913

Patriocetus ehrlichi (Van Beneden, 1865) (Late Oligocene, Linz/Donau, Austria)

Patriocetus kazakhstanicus Dubrovo and Sanders, 2000 (Late Oligocene, Mangyshlak Peninsula, Kazakhstan)

Subfamily Squalodontinae (Brandt, 1872) Rothausen, 1968

Eosqualodon Rothausen, 1968

Eosqualodon langewieschei Rothausen, 1968 (Late Oligocene, Westfalia, Germany)

Eosqualodon latirostris (Capellini 1904) (Early Miocene, Italy)

Phoberodon Cabrera, 1926

Phoberodon arctirostris Cabrera, 1926 (Early Miocene, Patagonia, Argentina)

Squalodon Grateloup, 1840

Squalodon bellunensis Dal Piaz, 1916 (Early Miocene, Italy)

Squalodon kelloggi (Rothausen, 1968 (Early Miocene, France)

Squalodon antverpiensis van Beneden, 1861 (Late Middle Miocene, Belgium)

Squalodon atlanticus Leidy, 1861 (Middle Miocene, New Jersey, Maryland

Squalodon bariensis (Jourdan, 1861) (Middle Miocene, France)

Squalodon calvertensis Kellogg, 1923 (Middle Miocene, Maryland)

Squalodon whitmorei Dooley, 2005 (Middle Miocene, Maryland, Virginia)

Kelloggia Mchedlidze, 1976

Kelloggia barbarus (Aslanova, 1976) (Late Oligocene, Azerbaidzhan)

Prosqualodon Lydekker, 1894

Prosqualodon australis Lydekker, 1892 (Early Miocene, Patagonia, Argentina)

Prosqualodon davidis Flynn, 1923 (Late Oligocene, New Zealand)

"*Prosqualodon" hamiltoni* Benham, 1937 (Late Oligocene, New Zealand)

Family Waipatiidae Fordyce, 1994

Waipatia Fordyce, 1994

Waipatia maerewhenua Fordyce, 1994 (Late Oligocene, New Zealand)
Sulakocetus Mchedlidze, 1976
Sulakocetus dagestanicus Mchedlidze, 1976 (Late Oligocene, Caucasus)
Sachalinocetus Dubrovo, 1971
Sachalinocetus cholmicus Dubrovo, 1971(Early or Middle Miocene, Sakhalin Island, Russia)
Family Squalodelphinidae (Dal Piaz, 1916) Rice, 1998 (emended name)
Squalodelphis Dal Piaz, 1916
Squalodelphis fabianii Dal Piaz, 1916 (Early Miocene, Italy)
Notocetus Moreno, 1892 (= *Diochotichus* Ameghino, 1892)
Notocetus vanbenedeni (Moreno, 1892) (Early Miocene, Patagonia, Argentina)
Notocetus marplesi (Dickson, 1964) (Late Oligocene, New Zealand)
"Microcetus" hectori (Benham, 1935) (Late Oligocene, New Zealand)
Phocageneus Leidy, 1869
Phocageneus venustus Leidy, 1869 (Middle Miocene, Virginia, Maryland, North Carolina)
Family Platanistidae (Gray, 1846) Gray, 1863
Subfamily *incertae sedis*
Araeodelphis Kellogg, 1957
Araeodelphis natator Kellogg, 1957 (Middle Miocene, Maryland, North Carolina)
Subfamily Pomatodelphininae Barnes, 2002
Prepomatopdelphis Barnes, 2002
Prepomatopdelphis korneubergensis Barnes, 2002 (late Early Miocene, Austria)
Zarhachis Cope, 1868
Zarhachis flagellator Cope, 1868 (early Middle Miocene, Maryland)
Pomatodelphis Allen, 1921
Pomatodelphis inaequalis Allen, 1921 (=*Schizodelphis depressus* Allen, 1921,
Pomatodelphis bobengi (Case, 1934) Morgan, 1994 (originally described as a species of *Schizodelphis*)
Pomatodelphis stenorhynchus (Holl, 1829)
Subfamily Platanistinae (Gray, 1863) Barnes, 2002
Genus and species undetermined (Nye Formation, Early Miocene, Oregon)
Platanista Wagler, 1830
Platanista gangetica (Roxburgh, 1801)
Platanista gangetica gangetica (Roxburgh, 1801) (Ganges River Dolphin, India, Bangladesh, Nepal)
Platanista gangetica minor Owen, 1853 (Indus Dolphin, Pakistan)

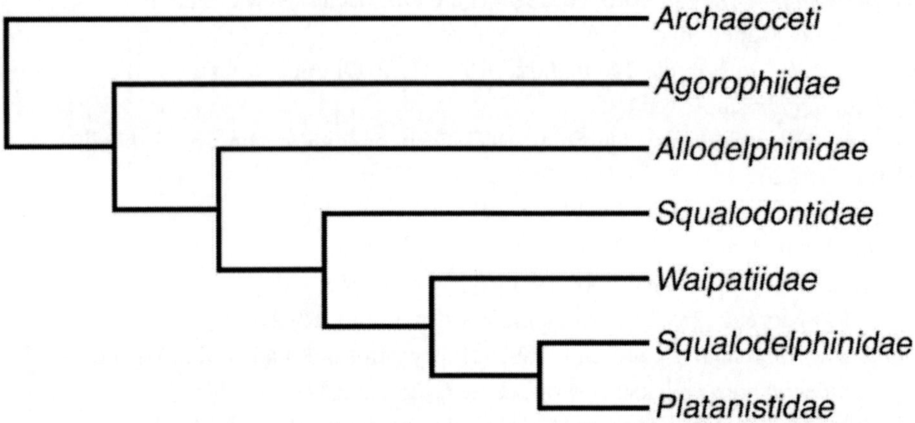

Figure 12. Postulated phylogenetic relationships between more primitive Cetacea and the families of the superfamily Platanistoidea: the Allodelphinidae, Squalodontidae, Waipatiidae, Squalodelphinidae, and Platanistidae. Although time is not implied in this image, the early diversification among the four extinct families of Platanistoidea, the Allodelphinidae, Squalodontidae, Waipatiidae, and Squalodelpinidae, is implied, occurring during the later part of the Oligocene and the early part of the Miocene. The platanistoids then gradually dwindled in diversity, so that only relict fresh water populations of the genus *Platanista* (family Platanistidae) survive now. The evolutionary diversification of the superfamily Platanistoidea has an inverse relationship with that of the superfamily Delphinoidea, whose members were small, not taxonomically diverse, and rare in Late Oligocene time, then diversified to become the most diverse group of living Cetacea (Modified from Barnes, 2006: Figure 9). This analysis used 64 characters among nine taxa of fossil Cetacea, including some described members of the superfamily Platanistoidea for which reasonably complete crania were available. The derived Late Eocene archaeocete, *Zygorhiza kochii* (Reichenbach, 1847) from Alabama, U. S. A., was used to root the tree, and the primitive odontocete, *Agorophius pygmaeus* (Müller, 1849) probably from South Carolina, U. S. A., was used as the outgroup for the Platanistoidea. This was the only resulting tree, having a tree length of 84, a Consistency Index of 0.76, a Retention Index of 0.75, and was obtained by manipulating a character matrix with MacClade Version 3.01, and using the Branch and Bound search option of PAUP Version 3.1.1.

Summary and Conclusions

1. The superfamily Platanistoidea is a clade of Odontoceti that was diverse during Oligocene and Miocene time, from approximately 30 to 10 or 12 million years ago. The group subsequently declined in diversity, so that now within this superfamily only the family Platanistidae survives, represented by the genus *Platanista,* living in rivers of south Asia. Most fossil platanistoids were marine, but fossil members of the family Platanistidae have been found in brackish and fresh water deposits as well as marine deposits. The evolutionary history of the Platanistoidea has an inverse relationship to the history of the superfamily Delphinoidea, which in the Oligocene were small and rare, and today are the most diverse group of Cetacea.
2. There are four extinct families within the Platanistoidea: the Allodelphinidae (Miocene), Squalodontidae (Oligocene and Miocene), Waipatiidae (Oligocene and Miocene), and Squalodelphinidae (Oligocene and Miocene). All fossils of these

families are from marine deposits. Fossils of the extant family Platanistidae are from marine, estuarine, and fresh water deposits.

3. Species in the extinct platanistoid family Allodelphinidae are known only from marine deposits around the margin of the North Pacific Ocean, their fossils having been found in California, Oregon, Washington, and Japan. They are known from the earliest Miocene, approximately 23 million years ago, to the Late Miocene, approximately 10 to 12 million years ago. Allodelphinids had skull lengths of approximately one meter or more, very long and slender rostra, with the lower jaw reaching the same length as the rostrum, many small and single-rooted teeth, and unusually long cervical vertebrae.
4. An extinct odontocete named *Squalodon errabundus* Kellogg, 1931, was named on the basis of isolated petrosal bones from the middle Middle Miocene age Sharktooth Hill Local Fauna in the Sharktooth Hill Bonebed, in Kern County, California, U. S. A. A complete skull, found with this same type of petrosal associated with its basicranium, demonstrates that this species is not a member of the Squalodontidae, but is actually an allodelphinid platanistoid, and the species therefore needs a new generic assignment. The new genus name *Zarhinocetus* is proposed here for this species, yielding the new combination *Zarhinocetus errabundus* (Kellogg, 1931).
5. The members of the extinct family Squalodontidae were large odontocetes with large teeth, and had extreme heterodonty, with the teeth differentiated as incisors, canines, premolars, and molars. They are known from Oligocene and Miocene deposits and from all of the world's major ocean basins.
6. The extinct family Waipatiidae includes marine odontocetes with relatively short and stout rostra and mandibles, symmetrical cranial vertices, and dentitions that are less prominently heterodont than those of squalodontids. Included are *Waipatia maerewhenua,* of Late Oligocene age from New Zealand, and possibly *Sulakocetus dagestanicus,* of Late Oligocene age from the Caucasus, and *Sachalinocetus cholmicus,* of Early or Middle Miocene age from Sakhalin Island, Russia.
7. A more derived family, the extinct Squalodelphinidae, includes platanistoids that have stout rostra and mandibles in which the cheek teeth have single roots (the result of the original two roots having coalesced), dorsoventrally thickened supraorbital processes, asymmetrical cranial vertices, and transversely compressed zygomatic processes of the squamosals. Members of this family include Early Miocene *Squalodelphis fabianii* from Italy, Early Miocene *Notocetus vanbenedeni* from Patagonia, Argentina, and probably *Notocetus marplesi* and *"Microcetus" hectori*, both of Late Oligocene age from New Zealand, and *Phocageneus venustus,* of Middle Miocene age from Virginia, Maryland, and North Carolina in the U. S. A.
8. The highly derived platanistoid family Platanistidae includes dolphins that have exceptionally narrow rostra and symphyseal portions of their mandibles, asymmetrical cranial vertices, and transversely compressed zygomatic processes of their squamosals. This is the family that includes the living, river-dwelling members of the genus *Platanista,* but fossil members of this group are found in deposits of marine, brackish, and fresh water origin. Fossils are known only from the Northern Hemisphere, and demonstrate that members of this family were much more widespread in the past than they are now. The most primitive named member of the

Platanistidae is *Araeodelphis natator,* an early Middle Miocene fossil from the western North Atlantic, whose more precise relationships are not yet known.

9. Two subfamilies are recognized in the Platanistidae, the fossil (Miocene only) subfamily Pomatodelphininae, and the extant (Miocene to Recent) subfamily Platanistinae.
10. The platanistid subfamily Pomatodelphininae includes late Early Miocene *Prepomatodelphis korneuburgensis* from shallow marine deposits of Austria, and more highly derived, long-snouted species of the genera *Zarhachis* and *Pomatodelphis* of Middle and Late Miocene age from the North Atlantic realm (France, Maryland, Virginia, North Carolina, Florida, and Alabama), all of which have very elongate snouts and dorsoventrally flattened rostra and symphyseal portions of their mandibles. Derived members of the Pomatodelphininae have dorsoventrally thickened frontal bones forming eminences dorsal to their orbits, and in at least one species, *Zarhachas flagellator,* these eminences were invaded by dorsal extensions of the supraorbital lobe of the pterygoid sinus, which extended dorsally through the anterior maxillary foramina.
11. The subfamily Platanistinae is extant, and includes the extant genus *Platanista,* whose living species are the fresh water susus of south Asia, *Platanista gangetica.* Fossils that appear to belong to the Platanistinae from the eastern margin of the North Pacific basin (Oregon, Washington) suggest that the Platanistinae formerly had a wider geographic range and were originally marine in habitat. The Platanistinae are comparatively highly derived odontocete taxa, having among their derived characters small size, enlarged and anteriorly extended zygomatic process of squamosal, atrophied eye, extreme left-skew asymmetry of the cranial vertex, reduced nasal bones, greatly enlarged supraorbital crests (formed by maxillary bones and pneumaticized by extensions from the middle ear air sinus system), extremely narrow supraoccipital-nuchal crest area, reduced lambdoidal crests, secondarily thickened zygomatic process of the jugal, transversely flattened rostrum and symphyseal part of the mandible, secondary heterodonty (crowns of anterior teeth greatly elongated, crowns of posterior teeth widened), and paedomorphism (which accounts for some of their derived characters).

ACKNOWLEDGMENTS

For many hours of discussion about platanistoid dolphins we are indebted to Mr. David J. Bohaska, Dr. Robert L. Brownell, Jr., Dr. R. Ewan Fordyce, Dr. Samuel A. McLeod, Dr. Gary S. Morgan, Dr. Christian de Muizon, Dr. Olivier Lambert, and Dr. Frank C. Whitmore, Jr. For access to collections under their care we thank the staff of the Natural History Museum in Vienna (Dr. Ortwin Schultz, Dr. Gudrun Daxner-Höck), United States National Museum of Natural History (Mr. David J. Bohaska, Dr. James G. Mead), Calvert Marine Museum, Yale Peabody Museum (the late Dr. John H. Ostrom), University of California Museum of Paleontology at Berkeley (Dr. J. Howard Hutchison, Dr. Patricia Holroyd), and the Muséum National d'Histoire Naturelle in Paris (Dr. Christian de Muizon). This study was made possible by support and assistance from the Natural History Museum of Los Angeles

County and its Foundation through the Fossil Marine Mammal Research Account, and by generous donations from Ronald and Judith Perlstein, Ms. Donna Matson, and by Mr. James E. Klein and Mrs. Sally Klein. Dr. Bruno Frolich (USNM Department of Anthropology) made the CT scan of the fossil skull of *Zarhachis flagellator* that appears in Figure 9C. John De Leon (former LACM staff photographer) made the photographs in Figures 2A-C and 3. Daniel N. Gabai (volunteer at LACM) made the photographs in Figures 1D, 3D, and 11, and helped format Figure 9. All other images were prepared by the authors.

REFERENCES

[1] Abel, O. (1905). Les Odontocètes du Boldérien (Miocène Supérieur) d'Anvers. *Mémoires du Musée Royal d'Histoire Naturelle de Belgique*, 3, 1-155.

[2] Addicott, W. O. (1972). Provincial middle and late Tertiary molluscan stages, Temblor Range, California. In E. H. Stinemeyer (Ed.), *Proceedings of the Pacific Coast Miocene Biostratigraphic Symposium (pp 1-26)*. Bakersfield, California: Society of Economic Paleontologists and Mineralogists.

[3] Allen, G. M. (1921). Fossil cetaceans from the Florida phosphate beds. *Journal of Mammalogy*, 2(3), 144-159, pls. 9-12.

[4] Au, W. L. (2002). Echolocation. In W. F. Perrin, H. Thewissen, & B. Wursig, (Eds.), *Encyclopedia of Marine Mammals* (pp. 358-367). San Diego, CA: Academic Press.

[5] Barnes, L. G. (1977). Outline of eastern North Pacific fossil cetacean assemblages. Charles A. Repenning, editor, Symposium: Advances in systematics of marine mammals. *Systematic Zoology*, 25(4), 321-343 [for December 1976].

[6] Barnes, L. G. (1978). A review of *Lophocetus* and *Liolithax* and their relationships to the delphinoid family Kentriodontidae (Cetacea: Odontoceti). *Natural History Museum of Los Angeles County Science Bulletin*, 28, 1-35.

[7] Barnes, L. G. (1985). The Late Miocene dolphin *Pithanodelphis* Abel, 1905 (Cetacea: Kentriodontidae) from California. *Contributions in Science, Natural History Museum of Los Angeles County*, 367, 1-27.

[8] Barnes, L. G. (1990). The fossil record and evolutionary relationships of the genus *Tursiops*. In S. Leatherwood & R. R. Reeves (Eds.), *The Bottlenose Dolphin* (pp. 3-26). San Diego, CA: Academic Press.

[9] Barnes, L. G. (2002a). Cetacea, Overview. In W. F. Perrin, H. Thewissen, & B. Wursig (Eds.), *Encyclopedia of Marine Mammals* (pp. 204-208) Sand Diego, CA: Academic Press.

[10] Barnes, L. G. (2002b). An Early Miocene long-snouted marine platanistid dolphin (Mammalia,Cetacea, Odontoceti) from the Korneuberg Basin (Austria). Das Karpat des Korneuburger Beckens, Volume 2, W. Sovis and B. Schmid, editors, *Beitrage zur Paläontologie*, 27, 407-418.

[11] Barnes, L. G. (2006). A phylogenetic analysis of the superfamily Platanistoidea (Mammalia, Cetacea, Odontoceti). *Beitrage zur Paläontologie*, 30, 25-42.

[12] Barnes, L. G., McLeod, S. A. Kearin, M. L., & Deering, M. A. (2003). The most primitive known platanistid (Cetacea; Odontoceti), a new Early Miocene species from southern California. *Journal of Vertebrate Paleontology*, 23(supplement to 3), 32A.

[13] Barnes, L. G., & Mitchell, E. D. (1984). *Kentriodon obscurus* (Kellogg, 1931), a fossil dolphin (Mammalia: Kentriodontidae) from the Miocene Sharktooth Hill Bonebed in California. *Contributions in Science, Natural History Museum of Los Angeles County*, 353, 1-23.

[14] Barnes, L. G., & Reynolds, R. E. (2007). A primitive Early Miocene platanistoid dolphin (Cetacea: Odontoceti) from Cajon Pass, San Bernardino County, California; p. 107. In R. E. Reynolds (Ed.), *Wild, scenic and rapid: a trip down the Colorado River Trough, Abstracts of Proceedings, The 2007 Desert Research Symposium, California State University Desert Studies Consortium and LSA Associates* (April, pp. 107) Zzyzx, CA: California State University.

[15] Berggren, W. A., Kent, D. V., Swisher, C. C., III, & Aubry, M. P. (1995). A revised Cenozoic geochronology and chronostratigraphy. *SEPM Special Publication*, 54, 129-212.

[16] Bohaska, D. J., Barnes, L. G., & Godfrey, S. J. (2007). *Araeodelphis natator* Kellogg,1957, the most primitive known member of the Platanistidae (Odontoceti, Cetacea) and a review of the Platanistidae from Calvert Cliffs (Miocene, Chesapeake Croup), Maryland, USA. In A. Warren (Ed.), *Geological Society of Australia, Abstracts, Number 85, 11th Conference on Australasian Vertebrate Evolution, Palaeontology, and Systematics, Melbourne, Australia, 10-13 April 2007, Programme Abstracts* (pp. 61-62): Sidney, Australia: Geological Society of Australia.

[17] Cabrera, A. (1926). Cétácéos fossiles del Museo de La Plata. *Revista Museo de La Plata*, 29, 363-411.

[18] Cassens, I., S. Vicario, V. G. Waddell, H. Balchowsky, D. Van Belle, W. Ding, F. Chen, R. S. L. Mohan, P. C. Simoes-Lopes, R. Bastida, A. Meye, M. J. Stanhope, & M. C. Milinkovitch. (2000). Independent adaptation to riverine habitats allowed survival of ancient cetacean lineages. *Proceedings of the National Academy of Science of the United States of America,* 97(21), 11343-11347.

[19] Cozzuol, M. A. (1996). The record of the aquatic mammals in southern South America. München Geowissenschaftliche Abhandlundgen. A. *Geologie und Paläontologie*, 30, 321-342.

[20] Crowley, B., Barnes, L. G., & Goedert. J. L. (1999). An Early Miocene cetacean assemblage from Washington State, and comparisons with other west coast assemblages. *Journal of Vertebrate Paleontology*, 19(supplement to 3), 40A.

[21] Deering, M. A., Kearin, M. L., Barnes, L. G., Heathcoat, E. M., Orlando, S. W., Reeves, C. E., & Calvano, G. (2003). Discovery of a new species of rare Early Miocene platanistid dolphin (Cetacea, Odontoceti) from the San Joaquin Hills, Orange County, southern California. Abstracts and Program, The Western Association of Vertebrate Paleontologists Annual Meeting, Mojave River Valley Museum, 14-16 February 2003, p. 8, Barstow, California

[22] Dooley, A. C., Jr. (2005). A new species of *Squalodon* (Mammalia, Cetacea) from the Middle Miocene of Virginia. *Virginia Museum of Natural History Memoir*, 8, 1-43.

[23] Dubrovo, I. A., & Sanders, A. E. (2000). A new species of *Patriocetus* (Mammalia, Cetacea) from the Late Oligocene of Kazakhstan. *Journal of Vertebrate Paleontology*, 20(3), 577-590.

[24] Fordyce, R. E. (1981). Systematics of the odontocete *Agorophius pygmaeus* and the family Agorophiidae (Mammalia: Cetacea). *Journal of Paleontology*, 55, 1028-1045.

[25] Fordyce, R. E. (1994). *Waipatia maerewhenua,* new genus and species (Waipatiidae, new family), an archaic Late Oligocene dolphin (Cetacea: Odontoceti: Platanistoidea) from New Zealand. *Proceedings of the San Diego Society of Natural History*, 29, 147-176.

[26] Fordyce, R. E. (2006). Chapter 33. A southern perspective on Cetacean evolution and zoogeography. In J. Merrick, M. Archer, G. M. Hickey, & M. S. Y. Lee (Eds.), *Evolution and Biogeography of Australasian Vertebrates* (pls. 116-124, pp 755-778). New South Wales, Australia: Australian Scientific Publishing.

[27] Fordyce, R. E, & Barnes, L. G. (1994). The evolutionary history of whales and dolphins, In G. W. Wetherill (Ed.), *1994 Annual Review of Earth and Planetary Sciences* (Volume 22, pp. 419-455). Palo Alto, California: Annual Reviews, Inc.

[28] Fordyce, R. E, Barnes, L. G., & Miyazaki, N. (1995). General aspects of the evolutionary history of whales and dolphins. In L.G. Barnes, Y. Hasegawa, & N. Inuzuka, (Eds.), *The Island Arc, Thematic Issue: Evolution and Biogeography of Fossil Marine Vertebrates in the Pacific Realm, Proceedings of the 29th International Geological Congress*, (3(4), pp 373-391). Kyoto, Japan: International Geological Congress.

[29] Fraser, F. C., & Purves, P. E. (1960). Hearing in cetaceans: evolution of the accessory air sacs and the structure of the outer and middle ear in Recent cetaceans. *Bulletin of the British Museum (Natural History), Zoology*, 7(1), 1-140, 53 pls.

[30] Geisler, J. H., & Sanders, A. E. (2003). Morphological evidence for the phylogeny of Cetacea. *Journal of Mammalian Evolution*, 10(1 & 2), 23-129

[31] Godfrey, S. J, Barnes, L. G., & Bohaska, D. J. (2006). *Araeodelphis natator* Kellogg, 1957, the most primitive known member of the Platanistidae (Odontoceti, Cetacea), and relationships to other clades within the Platanistoidea. *Journal of Vertebrate Paleontology*, 26(supplement to 3), 68A.

[32] Gottfried, M. D., Bohaska, D. J., & Whitmore, F. C. Jr. (1994). Miocene cetaceans of the Chesapeake Group. *Proceedings of the San Diego Society of Natural History*, 29, 229-238.

[33] Hamilton, H., Caballero, S., Collins, A. G., & Browell, R. L. Jr. 2001. Evolution of river dolphins. *Proceedings of the Royal Society of London, Biological Series*, 268, 549-556.

[34] Hershkovitz, P. (1966). Catalog of living whales. *Smithsonian Institution Bulletin*, 246, 1-259.

[35] Heyning, J. E. (1989). Comparative facial anatomy of beaked whales (Ziphiidae) and a systematic revision among the families of extant Odontoceti. *Contributions in Science, Natural History Museum of Los Angeles County*, 405, 1-64.

[36] Hulbert, R. C., Jr., & Whitmore, F. C. Jr. (2006). Late Miocene mammals from the Mauvilla Local Fauna, Alabama. *Bulletin of the Florida Museum of Natural History, University of Florida* 46(1), 1-28.

[37] Kellogg, A. R. (1923). Description of two squalodonts recently discovered in the Calvert Cliffs, Maryland; and notes on the shark-toothed cetaceans. *Proceedings of the United States National Museum*, 62(16), 1-69, pls. 1-20.

[38] Kellogg, A. R. (1924). A fossil porpoise from the Calvert Formation of Maryland. *Proceedings of the United States National Museum*, 63(14), 1-39, pls. 1-18.

[39] Kellogg, A. R. (1926). Supplementary observations on the skull of the fossil porpoise Zarhachis flagellator Cope. *Proceedings of the United States National Museum*, 67(28), 1-18, pls. 1-5.

[40] Kellogg, A. R. (1927). *Kentriodon pernix,* a Miocene porpoise from Maryland. *Proceedings of the United States National Museum*, 69(19), 1-55, pls. 1-14.

[41] Kellogg, A. R. (1928). History of whales - their adaptation to life in the water. *Quarterly Review of Biology*, 3, 29-76, 174-208.

[42] Kellogg, A. R. (1931). Pelagic mammals from the Temblor Formation of the Kern River region, California. *Proceedings of the California Academy of Sciences*, Series 4, 19(12), 217-397.

[43] Kellogg, A. R. (1936). A review of the Archaeoceti. *Carnegie Institution of Washington Publication*, 482, 1-366.

[44] Kellogg, A. R. (1944). Fossil cetaceans from the Florida Tertiary. *Bulletin of the Museum of Comparative Zoology at Harvard College*, 94(9), 433-471, pls. 1-6.

[45] Kellogg, A. R. (1957). Two additional Miocene porpoises from the Calvert Cliffs, Maryland. *Proceedings of the United States National Museum*, 107, 279-337, pls. 1-30.

[46] Kellogg, A. R. (1959). Description of the skull of *Pomatodelphis inaequalis* Allen. *Bulletin of the Museum of Comparative Zoology, Harvard College*, 121(1), 1-26, pls. 1-6.

[47] Kimura, T., & Ozawa, T. (2001). An Early Miocene odontocete from the Awa Group, central Japan. Abstract of the 150th Regular Meeting of the Paleontological Society of Japan, 92 [in Japanese].

[48] Lydekker, R. (1894). Contributions to a knowledge of the fossil vertebrates of Argentina. II. Cetacean skulls from Patagonia. *Anales del Museo de La Plata, Palaeontología Argentina*, 2(2), 1-13, pls. 1-6. [dated 1893, published 1894]; Buenos Aires.

[49] Maddison, W. P., & Maddison, D. R. (1992). MacClade, Version 3.01.

[50] McKenna, M. C., & Bell, S. K. (1997). *Classification of mammals above the species level*. New York, NY: Columbia University Press.

[51] Messenger, S. L. & McGuire, J. A. (1998). Morphology, molecules, and the phylogenetics of cetaceans. *Systematic Biology*, 47, 90-124.

[52] Miller, G. S. (1923). The telescoping of the cetacean skull. *Smithsonian Miscellaneous Collections*, 76(5), 1-70.

[53] Mitchell, E. D. (1965). *History of research at Sharktooth Hill, Kern County, California. Special Publication of the Kern County Historical Society*, Kern County, CA: Kern County Museum.

[54] Mitchell, E. D., & Tedford, R. H. (1973). The Enaliarctinae. A new group of extinct aquatic Carnivora and a consideration of the origin of the Otariidae. *Bulletin of the American.Museum of Natural History*, 151(3), 201-284.

[55] Morgan, G. S. (1994). Miocene and Pliocene marine mammal faunas from the Bone Valley Formation of central Florida. *Proceedings of the San Diego Society of Natural History*, 29, 239-268.

[56] Muizon, C. de. (1987). The affinities of *Notocetus vanbenedeni,* an Early Miocene platanistoid (Cetacea, Mammalia) from Patagonia, southern Argentina. *American Museum Novitates*, 2904, 1-27.

[57] Muizon, C. de. (1988a). Le polyphylétisme des Acrodelphidae, odontocètes longirostres des Miocène européen. *Bulletin du Muséum National d'Histoire Naturelle, Paris, Sciences de la Terre, Série 4*(10)1, 31-88.

[58] Muizon, C. de. (1988b). Les relations phylogénétiques des Delphinida (Cetacea, Mammalia). *Annales de Paléontologie*, 74(4), 159-227.

[59] Muizon, C. de. (1991). A new Ziphiidae (Cetacea) from the Early Miocene of Washington State (USA) and phylogenetic analysis of the major groups of odontocetes. *Bulletin du Muséum National d'Histoire Naturelle, Paris, Sciences de la Terre*, Série 4(12, C)3-4, 279-326.

[60] Muizon, C. de. (1994). Are the squalodonts related to the platanistoids? *Proceedings of the San Diego Society of Natural History*, 29, 135-146.

[61] Nikaido, M., Matsuno, F., Hamilton, H., Brownell, R. L. Jr., Cao, Y., Ding, W., Zuoyan, Z. Shedlock, A. M., Fordyce, R. E., Hasegawa, M., & Okada, N. (2001). Retroposon analysis of major cetacean lineages: The monophyly of toothed whales and the paraphyly of river dolphins. *Proceedings of the National Academy of Sciences of the United States of America*, 98, 7384-7389.

[62] Probst, J. (1886). Über die fossilen Reste von Zahnwalen (Cetodonten) aus der Molasse von Baltringen O. A. Laupheim. Jahreshefte des Vereins fur vaterländische Naturkunde in Württemberg, Stuttgart, 42, 102-145.

[63] Prothero, D. R., Liter, M.R., Barnes, L. G., Wang, X., Mitchell, E. D., McLeod, S. A., Whistler, D. P., Tedford, R. H., & Ray, C. E. (2008). Land mammals from the Middle Miocene Sharktooth Hill Bonebed, Kern County, California. In S. G. Lucas, G. S. Morgan, J. A. Spielman, & D. R. Prothero (Eds.), *Neogene Mammals, Bulletin of the New Mexico Museum of Natural History and Science* (44:i-iii, pp.299-314). New Mexico Museum of Natural History and Science

[64] Rice, D. W. (1998). Marine mammals of the world. Systematics and distribution. *The Society for Marine Mammalogy*, Special Publication, 4:i-ix, 1-231.

[65] Rommel, S. (1990). Chapter 2. Osteology of the bottlenose dolphin, In S. Leatherwood & R.R. Reeves (Eds.), *The Bottlenose Dolphin*, (pp. 26-49). San Diego, CA: Academic Press.

[66] Rothausen, K. (1968). Die systematische Stellung der europaischen Squalodontidae (Odontoceti:Mamm.). *Paläontologische Zeitschrift*, 42, 83-104.

[67] Simpson, G. G. (1945). The principles of classification and a classification of mammals. *Bulletin of the American Museum of Natural History*, 85, 1-350.

[68] Swofford, D. L. (2002). PAUP: Phylogenetic analysis using parsimony. Version 4.0b. Massachusetts: Sinauer Associates, Inc.

[69] Tedford, R. H., Skinner, M. F., Fields, R. W., Rensberger, J. M., Whistler, D. P., Galusha, T., Taylor, B. E., Macdonald, J. R., & Webb, S. D. (1987). Faunal succession and biochronology of the Arikareean through Hemphillian interval (Late Oligocene through earliest Pliocene epochs) in North America. In M. O. Woodburne (Ed.), *Cenozoic Mammals of North America. Geocrhonology and biostratigraphy* (pp.153-210).Berkeley CA, Los Angeles, CA, and London, United Kingdom: University of California Press.

[70] Tedford, R. H., Albright, L. B., III, Barnosky, A. D., Ferrusquía-Villafranca, I., Hunt, R. M., Jr., Storer, J. E., Swisher, III C. C., Voorhies, M. R., Webb, S. D., & Whistler D. P.. (2004). Mammalian biochronology of the Arikareean through Hemphillian interval

(Late Oligocene through early Pliocene epochs). In M. O. Woodburne (Ed.), *Late Cretaceous and Cenozoic Mammals of North America. Biostratigraphy and Geochronology* (pp.169-231). New York, NY: Columbia University Press.

[71] True, F. W. (1910). Description of a skull and some vertebrae of the fossil cetacean *Diochotichus vanbenedeni* from Santa Cruz, Patagonia. *Bulletin of the American Museum of Natural History*, 28(4), 19-32.

[72] Van Beneden, P. J., & Gervais, F. L. P. 1868-1880. (1880). Ostéographie des cétacés vivants et fossils, comprenant la description et l'iconographie, du squelette et du système dentaire de ces animaux ainsi que les documents relatifs à leur histoire naturelle. Paris, France: A.Bertrand, *Libraire de la Société de Géographie, text.*

[73] Whitmore, F. C., Jr., & Kaltenbach, J. A. (2007). Neogene Cetacea of the Lee Creek Phosphate Mine, North Carolina, In: C. E. Ray, D. J. Bohaska, I. A. Koretsky, L. W. Ward, & L. G. Barnes (Eds). *Geology and Paleontology of the Lee Creek Mine, North Carolina, IV, Special Publication of the Virginia Museum of Natural History* (Number 14, pp. 181-269). Martinsville VA: Virginia Museum of Natural History.

[74] Whitmore, F. C., & Sanders, A. E. (1977). Review of the Oligocene Cetacea. Charles A. Repenning, editor, Symposium: Advances in systematics of marine mammals. *Systematic Zoology*, 25(4), 304-320 [for December 1976].

[75] Wilson, L. E. (1935). Miocene marine mammals from the Bakersfield region, California *Bulletin of the Peabody Museum of Natural History, Yale University*, 4, 1-143, pl. 1.

INDEX

A

abatement, 438
abundance, 57, 66, 70, 295, 306, 316, 317, 355, 358, 375, 393
ACC, 167
accidental, 19, 164, 233, 238, 240, 242, 244, 384, 388, 420, 436
accounting, 38
accuracy, 288, 433
achievement, 368, 383
acid, 118, 121, 122, 124, 126, 154, 209, 396, 398, 399, 400, 401, 402, 404, 408, 411
acidic, 153
acidity, 181
acoustic, xvi, 62, 262, 331, 333, 343, 344, 345, 346, 347, 348, 349, 351, 353, 354, 355, 360, 367, 371, 380
acoustic signals, 262
acoustical, 164
acromion, 449
actin, 269
acute, 75, 76, 77
adaptation, 79, 113, 114, 119, 195, 201, 211, 213, 215, 302, 315, 396, 397, 423, 439, 484, 486
administration, 64
adult, 3, 20, 41, 42, 73, 84, 139, 149, 152, 228, 232, 308, 312, 367, 430, 433, 447, 450, 458, 459, 462
adults, xiv, 33, 152, 221, 222, 224, 226, 232, 257, 301, 307
aggregation, 133, 151, 268
aggression, 30, 33, 41, 48, 64, 151
aggressive behavior, 14
agility, 7
agricultural, 61, 379, 389, 438
air, xviii, 279, 392, 431, 446, 459, 462, 474, 482, 485
alcohol, 134, 166
algorithm, 104, 109, 115, 121, 289, 291, 405
alkaline, 288, 347
alkaline phosphatase, 288
allele, 121, 122, 123, 124, 125, 127, 139, 167, 168, 169, 170, 187, 189, 398, 399, 401, 408, 413, 414
alleles, xiii, xvii, 117, 118, 121, 124, 125, 126, 135, 139, 167, 168, 169, 190, 395, 396, 398, 399, 400, 401, 405, 408, 409, 410, 411, 412, 413, 416
allies, 201, 209
allopatric speciation, 182, 208
alpha, 414
alternative, 27, 44, 104, 112, 185, 270, 331, 383, 391, 442
alternative hypothesis, 44, 185
alternatives, xvi, 323, 330, 336
alters, 163
alveoli, 461
amino, 118, 121, 122, 126, 396, 398, 399, 400, 401, 404, 408, 411
amino acid, 118, 121, 122, 126, 396, 398, 399, 400, 401, 404, 408, 411
amino acids, 121
amplitude, 111
ampulla, 309
analysis of variance, 224
anatomy, 72, 76, 79, 209, 215, 440, 447, 485
Andes, 31, 154, 184, 188, 294
annealing, 103, 114, 166, 295
annual rate, 360
ANOVA, 224, 228
anoxic, 2
anthropogenic, xiv, xvi, 3, 14, 18, 20, 30, 56, 67, 113, 154, 221, 222, 272, 286, 323, 361, 379, 381, 386, 388, 389, 424
anthropological, 248
antibiotic, 134, 166

antigen, 124, 130, 396, 404, 411, 413, 414, 415, 416
aquaculture, 369
aquatic habitat, xii, 29, 30, 44, 46, 47, 61, 422
aquatic habitats, 422
arginine, 122
argument, 63, 182, 268
arsenic, 61
articulation, 197
Asia, xix, 21, 27, 50, 67, 68, 69, 129, 186, 187, 191, 192, 217, 296, 338, 372, 373, 390, 392, 421, 422, 437, 439, 441, 442, 443, 446, 447, 456, 462, 480, 482
Asian, 25, 63, 341, 363, 372, 373, 375, 381, 388, 391, 392, 394, 416, 435, 439, 442
assessment, 67, 114, 130, 229, 233, 246, 271, 272, 281, 324, 327, 337, 390, 391, 394, 420, 442
assignment, 202, 204, 455, 477, 481
asymmetry, 200, 450, 462, 482
asymptotic, xiii, 96, 131, 137, 138, 139, 143, 146, 147, 150, 309
Atlantic, xviii, 6, 98, 99, 130, 158, 181, 184, 191, 201, 211, 222, 230, 238, 262, 270, 273, 282, 303, 304, 307, 317, 318, 319, 320, 323, 324, 327, 333, 334, 335, 336, 337, 425, 446, 453, 456
Atlantic Ocean, 184, 238, 303, 318, 336
attachment, 314
attacks, 152
attitudes, 386
attractors, 63
autonomy, 232, 306
availability, xii, 17, 29, 30, 46, 57, 256, 258, 272, 325, 358, 420, 434, 436
averaging, 433
awareness, 2, 65, 331, 387, 420, 436, 438

B

BA1, 106, 111
babies, 364
back, 33, 162, 169, 195, 242, 301, 325, 344, 353, 354, 358, 367, 384, 429
bacteria, 2, 396
banking, 72, 77
banks, 154, 163, 330, 362
barrier, xiii, 58, 101, 109, 111, 118, 181, 184, 252, 257, 269
barriers, 64, 105, 108, 112, 113, 115, 163, 208, 353, 387, 426, 436
base pair, xiii, 101, 105, 132, 287, 399
basic needs, 331
Bayesian, 317
beaches, xv, 247, 251, 256, 258
beliefs, 42, 63
bending, xii, 71, 72, 74, 75, 77, 79
benefits, 118, 238, 266, 335
bias, 112, 168, 222, 224, 275, 312, 440
bifurcation, 169, 411, 471
binding, xiii, 28, 117, 121, 124, 217, 404, 411, 414, 425, 443
binomial distribution, 169, 170
bioaccumulation, 389
bioavailability, 334
biodiversity, 15, 25, 65, 370, 374, 388, 390, 438, 439
biogeography, 208
biological processes, 245
biomass, 30, 44, 111, 163, 325
biometric, xiii, 83, 134, 166
biomonitoring, 439
biopsies, 120, 134, 135, 166, 275
biopsy, 134, 166, 274, 275, 279
biota, 429, 443
biotic, 163, 228
birds, 206, 238
birth, 6, 8, 48, 231, 262, 273, 310, 311, 312, 314, 368, 430
birth rate, 312
births, 47, 48, 311, 366
bleeding, 42, 64
blocks, 399, 409, 411
blood, 368
boats, 62, 65, 66, 67, 154, 224, 240, 241, 243, 244, 326, 330, 346, 360, 361, 364, 378, 381, 385, 387, 431
body mass, 312
body size, 6, 8, 57, 202, 207, 430, 433, 449
body weight, 308, 311
bootstrap, 104, 110, 112, 136, 168, 188, 271, 290, 399, 405, 425
borderline, 174
bottleneck, xiv, 18, 124, 132, 158, 161, 162, 165, 167, 168, 171, 172, 174, 178, 179, 180, 182, 189, 344, 346
bottlenecks, xiv, 104, 161, 171, 187
brain, 181, 196, 198, 210, 450, 467
breathing, 42
breeding, xiii, xvii, 5, 14, 20, 65, 127, 131, 136, 150, 155, 311, 357, 363, 365, 367, 368, 369, 377, 383, 386, 387, 388, 434
bubble, 306
buffalo, 416
buffer, 103, 120, 166, 167, 398
bureaucracy, 389
bypass, 63

C

cadmium, 324, 334
calf, xv, 64, 84, 93, 247, 248, 252, 253, 256, 257, 275, 312, 365, 420, 433, 434
calving, 46, 47, 48, 260, 282, 305, 311
capillary, 288
cargo, 239, 241, 242, 243, 244, 347, 348, 349
cartilage, 465
CAS, 281, 359, 362, 366, 367, 370, 447, 458, 462, 464
case study, 186, 371
cast, 196, 199, 241, 312, 368
CAT, 120
catchments, 421
catfish, 84, 162, 164, 224, 391, 436
cats, 148, 159
cattle, 397, 410, 416
Caucasus, 455, 479, 481
CCC, 167
cell, 347
cement, 95, 98
cephalization, 2
cephalopods, 8, 307
channels, 13, 19, 30, 31, 38, 41, 44, 62, 63, 111, 148, 153, 163, 231, 292, 293, 379, 389, 429, 434
chemical properties, 113
chemicals, 2, 438
Chile, 101, 199, 201, 202, 211, 213, 214, 285
chimpanzee, 132
chlordane, 437
chlorine, 61
chloroform, 103, 135, 166
chordata, 5, 6, 8, 10, 15, 18, 20, 21
chromium, 362
chromosome, 57, 118, 124, 182
chromosomes, 165, 167, 169
CITES, 14, 61, 69, 102, 125, 186, 222, 238, 286, 435
classes, 163, 312
classical, 424
classification, xix, 48, 192, 194, 201, 206, 217, 270, 271, 281, 328, 424, 446, 447, 453, 477, 487
clay, 435
climate change, 435
cloning, 121, 159, 192, 398
close relationships, 448
clusters, 105, 133, 151, 267, 369, 392
Co, xiii, xvi, 14, 15, 16, 22, 27, 62, 64, 85, 98, 101, 106, 107, 108, 109, 110, 111, 120, 152, 154, 158, 166, 174, 186, 273, 285, 288, 290, 291, 293, 436
coastal areas, 2, 230, 258
coastal zone, 303
codes, 290
coding, 476
codon, 399
codons, 124, 400, 401
cohesion, 268
cohort, 275
collaboration, 96, 186, 233, 245, 359, 360
collisions, 20, 361, 379
colonization, xiv, 14, 112, 154, 161, 164, 165, 177, 178, 179, 273
commerce, 20, 63
commodity, 272
commons, 330, 335
communication, xvi, 48, 62, 112, 343, 353, 354, 362, 365, 426, 430, 433
communities, xv, 15, 51, 63, 64, 65, 96, 155, 164, 186, 222, 240, 247, 255, 293, 313, 330, 331, 336, 382, 386, 389, 436, 438
community, xvii, 3, 58, 63, 111, 157, 253, 358, 370, 377, 383, 384, 386, 387, 436
compatibility, 130
compensation, 331
competition, 230, 232, 275, 305, 307, 308, 311, 330, 384
competitor, 62, 84
compilation, 3
complement, 241
complexity, 77, 163, 238, 331
components, xv, 133, 163, 237, 238, 244
composites, 206, 460
composition, xii, xv, 29, 33, 67, 141, 163, 257, 261, 289
compounds, 25, 61, 334, 437, 440, 443
concentrates, 45
concentration, 118, 238, 324, 437
concussion, 164
confidence, 168, 171, 172, 174, 180, 203, 288, 292, 399, 433
confidence interval, 168, 171, 174, 399
confidence intervals, 168, 399
conflict, 238, 370, 388
connective tissue, 72, 430, 431
connectivity, xvi, 109, 286, 292, 293
consensus, 104, 119, 121, 122, 425, 433
constraints, xvi, 38, 306, 323
construction, xvi, 4, 6, 7, 14, 18, 20, 56, 62, 63, 102, 163, 164, 182, 199, 222, 243, 286, 343, 344, 353, 354, 380, 389, 420, 422, 436
consumption, 61, 63, 162, 308
contaminant, 62
contaminants, 56, 61, 276, 288
contamination, 7, 61, 62, 102, 133, 163, 164, 272, 324, 336, 412, 437
contingency, 249

Convention on International Trade in Endangered Species, 14, 102, 286
convergence, 266, 396, 399, 409, 411, 414, 416
conversion, 399, 409, 410, 416
convex, 197, 457, 463, 471
copper, 324
copulation, 48, 368
correction factors, 381
correlation, xv, 88, 89, 90, 93, 95, 105, 112, 133, 136, 137, 143, 148, 149, 150, 168, 180, 225, 229, 232, 237, 240, 241, 242, 244, 245, 311
correlation coefficient, 88, 150, 225, 229, 232, 241, 242, 244, 245
correlations, 88, 93, 94, 95, 159, 275
costs, 164, 245, 387
countermeasures, 371
couples, 126, 127, 128, 133
coupling, 275
covering, xiii, 18, 60, 101, 183, 306, 308
CPU, 347
cranial nerve, 466
cranium, 200, 266, 447, 450, 452, 456, 457, 459, 460, 463, 464, 465, 468, 469, 470, 473, 475, 477
CRC, 190
critical analysis, 20
critical value, 172
crocodile, 64
cross-validation, 88, 93
crown, 194, 425, 456, 472, 477
crustaceans, 8, 18, 307, 308, 420
crystalline, 430
CT scan, 459, 460, 483
CTA, 120
cultivation, 378
culture, 22, 245
cycles, 103, 120, 163, 166, 167, 287, 310, 368, 398, 423, 437
cycling, 103, 120, 397
cystic fibrosis, 190
cyt-b, 165
cytochrome, xvi, 28, 189, 217, 266, 267, 270, 271, 280, 281, 282, 285, 286, 287, 417, 423, 425, 443

D

danger, 62, 258
data analysis, 156, 224, 414
data set, 168, 171, 190, 289, 397
database, 126, 274
dating, 182, 270, 358
dead zones, 2
death, 33, 41, 62, 84, 273, 309, 311, 312, 365, 366, 379, 391
death rate, 273, 312
deaths, 62, 365, 366, 379
debt, 243, 330
debts, 243
decision makers, 438
decomposition, 40
definition, 48, 136, 311, 447
deforestation, 2, 18, 64, 162
deformation, 196
degradation, xvi, xvii, xviii, 2, 9, 19, 61, 65, 232, 272, 323, 325, 377, 379, 419, 420, 424, 437
degrading, 368
demersal fisheries, 25, 335
demographic change, 165, 167, 172, 181
demography, 162, 170, 276
denaturation, 120, 287, 397
density, 4, 44, 45, 46, 133, 152, 163, 184, 222, 231, 306, 313, 325, 328, 352, 353, 359, 361, 363, 367, 379
dependent variable, 88
deposition, 98
deposits, xviii, 190, 202, 206, 207, 208, 209, 211, 425, 434, 446, 450, 455, 456, 459, 480, 481, 482
depressed, 201, 463
depression, 163
derivatives, 61
desert, 127
destruction, 7, 162, 164, 286, 332, 435
detection, 44, 266, 269, 274, 349, 351, 353, 354, 355, 358, 370, 416
developed countries, 330, 331
developed nations, 330
developing countries, 330, 331
developing nations, 330, 331
deviation, 170, 172, 226, 227, 228, 292
diatoms, 308
diet, 23, 24, 38, 62, 67, 73, 95, 153, 163, 232, 233, 259, 307, 325, 332, 421, 437
dietary, 208, 325, 422
dietary habits, 208
differentiation, 27, 52, 69, 104, 105, 107, 108, 112, 113, 129, 156, 158, 179, 188, 191, 266, 267, 268, 269, 271, 273, 277, 292, 293, 305, 411
dimorphism, 8, 96, 151, 157, 274, 310, 320
diploid, 113
direct action, 61
disappointment, 386
discordance, 184
diseases, 125, 245
disenchantment, 53, 246
dispersion, 44, 112, 165, 184, 185
displacement, 258
disposition, 139, 243

dissolved oxygen, 153
divergence, 26, 56, 75, 113, 121, 122, 124, 170, 179, 182, 197, 210, 259, 270, 281, 400, 410, 411, 413, 425, 461
diversification, 182, 281, 480
DNA polymerase, 103, 397
dogmas, 328
dominance, 48, 124, 303
donations, 483
drainage, 184, 189, 280, 359, 378, 388, 422
drying, 462
duplication, 28, 410, 413, 416
duration, 349, 351

E

ears, 279, 361
Earth Science, 214, 216, 314
eating, 308
ecological, xiii, xv, 3, 50, 70, 101, 112, 113, 115, 118, 132, 153, 181, 190, 194, 201, 231, 238, 247, 258, 268, 275, 303, 305, 324, 331, 335, 383, 421, 437, 438
ecological systems, 154
ecologists, 3, 132
economic assistance, 125
economic change, 245
economic development, 344, 368, 369, 370, 383
ecosystem, 63, 65, 238, 303, 308, 330, 331, 371, 386, 396, 423, 437
ecosystem restoration, 331
ecosystems, 2, 51, 84, 374, 420, 423, 429, 435
Ecuador, 7, 13, 22, 28, 30, 39, 40, 53, 56, 67, 70, 125, 153, 155, 156, 159, 164, 165, 184, 186, 187, 262, 272, 276, 283, 422
EDGE, 390
education, 332, 413, 438
effluent, 133
effluents, 362, 437
election, 127, 313, 314, 374
electricity, 164
electrophoresis, 121
embryo, 127
employment, 65, 240
encouragement, 22
entanglement, xii, 7, 18, 29, 30, 42, 164, 186, 286, 325, 327, 336, 361, 420, 436
entrapment, 186
environment, xii, xiii, 2, 25, 29, 30, 52, 71, 101, 111, 112, 115, 118, 157, 202, 204, 206, 207, 208, 365, 368, 384, 386, 392, 396, 397, 422, 437
environmental change, 3, 118, 208, 303
environmental characteristics, 257
environmental effects, 266
environmental factors, 231
epidemiology, 130
equilibrium, 134, 135, 140, 141, 167, 168, 180, 191
erosion, 62
estimating, 306, 312, 314, 415
estimator, 93, 167
estuaries, 8, 255, 256, 262, 303, 360
estuarine, xviii, 24, 26, 27, 98, 126, 184, 201, 259, 264, 282, 303, 305, 316, 320, 422, 423, 446, 481
estuarine systems, 422
ethanol, 103, 286
eutrophication, 2, 163
evolution, xvii, 3, 72, 80, 104, 114, 115, 118, 119, 128, 129, 130, 157, 158, 160, 182, 185, 187, 188, 206, 210, 271, 279, 395, 396, 397, 412, 414, 415, 417, 419, 421, 423, 424, 425, 439, 447, 485
evolutionary process, xiv, 127, 193, 278
examinations, 368
exclusion, 157, 203, 448, 455
exercise, 180, 328
exonuclease, 288
expansions, xiv, 161, 162, 165, 173, 175, 180, 190
expenditures, 335
experimental design, 225, 248
exploitation, 61, 325, 330, 337, 378, 379, 421, 435
explosions, 164
explosives, 20, 164, 361, 379
exposure, 68, 196, 232, 362, 467
exposure, 67
extinction, 3, 20, 27, 28, 119, 192, 276, 283, 293, 303, 355, 358, 361, 362, 364, 373, 379, 380, 381, 385, 386, 387, 388, 389, 390, 392, 420, 421, 423, 424, 443
extraction, 102, 103, 134, 135, 166, 286, 288, 420
extrapolation, 306
eye, 262, 289, 430, 431, 462, 474, 482
eyes, 6, 8, 18, 20, 64, 162, 222, 242, 243, 286, 301, 430

F

failure, 130, 210, 367, 381, 388
false negative, 361
FAO, 25, 234, 238, 280, 315, 335
farming, 163
farms, 148
fascia, 431
fat, 14, 62, 63, 224, 243, 245
fauna, 62, 184, 203, 204, 214, 215, 216, 338, 378, 420, 440
feces, 368

feeding, xv, 73, 75, 76, 244, 247, 248, 252, 253, 254, 256, 257, 258, 260, 308, 324, 364, 368, 420, 421, 437
females, xiii, 3, 15, 18, 20, 33, 38, 85, 112, 113, 131, 136, 137, 139, 143, 147, 149, 150, 154, 162, 226, 228, 232, 274, 275, 276, 291, 309, 310, 312, 364, 365, 366, 368, 430
fertility, 119, 362, 379
fertilization, 126, 127
fertilizers, 163, 437
fetal, 33, 38
fetus, 430
fiber, 413
fibrosis, 190
fidelity, xv, 7, 13, 52, 68, 113, 148, 154, 233, 247, 259, 260, 275, 277, 282, 386
filters, 372
finance, 243
financial support, 113, 258
fingerprints, 416
fire, 14
firearms, 84, 164
fish production, 20, 163, 364, 369
fishers, 230
fission, 152, 281
flare, 467
flexibility, 72, 73, 74, 75, 78, 79
flight, 306
float, 225
floating, 163
flood, 31, 48, 111, 163, 384, 433, 437
flooding, 63, 118, 258, 433, 437
flora, 420
flora and fauna, 420
flow, xiii, xvi, 7, 13, 19, 57, 101, 102, 105, 107, 109, 111, 118, 141, 148, 150, 155, 156, 157, 159, 163, 181, 242, 269, 273, 276, 283, 285, 289, 292, 293, 305, 311, 373, 386, 389, 420, 434, 438
fluctuations, xii, 29, 30, 44, 65, 231, 308, 434
fluvial, 28, 188, 203, 206, 209, 248, 255, 303
FMA, 328, 329, 331
FMAs, 328
focusing, xv, 261
folklore, 280
food, 2, 17, 57, 61, 62, 63, 112, 133, 152, 163, 230, 256, 258, 303, 307, 324, 331, 358, 384, 423, 435, 436, 437, 438
foramen, 87, 266, 451, 456, 457, 466, 467, 471, 474, 475, 477
forests, 22, 31, 46, 57, 77
formamide, 398
fossil fuels, 2
founder effect, 310
fragmentation, 5, 63, 162, 267, 345, 353, 422
freedom, 254, 256
freezing, 243
frequency distribution, 189, 309
fresh water, xiii, xviii, 50, 101, 187, 204, 208, 248, 360, 445, 447, 456, 459, 480, 481, 482
fusion, 72, 79, 85, 152, 275, 281

G

garbage, 60
gas exploration, 18, 56
gastrointestinal, 309, 313
gauge, 223, 224, 225, 231
gel, 121, 167, 398
gels, 135, 167
GenBank, 289, 397, 399, 400, 401, 405
gender, xiv, 221, 224, 226, 227, 228, 232, 282
gene pool, 142, 148, 436
genealogy, xvi, 169, 286, 290, 291, 292
generation, 108, 126, 136, 137, 139, 148, 150, 179, 180, 289, 291
genes, xvii, 27, 57, 69, 113, 118, 119, 124, 128, 129, 134, 136, 158, 165, 191, 209, 269, 286, 395, 410, 411, 413, 414, 424
genetic diversity, 15, 22, 102, 125, 129, 134, 154, 156, 179, 185, 188, 273, 276, 282, 307, 410
genetic drift, 142, 156, 163, 182, 273
genetic marker, 287
genetics, xiii, 3, 8, 22, 57, 64, 104, 131, 132, 143, 149, 154, 155, 156, 157, 179, 181, 183, 268, 276, 277, 280, 281, 318, 414, 415
genome, 115, 116, 187, 210, 217, 400, 401, 443
genomes, 132
genomic, 103, 120
genomics, 127
genotype, 129, 130, 141, 275, 398
genotypes, 119, 125, 136, 266
genre, 50
geography, 113, 214
gestation, 262, 310, 311
gill, xv, 33, 42, 43, 45, 162, 164, 237, 308
girth, 33, 151
glaciation, 179
Global Positioning System, 431
global warming, 2
God, 2
gold, 61, 64, 163
goodness of fit, 88
government, 61, 113, 293, 330, 331, 363, 369, 370, 436, 438
GPS, 33
graduate students, 3

grassland, 30
group size, xii, xv, 29, 30, 33, 38, 46, 133, 151, 247, 248, 249, 252, 253, 254, 255, 257, 258, 344, 345, 349, 355, 427, 433
grouping, 104, 133, 194, 209, 270, 271, 424
growth, 2, 6, 9, 25, 38, 51, 84, 86, 96, 97, 98, 105, 115, 168, 171, 172, 178, 179, 180, 188, 190, 191, 222, 231, 248, 275, 305, 310, 312, 313, 314, 317, 320, 321, 330, 416, 437, 440
growth rate, 6, 9, 38, 179, 222, 231, 312, 313
guidelines, 65, 268
guilty, 382
gums, 38
guns, 84
Guyana, 13, 56, 154, 222, 273

H

HA1, 261
haplotype, xvi, 57, 104, 105, 106, 107, 109, 110, 111, 112, 170, 177, 178, 179, 268, 285, 289, 290, 291, 292, 293, 305, 344, 412
haplotype analysis, 344
haplotypes, xiii, xvi, 101, 102, 104, 105, 106, 107, 109, 110, 111, 112, 170, 177, 178, 181, 182, 183, 267, 268, 269, 273, 285, 289, 290, 291, 292, 361
harassment, 61
harbour, 52, 158, 191, 276, 282, 380
harm, 42, 229
harvest, 61
health, 2, 14, 65, 73, 324, 331, 362, 367, 368, 379, 384
health problems, 14, 73
health status, 324
heart, 2
heat loss, 310
heavy metal, 2, 7, 19, 26, 61, 163, 324, 335, 336
heavy metals, 2, 26, 61, 163, 324, 335, 336
height, 30, 87, 381, 433, 461, 468, 472, 473
hematology, 367
hemisphere, xviii, 209, 446, 454
hepatitis, 130
hepatitis C, 130
heterogeneity, 108, 109, 133, 148, 159, 244, 434
heterogeneous, 245
heterozygosity, 119, 126, 127, 139, 182
heterozygote, 126, 141, 146, 150
heuristic, 289
hexachlorobenzene, 437
hexachlorocyclohexane, 437
high resolution, 460
high-frequency, 345, 347, 372
histogram, 411
HIV, 126
HIV-1, 126
HLA, 126, 127, 128, 129, 404, 415, 416
Holocene, 181, 182, 183, 273, 378
homogeneity, 148, 273
homology, 121, 399
homozygosity, 119, 121, 167
homozygote, 139, 140, 141
hookworm, 126
horizon, 464
hormone, 368, 371
hormones, 368, 371
human activity, 6, 45, 358, 364
human genome, 187
human leukocyte antigen, 127, 404
humans, 2, 33, 42, 61, 63, 116, 118, 119, 130, 148, 166, 179, 180, 331, 363, 397, 415
humerus, 431
humidity, 181
hunting, 14, 19, 61, 257, 381, 420, 430
husbandry, 383
hybrid, 245
hybridization, 116, 269
hydro, 124, 429
hydrologic, 19, 163, 362
hydrological, 162, 185, 208, 420
hydrology, 116, 349, 353
hydrophilic, 124
hydrophobic, 122, 124
hydrophone, 347
hypothesis, 44, 48, 116, 181, 184, 187, 192, 195, 207, 208, 275, 304, 415, 423

I

ice, 225, 232, 240, 241, 242, 244, 365, 366, 398
identification, xii, 23, 25, 29, 31, 33, 37, 49, 52, 66, 68, 104, 126, 186, 196, 212, 272, 274, 275, 277, 280, 282, 321, 345, 361, 383, 416, 439
identity, 121, 269, 397, 412
illumination, 121
images, xii, 4, 19, 21, 44, 71, 73, 74, 75, 453, 460, 469, 475, 483
imagination, 53, 246
imaging, 72
immune response, 118, 128
immune system, 395
immunogenetics, 127, 397
immunological, 396
implementation, xii, xvi, 55, 323, 331, 332, 385, 387, 389
imports, 164
imprisonment, 382

in situ, xvii, 5, 134, 184, 345, 357, 362, 364, 368, 377, 382, 385, 386, 388
in utero, 311
in vitro, 126, 127
in vitro fertilization, 126, 127
in vivo, 124
inbreeding, 155, 163
incentive, 65
inclusion, 108, 193, 447
income, 62, 330, 384
increased access, 47
India, xvii, 3, 19, 25, 27, 191, 264, 389, 391, 419, 420, 421, 422, 425, 426, 428, 434, 436, 437, 438, 439, 440, 441, 442, 443, 462, 479
Indian, 7, 15, 20, 22, 84, 96, 134, 155, 164, 165, 186, 270, 295, 321, 358, 389, 419, 420, 421, 423, 424, 425, 431, 432, 435, 438, 439, 441
Indian Ocean, 270
Indians, 15, 18, 164
indication, 113, 141, 266, 275, 325, 411, 459
indicators, 2, 23, 186, 334
indices, 171
indigenous, 2, 14, 58, 62, 63, 66, 84, 222
indigenous, 15, 61
indigenous peoples, 2
indirect effect, 20, 222
Indo-Pacific, 270, 281, 396
industrial, 231, 238, 337, 362, 379, 386, 389, 437, 438
industrialization, 324, 379
industry, 2, 6, 231, 331, 354
infection, 127, 130, 309, 367
infections, 26, 48, 63
infectious, 119, 127
infectious disease, 119, 127
inferences, 158
infertility, 127
ingestion, 323, 336
inherited, 274, 275, 276
injuries, 365, 384
injury, 99, 257, 372
insertion, 95, 210, 441
insight, 48, 130
instability, 231
institutions, 96, 293, 370, 381
instruction, 398
instruments, 243
integration, 2, 182
integrity, 150, 437
interaction, xiv, xv, 33, 118, 187, 221, 222, 230, 231, 234, 237, 240, 244, 257
interactions, 27, 48, 60, 67, 73, 98, 119, 222, 233, 234, 238, 246, 249, 254, 256, 257, 293, 321, 335, 338, 384
interdisciplinary, 248
interface, 116, 130, 192, 248, 255
intermediaries, 268
interrelationships, 193, 217
interval, 168, 171, 174, 310, 311, 314, 347, 348, 431, 487
intervention, 256, 364
interviews, 31, 59, 62, 248, 326, 328
intestine, 309, 313, 319, 430
intrinsic, 312
intron, 27, 52, 57, 69, 124, 126, 129, 191
introns, 118, 165, 182, 269
invasive, 157
invertebrates, 203, 308, 422, 437
investigations, 26, 52, 68, 115, 129, 158, 159, 190, 192, 213, 216, 315, 391, 442
investment, 244, 311, 331
iron, 347
irrigation, 389, 436, 437
island, 158, 190, 289
isolation, 5, 56, 107, 113, 150, 157, 266, 273, 293, 305, 345, 420, 423
isopods, 308
isotopes, 421
isozyme, 155

J

Japan, 343, 346, 347, 354, 360, 383, 397, 398, 406, 417, 449, 450, 478, 481, 485, 486
Japanese, 187, 358, 400, 416, 486
jaw, 42, 62, 307, 308, 430, 450, 481
joining, 104, 106, 110, 114, 121, 170, 177, 186, 399, 405

K

kidney, 336, 338, 437
kidneys, 324
killing, xii, 7, 29, 30, 164, 382, 389, 420, 435, 436
Korea, 121

L

labeling, 172
lactating, 38, 152, 307
lactation, 33
lagoon, 14, 42, 134, 135, 139, 141, 142, 147, 148, 149, 150, 257

lakes, 16, 30, 31, 33, 34, 35, 36, 38, 43, 44, 45, 46, 111, 133, 152, 181, 231, 258, 344, 358, 359, 360, 362, 363, 368, 369, 379, 390, 422
lamina, 448, 451, 456, 460, 469
laminar, 449
laminated, 204
land, xii, 55, 68, 111, 198, 206, 371, 392, 420, 421
language, 311
large intestine, 309, 430
larval, 308
Latin America, 24, 51, 129, 158, 313, 314, 315, 316, 317, 318, 319, 320, 330, 331, 332, 334, 335, 336, 337, 338, 339
Latin American countries, 330
law, 18, 61, 183, 293, 330
law enforcement, 293, 330
laws, xii, xvii, 55, 60, 61, 64, 164, 238, 377
leadership, 243
leakage, 62
learning, 307
learning process, 307
legal protection, xii, 55
legislation, xvii, 164, 377, 381
lesions, 61, 64
leukocyte, 127, 130, 404
life expectancy, 311
life span, 6, 8, 231, 309, 311
lifespan, 210
lifetime, 311, 347
likelihood, 136, 190, 244, 267, 310, 327
limitations, 79, 267, 312, 354, 422
linear, xiii, xiv, xv, 13, 31, 83, 88, 90, 93, 95, 96, 122, 133, 153, 221, 224, 242, 247, 249, 361
linear model, 249
linear regression, xiv, 88, 90, 93, 95, 221, 224, 242
lingual, 207
linguistic, 115
liniment, 436
linkage, 369
links, 2
lithium, 347
liver, 336, 338, 437
location, 15, 17, 37, 44, 57, 72, 111, 153, 209, 240, 265, 266, 273, 288, 291, 307, 309, 311, 430, 433, 436, 449, 469, 475, 477
locomotion, xii, 71, 72
locus, xvii, 121, 122, 124, 128, 168, 169, 278, 289, 395, 398, 399, 400, 401, 408, 409, 410, 411, 412, 414, 415
long distance, xiii, 17, 101, 112, 344
long period, 313, 383, 412
love, 6, 7, 14, 22, 84, 162, 222, 272, 280, 282
lumbar, 71, 73, 74, 78, 79, 202
lung, 42, 63
lungs, 309
lupus, 132
lying, 152, 423

M

major histocompatibility complex (MHC), xiii, xvii, 14, 28, 117, 118, 119, 120, 124, 125, 126, 127, 128, 129, 130, 395, 397, 398, 399, 400, 401, 402, 403, 404, 405, 408, 409, 410, 411, 412, 413, 414, 415, 416, 417
malaria, 127
males, xiii, 3, 8, 14, 18, 20, 85, 112, 113, 119, 127, 130, 131, 136, 137, 143, 147, 149, 150, 154, 226, 228, 232, 274, 275, 276, 309, 310, 364, 365, 368, 430
mammal, 2, 22, 73, 80, 98, 119, 125, 130, 148, 159, 189, 198, 204, 215, 233, 238, 311, 330, 339, 358, 370, 378, 380, 387, 421, 440, 486
mammalian, 23, 24, 50, 155, 214, 260, 279, 295, 315, 317, 485, 487
management, xii, xv, xvi, 25, 29, 49, 98, 111, 113, 128, 164, 190, 238, 261, 276, 281, 282, 286, 318, 321, 327, 328, 330, 331, 334, 337, 338, 339, 345, 355, 364, 375, 383, 389, 417, 436, 440
mandible, xix, 198, 201, 206, 217, 446, 450, 454, 456, 457, 461, 463, 464, 465, 470, 472, 473, 477, 482
mandibular, 85, 87, 91, 197, 198, 201, 203, 206, 207, 302, 453, 465
mangroves, 251
manipulation, 134, 166, 475
manpower, 387
manufacturer, 397
marine environment, 193, 203, 204, 422
marine mammals, 81, 98, 114, 129, 159, 186, 187, 213, 280, 315, 325, 331, 333, 334, 336, 483, 488
market, 6, 15, 63, 84, 162, 222, 239, 240, 241, 243, 245, 272
market prices, 243
marketing, xv, 237
markets, 6
Markov chain, 67, 135, 139
Maryland, 200, 211, 214, 445, 453, 455, 457, 458, 460, 478, 479, 481, 482, 484, 485, 486
mask, 133, 293, 362
masseter, 95
mastication, 95
maternal, 111, 151
matrix, 88, 138, 312, 328, 405, 465, 475, 480
maxilla, 85, 197, 449, 459, 463, 466, 467, 468, 469, 470, 471, 472, 473, 477

maxillary, 201, 202, 206, 451, 456, 457, 458, 459, 460, 462, 463, 466, 472, 474, 477, 482
Maximum Likelihood, 290
measurement, 76
measures, xiii, xvi, xvii, 20, 83, 84, 85, 86, 96, 136, 170, 185, 208, 225, 275, 323, 357, 362, 363, 367, 368, 370, 381, 382, 386, 387, 388, 421
meat, 42, 63, 245, 378, 389, 435, 436
media, 422
median, 104, 121, 170, 177, 301
melon, 7, 262, 301, 433
melting, 433
membership, 281
memory, 347
men, 15, 225, 241, 243
merchandise, 382
mercury, 14, 18, 23, 24, 25, 26, 61, 65, 66, 67, 68, 102, 115, 163, 186, 324, 334, 338, 362
metabolic, 310, 325
metabolic rate, 310
metals, 2, 25, 26, 61, 163, 325, 335, 336
metric, 305, 437
microorganisms, 163
microsatellites, xiii, xiv, 113, 118, 124, 130, 131, 132, 135, 136, 139, 140, 141, 142, 156, 161, 162, 165, 166, 168, 171, 172, 177, 179, 187, 191, 269, 274, 275, 397
microscopy, 97
migrant, 147, 148, 149, 150
migrants, 104, 108, 142, 149, 289
migration, 63, 64, 111, 135, 138, 143, 147, 149, 150, 156, 158, 178, 184, 188, 273, 276, 293
milk, 38, 115, 243, 245, 368
mining, 18, 20, 56, 61, 362
mitochondrial DNA, xiii, 16, 56, 101, 115, 116, 118, 177, 188, 275, 286, 287, 293, 302, 305, 416, 417
mixing, 113, 245, 303, 423
MMT, 347
mobility, xii, 13, 71, 72, 73, 74, 75, 76, 77, 79
modeling, 328
models, xiii, 83, 88, 89, 93, 95, 104, 131, 136, 143, 148, 150, 168, 188, 191, 312
molasses, 181
molecular biology, 3
molecular dating, 182
molecular markers, xv, 22, 118, 122, 124, 125, 139, 154, 165, 167, 185, 261, 276
molecular weight, 135
molecules, 215, 397, 399, 400, 415, 439, 440, 486
monsoon, 19, 421, 429, 433
Morbillivirus, 24
morning, 42, 64
morphological, xi, xii, xvii, 1, 3, 7, 16, 57, 71, 72, 79, 118, 183, 194, 195, 199, 203, 207, 209, 262, 266, 268, 269, 270, 280, 302, 419, 421, 424, 429, 434, 477
morphology, 26, 59, 72, 79, 80, 81, 116, 163, 194, 195, 196, 198, 201, 206, 208, 209, 210, 211, 215, 246, 266, 268, 270, 276, 314, 319, 422, 457
morphometric, xiii, 14, 26, 59, 83, 84, 88, 89, 95, 96, 158, 248, 259, 262, 265, 268, 281, 305
mortality rate, 276
mosaic, 13, 414
mothers, 66
motors, 49, 245
mountains, 118, 154
mouse, 126
mouth, xvi, 7, 8, 15, 42, 111, 134, 184, 241, 248, 251, 253, 264, 267, 269, 306, 343, 344, 346, 353, 354, 363, 367, 368, 426, 430, 449
movement, xii, xvi, 29, 30, 37, 57, 63, 75, 149, 252, 253, 258, 305, 343, 344, 346, 353, 354, 355, 374
mtDNA, xiii, xiv, 69, 101, 102, 103, 104, 105, 106, 109, 110, 111, 112, 113, 124, 125, 129, 158, 161, 162, 165, 166, 170, 174, 177, 178, 181, 182, 183, 190, 191, 273, 280, 283, 291, 305, 321, 344, 361
multiple regression, xiii, 83, 88, 90, 91, 94, 95, 242
multiple regression analysis, 88
multiples, 147
multiplexing, 277
multiplication, 170
multivariate, 208, 305
muscle, 5, 95, 231, 430, 431, 437
muscle mass, 5, 231
mutation, 69, 106, 116, 124, 167, 168, 169, 170, 179, 190, 192, 409, 416
mutation rate, 69, 170, 179, 409, 416
mutations, 115, 121, 169, 172, 188, 414
mycology, 81

N

nares, xiii, xviii, 83, 87, 89, 91, 95, 183, 196, 199, 446, 453, 454, 463, 477
narratives, 252
nation, 330
National Academy of Sciences, 24, 114, 115, 126, 127, 129, 187, 190, 192, 213, 216, 372, 414, 415, 439, 441, 487
National Marine Fisheries Service, 187
National Research Council, 113
native species, xii, 55
natural, 3, 30, 48, 51, 58, 64, 65, 114, 124, 159, 183, 188, 238, 248, 275, 277, 283, 306, 308, 327, 329,

362, 363, 364, 368, 369, 383, 384, 385, 386, 388, 392, 417, 462
natural habitats, 369
natural resource management, 238
natural resources, 65, 248
natural selection, 124, 183, 417
NCA, 267, 268, 273
neck, xii, 71, 72, 73, 74, 75, 76, 77, 79, 112, 301, 450
nematode, 309
neonatal, 372
neonate, 38
neonates, 33, 38, 39, 62, 63, 434
Nepal, xvii, 19, 27, 52, 159, 419, 420, 421, 422, 426, 428, 429, 430, 434, 436, 442, 443, 462, 479
nesting, 128
network, 106, 111, 170, 177, 178, 268, 272, 289, 291, 370, 380, 423
newspapers, 245
next generation, 137
NGOs, 436
Ni, 390
Nielsen, 84, 98, 169, 190
NIH, 73
nitrate, 135, 167
NOAA, 129, 282
noise, xvi, 2, 20, 49, 164, 323, 347, 348, 353, 354, 362, 435, 437
non-human, 48, 415
non-random, 150
normal, 2, 224, 411, 473, 474
nuclear, 27, 28, 52, 69, 118, 124, 129, 158, 165, 177, 182, 191, 209, 210, 217, 269, 271, 276, 282, 302, 425, 443
nucleic acid, 209
nucleotide sequence, 121, 400, 401, 405, 411, 439
nucleotides, 287, 288, 399, 400
nursing, 152, 368
nutrient, 4, 31, 461
nylon, 65, 163, 164, 230, 245, 324, 436

O

obligate, 201, 378, 381, 421
observations, 8, 32, 38, 47, 72, 73, 76, 77, 115, 159, 183, 190, 192, 198, 252, 253, 255, 257, 259, 346, 347, 348, 352, 353, 359, 383, 441, 448, 477, 486
occipital regions, 75
oceans, 362, 462
offshore, 304, 330, 334, 384
oil, 2, 7, 15, 27, 42, 56, 62, 84, 133, 163, 286, 324, 378, 389, 391, 420, 435, 436, 442
oil spill, 133, 163, 324
oils, 391
olfactory, 196
olfactory nerve, 196
orbit, 87, 201, 202, 448, 456, 460, 469, 473, 474, 477
organ, 367, 459
organic, 118, 120, 324, 338
organochlorine compounds, 437
organometallic, 25, 440
oscillations, 270
osmotic, 303, 423
ossification, 465
osteology, 80, 217
otters, 77, 80
ovaries, 51, 311, 317
ovary, 311
overexploitation, 424
ovulation, 311, 368
oxygen, 2, 44, 153

P

Pakistan, 422, 424, 447, 458, 462, 479
palate, 204, 205, 207, 448, 456, 469
paleontology, 302
parameter, 27, 147, 169, 240, 313, 320, 328, 338, 399, 405, 408
parameter estimates, 328
parameter estimation, 313
parasite, 118, 119, 238
parasites, 48, 128, 129, 276, 308, 309, 318, 396
parents, 136
parietal, 197
participatory research, xv, 247
partition, 108
passive, xvi, 187, 234, 246, 343, 345, 355, 371, 375, 387
pastoral, 69
paternity, 132, 157, 275
pathogenic, 124, 125, 163
pathogens, xiii, 2, 14, 117, 119, 124, 125, 396, 413
pathways, 396
PCR, 103, 120, 132, 135, 166, 167, 287, 288, 397, 398, 412
penalty, 410
penguins, 324
penis, 6
peptide, xiii, 117, 119, 121, 396, 404, 416
peptides, 121, 124, 414
percentile, 170
perception, 119, 207, 248
perceptions, 63, 119
peritoneal, 430
peritonitis, 127

pesticide, 7, 286, 362
pesticides, 14, 28, 61, 64, 69, 102, 163, 437, 441
pH, 31, 57, 103, 112, 113, 118, 397
pH values, 118
phalanges, 16, 183
phalanx, 183
phenol, 103, 135, 166
phenotypic, 265, 266, 269, 328
phenotypic plasticity, 269
philosophical, 336
phosphate, 483, 488
phosphorus, 18, 61
photographs, 75, 76, 78, 212, 483
photoperiod, 30
phylogenetic tree, 267, 405, 475, 477
phylogeny, xvii, 25, 47, 102, 116, 189, 199, 203, 206, 208, 210, 214, 217, 266, 271, 278, 322, 390, 419, 424, 439, 443, 485
physicians, 127
physicochemical, 57, 124, 125
physicochemical properties, 124
physiological, 2, 3
physiology, 367, 368, 370
phytoplankton, 2
pig, 272
pinhole, 430
planning, 381
plants, 163, 393
plastic, 323
plasticity, 102, 266, 269
plastics, 2
platforms, 346
play, 2, 48, 232, 331
Pleistocene, 111, 181, 182, 207, 208, 216, 270, 303, 320, 390
Pliocene, 7, 56, 111, 181, 184, 189, 199, 201, 202, 203, 209, 211, 213, 215, 270, 303, 423, 440, 441, 486, 487, 488
pneumonia, 367
poison, 381
poisoning, 41, 43
poisonous, 63
Poisson distribution, 133, 168
polar bears, 157
polarity, 269, 311, 469, 475
pollutant, 383
pollutants, 324, 325, 336, 362, 379, 386, 389
pollution, xi, xvi, xvii, xviii, 1, 2, 4, 7, 9, 19, 25, 63, 64, 65, 68, 126, 164, 286, 323, 324, 344, 357, 362, 382, 383, 392, 419, 420, 435, 437, 438
polyacrylamide, 398
polychlorinated biphenyls (PCBs), 324, 333, 437
polymerase, 103, 120, 132, 160, 166, 282, 397
polymerase chain reaction, 103, 132, 160, 282
polymorphism, xiii, 14, 116, 117, 118, 119, 121, 122, 124, 126, 127, 128, 130, 192, 396, 398, 399, 411, 412, 413, 414, 415, 416
polymorphisms, 118, 119, 126, 132, 160, 415
pools, 19, 367, 368, 429, 434, 435
population density, 4, 133, 163, 222, 379
population group, 108
population growth, 2, 9, 105, 115, 168, 171, 172, 188, 190, 222, 312, 313
Population Growth Rate, 312
population size, xi, xii, 1, 18, 28, 55, 65, 113, 142, 154, 167, 168, 170, 179, 180, 185, 188, 192, 231, 273, 276, 279, 320, 344, 357, 359, 360, 365, 371, 374, 386, 390, 393, 396, 420, 421
precipitation, 30, 60
predators, 2, 3, 45, 64, 111, 308, 310, 421
prediction, 88, 97
preference, 57, 129, 258, 304, 434
pregnancy, 33, 130, 210, 310
pregnant, 38, 67, 245, 365, 366, 368, 380, 430, 436
press, 69, 98, 115, 129, 158, 191, 200, 201, 215, 316, 355
pressure, 118, 119, 124, 125, 325, 326, 347, 348, 370, 387, 396, 397, 435
primate, 118, 157, 396, 415
primates, 2, 126, 128, 148, 157, 158, 191
probability, xiv, 90, 131, 136, 140, 141, 147, 149, 150, 169, 276, 288, 306, 328, 354, 370
probability distribution, 328
probe, 354
production, 20, 163, 206, 243, 364, 369, 372, 382
productivity, 118, 362, 364
profit, 243, 330
progeny, 139
program, xvii, 20, 55, 56, 59, 104, 121, 139, 168, 170, 288, 289, 347, 357, 365, 367, 370, 377, 381, 385, 386, 387, 388, 389, 399
propagation, 384, 386
proposition, 64, 204
propulsion, 72
protected area, 18, 31, 49, 61, 64, 383
protected areas, 18, 49, 61, 64, 383
protection, xii, xvii, 20, 55, 60, 65, 119, 127, 164, 286, 332, 355, 357, 363, 365, 367, 368, 369, 371, 374, 377, 382, 383, 386, 387, 390, 421
protein, 28, 62, 126, 217, 416, 425, 443
proteins, 118, 395, 414
protocol, 103, 120, 167, 286, 288, 368, 397
protocols, 104, 121, 288
pseudogene, 399, 401
PSI, 126
public, 2, 65, 238, 370, 382, 387

public awareness, 65, 387
public education, 382
pulse, 311
pure line, 132
purification, 288
PVA, 388
P-value, 399

R

radiation, xviii, 182, 303, 423, 445, 446
radius, 77
rain, 13, 22, 52, 68, 163, 181, 228, 434
rain forest, 22
rainfall, 228
rainforest, 30, 31
random, 84, 133, 136, 148, 150, 257, 289, 411
RAPD, 148, 150, 182
rash, 270
rats, 160
reading, 96, 121
realism, 393
reality, 139, 382
recall, 182
recognition, 119, 124, 126, 128, 248, 268, 271, 277, 381, 387, 396, 411, 477
recombination, 132, 399, 409, 411, 416, 417
reconstruction, 3, 110, 188, 290, 405
recovery, xvii, 20, 192, 326, 332, 377, 381, 383, 385, 386, 387, 388, 389, 390, 392
refining, 388
reforms, 243
refuge, 45, 46, 50, 152, 435
regeneration, 386
regional, 66, 76, 292, 382, 384
regression, xiii, xiv, xv, 83, 88, 89, 90, 91, 93, 94, 95, 96, 97, 221, 224, 229, 237, 242, 303, 423
regression analysis, 88, 90, 91, 93, 224, 229
regression equation, xiii, 83, 90, 91, 92, 95
regression method, 96
regressions, 88, 89, 93, 94, 95, 229
regular, 63, 84, 243, 347, 359
regulation, 293
regulations, xvii, 187, 222, 331, 364, 369, 377, 382
relevance, xv, 261
relict species, 182, 303, 396
reproduction, xii, 25, 29, 30, 39, 40, 47, 48, 50, 51, 52, 53, 97, 128, 154, 232, 248, 311, 312, 315, 317, 321, 383, 439
reproductive activity, 41, 151
reproductive organs, 63
reptiles, 50, 164, 213, 293
reserves, xvii, 357, 359, 362, 363, 369, 370, 377, 382, 384
reservoir, 186, 354, 369, 388
reservoirs, 437
residential, 443
residues, xiii, 25, 117, 124, 324, 399, 400, 401, 404
resistance, 119, 126, 127
resolution, 188, 439, 460
resource availability, 358
resource management, 238
resources, 2, 65, 84, 113, 118, 125, 154, 163, 181, 185, 230, 248, 258, 293, 303, 305, 307, 330, 331, 332, 369, 382, 384, 386, 387, 423, 435, 438
respiratory, 48, 63
respiratory problems, 63
retention, 410
returns, 31
revenue, xv, 237
rice, 378
rings, 382
risk, 163, 164, 224, 257, 321, 328, 360, 362, 386, 393
risks, 384
river basins, xii, 29, 30, 50, 52, 56, 66, 80, 133, 165, 184, 303, 420, 424
river systems, xvii, 21, 31, 377, 381, 388, 419, 420, 421, 423, 424, 435
rocky, 426
rodent, 148
rodents, 119, 148
rolling, 20, 361, 365, 367, 379, 382, 384
runoff, 2

S

safety, 134, 165
saline, 248
salinity, 230, 256, 258
saliva, 368
salmon, 162, 187, 253
salt, xviii, 258, 445
saltwater, 423
sample, xiv, 34, 35, 37, 46, 47, 84, 85, 112, 121, 134, 136, 139, 141, 161, 168, 169, 171, 172, 174, 177, 223, 227, 266, 289, 291, 313, 398, 401, 412, 423
sample variance, 169
sampling, xiii, 38, 44, 46, 131, 132, 134, 135, 136, 139, 142, 148, 150, 153, 157, 158, 168, 223, 225, 226, 231, 239, 249, 271, 275, 287, 288, 291, 292, 347, 368
sand, 4, 163, 308, 344, 429, 434, 435, 437
SAP, 288
scapula, 449

scientific community, 3, 63, 358
scores, 288, 399
sea level, 58, 181, 184, 203, 270, 303, 423, 424, 439, 440
sea-level rise, 423
seals, 99, 280
search, 46, 57, 112, 121, 126, 154, 257, 258, 277, 289, 364, 396, 411, 480
searches, 389
searching, 165
seasonal pattern, xii, 29, 30, 43, 45
seasonality, xii, 29, 39, 44, 46, 47, 48, 50, 51, 52, 230, 231, 242, 262, 371
secretion, 368
sediment, 153, 163, 373, 420, 423, 434
sedimentation, 19, 437
sediments, 118, 163, 185, 198, 203, 206, 208, 437, 450
segregation, 25, 52, 113, 115, 126, 152, 154, 157, 436
selecting, 312
selenium, 324, 334, 338
SEM, 97
semi-natural, xvii, 362, 364, 365, 369, 375, 377, 382, 384, 386, 389, 393
senescence, 311
sense organs, 422
sensitivity, 347
separation, xiii, 56, 101, 106, 111, 124, 163, 177, 179, 182, 267, 345, 461
sequencing, 115, 287
series, xii, xvii, 3, 55, 56, 179, 208, 243, 337, 347, 354, 360, 377, 381, 386, 387
serum, 368, 371
services, 415
settlements, 64
settlers, 2, 42, 164
severity, 388
sewage, 2, 20, 60, 362, 364
sex, 33, 41, 147, 148, 150, 154, 158, 274, 275, 276, 277, 279, 281, 304, 333
sexual behavior, 48, 368, 374
sexual dimorphism, 151, 274, 320
shape, 26, 42, 75, 107, 170, 196, 197, 259, 266, 270, 281, 317, 474
shaping, 413
shares, 275, 344, 461
sharing, 96, 105, 106, 128, 396, 413
sheep, 127, 272, 397
Shell, 53, 278
shelter, 18, 152
shipping, xvi, 343, 344, 346, 353, 354, 361, 364, 369
shoot, 43
shores, 60
shrimp, 288, 328
siblings, 275
significance level, 142, 169, 170
signs, 33, 277, 389
silica, 287
silver, 121, 135, 167, 398
similarity, xvii, 47, 121, 204, 386, 395, 396, 397, 399, 408, 412, 416, 424, 430
simple linear regression, xiv, 221, 224
simulation, 143, 150, 172, 283
simulations, 139, 143, 146, 147, 148, 149, 150, 154, 168, 171, 179
sinus, xix, 128, 405, 406, 446, 449, 451, 458, 459, 462, 470, 474, 475, 482
sinuses, xviii, 446, 460
sites, xiii, 32, 43, 47, 48, 67, 104, 107, 110, 117, 121, 122, 124, 201, 223, 224, 232, 239, 289, 291, 369, 384, 399, 401, 404, 411, 416, 435
skeleton, 74, 80, 97, 194, 214, 422, 447, 450
skills, 387
skin, 102, 120, 286, 288, 301, 308, 318
SLA, 406
small intestine, 309
small mammals, 159
sociability, 133
social behavior, 14, 48, 151, 155
social behaviour, 272, 274
social group, 148, 157
social organization, xv, 64, 155, 156, 157, 261, 275
social structure, xv, 56, 131, 132, 148, 150, 155, 157, 248, 261, 262, 274, 275, 276
social welfare, 331
socioeconomic, 240, 335
software, xii, 71, 73, 104, 105, 108, 111, 115, 136, 156, 170, 289, 411, 414, 415, 416
sounds, xvi, 60, 343, 372
South Asia, 20, 55, 420
South Pacific, xviii, 446, 454
spatial, xviii, 60, 102, 113, 139, 142, 152, 168, 273, 289, 293, 346, 419, 434
spatial analysis, 273, 289
spawning, 44, 163, 437
specialization, 202
speciation, xv, 160, 182, 188, 190, 208, 261, 267, 270, 396, 412
speculation, 46
speed, 77, 79, 354, 369, 381, 427, 432, 433, 435
spelling, 448
sperm, 98, 210, 275, 311, 383, 425, 466
spills, 133, 163, 324, 379
spine, 183, 448, 465, 470
spiritual, 286

spontaneous abortion, 119
sporadic, 423
sports, 43
Sri Lanka, 162, 186
stability, 77, 79, 172, 222, 331, 358
stages, 40, 464, 483
standard deviation, 141, 170, 226, 227, 228, 292
standard error, 135, 399, 404, 408
standard of living, 331
stars, 382
statistical analysis, 88
statistics, xiii, 131, 134, 136, 137, 138, 139, 142, 143, 147, 150, 157, 168, 174, 179, 227, 228, 260, 289
stereotypical, 383
sternum, 16, 183, 202
stochastic, 142
stock, 208, 246, 256, 306, 308, 317, 329, 334
stomach, 33, 38, 253, 308, 309, 314, 316, 319, 323, 431, 437
strategies, xvii, 111, 116, 130, 192, 245, 257, 275, 327, 337, 377, 383, 388, 389, 421
streams, 25, 439
stress, 62, 63, 65, 126, 303
strikes, xii, 29, 30, 42, 48, 362
structuring, 276, 305, 328
subcutaneous tissue, 430, 431
summer, 48, 228, 229, 230, 231, 233, 248, 421
sunlight, 434
superimposition, 74
supply, 230, 238, 243
surface area, 44, 45
surface water, 392, 443
surveillance, 243
survival, 24, 26, 114, 119, 126, 187, 190, 213, 277, 303, 310, 311, 312, 313, 315, 327, 344, 358, 369, 386, 421, 437, 438, 439, 484
survival rate, 311, 312, 313, 327
surviving, 364, 380, 423
survivors, 195
sustainability, 28
suture, 87, 196, 469, 474
swelling, 42
synergistic effect, 232
synthesis, 317
systematics, xv, 28, 114, 115, 217, 261, 266, 268, 277, 278, 280, 282, 417, 443, 483, 488

T

tactics, 155, 321
tangible, 438, 464
tannins, 118
targets, 18, 326, 438
taxa, xiv, 55, 81, 183, 191, 193, 199, 202, 203, 209, 211, 215, 268, 271, 278, 388, 423, 425, 448, 449, 458, 465, 468, 470, 472, 475, 476, 480, 482
taxonomic, xv, 7, 22, 30, 56, 194, 199, 204, 209, 261, 266, 267, 269, 270, 271, 310, 424, 446, 477
taxonomy, 3, 47, 195, 199, 201, 248, 262, 266, 271, 276, 280, 282
technological change, 245
television, 272, 382
temperature, 30, 32, 48, 103, 166, 256, 286, 398, 431
temporal, xviii, 44, 87, 90, 91, 152, 159, 177, 180, 198, 325, 346, 349, 419, 434, 468, 471
territorial, 133, 275
territory, 61, 63
testes, 311, 430, 431
testis, 275, 430
thoracic, xii, 71, 74, 75, 77
thorax, 74, 76, 78
tides, 257, 258
timing, xv, 30, 261
tissue, 286, 287, 288, 430, 431, 459
tolerance, 242
topology, 267
tourism, 49, 62, 65, 251, 335
tourist, 59, 66
toxic, 56
toxicity, 325
toxins, 2, 61
trace elements, 324, 333, 336, 338, 393
trade, xv, 237, 239, 240, 245
tradition, 7
traffic, xvi, xvii, 18, 49, 62, 63, 164, 256, 293, 343, 344, 353, 354, 357, 362, 382, 383, 386, 436, 437
training, 61, 73, 75, 76, 368
traits, 85, 95, 96, 118, 130, 160, 204, 210, 262
transition, 45, 46, 107, 138, 228, 229, 243, 421, 423, 424
transitions, 105, 107, 136, 228
translation, 124
translocation, 111, 113, 365, 377, 384, 386, 387, 390
translocations, 113
transmission, 238
transparency, 304
transparent, 111, 112, 351, 352, 353
transport, 4, 20
transportation, 344, 364, 369
traps, 42, 43, 187, 234
trauma, 41
travel, 13, 17, 33, 37, 44, 46, 57, 65, 149, 347, 348, 349
travel time, 347, 348, 349
trawling, 327, 328

trees, 15, 112, 283, 289, 290, 291, 399, 400, 405, 425
Triassic, 440
tribes, 22
tropical areas, 40
trout, 73, 76
trucks, 239
tsunami, 22
turbulence, xvi, 58, 343, 352, 353, 354
turtle, 388
turtles, 84, 164, 206, 238
tyrosine, 122

U

ultrasound, 181
uncertainty, 27, 271, 313, 320, 328, 338, 349, 433
unemployment, 330
UNEP, 317, 319, 320, 333, 337, 438, 441
UNESCO, 438
uniform, xvi, 343, 354, 465, 471
United Nations (UN), 25, 317, 319, 320, 333, 335, 438
urbanization, 2
Uruguay, xvi, 8, 85, 98, 184, 211, 234, 301, 305, 306, 307, 308, 309, 310, 311, 312, 313, 314, 315, 319, 324, 325, 328, 330, 331, 337, 422
UV, 121

V

validation, 88, 93
variability, xvii, 44, 45, 108, 116, 117, 119, 121, 125, 136, 155, 157, 192, 226, 265, 276, 317, 395, 396, 397, 399, 400, 401, 411, 414, 416
variables, xv, 77, 84, 85, 87, 88, 89, 90, 91, 93, 95, 232, 237, 240, 241, 244, 247, 248, 249, 254, 256, 257, 258, 305
variance, 88, 104, 108, 134, 136, 142, 148, 167, 168, 169, 170, 171, 173, 179, 224, 273, 289, 291
vector, 137, 148, 398
vegetation, 30, 43, 44, 45, 47, 112, 422
velocity, 133, 420
vertebrae, xii, 8, 13, 71, 72, 73, 74, 77, 78, 79, 112, 202, 217, 378, 449, 450, 481, 488
vertebrates, 118, 119, 132, 203, 206, 208, 280, 309, 380, 415, 486
vessels, xiv, 42, 49, 61, 62, 221, 223, 224, 225, 230, 231, 232, 238, 240, 241, 242, 244, 252, 326, 364
virulence, 118
virus, 127, 130, 416
virus infection, 127, 130
viruses, 395
visible, 38, 48, 196, 353, 430, 469
vision, 112
vocalizations, 14
vulnerability, 44
vulva, 15

W

waste water, 133, 163
water quality, 2, 56, 61, 64, 65, 163, 362, 364
watershed, 20, 23, 118, 186
waterways, 31, 49, 424
weight ratio, 311
welfare, 331, 383
wetlands, 392
wild animals, 61, 345, 438
wildlife, 50, 63, 66, 279, 345, 386, 387, 388, 391
wind, 225, 346, 434
windows, 116, 130, 192, 306
winter, 48, 228, 229, 230, 231, 421
witchcraft, 222
women, 67, 245
wood, 152, 165, 241
workers, 266, 277, 429
writing, 370, 464

X

xylene, 398

Y

Y chromosome, 118, 124, 165, 182
yield, 330, 411

Z

zinc, 324, 334
Zn, 333
zoogeography, 485
zygomatic, xiii, xviii, 83, 87, 89, 90, 91, 95, 196, 197, 210, 446, 450, 451, 453, 455, 456, 460, 461, 462, 466, 468, 471, 473, 474, 477, 481, 482
zygomatic arch, 196, 450, 466